THE ELEMENTS*

	SYMBOL	ATOMIC NO.	ATOMIC MASS		SYMBOL	ATOMIC NO.	ATOMIC MASS
Actinium	Ac	89	[227]†	Mercury	Hg	80	200.59
Aluminum	Al	13	26.98154	Molybdenum	Mo	42	95.94
Americium	Am	95	[243]†	Neodymium	Nd	60	144.24
Antimony	Sb	51	121.75	Neon	Ne	10	20.179
Argon	Ar	18	39.948	Neptunium	Np	93	237.0482‡
Arsenic	As	33	74.9216	Nickel	Ni	28	58.71
Astatine	At	85	[210]†	Niobium	Nb	41	92.9064
Barium	Ba	56	137.34	Nitrogen	N	7	14.0067
Berkelium	Bk	97	[249]†	Nobelium	No	102	[254]†
Beryllium	Be	4	9.0128	Osmium	Os	76	190.2
Bismuth	Bi	83	208.9804	Oxygen	O	8	15.9994
Boron	B	5	10.81	Palladium	Pd	46	106.4
Bromine	Br	35	79.904	Phosphorus	P	15	30.97376
Cadmium	Cd	48	112.40	Platinum	Pt	78	195.09
Calcium	Ca	20	40.08	Plutonium	Pu	94	[242]†
Californium	Cf	98	[251]†	Polonium	Po	84	[210]†
Carbon	C	6	12.011	Potassium	K	19	39.098
Cerium	Ce	58	140.12	Praseodymium	Pr	59	140.9077
Cesium	Cs	55	132.9054	Promethium	Pm	61	[147]†
Chlorine	Cl	17	35.453	Protactinium	Pa	91	231.0359‡
Chromium	Cr	24	51.996	Radium	Ra	88	226.0254‡
Cobalt	Co	27	58.9332	Radon	Rn	86	[222]†
Copper	Cu	29	63.546	Rhenium	Re	75	186.2
Curium	Cm	96	[247]†	Rhodium	Rh	45	102.9055
Dysprosium	Dy	66	162.50	Rubidium	Rb	37	85.4678
Einsteinium	Es	99	[254]†	Ruthenium	Ru	44	101.07
Erbium	Er	68	167.26	Samarium	Sm	62	150.4
Europium	Eu	63	151.96	Scandium	Sc	21	44.9559
Fermium	Fm	100	[253]†	Selenium	Se	34	78.96
Fluorine	F	9	18.99840	Silicon	Si	14	28.086
Francium	Fr	87	[223]	Silver	Ag	47	107.868
Gadolinium	Gd	64	157.25	Sodium	Na	11	22.98977
Gallium	Ga	31	69.72	Strontium	Sr	38	87.62
Germanium	Ge	32	72.59	Sulfur	S	16	32.06
Gold	Au	79	196.9665	Tantalum	Ta	73	180.9479
Hafnium	Hf	72	178.49	Technetium	Tc	43	98.906‡
Helium	He	2	4.00260	Tellurium	Te	52	127.60
Holmium	Ho	67	164.9304	Terbium	Tb	65	158.9254
Hydrogen	H	1	1.0079	Thallium	Tl	81	204.37
Indium	In	49	114.82	Thorium	Th	90	232.0381‡
Iodine	I	53	126.9045	Thulium	Tm	69	168.9342
Iridium	Ir	77	192.22	Tin	Sn	50	118.69
Iron	Fe	26	55.847	Titanium	Ti	22	47.90
Krypton	Kr	36	83.80	Tungsten	W	74	183.85
Lanthanum	La	57	138.9055	Uranium	U	92	238.029
Lawrencium	Lr	103	[257]†	Vanadium	V	23	50.9414
Lead	Pb	82	207.2	Xenon	Xe	54	131.30
Lithium	Li	3	6.941	Ytterbium	Yb	70	173.04
Lutetium	Lu	71	174.97	Yttrium	Y	39	88.9059
Magnesium	Mg	12	24.305	Zinc	Zn	30	65.38
Manganese	Mn	25	54.9380	Zirconium	Zr	40	91.22
Mendelevium	Md	101	[256]†				

* Only 103 elements are listed, as there is no international agreement for the names of elements 104–109.

† Mass number of most stable or best known isotope.

‡ Mass of most commonly available, long-lived isotope.

World of Chemistry

World of Chemistry

MELVIN D. JOESTEN
Professor of Chemistry
Vanderbilt University
Nashville, Tennessee

DAVID O. JOHNSTON
Justin Potter Distinguished
Professor of Chemistry
David Lipscomb University
Nashville, Tennessee

JOHN T. NETTERVILLE
Guest Professor, Retired
David Lipscomb University
Nashville, Tennessee

JAMES L. WOOD
Resource Consultants, Inc.
Brentwood, Tennessee

Essays by Roald Hoffmann
Cornell University

Boxed features in each chapter provided by

Isidore Adler
University of Maryland

Nava Ben-Zvi
Hebrew University of Jerusalem
and Open University of Israel

Saunders Golden Sunburst Series
SAUNDERS COLLEGE PUBLISHING
Philadelphia   Ft. Worth   Chicago   San Francisco
Montreal   Toronto   London   Sydney   Tokyo

Text Typeface: Times Roman
Compositor: Progressive Typographers
Acquisitions Editor: John Vondeling
Associate Editor: Kate Pachuta
Managing Editor: Carol Field
Project Editor: Maureen Iannuzzi
Copy Editor: Linda Davoli
Manager of Art and Design: Carol Bleistine
Art and Design Coordinator: Doris Bruey
Text Designer: Ed Butler
Cover Designer: CIRCA 86
Text Artwork: J&R Technical Services
Layout Artist: Dorothy Chattin
Director of EDP: Tim Frelick
Production Manager: Charlene Squibb

Cover Credit: COMSTOCK, Inc.

Printed in the United States of America

World of Chemistry

0-03-030167-X

Library of Congress Catalog Card Number: 90-052734

0123 032 987654321

Preface

APPROACH AND SCOPE

World of Chemistry has been prepared for those students wishing a one- or two-semester course in college chemistry. This text is based on the fifth edition of *Chemistry and Society,* and the fundamental approach of this series of texts has been preserved—physical and chemical discoveries are presented along with the impact of these discoveries on our way of life. These human investigations of nature have produced modern chemical knowledge. Throughout the text the student is confronted with chemical applications that dramatically affect the quality of human life.

This text is presented as a stand alone course in chemistry for liberal arts students or as an integral part of a comprehensive package including a series of 26 thirty-minute video programs, a teachers' guide, a laboratory manual with teachers' guide, and a student study manual. The video series, entitled *The World of Chemistry,* was developed by Dr. Isidore Adler, University of Maryland, and Dr. Nava Ben-Zvi, Hebrew University of Jerusalem and Open University of Israel, and was sponsored by the Annenberg/CPB Project along with corporate sponsors. The video programs feature Nobel laureate and Priestley medalist Roald Hoffmann, and is a comprehensive survey of the field of chemistry and its impact on modern society. The series was produced jointly by The University of Maryland, College Park, Maryland and The Educational Film Center, Annandale, Virginia.

No previous knowledge of chemistry is assumed or required in this presentation. However, the approach is sufficiently different to challenge and interest the student with a background in high school chemistry.

To the beginning student there may be a mystery in chemistry, but to leave the workings of the chemist as a mystery argues that the liberally educated person must be dependent upon the chemist for those chemical decisions that affect society as a whole. *World of Chemistry* is based on the belief that the liberal arts student can see and appreciate the chain of events leading from chemical fact to chemical theory and the ingenious manipulation of materials based on the chemical theories. Thoughtful students will then see that the intellectual struggles in chemistry are closely akin to their own personal intellectual pursuits and will feel that each educated individual should and can have a say in how the applications of chemical knowledge are to affect the human experience.

The topics covered in this book have been selected based on what we have observed to be student interests. As a team of authors, and as individual teachers of chemistry at the collegiate level for many years, we have observed the following intense interests of our students:

1. Feeling the satisfaction of understanding the cause of natural phenomena.
2. Understanding the scientific bases for making the important personal choices demanded for the use of chemicals and chemical products.
3. Participating on a rational basis in the societal choices that will affect the quality of human life.
4. Helping to preserve and restore the quality of the environment along with a sensible approach to the recycling of natural resources.
5. Developing an insight into the perplexing problem of chemical dependency.
6. Sensing the balance involved in population control, the chemical control of disease, and the ability of the world to produce food.
7. Choosing personal habits in exercise programs and in nutritional selections compatible with healthful living.
8. Going places in vehicles that reflect the best uses of the materials used in transportation.
9. Using present energy reserves at a sensible rate as new energy sources are developed for the long term.

All of these paramount interests, as well as many of lesser note, are featured in this chemistry text on material substances and their uses.

World of Chemistry utilizes the common sense approach that is too often lost as the chemical community presents itself to the educated public at large. As in the total human experience, there is in chemical studies the fundamental relationship between cause and effect — structure causes function, chemical periodicity, and consequent material properties. We have selected carefully that thread of chemical history that shows chemistry to be the human endeavor it is. Much of the text has been written consistent with the essential chemical story to be told, the effectiveness of the communications with college students employed over the years, and with the critical reviews received from our peers. The philosophical setting for the presentation is made in Chapter 1, allowing the text and teacher to whet the appetite of the students for understandings of what may have previously been thought to belong only to the scientific elite. Chapters 2 through 15, while laying the groundwork for any intellectual consideration of the effect of chemistry on society, are replete with interesting applications to which the liberal arts students readily relate. The remainder of the text addresses problems about which there is intense interest in the general public, some of which are synthetic materials that dramatically alter the human environment; the nutritional basis of healthy living; medicines and drugs; pollution and the conservation of natural resources; consumer chemistry; and the agricultural production of food for a hungry world population.

A concerted effort has been made to inform the reader about certain vital and/or interesting matters that have bases in chemistry. Some examples are designer drugs, viral diseases in general and AIDS in particular, chemical treatment for major diseases and even hair loss; and the almost unbelievable applications that appear to be in the works for new materials — such as the electrolytic ceramics and superconducting materials.

ENHANCEMENTS

Full color throughout the text greatly enhances the effectiveness of the teaching aids that help the book communicate. Boldface for new terms and concepts along with marginal notes add emphasis to focus reader attention. Numerous interesting featurettes have been added throughout the text. Self-tests, which have proven so helpful in assisting the student in measuring retention and comprehension, are retained. Questions at the ends of the chapters provide for an additional measure of study and opportunities for extended research. Matching sets are included at the ends of the chapters to keep the necessary vocabulary in mind during chapter review. Numerous illustrations, which are a logical extension of the text, often communicate better than words. Boxed features in each chapter from the videos of *The World of Chemistry,* labelled with the icon 📹, facilitate the crossover excitement and understandings needed for interesting concepts that are only mentioned in the video presentation. Essays by Dr. Hoffmann offer a rare opportunity for the liberal arts student to sense that the human effort to know, appreciate, and understand is really common throughout all of the threads of intellectual activities. Dr. Hoffmann communicates to an appreciable degree with the casual student and offers a challenge and prod to the exceptional student to new and independent thought.

ACKNOWLEDGMENTS

We are deeply grateful to all who have contributed to the improvement of the manuscript and teaching aids for this book. We express a special measure of appreciation to Roald Hoffmann. Nava Ben-Zvi and Isidore Adler made it possible for us to bring two projects into one while retaining the individual usefulness of both the video series and the text. We sincerely appreciate the help of our colleagues who reviewed our manuscript and offered suggestions for improvement: Robert C. Belloli, California State University at Fullerton; John J. Fortman, Wright State University; J. W. Robinson, Louisiana State University; Robert E. Miller, Keene State College; Keith Kennedy, St. Cloud State University; Sheldon S. York, University of Denver; J. Leland Hollenberg, University of Redlands; Tamar Y. Susskind, Oakland Community College; Gilbert Castellan, University of Maryland; Rudolph S. Bottei, University of Notre Dame; Gary D. White, Cleveland State University; and Jerry A. Driscoll, University of Utah. Numerous other users of our texts have helped with corrections for which we are thankful. We also want to thank George Parks, Phillips Petroleum, for providing several computer-generated molecular structures for our book.

Lisa Ragsdale, Project Manager of *The World of Chemistry* video series was a delight from the beginning to the end. We appreciate the entire staff at Saunders College Publishing, professionals every one; they know how to get the job done! Kate Pachuta, as she has done before, has facilitated the flow of information and ideas, prompted the necessary decisions, pushed for meeting deadlines, and worked for excellence in every facet of editorial control. Maureen Iannuzzi, Project Editor, has been patient and helpful throughout the process of turning our manuscript into a book.

We give our most abundant thanks to John Vondeling, who has placed confidence in our series of chemistry texts for more than twenty years. The ambitious project in producing a liberal arts chemistry text in full color and, at the same time, coordinating with a complete video chemistry course of the caliber of the Annenberg/CPB Project is another first for John. We appreciate his leadership in the world of collegiate chemistry.

Much help has come our way, but, of course, the responsibility for the contents of the text rests entirely on us.

As in all of our previous works, we dedicate this effort to our spouses and gratefully acknowledge their support and understanding during the preparation of this manuscript.

Melvin D. Joesten
David O. Johnston
John T. Netterville
James L. Wood

June 1990

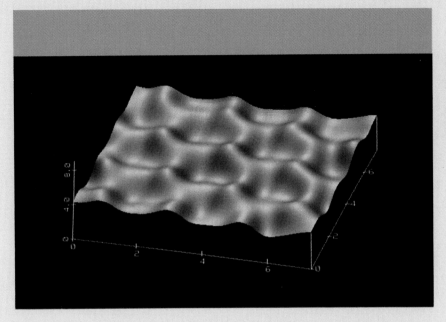

Support Package

◁ **The Study Guide** to Accompany The World of Chemistry. Provides a consistent method of incorporating the television programs with the textbook. To help the student focus on the content of the programs and to develop main points in greater detail, each unit of the study guide contains an overview, guidelines for viewing, an amplification of the various points covered, study questions, and a relevant "hands on" activity.

◁ **The Laboratory Manual** for The World of Chemistry. Offers 20 experiments, keyed to the video programs, which may be incorporated into the course. An alternate laboratory manual, *Laboratory Manual to Accompany Chemistry and Society,* 5th Edition, by Mark M. Jones *et al.,* offers 45 experiments which may be incorporated into the course.

◁ **The Faculty Telecourse Manual.** Contains general guidelines, teaching tips, and instructional resources for each study guide unit. This manual is creatively designed to combine student materials and faculty notes in one volume. Institutions licensing *The World of Chemistry* will receive one copy free.

◁ **Instructor's Manual & Testbank.** Includes answers to questions at the end of chapters, and multiple-choice test questions for each chapter.

◁ **Computerized Test Bank.** Available for IBM and Macintosh computers; free upon textbook adoption.

◁ **Overhead Transparencies.** 100 color images taken from the text. Available free upon textbook adoption. These acetates can be used to illustrate principles and concepts.

Contents Overview

Contents

CHAPTER 3   ATOMS   **53**

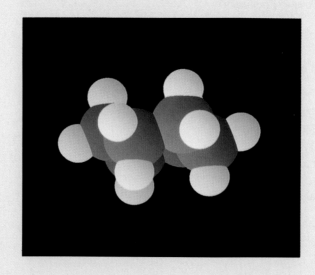

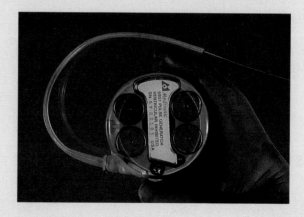

CHAPTER 6   SOME PRINCIPLES OF CHEMICAL REACTIVITY   146

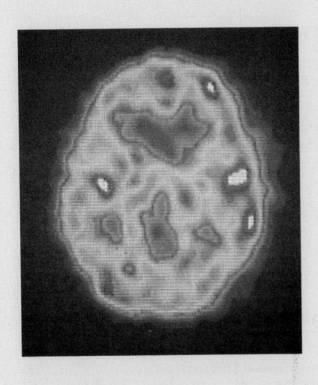

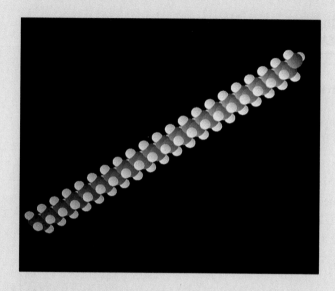

CHAPTER 13 ORGANIC CHEMICALS — ENERGY AND MATERIALS FOR SOCIETY 373

CHAPTER 14   GIANT MOLECULES — THE SYNTHETIC POLYMERS   409

CHAPTER 15   CHEMISTRY OF LIVING SYSTEMS   446

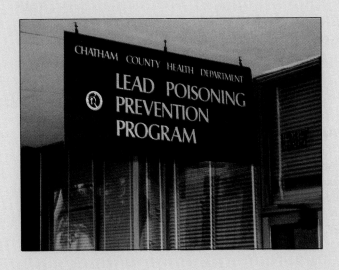

CHAPTER 16   TOXIC SUBSTANCES   490

CHAPTER 17   WATER—PLENTY OF IT, BUT OF WHAT QUALITY?   526

CHAPTER 18 CLEAN AIR—SHOULD IT BE TAKEN FOR GRANTED? 554

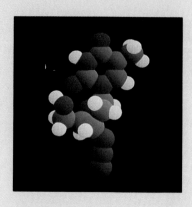

CHAPTER 21   CHEMISTRY AND MEDICINE   661

1

Impact of Science and Technology on Society

(United Nations Report on Environment and Development.)

WHAT IS SCIENCE?

Science can be defined in a number of ways. Perhaps the place to start is with the derivation of the word *science,* which comes from the Latin *scientia,* meaning "knowledge." Science is a human activity involved in the accumulation of knowledge about the universe around us. Pursuit of knowledge is common to all scholarly endeavors in the humanities, social sciences, and natural sciences. Historically, the natural sciences have been closely associated with our observations of nature — our physical and biological environment. Knowledge in the sciences, then, is more than a collection of facts; it involves comprehension, correlation, and an ability to explain established facts, usually in terms of a physical cause for an observed effect.

Figure 1–1 is a classification for the natural sciences. There is no sharp distinction between the two groups of sciences or among members within a group since new disciplines emerge that bridge areas at different levels. In the physical sciences, for example, there are biophysicists, geochemists, bioinorganic chemists, and chemical physicists. Some of these names define broad interdisciplinary fields, and others refer to more specialized subfields. The dynamic character of science is illustrated by the emergence of new disci-

Figure 1–1 Organizational chart for the natural sciences, with emphasis on chemistry.

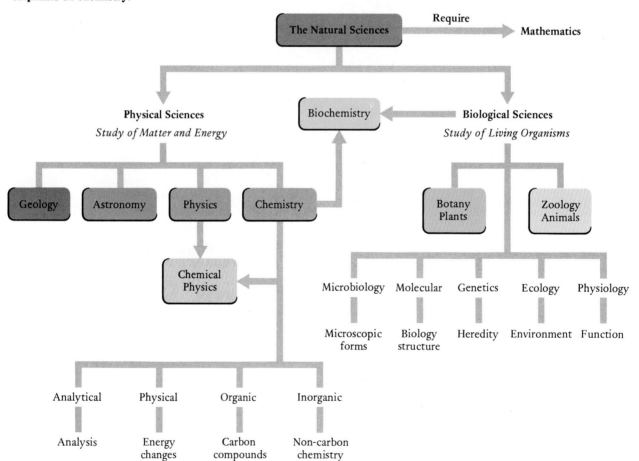

plines. Current chemical research is particularly active in those areas that either link subdisciplines of chemistry, such as organometallic chemistry, or that link chemistry with other sciences, such as bioinorganic chemistry or chemical physics.

WHAT IS CHEMISTRY?

There are many different subdivisions of science because there are many different ways to focus on the world around us. In this book we shall study the science of chemistry, which is one of the physical sciences. Chemistry is concerned with the study of matter and its changes. Since matter is the material of the universe, every object we see or use is part of the chemical story. Our body is a sophisticated chemical factory with hundreds of chemical reactions occurring even as you read this page. Because chemistry is so intimately involved in every aspect of our contact with the material world, chemistry can be regarded as the central science, an integral part of our culture, having an influence on almost every aspect of our lives.

WHAT IS THE DIFFERENCE BETWEEN TECHNOLOGY AND SCIENCE?

A deeper understanding of scientific knowledge comes when we distinguish among basic science, applied science, and technology. The difference between basic and applied science is determined by the motivation for doing the work. **Basic science** is the pursuit of knowledge about the universe with no short-term practical objectives for application. An example of **basic research** is seeking the answer to the question "What is penicillin?" by determining the molecular structure of penicillin. **Applied science** has well-defined, short-term goals related to solving a specific problem. For example, after the antibacterial action of penicillin was discovered, scientists conducted **applied research** on the effectiveness of penicillin against different

(a)

(b)

Reaction of aluminum with bromine to give aluminum bromide. (Charles D. Winters.)

(a) Basic research led to the discovery of Nylon by Dr. Wallace Carothers in 1935. (Charles D. Winters.) (b) Applied research and technology led to today's applications of the use of Nylon in rugs. (Courtesy of Du Pont de Nemours & Company.)

(a)

(b)

types of bacterial infections. Thus, both basic and applied scientific research produce new knowledge.

Technology is the use of scientific knowledge to manipulate nature. This may involve production of (1) new drugs, (2) better plastics, (3) safer automobiles, (4) nuclear weapons, or (5) chicken feed that causes eggs to have lower cholesterol. For example, engineers developed economical methods for large-scale production of penicillin from a knowledge of the results of basic and applied research on the chemistry and biochemistry of penicillin.

The important point is that technology, like science, is a human activity. Decisions about technological applications and priorities for technological developments are made by men and women; whether scientific knowledge is used to promote good or bad technological applications depends on those persons in industry and government who have the authority to make such decisions.

In a democratic society the voters may influence these decisions. Therefore, it is important to have an informed citizenry who can critically evaluate societal issues that are the consequences of technology.

The image most people have of science is strongly influenced by their familiarity with technological advances and, in most instances, is their only view of scientific progress. It is important to get beyond the everyday image of technology in order to recognize the symbiotic relationship of science and technology. Modern science depends on technological advances, especially in the development of more sophisticated instrumentation, to examine in greater depth the unanswered questions about the universe. Examples of this interrelationship are given throughout this book.

Perhaps more easily seen are the advances in technology that have occurred whenever new scientific discoveries are made. Regardless of the type of scientific discovery, there is a delay between a discovery and its

Symbiosis is the close association of two dissimilar things in a mutually beneficial relationship.

TABLE 1–1 How Long It Has Taken Some Fruitful Ideas to Be Technologically Realized

Innovation	Conception	Realization	Incubation Interval (Years)
Antibiotics	1910	1940	30
Cellophane	1900	1912	12
Cisplatin, anticancer drug	1964	1972	8
Heart pacemaker	1928	1960	32
Hybrid corn	1908	1933	25
Instant camera	1945	1947	2
Instant coffee	1934	1956	22
Nuclear energy	1919	1945	26
Nylon	1927	1939	12
Photography	1782	1838	56
Radar	1907	1939	32
Recombinant DNA drug synthesis	1972	1982	10
Roll-on deodorant	1948	1955	7
Self-winding wristwatch	1923	1939	16
Videotape recorder	1950	1956	6
Xerox copying	1935	1950	15
X rays in medicine	Dec. 1895	Jan. 1896	0.08
Zipper	1883	1913	30

technological application. The incubation period depends on (1) the rate of information transmittal, (2) the recognition of the applicability of the discovery, (3) the invention of a technological application for the new science, and (4) the large-scale manufacture of the new invention. The incubation times for several ideas or scientific discoveries are given in Table 1–1. Although there are exceptions, innovations based on applied research tend to happen faster than those developed from basic research.

HOW IS SCIENCE DONE?

Scientific Method

The methodology of science is often summarized by the term *scientific method*. A scientific method is a logical approach to solving scientific problems that may include (1) **observation,** or facts gathered by experiment, (2) **inductive reasoning** to interpret and classify facts by a general statement **(law),** (3) **hypothesis,** or speculation about how to explain facts or observations, (4) **deductive reasoning** to test a hypothesis with carefully designed experiments, and (5) **theory,** or a tested hypothesis or model to explain laws. Variations of this approach are often practiced in scientific research, and the imagination, creativity, and mental attitude of the scientist are often more important than the actual procedure.

The strictest intellectual honesty is required in the collection of observable facts and in the effort to arrange these facts into a pattern that reveals the

Inductive reasoning moves from specific facts to generalization. Deductive reasoning moves from generalization to specific facts.

Another outline of the scientific method: observe, generalize, theorize, test, and retest.

underlying cause of the *observed* behavior. The data normally must be collected under conditions that can be reproduced anywhere in the world. Then new data can be obtained to confirm or to refute the correctness of the suggested pattern. The results represent a unique type of objective truth that is ideally independent of differences in the language, culture, religion, or economic status of the various observers. Such established truth is appropriately referred to as **scientific fact.**

A scientific fact can be verified independently of any particular observer.

A **scientific fact** is an observation about nature that usually can be reproduced at will. For example, wood readily burns in the presence of air at a sufficiently high temperature. If you have any doubt about this fact, it is easy enough to set up an experiment that will readily demonstrate the fact anew. You only need some wood, air, and a source of heat. The repeatability of a scientific fact distinguishes it from a historical fact, which obviously cannot be reproduced. Of course, some scientific facts — such as the movement of heavenly bodies — are also historical facts and are not repeatable at will.

A scientific law summarizes a large number of related facts. A scientific law predicts what *will* happen. A governmental law describes what people *should* or *should not* do.

Often a large number of related scientific facts can be summarized into broad, sweeping statements called **natural, or scientific, laws.** The law of gravity is a classic example of a natural law. This law — all bodies in the universe have an attraction for all other bodies that is directly proportional to the product of their masses and inversely related to the square of their separation distance — summarizes in one sweeping statement an enormous number of facts. It implies that any object more dense than air that is lifted a short distance from the surface of the earth will fall back if released. Such a natural law can be established in our minds only by inductive reasoning; that is, you conclude that the law applies to all possible cases, since it applies in all of the cases studied or observed. A well-established law allows us to predict future events. When convinced of the generality of a scientific law, we may reason deductively, based on our belief that if the law holds for all observed situations, it will surely hold for any new related event.

The same procedure is used in the establishment of chemical laws, as can be seen from the following example. Suppose an experimenter carried out hundreds of different chemical changes in closed, leakproof containers, and suppose further that the containers and their contents were weighed before and after each of the chemical changes. Also, suppose that in every case the container and its contents weighed exactly the same before and after the chemical change had occurred. Finally, suppose that the same experiments were repeated over and over again, the same results being obtained each time, until the experimenter was absolutely sure that the facts were reproducible. It can be understood then that the experimenter would reasonably conclude that:

All chemical changes occur without any detectable loss or gain in weight.

This is indeed a basic chemical law and serves as one of the foundations of modern chemical theory.

After a natural law has been established, its explanation will be sought because chemists are usually not satisfied until they have explained chemical laws logically in terms of the submicroscopic structure of matter. This is indeed a difficult process, and until recently its progress had been painfully slow because of our lack of direct access into the submicroscopic structure of matter with our physical senses. All we can do is collect information in the macroscopic world in which we live and then try, by circumstantial reasoning, to visualize what the submicroscopic world must be like in order to explain our macroscopic world. Such a visualization of the submicroscopic world is called a **theoretical model.** If the theoretical model is successful in explaining a number of chemical laws, a major scientific theory is built around it. The atomic theory and the electron theory of chemical bonding are two such major theories, and both will be discussed in relation to chemical laws in later chapters.

Consider again the chemical law concerning the conservation of weight in chemical changes. What is a possible theoretical model that could explain this law? If we assume that matter is made up of atoms that are grouped in a particular way in a given pure substance, we can reason that a chemical change is simply the rearrangement of these atoms into new groupings without the loss or destruction of those atoms and, consequently, rearrangement into new substances. If the same atoms are still there, they should have the same individual characteristic weight, and hence the law of conservation of weight is explained. The set of boxes in Figure 1–2 summarize the relationship between facts, a chemical law, and a theory or model. Note that the chemical law, known as the **law of conservation of matter** for chemical reactions, summarizes the facts shown in the first set of boxes. In the last box is a version of the atomic theory model that explains the chemical law.

For a scientific theory to have lasting value, it must not only explain the pertinent facts and laws at hand but also be able to explain or accommodate new facts and laws that are obviously related. If the theory cannot consist-

The recent development of the Scanning Tunneling Electron Microscope (STM), described in Chapter 3, allows us to "see" individual atoms.

Theories are ideas or models used to explain facts and laws.

See Chapter 3 for a discussion of atoms.

Figure 1–2   Example of Relationship between Facts, Chemical Laws, and Theories

Some chemical facts	A chemical law	A model or theory to explain the law
(a) 2 units of hydrogen by weight react with 16 units of oxygen. Result: 18 units of water	All chemical reactions occur without any detectable loss or gain in weight.	H atoms—one unit of mass O atoms—16 units of mass N atoms—14 units of mass (a) $H + H + O \longrightarrow H_2O$ water
(b) 3 units of hydrogen by weight react with 14 units of nitrogen. Result: 17 units of ammonia		(b) $H + H + H + N \longrightarrow NH_3$ ammonia
(c) 14 units of nitrogen by weight react with 16 units of oxygen. Result: 30 units of nitric oxide		(c) $N + O \longrightarrow NO$ nitric oxide

Figure 1–3   Dr. Barnett Rosenberg holds in his left hand a mouse that will die of cancer in a few days. In his right hand is a mouse that has been infected with cancer but will survive because it has received cisplatin. (Barros Research Institute.)

The use of the term *serendipity* for accidental discoveries was first proposed in 1754 by Horace Walpole after he read a fairy tale titled "The Three Princes of Serendip." Serendip was the ancient name of Ceylon and the princes, according to Walpole, "were always making discoveries by accident, of things they were not in quest of."

Carcinogens are substances that cause cancer.

Figure 1–5 is an abstract of the first published report of this discovery.

ently perform in this manner, it must be revised until it is consistent, or, if this is not possible, it must be discarded completely. You must not allow yourself to think that this process of trying to understand nature's secret is nearing completion. The process is a continuing one.

The word *theory* is often used in a different sense from the one discussed previously. If a student is absent from the chemistry class, his neighbor may say, "I do not know why he is absent, but my theory is that he is sick and unable to come to class." The speculative guess of the student about his absent friend is what scientists call a **hypothesis** and is vastly different from the broad theoretical picture used to explain a number of laws. The reader should be alert for the considerable amount of confusion that has resulted from the different meanings associated with this word. In this book the word *hypothesis* is used when speaking of a speculation about a particular event or set of data, and the word *theory* is reserved for the broad imaginative concepts that have gained wide acceptance by withstanding scrutiny and by their ability to explain facts and laws over a long period.

Experimental Methods

Discoveries come about through the observation of nature or by experimentation that can be categorized as trial and error, planned research, or accidental discoveries (serendipity).

Discovery by trial and error begins when one has a problem to solve and does various experiments in the hope that something desirable will emerge. The next set of experiments then depends on the results obtained in the first set. The discovery of the Edison battery by Thomas Edison's group is an example of discovery by trial and error. Edison's group performed more than 2000 experiments, each guided by the previous one, before settling on the composition of Edison's battery.

Discovery by planned research comes from carrying out specific experiments to test a well-defined hypothesis. The carcinogenic nature of some compounds is determined by progressing through a set pattern of experimental tests.

Discovery by accident is really a misnomer. The investigator is usually actively involved in investigating nature through experimentation, but "accidentally" finds some phenomenon not originally imagined or conceived. Thus, the accident has an element of serendipity and is not likely seen unless the investigator is a trained observer. As Louis Pasteur said, "Chance favors the prepared mind."

The discovery of one of the leading anticancer drugs, cisplatin, is an example of such an accidental discovery. In 1964, Barnett Rosenberg (Fig. 1–3) and his co-workers at Michigan State University were studying the effects of an electric current on bacterial growth. They were using an electrical apparatus with platinum electrodes to pass a small alternating current through a live culture of *Escherichia coli* bacteria. After an hour, they examined the bacterial culture under a microscope and observed that cell division was no longer taking place. After thorough analysis of the culture medium and additional experimentation, they determined that traces of several different platinum compounds were produced during electrolysis from the reaction of the platinum electrodes with chemicals in the culture medium.

SERENDIPITY

The history of science is replete with examples of serendipity, simply described as a fortuitous and happy discovery or observation from which many important future developments have flowed. There is a special quality to serendipity. It is significant only where the discoverer or observer recognizes through a burst of intuition that there is something that needs further exploration and then proceeds to devote a serious effort to exploit the discovery. Much more often than not, of course, is the requirement of a prepared mind.

A classic example of serendipity is to be found in the story of W. H. Perkins as told by science historian John K. Smith of Lehigh University.

He was a brilliant young chemist, who, while working in his home laboratory in 1865 in an effort to synthesize badly needed quinine succeeded instead in creating the dye "mauve." Perkins recognized the importance of his accidental discovery and as a consequence of his efforts succeeded in establishing the beginning of the dye industry. The spinoffs were enormous. There are today as a consequence a large variety of materials such as drugs, explosives, fertilizers which play such an important role in the affairs of society. One of the most important consequences, for example, is aspirin, easily one of the most useful drugs in the history of pharmaceutical chemistry.

The World of Chemistry (Program 2) "Color."

W. H. Perkins. (*The World of Chemistry*, Program 2.)

Careful observation was essential since platinum electrodes are commonly regarded as inert or unreactive, and only a few parts of platinum compounds per million parts of culture medium were present. Additional testing indicated a compound known as cisplatin was responsible for inhibiting cell division in *E. coli*. Approximately two years after its initial discovery, the Rosenberg group had the answer to the question "What caused inhibition of cell division in *E. coli* bacteria?" At this point they hypothesized that cisplatin might inhibit cell division in rapidly growing cancer cells. The compound was tested as an anticancer drug, and in 1979 the U.S. Food and Drug Administration approved its use as such. The drug has now been proved to be effective alone or in combination with other drugs for the treatment of a variety of cancers.

Another interesting aspect of the story is that cisplatin, a compound that contains two chloride ions, two ammonia molecules, and platinum, was first prepared in 1845. Although its chemistry had been studied thoroughly since then, the biological effects of cisplatin and its inhibition of cell division were not discovered until "the accident" 120 years later.

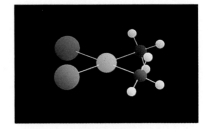

Model of cisplatin molecule. (Courtesy of Phillips Petroleum.)

An ion is a charged atom or group of atoms.

HOW DO SCIENTISTS COMMUNICATE?

Scientific knowledge is cumulative, and progress in science and technology depends on access to this body of knowledge. Since the earliest beginnings of science, this knowledge has been transmitted primarily by the written word.

Figure 1–4   Growth rate of papers abstracted by *Chemical Abstracts.*

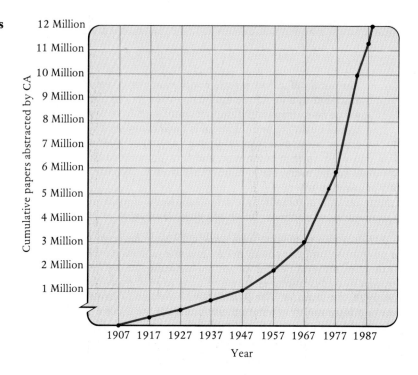

The invention of the printing press led to the development of scientific journals and other publications collectively known as the scientific literature. The explosive expansion of the scientific literature since the 1940s makes information management an essential part of modern science and technology.

Abstracting journals publish summaries of scientific publications.

The explosion of chemical literature can be illustrated dramatically by examining the growth rates of *Chemical Abstracts* (CA), which provides comprehensive coverage of the chemical literature worldwide. Figure 1–4 shows the growth rate of papers abstracted per year. Note the sharp increase since 1947. The number of journals monitored by CA has grown from 400 in 1907 to over 12,000 at the present time. In addition, CA monitors patent documents issued by 26 nations and two international bodies, conference and symposium proceedings, dissertations, government reports, and books. The publications abstracted by CA include leading physical, biological, med-

Figure 1–5   Example of abstract found in *Chemical Abstracts.* **This one was published in Volume 62, abstract 13543a, 1965. (This CA citation is copyrighted by The American Chemical Society and is reprinted by permission. No further copying is allowed.)**

> **Inhibition of cell division in Escherichia coli by electrolysis products from a platinum electrode.**   Barnett Rosenberg, Loretta Van Camp, and Thomas Krigas (Michigan State Univ., Lansing).   *Nature* 205(4972), 698–9(1965)(Eng).   *E. coli* cells grown in a continuous culture app. with an esp. designed chamber contg. 2 Pt electrodes were subjected to an elec. field in an O atm. With an applied voltage of 500 to 6000 cycles/sec., cell division ceased and filamentous growth occurred.   A similar effect was shown with a soln. of 10 ppm. $(NH_4)_2PtCl_6$.   Pt(IV) was found in the electrolyzed medium in concns. of 8 ppm.   Other transition metal ions, including Rh, in concns. of 1–10 ppm. inhibited cell division in *E. coli* while not interfering with growth.
> Gloria H. Cartan

ical, and other technical journals as well as chemistry-oriented journals. As a result, CA can be used for comprehensive searches of subject matter that crosses interdisciplinary lines.

An example of a CA abstract is shown in Figure 1–5. In 1988, CA published 474,545 abstracts, and 22% of these were patent abstracts. This is one measure of the ratio of technological application to scientific knowledge. CA indexes make this mountain of information accessible. Over 93 volumes make up the eleventh Collective Index of Chemical Abstracts for the period 1982 to 1986, and this collective index covers 2.3 million documents (Figure 1–6).

TECHNOLOGY AND THE INDUSTRIAL REVOLUTION

Over the last 200 years, accumulated scientific and technical knowledge has been put to use on an extensive scale in Europe and in those areas of the world that had the means and the will to follow the examples set by England, where the Industrial Revolution began. The result is the development of a society largely dependent upon and supported by a constantly changing technology. The first consequence of this technology was to increase the rate at which things can be produced. This, in turn, continually changed the occupational patterns of millions of human beings and brings forcefully to mind the persistence of change in our pattern of life.

These changes profoundly influenced the way people think about their material wants and the ways those desires can be satisfied. For example, there seems to be little argument with the statement, "If the number of human beings on the Earth could be stabilized, a much higher standard of living could prevail over most of the planet." A statement such as this would have been greeted with widespread derision 500 years ago. Although we are dependent on technology, people are today beginning to doubt its ability to solve both personal and social problems in the long run. It is obvious that confusion exists on this point since the cries about the curses of technology come from people who are highly dependent on it and who are even asking for more from technology.

Almost as soon as the Industrial Revolution began in England, the public realized that technological progress brought with it a series of problems. The first was the necessity for progress to be accompanied by changing patterns of employment.

It is obvious that if a machine makes as much thread as 100 workers can make, the workers will be replaced. The new opportunities that result from such a machine are rarely of benefit directly to the displaced workers, but the wealth of the country is increased since there are now 100 workers able to do other work. However, the initial reaction of the workers in 18th-century England was to riot and break up the machinery.

The increased use of fuels of all sorts, especially the introduction of coal and coke into metallurgical plants and then the use of coal to fuel engines, led to widespread problems with air pollution that were recognized and discussed over 200 years ago.

Figure 1–6   The 11th Collective Index of Chemical Abstracts is over 17 feet tall. *Chemical Abstracts* **have asked the** *Guinness Book of World Records* **to consider whether the 11th Collective Index is the world's largest index. (Courtesy of** *Chemical Abstracts.***)**

Technological unemployment results when people are displaced from their jobs by machines.

A modern version of change from technological progress is the use of robots for routine assembly work.

Automobile assembly line. (*The World of Chemistry*, Program 22.)

Fritz Haber (1898–1934).

Technological progress is always obtained at some cost, and the cost may not be obvious at the outset.

An important technological development was recognized as necessary in 1890 by Sir William Crookes, as he addressed the British Association for the Advancement of Science on the problem of the fixed-nitrogen supply (that is, nitrogen in a chemical form that plants can use). At the time, scientists recognized that the world's future food supply would be determined by the amount of nitrogen compounds made available for fertilizers. The source of these nitrogen supplies was then limited to rapidly depleting supplies of guano (bird droppings) in Peru and to sodium nitrate in Chile. It was realized that when these were exhausted, widespread famine would result unless an alternative supply could be developed. This problem was recognized first by English scientists as a potentially acute one, because by the 1890s England had become very dependent on imported food supplies. The Industrial Revolution allowed the population to grow rapidly, so the number of hungry people soon outstripped the domestic food supply.

Widespread interest in this problem led to research on a number of chemical reactions to obtain nitrogen from the relatively inexhaustible supply present in air. Air is 21% oxygen and 79% nitrogen. The nitrogen in the air is present as the rather unreactive molecule N_2, and in this form it can be used as a source of other nitrogen compounds by only a few kinds of bacteria. Some is also transformed into nitrogen oxide by lightning, and when this is washed into the soil by rain, the fixed nitrogen can be utilized by plants. The amount of nitrogen transformed by these processes into chemical compounds useful to plants is quite limited and cannot be increased easily.

Several chemical reactions were developed to form useful compounds from atmospheric nitrogen, but the best known and most widely used one has an ironic history. While England was interested in nitrogen for fertilizers, Germany was interested in nitrogen for explosives. The German General Staff realized that the British Navy could blockade German ports and cut them off from the sources of nitrogen compounds in South America. As a consequence, when the German chemist Fritz Haber showed the potential of an industrial process in which nitrogen reacts with hydrogen in the presence of a suitable catalyst to form ammonia (an essential ingredient in the preparation of explosives), the German General Staff was quite interested and furnished support through the German chemical industry for the study of the reaction and the development of industrial plants based on it. The first such plant was in operation by 1911, and by 1914 such plants were being built very rapidly.

When World War I broke out in August of 1914, many people thought that a shortage of explosives based on nitrogen compounds would force the war to end within a year. Unfortunately, by this time the nitrogen-fixing industry in Germany was capable of supplying the needed compounds in large amounts. This process thus prolonged the war considerably and resulted in an enormous increase in mortality. Subsequently, the ammonia process (Haber process) has been used on a huge scale to prepare fertilizers and is now largely responsible for the fact that Earth can support a population of more than 5 billion. Ammonia production by this process exceeds 40,000 tons per day in the United States alone.

The same type of problem seems to arise from the development of many technological processes. The utilization of nuclear energy brings with it the ability to make nuclear explosives from one of the byproducts. The development of rapid and convenient means of transportation, such as the automobile and the airplane, also brings forth new weapons of war and air pollution problems. We must learn to control our technology in such a manner as to maximize its benefits and minimize its disadvantages. These problems arise with all technological developments, even the most primitive. For example, the discovery of the techniques necessary to manufacture iron led first to the development of weapons (swords) by their discoverers, the Hittites, who then proceeded to conquer their neighbors and lead the first successful invasion of Egypt (ca. 1550 B.C.).

Now let's turn our attention to the present and see where chemistry and technology may allow us to go.

Application of ammonia to a field. (Courtesy Farmland Industries, Inc.)

THE CHIP AND THE SPLICE

Two major technological revolutions are currently under way: the microelectronic revolution and the biotechnology revolution.

The Microelectronic Revolution

The **chip,** a nickname for the integrated circuit, is a small slice of silicon that contains an intricate pattern of electronic switches (transistors) joined by "wires" etched from thin films of metal. Some are information storers called memory chips; others combine memory with logic function to produce computer or microprocessor chips. These two applications make the chip capable of almost infinite application. A microprocessor chip, for example, can provide a machine with decision-making ability, memory for instructions, and self-adjusting controls.

In everyday life we see many examples of the influence of the chip: digital watches; microwave oven controls; new cars with their carefully metered fuel-air mixtures; hand calculators; cash registers that total bills, post sales, and update inventories; and computers of a variety of sizes and capacity—all of these make use of the chip. By looking at Figure 1–7, you can appreciate the technological advancement represented by the chip. Small microprocessor chips such as those pictured in Figure 1–7 hold over 30,000 transistors.

The story of the chip starts with the invention of the transistor in 1947. The **transistor** is a semiconductor device that acts either as an amplifier or as a current switch. Although transistorized circuits were a tremendous improvement over vacuum tubes, large computer circuits using 50,000 or more transistors and similar numbers of diodes, capacitors, and resistors were difficult to build because computers had to be wired together in a continuous loop, and a circuit with 100,000 components could easily require a million soldered connections. The cost of labor for soldering and the chance for defects were high. In the late 1950s, the Navy's newest destroyers required 350,000 electronic components and millions of hand-soldered connections.

The small chip stores and processes vast amounts of information at amazing speeds.

Figure 1–7 A typical computer chip. (AT&T Bell Laboratories.)

Figure 1–8 (a) Original integrated circuit built by Jack Kilby. (Courtesy of Texas Instruments, Inc.). (b) Photomicrograph of a microprocessor chip with 200,000 transistors on one piece of silicon. (Courtesy of Motorola, Inc.)

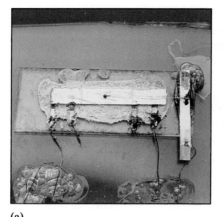

(a)

(b)

The transistor, smaller than the head of a match, replaced vacuum tubes ten or more cubic centimeters in volume.

It was clear that the limit to supercircuits using transistors was the number of individual connections required.

In 1958, Jack Kilby at Texas Instruments and Robert Noyce at Fairchild Semiconductor, working independently, came up with the solution. Make the semiconductor (silicon or germanium) in the transistor serve as its own circuit board. If all the transistors, capacitors, and resistors could be integrated on a single slice of silicon, connections could be made internally within the semiconductor and no wiring or soldering would be necessary. Figure 1–8(a) shows Kilby's hand-made chip and a modern chip (b) with 200,000 transistors on one piece of silicon.

The Biotechnology Revolution — Designer Genes

The biotechnology revolution began after the first successful gene-splicing and gene-cloning experiments produced **recombinant DNA** in the early 1970s. In Chapter 15 we shall discuss the biochemistry of DNA and the genetic code. The present discussion will focus on the potential of recombinant DNA technology for solving three of the world's greatest problems: hunger, sickness, and energy shortages.

Recombinant means capable of genetic recombination.

The process for forming and cloning recombinant DNA molecules is outlined in Figure 1–9. The basic idea is to use the rapidly dividing property of common bacteria, such as *E. coli,* as a microbe factory for producing recombinant DNA molecules that contain the genetic information for the desired product. Rings of DNA called **plasmids** are isolated from the *E. coli* cell. The ring is cut open with a cutting enzyme, which also cuts the appropriate gene segment from the desired human, animal, or viral DNA. The new gene segment is spliced into the cut ring by a splicing enzyme. The altered DNA ring (recombinant DNA) is reinserted into the host *E. coli* cell. Each plasmid is copied many times in a cell. When the *E. coli* cells divide, they pass on to their offspring the same genetic information contained in the parent cell.

"Cut" and "splice" are used figuratively, since we are really causing the breaking and making of chemical bonds; see Chapter 5.

The commercial potential of recombinant DNA technology was recognized very early, and several biotechnology companies were started by the

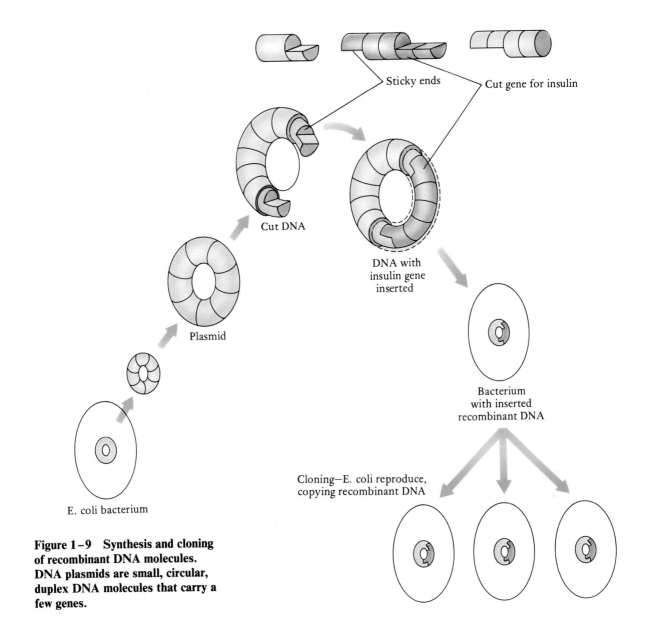

Sticky ends

Cut gene for insulin

Cut DNA

DNA with
insulin gene
inserted

Plasmid

Bacterium
with inserted
recombinant DNA

E. coli bacterium

Cloning—E. coli reproduce,
copying recombinant DNA

**Figure 1–9   Synthesis and cloning
of recombinant DNA molecules.
DNA plasmids are small, circular,
duplex DNA molecules that carry a
few genes.**

scientists who did key experiments in gene splicing and gene cloning. A
recent study by the U.S. Office of Technology Assessment gives a measure of
the growth of biotechnology. The Federal government and industrial companies are spending over $4.3 billion on biotechnology each year. There are
more than 403 biotechnology companies, but most of them are still doing
development work and have few products to sell. The risk is high, but the
payoff of successful ventures will be great. Some of the stronger biotechnology venture companies are beginning to report quarterly profits as their
products reach the market. Although the total annual sales in 1988 was

about $200 million, recent products have the potential to increase this figure to more than $1 billion by 1990. Some of the early applications illustrate how biotechnology can help to solve some of the world's problems in combating disease, hunger, and energy shortages.

One of the earliest benefits of recombinant DNA was the biosynthesis of **human insulin** in 1978. Millions of diabetics depend on the availability of insulin, but many are allergic to animal insulin, which was the only previous source. The biosynthesized human insulin is now being marketed by a firm called Genentech. Biotechnology firms are also producing **human growth hormone,** which is used in treating youth dwarfism, and **interferon,** which is a potential anticancer agent. Genentech received FDA approval in fall, 1987, to market **tissue-plasminogen-activator (TPA),** which dissolves the blood clots that cause heart attacks. If treatment with TPA begins soon enough, it could not only save lives but also reduce the damage caused by an attack.

The use of TPA is discussed in Chapter 21.

As the previous discussion indicates, most of the emphasis in biotechnology has been on the development of new drugs. Although progress is lagging in research on the use of biotechnology in agriculture and in the development of biological means of waste disposal, there is certainly recognition of the potential of biotechnology in these areas. Genetic engineers are trying to modify crops so they will make more-nutritious protein, resist disease and herbicides, and even provide their own nitrogen fertilizer. Another example is the use of recombinant DNA techniques to produce vaccines against diseases that attack livestock. Strains of bacteria are being developed that will convert garbage, plant waste material such as cornstalks, and industrial wastes into useful chemicals and fuels. This not only helps to solve the energy shortage but also provides a way to recycle wastes.

Although products such as human insulin and human growth hormone are currently manufactured using recombinant DNA techniques on *E. coli* bacteria, genetic engineers are now working on making genetic changes in animals. For example, in April, 1988, the first U.S. animal patent was awarded to Harvard University for a mouse that carries an activated cancer-causing gene. Mice carrying this gene can be used to test for possible carcinogens with greater sensitivity. These mice can also be subjected to substances such as vitamin E or beta-carotene, which are thought to protect against cancer.

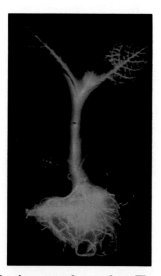

A luminescent tobacco plant. The gene of the firefly luciferase enzyme has been incorporated into the genetic material of the tobacco plant. (Courtesy of Dr. Marlene DeLuca, University of California at San Diego.)

Scientists at Genentech have created mice that produce human TPA in their milk. These mice can produce grams of TPA per liter of milk compared to a typical output from a bioreactor of *E. coli* of milligrams per liter—a thousandfold increase. Although the gene for making human TPA is present in all of the cells of the genetically altered (transgenic) mice, the protein is only synthesized in the mammary gland. The researchers accomplished this by fusing genes that normally control the production of mouse milk proteins to the gene for TPA. Since extracting milk from mice would not be simple, research is underway to produce transgenic cows, sheep, or goats that secrete TPA. Biogenetic engineers estimate that the world's supply of TPA could be obtained by milking a herd of 100 transgenic cows.

Not too far in the future, hereditary disorders caused by a defective gene will be treatable, if not curable, using gene therapy. Over 3,500 genetic diseases afflict as many as 1 out of every 20 newborn babies. They cause half of all miscarriages, nearly half of all infant deaths, and 80% of mental retardation. Other disorders influenced by defective genes are heart disease, diabetes, many forms of cancer, alcoholism, and mental illness. Most genetic diseases involve more than one defective gene, and research has only begun to identify the defective ones. However, some of the discoveries so far indicate the potential of gene therapy. Defective genes have been located that relate to Alzheimer's disease; Down's syndrome; Huntington's chorea; cystic fibrosis; elephant man disease; and cancer of the eye, colon, and lungs. However, there is a big difference between being able to predict a genetic disease and curing it. It may ultimately be possible to replace defective genes with healthy ones, or to otherwise compensate for the faulty gene. Animal testing is underway, but it will be many years before gene therapy includes effective cures.

A decade ago, at the beginning of the recombinant DNA era, many people, including the scientists working in the area, saw danger in biotechnology. Since *E. coli* is an intestinal bacterium, what if some of the genetically engineered *E. coli* escaped and found its way into people's intestines? These fears led to an 18-month moratorium on recombinant DNA research. However, the evidence to date shows that the *E. coli* used in recombinant DNA technology is too delicate to survive outside its environment. In addition, strict regulations are being followed in the experiments with genetically engineered bacteria to ensure against such problems.

But there is the deeper concern of ethics. Should a fetus found to be carrying the defective gene that causes Huntington's chorea (a degenerative brain disease) be aborted? If it were possible, should we raise IQs from 80 to 100 or from 120 to 160? Should we alter human genes to improve health, longevity, strength, and so forth? By altering life forms, whether plant, animal, or human, are we playing God? Debates on these questions have been going on since the beginning of biotechnology and will continue. The mechanism for resolving such important questions offers a new challenge to a free society.

The control of life, not just the environment, is at hand!

THE RISKS OF TECHNOLOGY

We have described some of the potential benefits of technology, but we also need to examine its risks. The public is well aware of the dangers of chemicals in the environment. Many persons have developed **chemophobia** (an unreasonable fear of chemicals) because of careless industrial practices such as improper disposal of hazardous wastes; environmental pollution of air, water, and earth; and catastrophic accidents. The names Bhopal, Chernobyl, and *Challenger* remind us of the influence of human error in increasing the risks associated with technology.

The chemical-plant accident that occurred in Bhopal, India on December 3, 1984, was the worst in history. Methyl isocyanate, a deadly gas used in the preparation of pesticides, escaped from a storage tank at the Union Carbide plant, killing over 2000 people and injuring tens of thousands. Numerous violations of safety procedures contributed to the disastrous leak. The explosion of the space shuttle *Challenger* on January 28,

The Chernobyl nuclear power plant after the accident. (Photography by V. Zufarov; courtesy of Fotokhronika Tass.)

1986, was caused by defective plastic O-rings between casing sections in the booster rocket. The explosion at the Chernobyl nuclear plant in the Soviet Union on April 26, 1986, which released large amounts of radioactive material into the atmosphere, was the result of a number of violations of operating regulations by the workers at the plant.

Careless disposal of hazardous wastes has been the cause of many problems. Love Canal in Niagara Falls, New York, the Times Beach community in Missouri, and the Minamata Bay and Jinzu River in Japan are just a few locations where serious problems have resulted from improper disposal of hazardous wastes.

The control of chemical hazards is essential for everyone's well-being.

Love Canal, the neighborhood that in 1977 discovered it was built on a toxic chemical dump, was the first publicized example of the problems of chemical waste dumps. In the mid-1970s heavy rains and snows seeped into the dump and pushed an oily black liquid to the surface. The liquid contained at least 82 chemicals, 12 of which were suspected carcinogens.

The entire community of Times Beach, Missouri, was bought by the U.S. Environmental Protection Agency (EPA) in 1983, and the 2200 residents were relocated because dioxins, a group of very toxic chemicals produced in small amounts during the synthesis of a herbicide, were found in the soil at concentrations as high as 1100 times the acceptable level.

Discarded waste barrels. (*The World of Chemistry,* Program 25.)

In the 1950s, tons of waste mercury were dumped into the bay at Minamata, Japan. In the next few years thousands of persons in the Minamata area suffered paralysis and mental disorders, and over 200 people died. Several years passed before it was determined that these people had been poisoned by methyl mercury compounds. Anaerobic bacteria in the sea bottom converted mercury to methyl mercury compounds, which were eaten by plankton. The methyl mercury compounds were carried up the food chain and eventually accumulated in the fatty tissue of fish. Since fish are a major part of the Japanese diet, intake of methyl mercury compounds reached levels that caused the sickness now known as **Minamata disease.**

Itai-Itai disease makes bones brittle and easily broken. It is caused by cadmium poisoning and was also first observed on a major scale in Japan. Itai means "it hurts" in Japanese and graphically illustrates the pain associated with this disease. Many cases were observed downstream from a zinc-refining plant on the Jinzu River in Japan. Cadmium is a byproduct of the zinc-refining industry and is used in various alloys and in nickel-cadmium rechargeable batteries.

Fir trees on Mount Mitchell in North Carolina killed by acid rain. (John Shaw/Tom Stack and Associates.)

What is an Acceptable Risk?

Risk assessment for individuals involves a consideration of the likelihood or probability of harm and the severity of the hazard. Assessment of societal risks combines probability and severity with the number of persons affected. The science of risk assessment is still evolving, but it is clear that the importance of public perception of risks needs to be recognized before risk assessment can be quantified. Often there is little correlation between the actual statistics of risk and the perception of risk by the public or by individuals. For example, we are all aware that the risk of injury or death is much lower from traveling in a commercial airplane than from traveling in an automobile, yet all of us know persons who avoid airplane flights because of their fear of a crash.

What factors influence public perception of risk? Catastrophic accidents such as those at Bhopal and Chernobyl obviously affect public perception of risk. In addition, people tend to judge involuntary exposure to activities or technologies (such as living near a hazardous dump site) as riskier than voluntary exposure (such as smoking). In other words, persons rate risks they can control lower than those they cannot control.

No absolute answer can be provided to the question "how safe is safe enough?" The determination of acceptable levels of risk requires value judgments that are difficult and complex, involving the consideration of scientific, social, and political factors. Over the years a number of laws designed to protect human health and the environment have been enacted to provide a basic framework for making decisions. The fact that three types of laws exist in this area adds to public confusion about risk assessment and its meaning.

The living must accept risk. The question is how much.

Risk-based laws are zero-risk laws that allow no balancing of health risks against possible benefits. The Delaney Clause of the Federal Food, Drug, and Cosmetic Act is such a law. It specifically bans the use of any intentional food additive that is shown to be a carcinogen in humans or animals, regardless of any potential benefits. The rationale for this law is the nonthreshold theory of carcinogenesis, which assumes that there is no safe level of exposure to any cancer-causing agent.

Chapter 20 discusses food additives.

The Safe Drinking Water Act, the Toxic Substances Control Act, and the Clear Air Act are **balancing laws;** they balance risks against benefits. The Environmental Protection Agency is required to balance regulatory costs and benefits in its decision-making activities. Risk assessments are used here. Chemicals are regulated or banned when they pose "unreasonable risks" to or have "adverse effects" on human health or the environment.

Technology-based laws impose technological controls to set standards. For example, parts of the Clean Air Act and the Clean Water Act impose pollution controls based on the best economically available technology or the best practical technology. Such laws assume that complete elimination of the discharge of human and industrial wastes into water or air is not feasible. Controls are imposed to reduce exposure, but true balancing is not attempted; the goal is to provide an "ample margin of safety" to protect public health and safety.

Clean air is discussed in Chapter 18. Clean water is discussed in Chapter 17.

RISK MANAGEMENT

Those in responsible positions in business and government now have a greater awareness of the need to solve the environmental problems associated with technological production and to assess the risks of technology. The important point is that our present environmental problems often stem from decades of neglect. The Industrial Revolution brought prosperity, and little thought was given to the possible harmful effects of the technology that was providing so many visible benefits. However, as shown by the Bhopal, *Challenger,* and Chernobyl incidents, serious accidents still occur despite the development of safety procedures that, if enforced, would prevent such accidents.

Government must and the chemical industry should take the lead in demonstrating their willingness to help solve the problems caused by previous lack of foresight. This should be done through cooperation and not confrontation. We need a science policy that is based on input from responsive leaders, from both industry and government, who provide a forum to examine the facts and then reach responsible decisions that lead to prompt action.

Is this possible? You may doubt it, but you have a responsibility to future generations to do your part in seeing that responsible action is taken. We cannot and should not "turn off" science and technology. Those who long for the "good old days" should know that diseases such as malaria, smallpox, and polio took many lives in the past. Without the antibiotics and vaccines discovered by science and developed by technology, the death toll would still be great from these. In addition, none of the modern fertilizers would have been discovered to increase crop yields, and worldwide famine would have occurred. You could add to this list many things that are of a humanitarian nature before you even start listing the technical advances that have raised the comfort level of our lives.

Risk management requires value judgments that integrate social, economic, and political issues with the scientific assessment of the risk. The determination of the acceptability of the risk is a societal issue, not a scientific one. It is up to all of us to weigh the benefits against the risks in an intelligent and competent manner. The assumption of this text is that the wit to deal with environmental problems caused by uncontrolled technology is to be found in the educated public at large, not in the select group that stands to make short-term financial or political profit. Always keep in mind that, except in the case of some radioactive wastes, the knowledge is available to "clean up" after any industrial operation; it is just a matter of cost, energy, and values.

It is apparent that we need citizens to take responsibility for being informed about the technological issues that affect society. Albert Gore, Jr., U.S. senator from Tennessee, said in support of better science education,

Science and technology are integral parts of today's world. Technology, which grows out of scientific discovery, has changed and will continue to change our society. Utilization of science in the solution of practical problems has resulted in complex social issues that must be intelligently addressed by all citizens. Students must be prepared to understand technological innovation, the productivity of technology, the impact of the products of technology on the quality of life, and the need for critical evaluation of societal matters involving the consequences of technology.

WHAT IS YOUR ATTITUDE TOWARD CHEMISTRY?

Before beginning this study of chemistry and its relationship to our culture, each of us needs to examine our prejudices (if any) and attitudes about chemistry, science, and technology. Many nonscientists regard science and its various branches as a mystery and have the attitude that they cannot possibly comprehend the basic concepts and consequent societal issues. Many also have chemophobia and a feeling of hopelessness about the environment. Many of these attitudes are the result of reading about the harmful effects of technology. Some of these harmful effects are indeed tragic. However, what is needed is a full realization of both the benefits and the harmful effects that can be attributed to science and technology. In the analysis of these pluses and minuses, we need to determine why the harmful effects occurred and whether the risk can be reduced for future generations. This book will give you the basics in chemistry, which we hope will afford you a healthier and more satisfying life by allowing you to make wise decisions about personal problems and problems that concern our world.

SUMMARY

1. We need an informed citizenry to use and evaluate scientific and technological advances.
2. To be informed about chemical problems requires basic knowledge about what matter is and what matter does.
3. More sophisticated chemical problems require a deeper understanding of the workings (facts and theories) of chemistry.
4. You should be involved. As an educated person, you have a responsibility and a privilege.

SELF-TEST 1–A*

1. The ultimate test of a scientific theory is its agreement with
   _____ .
2. Different workers, in different countries, who carry out a particular laboratory experiment in exactly the same way should get
   _____ result.
3. The chip is the nickname for _____ .
4. The common bacterium used as the microbe factory in recombinant DNA technology is _____ .
5. Arrange from most abstract to general to specific: laws, facts, theories.
6. Itai-Itai disease is caused by _____ poisoning.
7. An example of a technology-based law is
   _____ .
8. The letters in TPA, a genetic engineering product for treatment of heart attacks, stand for _____ .

---

*Use these self-tests as a measure of how well you understand the material. Take a test only after careful reading of the material preceding it. Do not return to the text during the self-test, but reread entire sections carefully if you do poorly on the self-test on those sections. The answers to the self-tests are in Answers to Self-Test Questions and Matching Sets at the end of the text.

QUESTIONS

1. Distinguish between theory and law in chemistry.
2. Give an example of a chemical fact.
3. Give an example of a chemical law.
4. How many times do you think a given experiment should yield the same result before a scientific fact is considered to have been established?
5. Distinguish between basic science, applied science, and technology.
6. Persons often confuse science with scientism. Look up the definition of *scientism* in a dictionary and discuss why it is important to society that science not be confused with scientism.

ESSAY by ROALD HOFFMANN

At Millesgården, near Stockholm, the work of the great Swedish sculptor Carl Milles is splendidly displayed. During a recent visit there I had a new insight into one sculpture group, the Aganippe fountain. Its theme is classical in origin, but Milles's interpretation is unusual. The spring of Aganippe, on the slopes of Mount Helicon in Greece, was said to inspire artists and poets. Milles portrays Aganippe as a female figure at the edge of a pool. From the pool rise several dolphins, arched in mid-leap. Three of the dolphins carry men, who represent Music, Painting, and Sculpture. Water rises from the beaks of the dolphins; this is after all a fountain, and Milles was a master designer of fountains.

What is natural and what is unnatural about this work of art? Like all fountains, it is obviously artificial. Someone has thought up a clever device that combines beauty and hydraulic engineering, to manipulate a natural substance, water, for artistic purposes. Water doesn't "want" to run up, nor does it "want" to run in controlled channels, much less through dolphin beaks. We make complicated machines to channel water and pump it up so that it can flow down naturally. Pumps, meters, valves—all those hidden artificial mechanisms! What could be more synthetic than a fountain?

The fountain's figures are cast in bronze, and their mechanical parts are made from other metals. The bronze itself is an artificial alloy of copper and tin, perhaps with a little lead and zinc. Indeed, the discovery of how to make bronze was important enough in human history that the Bronze Age was named after it. The elements in the alloy are smelted from their ores and refined in a remarkable process by artificial machines. However, the ores of copper and tin—covellite, cuprite, cassiterite, and others—are certainly natural.

Milles's fountain thus began with natural ores that were subjected to unnatural smelting and alloying technology. The resulting metal was used by natural man in the obviously unnatural act of sculpture to manipulate the most natural of substances, water, and to construct images of natural man and horse and dolphin. Any imagined separation of the natural and unnatural can be proved false by the careful analysis of any object in our world, such as Milles's fountain.

Most scientists, especially chemists, are sympathetic to this argument. They feel attacked by society as producers of unnatural, frequently dangerous materials. A survey of the media shows a consistent use of negative terms whenever chemistry is mentioned. Adjectives such as explosive, poisonous, toxic, and polluting are so often joined with chemical names that they have become stock phrases. Natural products are given positive associations—organically grown, unadulterated, preservative-free—while synthetics seem undesirable. Yet many manmade items are economically successful, for they shelter us, heal us, and make life easier, more interesting, more colorful.

The scientist understands that in any activity in which humans are involved—art, science, business, child-rearing—it makes little sense to separate the natural and unnatural. However, people continue to distinguish natural from unnatural, and to fear the artificial, for important reasons. No amount of "rationality" is going to make these real intellectual concerns go away. And those concerns are as valid for scientists as they are for other people. Even the people

with the greatest stake in producing and selling synthetic things take pleasure in natural surroundings. Their homes are decorated with real plants, not plastic and fabric imitations. No tanning salon will substitute for their real Hawaiian tan; they will avoid plastic shingles on their roofs and woodgrain imitations in their furniture. It seems clear to me that even the scientist—complaining about "unreasonable" people who cannot see the impossibility of separating the natural and unnatural—proves the attractiveness of such a separation in his or her own daily life.

Let us think about why we do prefer the natural, no matter who we are. I see many interconnected emotional forces at work, including six that I can label: romance, status, alienation, pretense, scale, and spirit. **Romance:** Romantic fiction, in literature, music, and other arts, is based on an unrealistic striving for what no longer is (or perhaps never could be). An idyllic farm is one favorite location. The irony of these unreal but attractive constructions, which are supposedly about the natural, is that such stories are fine for everyone except the people who had to live in their settings. It doesn't matter that the real stable smelled bad; my mind's stable smells just right. A reaching out for nature, for real wood, the smell of hay, the feel of wind in the sails, still determines our *desires*.
Status: Synthetics are successful because of some combination of lower cost, greater durability, more versatility, and perhaps new capabilities, compared with some natural materials. The replacement makes a wider range of materials available more cheaply to a larger group of people. Sanitary water delivery and waste disposal, better shelter, and reduction of

The Aganippe fountain sculptures by Carl Milles. (Roald Hoffmann.)

the death rate in infancy and childhood are now available to many more people than a hundred years ago. Although we still have a long way to go, this is what chemists and engineers can be proud of having achieved.

But human beings are (nicely) strange. When the synthetic becomes inexpensive and available to all, a curious inversion of taste occurs: the fashion makers decree that the "natural" has more status. A cotton shirt is supposed to feel more luxurious than a "permanent press" blend that includes synthetic fibers. A wood floor is seen as nicer than plastic tile, and the rarer the wood the better.

Perhaps what we want is not so much to be superior to others, but to be somewhat (not too much!) different. The natural provides, in its infinite variations, that opportunity to be slightly different.

Alienation: In routine work on an assembly line, in selling lingerie, and even in scientific research, we often feel dehumanized. We work on a piece of something, not on the whole. To be efficient we work repetitiously. All around us are machines whose workings we don't understand.

The synthetic and unnatural product is almost always inexpensive because it is mass-produced. The objects made in this way appear identical, and tend to reinforce the dehumanization that we feel. The typical mass-produced object shows little history of its making, so we don't see the signature of a human hand on what is around us.

Pretense: "False" is equivalent to "bad" in all things that are significant to human beings. To tell a lie, to pretend to be what one isn't, is not to be good. Much of the synthetic world of chemicals is not only unnatural in the sense of being manmade, but it often pretends to be what it is not. Plastic plates carry the patterns of porcelain, and plastic surfaces on furniture imitate wood grain. Paper napkins simulate linen, lace, and embroidery. Some of this is fine, but too much imitation always leads to disgust. One soon longs for the authentic.

Scale: There can be too many of one thing. The first plastic ashtray looks interesting, but as more and more of them appear, they quickly begin to bore us. The repetitive nature of mass-produced objects is often the only feature that impresses us.

Sometimes it is the abundance of artificial objects in our environment, rather than the repetition of one, that repels us. The typical American motel room, for instance, offers little that is not artificial. The variety of plastics and synthetic fibers in the furnishings of the room is astonishing, and even intellectually interesting as an example for a course about polymers. However, one is hardly attracted to that setting.

Spirit: What makes all of us seek out the natural? No simple psychological or sociological argument explains this. I believe that the soul has an innate need for the unique and growing that is life. I see a fir tree trying to grow in the apparent absence of topsoil, in a cleft of a cliffside, and I think how it will eventually split the rock. The plants trying to grow in my office remind me of that tree. Even the grain in the wood of my desk, though it tells me of death, also tells me of that tree. I see a baby satisfied after feeding, and its smile unlocks a memory of the smiles of my children when they were small, a line of ducklings forming after their mother, or that fir tree.

Let us recognize the ambiguity of our feelings toward both manmade and natural objects. One part of us knows that the natural and unnatural are inseparable. In science and art and commerce, everything made by human hands, tools, and minds is artificial—and that can be good or bad. At the same time, each of us is drawn to the natural, for complex reasons. To be inconsistent in this way is to be nothing more nor less than human.

2

The Chemical View of Matter — Order in Disorder

The material of which the universe is composed is generally found in a highly mixed composition of matter as illustrated by this vein of gold ore. It would appear that matter, or the elements, originated in a totally mixed state and that it has separated to some extent through natural processes. Exceptions occur as illustrated by rather large nuggets of pure gold. (Brian Parker/Tom Stack and Associates.)

Matter occupies space and has weight.

D o you enjoy the material things around you? Sometimes yes and sometimes no, right? Think beyond your immediate setting: how many materials and things are in this universe of ours? Too many to count? These things—all of them—that we can see, touch, and weigh are made of matter. Although an uncountable number of materials and objects exist, is there any order and simplicity in the makeup of **matter**? Can we reasonably hope to control matter to make life more pleasant for the human race? The science of **chemistry** addresses these fundamental questions.

Most of the things we use in our daily life are very different from the materials that are an obvious part of our natural surroundings. Practically everything we use has been changed from a natural state of little or no utility to one of very different appearance and much greater utility. The processes by which the materials found in nature can be changed and a detailed description of such changes are highly intriguing. This is a basic dimension of the science of chemistry: the **changes** in matter.

Chemical change alters the kind of matter without changing the amount of matter. Examples: a. Chemical change—burning, rusting, souring; b. Physical change—melting, boiling, cutting; c. Nuclear change—producing nuclear energy and radioisotopes.

Our attention will be focused on a particular kind of change—**chemical change** (Fig. 2-1). In any chemical change, the starting material is changed into a different kind of matter. Matter may also undergo other kinds of changes: **physical changes,** which result in new forms of the same material, and **nuclear changes,** which often change matter into energy while producing new substances. More complete definitions of these types of changes will follow. Also, as you study further about physical and chemical changes, you will come to realize that the categorization of material change is not as clear-cut as it first appears.

In physical terms, energy is the ability to do work and work is a physical force exerted through a distance.

What causes changes to occur in matter? It is **energy!** Examples of energy are heat, light, sound, and electricity. Energy and matter are not the same even though they are closely related. Energy has the ability to move matter (engines, eardrums, and motors, for example). Matter is converted into energy in nuclear reactors and nuclear bombs. Energy can infiltrate matter and manifest itself through the actions of the matter. A sample of hot water contains more energy than the same sample when it is cold. Some forms of energy can exist apart from matter; examples are light and radiant heat. It appears that all forms of energy are generated by changes in matter, and that matter, in turn, can absorb energy to produce other physical and chemical changes. Indeed, energy by definition is that which can produce change in matter. It follows, then, that a study of chemistry involves still another dimension: the *energy* associated with chemical changes.

It is difficult, in a few words, to establish the exact bounds of chemistry. Even so, it will be helpful to think of chemistry as

> the study of the kinds of matter and the changes of one kind of matter into another with the associated energy changes.

Since the feature used to recognize a chemical change is the production of a different kind of matter, recognition of a chemical change requires a recognition of different kinds of matter. In a natural state the kinds of matter are usually mixed together, and the separation of such mixtures has to precede their systematic classification. After an examination of the methods

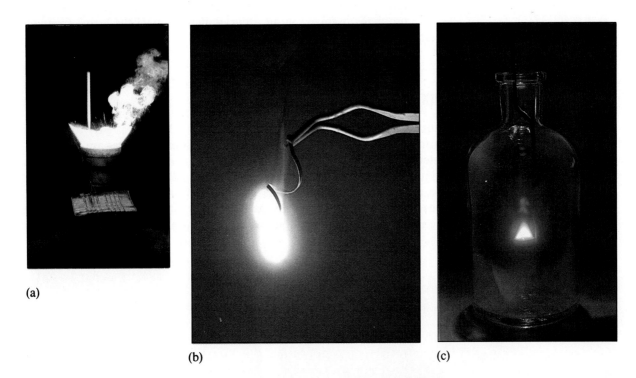

(a)

(b)   (c)

(d)

Figure 2–1   Chemical change results in different kinds of matter. (a) Aluminum powder reacts with iron oxide in the thermite reaction, a reaction that railroaders used to weld rails with the resulting molten iron. (Charles D. Winters.) (b) Metallic magnesium burns in air to produce a white solid, magnesium oxide. (c) Yellow sulfur, a solid, reacts with oxygen, a gas, to produce another gas, poisonous sulfur dioxide. (d) Photographs of zinc powder, sulfur, a mixture of these two powders and a sample of the compound zinc sulfide. Note that a chemical change, in contrast to a physical mixing, produces a new substance (Beverly March.)

of separating such mixtures into their components, we can appreciate some of the problems involved in an accurate definition of the terms *kinds of matter* and *chemical change.*

MIXTURES AND PURE SUBSTANCES

Most natural samples of matter are mixtures. Often, it is easy to see the various ingredients in a mixture (Fig. 2–2). Some mixtures are obviously heterogeneous, as the uneven texture of the material is clearly visible. Some mixtures appear to be homogeneous when actually they are not. For example, the air in your room appears homogeneous until a beam of light enters the room, revealing floating dust particles. Milk appears smooth in texture to the eye, whereas magnification reveals an uneven distribution of materials.

Homogeneous means of smooth texture, the same throughout. Heterogeneous means of nonuniform texture, not the same at every observed point.

(a)

(b)

Figure 2–2   (a) This NASA photograph of a moon rock and many similar ones show that lunar materials, like the solid formation in the crust of the earth, tend to be mixtures of more basic substances. It is likely that this is characteristic of crust materials in the universe. (b) The isolation of pure carbon in nature, the chemical makeup of diamond, is the exception, as there are relatively few pure substances isolated in natural formations. Typical rough and cut diamond of jewel quality is shown here. (c) Artificial diamond and graphite in a pencil and carbon powder. (Courtesy of General Electric Company.)

(c)

Colloids are mixtures that appear to be homogeneous in normal lighting but actually are heterogeneous (Fig. 2–3). Homogeneous mixtures do exist; such mixtures are **solutions.** No amount of optical magnification will reveal a solution to be heterogeneous, for heterogeneity in solutions exists only at atomic and molecular levels where the individual particles are too small to be seen with ordinary light. Examples of solutions are clean air (mostly nitrogen and oxygen), sugar-water, and some brass alloys (which are homogeneous mixtures of copper and zinc).

When a mixture is separated into its components, the components are said to be *purified.* However, most efforts at separation are incomplete in a single operation or step, and repetition of the process is necessary to produce a purer substance. Ultimately in such a procedure the experimenter may arrive at **pure substances,** samples of matter that cannot be purified further. For example, if sulfur and iron powder are ground together to form a mixture, the iron can be separated from the sulfur by repeated stirrings of the

Purification separates the kinds of matter.

Figure 2–3   The Tyndall Effect. A colloid in a clear liquid or gas will scatter light because of the relatively large size of the dispersed particles in contrast to the much smaller sizes of atoms and small molecules. A solution, which is a dispersion at the molecular level, will pass light with no scatter. (a) A laser light show works best in relatively "dirty" air as the beams can be seen in every direction because of the scatter (Fritz Goro). (b) Sunbeams in a forest can be seen only if there is colloidal moisture or dust. In clean air neither the laser light nor the sunbeam can be observed at any angle of view, the light traveling straight through with no scatter. (H. Armstrong Roberts.)

(a)

(b)

Figure 2–4   Mixed powdered iron and sulfur illustrate heterogeneous mixtures. The magnetic property of iron allows a physical separation of the two pure substances. (Charles Steele.)

mixture with a magnet (Fig. 2–4). When the mixture is stirred the first time and the magnet removed, much of the iron is removed with it, leaving the sulfur in a higher state of purity. However, after just one stirring the sulfur may still have a dirty appearance due to a small amount of iron that remains. Repeated stirring with the magnet, or perhaps the use of a very strong magnet, will finally leave a bright yellow sample of sulfur that apparently cannot be purified further by this technique. In this purification process a property of the mixture, its color, is a measure of the extent of purification. After the bright yellow color is obtained, it could be assumed that the sulfur has been purified.

Drawing a conclusion based on one property of the mixture may be misleading because other methods of purification might change some other properties of the sample. It is safe to call the sulfur a pure substance only when all possible methods of purification fail to change its properties. This assumes that all pure substances have a set of properties by which they can be recognized, just as a person can be recognized by a set of characteristics.

A pure substance, then, is a kind of matter with properties that cannot be changed by further purification.

Most materials in nature are mixtures; a few are relatively pure substances.

There are some naturally occurring pure substances. Rain is very nearly pure water, except for small amounts of dust, dissolved air, and various pollutants. Gold, diamond, and sulfur are also found in very pure form. These substances are special cases. The human, a complex assemblage of mixtures, lives in a world of mixtures—eating them, wearing them, living in houses made of them, and using tools made of them.

Although naturally occurring pure substances are not common, it is possible to produce many pure substances from natural mixtures. Relatively pure substances are now very common as a consequence of the development of modern purification techniques. Common examples are refined sugar, table salt (sodium chloride), copper, sodium bicarbonate, nitrogen, dextrose, ammonia, uranium, and carbon dioxide—to mention just a few. In all, over 10 million pure substances have been identified and cataloged.

Over 10 million pure substances have been identified!

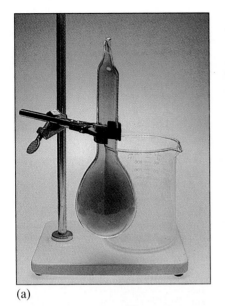
(a)

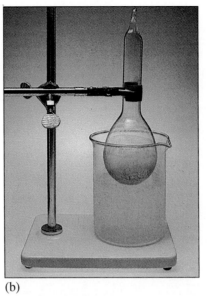
(b)

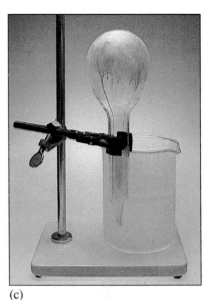
(c)

Figure 2–5 Three phases of matter for nitrogen dioxide. The gas (a) can be frozen to a solid (b) by cooling in liquid nitrogen, which on melting results in the liquid (c) running down the inside of the tube. (Charles D. Winters.)

STATES OF MATTER

As we examine mixtures and pure substances, it is easy to recognize three of the four states of matter — **solids, liquids,** and **gases** — and to note that both mixtures and pure substances can exist in the different states. Solids have definite shapes and volumes, liquids have definite volumes but indefinite shapes, and gases will take any shape or volume imposed by the containing vessel (Fig. 2–5). **Plasmas** constitute a fourth state of matter, which is not as common in our everyday experience. However, plasma is probably the most

A solid maintains both shape and volume, in contrast to a liquid which assumes any confining shape for a given volume. (Charles D. Winters.)

Native gold. (© Gemological Institute of America.)

**Figure 2–6   The Rosette Nebula.
Light from such nebulas is produced
by highly energized atoms, molecules, and ions. (Courtesy of The
California Institute of Technology.)**

An ion is a charged atom or group of
atoms.

prevalent state of matter in the universe (Fig. 2–6). A plasma is much like a
gas except it is composed of charged particles called ions, which dramatically
respond to electric and magnetic forces. Natural materials in the plasma state
include flames, the outer portion of the earth's atmosphere, the atmosphere
of the stars, much of the material in nebular space, and part of a comet's tail.
The aurora borealis offers a dazzling display of matter in the plasma state
streaming through a magnetic field.

DEFINITIONS: OPERATIONAL AND THEORETICAL

Chemistry begins with observation and
experiments.

A pure substance is identified **operationally,** meaning through specific experiments or operations. When further purification efforts are unsuccessful
in changing the properties of a substance, it is said to be a pure substance. It is
evident, then, that operational definitions result from performing operations
or tests on matter and summarizing the results in a statement. For example,
iron is a magnetic metal that melts at 1535°C, boils at 3000°C, and is 7.86
times more dense than water. These properties come from the operations of
applying heat to the pure substance and measuring temperatures, weights,
and volumes. When all of the properties of pure iron have been listed, we find
that the pure substance has been characterized in a way that distinguishes it
from all other pure substances. For all the millions of pure substances, no
two of them have exactly the same set of properties.

A pure substance also can be defined in theoretical terms, that is, in
terms of the molecules, atoms, and subatomic particles that compose it. Both

types of definitions are important in the study of chemistry, and both are used in this text. **Theoretical** definitions follow the development of the theory on which they are based.

Pure substances are defined in terms of atoms and molecules in Chapter 3.

SELF-TEST 2-A

1. Four common materials that cannot be pure substances are:
   a. _____   c. _____
   b. _____   d. _____
2. In the human experience, which usually comes first, the operational definition or the theoretical definition?
3. Four common materials that are very nearly pure substances are:
   a. _____   c. _____
   b. _____   d. _____
4. All of the properties of two different pure substances could be identical. True ( ) False ( )
5. Three types of changes that are fundamental to nature are:
   a. _____   b. _____   c. _____
6. A homogeneous mixture is a _____ .
7. Solutions may exist in solid, liquid, or gaseous states. True ( ) False ( )

MATCHING SET

_____ 1. Definite shape
         and volume
_____ 2. Indefinite shape
         and volume
_____ 3. Composed of
         charged particles
_____ 4. Indefinite shape,
         definite volume

a. gas
b. liquid
c. solid
d. plasma

SEPARATION OF MIXTURES INTO PURE SUBSTANCES

The separation of mixtures is usually more difficult than the magnetic separation of iron and sulfur described previously. Most beginning chemistry students would find it bewildering to separate a piece of granite into pure substances; indeed, a trained chemist might find this difficult. Since each of the pure substances in granite has a set of properties unlike those of any other pure substance, it should be possible to use these properties to separate the pure substances, just as the attraction of iron to a magnet is used to separate iron from sulfur.

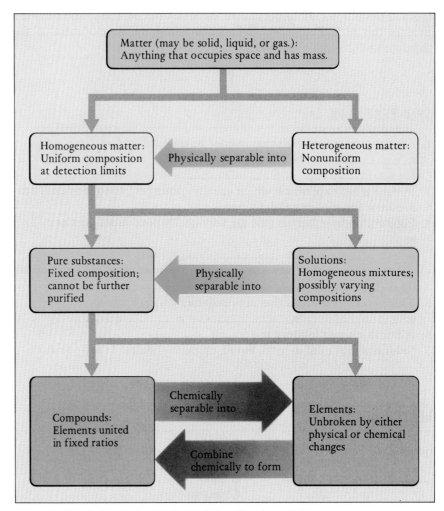

Figure 2–7   A chemical and physical classification of matter.

Refer to Figure 2–7 as you consider the operational definitions of the classifications of matter. A homogeneous sample of matter may be a pure substance or a mixture as in the case of a solution. Note that a pure substance is considered *elemental* only if all attempts to reduce it to two or more pure substances fail as illustrated in the following section. Also take note whether a physical change or a chemical change is required to go from one type of matter to another.

Many different methods have been devised to separate the pure substances in a mixture. In each case, differing properties of the pure substances are exploited to effect the separation. Figure 2–8 illustrates four commonly used methods: chromatography, distillation, recrystallization, and filtration. Figure 2–11 illustrates a separation by distillation that has been of interest to many.

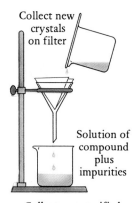

a. Dissolve solid in minimum quantity of hot solvent.

b. Cool solution (generally in ice + water). New crystals form.

Collect new crystals on filter

Solution of compound plus impurities

c. Collect new purified crystals on filter.
d. Repeat process if necessary.

Recrystallization

(b)

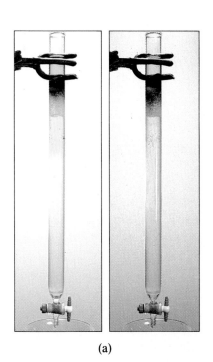

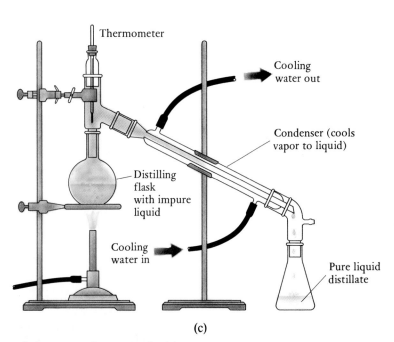

Thermometer

Cooling water out

Condenser (cools vapor to liquid)

Distilling flask with impure liquid

Cooling water in

Pure liquid distillate

(a)

(c)

Figure 2–8 Four methods of purifying mixtures of elements and compounds. (a) Chromatography. Owing to the absorbent character of paper or powdered silica, water moves along the medium and carries the ink dyes or spinach pigments at different rates, depending on the different attractions of the dyes or pigments for the paper or powdered silica; hence, the colored materials are separated. Chromatography is now widely applied to the separation of colorless materials with different "seeing" techniques. (Charles Steele.) (b) Recrystallization can be used to separate some solid mixtures based on different degrees of solubility in a solvent, and filtration isolates solids from liquids. (c) Distillation. Sodium chloride dissolves in water to form a clear solution. When heated above the boiling point (indicated by thermometer), water will vaporize and pass into the cool condenser where it liquifies as pure water.

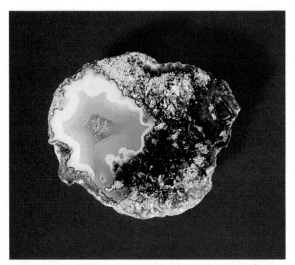

Figure 2–9   This broken geode shows the natural separation of crystalline rock materials. The layers of different chemicals form through recrystallization processes. (Ray Simons.)

Figure 2–10   Scanning electron micrograph of particles of asbestos filtered from a sample of air by a small-pore filter.

Figure 2–11   Distillation. Some of the most useful purification techniques copy processes in nature and date back to alchemical times. Distillation allows a more volatile substance to be separated from a less volatile one. In this case, alcohol is partially separated by evaporation from water and other ingredients. One distillation can produce a mixture that is 40% alcohol from one that is only 12%. Further distillations would produce an even better separation.

ELEMENTS AND COMPOUNDS

Experimentally, pure substances can be classified into two categories: those that can be broken down by chemical change into simpler pure substances and those that cannot. Table sugar (sucrose), a pure substance, will decompose when heated in the oven, leaving carbon, another pure substance, and evolving water. No chemical operation has ever been devised that will decompose carbon into simpler pure substances. Obviously sucrose and carbon belong to two different categories of pure substances. Only 89 substances found in nature cannot be reduced chemically to simpler substances; 20 others are available artificially. These 109 substances are called **elements.** Pure substances that can be decomposed into two or more different pure substances are referred to as **compounds.** Even though there are presently only 109 known elements, there appears to be no practical limit to the number of compounds that can be made from the 109 elements.

These are operational definitions of element and compound.

Figure 2–12 Relative abundance (by mass) of the most common elements in Earth's crust, the whole Earth, and the universe. Note that Earth's crust differs significantly from the cosmic array of elements.

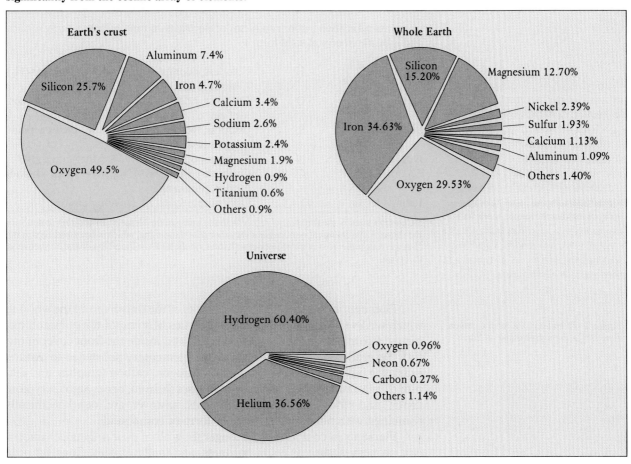

TABLE 2–1 Some Common Elements*

Name	Symbol	Properties of Pure Elements
METALS		

(A metal is a good conductor of electricity, can have a shiny or lustrous surface, and in the solid form usually can be deformed without breaking.)

Name	Symbol	Properties of Pure Elements
Iron Latin, *ferrum*	Fe	Strong, malleable, corrodible
Copper Latin, *cuprum*	Cu	Soft, reddish-colored, ductile
Sodium Latin, *natrium*	Na	Soft, light metal, very reactive, low melting point
Silver Latin, *argentum*	Ag	Shiny, white metal, relatively unreactive, good conductor of electricity and heat
Gold Latin, *aurum*	Au	Heavy, yellow metal, very unreactive, ductile, good conductor
Chromium	Cr	Resistant to corrosion, hard, bluish-gray, brittle

NONMETALS

(A nonmetal is often a poor conductor of electricity, normally lacks a shiny surface, and is brittle in crystal-solid form.)

Name	Symbol	Properties of Pure Elements
Hydrogen	H	Colorless, odorless, occurs as a very light gas (H_2), burns in air
Oxygen	O	Colorless, odorless gas (O_2), reactive, constituent of air
Sulfur	S	Odorless, yellow solid (S_8), low melting point, burns in air
Nitrogen	N	Colorless, odorless gas (N_2), rather unreactive
Chlorine	Cl	Greenish-yellow gas (Cl_2), very sharp choking odor, poisonous
Iodine	I	Dark purple solid (I_2), sublimes easily

Solids sublime if they readily evaporate without melting. Example: moth balls.

* Chemists usually use the symbol rather than the name of the element. In addition to denoting the element, the chemical symbol has a very specialized meaning, which is described later in this chapter. A complete list of the elements with their symbols can be found inside the front cover of this book.

Figure 2–12 displays the most common elements in the universe and on Earth.

Elements are the basic building blocks of the universe and the world in which we live. Table 2–1 lists the properties of some of the common elements; a complete list of the elements is found inside the front cover of this text. Several elements are found as the elementary substance in nature; examples include gold, silver, oxygen, nitrogen, carbon (graphite and diamond), platinum, sulfur, and the noble gases (helium, neon, argon, krypton, xenon, and radon). Many more elements, however, are found chemically combined with other elements in the form of compounds.

Elements in compounds no longer show all of their original, characteristic properties, such as color, hardness, and melting point. Consider ordinary sugar as an example. It is made up of three elements: carbon (which is

(a)

(b)

(c)

(d)

Copper sulfate can be separated from sand (a) by dissolving the soluble copper sulfate in water (b), followed by filtering the sand from the solution (c). Solid copper sulfate can be recovered from the solution (d) by evaporating the water from the nonvolatile copper sulfate. (Charles D. Winters.)

usually a black powder), hydrogen (the lightest gas known), and oxygen (a gas necessary for respiration). The compound sucrose is completely unlike any of the three elements of which it is composed; it is a white crystalline powder that, unlike carbon, is readily soluble in water.

A careful distinction should be made between a compound of two or more elements and a *mixture* of the same elements. The two gases hydrogen and oxygen can be mixed in all proportions. However, these two elements can and do react chemically to form the compound water. Not only does water exhibit properties peculiar to itself and different from those of hydrogen and oxygen, but it also has a definite percentage composition by weight (88.8% oxygen and 11.2% hydrogen). In addition to the distinctly different properties between compounds and their parent elements, there is this second distinct difference between compounds and mixtures:

> **Compounds have a definite percentage composition by weight of the combining elements.**

Compounds have a fixed composition of the elements they contain.

WHY STUDY PURE SUBSTANCES AND THEIR CHANGES? COMFORT, PROFIT, AND CURIOSITY!

Perhaps by now you are wondering why we should be interested in elements and compounds and their chemical properties. There are two basic reasons. The first is the belief that the knowledge of chemical substances and chemical changes will allow us to bring about desired changes in the nature of everyday life. Two hundred years ago most of the materials surrounding a normal person could be changed only by physical means. Only a few useful materials, such as iron and pottery, were the product of chemical change. By contrast, today's synthetic fibers, plastics, drugs, latex paints, detergents, new and better fuels, photographic films, and audio and video tapes are but a few of the materials produced by controlled chemical change. (We shall return to examine the chemistry of many of these later.) You will find it difficult to find more than a few objects in your home that have not been altered by a desirable chemical change. Not only is it important to bring about desirable changes, but also in the areas of toxicity and pollution it is important to avoid undesirable changes.

A second driving reason for the study of elements, compounds, and their properties is simple curiosity. Chemicals and chemical change are a part of nature that is open to investigation, and, like the mountain climber, we shall find this task both interesting and challenging simply because it is there. If we hope to understand matter through basic research, the first steps are to discover the simplest forms of matter and to study their interactions. Curiosity draws many chemists toward these basic research activities.

THE STRUCTURE OF MATTER EXPLAINS CHEMICAL AND PHYSICAL PROPERTIES

For reasons that are partly theoretical and partly practical, we are deeply interested in the **structure of matter** — that is, the minute parts of matter and how these parts are fitted together to make larger units. Why does an element or compound have the properties it has? Why does one element or compound undergo a change that another element or compound will not undergo? Inanimate matter is the way it is because of the nature of its parts. A watch is what it is because of the nature of its individual parts. So is a car, a refrigerator, and the salt in your salt shaker. The individual parts (smaller than the whole) are what determine the nature (actions and properties) of the whole. The most basic parts of matter, as we shall see, are very small. If we even hope to understand the nature of matter, it is absolutely necessary that we have some understanding of these minute parts and how they are related to each other.

The causes of the properties of matter lie in the structure and composition of its parts.

A very large portion of today's research in chemistry is aimed at sorting out and elucidating the structure of matter. Indeed,

> the basic theme of this text is the relationship between the structure of matter and its properties.

This theme of structure and related properties is of great interest because if we know exactly how the minute parts of matter are put together and how tightly they are bound, we can discover exact relationships between structure and properties. Armed with this understanding, we can make changes that result in new substances and predict the properties of these substances. Such knowledge can save many months of trial and error that otherwise may be required to prepare a product with the desired qualities. Although this day of predicting chemical changes based on structural characteristics has not arrived completely, such significant advances have been made that the practice of modern chemistry would not be possible without such knowledge.

Submicroscopic structures help to explain chemistry.

Samples of matter large enough to be seen and felt and handled, and thus large enough for ordinary laboratory experiments, are called **macroscopic** samples, in contrast to **microscopic** samples, which are so small that they have to be viewed with the aid of a microscope. The structure of matter that really interests us, however, is at the **submicroscopic** level. Our senses have very limited access into this small world of structure, and any conclusions about it will have to be based largely on circumstantial evidence gathered in the macroscopic and microscopic worlds (Fig. 2–13).

(b)

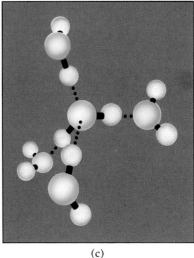

(c)

Figure 2–13 Direct observation stops at the microscopic level. Convinced of structure beyond the microscopic level, the chemist employs circumstantial evidence to construct the world of molecules, atoms, and subatomic parts in the mind's eye. (Snowflakes and water molecule models: *The World of Chemistry,* Program 12.)

MODELS AND MODELING

The history of science is filled with examples of models—mental concepts that scientists and investigators use to explain the unseen. Many of these models are triumphs of the human mind and represent extraordinary flights of the imagination. Among the classic examples are models of the atom, the kinetic molecular model of gases, and the replication of the molecules of DNA. Models must agree with facts, be modified to agree with facts, or be abandoned.

A remarkable example of the use of models to explain observed phenomena is found in the recent observations of the Galilean moon (Jupiter's Io) by that astonishing spacecraft Voyager (Fig. 2–14). A recent encounter with this moon showed it to be, to eveyone's surprise, one of the most bizarre and active bodies in the solar system. Dr. Torrence Johnson, director of the Voyager Project, points out,

Prior to the Jupiter visit, the model of Io, based on terrestrial observations, was of a moon similar to our own in age, composition, etc. The only inconsistency was that the surface was too bright. The model proposed at the time attributed its high reflectivity to the possibility that Io was covered by a layer of ice which had long since evaporated and left behind deposits of highly reflecting salt.

We now know that this model is invalid. Io is unique in the solar system in being one of the most active volcanic bodies, showing at the time of the encounter at least seven active volcanoes. Enormous quantities of sulfur and sulfur dioxide were being spewed into space. Where does the energy come from? The existing Ionian model proposes that the energy comes from the gravitational tugging from Jupiter and the other moons of this planet, which together churn the contents of Io as the orbital configurations change.

The World of Chemistry (Program 4) "Modeling the Unseen."

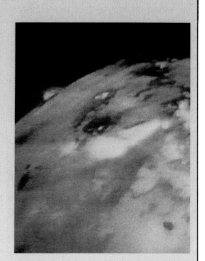

Figure 2–14 Voyager photograph of Io, a moon of Jupiter. Note the volcanic activity in the upper left. The cold-moon model of this planet had to be changed as a result of these observations. (*The World of Chemistry,* Program 4.)

Scope of chemistry.

We can now extend our concept of the science of chemistry. It is science that investigates the properties and changes of pure substances, but chemistry is also deeply concerned with structure, both macrostructure and submicrostructure, in an effort to give plausible reasons for properties and change, with emphasis on chemical change.

CHEMICAL, PHYSICAL, AND NUCLEAR CHANGES—A CLOSER LOOK

Chemical and physical changes were originally defined in an operational sense, whereas nuclear changes were first conceived in theoretical terms. We are now prepared to consider these important transformations in matter from both points of view, even though theoretical understanding will develop further as you study the following chapters.

A chemical change involves the disappearance of one or more pure substances and the appearance of one or more other pure substances. In theoretical terms, a chemical change produces a new arrangement of the atoms involved without a loss or gain in the number of atoms. In a physical change, the pure substance is preserved from an operational point of view even though it may have changed its physical state or the gross size and shape of its pieces. In theory, the physical change may break the gross arrangement of the overall structure of the theoretical particles (cutting a diamond) or even separate the particles (melting and vaporizing a metal), but the fundamental particles remain and their tendency to interact with each other is retained. A nuclear change occurs when atoms of one type are changed into atoms of another type. Physically, the nuclear change produces new pure substances, as does the chemical change; however, nuclear changes are generally associated with conversion of elements into other elements, radioactive emissions, and very large energy transformations relative to those involved in chemical and physical changes. Also, matter is destroyed in many nuclear changes and converted into energy.

Nitrogen, a liquid at −196°C, is poured into a beaker (a) where it is used to freeze a rose (b). The flower tissue has different properties after undergoing this physical change (c). (Charles D. Winters.)

(a)

(b)

(c)

Consider the following examples of the three types of change:

1. *Chemical changes.* Rusting of iron, burning of gasoline in an automobile engine, preparation of caramel by heating sugar, preparation of iron from its ores, solution of copper in nitric acid, ripening and decay of fruit.
2. *Physical changes.* Evaporation of a liquid, melting of iron, drawing of metal wire, freezing of water, crystallization of salt from sea water, grinding or pulverization of a solid, cutting and shaping of wood, bending, breaking, molding.
3. *Nuclear changes.* Production of the elements in stars, splitting of uranium atoms in an atomic bomb, fusion of light atoms in thermonuclear bomb, production of radioisotopes in a nuclear reactor, natural radiation from uranium, radium, and radon.

SELF-TEST 2–B

1. The two large categories into which elements can be divided are:
   a. _____ and b. _____
   A half-dozen examples of each are:
   a. _____   a. _____
   b. _____   b. _____
   c. _____   c. _____
   d. _____   d. _____
   e. _____   e. _____
   f. _____   f. _____
2. How many elements are presently known?
3. A compound has properties that are combinations of the elemental properties. True (　) False (　)
4. Four physical changes not listed in the chapter are:
   a. _____
   b. _____
   c. _____
   d. _____
5. Four chemical changes not listed in the chapter are:
   a. _____
   b. _____
   c. _____
   d. _____
6. A chemical change always produces a new
   _____.
7. _____ structures explain chemical properties.
8. Put the following in order of decreasing size: 1. microscopic 2. molecular 3. macroscopic.
9. In what type of natural change are atoms altered in their internal structure?
10. Which element is most abundant:
    a. In the crust of the Earth?

the oxidation half-reaction is
the hydrogen ions. H_2O_2 to form bo
Permagnanate ion as an oxidizing agen
$H_2O_2 + 2H^+ + 2e^-$ -------- $2H_2O$ Hydr
a mole of atoms would be 602,000,000
a chemical reaction, the sum of a
ury + oxygen-------- mercuric oxy
Thermo-nuclear compound

"chemistry"

The shorthand used in chemical expressions are merely time-saving devices, a technique used in the recording of most human enterprises. Consider R, H, E, HR, SO, ERA, X, etc., used in baseball jargon.

Chemical symbols are abbreviations for the different elements.

A mole contains 6.02×10^{23} particles.

A mole of water molecules in the liquid state occupies only about four teaspoonfuls. Are molecules small . . . or are they small?

b. In the bulk of the Earth?

c. In the universe?

11. Name four separation techniques (or purification methods).

12. List as many elements as you can recall without looking at a reference.

THE LANGUAGE OF CHEMISTRY

Symbols, formulas, and equations are used in chemistry to convey ideas quickly and concisely. These shorthand notations are merely a convenience and contan no mysterious concepts that cannot be expressed in words. Certain characters are used often, and a general familiarity with them will help in reading a chemical text.

A **chemical symbol** for an element is composed of one, two, or three letters—the first letter a capital and the second and third are lowercase letters. The symbol represents three concepts. First, it stands for the element in general. H, O, N, Cl, Fe, and Pt are shorthand notations for the elements hydrogen, oxygen, nitrogen, chlorine, iron, and platinum, respectively. It is often useful and timesaving to substitute these symbols for the words themselves in describing chemical changes. Some symbols originate from Latin words (such as Fe, from *ferrum,* the Latin word for iron); others come from English, French, and German names. Second, the chemical symbol stands for a single atom of the element. The **atom** is the smallest particle of the element that can enter into chemical combinations. Third, the elemental symbol stands for a mole of the atoms of the element. The **mole** is a term in chemical usage (derived from the Latin for "a pile of" or "a quantity of") that means the quantity of substance that contains 602 sextillion identical particles. Just as a dozen apples would be 12 apples, a mole of atoms would be 602,200,000,000,000,000,000,000 atoms or 6.022×10^{23} atoms. How big is this number? A mole of textbooks like this one would cover the entire surface of the continental United States to a height of 190 miles! To match the population density on Earth, a mole of people would require 150 trillion planets. It takes 134,000 yr (years) for a mole of water drops (0.05 mL [milliliter] each) to flow over Niagara Falls at a flow rate of 112,500,000 gal/min (gallons per minute). A mole is a *very* large number. Yet, it turns out that a mole of very small atoms is usually a convenient amount for laboratory work. Thus, the symbol Ca can stand for the element calcium, or a single calcium atom, or a mole of calcium atoms. It will be evident from the context which of these meanings is implied.

Atoms can bond together to form molecules. A **molecule** is the smallest particle of an element or a compound that can have a stable existence in the close presence of like molecules. One or more of the same kind of atom can make a molecule of an element. For example, two atoms of hydrogen will bond together to form a molecule of ordinary hydrogen, four atoms of phosphorus form a molecule of phosphorus, and eight sulfur atoms will form a single molecule. Subscripts in a chemical formula show the number of atoms involved: H_2 means a hydrogen molecule is composed of two atoms, P_4 means a phosphorus molecule is composed of four atoms, and S_8 means a

(a)

(b)

Figure 2–15   Molar amounts, one mole, of (a, back row) bromine, aluminum, mercury, copper (front row) sulfur, zinc, and iron. (Charles Steele.) (b) White sodium chloride, orange potassium dichromate, red cobalt chloride hexahydrate, green nickel chloride hexahydrate, and blue copper sulfate pentahydrate. (Richard Roese.)

sulfur molecule is composed of eight atoms. The noble gases, such as He, have monatomic molecules (meaning they have one atom).

   When unlike atoms combine, as in the case of water (H_2O) or sulfuric acid (H_2SO_4), the formulas tell what atoms and how many of each are present in a molecule of the compound. For example, an H_2SO_4 molecule is composed of two hydrogen atoms, one sulfur atom, and four oxygen atoms. A **formula** of a molecular substance can stand not only for the substance itself but also for one molecule of the substance or for a mole of such molecules, depending on the context.

H_2SO_4: 2 atoms of hydrogen, 1 atom of sulfur, 4 atoms of oxygen

Some compounds are composed of charged atoms or groups of atoms held together by the attraction between positive and negative ions. Such compounds contain no molecules at all. See Chapter 5.

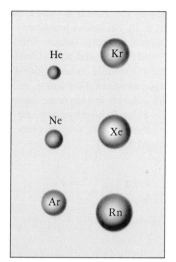

Figure 2–16　Atomic models showing relative sizes for a family of the elements, the noble gases (helium, neon, argon, krypton, xenon, and radon).

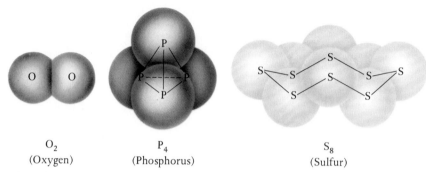

O_2
(Oxygen)

P_4
(Phosphorus)

S_8
(Sulfur)

Figure 2–17　Elemental atoms tend to group into stable arrangements. A molecule is the smallest particle of an element or a compound that can have a stable existence in the close presence of like molecules.

When elements or compounds undergo a chemical change, the formulas, arranged in the form of a **chemical equation,** can present the information in a very concise fashion. For example, carbon (C) can react with oxygen (O_2) to form carbon monoxide (CO). Like most solid elements, carbon is written as though it had one atom per molecule; oxygen exists as diatomic (two-atom) molecules, and carbon monoxide molecules contain two atoms, one each of carbon and oxygen. Furthermore, one oxygen molecule will combine with two carbon atoms to form two carbon monoxide molecules. All of this information is contained in the equation

$$2\,C + O_2 \longrightarrow 2\,CO$$

The arrow ($\rightarrow$) is often read "yields"; the equation then states the following information:

1. Carbon plus oxygen yields carbon monoxide.
2. Two atoms of carbon plus one diatomic molecule of oxygen yield two molecules of carbon monoxide.
3. Two moles of carbon atoms plus one mole of diatomic oxygen molecules yield two moles of carbon monoxide molecules.

The number written before a formula, the **coefficient,** gives the amount of the substance involved, and the **subscript** is a part of the composition of the pure substance itself. Changing the coefficient only changes the amount of

Chemical equations summarize information on chemical reactions in a concise fashion.

Figure 2–18　Since a model cannot be the reality, a model maker is likely to emphasize certain aspects of that reality in a particular model. The earlier ball-and-stick model for the water molecule gives the geometry of atomic attachment, but the space-filling model developed later depicts the overall geometry as well as the relative sizes of and the distances between the atoms. (Charles Steele.)

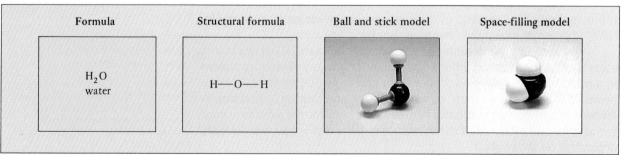

Formula	Structural formula	Ball and stick model	Space-filling model
H_2O water	H—O—H		

the element or compound involved, whereas changing the subscript would necessarily involve changing from one substance to another. For example, 2 CO means either two molecules of carbon monoxide or two moles of these molecules, whereas CO_2 would mean a molecule or a mole of carbon dioxide, a very different substance.

STANDARDS AND MEASUREMENT

The establishment of natural facts and laws is obviously dependent on accurate observations and measurements. Although one can be as accurate in one language as in another or in one system of units as in another, there has been an effort since the French Revolution to have all scientists embrace one simple system of measure. The hope was and is to facilitate communication in science. The metric system, which was born of this effort, has two advantages. First, it is easy to convert from one unit to another since subunits and multiple units differ only by factors of ten. Consequently, to change millimeters to meters, one has only to shift the decimal three places to the left. Compare the difficulty of the decimal shift to the problem of changing inches to miles. The second advantage is that standards for measurements are defined by reproducible phenomena of nature rather than by the length of the king's foot or some other such changeable standard. For example, length is now defined in terms of the length of a particular wavelength of light, a number we believe to be invariant.

The International System of Units, abbreviated SI (from the French *Système International*), was adopted by the International Bureau of Weights and Measures in 1960. It, by international agreement, sets the standards (units) against which essentially all scientific measurements are made. The SI system is an extension of the metric system, retaining its ease of unit conversion but doing a better job in the definition of units based on physical phenomena.

Seven fundamental units are required to describe what is now known about the universe.

Physical Quantity	Name of Unit	Symbol
1. Length	meter	m
2. Mass	kilogram	kg
3. Time	second	s
4. Thermodynamic temperature	kelvin	K
5. Luminous intensity	candela	cd
6. Electric current	ampere	A
7. Amount of a substance	mole	mol

The international spelling of the unit of length is metre.

Other necessary units are derived from these seven. For example, volume is defined in terms of cubic length (cubic centimeters [cm^3]).

In this book, we shall frequently employ five units and a sixth one to a lesser degree. Along with a suitable common equivalent, the units are:

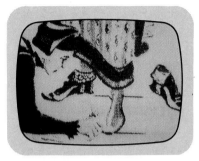

Figure 2–19   The length of the King's foot, while serving vanity well, was short lived as a standard for the measurement of length if there is a desire for accuracy and reproducibility. (*The World of Chemistry,* Program 3.)

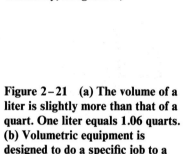

Figure 2–20   One inch is equal to 2.54 centimeters (cm). (**Charles Steele.**)

Name of Unit	Symbol	Common Equivalent
1. meter	m	39.4 inches
2. liter	L	1.06 quarts
3. gram	g	0.0352 ounce
4. degrees Celsius	°C	Water boils at 100°C and freezes at 0°C.
5. calorie	cal	Energy required to heat 1 g of water 1°C
6. mole	mol	Number of atoms in 12 g of carbon-12 isotope

Figure 2–21   (a) The volume of a liter is slightly more than that of a quart. One liter equals 1.06 quarts. (b) Volumetric equipment is designed to do a specific job to a defined level of precision in the easiest possible way; hence, such equipment takes many forms. (**Charles D. Winters.**)

Figures 2–19 through 2–23 along with Table 2–2 will aid the student in developing a "feel" for these units. Appendix A gives further information about the SI system and precise definitions for some of the units commonly used in elementary chemistry courses.

Although it is not the purpose of this text to require students to master a number of quantitative calculations, the authors do hope the reader will

(a)

(b)

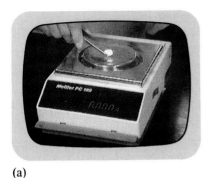 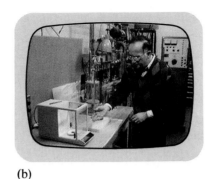

(a)   (b)

Figure 2–22 (a) A sample is weighed on a laboratory balance. (b) Dr. Hoffmann at an analytical balance. (*The World of Chemistry,* Program 3.)

realize that the SI system requires no higher level of thinking than the English system and that problem solving is usually easier in the newer system. Consider the following examples:

EXAMPLE 1

1. English: How many feet are in 0.5 miles?

    ? feet = 0.5 miles × 5280 ft/mile = 2640 ft

2. SI: How many meters are in 2 km?
   The prefix kilo- means 1000 times, so

    ? m = 2 km × 1000 m/km = 2000 m

   It is hardly worth the trouble to write anything down in this solution. One just thinks 2000 meters as one thinks one dollar for ten dimes.

EXAMPLE 2

1. English: How many ounces are in 1.50 gal?
    You might remember there are 32 oz/qt and 4 qt/gal. Your solution then would be:

    ? ounces = 1.50 gal × 4 qt/gal × 32 oz/qt = 192 oz

TABLE 2–2 Recipe for Perfect Brownies in English and Metric Measurements

Ingredient	English	Metric (SI)
Unsweetened chocolate squares	2 oz	60 g
Butter or margarine	½ c	120 mL
Sugar	1 c	240 mL
Eggs	2	2
Vanilla	1 tsp	5 mL
Sifted enriched flour	½ c	120 mL
Chopped walnuts	½ c	120 mL
Oven	325°F	163°C
Pan	8 × 8 × 2 in.	20 × 20 × 5 cm

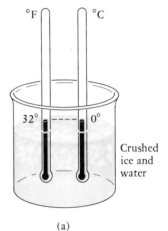

(a)

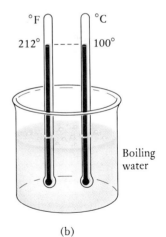

(b)

Figure 2–23 Temperature scales are defined in terms of the expansion of common materials such as mercury and in terms of fixed reference points such as the changes of state of water (a and b) and other common materials. It is only a matter of preference and convenience whether one scale or another is used.

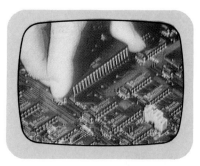

Figure 2–24 Units of measurement are selected for convenience: the millimeter is suitable for the distances between the small parts of this electronic circuit board, and the kilometer is better for measuring intercontinental distances. The SI system simply adds the ease of using multiples of 10 (*The World of Chemistry,* Program 5.)

2. SI: How many milligrams are in a coin that weighs 5 g?

The prefix milli- means one-thousandth of, so there are 1000 mg in a gram. Five grams then would be 5000 mg.

EXAMPLE 3

A typical piece of white bread contains 70 dietary calories (Cal). One dietary calorie is equal to 1000 small (scientific) calories (cal). How many small calories are in a typical piece of white bread? 70,000 small calories.

(Note: You can see why the dietitians like the larger unit; with it, they can use smaller numbers in their notations.)

ACCURACY AND PRECISION

A measured value is accurate if it is in agreement with the true value. The level of **accuracy** increases as the difference between the measured value and the accepted value becomes smaller. A group of measured values are precisely measured if they are in close agreement with each other; the **precision** of the measurements increases as the scatter of the measurements decreases.

It is quite possible to be precise but inaccurate, as one would expect if measuring with a faulty instrument. For example, a stopwatch designed to measure to the nearest thousandth of a second would always measure times too large if the watch is regularly running too fast even though it is reproducing its measurements precisely. The results from three dart players are presented in Figure 2–25 to illustrate accuracy and precision. Player **a** is precise but off the mark. Player **b** has relatively poor precision but, if the average of the throws could be used as the result, the accuracy would be better than the precision. Player **c** has both good accuracy and precision. The goal in scientific measurement is to be precise and accurate.

SELF-TEST 2–C

1. Consider the chemical equation: $CH_4 + 2 O_2 \rightarrow CO_2 + 2 H_2O$. Explain what is meant by the symbols:
   a. O _____
   b. $2 O_2$ _____
   c. CH_4 _____
   d. $\rightarrow$ _____
   e. H_2O _____
   f. $2 H_2O$ _____
2. Name three concepts that a chemical symbol can represent.
3. A chemical formula gives what two pieces of information?
4. Which is proper to change when you balance a chemical equation: a coefficient or a subscript?
5. The SI system is more accurate than the English system of weights and measures. True ( ) False ( )
6. How many fundamental physical units are required to express our present knowledge of the universe?

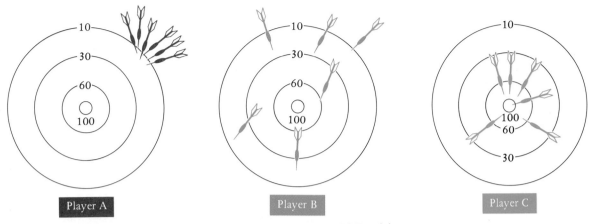

Figure 2–25 **(a) Precision may be excellent with poor accuracy. (b) Precision may be poor with relatively good or relatively poor accuracy. (c) Precision may be good with good accuracy.**

MATCHING SET

_____ **1.** Produces a new type of matter

_____ **2.** Air

_____ **3.** Unchanged by further purification

_____ **4.** Used to separate a solid from a liquid

_____ **5.** Cannot be reduced to simpler substances

_____ **6.** SI system

_____ **7.** Symbol for iron

_____ **8.** Mole of atoms

_____ **9.** Molecule containing three oxygen atoms

_____ **10.** Carbon monoxide

_____ **11.** 100 cm

_____ **12.** Element symbol

_____ **13.** Theoretical definition

_____ **14.** Basic research

_____ **15.** Plasma

_____ **16.** Solution

a. Filtration
b. Chemical change
c. Element
d. Properties of pure substance
e. Mixture
f. Uses multiples of ten
g. O_3
h. CO
i. Fe
j. 1 m
k. 6.022×10^{23} atoms
l. Based on idea of atoms
m. Information oriented
n. Made of charged particles
o. The element, an atom or a mole
p. Homogeneous mixture

QUESTIONS

1. Name as many materials as you can that you have used during the past day that were not chemically changed by artificial means.
2. Identify the following as physical or chemical changes. Justify your answer in terms of the operational definitions for these types of changes.
   a. Formation of snowflakes
   b. Rusting of a piece of iron
   c. Ripening of fruit
   d. Fashioning a table leg from a piece of wood.
   e. Fermenting grapes
   f. Boiling a potato
3. If physics is the study of matter and energy, why can it be said that the study of chemistry is a special case within the general study of physics?
4. Would it be possible for two pure substances to have exactly the same set of properties? Give reasons for your answer.
5. Chemical changes can be both useful and destructive to humanity's purposes. Cite a few examples of each kind of change from your own experience. Also give evidence from observation that each is indeed a chemical change and not a physical change.
6. For many years water was thought to be an element. Explain how this could be in keeping with the operational definitions of elements and compounds.
7. Classify each of the following as a physical property or a chemical property. Justify your answer in terms of operational definitions.
   a. Density
   b. Melting temperature
   c. Substance that decomposes into two elements upon heating
   d. Electric conductivity
   e. A substance that does not react with sulfur
   f. Ignition temperature of a piece of paper
8. Classify each of the following as an element, a compound, or a mixture. Justify each answer.
   a. Mercury
   b. Milk
   c. Pure water
   d. A tree
   e. Ink
   f. Iced tea
   g. Pure ice
   h. Carbon
9. Which of the materials listed in Question 8 can be pure substances?
10. Explain how the operational definition of a pure substance allows for the possibility that it is not actually pure.

11. Why do theoretical definitions come after operational definitions in a particular concept?
12. Is it possible for the properties of iron to change? What about the properties of steel? Explain your answer.
13. Suggest a method for purifying water slightly contaminated with a dissolved solid.
14. Define a solution as a particular kind of mixture.
15. Given the following sentence, write a chemical reaction using chemical symbols that conveys the same information. "One nitrogen molecule containing two nitrogen atoms per molecule reacts with three hydrogen molecules, each containing two hydrogen atoms, to produce two ammonia molecules, each containing one nitrogen and three hydrogen atoms."
16. Aspirin is a pure substance, a compound of carbon, hydrogen, and oxygen. If two manufacturers produce equally pure aspirin samples, what can be said of the relative worth of the two products?
17. Is it possible to have a mixture of two elements and also to have a compound of the same two elements? Explain. Can you think of an example?
18. Name four forms of energy.
19. What is the difference between a chemical change and a nuclear change?
20. Describe in words the chemical process that is summarized in the following equation:

$$2 \, Na + Cl_2 \longrightarrow 2 \, NaCl$$

21. How many *atoms* are present in each of the following:
    a. One mole of He
    b. One mole of Cl_2
    c. One mole of O_3
22. Would you think that tea in tea bags is a pure substance? Use the process of making tea to make an argument for your answer. How would your argument apply to instant tea?
23. Which contains more atoms?
    a. One mole of water, H_2O
    b. One mole of hydrogen, H_2
    c. One mole of oxygen, O_2
24. Find as many pure substances as you can in a kitchen.
25. The number 12 is to a dozen, and 144 is to a gross, as 6.022×10^{23} is to a(n) _____.
26. If you had a mole of elephants, how many moles of elephant ears would you have? A mole of O_3 molecules contains how many moles of oxygen atoms?
27. How tall are you in meters?
28. Which is colder: $0°C$ or $0°F$?

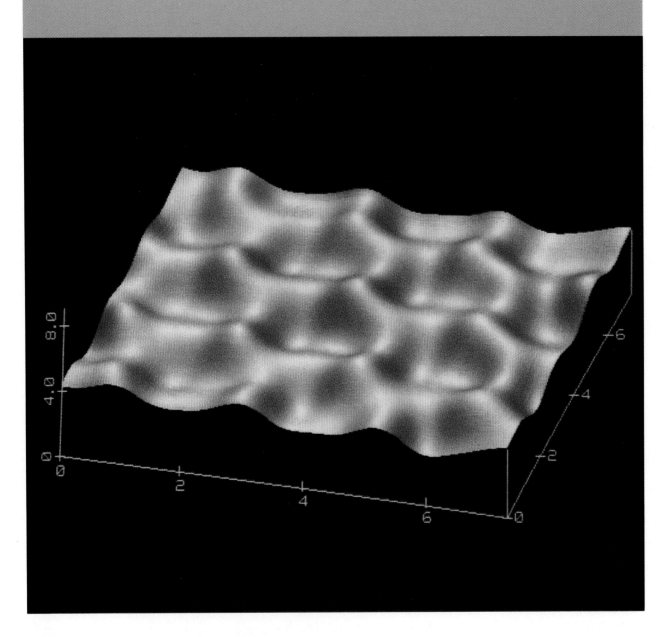

3

Atoms

The position of individual atoms on the surface of a graphite crystal is shown in this Scanning Tunnelling Microscope (STM) image. (A & I Color.)

As you focus on a dot over an "i," can you visualize the thousands and thousands of individual, very small atoms in the dot? Trying to fathom the minuteness of the atom is as deeply challenging to the human mind as trying to fathom the wholeness of the universe. One lures the mind to unseen smallness; the other to unseen largeness.

Beginning almost 200 years ago, scientists began to probe the atom. We now understand something of what composes atoms, and as of just a few years ago, we can now actually see the general outline of individual atoms with a special microscope called the scanning tunneling microscope (STM) (see Fig. 3–20).

The nature of atoms is the subject of this chapter. Assembling the pieces of this puzzle has taken more than 2000 years of human endeavor, but much of this work has been done in the past 190 years. What difference does it make what is inside atoms, or even if they exist? Simply, it is thrilling to discover what makes nature work, and this has been a strong driving force for many individuals as they have studied the atom.

As we shall find out, understanding atoms helps us understand such intriguing phenomena as bonding, chemical reactivity, the nature of light, and nuclear reactions. Perhaps the most powerful outcome of a knowledge about atoms is the ability to predict accurately the properties of matter that lead to new medicines to combat diseases, new fibers and plastics, building materials, and other things that make our lives enjoyable and productive.

THE GREEK INFLUENCE

Because the writings of Leucippus and Democritus have been destroyed, we know about their ideas only from recorded opposition to atoms and from a lengthy poem (55 B.C.) by the Roman poet, Lucretius.

The ancient Greeks recorded the first theory of atoms. Leucippus and his student, Democritus (460–370 B.C.), argued for the concept of atoms. Democritus used the word **atom** (literally means "uncuttable") to describe the ultimate particles of matter, particles that could not be divided further. He reasoned that in the division of a piece of matter, such as gold, into smaller and smaller pieces, one would ultimately arrive at a tiny particle of gold that could not be further divided and still retain the properties of gold. The atoms that Democritus envisioned representing different substances were all made of the same basic material. His atoms differed only in shape and size.

Democritus used his concept of atoms to explain the properties of substances. For example, the high density and softness of lead could be caused by lead atoms packed very closely together like marbles in a box and moving easily one over another. Iron was known to be a less dense metal that is quite hard. Democritus argued that the properties of iron resulted from atoms shaped like corkscrews, atoms that would entangle in a rigid but relatively lightweight structure. Although his concept of the atom was limited, Democritus did explain in a simple way some well-known phenomena, such as the drying of clothes, how moisture appears on the outside of a vessel of cold water, how an odor moves through a room, and how crystals grow from a solution. He imagined the scattering or collecting of atoms as needed to explain the events he saw. All atomic theory has been built on the assumption of Leucippus and Democritus: atoms are the cause of the phenomena that we can see.

Plato (427–347 B.C.) and Aristotle (384–322 B.C.) led the arguments against the atom by asking to be shown atoms. They also argued that the idea of atoms was a challenge to God. If atoms could be used to explain nature, there would be no need for God. For centuries most of those in the mainstream of enlightened thought rejected or ignored the atoms of Democritus.

Ideas about atoms drifted in and out of philosophical discussions for about 2200 years without generally affecting how nature was understood. Galileo (1564–1642) reasoned that the appearance of a new substance through chemical change involved a rearrangement of parts too small to be seen. Francis Bacon (1561–1626) speculated that heat might be a form of motion by very small particles. Robert Boyle (1627–1691) and Isaac Newton (1642–1727) used atomic concepts to interpret physical phenomena.

However, it was John Dalton (1766–1844), an English schoolteacher, who forcefully revived the idea of the atom. By Dalton's time experimental results had gained a position of greater respect than authoritative opinions. More clearly than any before him, Dalton was able to explain general observations, experimental results, and laws relative to the composition of matter. Dalton was particularly influenced by the experiments of two Frenchmen, Antoine Lavoisier (1743–1794) and Joseph Louis Proust (1754–1826). We shall look at the major contributions of these two experimentalists before we examine Dalton's theory.

The Parthenon is contemporary with early ideas about atoms. (*The World of Chemistry,* Program 6.)

Although the idea was proposed three and a half centuries earlier by Roger Bacon, it was not until 1620 that Francis Bacon wrote his book, *New Organon,* which put experimental science in the most refined and scholarly terms and made it possible for other scholars to accept it.

ANTOINE-LAURENT LAVOISIER: THE LAW OF CONSERVATION OF MATTER IN CHEMICAL CHANGE

There are many reasons why Antoine-Laurent Lavoisier has been acclaimed the father of chemistry. He clarified the confusion over the cause of burning. He wrote an important textbook of chemistry, *Elementary Treatise on Chemistry.* He was the first to use systematic names for the elements and a few of their compounds. Although he made still other contributions, his most notable achievement was to show the importance of very accurate weight measurements of chemical changes. His work began the process of establishing chemistry as a **quantitative** science.

Lavoisier weighed the chemicals in such changes as the decomposition of mercury oxide by heat into mercury and oxygen.

$$2\,HgO \longrightarrow 2\,Hg + O_2$$
Mercury Oxide   Mercury   Oxygen

Very accurate measurements showed that the total weight of all the chemicals involved remained constant during the course of the chemical change. Similar measurements on many other chemical reactions led Lavoisier to the summarizing statement now known as the **law of conservation of matter:**

Matter is neither lost nor gained during a chemical reaction.

Lavoisier used quantitative analysis (measured quantities) early in the development of modern chemistry. Today quantitative analysis is used routinely in chemistry.

(a)

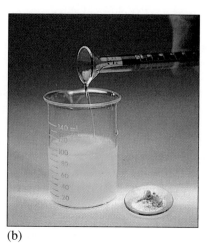
(b)

(c)

Figure 3–1   (a) A weighed sample containing a barium salt is dissolved in water. (b) Excess sulfuric acid is added to react with the barium and form barium sulfate, which precipitates. (c) The barium sulfate is collected on a filter paper. After drying and weighing the barium sulfate, the amount of barium in the original barium sample can be calculated. (Charles D. Winters.)

In other words, if one weighed all of the products of a chemical reaction—solids, liquids, and gases—the total would be the same as the weight of the reactants. Substances can be destroyed or created in a chemical reaction, but matter cannot. In an atomic view, a chemical reaction was just a recombination of atoms. As a further example of the law of conservation of matter, consider Figure 3–1.

The Personal Side

With all of his success, Lavoisier had his problems and disappointments. His highest goal, that of discovering a new element, was never achieved. He lost some of the esteem of his colleagues when he was accused of saying the work of someone else was his own. In 1768, he invested half a million francs in a private firm retained by the French government to collect taxes. He used the earnings (about 100,000 francs a year) to support his research. Although Lavoisier was not actively engaged in tax collecting, he was brought to trial as a "tax-farmer" during the French Revolution. Lavoisier, along with his father-in-law and other tax-farmers, was guillotined on May 8, 1794, just two months before the end of the revolution.

A. L. Lavoisier and wife by David. (Metropolitan Museum of Art.)

JOSEPH LOUIS PROUST: THE LAW OF CONSTANT COMPOSITION

Following the lead of Lavoisier, several chemists investigated the quantitative aspects of compound formation. One such study, made by Proust in 1799, involved copper carbonate. Proust discovered that, regardless of how copper carbonate was prepared in the laboratory or how it was isolated from nature, it always contained five parts of copper, four parts of oxygen, and one part of carbon by weight. His careful analyses of this and other compounds led to the belief that a given compound has an unvarying composition. These and similar discoveries are summarized by the **law of constant composition:**

Compounds have constant composition, whereas mixtures may have variable composition.

> **In a compound, the constituent elements are always present in a definite proportion by weight.**

Pure water, a compound, is always made up of 11.2% hydrogen and 88.8% oxygen by weight. Pure table sugar, another compound, always contains 42.11% carbon by weight. Contrast these with 14-carat gold, a mixture that should be at least 58% gold, from 14% to 28% copper, and 4% to 28% silver by weight. This mixture can vary in composition and still be properly called 14-carat gold, but a compound that is not 11.2% hydrogen and 88.8% oxygen is not water.

The Personal Side

Proust's generalization has been verified many times for many compounds since its formulation, but its acceptance was delayed by controversy. Comte Claude Louis Berthollet (1748–1822), an eminent French chemist and physician, believed and strongly argued that the nature of the final product was determined by the amount of reacting materials one had at the beginning of the reaction. The running controversy between Proust and Berthollet reached major proportions, but more careful measurements supported Proust. Proust showed that Berthollet had made inaccurate analyses and had purified his compounds insufficiently—two great errors in chemistry.

Unlike Lavoisier, Proust saved his head during the French Revolution by fleeing to Spain, where he lived in Madrid and worked as a chemist under the sponsorship of Charles IV, King of Spain. When Napoleon's army ousted Charles IV, Proust's laboratory was looted and his work came to an end. Later, Proust returned to his homeland, where he lived out his life in retirement.

JOHN DALTON: THE LAW OF MULTIPLE PROPORTIONS

John Dalton made a quantitative study of different compounds made from the same elements. Such compounds differed in composition from each other, but each obeyed the law of constant composition. Examples of this concept are the compounds carbon monoxide, a poisonous gas, and carbon dioxide, a product of respiration. Both compounds contain only carbon and oxygen. Carbon monoxide is made up of carbon and oxygen in proportions

The Personal Side

Whereas Lavoisier is considered the father of chemical measurement, John Dalton is considered the father of chemical theory. Dalton, a gentle man and a devout Quaker, gained acclaim because of his work. He made careful measurements, kept detailed records of his research, and expressed them convincingly in his writings. However, he was a very poor speaker and was not well received as a lecturer. When Dalton was 66 years old, some of his admirers sought to present him to King William IV. Dalton resisted because he would not wear the court dress. Since he had a doctor's degree from Oxford University, the scarlet robes of Oxford were deemed suitable, but a Quaker could not wear scarlet. Dalton, being color-blind, saw scarlet as gray, so he was presented in scarlet to the court but in gray to himself. This remarkable man was, in fact, the first to describe color blindness. He began teaching in a Quaker school when only 12 years old, discovered a basic law of physics, the law of partial pressure of gases, and helped found the British Association for the Advancement of Science. He kept over 200,000 notes on meteorology. Despite his accomplishments he shunned glory and maintained he could never find time for marriage.

by weight of 3 to 4. Carbon dioxide is made up of carbon and oxygen in proportions by weight of 3 to 8. Note that for equal amounts of carbon (three parts), the ratio of oxygen in the two compounds is 8 to 4, or 2 to 1.

Methane is the main component of natural gas. Ethylene is the only component of polyethylene.

In 1803, after analyzing compounds of carbon and hydrogen such as methane (in which the ratio of carbon to hydrogen is 3 to 1 by weight) and ethylene (in which the ratio of carbon to hydrogen is 6 to 1 by weight) and compounds of nitrogen and oxygen, Dalton first clearly enunciated the **law of multiple proportions:**

> **In the formation of two or more compounds from the same elements, the weights of one element that combine with a fixed weight of a second element are in a ratio of small whole numbers (integers) such as 2 to 1, 3 to 1, 3 to 2, or 4 to 3.**

DALTON'S ATOMIC THEORY

Why do the laws of conservation of matter, constant composition, and multiple proportions exist? How can they be explained? John Dalton employed the idea of atoms and endowed them with properties that enabled him to explain these chemical laws (Fig. 3–2).

The major points of Dalton's theory, presented in modernized statements, are:

1. Matter is composed of indestructible* particles called atoms.
2. All atoms of a given element have the same properties such as size, shape, and weight,† which differ from the properties of atoms of other elements.

---

* Radioactive atoms are self-destructive. Dalton had no knowledge of this phenomenon.
† We now know that all of the atoms of the same element do not necessarily have the same weight. The idea of isotopes is introduced later in this chapter.

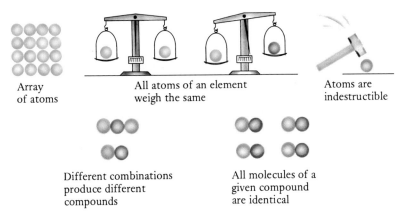

Figure 3–2 Features of Dalton's atomic theory. Isotopes have been ignored.

Array
of atoms

All atoms of an element
weigh the same

Atoms are
indestructible

Different combinations
produce different
compounds

All molecules of a
given compound
are identical

3. Elements and compounds are composed of definite arrangements of atoms, and chemical change occurs when the atomic arrays are rearranged.

Dalton's theory was successful in explaining the three laws of chemical composition and reaction. See Figure 3–3.

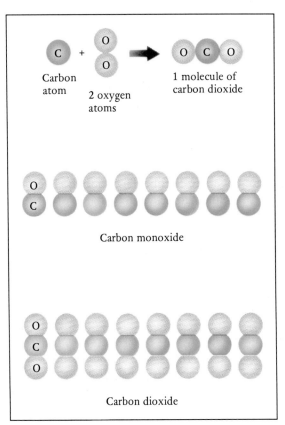

Figure 3–3 John Dalton's explanation of three laws of chemistry in terms of atoms.

C + O O → O C O

Carbon
atom

2 oxygen
atoms

1 molecule of
carbon dioxide

O
C

Carbon monoxide

O
C
O

Carbon dioxide

THE BEGINNING OF THE CONCEPT OF ATOMIC WEIGHTS

John Dalton's idea about unique atomic weights for the atoms of the different elements naturally generated interest in searching for the atomic weight characteristic of each element. It was not until after 1860 that chemists developed a consistent set of atomic weights, although several notable attempts were made before that, including an early attempt by Dalton himself. In September 1860, many of the most brilliant minds in chemistry met in Karlsruhe, Germany, to discuss the inconsistencies in the atomic weights proposed at that time. There were differences of opinion on whether the formula of water was HO or H_2O, whether hydrogen gas was H_2 or H, and whether oxygen gas was O_2 or O. Water is 88.8% oxygen and 11.2% hydrogen by weight—a firmly established experimental fact at that time. If water is HO, as Dalton argued (based on his belief that the simplest formula is likely to be the correct one), then the weight of an oxygen atom should be about eight times that of a hydrogen atom:

$$\frac{\text{weight of an oxygen atom}}{\text{weight of a hydrogen atom}} = \frac{88.8}{11.2} = \frac{7.9}{1}$$

If the formula for water is H_2O, as the scientist Amedeo Avogadro (1776–1856) had proposed in 1811, then one oxygen atom is 88.8% of the molecule but two hydrogen atoms are 11.2%. Each hydrogen atom would be ½(11.2%), or 5.6%. With the formula H_2O, then, an oxygen atom would be about 16 times heavier than a hydrogen atom:

$$\frac{\text{weight of an oxygen atom}}{\text{weight of a hydrogen atom}} = \frac{88.8}{5.6} = \frac{15.9}{1}$$

Near the end of the meeting at Karlsruhe, Stanislao Cannizzaro (1826–1910) argued for the ideas of Avogadro. In spite of his arguments, which later proved to be correct, the confusion about atomic weights was not resolved during the conference. Several years later, chemists finally accepted H_2O as the formula of water, and a consistent set of atomic weights was agreed on and used.

The fact that an oxygen atom is about 16 times heavier than a hydrogen atom does not tell us the weight of either atom. These are relative weights in the same way that a grapefruit may weigh twice as much as an orange. This information gives neither the weight of the grapefruit nor that of the orange. However, if a specific number is *assigned* as the weight of any particular atom, this fixes the numbers assigned to the weights of all other atoms. The standard for comparison of relative atomic weights was for many years the weight of the oxygen atom, which was taken as 16.0000 atomic weight units. This allowed the lightest atom, hydrogen, to have an atomic weight of 1.008, or approximately 1.

The modern set of atomic weights (inside the front cover) is an outgrowth of the set of weights begun in the 1860s. The present atomic weight scale, adopted by scientists worldwide in 1961, is based on assigning the weight of a particular kind of carbon atom, the carbon-12 atom, as exactly 12 atomic weight units. On this scale, an atom of magnesium (Mg) with an

Stanislao Cannizzaro (1826–1910), Italian chemist whose work resulted in the clarification of the atomic weight scale.

Atomic weights are relative weights with the carbon-12 atom being the standard.

atomic weight of about 24 has twice the weight of a carbon-12 atom. An atom of titanium (Ti) with an atomic weight of 48 has four times the weight of a carbon-12 atom.*

ATOMS ARE DIVISIBLE — DALTON AND THE GREEKS WERE WRONG

Dalton's concept of the indivisibility of atoms was severely challenged by the subsequent discoveries of radioactivity and cathode rays and was even in conflict with some previously known electrical phenomena such as static electric charges.

Electric charge was first observed and recorded by the ancient Egyptians, who noted that amber, when rubbed, attracted light objects. A bolt of lightning, a spark between a comb and hair in dry weather, and a shock on touching a doorknob are all results of the discharge of a buildup of electric charge.

The two types of electric charge had been discovered by the time of Benjamin Franklin (1706–1790). He named them positive (+) and negative (−) because they appear as opposites, in that they neutralize each other. The existence and nature of the two kinds of charge, and their effects on each other, can be shown with a simple electroscope (Fig. 3–4). When a hard rubber rod is rubbed vigorously with silk and allowed to touch the lightweight balls, the balls spring apart immediately. The touching allowed the rod and the balls to share the same type of charge (positive). If the rod is then brought near one of the balls, the ball moves away from the rod. This movement indicates that *like charges repel.*

If the same rod is now rubbed vigorously with wool and brought near the charged balls, they move toward the rod. The opposite type of charge is now on the rod. The generalization is:

Unlike charges attract, and like charges repel.

The discovery of natural radioactivity, a spontaneous process in which some natural materials give off very penetrating radiations, indicated that atoms must have some kind of internal structure. Henri Becquerel (1852–1908) discovered this property in natural uranium and radium ores in 1896. His student, Marie Curie (1867–1934), isolated the radioactive element radium and some of its pure compounds. It turns out that radioactivity is characteristic of the elements, not the compounds, and that about 25 elements are naturally radioactive.

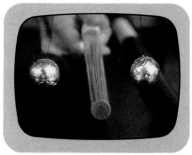

(a)

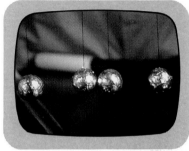

(b)

Figure 3–4 (a) Like-charged aluminum-foil-covered balls repel each other. (b) Unlike-charged balls attract. (*The World of Chemistry,* Program 6).

---

* What we are talking about here is really atomic mass. Atomic weight, an uncorrected misnomer from the past, persists today, for example, in the "Tables of Atomic Weights" published in most chemistry textbooks. If they were really atomic weights, they would change value wherever the force of gravity changes on earth (less at the equator, more at the poles). Instead, we have only one table for the whole world, which means atomic weights are really atomic masses. In this text, we shall use both terms: *atomic weight,* because it is a practice of chemistry and a term possibly more familiar to students, and *atomic mass,* where it seems necessary to clarify the thought.

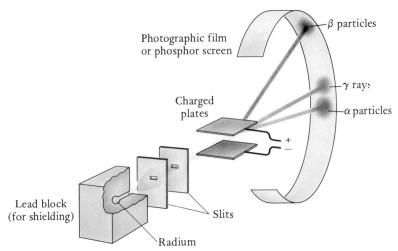

Figure 3–5   Separation of alpha, beta, and gamma rays by an electrical field.

Radioactive elements commonly emit alpha, beta, and gamma rays, as shown in Figure 3–5. Alpha and beta rays are composed of particles with charges and masses, but gamma rays have no detectable mass and are more like light. Alpha particles have a mass of 4 on the carbon-12 atomic weight scale, positive charge, and low penetrating power (they will not penetrate skin, for example). In the arrangement shown in Figure 3–5, alpha particles are attracted toward the negatively charged plate. Beta particles have a mass of 0.0005 on the carbon-12 atomic weight scale, negative charge (they are attracted toward the positive plate), and enough penetrating power to go through kitchen-strength aluminum foil. Gamma rays are a type of electromagnetic energy like light and X rays, but more penetrating. Gamma rays can penetrate a considerable thickness of aluminum and even thin sheets of lead. They are not deflected at all by charged plates.

Cathode rays, as we shall see later, are similar to beta rays in that both are composed of negatively charged particles, with identical charges and masses.

The discoveries of natural radiation, cathode rays, and electric charge are evidence that atoms can be divided and some of them divide spontaneously. The smallest atom is 1836 times more massive than the beta or cathode-ray particle. Therefore, beta (and cathode-ray) particles must be

TABLE 3–1   Summary of Properties of Alpha Particles, Beta Particles, and Gamma Rays

	Charge	Relative Mass	Symbols
Alpha particle	Positive ($+2$)	4	α, $^4_2\alpha$, $^4_2\mathrm{He}$
Beta particle	Negative (-1)	0.0005	β, $^0_{-1}\beta$
Gamma ray	Neutral (0)	0	γ, $^0_0\gamma$

subatomic in origin. The properties of the three fundamental subatomic particles are summarized in Table 3–1.

SELF-TEST 3-A

1. Two Greek philosophers who were influential in advocating the concept of atoms were _____ and _____ .
2. The Greek approach to the "discovery" of atoms can best be described as:
   a. Experimentation
   b. Philosophy (use of logic)
   c. Direct observation of atoms
   d. Consistent explanation of well-known, established laws of nature
   e. Deductive reasoning
3. The law of conservation of matter states that matter is neither lost nor _____ in a _____ reaction.
4. The law of multiple proportions explains the existence of compounds like _____ and _____ .
5. a. Assume that you are a chemist of many years ago. Your field of study is compounds composed of nitrogen and oxygen only. You know about several. One contains 16 g of oxygen for every 14 g of nitrogen; another contains 32 g of oxygen for every 14 g of nitrogen. Your assistant discovers what he claims is a new compound of nitrogen and oxygen. On analysis, the compound is found to contain 8 g of O for every 14 g of N. Has your assistant discovered a new compound, or is it one of the others?
   b. What is the ratio by weight of O in these compounds for a given weight of N?
   c. What fundamental law of chemistry is illustrated by a comparison of the compounds?
6. According to Dalton's atomic theory, what happens to atoms during a chemical change?
   a. Atoms are made into new and different kinds of atoms.
   b. Atoms are lost.
   c. Atoms are gained.
   d. Atoms are recombined into different arrangements.
7. According to Dalton's atomic theory, a compound has a definite percentage by weight of each element because
   a. All atoms of a given element weigh _____ .
   b. All molecules of a given compound contain a definite number and kind of _____ .
8. Like charges _____ ; unlike charges _____ .
9. The three types of radiation from a radioactive element such as radium are _____ , _____ , and _____ , of which _____ pass through an electric field without being deflected.
10. The closest display of real "pictures" of individual atoms is provided by the _____ _____ microscope.

THE ELECTRON—THE FIRST SUBATOMIC PARTICLE DISCOVERED

The first ideas about electrons came from experiments with cathode-ray tubes. A forerunner of neon signs, fluorescent lights, and TV picture tubes, a typical cathode-ray tube is a partially evacuated glass tube with a piece of metal sealed in each end (Fig. 3–6). The pieces of metal are called electrodes; the one given a negative charge is the **cathode,** and the one given a positive charge is the **anode.**

If a sufficiently high voltage is applied to the electrodes, an electric discharge can be created between them. This discharge appears to be a stream of particles emanating from the cathode. This **cathode ray** will cause gases and fluorescent materials to glow and will heat metal objects in its path to red heat. Cathode rays travel in straight lines and cast sharp shadows. Unlike light, however, cathode rays are attracted to a positively charged plate. This led to the conclusion that cathode rays are a stream of negatively charged particles.

Careful microscopic study of a screen that emits light when struck by cathode rays shows that the light is emitted in tiny, random flashes. Thus, not only are cathode rays negatively charged, but they also appear to be composed of particles, each one of which produces a flash of light upon collision with the material of the screen. The cathode-ray particles became known as **electrons.**

Cathode rays are streams of the negatively charged particles called *electrons.*

Figure 3–6 Deflection of a cathode ray by an electric field and by a magnetic field. When an external electric field is applied, the cathode ray is deflected toward the positive pole. When a magnetic field is applied, the cathode ray is deflected from its normal straight path into a curved path.

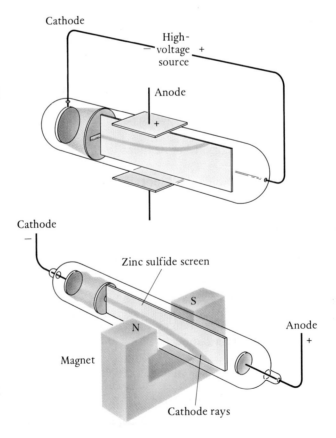

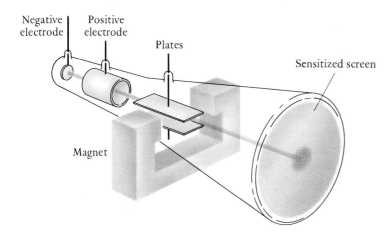

Negative electrode
Positive electrode
Plates
Sensitized screen
Magnet

Figure 3–7 J. J. Thomson experiment. Electric field, applied by plates, and magnetic field, applied by magnet, cancel each other's effects to allow cathode ray (electron beam) to travel in straight line.

The charge and mass of an electron were determined by a combination of experiments by Sir Joseph John Thomson in 1897 and by Robert Andrews Millikan in 1911. Both scientists were awarded Nobel Prizes, Thomson in 1906 and Millikan in 1923.

By using a specially designed cathode-ray tube (Fig. 3–7), Thomson applied electric and magnetic fields to the rays. Using basic laws of electricity and magnetism, he determined the **charge-to-mass ratio** of the electron. He was able to measure neither the absolute charge nor the absolute mass of the electron, but he established the ratio between the two numbers and made it possible to calculate either one if the other could ever be measured. What Thomson did for the concept of the electron is like showing that a peach weighs 40 times more than its seed. What is the weight of the peach? What is the weight of the seed? Neither is known, but if it can be determined by other means that the peach weighs 120 g, then the weight of the seed, by ratio, must be 3 g.

Thomson discovered the charge-to-mass ratio of the electron.

The Personal Side

Sir Joseph John Thomson (1856–1940) was a scientist chiefly working in mathematics until he was elected Cavendish Professor at Cambridge University in 1884. He was not skilled at experimental techniques, but his ability to suggest experiments and interpret their results led to the discovery of the electron. In 1897, Thomson wrote in the *Philosophical Magazine,* "We have in the cathode rays a new state, a state in which the subdivision of matter is carried very much further than in the ordinary gaseous state—this matter being the substance from which the chemical elements are built up." Thomson won the Nobel Prize in 1906 and was knighted in 1908. Seven of his research assistants later won Nobel Prizes for their own research work. Thomson was buried in Westminster Abbey near the grave of Sir Isaac Newton.

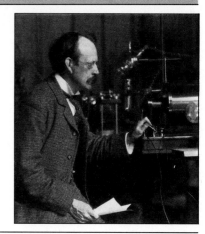

Sir J. J. Thompson (1856–1909) in his laboratory at Cambridge University. (University of Cambridge, Cavendish Laboratory.)

An important part of Thomson's experimentation was his use of 20 different metals for cathodes and of several gases to conduct the discharge. Every combination of metals and gases yielded the same charge-to-mass ratio for the cathode rays. This led to the belief that electrons are common to all of the metals and gases used in the experiments, and probably to all atoms in general. Thus, it appeared that the electron was one of the atomic building blocks of which atoms are made.

Millikan measured the fundamental charge of matter—the charge on an electron. A simplified drawing of his apparatus is shown in Figure 3–8. The experiment consisted of measuring the electric charge carried by tiny drops of oil that are suspended in an electric field. By means of an atomizer, oil droplets were sprayed into the test chamber. As the droplets settled slowly through the air, high-energy X rays were passed through the chamber to charge the droplets negatively (the X rays caused air molecules to give up electrons to the oil). By using a beam of light and a small telescope, Millikan could study the motion of a single droplet. When the electric charge on the plates was increased enough to balance the effect of gravity, a droplet could be suspended motionless. At this point, the gravitational force would equal the electric force. Measurements made in the motionless state, when inserted into equations for the forces acting on the droplet, enabled Millikan to calculate the charge carried by the droplet.

Millikan found different amounts of negative charge on different drops, but the charge measured each time was always a whole-number multiple of a very small basic unit of charge. The *largest* common divisor of all charges measured by this experiment was 1.60×10^{-19} C. (The coulomb, abbreviated C, is a charge unit). Millikan assumed this to be the fundamental charge, which is the charge on the electron.

With a good estimate of the charge on an electron and the ratio of charge to mass as determined by Thomson, the very small mass of the electron could be calculated. The mass of an electron is 9.11×10^{-28} g. On the carbon-12 relative scale, the electron would have a weight of 0.000549 atomic weight units. The negative charge on an electron of -1.60×10^{-19} C is set as the standard charge of -1.

Electrons are present in all of the elements.

Millikan measured the charge on an electron.

Only a whole number of electrons may be present in a sample of matter.

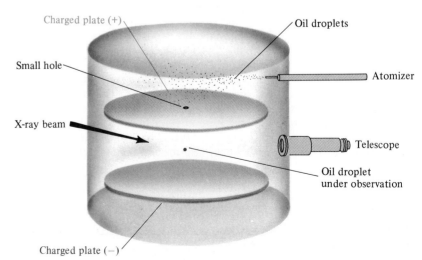

Figure 3–8   The Millikan oil drop experiment. Tiny oil droplets, produced by an atomizer, fall through a hole in the upper plate. X rays give some of these droplets a negative charge. By adjusting the voltage between the plates, the negatively charged droplets fall more slowly due to attraction by the positively charged upper plate. When a droplet is sighted in the telescope and the voltage is adjusted just enough to hold the droplet stationary, its charge can be calculated when the mass of the droplet and the voltage are known.

Charged plate (+)

Oil droplets

Small hole

Atomizer

X-ray beam

Telescope

Oil droplet under observation

Charged plate (−)

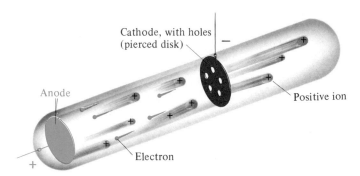

Figure 3-9 Cathode ray tube with a perforated cathode. Electrons collide with gas molecules and produce positive ions, which are attracted to the negative cathode. Some of the positive ions pass through the holes and form a positive ray. Like cathode rays, positive rays are deflected by electric and magnetic fields.

PROTONS — THE ATOM'S POSITIVE CHARGE

The first experimental evidence of a fundamental positive particle came from the study of canal rays. A special type of cathode-ray tube produces canal rays (Fig. 3-9). The cathode is perforated, and the tube contains a gas at very low pressure. When high voltage is applied to the tube, cathode rays can be observed between the electrodes as in any cathode-ray tube. On the other side of the perforated cathode, a different kind of a ray is observed. These rays are attracted to a negative plate brought alongside the rays. The rays must therefore be composed of positively charged particles. Each gas used in the tube gives a different charge-to-mass ratio for the positively charged particles. When hydrogen gas was used, the largest charge-to-mass ratio was obtained, indicating that hydrogen provides the positive particles with the smallest mass. This particle was considered to be the fundamental positively charged particle of atomic structure, and was called a **proton** (from Greek for "the primary one").

For a numerator of fixed size, the larger the ratio, the smaller the denominator.

Experiments on canal rays were begun in 1886 by E. Goldstein and further work was done later by W. Wien. The production of canal rays is caused by high-energy electrons moving from the negative cathode to the positive anode, hitting the molecules of gases occupying the tube. Electrons are knocked from some atoms by the high-energy electrons, leaving each molecule with a positive charge. The positively charged molecules are then attracted to the negative electrode. Since the electrode is perforated, some of the positive particles go through the holes or channels (hence the name *canal rays*).

Canal rays are streams of positive ions derived from the gases present in the discharge tube.

The mass of the proton is 1.67261×10^{-24} g, which is 1.00727 relative weight on the carbon-12 scale. The charge of $+1$ on the proton is equal in size but opposite in effect to the charge on the electron.

An ion *is an atom or a group of atoms carrying a charge.*

NEUTRONS — NEUTRAL PARTICLES FOUND IN MOST ATOMS

The masses calculated for atoms indicated that neutral particles with about the mass of the proton must be present in the atom in addition to the protons and electrons. This third type of particle in the atom proved hard to find. Since the particle has no charge, the usual methods of detecting small individual particles could not be used.

It remained for James Chadwick, in 1932, to devise a clever experiment that produced neutrons by a nuclear reaction and then detected them by having the neutrons knock hydrogen ions, a detectable species, out of paraffin.

A neutron has no electric charge and has a mass of 1.67492×10^{-24} g, which is a relative weight of 1.00867 on the carbon-12 scale.

Paraffin is the hydrocarbon that seals home-canned strawberry preserves.

THE NUCLEUS—AN AMAZING ATOMIC CONCEPT

Alpha particles are scattered by the nuclei of the gold atoms.

When Ernest Rutherford and his students directed alpha particles toward a very thin sheet of gold foil in 1909, they were amazed to find a totally unexpected result (Fig. 3–10). As they had expected, the paths of most of the alpha particles were only slightly changed as they passed through the gold foil. The extreme deflection of a few of the alpha particles was a surprise. Some even "bounced" back toward the source. Rutherford expressed his astonishment by stating that he would have been no more surprised if someone had fired a 15-inch artillery shell into tissue paper and then found it in flight back toward the cannon.

What allowed most of the alpha particles to pass through the gold foil in a rather straight path? According to Rutherford's interpretation, the atom is mostly *empty space* and, therefore, offers little resistance to the alpha particles (Fig. 3–11).

What caused a few alpha particles to be deflected? According to Rutherford's interpretation, concentrated at the center of the atom is a **nucleus**

Figure 3–10 Rutherford's gold foil experiment. A cylindrical scintillation screen is shown for simplicity; actually, a movable screen was employed. Most of the alpha particles pass straight through the foil to strike the screen at point A. Some alpha particles are deflected to points B, and some are even "bounced" backward to points such as C.

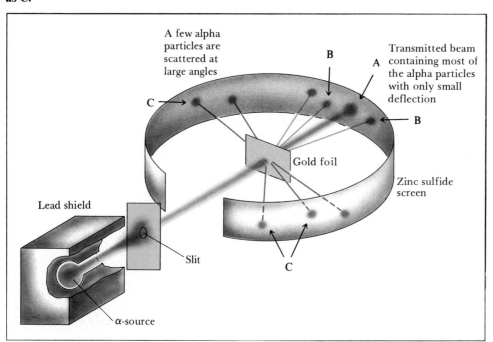

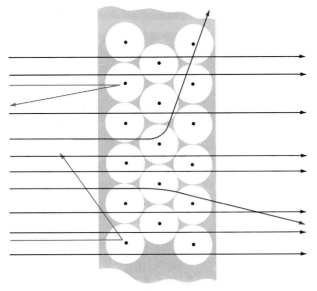

Figure 3–11 Rutherford's interpretation of how alpha particles interact with atoms in a thin gold foil. Actually, the gold foil was about 1000 atoms thick. For illustration purposes, points are used to represent the gold nuclei, and the path widths of the alpha particles are drawn much larger than scale.

containing most of the mass of the atom and all of the positive charge. When an alpha particle passes near the nucleus, the positive charge of the nucleus repels the positive charge of the alpha particle; the path of the smaller alpha particle is consequently deflected. The closer an alpha particle comes to a target nucleus, the more it is deflected. Those alpha particles that meet a nucleus head on bounce back toward the source as a result of the strong positive-positive repulsion, since the alpha particles do not have enough energy to penetrate the nucleus.

Rutherford's calculations, based on the observed deflections, indicate that the nucleus is a very small part of an atom. An atom occupies about a million million times more space than does a nucleus; the radius of an atom is about 10,000 times greater than the radius of its nucleus. Thus, if a nucleus were the size of a baseball, then the edges of the atom would be about one third of a mile away. And most of the space in between would be absolutely empty.

Alpha-particle scattering can be explained if the nucleus occupies a very small volume of the atom.

Since the nucleus contains most of the mass and all of the positive charge of an atom, the nucleus must be composed of the most massive atomic particles, the protons and neutrons. The electrons are distributed in the near-emptiness outside the nucleus.

Truly, Rutherford's model of the atom was one of the most dramatic interpretations of experimental evidence to come out of this period of significant discoveries.

ATOMIC NUMBER—EACH ELEMENT HAS A NUMBER

The **atomic number** of an element indicates the number of protons in the nucleus of the atom, which is the same as the number of electrons outside the nucleus. The two types of particles must be present in equal numbers for the

Ernest Rutherford.

TABLE 3–2 Summary of Properties of Electrons, Protons, and Neutrons

	Relative Charge	Relative Mass	Location
Electron	−1	0.00055	Outside the nucleus
Proton	+1	1.00727	Nucleus
Neutron	0	1.00867	Nucleus

atom to be neutral in charge. Note that the periodic table of the elements, inside the back cover, is an arrangement of the elements consecutively according to atomic number. Beginning with the atomic number 1 for hydrogen, there is a different atomic number for each element.

$$\begin{pmatrix} \text{Number of} \\ \text{electrons} \\ \text{per atom} \end{pmatrix} = \begin{pmatrix} \text{Number of} \\ \text{protons} \\ \text{per atom} \end{pmatrix} = \begin{pmatrix} \text{Atomic number} \\ \text{of the} \\ \text{element} \end{pmatrix}$$

The **atomic mass** of a particular atom is the sum of the masses of the protons, neutrons, and electrons in that atom. Since an electron has such a small mass, the atomic mass is very nearly the sum of the masses of the protons and neutrons in the nucleus. Both protons and neutrons have masses of approximately 1.0 on the atomic weight scale (see Table 3–2). Hydrogen, with an atomic weight of 1, must be composed of one proton (and no neutrons) in the nucleus and one electron somewhere outside the nucleus. Helium has an atomic number of 2 and an atomic weight of 4. The atomic number of 2 indicates two protons and two electrons per atom of helium. The atomic weight of 4 means that, in addition to the two protons in the nucleus, there are two neutrons.

$$\begin{pmatrix} \text{Approximate} \\ \text{number of} \\ \text{neutrons} \\ \text{per atom} \end{pmatrix} = \begin{pmatrix} \text{Atomic weight} \\ \text{of the} \\ \text{element} \end{pmatrix} - \begin{pmatrix} \text{Atomic number} \\ \text{of the} \\ \text{element} \end{pmatrix}$$

A notation frequently used to show the atomic mass (also called **mass number**) and atomic number of an atom uses subscripts and superscripts to the left of the symbol:

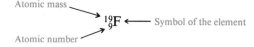
Atomic mass ⟶ $^{19}_{9}F$ ⟵ Symbol of the element
Atomic number ⟶

For an atom of fluorine, $^{19}_{9}F$, the number of protons is 9, the number of electrons is also 9, and the number of neutrons is $19 − 9 = 10$.

ISOTOPES — DALTON NEVER GUESSED!

Many of the elements, when analyzed by a special type of canal-ray tube called a *mass spectrometer* (Fig. 3–12), are found to be composed of atoms of different masses. Atoms of the same element having different atomic masses are called **isotopes** of that element.

The lightest atom is the hydrogen atom.

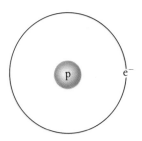

Hydrogen atom

Notations such as $^{19}_{9}F$ are used to represent isotopes of an element.

Some really heavy atoms have over 155 neutrons in the nucleus.

Isotopes are atoms of the same element having different numbers of neutrons.

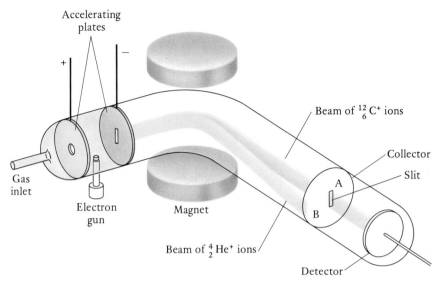

Figure 3–12 Mass spectrometer. Sample to be studied is injected near filament. Electrodes (not shown) subject sample to electron beam that ionizes a part of the sample by knocking electrons from neutral atoms or molecules. Electrodes are arranged to accelerate positive ions toward first slit. The positive ions that pass the first slit are immediately put into a magnetic field perpendicular to their path and follow a curved path determined by the charge-to-mass ratio of the ion. A collector plate, behind the second slit, detects charged particles passing through the second slit. The relative magnitudes of the electrical signals are a measure of the numbers of the different kinds of positive ions.

The element neon is a good example to consider (Fig. 3–13). A natural sample of neon gas is found to be a mixture of three isotopes of neon:

$$^{20}_{10}\text{Ne} \qquad ^{21}_{10}\text{Ne} \qquad ^{22}_{10}\text{Ne}$$

The fundamental difference between isotopes is the different number of neutrons per atom. All atoms of neon have 10 electrons and 10 protons; about 90% of the atoms have 10 neutrons, some have 11, and others have 12. Because they have different numbers of neutrons, they must have different masses. Note that all the isotopes have the same atomic number. They are all neon.

There are only 109 known or claimed elements, yet more than 1000 isotopes have been identified, many of them produced artificially (see Chap-

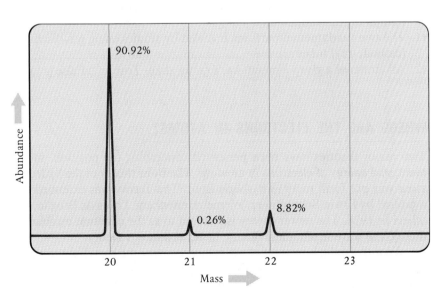

Figure 3–13 Mass spectrum of neon (+ 1 ions only). The principal peak corresponds to the most abundant isotope, neon-20. Percent relative abundance is shown.

ter 7). Some elements have many isotopes; tin, for example, has 10 natural isotopes. Hydrogen has three isotopes, and they are the only three that are generally referred to by different names: ^1_1H is called protium, ^2_1H is called deuterium, and ^3_1H is called tritium. Tritium is radioactive. The natural assortment of isotopes, each having its own distinctive atomic mass, results in fractional atomic weights for many elements.

The weighted average of the atomic weights of the isotopes in a natural mixture is the noninteger atomic weight of the element.

SELF-TEST 3-B

1. Isotopes of an element are atoms that have nuclei with the same number of _____ but different numbers of _____ .

2. The nucleus of an atom occupies a relatively large ( ) or small ( ) fraction of the volume of the atom.

3. The positive charges in an atom are concentrated in its _____ .

4. The negatively charged particles in an atom are _____; the positively charged particles are _____; and the neutral particles are _____ .

5. In a neutral atom there are equal numbers of _____ and _____ .

6. The number of protons per atom is called the _____ number of the element.

7. The mass of the proton is _____ times the mass of the electron.

8. An atom of arsenic, $^{75}_{33}As$, has _____ electrons, _____ protons, and _____ neutrons.

9. Positive (canal) rays obtained with different gases are (different/identical), and the cathode rays obtained using different cathodes are (different/identical).

10. Cathode rays are composed of a universal constituent of matter named _____ .

11. The two fundamental particles revealed by studies using gas-discharge (cathode-ray) tubes are the _____ and _____ .

12. All atoms of a given element are exactly alike. True ( ) False ( )

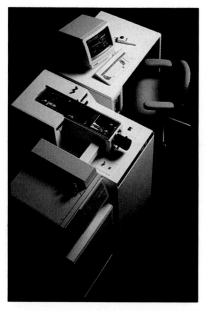

A modern mass spectrometer. (Courtesy of Finnigan Corporation.)

WHERE ARE THE ELECTRONS IN ATOMS?

Two major theories have been presented concerning the position, movement, and energy of electrons in an atom. The **Bohr theory** of the hydrogen atom was put forth in 1913 by Niels Bohr. This theory was extended and modified by Erwin Schrödinger, Werner Heisenberg, Louis de Broglie, and others in 1926. The newer theory is referred to as the **quantum mechanical theory,** the wave mechanical theory, or Schrödinger's theory.

To understand how these theories have become the basis for our understanding of the behavior of electrons in atoms, we first need to look at the nature of light and how it is related to matter.

Electromagnetic Radiation

Electromagnetic radiation is familiar to all of us—sunlight, headlights on automobiles, light from camera flash attachments, dentist's X rays, and radio waves that we use for communications are a few examples (Fig. 3–14). All of these kinds of radiation seem very different, but are actually very similar in some properties. All electromagnetic radiation travels through

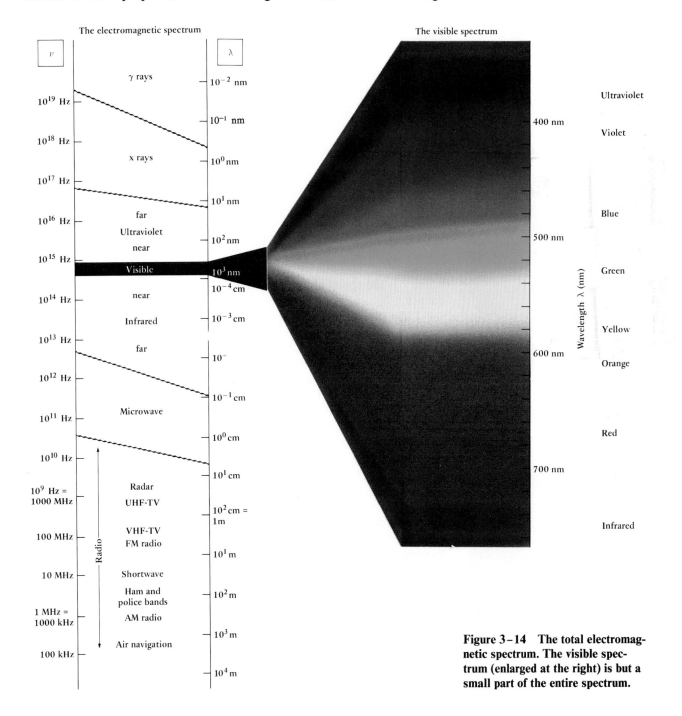

Figure 3-14 **The total electromagnetic spectrum. The visible spectrum (enlarged at the right) is but a small part of the entire spectrum.**

Figure 3–15 Illustrations of wavelength and frequency with water waves. The waves are moving toward the post. (a) The wave has a long wavelength, large λ, and low frequency (the number of times its peak hits the post). (b) The wave has a shorter wavelength and a higher frequency. It hits the post more often per unit of time.

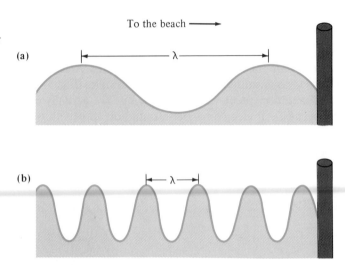

To the beach ⟶

(a) λ

(b) λ

Infrared radiation used for night vision. (*The World of Chemistry*, Program 10)

Max Planck (1858–1947) was awarded the Nobel prize in 1918.

space at the same rate (186,000 miles/s [miles per second] or 3.00×10^8 m/s [meters per second] in a vacuum) and can be described in terms of a frequency (v) and a wavelength (λ). The frequency of all electromagnetic radiation is related to its wavelength by

$$v\lambda = c$$

where c is the speed of light. Figure 3–15 illustrates wavelength and frequency.

Light (the kind we can see, visible light; the kind we can feel, infrared radiation; and the kind that burns us when we stay out in the sun too long, ultraviolet radiation) plays a role in understanding electrons in atoms.

Have you ever sat near an electric resistance heater as it heats up? Of course, we cannot see the metal atoms in the heater wire, but as electric energy flows through the wire, the atoms gain some of this energy. First the wire emits a slight amount of heat you can feel (infrared). As the wire heats further, it begins to glow and emit visible light. First red light is emitted, then orange. If the heater gets really hot, the wire appears almost white (white hot). At the close of the 19th century, scientists were trying to explain light and heat emissions from hot objects. The wave theory predicted that as the object got hotter, its color should shift all the way to the blue and finally to the violet, but no object ever did this.

In 1900, Max Planck (1858–1947) suggested a revolutionary idea. He reasoned that the energy that can be gained or lost by an atom (such as an iron atom in a hot wire) must be some whole number (integer) multiple of a minimum energy expressed as **nhv**, where n is a whole number and h is a constant. The lowest frequency used to describe the atom is v. The value of h is 6.626×10^{-34} and is now called **Planck's constant.** Planck reasoned that an atom could have energies of one hv, two hv, three hv, four hv, and so on, but never an energy like 3.3 hv. The minimum energy hv was called a **quantum.** Using his quantum hypothesis, Planck was able to calculate a spectrum for a hot object that agreed with its experimental spectrum. A

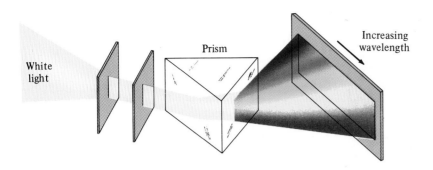

Figure 3–16 Spectrum from white light produced by refraction in a glass prism. The different colors blend into one another smoothly.

White light

Prism

Increasing wavelength

spectrum is what is observed when light of several different colors is separated or dispersed by some type of transparent matter such as glass or a droplet of water.

Dispersed light produces a spectrum.

The spectrum of white light is the rainbow display of separated colors shown in Figure 3–16. An **emission** spectrum is observed when the light emitted by atoms energized by a flame or an electric arc is allowed to pass through a narrow vertical slit and then through a prism of glass or quartz. If sunlight or light from a white-hot solid is dispersed, all of the colors of the rainbow are seen. This is a **continuous emission spectrum.** However, if the light from an energized gaseous element is dispersed, only a few colored lines are produced, the lines being separated by black spaces. This is a **bright-line emission spectrum** (Fig. 3–17). Each line is a pure color and is really an image of the slit in that particular color. A quantum of light (photon) of any given color has a characteristic energy that is different from the energy of a quantum of any other color.

When a theory can accurately predict experimental results, the theory is usually regarded as useful. At first, however, Planck's quantum theory was not widely accepted due to its radical nature. But when Planck's quanta were used to explain another phenomenon called the **photoelectric effect,** quanta of electromagnetic energy were firmly accepted. In the early 1900s it was known that certain metals exhibited a photoelectric effect, that is, they emitted electrons when illuminated by light of certain wavelengths. The metal cesium (Cs) will emit electrons when illuminated by red light, whereas

Figure 3–17 Atomic emission spectra for hydrogen, mercury, and neon. Excited gaseous elements will produce characteristic spectra which can be used to identify as well as quantify.

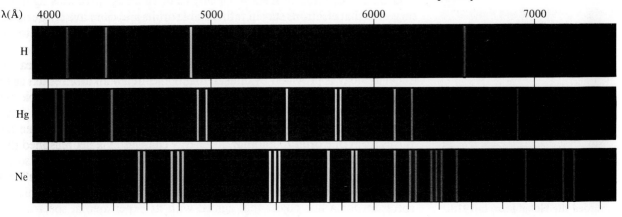

Figure 3-18   The photoelectric
effect. (a) A photocell's metallic
cathode is struck by photons. Light
above a certain frequency causes
electrons to be ejected. These are attracted to the anode, causing an
electrical flow through the cell. (b)
As the frequency of light striking a
metal cathode is increased, some
frequency causes electrons to be
ejected. The number of electrons is
dependent on the light intensity, not
the frequency.

Figure 3–18 The photoelectric effect. (a) A photocell's metallic cathode is struck by photons. Light above a certain frequency causes electrons to be ejected. These are attracted to the anode, causing an electrical flow through the cell. (b) As the frequency of light striking a metal cathode is increased, some frequency causes electrons to be ejected. The number of electrons is dependent on the light intensity, not the frequency.

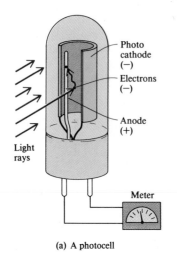

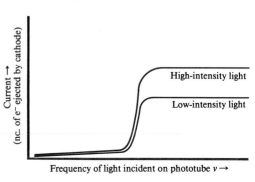

(a) A photocell

(b) Current in a photocell as a function of frequency and intensity of incident light.

other metals require yellow light and others require ultraviolet light. Figure 3–18 shows how an electric current suddenly increases when light of a frequency above a certain value shines on a photosensitive metal. In view of Planck's quantum theory, only photons above a certain energy (frequency) are capable of causing electrons to be ejected from the metal. This idea was first suggested by Albert Einstein in 1905. Einstein received a Nobel Prize in 1921 for this idea involving quantum theory. Now that the existence of electromagnetic quanta was firmly established, quanta could be applied to atomic structure. The first to do this was Niels Bohr, a Danish physicist.

The Bohr Model of the Atom

In Bohr's concept, electrons revolve around a nucleus in definite orbits, much as planets revolve around the sun. He equated classical mathematical expressions for the force tending to keep the electron traveling in a straight line and the force tending to pull the electron inward (the positive-to-negative attraction between a proton and an electron). The total energy of the atom is the kinetic energy of the electrons plus the potential energy due to the electron's separation from the nucleus. In a revolutionary sort of way, Bohr suggested that electrons stay in rather stable orbits and can have only certain energies within a given atom. According to Bohr, an electron can travel in one orbit for a long period or in another orbit some distance away for a long period, but it cannot stay for any measurable time between the two orbits. A rough analogy is provided by considering books in a bookcase. Books may rest on one shelf or on another shelf for very long periods but cannot rest between shelves. In moving a book from one shelf to another shelf, the potential energy of the book changes by a definite amount. When an electron moves from one orbit to another, its energy changes by a definite amount, called a **quantum** of energy.

Energy is required to separate objects attracted to each other. For example, energy is required to lift a rock from the earth, or to separate two

Energy of matter in motion is *kinetic* energy.

Energy stored in matter is *potential* energy.

Bohr assumed that atoms can exist only in certain energy states.

The Personal Side

Bohr had a close call in his escape from Denmark when Hitler's forces ravaged the country. Having done all he could to get Jewish physicists to safety, Bohr was still in Denmark when Hitler's army suddenly occupied the country in 1940. In 1943, to avoid imprisonment, he escaped to Sweden. There he helped to arrange the escape of nearly every Danish Jew from Hitler's gas chambers. He was later flown to England in a tiny plane, in which he passed into a coma and nearly died from lack of oxygen.

He went on to the United States, where until 1945 he worked with other physicists on the atomic bomb development at Los Alamos. His insistence upon sharing the secret of the atomic bomb with other allies, in order to have international control over nuclear energy, so angered Winston Churchill that he had to be restrained from ordering Bohr's arrest. Bohr worked hard and long on behalf of the development and use of atomic energy for peaceful purposes. For his efforts, he was awarded the first Atoms for Peace Prize in 1957. He died in Copenhagen on November 18, 1962.

magnets, or to pull a positive charge away from a negative charge. Bohr suggested that a quantum of energy is required to move an electron from one energy level to another. The energy added to move an electron farther from its nucleus is stored in the system as potential energy (energy of position). Thus, the electron has more energy when it is in an orbit farther from the nucleus and less energy when it is in an orbit close to the nucleus. When the electron passes from an outer orbit to an inner orbit, energy is emitted from the atom, generally in the form of light.

Bohr used the idea of electrons moving up and down a "bookcase" of energy levels corresponding to orbits to explain the observable bright-line emission spectrum of hydrogen.

According to Bohr, the light forming the lines in the bright-line emission spectrum of hydrogen comes from electrons moving toward the nucleus after having first been energized and pushed to orbits farther from the nucleus. A movement between two particular orbits involves a definite quantum of energy. Each time an electron moves from one orbit to another orbit closer to the nucleus, energy loss occurs and a quantum of light having a characteristic energy (and color) is emitted. Transitions toward the nucleus between two outer orbits emit quanta having smaller characteristic energies. Transitions from an outer orbit to orbits near the nucleus emit quanta having larger characteristic energies. Each line corresponds to its own particular energy of the same-sized quanta of light.

Not only could Bohr explain the cause of the lines in the bright-line emission spectrum of hydrogen, but he also calculated the expected wavelengths of the lines. He expressed the results of his calculations in the alternative view of the nature of light, its wave nature.

With brilliant imagination, Bohr applied a little algebra and some classical mathematical equations of physics to his tiny solar-system model of the hydrogen atom. The unprecedented requirement was that only a few allowable paths (quantized orbits) are available in which electrons can move stably around the nucleus. A further requirement was that energy differences

Bohr used orbits, but they are now called shells or energy levels.

TABLE 3–3 Agreement Between Bohr's Theory and the Lines of the Hydrogen Spectrum*

Changes in Energy Levels	Wavelength Predicted by Bohr's Theory (nm)	Wavelength Determined from Laboratory Measurement (nm)	Spectral Region
$2 \rightarrow 1$	121.6	121.7	Ultraviolet
$3 \rightarrow 1$	102.6	102.6	Ultraviolet
$4 \rightarrow 1$	97.28	97.32	Ultraviolet
$3 \rightarrow 2$	656.6	656.7	Visible red
$4 \rightarrow 2$	486.5	486.5	Visible blue-green
$5 \rightarrow 2$	434.3	434.4	Visible blue
$4 \rightarrow 3$	1876	1876	Infrared

* These lines are typical; other lines could be cited as well, with equally good agreement between theory and experiment. The unit of wavelength is the nanometer (nm), 10^{-9} m.

(quanta) existed between any two orbits. Bohr was able to calculate the wavelengths of the lines in the hydrogen spectrum, some of which are shown in Table 3–3. Note the close agreement between the measured values and the values predicted by the calculations of the Bohr theory. Niels Bohr had tied the unseen (the interior of the atom) with the seen (the observable lines in the hydrogen spectrum)—a fantastic achievement.

The Bohr theory was accepted almost immediately after its presentation, and Bohr was awarded the Nobel Prize in physics in 1922 for his contribution to the understanding of the hydrogen atom.

One way to describe the interior of atoms other than hydrogen is to use the orbits devised by Bohr for hydrogen and insert the proper number of electrons. Although this is at best a rough approximation, we shall find in Chapter 5 that it is adequate for explaining such phenomena as some kinds of bonding and the formation of ions.

Bohr's theoretical calculations for hydrogen spectral lines agreed amazingly well with experimental data.

Atom Building Using the Bohr Model

Let us build up some atoms in Tinker Toy fashion. First, we need a few rules in order to play the game. Recall that the atomic number is the number of electrons (and protons) per atom of the element. Consistent with the ionization energies (discussed in Chapter 5), the maximum number of electrons per energy level is $2n^2$, where n is the number of the energy level. Energy levels are numbered with integers, beginning with 1 for the one closest to the nucleus. As practice, use the formula to check these numbers.

Orbit	Maximum Number of Electrons
1	2
2	8
3	18
4	32
5	50

Placement of Electrons in Ground State

Element	Atomic Number	Bohr Model	Wave Mechanical Model
Hydrogen (H)	1	(1p) 1)e	$1s^1$
Helium (He)	2	(2p 2n) 2)e	$1s^2$
Lithium (Li)	3	(3p 4n) 2)e 1)e	$1s^2 2s^1$
Beryllium (Be)	4	(4p 5n) 2)e 2)e	$1s^2 2s^2$
Boron (B)	5	(5p 6n) 2)e 3)e	$1s^2 2s^2 2p^1$
Carbon (C)	6	(6p 6n) 2)e 4)e	$1s^2 2s^2 2p^2$ (or $2p_x^1 2p_y^1$)
Nitrogen (N)	7	(7p 7n) 2)e 5)e	$1s^2 2s^2 2p^3$ (or $2p_x^1 2p_y^1 2p_z^1$)
Oxygen (O)	8	(8p 8n) 2)e 6)e	$1s^2 2s^2 2p^4$ (or $2p_x^2 2p_y^1 2p_z^1$)
Fluorine (F)	9	(9p 10n) 2)e 7)e	$1s^2 2s^2 2p^5$
Neon (Ne)	10	(10p 10n) 2)e 8)e	$1s^2 2s^2 2p^6$
Sodium (Na)	11	(11p 12p) 2)e 8)e 1)e	$1s^2 2s^2 2p^6 3s^1$
Magnesium (Mg)	12	(12p 12n) 2)e 8)e 2)e	$1s^2 2s^2 2p^6 3s^2$
Aluminum (Al)	13	(13p 14n) 2)e 8)e 3)e	$1s^2 2s^2 2p^6 3s^2 3p^1$
Silicon (Si)	14	(14p 14n) 2)e 8)e 4)e	$1s^2 2s^2 2p^6 3s^2 3p^2$ (or $3p_x^1 3p_y^1$)
Phosphorus (P)	15	(15p 16n) 2)e 8)e 5)e	$1s^2 2s^2 2p^6 3s^2 3p^3$ (or $3p_x^1 3p_y^1 3p_z^1$)
Sulfur (S)	16	(16p 16n) 2)e 8)e 6)e	$1s^2 2s^2 2p^6 3s^2 3p^4$ (or $3p_x^2 3p_y^1 3p_z^1$)
Chlorine (Cl)	17	(17p 18p) 2)e 8)e 7)e	$1s^2 2s^2 2p^6 3s^2 3p^5$
Argon (Ar)	18	(18p 22n) 2)e 8)e 8)e	$1s^2 2s^2 2p^6 3s^2 3p^6$
Potassium (K)	19	(19p 20n) 2)e 8)e 8)e 1)e	$1s^2 2s^2 2p^6 3s^2 3p^6 4s^1$
Calcium (Ca)	20	(20p 20n) 2)e 8)e 8)e 2)e	$1s^2 2s^2 2p^6 3s^2 3p^6 4s^2$

Figure 3–19 **Electron arrangements of the first 20 elements. The nuclear contents of a typical isotope are shown.**

A general, overriding rule to the preceding numbers is that the outside energy level can have no more than eight electrons. When electrons are placed in shells as close to the nucleus as possible, the electrons are said to be in their **ground state.**

The ground state is the lowest energy state for all of the electrons in an atom.

You might like to follow along in Figure 3–19 (the Bohr model column) as the building-up process is described. Hydrogen (H), with atomic number 1, has one electron. In its ground state, this electron is in the first energy level.

The two electrons of helium (He) are in its first energy level since the first shell can have a maximum of two electrons.

For all atoms of other elements, two electrons are in the first energy level, and the other electrons of the atoms are assorted into higher numbered energy levels. In atomic-number order, lithium (Li) through neon (Ne), two electrons are placed in the first energy level (which fill it), and into the second energy level are placed one, two, three, and so on to eight electrons (for Ne). Eight electrons fill the second energy level.

Sodium (Na), with 11 electrons, has the first two energy levels filled with 2 and 8 electrons, respectively, and has 1 electron in the third energy level. Each succeeding element in atomic-number order, magnesium (Mg) through argon (Ar), adds 1 more electron to the third energy level of its atoms.

At Ar, the maximum of eight electrons in the outside energy level comes into play. When 19 electrons are present, as in an atom of potassium (K), the first energy level has 2 electrons, the second energy level has 8 electrons, and the third energy level could have the other 9 electrons (maximum of 18 electrons) if it were not the outside energy level. So to accommodate 19 electrons, there are two choices: 2-8-9 or 2-8-8-1. The first choice violates the requirement of no more than 8 in the outside energy level. The second is the proper choice. Calcium (Ca) with 20 electrons per atom has an electronic arrangement of 2-8-8-2.

Beginning with scandium (Sc), atomic number 21, and continuing through zinc (Zn), atomic number 30, 10 electrons are added to the third energy level to complete its maximum of 18. Zinc has the electronic arrangement 2-8-18-2.

You might pause in your reading here and predict the ground-state electronic arrangement of gallium (Ga), atomic number 31, and rubidium (Rb), atomic number 37, using this system.

THE WAVE THEORY OF THE ATOM

The Bohr model failed when applied to elements other than hydrogen because it could not account exactly for the line spectra of atoms with more than one electron. It was also weak in explaining why the periods (the horizontal rows) of the periodic table vary considerably in length.

After Bohr's work, a more modern, highly sophisticated mathematical theory of the atom was developed by Schrödinger, Heisenberg, Dirac, and others. In this theory, electrons are treated as having both a particle and a wave nature. The locations of the electrons are treated as **probabilities,** without seeking to locate the exact spot for an electron at a given time. This approach suggested that the Bohr theory describing the electrons with fixed orbits sought more precision than nature would allow.

A Frenchman, Louis de Broglie, was the first to suggest (in 1924) that electrons and other small particles should have wave properties. In this respect, he said, electrons should behave like light, a suggestion that scientists of the time found hard to accept. However, in a few years separate experi-

Electrons are described by both particle and wave theories.

ments by George Thomson (son of J. J. Thomson) in England and Clinton Davisson in the United States justified de Broglie's hypothesis. The **electron microscopes** found in many research laboratories today are built and operated on our understanding of the wave nature of the electron.

It should not be surprising to find that matter can be treated by both wave and particle theories (the duality of matter), since its convertible counterpart — light — has been treated successfully by both theories for a long time. Keep in mind that we do not really know if matter or light is a wave or a particle. However, because there are limits on what we can visualize in our physical world, in talking about something like subatomic behavior we are forced to use physical models based on known behavior, rather than more sophisticated models that would describe some type of intermediate behavior with which we are unfamiliar in our macroscopic world.

Electron microscope. (*The World of Chemistry,* **Program 15**)

The wave theory of the atom was developed in the 1920s, principally by Erwin Schrödinger (1887–1961). The most fundamental aspects of the theory are the mathematical wave equations used to describe the electrons in atoms. Solutions to the equations are called wave functions, or **orbitals.** Calculations involving the wave equations are complicated and time-consuming, but we do not need to do the elaborate calculations in order to use the results.

The principal result of the wave equation for an atom is a series of orbitals. Orbitals are different in their type, energy, and likely configuration in space (related to the probability of finding the electron there). The types of orbitals are distinguished by the letters *s, p, d,* and *f.* These letters were derived from terms in spectroscopy (sharp, principal, diffuse, and fundamental, respectively) and emphasize again that atomic theory developed very closely with atomic spectra.

The orbitals of the wave theory are actually subdivisions of the Bohr energy levels or *shells.** Each energy level has an *s* orbital. Beginning with the second energy level, each shell also has a set of three *p* orbitals. The third energy level and all energy levels thereafter also have a set of five *d* orbitals. The fourth energy level and all energy levels thereafter also have a set of seven *f* orbitals. We shall need only *s* and *p* orbitals for the explanations given in this book, since the electrons in the outer energy levels of the various atoms are in the *s* and *p* orbitals of those atoms. These outer *s* and *p* electrons are the ones most involved with other atoms.

Only the **probability** of finding an electron in a given volume of space around the nucleus can be calculated from the orbital resulting from the Schrödinger equation. In order to portray the probabilities of finding an electron, usually the surface of a region in space (similar to the surface of a balloon) is plotted that will enclose the volume where the electron will be expected to be found 90% of the time. Actually, the electron structure of the atom is rather "messy" to describe in any definite way because of the uncertainty of locating its electron.

---

* This shows that both the Bohr and Schrödinger theories can yield the same result in simple cases, as they should, since they are describing the same thing.

THE SCANNING TUNNELING MICROSCOPE

Chemists and physicists had more than ample evidence of the existence of atoms before the invention of the scanning tunneling microscope (STM). They were able to "see" atoms through a large variety of phenomena, but to say that they saw atoms had a special meaning. What they were seeing by such techniques as X-ray diffraction was a manifestation of many atoms and a composite picture due to the scattering of X rays from many planes of a crystal.

Yet chemists and physicists have always dreamed of being able to see individual atoms directly, that is, of being able to produce images with a direct correspondence to the atom's actual position in the sample.

These dreams began to be realized in the 1950s. An early and spectacular effort was reported by Erwin Mueller using a field ion microscope that he invented, which made it possible to image individual atoms on a crystal's surface. The even more remarkable **STM** not only makes it possible to see individual atoms and how they are arranged on a surface, but also permits the study of atom migration and atomic dislocations on surfaces. The development of the STM is considered an event of such magnitude that its developers, Gerd Binnig and Heinrich Rohrer of IBM's Zurich Research Laboratory in Switzerland, received the Nobel Prize in physics in 1986. The STM is an astonishing device because of its inherent simplicity. It consists of a tungsten needle, hardly more than a single atom wide at the end. When this needle is lowered to within a few atoms thickness of the surface to be imaged and a small voltage is applied, electrons tunnel, that is, they pass from the tungsten atom into the electron clouds of the atoms on the surface and produce a measurable current. By adjusting the up–down position of the tungsten needle as it moves across the surface, a constant tunneling current is maintained. As this takes place, however, the positions of the atoms are actually measured giving a picture of the atomic landscape.

The potential for studying materials using the STM is great, particularly in the area of catalysts. Recently, STM studies have also been able to see actual amino acid molecules on the surfaces of crystals. Amino acids are the basic building blocks of all living matter.

The World of Chemistry (Program 6), "The Atom."

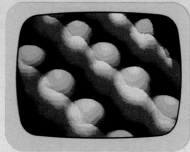

STM image of gallium arsenide.
(*The World of Chemistry*,
Program 6)

As shown in Figure 3–20 (a), an *s* orbital is always spherical. A *p* orbital is shaped like a dumbbell (b). Three different spatial orientations are possible for the *p* orbital, hence the designations p_x, p_y, and p_z. The *d* and *f* orbitals have more complicated shapes (c). The shapes of the orbitals are about the best that we can do to relate the wave theory of electron probabilities in a pictorial way, for we are trying to visualize where in space a given electron will be. The geometries associated with the orbitals help to explain the structures of molecules that result from the interaction and combinations of atoms.

The first Bohr energy level is synonymous with the 1*s* orbital. (The 1 indicates the shell; the *s* indicates the type of orbital.) According to the wave

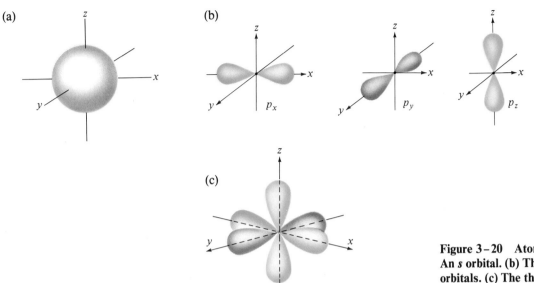

Figure 3–20   Atomic orbitals. (a) An *s* orbital. (b) The three *p* orbitals. (c) The three *p* orbitals shown together.

theory, an orbital may be occupied by a *maximum* of two electrons. Hence, the $1s$ orbital (and also the first shell) can have a maximum of two electrons. When the electron of a hydrogen atom is as close to its nucleus as it can stably be (in its ground state), the electron is in a $1s$ orbital, expressed as $1s^1$. The two electrons of helium can be in the same orbital and are in the $1s$ orbital in their ground state, $1s^2$.

Since the second Bohr energy level can have a maximum of eight electrons, four orbitals are required to accommodate the eight electrons. The four orbitals are a $2s$ orbital and a set of three $2p$ orbitals designated $2p_x, 2p_y$, and $2p_z$ (Fig. 3–20). When the three electrons of a lithium atom are in their ground state, two electrons are in the $1s$ orbital and the other electron is in the $2s$ orbital, $1s^2 2s^1$ (Fig. 3–19). The five electrons of a boron atom are distributed with two in the $1s$ orbital, two in the $2s$ orbital, and one in the $2p$ orbital (either $2p_x, 2p_y$, or $2p_z$, since these orbitals have the same energy unless the atom is in a strong magnetic field), $1s^2 2s^2 2p^1$.

If there are two, three, or four electrons in a set of p orbitals, the electrons are spread among the orbitals as much as possible. This gives a more stable arrangement of electrons from spin and angular momentum considerations. What this means is that no orbital in a set will have two electrons until each orbital in the set has one electron. Thus, the six electrons of carbon distribute as $1s^2 2s^2 2p_x^1 2p_y^1$. The seven electrons of nitrogen have the configuration $1s^2 2s^2 2p_x^1 2p_y^1 2p_z^1$, and the eight electrons of oxygen are arranged in the ground state as $1s^2 2s^2 2p_x^2 2p_y^1 2p_z^1$.

In Figure 3–19, wave-mechanical electronic arrangements are shown through atomic number 20. Note the ground-state order of energy for the orbitals: $1s\,2s\,2p\,3s\,3p\,4s$. How does the number of electrons in the $2s$ and $2p$ orbitals compare with the number of electrons in the second shell? The $3s$ and $3p$ and the third shell?

Erwin Schrödinger (1887–1961). He shared the Nobel Prize in Physics in 1933. (AIP Niels Bohr Library.)

What, then, is an atom really like? The atomic concepts have changed over a long period of time. We have Dalton's concept of an atom as a hard sphere smilar to a small billiard ball. We have Bohr's concept of the atom as a small three-dimensional solar system with a nucleus and electrons in paths called energy levels, or shells. In the moden theory, we have more detail in that energy levels are now sub-levels called orbitals, and we are given approximate spaces where electrons exert their greatest influence in an atom. Why present all three theories? First, an understanding of the simpler Dalton and Bohr theories helps us to understand the more complicated, more detailed modern theory of the atom. Second, all three theories help us to understand the phenomena we observe. We simply use whatever detail is necessary to explain what we see. For example, the simpler Dalton concept adequately explains many properties of the gaseous, liquid, and solid states. Most bonding between atoms of the light elements can be explained by application of the orbits of Bohr. The shapes of molecules and the arrangement of atoms with respect to each other can best be explained by the orbital representations of the modern theory. In the explanations given in this text, we shall follow the principle that simplest is best.

SELF-TEST 3–C

1. Under some conditions light has properties of _____ and under other conditions exhibits the properties of _____.
2. When light is dispersed into the different colors composing the light, a _____ is produced.
3. According to Bohr's theory, light of characteristic wavelength is produced as an electron passes from an orbit ( ) closer to or ( ) farther from the nucleus to an orbit ( ) closer to or ( ) farther from the nucleus.
4. Which of the following led to the moden theory of the atom and was not included in the Bohr theory?
   a. Concept of the nucleus
   b. Quantum theory
   c. Particle nature of the electron
   d. Wave nature of the electron
5. According to de Broglie, every moving particle has not only mass and velocity but also a characteristic _____.
6. The maximum number of electrons in the $n = 3$ energy level is _____, and the maximum number in any orbital is _____.
7 Consider the meaning of the representations of the orbitals shown in Figure 3–21.
   a. Are the representations those of the paths of electrons?
   b. Are the representations the containers of electrons?
   c. Do the representations show where an electron is most likely to be found?

MATCHING SET

_____ 1. Atomic mass
_____ 2. Unlike electric charges
_____ 3. $2n^2$
_____ 4. Nucleus
_____ 5. Electron
_____ 6. ^{22}Ne and ^{20}Ne
_____ 7. Atomic number
_____ 8. Quantum
_____ 9. Gamma ray
_____ 10. Particles in an H atom
_____ 11. Neutron
_____ 12. G. P. Thomson C. J. Davisson
_____ 13. de Broglie
_____ 14. Orbital
_____ 15. Photoelectric effect

a. Attract
b. Equal to number of protons in nucleus
c. Demonstrated wave nature of electron
d. Cathode-ray particle
e. Neutrons plus protons
f. A small, definite amount of energy
g. Proton and an electron
h. Uncharged elementary particle
i. Contains most of the mass in an atom
j. Maximum number of electrons in an energy level
k. Predicted wave nature of electron
l. Probable location for electrons in an atom
m. A form of radiant energy
n. Isotopes
o. Planck

QUESTIONS

1. What kinds of evidence did Dalton have for atoms that the early Greeks (Democritus, Leucippus) did not have?
2. How does Dalton's atomic theory explain:
   a. The law of conservation of matter?
   b. The law of constant composition?
   c. The law of multiple proportions?
3. Although there may not be a very reliable way to check the conservation of matter in a large explosion of dynamite, what leads us to believe that the law of conservation of matter is obeyed?
4. Describe in detail Rutherford's gold-foil experiment under the following headings:
   a. Experimental setup
   b. Observations
   c. Interpretations
5. Why was Thomson's charge-to-mass ratio determination for electrons very significant although he did not determine either the charge or the mass of the electron?
6. How do the following discoveries indicate that the Daltonian model of atoms is inadequate?
   a. Cathode rays
   b. Positive rays
   c. Nucleus
   d. Natural radioactivity
   e. Isotopes
7. Characterize the three types of emissions from naturally radioactive substances as to charge, relative mass, and relative penetrating power.
8. Explain what the following terms mean:
   a. Isotopes of an element
   b. Atomic number
   c. An alpha emitter
9. If electrons are a part of all matter, why are we not electrically shocked continually by the abundance of electrons about and in us?
10. There are more than 1000 kinds of atoms, each with a different weight. Yet there are only 109 elements. How does one explain this in terms of subatomic particles?
11. What is a practical application of cathode-ray tubes?
12. A common isotope of Li has a mass of 7. The atomic number of Li is 3. What are the constituent particles in its nucleus?

13. The element iodine (I) occurs naturally as a single isotope of atomic mass 127; its atomic number is 53. How many protons and how many neutrons does it have in its nucleus?

14. Suppose Millikan had determined the following charges on his oil drops:

1.33×10^{-19} C
2.66×10^{-19} C
3.33×10^{-19} C
4.66×10^{-19} C
7.92×10^{-19} C

What do you think his value for the electron's charge would have been?

15. What is a quantum? What is a photon?

16. Discuss, in quantum terms, how a ladder works.

17. Distinguish between atomic number and atomic weight.

18. Distinguish between a continuous spectrum and a bright-line spectrum under the two headings:
    a. General appearance
    b. Source

19. How does the Bohr theory explain the many lines in the spectrum of hydrogen although the hydrogen atom contains only one electron?

20. Helium, neon, argon, krypton, xenon, and radon form a group of similar elements in that they form very, very few compounds. From their atomic structures, suggest a reason for this similarity in relative inactivity.

21. Compare your view of a valley as you walk from a mountain top to the floor of the valley with the progression of atomic theory.

22. What is constant about a compound?
    a. The weight of a sample of the compound
    b. The weight of one of the elements in samples of the compound
    c. The ratio by weight of the elements in the compound

23. If pure water is 88.8% oxygen and 11.2% hydrogen by weight,
    a. Is it likely to have *only* 88.8 g of oxygen in 100 g of water?
    b. Is it likely to have *exactly* 22.2 g of oxygen in 25.0 g of water?

24. If you found the number of wheels received by an assembly plant to be twice the number of motors, what type of vehicle would you assume to be assembled there? Of what chemical law does this remind you?

25. In recent years we have found that pure substances, such as some plastics, do vary in composition and that some elements can be decomposed (nuclear fission). What does this say to you about concepts and progress in science?

26. Why is it impossible to produce a positive charge without producing a negative charge at the same time?

27. Krypton is the name of Superman's home planet and also that of an element. Look up the element krypton, and list its symbol, atomic number, atomic weight, and electronic arrangement.

28. Explain in your own words why alpha particles are deflected in one direction in an electric field while beta particles are deflected in the opposite direction.

29. Read about lasers, and seek similarities between the explanation for the generation of laser light and the explanation for the production of bright-line elemental spectra.

30. Without looking at Figure 3–19 (except for checking later), write out the placement of electrons in their ground state
    a. Into energy levels according to the Bohr theory for atoms having 6, 10, 13, and 20 electrons.
    b. Into orbitals according to the wave-mechanical theory for atoms having 7, 13, 16, and 20 electrons.

4

Elements in Useful Order — The Periodic Table

Busy little bees make each cell a copy of each other cell so the pattern is repetitive or periodic.
(Charles D. Winters.)

A t this point in the story of chemistry, we have the first meaningful opportunity to discuss the **periodic table** (or **chart**), which you have probably noticed on the classroom wall. The careful study of the properties of the elements (defined and partially described in Chapter 2) led to the formulation of the periodic table, and the established periodic table furnishes evidence for atomic theory (Chapter 3).

What is there to know about the periodic table? Why is a so-called periodic table on a wall of most science classrooms and labs? Is it just a portrait of chemistry to adorn a wall, or is it useful? Why is the name "periodic" appropriate? Why is the table so arranged, and what are its important features? Does the table give order to the 109 known elements?

We shall discover in this chapter that the periodic table is important because it summarizes, correlates, and predicts a wealth of chemical information. In essence, the periodic table does bring order to 109 individual elements. Elements in an orderly arrangement provide the same benefits as your class notes arranged in a logical order, your room neatly and orderly arranged, or the goods arranged into departments in a store—ease of use and facilitation of understanding. From the standpoint of its logic, the periodic table can be of great help to a student of chemistry. As you read and study this chapter, look not only for what chemical information is summarized but how it is summarized, correlated, simplified, and predicted by the orderly arrangement of the elements in the periodic table.

ELEMENTS DESCRIBED

If we are to find and see order among the elements, we must have some general acquaintance with them. A few of the chemical and physical properties of 20 of the elements are summarized in Table 4–1. This format has been chosen so you can compare the properties more easily. The properties chosen for comparison are density, hardness, and relative reactivity.

The mass of a substance in a given volume is its **density**. For example, 1 mL of water at room temperature contains 1 g of matter; its density is 1 gram per milliliter (1 g/mL), whereas lithium (Li) has a density of 0.534 g/mL. A piece of lithium will float on water as it reacts with the water. Most gases have very low densities, about 0.001 g/mL, whereas many metals have densities much greater than that of water (Fig. 4–1).

Hardness is a relative term used to describe solids. Diamond, a form of the element carbon, is one of the hardest substances known; however, talc, like that used in talcum powder, is among the softest of the solids. **Reactivity** refers to the ease and intensity with which an element reacts with other elements. An element that is very reactive may react vigorously with air upon exposure, with moisture on your fingertips if you touch it, or with other elements. An element that is unreactive may form no compounds at all!

Metals usually have high reflectivity (known as metallic luster), the ability to be bent and drawn into wire without shattering, and higher densities than nonmetals. Most metals conduct heat and electricity well and react with *nonmetals.* Because most metals conduct heat so well, in a cool environ-

Figure 4–1 Density is mass per unit volume. Objects with less density float on liquids with greater density, and liquids with less density float on liquids with greater density if the two liquids do not mix. Cork floats on gasoline, the top liquid layer. Oak wood sinks in gasoline but floats on water, the middle layer. Brass sinks in gasoline and water but floats on mercury, the densest of the three liquid layers.

TABLE 4–1 Some Properties of 20 Elements

Element	Atomic Number	Description	Compound Formation *	
			With Cl (or Na)	With O (or Mg)
Hydrogen (H)	1	Colorless gas; reactive	HCl	H_2O
Helium (He)	2	Colorless gas; unreactive	None	None
Lithium (Li)	3	Soft metal; low density; very reactive	LiCl	Li_2O
Beryllium (Be)	4	Harder metal than Li; low density; less reactive than Li	$BeCl_2$	BeO
Boron (B)	5	Both metallic and nonmetallic; very hard; not very reactive	BCl_3	B_2O_3
Carbon (C)	6	Brittle nonmetal; unreactive at room temperature	CCl_4	CO_2
Nitrogen (N)	7	Colorless gas; nonmetallic; not very reactive	NCl_3	N_2O_5
Oxygen (O)	8	Colorless gas; nonmetallic; reactive	Na_2O, Cl_2O	MgO
Fluorine (F)	9	Greenish-yellow gas; nonmetallic; extremely reactive	NaF, ClF	MgF_2, OF_2
Neon (Ne)	10	Colorless gas; unreactive	None	None
Sodium (Na)	11	Soft metal; low density; very reactive	NaCl	Na_2O
Magnesium (Mg)	12	Harder metal than Na; low density; less reactive than Na	$MgCl_2$	MgO
Aluminum (Al)	13	Metal as hard as Mg; less reactive than Mg	$AlCl_3$	Al_2O_3
Silicon (Si)	14	Brittle nonmetal; not very reactive	$SiCl_4$	SiO_2
Phosphorus (P)	15	Nonmetal; low melting point; white solid; reactive	PCl_3	P_2O_5
Sulfur (S)	16	Yellow solid; nonmetallic; low melting point; moderately reactive	Na_2S, SCl_2	MgS
Chlorine (Cl)	17	Green gas; nonmetallic; extremely reactive	NaCl	$MgCl_2$, Cl_2O
Argon (Ar)	18	Colorless gas; unreactive	None	None
Potassium (K)	19	Soft metal; low density; very reactive	KCl	K_2O
Calcium (Ca)	20	Harder metal than K; low density; less reactive than K	$CaCl_2$	CaO

*The chemical formulas shown are lowest ratios. The molecular formula for $AlCl_3$ is Al_2Cl_6, and for P_2O_5 is P_4O_{10}.

ment metals feel colder to the touch than do most nonmetals. The metal is conducting heat from your skin and you feel cooler.

Nonmetals are insulators; that is, they are extremely poor conductors of heat and electricity. Their crystals are brittle and tend to shatter easily. Therefore nonmetals cannot be drawn into wire or beaten into shapes like metals. Many nonmetals are gases at room temperature. They are usually less dense than metals, and they react readily with metals and other non-metals (Fig. 4–2).

Although you probably know the properties of some of the elements listed in Table 4–1, our primary purpose is not to ask you to learn these properties, but rather to use them to search out any trends and similarities among the elements. Do you see any trends or similarities among the elements listed in the table? The elements in Table 4–1 are listed in atomic number order, but is there another, better arrangement for them?

(a)

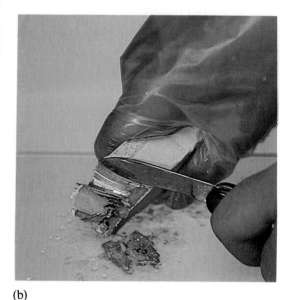

(b)

Figure 4–2 (a) Sulfur is a dull and brittle nonmetal. (Courtesy of Pennzoil Co.) (b) Sodium metal has the metallic sheen characteristic of metals. (Charles D. Winters.)

A helix is similar to a coiled spring. An early periodic table took this form.

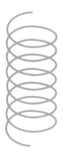

THE PERIODIC LAW—THE BASIS OF THE PERIODIC TABLE

The periodic law did not occur to anyone until 1869, although considerable information was available concerning the then-known elements. Parts of the complete idea had occurred as early as 1817 when Johann Wolfgang Döbereiner saw trends and similarities among several groups of three elements each, which he called *triads*. By 1862, A. Beguyer de Chancourtois saw similarities in elements along vertical lines when the elements were arranged in order of their atomic weights along a helix. A most interesting insight occurred in 1866 when John Newlands arranged elements in the order of their atomic weights and observed that every eighth element had similar properties. Newlands coined the "Law of Octaves" for which he was harshly ridiculed by his peers. All of these early ideas were incomplete and gained no lasting support.

On the evening of February 17, 1869, at the University of St. Petersburg (now Leningrad) in Russia, a 35-year-old professor of general chemistry, Dmitri Ivanovich Mendeleev (1834–1907), was writing a chapter for his soon-to-be-famous textbook on chemistry. He had the properties of each element written on a separate card for each element. While he was shuffling the cards trying to gather his thoughts before writing his manuscript, Mendeleev realized that if the elements were arranged in the order of their atomic weights, there was a trend in properties that repeated itself several times! Thus the periodic law and table were born, although only 63 elements had been discovered by 1869 (for example, the noble gases, He, Ne, Ar, Kr (krypton), Xe (xenon), and Rn (radon), were not discovered until after 1893), and the clarifying concept of the atomic number was not known until 1913.

	Group I R_2O RCl	Group II RO RCl_2	Group III R_2O_3 RCl_3	Group IV RO_2 RCl_4	Group V R_2O_5 RH_3	Group VI RO_3 RH_2	Group VII R_2O_7 RH	Group VIII RO_4
1	H = 1							
2	Li = 7	Be = 9.4	B = 11	C = 12	N = 14	O = 16	F = 19	
3	Na = 23	Mg = 24	Al = 27.3	Si = 28	P = 31	S = 32	Cl = 35.5	
4	K = 39	Ca = 40	— = 44	Ti = 48	V = 51	Cr = 52	Mn = 55	Fe = 56, Co = 59 Ni = 59, Cu = 63
5	(Cu = 63)	Zn = 65	— = 68	— = 72	As = 75	Se = 78	Br = 80	
6	Rb = 85	Sr = 87	?Yt = 88	Zr = 90	Nb = 94	Mo = 96	— = 100	Ru = 104, Rh = 104 Pd = 106, Ag = 108
7	(Ag = 108)	Cd = 112	In = 113	Sn = 118	Sb = 122	Te = 125	I = 127	
8	Cs = 133	Ba = 137	?Di = 138	?Ce = 140	—	—	—	— — — —
9	(—)	—	—	—	—	—		
10	—	—	?Er = 178	?La = 180	Ta = 182	W = 184	—	Os = 195, Ir = 197 Pt = 198, Au = 199
11	(Au = 199)	Hg = 200	Tl = 204	Pb = 207	Bi = 208	—	—	
12	—	—	—	Th = 231	—	U = 240	—	— — — —

Figure 4–3 An 1871 version of Mendeleev's periodic table. The formulas for simple oxides, chlorides, and hydrides are shown under each group heading. R represents the element in each group.

Within a month, Mendeleev had prepared a paper and had delivered it before the Russian Chemical Society. His idea and textbook achieved great success, and he rose to a position of prestige and fame as he continued to teach at St. Petersburg. In 1890, he resigned from the university during an episode of student unrest against the government, in which he sided with the students.

By 1871, Mendeleev published a more elaborate periodic table (Fig. 4–3). This version was the forerunner of the modern table currently seen in classrooms and textbooks.

Two features of the 1871 version were especially interesting. Empty spaces were left in the table, and there was a problem with the positions of tellurium (Te) and iodine (I).

The empty spaces showed the genius and daring of Mendeleev. He left the empty spaces to retain the rationale of ordered arrangement based on periodic recurrence of the properties. For example, in atomic weight order are copper (Cu), zinc (Zn), and then arsenic (As). If As had been placed next

**Dmitri Mendeleev (1834–1907).
Born in Siberia, Mendeleev rose to
Professor of Chemistry at St.
Petersburg (now Leningrad) and
then to director of the Russian
Bureau of Weights and Measures.
Although a prolific writer, a
versatile chemist and inventor, and a
popular teacher, the fame of this
brilliant scientist rests on his
discovery of the periodic law. (The
World of Chemistry, Program 7.)**

to Zn, As would have fallen under Al. But As forms compounds similar to those formed by P and antimony (Sb), not Al. Mendeleev reasoned that two as yet undiscovered elements existed and moved As over two spaces to the position below P. The two missing elements were soon discovered: gallium (Ga) in 1875 and germanium (Ge) in 1886. In later years the gaps in this 1871 periodic table were filled as the predicted elements were discovered.

Mendeleev aided the discovery of the new elements by predicting their properties with remarkable accuracy, and he even suggested the geographical regions in which minerals containing the elements could be found. The properties of a missing element were predicted by consideration of the properties of its neighboring elements in the table. He had learned from Döbereiner, perhaps, that the density of an element is approximately equal to the arithmetical average of the density of the lighter element above the missing element and the density of the heavier element just below. An example of Mendeleev's prediction of the properties of an undiscovered element is shown in Table 4–2. The term *eka* comes from Sanskrit, and means "one"; thus, *ekasilicon* means "one place away from silicon." He also predicted the properties of ekaboron (scandium) and ekaaluminum (gallium).

The empty spaces in the table and Mendeleev's predictions of the properties of missing elements stimulated a flurry of prospecting for elements in the 1870s and 1880s. As a result, Ga was discovered in 1875; scandium (Sc), samarium (Sm), holmium (Ho), and thulium (Tm) in 1879; gadolinium (Gd) in 1880; neodymium (Nd) and praseodymium (Pr) in 1885; and Ge and dysprosium (Dy) in 1886. Many of these elements are not even common today, yet they are important as ingredients in catalysts and color television screens.

If Mendeleev had followed the atomic weight order precisely, some elements with similar properties would not have been in the same column, or group. In the 1869 table, Te with an atomic weight of 128 was placed one position ahead of iodine, which has a lower atomic weight of 127. On the basis of its chemical properties, Te belonged with Sb, S, and O, and I belonged with F, Cl, and Br (bromine).

Mendeleev believed the atomic weight of tellurium was in error, but this was later shown not to be the case. In the 1871 table, the weight of Te had been changed from 128 to 125—an example of the unwise practice of

TABLE 4–2 Some of Mendeleev's Predicted Properties of Ekasilicon and the Corresponding Observed Properties of Germanium

	Ekasilicon (Es)	Germanium (Ge)
Atomic weight	72	72.6
Color of element	Gray	Gray
Density of element (g/mL)	5.5	5.36
Formula of oxide	EsO_2	GeO_2
Density of oxide (g/mL)	4.7	4.228
Formula of chloride	$EsCl_4$	$GeCl_4$
Density of chloride (g/mL)	1.9	1.844
Boiling point of chloride (°C)	Under 100	84

PERIODIC CHART

Among the most significant contributions to the modern periodic chart is that made by Nobel Laureate Glenn Seaborg. Among other things he demonstrated the importance of maintaining the courage of one's convictions.

Thanks to his insights it is now very well established that the transuranium elements (atomic numbers greater than 92), a number of which he either discovered or helped to discover during the Manhattan Project, are members of the actinide series. Actinides are the elements following actinium and belonging in a grouping off the main periodic chart.

Until Seaborg offered his version of the periodic table, chemists were convinced that Th, Pa, and U belonged in the main body of the table, Th under Hf, Pa under Ta, and U under W. When Seaborg proposed that Th was the beginning of the actinides and that the transuranium elements belonged as a group under the rare earths, some prominent and famous inorganic chemists, many of them Seaborg's friends, tried to discourage his publication of this finding in the open literature. One very prominent inorganic chemist felt that he would ruin his scientific reputation. Nevertheless Seaborg, strongly convinced, persisted and as a result, properly placed this most important class of elements where they are today. Based on Seaborg's expansion of the periodic table it was possible to predict accurately the properties of many of the as yet undiscovered transuranium elements. Subsequent preparation in atomic accelerators of these elements proved him right and it was fitting that he was awarded the Nobel Prize in 1951 for his outstanding work.

The World of Chemistry (Program 7) "The Periodic Table."

Glenn Theodore Seaborg (1912–) began his college education as a literature major, but changed to science in his junior year at the University of California. For his preparation and discovery of several transuranium elements, he shared the 1951 Nobel prize in chemistry with E. M. McMillan (1907–), who started Seaborg in this area of research. (*The World of Chemistry, Program 7.*)

---

changing data to fit a theory. The record is not clear as to why he changed the value.

Other reversed pairs in the modern periodic table are uranium (U) before neptunium (Np), Ar before K, cobalt (Co) before nickel (Ni), and thorium (Th) before protactinium (Pa). Upon realization of the atomic number concept in 1913, the question was resolved.

About nine months after Mendeleev delivered his paper before the Russian Chemical Society, Julius Lothar Meyer (1830–1895), a German physician and professor of chemistry at the University of Tübingen, prepared a table very similar to Mendeleev's. Apparently, both men were unaware of each other's work, yet both had left gaps for undiscovered elements. Meyer's table was based primarily on the repeatable trends in physical properties as the property is plotted against the atomic weight. Meyer grouped elements in subfamilies so, for example, zinc (Zn), cadmium (Cd), and mercury (Hg) with similar chemical properties could be separated from their chemical cousins Mg, Ca, strontium (Sr), and barium (Ba). Mendeleev's table was superior because Meyer did not predict the properties of the undiscovered elements, and he did not rectify atomic-weight position errors.

Building on the work of Mendeleev and Meyer, and others, and using the clarifying concept of the atomic number, we are now able to state the modern periodic law: *When elements are arranged in the order of their atomic numbers, their chemical and physical properties show repeatable trends.*

Refer again to Table 4–1 and note how the trend in properties from Li to Ne matches the trend from Na to Ar. The pattern in the properties of the elements, then, is *periodic;* hence the name *periodic law* or *table.* Other familiar periodic phenomena include the average daily temperature, which is periodic with time in a temperate climate. Low temperatures in January give way to high temperatures in July and low temperatures again in December. The trend repeats each year, not with exactly the same numbers but with the same pattern of change. Drowsiness follows a trend each 24 h. Cash flow follows a cycle related to pay day. Hunger pains may be periodic several times each day. A shingled roof has the same pattern over and over and is, therefore, periodic (Fig. 4–4).

So, to build up a periodic table according to the periodic law, line up the elements in a horizontal row in the order of their atomic numbers. Every time you come to an element with similar properties to one already in the row, start a new row. The columns, then, will contain elements with similar properties.

The properties of the elements are periodic functions of their atomic numbers.

Figure 4–4   A shingled roof with a repeatable pattern is one of many periodic phenomena.

FEATURES OF THE MODERN PERIODIC TABLE

A modern, popular version of the periodic table is shown in Figure 4–5. Note the following features.

The vertical columns are called **groups.**

The horizontal rows are called **periods.**

The letters A and B distinguish **families** of elements. For example, Group IA is the alkali metal family, and the closely related Group IB is sometimes called the coinage metal family.

The groups of elements are catalogued into four categories. The A groups are the **representative** elements. As we shall see, simple atomic theory represents these elements well. The B groups and Group VIII are the **transition** elements that link the two areas of representative elements. The **inner transition** elements are the lanthanide series, which fits between lanthanum (La) and hafnium (Hf), and the actinide series, which fits between actinium (Ac) and element 104. The **noble gases** are unique and comprise a group to themselves.

SELF-TEST 4–A

1. In which group of the periodic table are Mg, Pd, Cl, Ga, Ag?
2. In which period of the periodic table are Li, Mo, Nd, U, Br?
3. Mark the following as transition elements (T), representative elements (R), noble gases (N), or inner transition elements (I): Be, P, Cr, Kr, Am.

Atomic number — 11; Name — Sodium; Symbol — Na; Atomic weight — 22.9898; Electron structure — 2 8 1

Atomic weights are based on Carbon-12. Atomic weights in parentheses indicate the most stable or best-known isotope. Slight disagreement exists as to the exact electronic configuration of several of the high atomic-number elements.

Period	Group IA	IIA	IIIB	IVB	VB	VIB	VIIB	VIII	VIII	VIII	IB	IIB	IIIA	IVA	VA	VIA	VIIA	Noble Gases
1	1 Hydrogen H 1.0079 (1)																	2 Helium He 4.0026 (2)
2	3 Lithium Li 6.939 (2 1)	4 Beryllium Be 9.0122 (2 2)											5 Boron B 10.811 (2 3)	6 Carbon C 12.0112 (2 4)	7 Nitrogen N 14.0067 (2 5)	8 Oxygen O 15.9994 (2 6)	9 Fluorine F 18.9984 (2 7)	10 Neon Ne 20.183 (2 8)
3	11 Sodium Na 22.9898 (2 8 1)	12 Mag-nesium Mg 24.312 (2 8 2)											13 Aluminum Al 26.9815 (2 8 3)	14 Silicon Si 28.086 (2 8 4)	15 Phos-phorous P 30.9738 (2 8 5)	16 Sulfur S 32.064 (2 8 6)	17 Chlorine Cl 35.453 (2 8 7)	18 Argon Ar 39.948 (2 8 8)
4	19 Potassium K 39.098 (2 8 8 1)	20 Calcium Ca 40.08 (2 8 8 2)	21 Scandium Sc 44.956 (2 8 9 2)	22 Titanium Ti 47.90 (2 8 10 2)	23 Vanadium V 50.942 (2 8 11 2)	24 Chromium Cr 51.996 (2 8 13 1)	25 Manga-nese Mn 54.938 (2 8 13 2)	26 Iron Fe 55.847 (2 8 14 2)	27 Cobalt Co 58.933 (2 8 15 2)	28 Nickel Ni 58.71 (2 8 16 2)	29 Copper Cu 63.546 (2 8 18 1)	30 Zinc Zn 65.38 (2 8 18 2)	31 Gallium Ga 69.72 (2 8 18 3)	32 Germa-nium Ge 72.59 (2 8 18 4)	33 Arsenic As 74.922 (2 8 18 5)	34 Selenium Se 78.96 (2 8 18 6)	35 Bromine Br 79.904 (2 8 18 7)	36 Krypton Kr 83.80 (2 8 18 8)
5	37 Rubidium Rb 85.47 (2 8 18 8 1)	38 Strontium Sr 87.62 (2 8 18 8 2)	39 Yttrium Y 88.905 (2 8 18 9 2)	40 Zirconium Zr 91.22 (2 8 18 10 2)	41 Niobium Nb 92.906 (2 8 18 12 1)	42 Molyb-denum Mo 95.94 (2 8 18 13 1)	43 Techne-tium Tc (99) (2 8 18 14 1)	44 Ruthe-nium Ru 101.07 (2 8 18 15 1)	45 Rhodium Rh 102.905 (2 8 18 16 1)	46 Palladium Pd 106.4 (2 8 18 18 0)	47 Silver Ag 107.868 (2 8 18 18 1)	48 Cadmium Cd 112.40 (2 8 18 18 2)	49 Indium In 114.82 (2 8 18 18 3)	50 Tin Sn 118.69 (2 8 18 18 4)	51 Antimony Sb 121.75 (2 8 18 18 5)	52 Tellurium Te 127.60 (2 8 18 18 6)	53 Iodine I 126.904 (2 8 18 18 7)	54 Xenon Xe 131.30 (2 8 18 18 8)
6	55 Cesium Cs 132.905 (2 8 18 18 8 1)	56 Barium Ba 137.34 (2 8 18 18 8 2)	*57 Lan-thanum La 138.91 (2 8 18 18 9 2)	72 Hafnium Hf 178.49 (2 8 18 32 10 2)	73 Tantalum Ta 180.948 (2 8 18 32 11 2)	74 Wolfram (Tungsten) W 183.85 (2 8 18 32 12 2)	75 Rhenium Re 186.2 (2 8 18 32 13 2)	76 Osmium Os 190.2 (2 8 18 32 14 2)	77 Iridium Ir 192.2 (2 8 18 32 17 0)	78 Platinum Pt 195.09 (2 8 18 32 17 1)	79 Gold Au 196.967 (2 8 18 32 18 1)	80 Mercury Hg 200.59 (2 8 18 32 18 2)	81 Thalium Tl 204.37 (2 8 18 32 18 3)	82 Lead Pb 207.19 (2 8 18 32 18 4)	83 Bismuth Bi 208.980 (2 8 18 32 18 5)	84 Polonium Po (210) (2 8 18 32 18 6)	85 Astatine At (210) (2 8 18 32 18 7)	86 Radon Rn (222) (2 8 18 32 18 8)
7	87 Francium Fr (223) (2 8 18 32 18 8 1)	88 Radium Ra (226) (2 8 18 32 18 8 2)	**89 Actinium Ac (227) (2 8 18 32 18 9 2)	104	105	106	107	108	109									

Transition elements: groups IIIB through IIB.

Inner transition elements

Series														
* Lanthanide series 6	58 Cerium Ce 140.12 (2 8 18 20 8 2)	59 Praseo-dymium Pr 140.907 (2 8 18 21 8 2)	60 Neo-dymium Nd 144.24 (2 8 18 22 8 2)	61 Prome-thium Pm (147) (2 8 18 23 8 2)	62 Samarium Sm 150.35 (2 8 18 24 8 2)	63 Europium Eu 151.96 (2 8 18 25 8 2)	64 Gado-linium Gd 157.25 (2 8 18 25 9 2)	65 Terbium Tb 158.924 (2 8 18 27 8 2)	66 Dyspro-sium Dy 162.50 (2 8 18 28 8 2)	67 Holmium Ho 164.930 (2 8 18 29 8 2)	68 Erbium Er 167.26 (2 8 18 30 8 2)	69 Thulium Tm 168.934 (2 8 18 31 8 2)	70 Ytter-bium Yb 173.04 (2 8 18 32 8 2)	71 Lutetium Lu 174.97 (2 8 18 32 9 2)
** Actinide series 7	90 Thorium Th 232.038 (2 8 18 32 18 10 2)	91 Protac-tinium Pa (231) (2 8 18 32 20 9 2)	92 Uranium U 238.03 (2 8 18 32 21 9 2)	93 Nep-tunium Np (237) (2 8 18 32 22 9 2)	94 Plu-tonium Pu (242) (2 8 18 32 23 9 2)	95 Ameri-cium Am (243) (2 8 18 32 25 8 2)	96 Curium Cm (247) (2 8 18 32 25 9 2)	97 Berkelium Bk (247) (2 8 18 32 26 9 2)	98 Califor-nium Cf (251) (2 8 18 32 27 9 2)	99 Einstein-ium Es (254) (2 8 18 32 28 9 2)	100 Fermium Fm (253) (2 8 18 32 29 9 2)	101 Mende-levium Md (256) (2 8 18 32 30 9 2)	102 Nobelium No (254) (2 8 18 32 31 9 2)	103 Lawren-cium Lr (257) (2 8 18 32 32 9 2)

Figure 4–5 Periodic table of the elements. (1) Names and symbols are based on IUPAC names: 104, Unnilquadium; 105, Unnilpentium; 106, Unnilhexium; 107, Unnilseptium; and 109, Unnilennium. (2) Element 108 is claimed, but not confirmed. (3) Element 109 was reportedly discovered in 1982. (Modified from Morris Hein, et al., *Foundations of Chemistry in the Laboratory,* 4th ed. Belmont, CA, Dickenson Publishing Co., 1977.)

4. Who was primarily responsible for formulating the period table?
5. According to the periodic law, when the elements are arranged in the order of their _____, their properties show periodicity.
6. When a phenomenon shows the same pattern over and over, we say the pattern is _____.
7. Elements that conduct heat and electricity well are classified as _____.
8. Elements that do not conduct heat and electricity well are gases and brittle solids. These are classified as _____.

USES OF THE PERIODIC TABLE

The periodic table is useful to chemists and students of chemistry in many ways. In addition to being a handy reference for atomic weights and numbers, among other information, the arrangement of the elements—the position of an element in the table—presents a wealth of useful information.

In the following sections, we shall describe three major uses of the periodic table: (1) elements in a group have similar properties, (2) elements in successive periods show repeating trends in their properties, and (3) the periodic table supports theory and relates theoretical concepts. These memory aids will help you as you study chemistry.

Elements in a Group Have Similar Properties

Refer back to Table 4–1 and note the properties of Li, Na, and K. They are soft metals, have low density, are very reactive, and form chlorides and oxides with the formulas MCl and M_2O. Now notice on the modern periodic table (Fig. 4–5) that rubidium (Rb), cesium (Cs), and francium (Fr) are also in the same group, Group IA. What properties would you expect Rb, Cs, and Fr to have? If you predicted soft metals, low density, very reactive, MCl, and M_2O, you are right. Elements in a group have similar properties, but not the same properties.

In formulas MCl and M_2O, M represents an alkali metal.

Some properties of elements in a group differ by degree in a regular pattern. For example, the melting points beginning with Li and going down the column through Cs are (in °C) 179, 98, 64, 39, and 28, respectively. Lithium reacts slowly with water, sodium reacts faster, potassium still faster, and for the elements at the bottom of the group, just exposure to moist air produces a vigorous explosion (Fig. 4–6).

Some properties differ by degree but not in a regular pattern. For example, the densities (in g/mL) of the solids Li through Cs are 0.53, 0.97, 0.86, 1.53, and 1.87, respectively.

Some properties are the same for every member of a group. Elements in a group generally react with other elements to form similar compounds. This is the most useful and powerful inference that can be made from the periodic table. For example, if the formula for the compound composed of Li and Cl is LiCl, then probably there is a compound of Rb and Cl with the formula

(a)

(b)

Figure 4-6   (a) Sodium reacts with water. (b) Potassium is a more active metal, reacting quickly with either air or water. (*The World of Chemistry,* Program 7.)

RbCl; the compound for Rb and Br (in Group VIIIA with Cl) would probably have the formula of RbBr. Likewise, if the formula Na_2O is known, then a compound with the formula K_2S predictably exists. This ability to predict formulas from the periodic table has limitations. For example, Na, K, Rb, and Cs all form superoxides (formula MO_2), but no superoxide with Li is known. However, the limitations do not prohibit the use of the periodic table for predicting formulas. In general, elements in the same group of the periodic table form some of the same types of compounds!

Hydrogen probably should be in a group by itself, although you may see H in both Group IA and Group VIIA in some periodic tables. Hydrogen forms compounds with formulas similar to those of the alkali metals, but with vastly different properties, such as those between NaCl and HCl; Na_2O and H_2O. Hydrogen also forms compounds similar to those of the halogens: NaCl and NaH (sodium hydride); $CaBr_2$ and CaH_2 (calcium hydride).

Hydrogen—the element without a home on the periodic table.

Four groups of elements in the periodic table are referred to by the name of the group. These four groups and some of their general properties are described below.

The **alkali** metals are Group IA (Li, Na, K, Rb, Cs, and Fr). The name *alkali* derives from an old word meaning "ashes of burned plants." When alkali metal oxides react with water, alkalis are formed. Alkalis (later in this book to be called bases) taste bitter (but don't try it!), feel slick when in solution, and neutralize or destroy the effects of acids. A common alkali is sodium hydroxide (NaOH), known commercially as lye. It is formed when sodium oxide (Na_2O) reacts with water.

$$Na_2O + H_2O \longrightarrow 2\ NaOH$$

All of the 21 isotopes of the alkali metal francium are naturally radioactive.

Radioactivity is discussed in Chapter 7.

The **alkaline earths** are Group IIA (Be, Mg, Ca, Sr [strontium], Ba [barium], and Ra [radium]). They are metals harder and less reactive than the alkali metals. They form MCl_2 and MO. As oxides, some react with water to form alkalis. For example, CaO (lime) reacts with water to form $Ca(OH)_2$ (slaked lime), a widely used alkali because of its low cost. Theatrical "limelight," a brilliant white light, got its name from calcium oxide being placed in

Figure 4-7   Why the halogens are in the same chemical group is not obvious from the color and state of the elements. At room temperature, chlorine is a greenish-yellow gas, bromine is a red liquid, and iodine is a grayish-purple solid. Formulas of compounds formed by metals with the halogens confirm the similarities among Cl_2, Br_2, and I_2, as well as with F_2 and At_2. The same number of valence electrons in each atom of the halogens (seven) is the cause of compound formation similarity among the halogens. (Charles Steele.)

Aliases: noble gases, inert gases, rare gases.

an electric arc. In the Middle Ages, an *earth* was any solid substance that did not melt and was not changed by fire into some other substance. Under these conditions, many of the alkaline-earth compounds change into oxides, having high melting temperatures (in excess of 1900°C) and the same general white appearances of the original compounds. These early investigators could not attain temperatures high enough to melt the oxides; hence, the name *earth* was applied and has stuck to this day.

The **halogens** are Group VIIA (F, Cl, Br, I, and At [astatine]; Fig. 4-7). Fluorine and Cl are gases at room temperature, whereas Br is a liquid and I is a solid. In the elemental state, each of these elements exists as diatomic molecules (X_2). All isotopes of At are naturally radioactive and disintegrate quickly. If you could accumulate enough astatine, it would be a solid at room temperature. The name *halogen* comes from a Greek word and means "salt-producing." The most famous salt involving a halogen is sodium chloride (NaCl), table salt. But there are many other halogen salts such as calcium fluoride (CaF_2), a natural source of fluorine; potassium iodide (KI), an additive to table salt that prevents goiter; and silver bromide (AgBr), the active photosensitive component of photographic film.

The **noble gases** are He, Ne, Ar, Kr, Xe, and Rn. All are colorless monatomic gases at room temperature. They are referred to as noble because they generally lack chemical reactivity. The derivations of some of the names of these elements are consistent with their inactivity: argon (from the Greek, *argon,* meaning "inactive"); xenon (from the Greek, *xenon,* meaning "stranger"). Helium (Greek, *helios,* meaning "the sun") was discovered by analysis of the sun's light; later it was found on Earth. Neon (Greek, *neos,* meaning "new") is the common gas that glows in the tubes of "neon" lights. Neon glows red when excited in a discharge lamp and argon glows blue (Fig.

Figure 4–8 Noble gases in cathode ray discharge tubes give characteristic glows. Neon glows red; argon glows blue. (*The World of Chemistry,* Program 8.)

4–8). Other gases and painted tubes are used to give different colors. Radon is naturally radioactive.

Until 1962, it was thought that all of the noble gases had absolutely no chemical reactivity. On some older periodic tables, the noble gas column was called "inert gases." Many reasons were presented to explain why the noble gases were inactive and why they never would react. Beginning in 1962, however, the situation began to change. A Canadian, Neil Bartlett, prepared the compound O_2PtF_6. Realizing that xenon might also form a similar compound, he discovered the first noble-gas compound, $XePtF_6$. His discovery was followed quickly by the work of scientists at Argonne National Laboratory, who made some 30 compounds involving the heavier members of the noble gases combined with fluorine or oxygen. Some of the first prepared compounds were KrF_2, KrF_4, XeF_2, XeF_4, XeF_6, XeO_3, XeO_4, and RnF_4. No compounds with He, Ne, or Ar have yet been reported.

The name *radon* comes from "radium" (Latin, *radius,* meaning "ray"). At first, radon was called niton (Latin, *nitens,* meaning "shining"). The connection between radon, cancer, and basements is discussed in Chapter 7.

Xenon and oxygen have similar ionization energies (defined later in this chapter).

Similar Trends Occur in Successive Periods

What should be remembered about the periodic nature of certain properties is where the low values and high values occur. For example, from Table 4–1, the number of chlorine atoms combining with elements of the period Li through Ne is 1-2-3-4-3-2-1-0, respectively. The same trend occurs for Na through Ar (Fig. 4–9). These numbers correspond to the old term *combining*

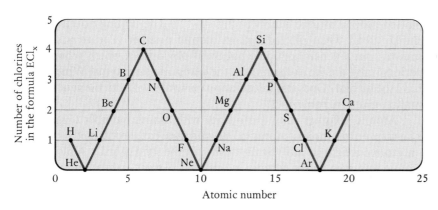

Figure 4–9 The number of chlorine atoms in the formula ECl_x, where E represents the element and x is the subscript on chlorine. The number of chlorine atoms is the combining power or valence of E.

Figure 4-10   Metals are the light-colored blocks in the center and on the left. The nonmetals are dark-colored blocks on the right. Metalloids are the few elements between the metals and nonmetals. (*The World of Chemistry,* Program 19.)

Valence electrons are the electrons in the shell most distant from the nucleus.

power, or **valence** (Latin, *valens,* meaning "strength"). If each chlorine has a combining power of one, then Li has a combining power of one (LiCl, one Li to one Cl), Be a combining power of two ($BeCl_2$), B a combining power of three (BCl_3), and so on. A modern, atomic-theory interpretation of combining power is discussed in the next chapter.

From left to right across each period, metallic character gives way to nonmetallic character (Fig. 4-10). The elements with the most metallic character are at the lower left part of the period table near Fr. The elements with the most nonmetallic character are found near the upper right portion of the periodic table near F.

Metallic character means having properties of metals (see page 88 and bonding).

The heavy line on the periodic table that begins at B and staircases down to At roughly separates the metals and the nonmetals. Most of the elements (about 80%) lie to the left of this line and are considered metals. The elements positioned along the line are considered **semimetals** or **metalloids.** Their properties are intermediate between those of metals and nonmetals. For example, Si, Ge, and As are **semiconductors,** and B conducts electricity well only at high temperatures. It is these semiconductor elements that form the basic components of memory chips and computer logic.

Semiconductors conduct electricity less than metals such as silver and copper, but more than insulators such as sulfur; semiconductors are components of transistors.

Notice the periodic patterns of melting points of the elements when plotted against atomic number (Fig. 4-11). The boiling points follow a similar trend when plotted against atomic number. The trends are not smooth, but a general periodic pattern is obvious. In which groups of the periodic table are the elements with the lowest melting points? Which groups have the highest? Does this information correlate well with the general information given in Table 4-1?

Atomic volumes show periodicity with atomic number (Fig. 4-12). The volume of a mole of atoms in the solid state can be obtained by dividing the atomic weight (g/mol) by the density of the solid (g/mL). A plot of such atomic volumes versus atomic weight was a main exhibit in support of Meyer's periodic table. Why do atoms get larger from top to bottom of a group? Do you suppose it has something to do with more layers making a larger onion? Yes, by analogy, the larger atoms simply have more energy levels (orbitals) inhabited by electrons than do the smaller atoms.

Larger atoms have more shells occupied by electrons.

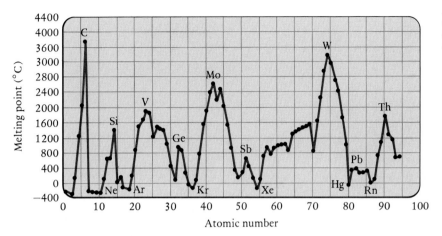

Figure 4–11 The periodic nature of the melting points of the elements when plotted versus atomic number.

Atomic volumes decrease across a period from left to right. You may see a paradox in adding electrons and getting smaller atoms, but you are adding protons, too. The greater nuclear charge pulls electrons in similar orbitals (same shell) closer to the nucleus and causes contraction of the atomic volume.

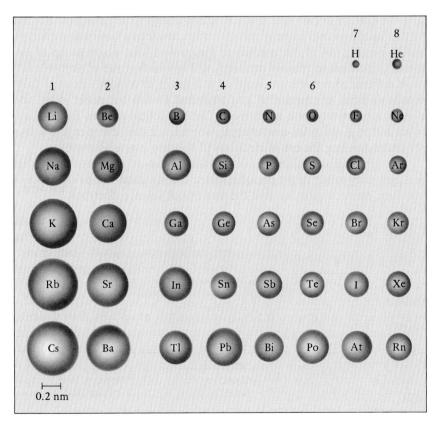

Figure 4–12 Relative sizes of atoms of the A group elements. Atoms increase in size as one goes down a group and, in general, decrease in going across a row in the periodic table. Hydrogen has the smallest atom, cesium the largest.

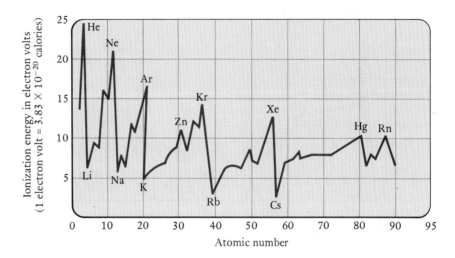

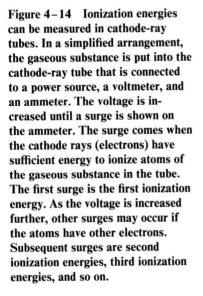

Figure 4–14 Ionization energies can be measured in cathode-ray tubes. In a simplified arrangement, the gaseous substance is put into the cathode-ray tube that is connected to a power source, a voltmeter, and an ammeter. The voltage is increased until a surge is shown on the ammeter. The surge comes when the cathode rays (electrons) have sufficient energy to ionize atoms of the gaseous substance in the tube. The first surge is the first ionization energy. As the voltage is increased further, other surges may occur if the atoms have other electrons. Subsequent surges are second ionization energies, third ionization energies, and so on.

Periodic relationships are also seen when the first ionization energies of the elements are plotted against atomic number (Fig. 4–13). The **first ionization energy** is the energy required to remove the first electron from an atom; the energy required to remove the second electron is the second ionization energy, and so forth for the third and subsequent electrons. Ionization energies can be determined experimentally for some elements by inserting the gaseous element into a cathode-ray tube and increasing the voltage until a surge of current occurs (the first ionization energy), increasing the voltage further until a second surge of current occurs (the second ionization energy), and so on (Fig. 4–14). According to Figure 4–13 for which group of elements is it easiest to remove an electron? For which group of elements is it most difficult to remove an electron? Is it easier to remove electrons from metals or from nonmetals? If you answered Group IA, noble gases, and metals, respectively, you are correct. The theoretical reasons for your answers will be given in the next chapter, where ionization energies give a basis for understanding the use of electrons in bonding atoms to atoms.

In subsequent chapters, look for other trends in elemental properties and how they relate to the periodic table. Among these will be the types of bonding, electronegativity, and the number of valence electrons.

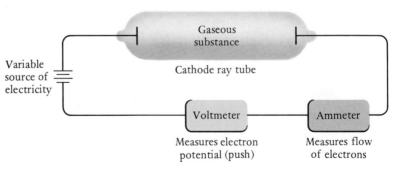

The Periodic Table and Atomic Theory Support Each Other

Atomic theory and the periodic table support the validity of each other. Atomic theory justifies the arrangement of the elements in the periodic table. The periodic table provides observational evidence for the growing understanding of trends in atomic structure. Chronologically, the periodic table was established prior to the development of our modern atomic theory and helped to make the general acceptance of modern atomic theory possible.

Why do elements in the same group in the periodic table have similar chemical behavior? The answer is because all of the elements in a group (particularly the representative elements, the A groups, and the noble gases) have atoms with similar structural features. Figure 4–5 gives the ground-state positions of electrons in orbitals for atoms with atomic numbers through 103. The electronic structures for the first 20 atoms are repeated in Figure 4–15. Note that in Group IA each element has atoms with one and only one electron in the outermost occupied shell (called the **valence shell**). Group IIA elements all have two electrons in the valence shells of their atoms. Group IIIA elements and atoms all have three electrons in the valence shell, and the pattern continues through Group VIIA. Note that the group number is the number of electrons in the valence shell of each atom in the group.

What is the structural feature of the noble gases that results in their having little or no chemical reactivity? The noble gases have eight electrons in the valence shell of each atom (except for He, which has a total of only two electrons). Eight electrons in the valence shell seem to provide a balanced, stable, structural arrangement that minimizes the tendency of an atom to react with other atoms.

The periods are not equal in size. Periods one through seven have 2, 8, 8, 18, 18, 32, and 23 (incomplete) elements, respectively. Since the larger elements have larger shells in which to place electrons, it takes more electrons (added to elements one by one) before the larger shells are filled and the few valence electrons are added to a new shell. Larger periods as the atoms of elements get larger are similar to longer rows and more seats per row in a stadium as you proceed from the field to higher in the stands (Fig. 4–16).

The valence shell is the outermost occupied orbit.

Eight electrons in the valence shell provide a stable electronic arrangement.

Figure 4–15 Electron arrangements of the first 20 elements. Above each column, or group, is the group number. The nuclear contents of a typical isotope are shown.

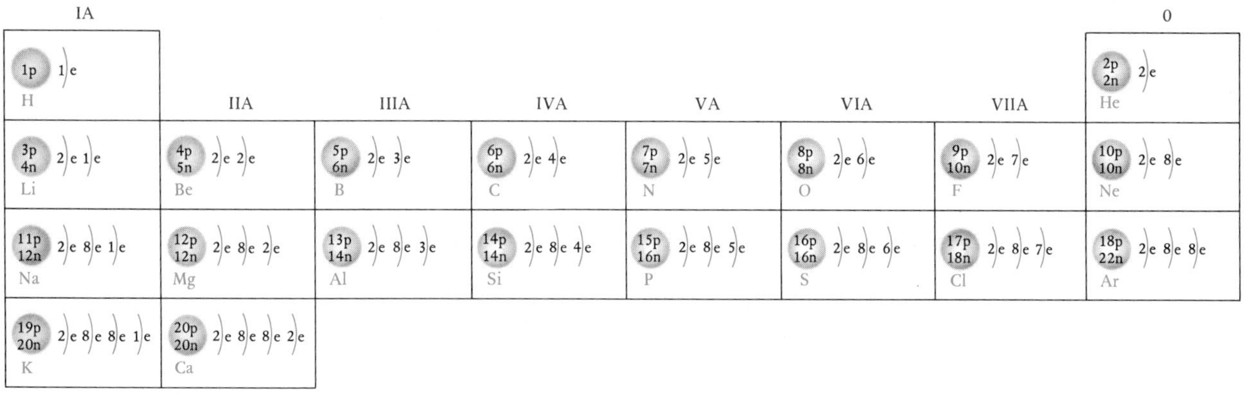

Figure 4-16   Periods in the periodic table and number of seats in rows in a stadium increase as the circles get larger. The circles (shells) get larger as shells of the elements are located further from the nucleus, similar to the length of the rows increasing as the rows are further from the field of play. See the text for an explanation. (Courtesy of Department of Athletics, Vanderbilt University.)

Nuclear accelerators are discussed in Chapter 7.

Why are there repeatable patterns of properties across the periods in the periodic table? Again, it is because there is a repeatable pattern in atomic structure. Each period begins with one electron in the valence shell of the atoms of the elements in Group IA. Each period builds up to eight electrons in the valence shell, and the period ends. This pattern repeats across periods two through six. As more elements are made by nuclear accelerators, period seven may be completed someday. When it does, the periodic table and atomic theory predict the last element in period seven (the final member of the noble gases) will be element number 118, with eight electrons in the valence shells of its atoms.

Atomic theory and the periodic table complement each other perfectly. One verifies the other, and vice versa.

THE PERIODIC TABLE IN THE FUTURE

The periodic table ties together well what is known about familiar elements, and it predicts accurately properties of unfamiliar elements. It is an indispensable memory aid. It makes intelligent and informal guessing easy, especially when it comes to predicting chemical formulas. All of these benefits are very important to students of chemistry.

Beyond these benefits, perhaps the most elegant contribution of the periodic table toward understanding nature is its stimulation of research. We are already aware of how Mendeleev's gaps stimulated the search for new elements. Going a step further, there is now active research under way to make elements beyond element 109. (There is a paradox here in using huge nuclear accelerators that cover many acres to try to shoot alpha particles and other very small particles into unseen, very small nuclei.) Part of the stimulus for this research is to see whether the prepared elements have the properties predicted by the periodic table. Without the periodic table, this reaching out would not be occurring. At least 20 elements have already been produced or discovered since 1939, bringing the number of known or claimed elements to 109. Predictions on elements 110 through 118 are that they will be very stable but still radioactive. Element 118 is expected to be a noble gas.

The periodic table will continue to stimulate the making of new compounds as it has in the past. An example from the past is sodium perbromate ($NaBrO_4$). Sodium perchlorate ($NaClO_4$) and sodium periodate ($NaIO_4$) had been known for some time. The fact that $NaBrO_4$ could not be prepared by the same methods was puzzling since Br is between Cl and I in Group VIIA of the periodic table, and all three elements should form similar compounds. All attempts failed (at least seven papers appeared in the chemical literature detailing why $NaBrO_4$ would never be made) until 1968, when another new compound, XeF_2, reacted with sodium bromate ($NaBrO_3$) in water to produce sodium perbromate for the first time. Faith in the correctness and predictive powers of the periodic table stimulated this research.

In the future, the groups in the periodic table will likely be labeled differently. Since 1959, the International Union of Pure and Applied Chemistry (IUPAC) has been considering the differences between the European and American practices of labeling A and B groups. For example, Group IIIA in European usage is Group IIIB in American practice. The IUPAC has decided to avoid the issue by recommending that the groups be labeled 1 through 18 consecutively from left to right.

IUPAC resolves subjective issues related to chemistry.

"Similarities within groups" are "similarities within groups"; new labels will not change the integrity of the form of the table. This text uses the American version. The preference for this version for introductory chemistry courses is to use "A" for all "representative" elements and "B" for all "transition" elements. As a result, the number of the "A" column is the number of outer (valence) electrons in the atoms.

"A rose by any other name . . . " is still a rose.

The periodic table is not a panacea for the chemist, but it is an important method for tying the properties and relationships of the elements together. For its place in *your* future, we propose this hypothesis: the periodic table will be your most lasting memory of the chemistry you study in this course. As you look at that portrait of chemistry on your classroom wall, what do you see now that you did not see before you read this chapter? You now see what you may not have seen before—a requirement for the appreciation of art—and of science.

SELF-TEST 4-B

1. What are the combining power and formula for the chloride compound of: Ga? Ba? Se? I?
2. Classify each of the following as a metal, nonmetal, or metalloid: Si, Ce, Cl, Cs, Ca, O, H, Ge.
3. How many electrons are in the valence shell of Na, Ca, F, Cl, O, Al, C?
4. The amount of energy required to remove an electron from an atom is called the _____ energy.
5. Which element in each pair has the greater ionization energy? He or O, Na or F, Ca or Br, K or S.
6. Which element in each pair has the larger atoms? Li or K, F or Br, Na or S, B or In.
7. In which groups of the periodic table would elements with the following electron configurations be found? 2-8-1, 2-8-4, 2-8-8-2.

MATCHING SET

____ 1. Periodic
____ 2. Ionization energy
____ 3. Larger atoms
____ 4. Two valence electrons
____ 5. A noble gas
____ 6. A metal
____ 7. A nonmetal
____ 8. A halogen
____ 9. An inner transition element
____ 10. Valence shell

a. Electron arrangement 2-8-2
b. Generally a gas or a brittle solid
c. Greater for Group VIIA than for Group IA
d. Eight valence electrons
e. At the bottom of a group
f. Praseodymium (Pr)
g. Electron arrangement 2-8-1
h. Seven valence electrons
i. Repeated pattern
j. Ruthenium (Ru)
k. Outermost occupied shell

QUESTIONS

1. State the periodic law.
2. How did the discovery of the periodic law lead to the discovery of elements?
3. Omitting argon, write the formulas for a bromide of each of the elements with atomic numbers 11 through 20.
4. From their position in the periodic table, predict which will be more metallic: (a) Be or B; (b) Be or Ca; (c) Ca or K; (d) As or Ge; (e) As or bismuth (Bi).
5. Use the information on the periodic chart to supply the following:
   a. The nuclear charge on cadmium (Cd)
   b. The atomic number of As
   c. The atomic mass (or mass number) of an isotope of Br having 46 neutrons
   d. The number of electrons in an atom of Ba
   e. The number of protons in an isotope of Zn
   f. The number of protons and neutrons in an isotope of Sr, atomic mass (or mass number) of 88
   g. An element forming compounds similar to those of Ga
6. In a general way, how do average daily temperatures over the past three years at your location relate to properties and electron structures of the elements when the elements are taken in the order of their atomic numbers? List other common occurrences you experience that are periodic.
7. Given one formula, based on the positions of the elements in the periodic table, predict the other formula
   a. $BaCl_2$; formula for Sr and Br
   b. Na_2S; formula for K and Se
   c. Al_2O_3; formula for Ga and S
   d. NCl_3; formula for P and Br

8. Sodium reacts violently with water and forms hydrogen in the process. Magnesium will react with water only when the water is very hot. Copper does not react with water. Suppose you find a bottle containing a lump of metal in a liquid and a label, "Cesium (Cs)." Based on your knowledge of the perioidic table, what danger is there, if any, of disposing of the metal by throwing it into a barrel of water?
9. Write the symbols of the halogen family in the order of increasing size of their atoms.
10. Why does Cs have larger atoms than Li?
11. What general electron arrangement is conducive to chemical inactivity?
12. How are the elements in a group related to each other?
13. Write the names and symbols of the alkaline-earth elements.
14. Write the symbols for the family of elements that have three electrons in the valence shells of their atoms.
15. What is common about the electron structures of the alkali metals?
16. Pick the electron structures below that represent elements in the same chemical family.
   a. $1s^2 2s^1$
   b. $1s^2 2s^2 2p^4$
   c. $1s^2 2s^2 2p^2$
   d. $1s^2 2s^2 2p^6 3s^2 3p^4$
   e. $1s^2 2s^2 2p^6 3s^2 3p^6$
   f. $1s^2 2s^2 2p^6 3s^2 3p^6 4s^2$
   g. $1s^2 2s^2 2p^6 3s^2 3p^6 4s^1$
   h. $1s^2 2s^2 2p^6 3s^2 3p^6 3d^1 4s^2$
17. Complete the following table:

Atomic No.	Name of Element	Electron Structure	Period	Metal or Nonmetal
6				
12				
17				
37				
42				
54				

18. How many electrons does the last element in each period have in its valence shell?

19. Answer this question without referring to the periodic table. Element number 55 is in Group IA, period 6. Describe its valence shell when all electrons are in the ground state. In how many shells are there electrons?

20. If element 36 is a noble gas, in what groups would you expect elements 35 and 37 to occur?

21. Oxygen and sulfur are very different elements in that one is a colorless gas and the other a yellow crystalline solid. Why, then, are they both in Group VIA?

22. Suppose the popular press reports the discovery of a large deposit of pure sodium in northern Canada. What is your reaction as an informed citizen?

23. True or False. There are more nonmetallic elements than metallic elements.

24. Many elements are known to form compounds with hydrogen. Letting E be an element in any group, the following table represents the possible formulas of such compounds.

Group	IA	IIA	IIIA	IVA	VA	VIA	VIIA
	EH	EH_2	EH_3	EH_4	EH_3	H_2E	HE

Following the pattern in the table, write the formulas for the hydrogen compounds of (a) Na, (b) Mg, (c) Ga, (d) Ge, (e) As, (f) Cl.

25. Write the symbol for an alkali metal, a lanthanide, an alkaline earth, a halogen, an actinide, and a transition metal (first series).

26. Below are some selected properties of Li and K. Before looking up the number, estimate values for the corresponding properties of Na.

	Lithium	Sodium	Potassium
Atomic weight	6.9	—	39.1
Density (g/cm³)	0.53	—	0.86
Melting point (°C)	180	—	63.4
Boiling point (°C)	1330	—	757

27. Give the names and symbols for two elements most like selenium (Se), atomic number 34.

28. What is the likelihood of discovering another family of elements such as the noble gases?

29. Complete the following by writing the predicted formulas.

Element	F	O	Cl	S	Br	Se
Na						
K						
B						
Al						
Ga						
C						
Si						

30. If element 118 is ever produced, what will be its position in the periodic table?

5

Chemical Bonds: The Ultimate Glue

Calcite crystal, calcium carbonate. (Brian Parker/Tom Stack and Associates.)

What holds matter together? Why does glue stick? What causes pieces of hard candy to stick together? Why is diamond so hard; why is wax soft? Or, in reverse, why do things break or fall apart? Why is table salt so brittle? Why does some paint peel? Why do some substances melt at a rather low temperature, and others melt at higher temperatures?

These and similar questions can be answered logically, and the answers will be consistent with experimental evidence if we think of matter as atoms bound to each other. Granted, it is a little hard to consider the Sears Tower or the Washington Monument or a living organism as a conglomeration of atoms bonded one to the other. But large pieces of matter, even the Rocky Mountains, conform to the same fundamental principles of nature as a small crystal of sugar or salt.

Chemical bonds hold atoms, molecules, and ions together.

Most of the reasons for matter bonding to matter (or atom bonding to atom) can be summarized by two concise notions:

1. Unlike charges attract.
2. Electrons tend to exist in pairs.

Couple these two ideas (one empirical, one theoretical) with the proximity requirement that only the outer electrons of the atoms (the *valence electrons*) interact, and you have the basic concepts that explain how atoms in over 10 million compounds bond to each other and how the surface of one material might be attracted or indifferent to another substance (water wets glass but not wax). Just how the different atoms use these principles to bond atom to atom is the subject of this chapter. We shall see that the action of an atom in the formation of a bond is dictated by its atomic structure, governed generally by its position in the periodic table.

Various interactions of the atoms cause the formation of five major types of chemical bonds (Fig. 5 – 1). These theoretical types of bonds, along with some common materials in which they occur, are:

1. **Ionic bonding:** Salts, such as table salt (sodium chloride); and metal oxides, such as lime, iron rust, ruby, and sapphire
2. **Covalent bonding:** Molecular compounds, such as water, methane, and sugar; and polymers, such as polyethylene
3. **Intermolecular bonding**
   a. Hydrogen bonding: water, ammonia, DNA, and proteins
   b. London forces: Liquid helium and solid CO_2 (dry ice)
4. **Metallic bonding:** Metals and alloys

As always, it is the properties of the substances that dictate and verify the related theories. Properties such as chemical reactivity, volatility (ability to pass into the gaseous state), melting point, electrical conductivity, and color often give some indication of how atoms are bonded to each other. For example, since melting involves atoms or molecules becoming less firmly bound to their neighbors, a high melting point implies that a solid is held together by very stable chemical bonds. As we shall see shortly, compounds composed of a network of tightly bound ions or atoms tend to have relatively high melting points. The volatility of a substance also indicates how strongly molecules are attracted to each other. For example, in the case of carbon

Bonding theories must explain the observed behavior of chemicals.

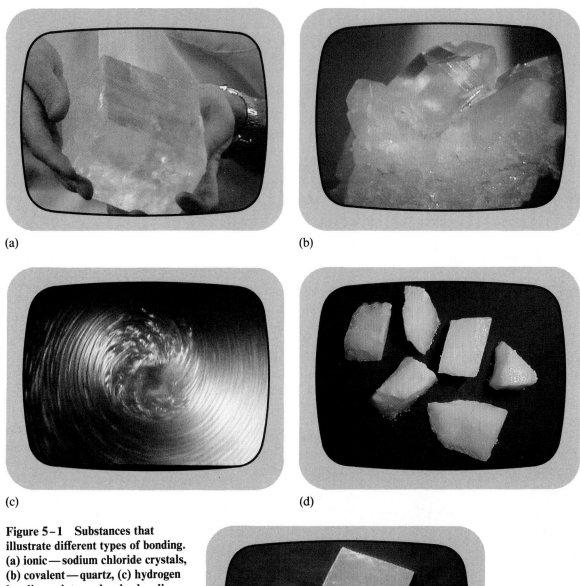

(a)

(b)

(c)

(d)

Figure 5-1 Substances that illustrate different types of bonding. (a) ionic—sodium chloride crystals, (b) covalent—quartz, (c) hydrogen bonding—intermolecular bonding between water molecules in liquid water, (d) London forces—intermolecular bonding between carbon dioxide molecules in solid carbon dioxide (dry ice), (e) metallic bonding in lithium metal. [(a), (b), (c), and (e) from *The World of Chemistry,* Programs 8, 6, 12, 15, respectively: "Chemical Bonds," "The Atom," "Water," and "The Busy Electron."]

(e)

dioxide (CO_2) we must assume that the bonding between molecules (intermolecular bonding) is slight, since it takes relatively little energy to break up solid CO_2 (dry ice). However, the individual CO_2 molecules are very stable as evidenced by the use of CO_2 gas to put out fires. The atoms in the carbon dioxide molecule must be strongly held together or the oxygen in it would keep the fire burning rather than put it out.

In the ensuing discussion of chemical bonds, major emphasis is placed on accounting for the properties of a given substance by the bonds that hold that substance together.

CO_2 changes readily from a solid to a gas (sublimes).

IONIC BONDS

A large number of compounds form hard, brittle crystalline solids with relatively high melting points. When melted or in solution, these compounds conduct electricity well, but when solid they do not conduct (Fig. 5–2). If in solution or melted, they are more likely to react quickly with each other. Compounds with these properties are known as **ionic compounds**. Examples of ionic compounds are sodium chloride (NaCl), magnesium fluoride (MgF_2), and calcium oxide (CaO).

All the properties of these compounds can be explained if the compounds are assumed to be composed of charged **ions** rather than neutral atoms. X-ray and mass spectrographic studies of these kinds of compounds strongly indicate that ions exist.

Recall that an ion is a charged atom or a charged group of atoms.

(a)

(b)

Figure 5–2 (a) Solid sodium chloride does not conduct electricity. (b) A solution of sodium chloride conducts electricity because the mobility of the ions and the attraction of the ions to the oppositely charged electrode carry the current through the liquid.

Metals lose electrons to form positive ions.

Nonmetals gain electrons to form negative ions.

Ionization energies are measured experimentally and correspond to the reaction:

atom ⟶ positive ion + electron.

Neon sign. (Grant Heilman, Runk/Schoenberger.)

Since electrons constitute the outermost parts of the atom, it is reasonable to assume that electrons—and only electrons—are manipulated to form ions. Electrons can be *removed* from an atom, with the result that part of the positive nuclear charge is not neutralized and the atom becomes an ion with a positive charge. Electrons can also be *gained* by an atom, with the result that excess negative charge has been added to the atom, making it a negative ion. Metals form positive ions in the presence of nonmetals, and nonmetals form negative ions in the presence of metals.

Recall that representative metals constitute the left side of the periodic table and that metals have one, two, or three valence electrons. Energy is required to remove these electrons from atoms of metals. This is the ionization energy that was shown to be periodic in Chapter 4 (lower ionization energies on the left of the periodic table, higher on the right; see Fig. 4–13 for review). Ionization energy must be added to an atom, and the energy can come either from an outside source (heat or electricity, for example) or from the energy given off when nonmetals receive and pair electrons. Since metals have the lower ionization energies, their electrons are easier to remove than the electrons of nonmetals.

A guiding principle that will help to predict which ions are most likely to form from atoms of metals and nonmetals becomes apparent when we examine a group of relatively inert gaseous elements: the noble gases. You are probably familiar with helium (He) and neon (Ne) in this group. Helium is used to fill weather balloons and blimps, and Ne is used in "neon" lights. Argon (Ar), krypton (Kr), xenon (Xe), and radon (Rn) are less well known members of the noble-gas family of elements. Until 1962, compounds of noble gases were unknown. Now, a number of compounds have been made, including XeF_4, XeF_6, $XeOF_4$, XeO_3, and KrF_2, but attempts to prepare stable compounds with He, Ne, and Ar have been unsuccessful so far. By any criterion, this family of elements is relatively inert compared with the other 83 natural elements.

The inertness of the noble gases must be related to their electron structure. When the electron arrangements of the noble gases are written out for the major energy levels, two features are apparent:

He	2				
Ne	2	8			
Ar	2	8	8		
Kr	2	8	18	8	
Xe	2	8	18	18	8

First, all except Xe have the lower energy levels filled to capacity. Second, except for He, the highest numbered energy level contains eight electrons. These electron arrangements appear to be particularly stable. Perhaps if other elements achieved these electron structures, they would also be stable chemically. This seems to be the case for a large group of elements.

Atoms of the group IA (or alkali metal) elements, lithium (Li), sodium (Na), potassium (K), rubidium (Rb), and cesium (Cs), have one more elec-

The electrons easiest to remove are the outermost ones, the valence electrons. Which is easier to remove, the peel or the seed of an orange?

tron than do atoms of the respective noble gases He, Ne, Ar, Kr, and Xe. If each of these metals loses an electron from each atom, the species would have the noble-gas electron arrangement. For example, an atom of Li has one valence electron in the ground-state arrangement. The loss of one electron gives the Li atom the electron structure of a He atom. A Li ion with only two electrons and three protons has a charge of 1^+.

Atoms of elements in group IIA of the periodic table (the alkaline-earth metals) have two valence electrons. Thus, for atoms of magnesium (Mg), calcium (Ca), strontium (Sr), and barium (Ba) to take on the noble-gas structure, each atom must lose two electrons. The loss of two electrons would leave two protons in the nucleus unneutralized, so each ion would have a charge of 2^+. To remove the third electron requires the breakup of pairs of electrons in a lower, main energy level. This takes considerably more energy.

The removal of electrons from atoms of metals with the consequent formation of positive ions can be depicted in varying degrees of detail. Figure 5–3 illustrates the formation of positive ions of Na, Mg, and aluminum (Al). It becomes more difficult to remove each succeeding electron from a given atom since the loss of each electron results in an increase in the positive charge, and this charge helps hold the remaining electrons more securely.

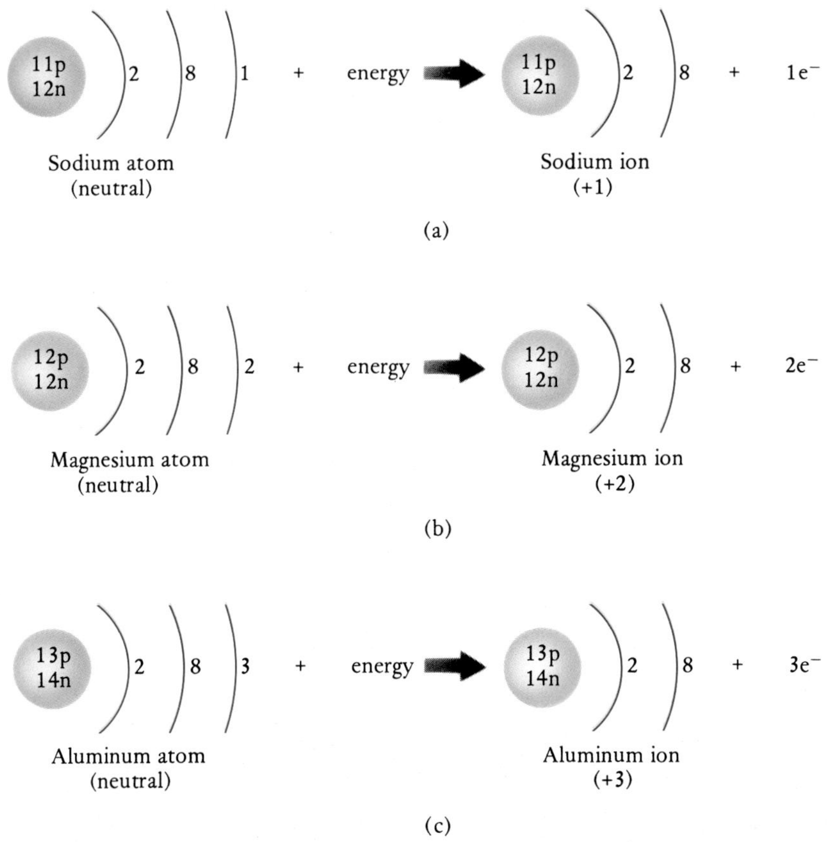

Valence electrons are in the highest stable energy level.

Sodium atom
(neutral)

Sodium ion
(+1)

(a)

Magnesium atom
(neutral)

Magnesium ion
(+2)

(b)

Aluminum atom
(neutral)

Aluminum ion
(+3)

(c)

Figure 5–3   The formation of positive ions from atoms of metals: (a) sodium, (b) magnesium, (c) aluminum.

TABLE 5–1 Electron Configurations of the Noble Gases and Ions with Identical Configurations

Species	Configuration
He, Li$^+$, Be^{2+}, H$^-$	2
Ne, Na$^+$, Mg^{2+}, F$^-$, O^{2-}	2-8
Ar, K$^+$, Ca^{2+}, Cl$^-$, S^{2-}	2-8-8
Kr, Rb$^+$, Sr^{2+}, Br$^-$, Se^{2-}	2-8-18-8
Xe, Cs$^+$, Ba^{2+}, I$^-$, Te^{2-}	2-8-18-18-8

Some simple, stable positive ions with a noble-gas electron configuration are listed in Table 5–1. Stable positive ions that do not conform to the noble-gas electron arrangement include most of the transition metal ions and some positive ions of elements to the right of the transition elements in the periodic table.

In summary,

positive ions are formed when metal atoms lose one electron (group IA), two electrons (group IIA), or three electrons (group IIIA) to nonmetal atoms. The resulting ions have the same electron arrangement as a noble gas.

Positive ions are stabilized by the presence of negative ions. The neutralization of charge stabilizes the charge on both types of ions. Stable negative ions can be produced by atoms that have six or seven valence electrons. These atoms may gain enough electrons to achieve the noble-gas structure. For example, atoms of group VIIA elements (the halogens) have seven valence electrons and need one each to have the electron arrangement of a noble gas. If atoms of fluorine (F), chlorine (Cl), bromine (Br), and iodine (I) gain one electron each, the resulting ions, F$^-$, Cl$^-$, Br$^-$, and I$^-$, have the same electron arrangement as Ne, Ar, Kr, and Xe, respectively. The group VIA elements (oxygen [O], sulfur [S], and selenium [Se]) need to gain two electrons for each atom to achieve the electron structure of a noble gas. The excess of two electrons per ion produces a charge of 2$^-$.

The gain of electrons by the nonmetals, like the loss of electrons by the metals, can be depicted in various degrees of detail (Fig. 5–4), Consider an atom of F (atomic number 9). Its electron arrangement is 2-7 (or

Figure 5–4 Different ways to depict the formation of negative ions from atoms of nonmetals: (a) fluorine, (b) oxygen.

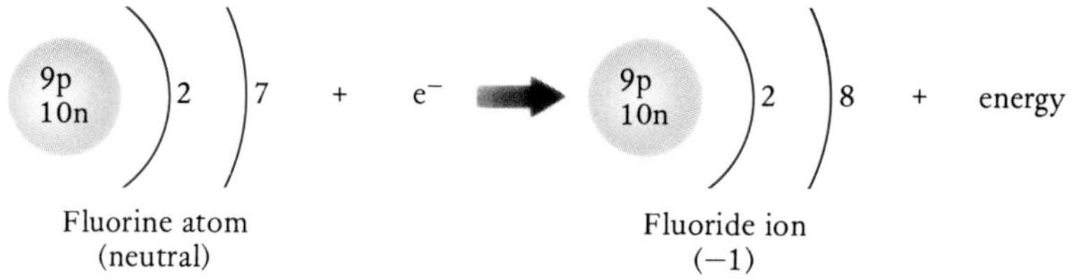

$$F + e^- \rightarrow F^- + \text{energy}$$
$$:\!\overset{\cdot\cdot}{F}\!\cdot\, + e^- \rightarrow\, :\!\overset{\cdot\cdot}{F}\!:^-$$

(a)

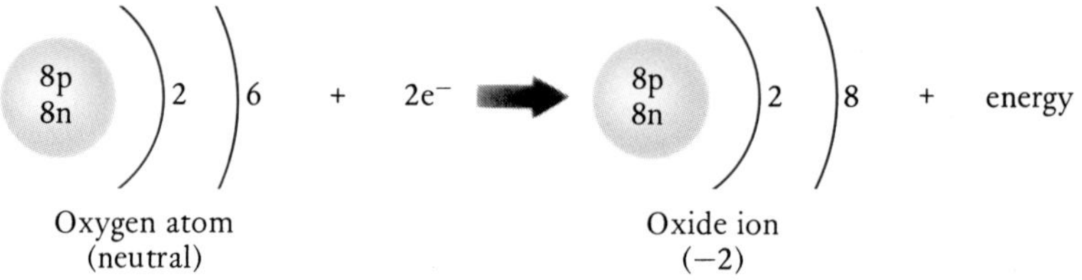

$$O + 2e^- \rightarrow O^{2-} + \text{energy}$$
$$\cdot\overset{\cdot\cdot}{O}: + 2e^- \rightarrow\, :\!\overset{\cdot\cdot}{O}\!:^{2-}$$

(b)

$1s^2 2s^2 2p_x^2 2p_y^2 2p_z^1$). A F atom needs one electron to pair with its one unpaired electron, and this single electron completes the stable eight valence electrons. Note in Figure 5–4a that an electron is on the left side of each equation, and is therefore gained.

The third depiction shown in Figure 5–4a for F is the informative **electron dot formula** (sometimes called Lewis dots), suggested by G. N. Lewis in 1916. Only the valence electrons are represented around the symbol for the element. The electrons are placed at north, south, east, and west positions adjacent to the symbol. Note that the method shows the pairing of electrons as well as the attainment of eight valence electrons when the negative ion is formed.

Figure 5–4b depicts how an oxygen atom becomes an oxide ion. The electron arrangement of O (atomic number 8) is 2-6 (or $1s^2 2s^2 2p_x^2 2p_y^1 2p_z^1$). Note that the electron dot formula shown in Figure 5–4b for the O atom has two pairs of electrons and two electrons that are not paired, in agreement with the electron arrangement of the valence electrons. Two electrons are added to an oxygen atom to form an oxide ion. Other common negative ions are listed in Table 5–1. As the number of electrons added to an atom is increased, it becomes more difficult for each succeeding electron to enter. Each electron that comes into an atom after the first must enter against an existing charge that is net negative.

In summary,

> **nonmetals in the presence of metals tend to gain one, two, or three electrons from metal atoms to form negative ions that have all valence electrons paired and have the stable eight-electron arrangement of noble gases.**

A memory aid tied to the periodic table is given in Figure 5–5 for ions formed by the elements.

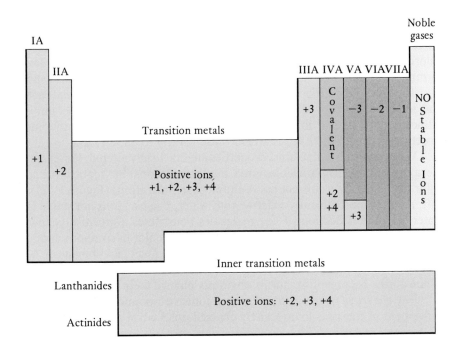

Figure 5–5 The periodic table and the formation of ions.

Figure 5–6   (a) Model of sodium chloride crystalline lattice. (b) Each Na^+ is surrounded by 6 Cl^-. (c) Each Cl^- is surrounded by 6 Na^+.

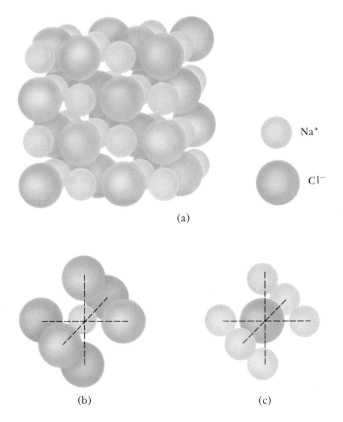

(a)

(b)   (c)

Ionic bonds are the attractions between ions of opposite charge.

Ionic compounds have no molecules.

An ionic structure is a regular geometrical array of ions.

A chemical formula of a compound is electrically neutral. Electrical neutrality requires an equal number of positive and negative charges in the crystal of the compound: two F^- for each Ca^{2+}; three O^{2-} for two Al^{3+}. In a crystal of table salt the Na^+ ions and Cl^- ions are held in place by electrical attraction between unlike charges. Furthermore, the ratio of sodium ions to chloride ions must be 1 to 1 if the compound is to be neutral. The simplest formula is found to be NaCl, so the theory is consistent with observations.

The crystalline structure of NaCl is shown in Figure 5–6. Note that each Na^+ ion is attracted by all six Cl^- ions around it. Similarly, each Cl^- ion is attracted by all six Na^+ ions around it. There is *no unique molecule* in ionic structures; no particular ion is attached exclusively to another ion.

When atoms become ions, properties are drastically altered. For example, a collection of Br_2 molecules is red, but bromide ions (Br^-) contribute no color to a crystal of a compound. A chunk of sodium atoms (Figure 5–7) is soft, metallic, and violently reactive with water, but Na^+ ions are stable in water. A large collection of Cl_2 molecules constitutes a greenish-yellow, poisonous gas, but chloride ions (Cl^-) produce no color in compounds and are not poisonous. In fact, Na^+ and Cl^- ions in the form of table salt can be put on tomatoes without fear of a violent reaction or of being poisoned. When atoms become ions, atoms obviously change their nature.

The *electrical conductivity* of melted ionic compounds is based on the movement of free ions to oppositely charged poles when an electric field is imposed (Fig. 5–2). The movements of the ions transport charge, or electric current, from one place to another. In a rigid solid, the immobile ions are not free to move, and the solid does not conduct electricity.

The *hardness* of ionic compounds is caused by the strong bonding between ions of unlike charge. The strong bonds require much energy to separate the ions and allow the freer movement of the melted state. Much energy means *higher melting points,* which are characteristic of ionic compounds.

Ionic compounds are *brittle* because the structure of the solid is a regular array of ions. Consider the structure of NaCl in Figure 5–6. If a plane of ions is shifted just one ion's distance in any direction, identically charged ions are now next to each other. This causes repulsion—there is no attraction—and the crystalline solid breaks. Sodium chloride cannot be hammered into a thin sheet; it shatters instead.

Ion sizes are different from parent atom sizes. Positive ions are smaller than the atoms from which they were made; negative ions are larger than their atoms (Fig. 5–8). A sodium atom with its single outer electron has a radius of 0.186 nm (nanometer). One would expect that when this electron is removed (forming the Na^+ ion) the resulting ion would be smaller. The Na^+ ion radius is 0.095 nm. This decrease in size results because there are now only ten electrons attracted to a charge of $+11$ on the nucleus, and these

Figure 5–7   A piece of sodium reacting with water. (Beverly March.)

One nanometer (nm) is 10^{-9} meter.

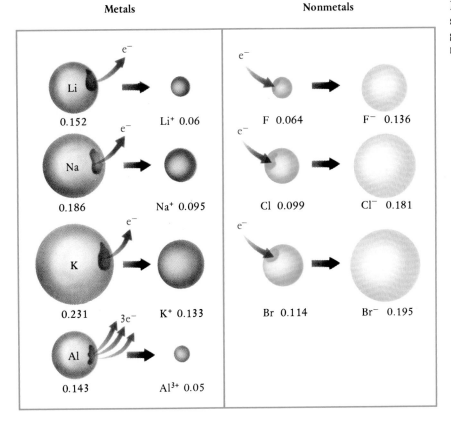

Metals   Nonmetals

Li   0.152   Li$^+$ 0.06   F 0.064   F$^-$ 0.136

Na   0.186   Na$^+$ 0.095   Cl 0.099   Cl$^-$ 0.181

K   0.231   K$^+$ 0.133   Br 0.114   Br$^-$ 0.195

Al   0.143   Al^{3+} 0.05

Figure 5–8   Relative sizes of selected atoms and ions. Numbers given are atomic or ionic radii in nanometers.

electrons are pulled closer to the nucleus by the charge imbalance. The same type of phenomenon is observed for all metal ions.

The metal ions with multiple charges (Al^{3+}, Fe^{2+}, etc.) are much smaller than the corresponding metal atom, because of still greater unneutralized positive charge on the nucleus.

Nonmetals gain electrons to form negative ions that are *larger* than the corresponding atoms. This phenomenon results from the addition of electrons to the outer shell of an atom without increasing the charge on the nucleus. The repulsion of the electrons and the lack of sufficient charge on the nucleus cause the expansion.

The sizes of ions are important because the strength of the forces that hold ions together in ionic compounds depends on the sizes (and charges) of the ions involved. If two ions have the same charge, the smaller ion would have a more concentrated charge and get closer to another ion to form a stronger bond.

SELF-TEST 5–A

1. Charged atoms are called _____ .
2. The attraction between positive and negative ions produces a(an) _____ bond.
3. A sodium atom loses _____ electron(s) in achieving a noble-gas configuration.
4. What is the correct formula for calcium iodide (Ca^{2+} and I^-)?
5. Which ion gained an electron in its formation: Na^+ or Cl^-?
6. Electrons in the outer shell may be called _____ electrons.
7. Positive ions are formed from neutral atoms by (   ) losing (   ) gaining electrons.
8. Positive ions are (   ) smaller (   ) larger than their parent atoms.
9. Negative ions are formed from neutral atoms by (   ) losing (   ) gaining electrons.
10. Negative ions are (   ) smaller (   ) larger than their parent atoms.
11. Predict the number of electrons lost or gained by the following atoms in forming ions. Indicate whether the electrons are gained or lost.
    Rb _____   _____
    Ca _____   _____
    K _____   _____
    S _____   _____
    Mg _____   _____
    Br _____   _____

G. N. Lewis (1875–1946) proposed in 1916 that a *covalent bond* results from sharing an electron pair. He was a professor of chemistry at University of California, Berkeley. Many of his ideas about bonding are still applicable today.

COVALENT BONDS

What holds together the atoms in molecules of carbon monoxide (CO), methane (CH_4), water (H_2O), quartz (SiO_2), ammonia (NH_3), carbon tetrachloride (CCl_4), and molecules of about 9 million other compounds in which

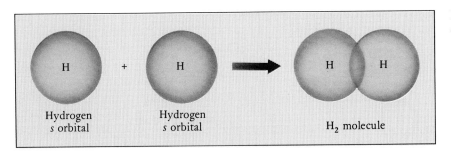

all of the elements are nonmetals? Compounds of only nonmetals are very poor conductors of electricity in the melted state. Remember that all non-metals have higher ionization energies than those of metals, and none are prone to form positive ions to balance possible negative ions.

The driving forces of electron pairing and the stable eight-electron (**octet**) arrangement of the noble gases can be accommodated by *sharing* pairs of electrons between atoms of elements in groups IVA, VA, VIA, and VIIA. The sharing of electrons between two atoms produces a **covalent bond**. The strength of the bond comes from interaction of an orbital of one atom with an orbital of another atom (Fig. 5–9). The shared electrons are held to each other by pairing forces, and the electron pairs are held to the two nuclei by attractions between unlike charges.

The drive to attain the stable eight valence electrons of a noble gas is known as the octet rule.

Single Covalent Bonds

A single covalent bond is formed when two atoms share a single pair of electrons. The simplest examples are diatomic (two-atom) molecules such as H_2 (hydrogen), F_2 (fluorine), and Cl_2 (chlorine).

A H atom has one electron. If a H atom could share its electron with another atom that has an unpaired valence electron of opposite spin, a stable pairing of the two electrons can be achieved and the H atom can then have the electron structure of He, a noble gas. This arrangement can be achieved by two H atoms sharing their single electrons. The electron dot formula for the H_2 molecule is

$$2\,\text{H} \cdot \longrightarrow \text{H} \mathbin{:} \text{H} + \text{energy}$$
$$\underset{\text{Atoms}}{\phantom{2\,\text{H}\cdot}} \quad \underset{\text{Molecule}}{\phantom{\text{H}\mathbin{:}\text{H}}}$$

To break a bond requires energy; when bonds are formed, energy is released.

Since each F atom has one unpaired electron $(\overset{..}{\underset{..}{:\text{F}}}\cdot)\,(1s^2 2s^2 2p_x^2 2p_y^2 2p_z^1)$, two F atoms also can share an electron each and form a single covalent bond and a F_2 molecule.

$$2\,\overset{..}{\underset{..}{:\text{F}}}\cdot \longrightarrow \overset{..}{\underset{..}{:\text{F}}}\mathbin{:}\overset{..}{\underset{..}{\text{F}:}} + \text{energy}$$
$$\underset{\text{Atoms}}{} \qquad \underset{\text{Molecule}}{}$$

Only the pair of electrons represented between the two symbols (the two F's) are bonding electrons. The other six pairs of electrons are called **nonbonding** valence electrons, and they repel each other.

Before reading any further, you might draw the electron dot structures of Cl_2, Br_2, and I_2.

Figure 5-10   Single bond formation in HF. (a) The electron dot representation; (b) the orbital interaction representation.

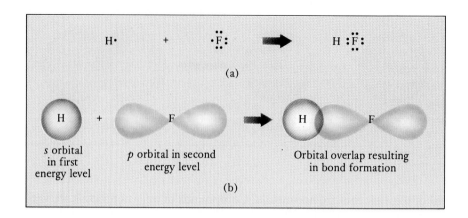

(a)

| s orbital in first energy level | p orbital in second energy level | Orbital overlap resulting in bond formation |

(b)

Electrons are all the same. The ×'s and dots distinguish the sources of the identical electrons.

When hydrogen (H×) and fluorine ($\cdot \ddot{\ddot{F}}$:) combine to form HF (H$\ddot{\ddot{F}}$:), the s orbital of H interacts with a half-filled p orbital of F (Fig. 5-10) to form a single covalent bond.

In a water molecule, two O—H single covalent bonds are formed. An oxygen atom has six valence electrons, of which two are unpaired ($\cdot \ddot{O}$:). It needs two more electrons to pair up its electrons and produce the stable octet of a noble gas. Two hydrogen atoms supply the two electrons.

$$:\overset{..}{\underset{.}{O}}\cdot + 2H^\times \longrightarrow :\overset{..}{\underset{\overset{\times}{H}}{O}}\!\!{}^\times\!H + energy$$

Water

The choice of H$:\ddot{O}$: over H$:\ddot{O}$:H will be explained later in this chapter.

An ammonia (NH$_3$) molecule has three N—H single covalent bonds. A nitrogen atom has five valence electrons, of which three are unpaired ($\cdot \overset{..}{N} \cdot$). The atom needs three more electrons to pair up its electrons and give it the stable eight. Three hydrogen atoms supply the three electrons to form the NH$_3$ molecule.

$$\cdot \overset{..}{\underset{.}{N}} \cdot + 3H^\times \longrightarrow H \overset{..}{\underset{\overset{\times}{H}}{N}}{}^\times H + energy$$

Ammonia

The rule that only eight electrons are in the valence shell of a bonded atom (octet rule) is not hard and fast; indeed, many compounds do not have an octet of electrons around one of the atoms in the molecule. For example, boron trifluoride, BF$_3$, is a stable compound with three B—F single covalent bonds. The boron atom has only three valence electrons ($^\times$B$^\times$), and each fluorine atom has seven valence electrons, one unpaired ($\cdot \ddot{F}$:). When a B atom and three F atoms share electrons to form a molecule of BF$_3$, the B has all of its electrons paired, but it has only six electrons (three pairs) in its valence shell.

$$^\times \overset{\times}{B}{}^\times + 3 \cdot \overset{..}{\underset{..}{F}}: \longrightarrow :\overset{..}{\underset{..}{F}} \overset{\overset{..}{F}:}{\underset{\times}{\overset{\times}{B}}} \overset{..}{\underset{..}{F}}: + energy$$

Since BF_3 gas can be isolated, the requirement to have all valence electrons paired must supersede the requirement to have eight electrons in the valence shell. In fact, G. N. Lewis, who is generally credited with developing the octet rule, realized the limitations of this theory. He wrote in 1923, "The electron pair, especially when it is held conjointly by two atoms, and thus constitutes the chemical bond, is the essential element in chemical structure." Yes, pairing sometimes supersedes the octet rule, but the molecular examples are few. Most covalent compounds of nonmetals follow the octet rule; and the simplicity of the octet rule justifies its use, remembering that occasionally there are exceptions.

Before reading on, draw the electron dot structures for methane (CH_4) and carbon tetrachloride (CCl_4) remembering that the four valence electrons of carbon are unpaired ($\cdot \overset{\cdot}{\underset{\cdot}{C}} \cdot$).

Have you ever considered the chemical formula for trisodium phosphate (Na_3PO_4) or for copper(II) sulfate ($CuSO_4$)? These are common substances sold in hardware stores and elsewhere for cleaning floors (Na_3PO_4) and killing algae in ponds ($CuSO_4$). Both of these substances have the properties of ionic compounds. When Na_3PO_4 is dissolved in water, sodium ions (Na^+) and a mixture of dihydrogen phosphate ($H_2PO_4^-$), hydrogen phosphate (HPO_4^{2-}), and phosphate (PO_4^{3-}) ions are formed. Copper sulfate forms copper ions (Cu^{2+}) and sulfate ions (SO_4^{2-}) in water. Since the PO_4^{3-} and SO_4^{2-} ions are composed of nonmetal atoms only, the P—O and S—O bonds are covalent. If the P and S atoms are surrounded by the oxygen atoms in electron dot structures, the experimentally correct arrangement is represented. Circles are used to represent the electrons transferred from the metal (Na, Cu) atoms to the PO_4^{3-} (addition of three electrons) and SO_4^{2-} (addition of two electrons). The bonds marked **coordinate covalent** are formed by one atom supplying both electrons for the shared bond.

The bonds between Na^+ and PO_4^{3-} and between Cu^{2+} and SO_4^{2-} are ionic bonds.

The structure of the $H_2PO_4^-$ and HPO_4^{2-} ions are drawn by forming a coordinate covalent bond between H^+ ions and oxygen atoms on PO_4^{3-}.

The phosphate ion and the sulfate ion are examples of **polyatomic (many-atom) ions,** which are held intact by covalent bonds. A few common examples are listed in Table 5–2.

Before continuing, reinforce your understanding of bonding by drawing the electron dot structure for the perchlorate ion ClO_4^-.

TABLE 5–2 A Few Polyatomic Ions

Ammonium	NH_4^+	Hypochlorite	ClO^-	Chromate	CrO_4^{2-}
Acetate	$CH_3CO_2^-$	Chlorate	ClO_3^-	Silicate	SiO_3^{2-}
Nitrate	NO_3^-	Perchlorate	ClO_4^-	Phosphate	PO_4^{3-}
Nitrite	NO_2^-	Carbonate	CO_3^{2-}	Arsenate	AsO_4^{3-}
Hydroxide	OH^-	Sulfate	SO_4^{2-}		

Figure 5-11 Electron dot structures of some molecules containing multiple bonds. Line structures are shown for comparison.

Formula	Name	Electron dot structure	Line structure
Double bonds:			
CO_2	Carbon dioxide	$:\overset{..}{O}::C::\overset{..}{O}:$	$O{=}C{=}O$
C_2H_4	Ethylene	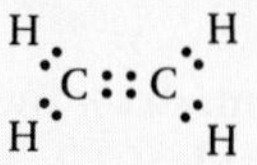 H C :: C H (H H)	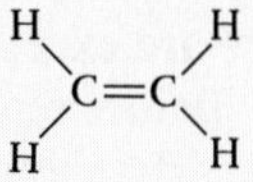 H—C=C—H
SO_3	Sulfur trioxide	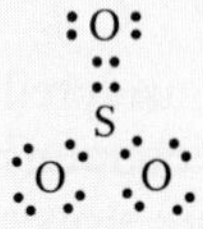 :O: S O O	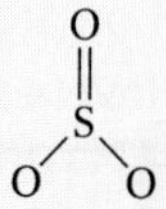 O S O O
Triple bonds:			
N_2	Nitrogen	$:N:::N:$	$N{\equiv}N$
CO	Carbon monoxide	$:C:::O:$	$C{\equiv}O$
C_2H_2	Acetylene	$H:C:::C:H$	$H{-}C{\equiv}C{-}H$

Multiple Bonding

An atom with fewer than seven electrons in its valence shell can form covalent bonds in two ways. The atom may share a single electron with each of several other atoms that can each contribute a single electron. This leads to **single** covalent bonds. But the atom can also share two (or three) pairs of electrons with a single other atom. In this case there will be two (or three) bonds between these two atoms. When two shared pairs of electrons join together the same two atoms, we speak of a **double bond,** and when three shared pairs are involved, the bond is called a **triple bond.** Examples of these bonds are found in many compounds such as those shown in Figure 5-11.

As we can see from these structures, molecules may contain several types of bonds. Thus, ethylene (Fig. 5-11) contains a double bond between the carbon atoms and single bonds between the hydrogen atoms and the carbon atoms. For convenience, an electron pair bond is often indicated by a dash as follows:

One pair of electrons is a *single* covalent bond.

A *double bond* consists of two electron pairs shared between two atoms.

A line between two atoms, as in H—H, represents a bonding pair of electrons.

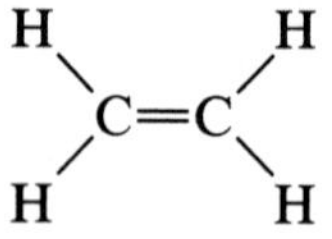

TABLE 5-3 Some Bond Lengths and Bond Energies

Bond type	C—C	C=C	C≡C	N—N	N=N	N≡N
Bond length (nm)	0.154	0.134	0.120	0.140	0.124	0.109
Bond energy (kcal/mol)	83	146	200	40	100	225

kcal/mol (kilocalories per mole) = thousands of calories necessary to break 6.02×10^{23} bonds.

The H_2 molecule with a single bond is shown as H—H; ethylene, with a double bond, is shown as $H_2C=CH_2$; and diatomic nitrogen, with a triple bond, is shown as N≡N. Note that in each of these cases the stable octet rule is obeyed if the shared electrons can be counted as belonging to both atoms.

Single, double, and triple bonds differ in length and strength. Triple bonds are shorter than double bonds, which in turn are shorter than single bonds. Bond energies normally increase with decreasing bond length as a result of greater orbital interaction. **Bond energy** is the amount of energy required to break a mole of the bonds. Some typical bond lengths and energies are listed in Table 5-3.

Polar Bonds

In a molecule like H_2 or F_2, where both atoms are alike, there is equal sharing of the electron pair. Where two unlike atoms are bonded, however, the sharing of the electron pair is unequal and results in a shift of electric charge toward one partner. Recall that the more nonmetallic an element is, the more that element attracts electrons. This is due to the relatively large ratio of nuclear charge to atomic size. In effect, the larger this ratio, the more strongly an atom attracts its electrons and those shared with other atoms in covalent bonding.

The attraction for the electrons in a chemical bond can be expressed on a quantitative basis and is called **electronegativity**. Nonmetallic character increases across and up the periodic table toward F, which has the largest electronegativity of the nonmetals (4). In 1932, Linus Pauling first proposed the concept of electronegativity of the nonmetals. The currently accepted values for electronegativities are shown in Figure 5-12. (Electronegativity values are relative numbers with an arbitrary value of 4.0 for the most electronegative element.) The electronegativities generally increase along a diagonal line drawn from francium (Fr) to F. The values for other elements are between these two extremes. Although electronegativities show a periodic trend (Fig. 5-13), the pattern is not as regular as that for ionization energies (Fig. 4-13) or atomic radii (Fig. 4-12).

When two atoms are bonded covalently and the electronegativities of the two atoms are the same, there is an equal sharing of the bonding electrons, and the bond is a **nonpolar** covalent bond. The bonds in H_2, F_2, and NBr_3 (nitrogen [N] and Br have the same electronegativity, 3.0) are nonpolar.

Two atoms with different electronegativities bonded covalently form a **polar** covalent bond. The bonds in HF, NO, SO_2, H_2O, CCl_4, and BeF_2 are polar.

In a *polar* bond, there is unequal sharing of bonding electrons.

The electronegativity of an atom is a measure of its ability to attract electrons in a covalent bond to itself. The most electronegative atom is fluorine.

Linus Pauling (1901–   ) has been awarded two Nobel prizes: in 1954 for his work on molecular structure and in 1963 for his efforts on nuclear disarmament. Only two other persons, Frederick Sanger and Marie Curie, have received two Nobel prizes.

Electronegativity increases

Most electronegative element

IA												IIIA	IVA	VA	VIA	VIIA	He —
H 2.2	IIA																
Li 1.0	Be 1.6											B 1.8	C 2.5	N 3.0	O 3.4	F 4.0	Ne —
Na 0.9	Mg 1.3	IIIB	IVB	VB	VIB	VIIB		VIII		IB	IIB	Al 1.6	Si 1.9	P 2.2	S 2.6	Cl 3.2	Ar —
K 0.8	Ca 1.0	Sc 1.4	Ti 1.5	V 1.6	Cr 1.7	Mn 1.6	Fe 1.8	Co 1.9	Ni 1.9	Cu 2.0	Zn 1.6	Ga 1.8	Ge 2.0	As 2.2	Se 2.6	Br 3.0	Kr —
Rb 0.8	Sr 0.9	Y 1.2	Zr 1.3	Nb 1.6	Mo 2.2	Tc 1.9	Ru 2.2	Rh 2.3	Pd 2.2	Ag 1.9	Cd 1.7	In 1.8	Sn 1.8	Sb 2.0	Te 2.1	I 2.7	Rn —
Cs 0.8	Ba 0.9	La 1.1	Hf 1.3	Ta 1.5	W 1.7	Re 1.9	Os 2.2	Ir 2.2	Pt 2.2	Au 2.5	Hg 2.0	Tl 2.0	Pb 2.3	Bi 2.0	Po 2.0	At 2.2	Xe —
Fr 0.8	Ra 0.9	Ac 1.1	104 —	105 —	106 —	107 —	108 —	109 —									

Electronegativity decreases

Most electropositive elements
(least electronegative)

☐ Metals ☐ Metalloids ☐ Nonmetals

Figure 5–12 Some electronegativity values in a periodic-table arrangement.

In a molecule of HF, for example, the bonding pair of electrons is more under the control of the highly electronegative fluorine atom than of the less electronegative hydrogen atom (Fig. 5–14).

When covalent bonds join different atoms, the bonds are generally polar because one of the atoms has distorted the electron distribution toward itself as a result of its greater electronegativity. Thus, polar covalent bonds occur in practically every molecule that has different kinds of covalently bonded atoms.

With regard to separation of charge, polar bonds fall between the extremes of pure covalent and ionic bonds. In a pure covalent bond, there is no charge separation; in ionic bonds there is complete separation; and in polar bonds the separation falls somewhere in between.

Figure 5–13 Periodic nature of electronegativities when plotted versus atomic number.

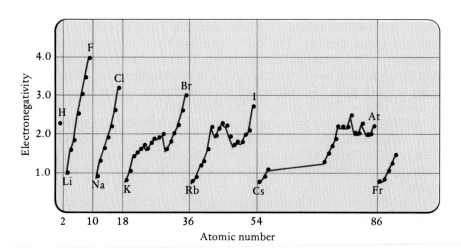

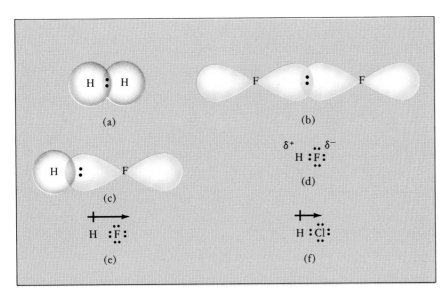

Figure 5–14 Polar bonds in HF and HCl. (a) Symmetrical distribution of electrons in H_2 results in the center of negative charge being identical with the center of positive charge. This is symbolized by the electron dots placed in the overlap area. (b) Overlap of p orbitals in F_2 also results in symmetrical distribution of charge. (c) In HF, the electron pair is displaced toward the fluorine nucleus since fluorine is more electronegative than hydrogen. Note the electron dots have been placed to the right of the overlap area to convey the idea of polarity (separation of charge). (d) δ^+ (delta positive, meaning fractional positive charge) and δ^- (delta negative, meaning fractional negative charge) are used to indicate poles of charge. In (e) and (f) an arrow is used to indicate electron shift, the arrow having a "plus" tail to indicate partial positive charge on the hydrogen atom. Note that the longer arrow in the HF structure indicates a greater degree of polarity than in HCl. This should not be confused with the greater bond length in HCl.

An experimental method for detecting polar molecules is represented in Figure 5–15. The polar bonds in beryllium difluoride (BeF_2), H_2O, CCl_4, and chloroform ($CHCl_3$) are indicated in Figure 5–16 by arrows in the direction of the electron shift, pointing toward the more electronegative atom.

A very common bond that we shall discuss frequently in the remainder of this text is the C—H bond. Since carbon (C) has an electronegativity of 2.5 and H of 2.2, the C—H bond is only slightly polar. The arrangement of C—H bonds around a C atom generally makes —CH_2— and —CH_3 groups nonpolar.

If a substance has polar bonds, it may have polar molecules or it may have nonpolar molecules; it all depends on the three-dimensional geometric shape of the molecule, discussed later in this chapter. If the electron shifts within the molecule balance out (are symmetrical), the substance has **non-**

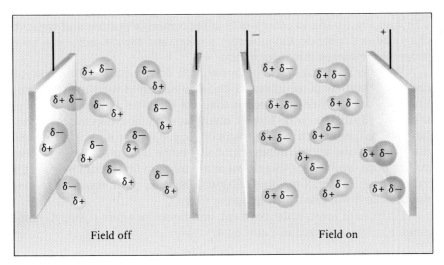

Field off   Field on

Figure 5–15 Physical evidence for both the existence of polar molecules and the degree of polarity is provided by a simple electric capacitor. The capacitor is composed of two electrically conducting plates with nonconducting material (an electrical insulator) between the plates. The storage of charge by the capacitor is least when there is a vacuum between the plates; charge storage is improved when nonpolar substances are placed between the plates and is most effective with polar substances between the plates. Energy is required to orient the molecules, which store the energy until the field is turned off. Then the energy is released as the dipoles become random again.

Figure 5–16 Polar bonds may or may not result in polar molecules. The polar bonds in BeF_2 and CCl_4 are arranged about the center atom in such a way as to cancel out the polar effect. In contrast, the polar bonds in H_2O and $CHCl_3$ molecules do not cancel as a result of the molecular shape but combine to give a polar molecule.

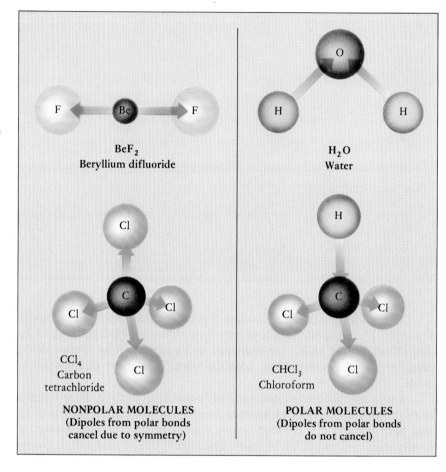

BeF₂
Beryllium difluoride

H₂O
Water

CCl₄
Carbon
tetrachloride

CHCl₃
Chloroform

NONPOLAR MOLECULES
(Dipoles from polar bonds
cancel due to symmetry)

POLAR MOLECULES
(Dipoles from polar bonds
do not cancel)

polar molecules. (See BeF_2 and CCl_4 in Fig. 5–16.) Or, to say it another way, if the centers of positive and negative charge coincide, the molecule is non-polar. On the other hand, if the electron shifts within the molecule do not balance out (are asymmetrical), the substance has **polar molecules** (H_2O and $CHCl_3$ in Fig. 5–16). In other words, if the centers of positive and negative charge do not coincide, the molecule is polar, and the substance will have properties reflecting this polar nature.

Whether a substance is polar or nonpolar can have a great effect on the chemical reactivity of the substance and its solubility in various liquids. For example, a rough rule of thumb is that **like dissolves like:** polar or ionic substances dissolve in polar liquids; nonpolar substances dissolve in nonpo-lar liquids. Therefore, if rubbing alcohol (2-propanol or isopropyl alcohol) dissolves in polar water, rubbing alcohol must be a polar molecule or have polar groups (such as —OH groups) in its structure that attract polar water molecules. Likewise, if gasoline will not dissolve in polar water, gasoline is nonpolar and will dissolve in nonpolar CCl_4.

Features of ionic, polar covalent, and nonpolar (pure covalent) com-pounds are summarized in Table 5–4.

TABLE 5-4 Characteristics of Ionic, Polar Covalent, and Pure Covalent Compounds

Type of Bond	Ionic	Polar Covalent	Pure Covalent
Disposition of the electrons	Transferred from metal to nonmetal	Partially transferred	Shared
Elements involved	Groups IA, IIA, transition, inner transition metals with groups VA, VIA, VIIA nonmetals	Nonmetals with nonmetals: IVA, VA, VIA, VIIA	Nonmetals with nonmetals: IVA, VA, VIA, VIIA
Electronegativity difference	Great (more than 2)	Small	None
Conductance of electricity as a solid	No	No	No
Conductance of electricity as a liquid (melted solid)	Yes	No	No
Molecules	No	Yes	Yes
Ions	Yes	No	No

SELF-TEST 5-B

1. **a.** An example of a molecule containing covalent bonding where the electrons are equally shared between the atoms is _____ ;
   b. one where electrons are unequally shared is _____ .
2. The number of electrons shared in a triple covalent bond is _____ .
3. There are _____ covalent bonds in an ammonia (NH_3) molecule.
4. Which atom cannot form a double bond, fluorine or sulfur?
5. **a.** How many valence (bonding) electrons are thought to be involved in molecules containing covalently bonded atoms of period 2 and 3 elements?
   b. This is known as the _____ rule.
   c. Is it true most of the time or all of the time?
6. Which is the most electronegative of all elements?
7. Does the electronegativity of the elements increase or decrease as you move down a group in the periodic chart?
8. Which is a polar molecule (H_2O, H_2, O_2, CCl_4)?

PREDICTING SHAPES OF MOLECULES

Valence-Shell Electron-Pair Repulsion Theory

A simple, reliable method for predicting the shapes of covalent molecules is the **valence-shell electron-pair repulsion theory (VSEPR),** which is based on the idea that electron pairs in the valence shell repel each other. In fact, this theory assumes that electron pairs in the valence shell of a central atom

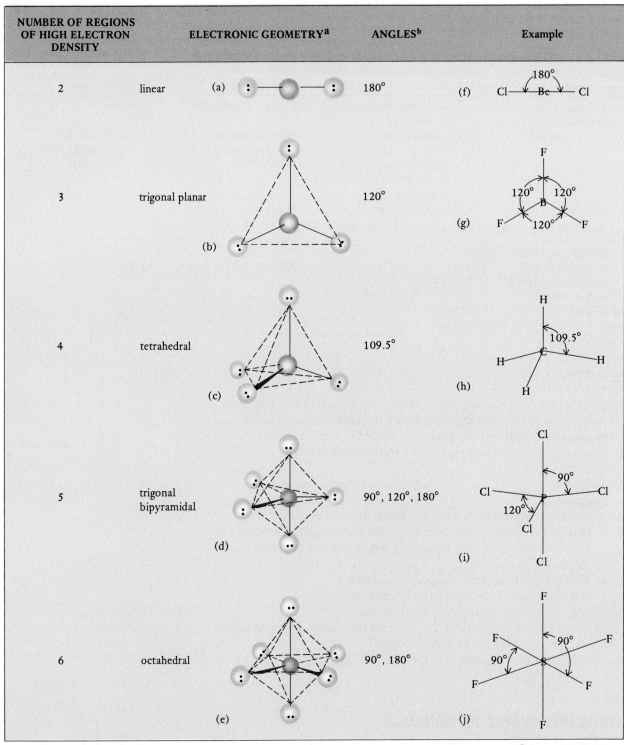

NUMBER OF REGIONS OF HIGH ELECTRON DENSITY	ELECTRONIC GEOMETRY[a]	ANGLES[b]	Example
2	linear (a)	180°	(f)
3	trigonal planar (b)	120°	(g)
4	tetrahedral (c)	109.5°	(h)
5	trigonal bipyramidal (d)	90°, 120°, 180°	(i)
6	octahedral (e)	90°, 180°	(j)

[a] Electronic geometries are illustrated here using only single pairs of electrons as regions of high electron density. The symbols ⦂ represent the regions of high electron density about the central atom ●. By convention, a line in the plane of the drawing is represented by a solid line ——, a line behind this plane is shown as a dashed line - - - -, and a line in front of this plane is shown as a wedge ◢ with the fat end of the wedge nearest the viewer. Each shape is outlined in blue dashed lines to help you visualize it.

[b] Angles made by imaginary lines through the nucleus and the centers of regions of high electron density.

Figure 5–17   Arrangements of electron pairs according to the valence-shell electron-pair repulsion theory.

Balloon models of VSEPR geometries for two to six electron pairs. (Charles D. Winters.)

behave like a group of electrically charged balloons that are connected to a central point. If similarly charged, the balloons would tend to be as far apart as possible. The geometric shapes that give the maximum distance for two, three, four, five, and six electron pairs are linear, trigonal planar, tetrahedral, trigonal bipyramidal, and octahedral, respectively (Fig. 5–17).

Since the noble-gas configuration of eight valence electrons gives a stable configuration for many common nonmetals, the most common geometry is tetrahedral, with four electron pairs at the corners of a tetrahedron. An example of a molecule with four bonding pairs of electrons is methane (CH_4 [Fig. 5–18]).

One of the advantages of VSEPR theory is the ability to predict shapes of molecules that contain both bonding and nonbonding pairs. Ammonia and water are two important examples. In both cases the central atom is surrounded by four pairs of electrons. Nitrogen in ammonia has one nonbonding pair and three bonding pairs, whereas oxygen in water has two nonbonding pairs and two bonding pairs.

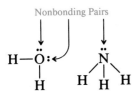

Figure 5–18 shows the tetrahedral representation of four electron pairs in CH_4, NH_3, and H_2O. The bonding angles in ammonia and water are pre-

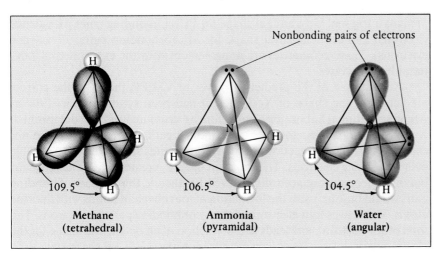

Methane
(tetrahedral)

Ammonia
(pyramidal)

Water
(angular)

Figure 5–18 Methane molecules have only bonding pairs of valence electrons; water and ammonia molecules have both bonding and nonbonding pairs. All three structures have a tetrahedral arrangement of valence electron pairs around the central atom. Since only atoms (not electron pairs) are represented in molecular structures, the different molecules have different molecular structures.

BONDING

Roald Hoffmann, John A. Newman Professor of Physical Science at Cornell University, shared the Nobel Prize in 1981 for his development of a set of rules governing the course of chemical reactions. Hoffmann and his coworkers recognized there was a significant relationship between the way electrons moved in a molecule, the nature of the bonds formed, and the ease or difficulty of the reactions that molecules undergo. The rules they developed allow chemists to predict reliably whether heat or light will cause a particular reaction to occur. The rules also can be used to understand the shape of molecules, especially the shapes of those which have a metal atom bonded to an organic fragment.

The World of Chemistry (Program 8) "Chemical Bonds."

Professor Roald Hoffmann, Cornell University. (*The World of Chemistry,* Program 8, "Chemical Bonds.")

Tetrahedral methane model. (Charles D. Winters.)

Pyramidal ammonia model. (Charles D. Winters.)

Angular water model. (Charles D. Winters.)

dicted to be slightly smaller than the normal tetrahedral angle of 109.5°. VSEPR theory attributes this to the larger volume occupied by nonbonding pairs compared to that occupied by bonding pairs. The increased volume spreads the nonbonding pairs farther apart and squeezes the bonding pairs closer together. Hence, the repulsions between nonbonding pairs are larger than those between bonding pairs.

Regarding the shapes of molecules with nonbonding electrons, only the shape formed by the atoms in the molecule is stated since X-ray structure studies locate atoms, not electron pairs. You can visualize the shape of the molecule by simply ignoring the nonbonding pairs. For example, NH_3 is pyramidal and H_2O is angular.

Often it is enough to know the formula of the molecule and the number of valence electrons of the central atom to predict shapes of other molecules even when the octet rule is not followed. For example, S, phosphorus (P), and other nonmetals in the third or higher periods form several covalent molecules that have more than eight electrons in the valence shell. VSEPR predicts a trigonal bipyramidal shape for PCl_5 (5 bonding pairs, 10 valence electrons) and an octahedral shape for SF_6 (6 bonding pairs, 12 valence electrons). These geometries are in agreement with the experimentally determined structures.

An example of the predictive power of VSEPR theory is the correct prediction of the shape of XeF_4 after it had been synthesized in 1962 at Argonne National Laboratory, but before its structure had been determined. Xenon is a noble gas, and as we mentioned earlier, noble gases were not expected to undergo reactions to form compounds since they have a stable octet of valence electrons. The isolation of XeF_4 created a real challenge for bonding theorists, but according to VSEPR theory, the number of bonding pairs would be four, from the formation of four covalent bonds with fluorine atoms. This leaves four electrons, or two nonbonding pairs of electrons. The total of six electron pairs leads to a prediction of an octahedral shape for the electron pairs, but where do you put the lone pairs? VSEPR theory tells us to place them as far from one another as possible, so placing them at opposite corners of an octahedron gives the maximum distance possible. This leaves a square planar shape for the XeF_4 molecule (cover up the nonbonding pairs to see this), and this was later shown by experiment to be the correct shape.

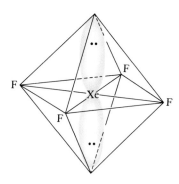

VSEPR theory can also be used to predict the shapes of molecules or ions that contain double or triple bonds. The rule is that if you treat double or triple bonds as though they were a single bond, your prediction of the geometry will be correct. For example, formaldehyde has the Lewis structure

Since the double bond would be treated the same as a single bond, the molecule would be predicted to have a trigonal planar shape based on three electron pairs. This is the correct shape of the formaldehyde molecule.

Crystals of xenon tetrafluoride, XeF_4. (Courtesy of Argonne National Laboratory.)

Molecular Structure Determination by Spectroscopy

How do we determine the structure of molecules? Many methods that are used to determine the structure of molecules or to identify molecules involve the interaction of matter with electromagnetic radiation. Probing matter with electromagnetic radiation is called **spectroscopy,** and each area of the electromagnetic spectrum (see Fig. 3–14) can be used as the basis for a particular spectroscopic method. Recall from Chapter 3 that electromagnetic radiation is emitted or absorbed in quantized packets of energy called **photons** with the energy of the photon represented by $E = h\nu$, where ν is the frequency of light. Molecules may absorb several different electromagnetic radiation frequencies depending on possible changes between their allowed energy levels. Each frequency must provide the exact package of energy needed to lift a molecule from one energy level to the next. For example, the energy of photons in the ultraviolet or visible region matches the energy needed to promote electrons from a lower energy level to a higher one.

Figure 5–19   Model of water molecule with springs for bonds. (After *The World of Chemistry*, Program 10, "Signals from Within.")

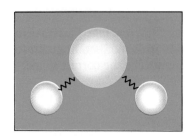

Infrared Spectroscopy

Infrared spectroscopy is of particular relevance for learning about covalent molecules because the energy required for causing the internal motions of molecules is similar to that of photons whose frequency is in the infrared region. Imagine that covalent bonds between atoms in a molecule are like springs that can only bend or stretch in specified amounts (Fig. 5–19). Bending or stretching the bonds or rotating the water molecule occur at discrete energies, that is, they are at specific energy levels. Each molecule has only specific sets of energy states, and the strength of the covalent bonds will determine what frequency of infrared light is necessary for changing one stretching, bending, or rotating energy level to another. In order to get "signals from within" the molecule, the molecule must be excited from one of these allowed energy states to another. An exact amount of energy must be added—a specific packet of energy. It is somewhat analogous to the sound frequency needed to break a glass. The vibration that will crack the glass can be determined by matching the sound frequency one gets when tapping the glass. A lower or a higher frequency will not crack the glass—the frequency must match the tone obtained when the glass is tapped. For example, a vibrating HCl molecule (Fig. 5–20) vibrates at a specific energy level. A package of energy representing the energy difference between that vibrational energy level and the next higher vibrational energy level must be provided in order to make the molecule vibrate at the next higher energy level. Electromagnetic radiation at the right frequency will be absorbed by the molecule. Photons of this energy are just right to make the molecule vibrate at the next higher energy level. Photons with too low or too high an energy do not cause vibration at the next higher energy level to occur, and the radiation passes through the molecule without being absorbed.

Since the covalent bonds in molecules differ in strength and number, the molecular motions and the number of vibrational energy levels will vary;

Unlike a child's swing which can have zero to maximum energy in any amount, the vibration in a water molecule apparently can have only definite frequencies, the quantum effect.

Imagine the matching of frequencies as being similar to jumping rope. To jump a rotating rope a person's frequency of jumping must match that of the rope rotation.

Figure 5–20   (a) Vibration of a hydrogen chloride molecule at a specific frequency. (b) Electromagnetic radiation at the right frequency will be absorbed by the molecule. (After *The World of Chemistry*, Program 10, "Signals from Within.")

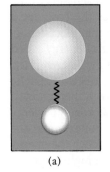

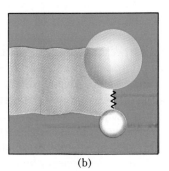

(a)   (b)

SIGNALS FROM WITHIN

The applications of electromagnetic spectroscopy in our everyday lives and in the lives of research chemists are so numerous that one can indeed fill volumes describing them. However, there are some particularly exciting examples that are used in the laboratories of the Drug Enforcement Agency (DEA), and others in the research of the Department of Agriculture.

In the laboratories of the DEA widespread use is made of infrared spectroscopy. These techniques make it possible not only to rapidly identify existing drugs but to actually develop a library of important signature spectra, or fingerprints of new drugs, so necessary for future identification. Thus, infrared spectroscopy becomes one of the most powerful weapons in the arsenal for fighting the scourge of drugs.

In the laboratories of the Department of Agriculture infrared spectroscopy has made it possible to identify readily the composition of the sex pheromones, or sex attractants, upon which insects depend for survival as a species. Once identified, it becomes possible to synthesize these substances and to use them in a powerful way to keep insects from proliferating.

The World of Chemistry (program 10) "Signals from Within."

Pheromones are discussed in Chapter 19.

hence, the infrared radiation absorbed by the molecules will differ. As a result, infrared spectroscopy can be used to learn about the structure of covalent molecules and even to analyze an unknown material by matching its infrared spectrum with that of a known compound. In fact, the infrared frequencies absorbed by a molecule are so characteristic that the infrared spectrum of a molecule can be regarded as its "fingerprint." An example of the use of infrared spectroscopy in the identification of vitamin C is shown in Figure 5–21.

Understanding how molecules absorb infrared light is also important to understanding the "greenhouse effect" described in Chapter 18. Although the primary concern has been with the increasing concentration of CO_2 in the atmosphere from the combustion of fossil fuels, water vapor, and the

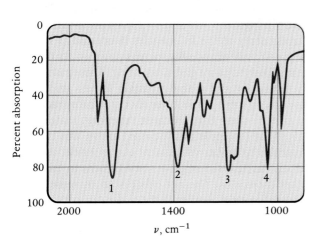

Figure 5–21 Infrared absorption spectrum of vitamin C (ascorbic acid). Chemists often use the reciprocal of the wavelength (cm^{-1}) for infrared absorption peaks. The strong absorptions at 1 (1640 cm^{-1}), 2 (1316 cm^{-1}), 3 (1136 cm^{-1}), and 4 (1042 cm^{-1}) help to identify vitamin C.

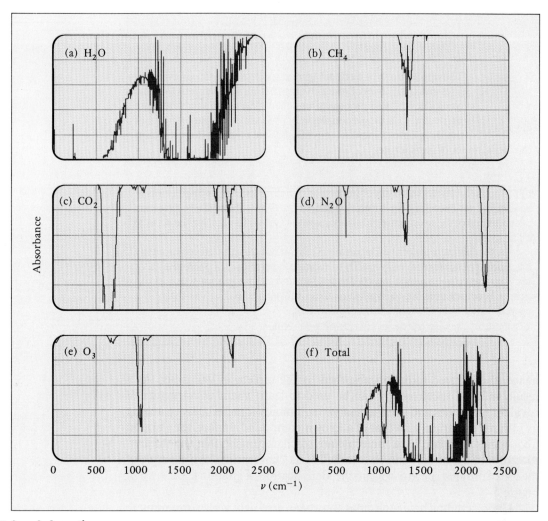

Figure 5–22   Infrared absorption regions for (a) H$_2$O, (b) CH$_4$, (c) CO$_2$, (d) N$_2$O, (e) O$_3$ and (f) total of (a) through (e). Note that the total shown in (f) amounts to considerable absorption of radiation in the infrared region.

Infrared frequencies are expressed in cycles/cm or reciprocal centimeters (cm^{-1}), which equals frequency v divided by the speed of light c.

trace gases—chlorofluorocarbons (CFCs), N$_2$O, CH$_4$, and O$_3$ also contribute to the effect. The characteristic infrared absorption frequencies for several of these gases are given in Figures 5–22 and 5–23. An understanding of the absorption of infrared energy at the frequencies illustrated in Figures 5–22 and 5–23 will help you understand why increasing amounts of these gases in the atmosphere are likely to cause global warming. These gases allow the sun's higher energy ultraviolet and visible radiation to pass through to warm the Earth, but they absorb some of the lower energy infrared radiation that is emitted from the surface of the Earth because the energy matches the vibrational energy levels of the molecule. As discussed above, this infrared radiation raises the molecules to higher vibrational energy levels, and this energy is eventually emitted as heat radiation—some returns to the Earth and some energizes other molecules. The net effect is an increase in the temperature of the lower atmosphere, similar to the rise in temperature in a greenhouse due to its glass enclosure. Of course, it is important not to lose

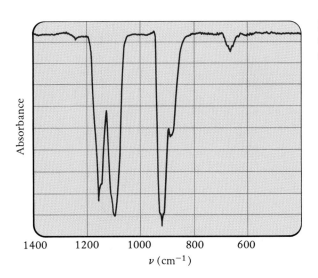

Figure 5–23   Infrared absorption spectrum of dichlorodifluoromethane (CFC-12 or Freon-12.)

sight of the fact that the insulating effect of water vapor and carbon dioxide is important to the existence of life on this planet. Without it, the daily temperature changes would be much more extreme.

INTERMOLECULAR FORCES

The attractive forces between molecules are collectively known as **van der Waals forces,** named after the Dutch physicist who won the Nobel Prize in 1910 for his studies of intermolecular forces in gases and liquids. They include dipole–dipole forces and temporary dipole forces **(London forces).** Hydrogen bonding is a special case of dipole–dipole interactions in which the hydrogen atom in a polar molecule interacts with an electronegative atom in either an adjacent or the same molecule.

Intermolecular bonds are bonds between molecules.

Hydrogen Bonding

When a hydrogen atom is covalently bonded to a highly electronegative atom like F, O, or N, the conditions are right for a very important type of intermolecular, positive-to-negative attraction called **hydrogen bonding.** The bond is produced by the attraction arising between a slightly positive hydrogen atom on one molecule and a very electronegative atom (N, O, or F) on another molecule (or at another location on the same molecule if the molecule is big enough to bend back on itself). The shifting of an electron pair toward very electronegative N, O, or F causes these atoms to take on a partial negative charge. The hydrogen bond, then, is a "bridge" between two highly electronegative atoms with a hydrogen atom bonded covalently to one of the electronegative atoms and electrostatically (positive-to-negative attraction) to the other electronegative atom. Hydrogen bonds have a strength of about one-tenth to one-fifteenth that of an average single covalent bond.

Hydrogen bonds can form between molecules or within a molecule. These are known as intermolecular and intramolecular hydrogen bonds, respectively.

Bond energies of hydrogen bonds are about 10% of normal covalent bond energies.

Figure 5–24   Boiling points of F_2, Cl_2, Br_2, and I_2, as a function of molecular weight.

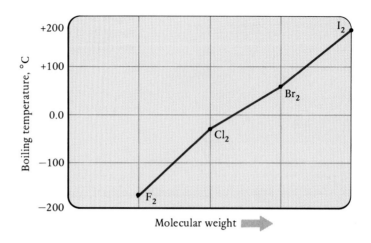

For a substance to evaporate or boil, its molecules must gain enough energy to break loose from each other.

Hydrogen bonds are found in many substances. They are responsible for such phenomena as hard candy getting sticky, cotton fabrics taking longer to dry than nylons, lanolin softening skin; and more importantly, the ultimate shape of proteins and enzymes; the mechanism for transmitting the genetic code; and a host of apparent anomalies in the nature of water.

In the absence of hydrogen bonding, the boiling points of molecular substances normally increase as the molecular weights increase. For example, the boiling points of F_2, Cl_2, Br_2, and I_2 increase rather regularly with increasing molecular weight (Fig. 5–24).

The general relationship between boiling points and molecular weights also holds for hydrogen chloride (HCl), hydrogen bromide (HBr), and hydrogen iodide (HI). However, the boiling point of hydrogen fluoride (HF), the lightest member of this series of compounds, is abnormally high (Fig. 5–25) because of hydrogen bonding. The association between HF molecules is illustrated in Figure 5–26. The increased association between HF molecules compared with that found between HCl molecules offers a ready explanation for the unusually high boiling point of HF.

Water provides a very common example of hydrogen bonding. Hydrogen compounds of oxygen's neighbors and family members are gases at

Figure 5–25   Boiling points of group VIA and group VIIA hydrides are plotted against molecular weights.

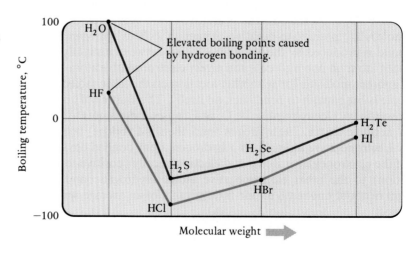

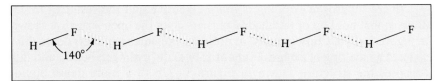

Figure 5–26   Hydrogen bonding in HF.

room temperature: CH_4, NH_3, H_2S, H_2Se, H_2Te, PH_3, HCl. But H_2O is a liquid at room temperature. Figure 5–25 shows that the boiling point of H_2O is about 200° higher than would be predicted if hydrogen bonding were not present. Water molecules are not linear, but are angular. The angular structure of the water molecule comes from the tetrahedral arrangement of four pairs of electrons around the oxygen atom (Fig. 5–18). Two nonbonding pairs of electrons are located on the oxygen atom, and the two bonding pairs are shifted toward the oxygen atom because of its electronegativity. As a result, the two hydrogen atoms have a partial positive charge, and the polar bonds in this geometry result in a polar molecule.

Water has an abnormally high boiling point because of hydrogen bonding between molecules.

(THE δ^- AND δ^+ REPRESENT PARTIAL CHARGES.)

In liquid and solid water, where the molecules are close enough to interact, the hydrogen atom on one of the water molecules is attracted to the **nonbonding electrons** on the oxygen atom of an adjacent water molecule. This is possible because of the small size of the hydrogen atom. Since each hydrogen atom can form a hydrogen bond to an oxygen atom in another water molecule, each water molecule can form a maximum of four hydrogen bonds to four other water molecules (Fig. 5–27a). The result is a tetrahedral cluster of water molecules around the central water molecule.

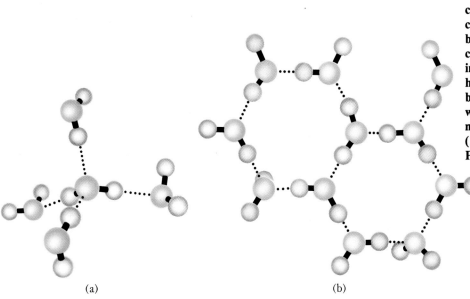

(a)   (b)

Figure 5–27   (a) Tetrahedral cluster of four water molecules around a fully hydrogen-bonded water molecule in the center. (b) Hydrogen bonding in the structure of ice. The hydrogen bonds are indicated by the dashed lines. In liquid water the hydrogen bonding is not as extensive as it is in ice. (*The World of Chemistry*, Program 12, "Water.")

Why does ice float?

In ice, hydrogen bonding is more extensive and the resulting three-dimensional hookup of tetrahedral clusters gives the open structure shown in Figure 5–27b. Consequently, at ordinary pressures ice is *less* dense than water. The melting of ice breaks about 15% of the hydrogen bonds, and this collapses the structure shown in Figure 5–27b to a more dense liquid. Imagine that water is made up of hydrogen-bonded clusters of water molecules that are in constant motion. The number of water molecules per cluster, and the speed of cluster motion is temperature-dependent. As the liquid is heated, more hydrogen bonds are broken and the clusters become smaller. Not all the hydrogen bonds are broken, however, and large aggregates of water molecules exist in liquid water, even near 100°C. As water is heated, thermal agitation disrupts the hydrogen bonding until, in water vapor, there is only a small fraction of the number of hydrogen bonds that are found in liquid or solid water.

Although hydrogen bonds are much weaker than ordinary covalent bonds, hydrogen bonding plays a key role in the chemistry of life. Later chapters in the text will discuss hydrogen bonding in connection with the properties of a number of substances such as water, DNA, proteins, alcohols, cotton, and hair.

Hydrogen bonding plays a key role in the chemistry of life.

London Forces

The forces between molecules that cannot be explained by dipole–dipole attractions or hydrogen bonds are often due to a weak attraction known as **London forces,** named after Fritz London who proposed the temporary dipole (meaning "two poles") concept in 1937.

In He, for example, the two electrons are nonbonding, and yet there is a slight attraction between two He atoms. This slight attraction allows He to become a solid at −272°C under high pressure.

London forces can arise from a variety of causes. For instance, as two atoms or molecules approach each other, intermolecular interaction will cause a temporary shifting of their electron clouds. An uneven electron distribution in an atom makes the atom itself temporarily polar, producing a dipole. The temporary poles on two adjacent atoms can interact with each other, resulting in a momentary attractive force (Fig. 5–28). The existence of the solid state of many nonpolar molecular substances (such as O_2, N_2, and He) can be explained by invoking London forces.

Figure 5–28 An illustration of London forces. One instantaneous dipole interacts with another in a neighboring atom.

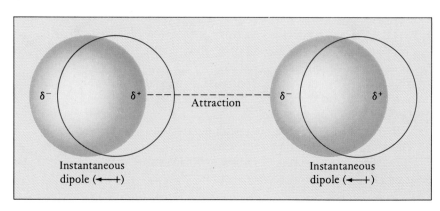

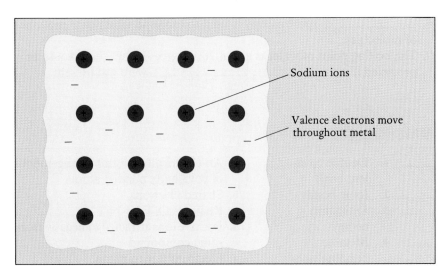

Figure 5-29   Model for metallic bonding.

Sodium ions

Valence electrons move throughout metal

METALLIC BONDING

Metals have some properties totally unlike those of other substances. For example, most metals are good conductors of heat and electricity; they are shiny solids and have relatively high melting points (with a few notable exceptions such as mercury [Hg]). Any theory of the bonding of metal atoms must be consistent with these properties. Structural investigations of metals have led to the conclusion that metals are composed of regular arrays, or lattices, of metal ions in which the bonding electrons are loosely held.

Figure 5-29 illustrates one model for metallic bonding in which the regular array, or lattice, of positively charged metal ions is embedded in a "sea" of mobile electrons. These mobile valence electrons are delocalized over the entire metal crystal, and the freedom of these electrons to move throughout the crystal is responsible for the properties associated with metals.

Loosely held electrons are found in metals. They can move freely and are not confined to the area between any particular pair of atoms.

SELF-TEST 5-C

1. ( ) Bonding ( ) nonbonding pairs of electrons repel more than
   ( ) bonding ( ) nonbonding pairs of electrons.
2. Give the molecular shapes expected for the hypothetical molecules AX_2, BZ_3, CY_4, and DQ_6. (There are no nonbonding valence electrons in these structures, and all of the bonds are single.)
3. The water molecule has _____ bonding pairs and _____ nonbonding pairs of electrons.
4. The ammonia molecule has _____ bonding pair(s) and _____ nonbonding pair(s) of electrons.
5. Internal motions of molecules include _____ , _____ , and _____ .

6. _____ spectroscopy is used to examine vibrational motions of molecules.

7. The boiling point of water is about 200 degrees higher than would be predicted if _____ _____ were not present.

MATCHING SET I

_____ 1. Double covalent bond

_____ 2. Ionic bonds

_____ 3. Ionization energy

_____ 4. Metallic bonding

_____ 5. London forces

_____ 6. Noble gas

_____ 7. Covalent bonds

_____ 8. NaCl

_____ 9. Metal ion

_____ 10. Hydrogen bonds

_____ 11. NH_3

_____ 12. Molecule

_____ 13. Single covalent bond

a. An electrically neutral arrangement of covalently bonded atoms

b. Shared electrons

c. Requires O, F, or N

d. Positive ions attracted to negative ions

e. Ionic compound

f. Electrons free to move

g. Element with eight valence-shell electrons

h. Covalent compound

i. Attraction between nonpolar neutral particles

j. Measures gaseous atom's hold on electron

k. Smaller than parent atom

l. Four electrons shared

m. Two electrons shared

MATCHING SET II (Shapes of Molecules)

_____ 1. $BeCl_2$ (no nonbonding valence electrons)

_____ 2. H_2O

_____ 3. $SiCl_4$

_____ 4. SF_6

_____ 5. BCl_3

_____ 6. XeF_4

_____ 7. NH_3

a. Linear

b. Trigonal planar

c. Bent

d. Tetrahedral

e. Octahedral

f. Pyramidal

g. Square planar

QUESTIONS

1. Diamond (a form of carbon) has a melting point of 3500°C, whereas carbon monoxide (CO) has a melting point of −207°C. What does this suggest about the kinds of bonding found in these two substances?

2. Write the electron configuration for the element potassium (atomic number 19). What will be the electron configuration when a K^+ ion is formed?

3. Is Ca^{3+} a possible ion under normal chemical conditions? Why?

4. Match the electron configurations that would be expected to lead to similar chemical behavior. The numbers denote the numbers of electrons in the shells.
   a. 2-2   d. 2-8-2
   b. 2-5   e. 2-1
   c. 2-8-8-1   f. 2-8-5

5. Fluorine (atomic number 9) has an electron configuration of 2-7. How many electrons will be involved in the formation of a single covalent bond?

6. What kind of bond (ionic, pure covalent, polar covalent) is likely to be formed by the following pairs of atoms?
   a. A group IA element with a group VIIA element
   b. A group VIA element with a group VIIA element
   c. Two Cl atoms
   d. Two elements with about the same electronegativity

7. Write the electron dot structures for the fluoride ion, F^-, the chloride ion, Cl^-, and the bromide ion, Br^-.

8. Draw the electron dot structure for water. Based on bonding theory, why is water's formula not H_3O?

9. Define the term *bond energy*.

10. Draw electron dot structures for the following molecules and/or ions:
    a. NF_3 h. CH_3OH o. ClO_3^-
    b. CCl_4 i. Br_2 p. SO_3^{2-}
    c. C_2Cl_2 j. HCl q. NH_4^+
    d. OF_2 k. BCl_3 r. OH^-
    e. H_2S l. PH_3 s. AsO_4^{3-}
    f. CO m. SiH_4
    g. N_2H_4 n. IBr

11. The members of the nitrogen family, N, P, As, and Sb, form compounds with hydrogen: NH_3, PH_3, AsH_3, and SbH_3. The boiling points of these compounds are
    SbH_3 $-17°C$
    AsH_3 $-55°C$
    PH_3 $-87.4°C$
    NH_3 $-33.4°C$
    Comment on why NH_3 doesn't follow the downward trend of boiling points.

12. Match the following substances with the type of bonding responsible for holding units in the solid together.
    solid Kr   ionic
    ice   covalent
    diamond   metallic
    CaF_2 hydrogen bonding
    iron   London forces

13. Predict the general kind of chemical behavior (i.e., loss, gain, or sharing of electrons) you would expect from atoms with the following electron arrangements:
    a. 2-8-1
    b. 2-7
    c. 2-4

14. Show how two fluorine atoms can form a bond by the interaction of their half-filled *p* orbitals.

15. Select the *polar* molecules from the following list and explain why they are polar:
    N_2, HCl, CO, NO

16. Boron trichloride has the electron dot formula

    $$:\overset{..}{Cl}:$$
    $$:\overset{..}{Cl}\overset{x}{\underset{..}{:}}B\overset{x}{\underset{..}{:}}\overset{..}{Cl}:$$

    What does this tell you about the octet rule even for period 2 elements?

17. Use your chemical intuition and suggest a reaction that might occur between boron trichloride (Question 16) and ammonia,

    $$H\overset{x}{:}\overset{..}{N}\overset{x}{:}H$$
    $$\overset{.x}{H}$$

18. Give the basic points of the valence-shell electron-pair repulsion theory.

19. Explain the 106.5° H—N—H bond angles of NH_3, the ammonia molecule.

20. $BeCl_2$ is a compound known to contain polar Be—Cl bonds, yet the $BeCl_2$ molecule is not polar. Explain.

21. How is an ionic bond formed? How is a covalent bond formed?

22. A compound will not conduct electricity when melted, and it melts at 46°C, a low melting point. What type of bond holds atom to atom in this compound?

23. What ions would probably be formed by Br, Al, Ba, Na, Ca, Ga, I, S, O, Mg, K, At, Fr, all group IA metals, all group VIIA nonmetals?

24. How are ionic solids held together?

25. A compound will conduct electricity when melted, but it is rather hard to melt. What type of bonds are in this compound?

26. In which case would hydrogen bonding be most extensive: (a) liquid water, (b) water vapor, or (c) ice?

27. Why is water a liquid at room temperature?

28. Which is harder to break, an ordinary covalent bond or a hydrogen bond?

29. What is the direction of energy transfer in a
    a. bond-making process?
    b. bond-breaking process?

30. Liquid water consists of water molecules held together by covalent O—H bonds, and the water molecules are held together loosely by hydrogen bonds. When water boils, which type of bond breaks first?

31. Explain why the infrared spectrum of a molecule is referred to as its "fingerprint."

32. For infrared energy to be absorbed by a molecule, what frequency of motion must the molecule have?

MOLECULAR BEAUTY I: STRUCTURE*

ESSAY by ROALD HOFFMANN

What makes a molecule beautiful? Let's begin with the obvious, its *structure*. Molecules have shapes, although until this century we could only guess what they were.

Those shapes can be simple, or they can be exquisitely intricate. All the small covalent molecules discussed in this chapter have very simple shapes. Structure I shows dodecahedrane ($C_{20}H_{20}$), a molecule with a somewhat more complex shape. (In this kind of shorthand diagram, the lines represent bonds between carbon atoms, one of which sits at each intersection or *vertex*. There is also an unshown hydrogen atom attached to each carbon atom.)

Dodecahedrane was first made in 1982 by Leo Paquette and his co-

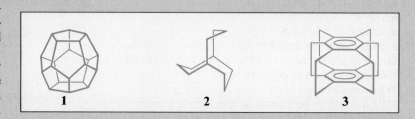

workers. It was a major synthetic achievement, many years in the making. It encloses a space shaped like a dodecahedron—a perfect solid with twelve symmetric, pentagonal faces —that is simply beautiful and beautifully simple.

Molecule 2 was named manxane (after the emblem of the Isle of Man)

by its makers, William Parker and his co-workers. Molecule 3, called superphane, was synthesized by Virgil Boekelheide's group. The hexagonal top and bottom "plates" in this structure are benzene rings (each containing six carbon atoms). All three molecules are simple, symmetrical, and devilishly hard to make.

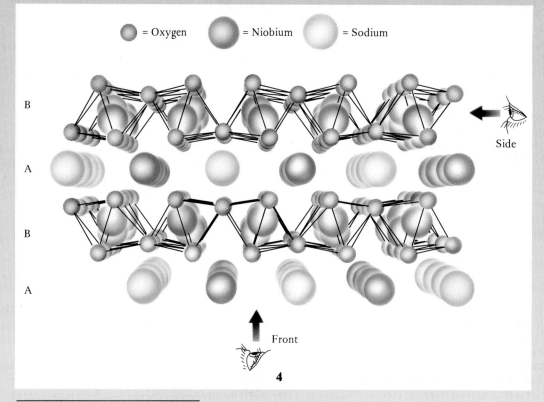

* Reprinted in part from **American Scientist,** Vol. 76, pp. 389–391, 1988.

Let's study a structure whose beauty is a bit harder to appreciate. Arndt Simon in Germany, Tony Cheetham in England, and their co-workers have recently made some inorganic compounds with formulas such as $NaNb_3O_6$, $NaNb_3O_5F$, and $Ca_{0.75}Nb_3O_6$. These are not discrete molecules like the previous group, but extended crystals in which atoms occur in repeating patterns. Structure 4 is one view of $NaNb_3O_6$.

The red balls represent oxygen (O), the dark blue ones niobium (Nb), and the light blue ones sodium (Na). This view chops out a chunk of the extended crystal. You should imagine it repeated thousands of times in three dimensions.

Let's take the structure apart to reveal its incredible beauty. In drawing 4 we clearly see layers or slabs. One kind of layer, marked A, contains only niobium and sodium atoms. The other kind of layer, B, is made of niobium and oxygen atoms. We'll examine this layer first.

The building block of the layer is an octahedron of oxygens around a niobium. One such unit is shown in drawing 5, in two views. In 5a, lines (bonds) are drawn from the niobium to each oxygen. In 5b, these lines are omitted but the oxygens are connected to form an octahedron. Which view is right? In a way, both are right —and both are incomplete. Three-dimensional molecular models like these are just our attempts to picture reality. There can be many representations, each designed to capture some aspect of the real molecule. In 5a the important feature is the arrangement of the chemical bonds, a pretty good choice. These are Nb—O bonds; there are no O—O bonds. Picture 5b emphasizes the octahedral shapes hiding in the structure.

You may wonder where these octahedra are in the complex structure of $NaNb_3O_6$. Let's take the models of drawing 5 and rotate them in space, to the position shown in drawing 6. If

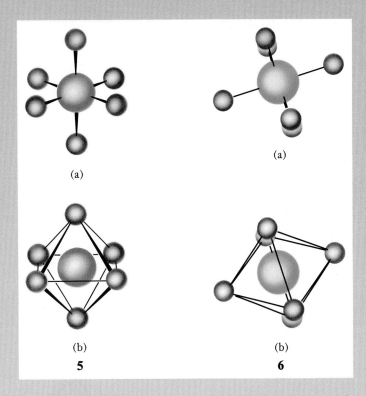

(a)

(a)

(b)

(b)

5   6

you compare 6b to the middle of layer B in 4, you will see the same shapes.

Now consider how the structure of layer B is built up. First, a large number of niobium-oxygen octahedra are linked into a chain by sharing opposite edges. Three views of such a chain appear in drawing 7. One of the views is from the same orientation as in 4. Let's call that the "top." The other two views are roughly from the "front" and "side" as marked in the original drawing, 4. The shared edges are emphasized by darker lines in the side view.

If you compare the top view of this chain with the view in 4, you will see a difference—the niobium atoms are in a neat straight line in 7, but are "staggered" in pairs in 4. Drawing 7 is just a suggestion of how the atoms might be connected, but 4 represents reality. For the moment let's accept this symmetrical asymmetry as one of those unexpected things that make life interesting.

Next, the one-dimensional chains of octahedra combine to form the complete B layer by sharing vertices with other identical chains. We imagine at first that they might do so in a nice straight way (see 8, which is a top view of such a layer). However, they don't; they "kink" (as in 9) in a more complicated but still symmetrical way. You might get the feeling that nature is playing tricks. This is a result of our tendency to believe at first in the most symmetrical suggestion of how things *might* be, without accounting for all the forces that must be balanced.

We now have layer B, this fantastic two-dimensional slab (repeated over and over in the crystal), consisting of edge-sharing one-dimensional chains of octahedra that are stitched together by sharing vertices. What about layer A?

Drawing 4 shows that layer A is made of needle-like lines of sodium and niobium. We might assume that

143

Top   Front   Side

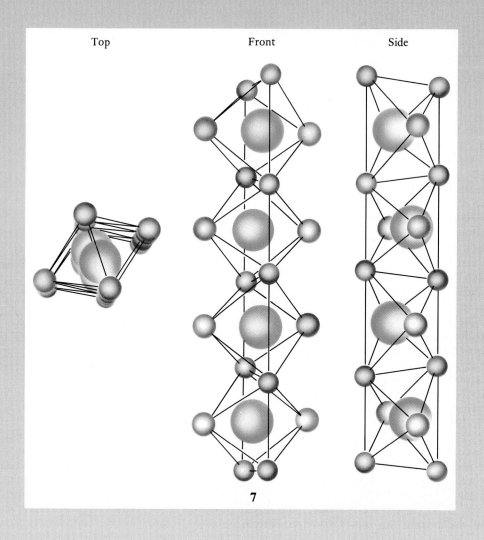

7

Top

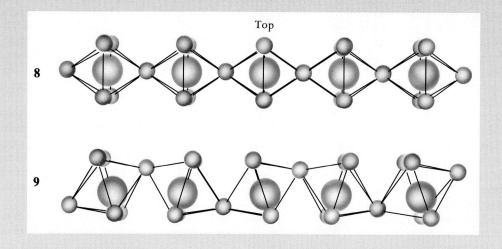

8

9

144

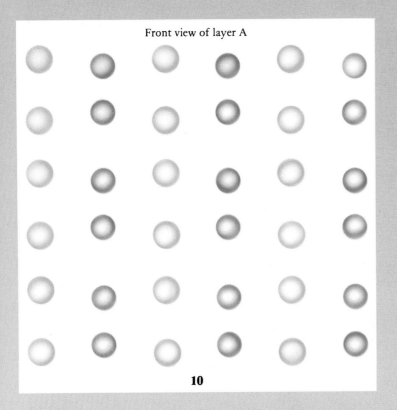

Front view of layer A

10

these atoms are equally spaced, but this crystal has another surprise in store.

Picture 10 is a front view of layer A. Although the sodium atoms are almost equally spaced, the niobium atoms clearly are not. They pair along the vertical direction (this pairing is hidden in the "top" view of 4) so that there are short and long separations. The short distance is considerably

shorter than that in pure niobium metal.

Why do niobium atoms pair? In the study we've done of the way electrons behave in these compounds, we found that the pairing is caused by the formation of Nb—Nb bonds along the needle. There are also bonds, not shown here, between the niobium atoms of layers A and B. Finally, there are links that connect the layers.

For instance, the niobiums in layer A are bonded to the oxygens in the two nearest B slabs.

$NaNb_3O_6$, an incredible material, is at the same time symmetrical and unsymmetrical and assembles itself into small black crystals. It is a testimonial to the natural forces that shape molecules, and to the human minds and hands that brought the structure into being.

6

Some Principles of Chemical Reactivity

Chemical reactions are sometimes spectacular and sometimes almost imperceptible, but they all have some features in common. This chapter explores some of the features that are important to all of us. (John Gillmoure.)

Literally millions of chemical changes have been observed and described in the literature. A typical premedical student taking organic chemistry will study the chemical changes involving approximately 2000 organic compounds in a two-semester course and, during this time, chemists will add many more than 2000 new organic compounds to the known list. Is there any hope, then, that a liberal-arts student like yourself, or even a chemist for that matter, can learn enough chemistry to grasp the sensibility that surrounds this science and discern its usefulness in meeting individual and societal needs? One possibility is to look for principles that are related to all of the chemical changes. Happily, there are only a few such generalizations, and these bring order to a multiplicity of facts. It is even more impressive when chemical principles, coupled with related theory, help predict new chemical information. Some of the more important principles are presented in this chapter.

REACTANTS BECOME PRODUCTS

In all chemical reactions some pure substances disappear and others appear. A familiar example is the change from shiny steel to a red-brown rust (Fig. 6–1). This is a very important reaction since losses from the corrosion of iron and steel in the United States cost slightly over $100/person every year, or tens of billions of dollars each year.

Some other chemical changes are given in the following equations:

REACTANTS		PRODUCTS
CaO + H_2O	$\longrightarrow$	$Ca(OH)_2$
Calcium Oxide (Quicklime) — Water		Calcium Hydroxide (Slaked Lime)
$2\,Na$ + $Cl_2(gas)$	$\longrightarrow$	$2\,NaCl$
Sodium — Chlorine		Sodium Chloride (Table Salt)
$H_2(gas)$ + I_2	$\longrightarrow$	$2\,HI$
Hydrogen — Iodine		Hydrogen Iodide

Sulfur being produced at a volcanic vent. (David Cavagnaro.)

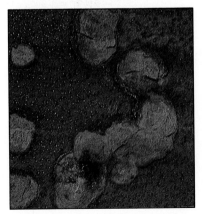

Figure 6–1   Underwater rust on a shipwreck.

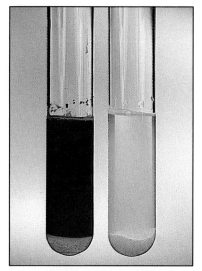

Figure 6–2   Though calcium oxide (lime) is not very soluble in water, as shown in the right tube, enough does dissolve to make the water solution very alkaline, as indicated by the acid-base indicator phenolphthalein which turns from colorless to red in strongly alkaline solutions. No indicator was added to the tube on the right. (Charles D. Winters.)

Figure 6–3   All chemical change is explained by the rearranging of atoms through the breaking and making of the bonds between atoms.

From these reactions we can note our first important point concerning chemical reactions:

> In chemical reactions, reactants become products; some substances are consumed and new ones appear.

Figure 6–4   Colorless hydrogen gas will burn in an atmosphere of red-brown bromine to yield hydrogen bromide, another colorless gas. The yellow of the hydrogen bottle is reflected from the glass surface. (CHEM Study Films.)

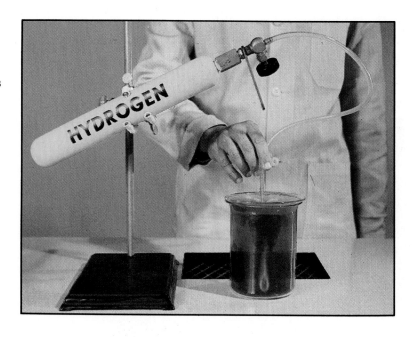

Figure 6–5   Steel wool burns explosively when heated to red heat and inserted into a container of oxygen gas. (Charles Steele.)

WEIGHT RELATIONSHIPS IN CHEMICAL REACTIONS

As was pointed out in Chapter 2, one of the first laws established in chemistry is the conservation of matter in chemical change. Pure substances are destroyed and created, but the matter involved is preserved. The practicality of the conserved quantitative relationship between the amount of reactants consumed and products produced enables the chemist to calculate the reaction yields easily.

All calculations involving chemical equations are based on the law of conservation of matter.

An important question concerning weight relationships in chemical reactions could be: What weight of aluminum (Al) can be produced from one ton of aluminum oxide (Al_2O_3)? Once the atomic weight scale was established, such important calculations could be accomplished readily in view of the conservation of matter involved. We shall return to this question in Example 4 after looking at simple relationships based on the balanced chemical equation.

Hydrogen (H_2) reacts with chlorine (Cl_2) to form hydrogen chloride (HCl). The weight of the reactants, H_2 and Cl_2, used in the reaction must equal the weight of the product, HCl. Accounting for atoms requires balancing an equation. To balance a chemical equation:

1. Place numbers (coefficients) only *before* formulas. (2 H_2 means 2 molecules of H_2 and $2 \times 2 = 4$ atoms of H)
2. Have the same number of each kind of atom on each side of the arrow. ($\longrightarrow$)
3. Do not change subscripts. Each symbol represents one atom. (H_2 means a molecule composed of two atoms.)

The reaction can be written more concisely by using the chemical equation:

$$H_2 + Cl_2 \longrightarrow 2\ HCl$$

Note that a coefficient, 2, has been placed in front of the hydrogen chloride so that two atoms of H and two atoms of Cl are represented in both the reactants and the products; that is, none are gained or lost. What is the meaning of the symbolism of the equation? There are two alternative but equally meaningful ways to interpret the equation:

> One molecule of H_2 reacts with one molecule of Cl_2 to form two molecules of HCl;

or,

> One mole of H_2 molecules reacts with 1 mole of Cl_2 molecules to form 2 moles of HCl molecules.

Once the equation is balanced, the relative number of moles for each substance involved is given by the respective coefficients. Now, recall that one mole of molecules weighs the molecular weight in grams. Remember that a single H_2 molecule weighs 2 on the atomic weight scale, but a mole of H_2 molecules weighs 2 g. As a result, a set of weights for all substances involved in the reaction can be obtained easily by adding up the atomic weights of the atoms represented by the formula. Since the coefficients denote the number of atoms or molecules, one can add up the atomic weights for the reactants or products to establish the ratio of weights of the chemicals involved in the reaction. The weight ratios are the same if we are referring to the individual atoms or molecules or the laboratory molar amounts of them. An analogy may be helpful: If a grain of corn weighs five times more than a grain of wheat, a dozen grains of corn should weigh five times more than a dozen grains of the wheat, and a mole of corn grains would weigh five times as much as a mole of wheat grains. This, of course, assumes that all of the corn grains are identical and all of the wheat grains are identical as well.

For example, the molecular weights of the molecules of H_2, Cl_2 and HCl involved in the previous reaction are:

	Atomic weight	×	No. of atoms	=	Total weight
H_2,	1	×	2	=	2
Cl_2,	35.5	×	2	=	71
HCl (two elements)					
H	1	×	1	=	1
Cl	35.5	×	1	=	35.5
		Total for HCl molecule 36.5			

Note that one molecule of HCl has a molecular weight (36.5) that is one half of the sum of the molecular weights for the reactants, H_2 (2), and Cl_2 (71.0). When you realize that two molecules of HCl are produced for each molecule of H_2 and/or Cl_2 that react, the conservation-of-weight relationship is understood properly.

$$
\begin{array}{ccccc}
H_2 & + & Cl_2 & \rightarrow & 2\,HCl \\
(1 + 1) & + & (35.5 + 35.5) & = & 2(1 + 35.5) \\
2 & + & 71 & = & 73 \\
& & 73 & = & 73
\end{array}
$$

Consider another example to illustrate the calculation of molecular weight. One of the components of gasoline is octane, C_8H_{18}. The molecular weight is calculated as follows:

Atomic weight	× No. of atoms	= Total weight
C	12.0 × 8	= 96.0
H	1.0 × 18	= 18.0
Molecular weight of C_8H_{18} =		114.0

The molecular weight is simply the sum of all of the atomic weights represented by the formula. Now, recall that the weight of one mole of molecules is the molecular weight in grams. The molecular weight of C_8H_{18} is 114.0, and 114.0 g is the weight of 6.02×10^{23} molecules or 1 mole of C_8H_{18}. It follows that 2 moles of C_8H_{18} molecules weigh 228 g, that is, 2 moles × 114 g/mole.

These facts allow several types of mole calculations to be made for chemical reactions. The following examples will illustrate some of the calculations.

EXAMPLE 1

How many moles of nitrogen (N_2) are required to react with 6 mol of H_2 in the formation of ammonia (NH_3)?

Ammonia is used as a crop fertilizer.

1. Write and balance the equation:

$$N_2 + 3 H_2 \longrightarrow 2 NH_3$$

2. Since 1 mol of N_2 reacts with 3 mol of H_2, how many moles of N_2 will react with 6 mol of H_2? The answer is 2 mol of N_2. If the number of moles of H_2 is doubled, then the number of moles of N_2 must be doubled to *keep the same ratio of N and H that react with each other.*

Chemists use "mol" as an abbreviation for "mole."

Figure 6–6 An industrial plant for the manufacture of ammonia. The many industrial uses of ammonia make it the third most abundant chemical manufactured in the United States. (Courtesy of W. M. Kellogg Company.)

Figure 6–7   Nitrogen dioxide, an air pollutant from automobile engine exhaust, is a heavy brown gas. It can also be produced in the reaction of copper and concentrated nitric acid as shown in this photograph. The blue-green color results from the copper ions going into solution. (Charles D. Winters.)

Nitrogen dioxide is a major air pollutant.

Figure 6–8   An organic sample can be analyzed for its H and carbon (C) contents because of the conservation of mass in chemical change. When a weighed sample is burned in pure O_2, water (H_2O) and CO_2 are formed. The H_2O can be absorbed in the first tube and the CO_2 in the second tube by using appropriate absorbers. The increases in weight in these two tubes measure the amounts of H_2O and CO_2 absorbed. Since we know the percentage of H in H_2O and C in CO_2, we can calculate the amounts of H and C that came from the burned sample.

EXAMPLE 2

How many moles of nitrogen dioxide (NO_2) will be produced by 4 mol of O_2 reacting with sufficient nitrogen oxide (NO)?

1. Write and balance the equation:

$$2\,NO + O_2 \longrightarrow 2\,NO_2$$

2. From the balanced equation we see that the number of moles of NO_2 produced is twice that of the O_2 reacting. Therefore, 4 mol of O_2 would produce 8 mol of NO_2.

EXAMPLE 3

How many grams of carbon dioxide (CO_2) can be produced by burning 2650 g of gasoline (C_8H_{18})? How many tons of CO_2 would be produced by burning 2650 tons of the same gasoline?

1. Write and balance the equation:

$$2\,C_8H_{18} + 25\,O_2 \longrightarrow 16\,CO_2 + 18\,H_2O$$

2. The balanced equation states that 2 mol of gasoline produce 16 mol of CO_2. Since the molecular weight of C_8H_{18} is 114, 2 mol would weigh 228 g. The molecular

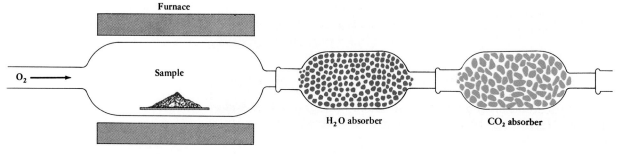

Furnace

$O_2 \longrightarrow$

Sample

H_2O absorber CO_2 absorber

weight of CO_2 is 44, that is, $(1 \times 12) + (2 \times 16)$, so 16 mol would weigh 16 mol $\times$ 44 g/mol = 704 g. Thus, 228 g of gasoline would produce 704 g of CO_2; that is, the weight of CO_2 produced is about three times the weight of gasoline burned. This means that our 2650 g of gasoline should produce about 8000 g of CO_2. To be more exact:

$$\text{grams of } CO_2 = 2650 \text{ g } C_8H_{18} \times \frac{704 \text{ g } CO_2}{228 \text{ g } C_8H_{18}} = 8182 \text{ g } CO_2$$

Since the tonnage ratio is the same as the gram ratio, 8182 tons of CO_2 would be produced from 2650 tons of the gasoline.

EXAMPLE 4

How many pounds of Al can be obtained from 1.00 ton of pure aluminum oxide (Al_2O_3)?

1. Write and balance the chemical equation:

$$2\ Al_2O_3 \longrightarrow 4\ Al + 3\ O_2$$

2. The balanced equation states that 2 mol of Al_2O_3 produce 4 mol of Al. Add up the formula weight of Al_2O_3.

Figure 6–9 Rolls of sheet aluminum obtained from the reduction of aluminum oxide. (Aluminum Association of America.)

$$\left(\frac{27.0 \text{ g}}{\text{mol Al}} \times \frac{2 \text{ mol Al}}{\text{mol } Al_2O_3} \right) + \left(\frac{16.0 \text{ g}}{\text{mol O}} \times \frac{3 \text{ mol O}}{\text{mol } Al_2O_3} \right) = \frac{102 \text{ g}}{\text{mol } Al_2O_3}$$

Thus, 2 mol of Al_2O_3 would weigh 204 g (2 mol $\times$ 102 g/mol). Four moles of Al weigh 4 mol $\times$ 27.0 g/mol = 108 g. Since 108 is about half of 204, a ton of Al_2O_3 should produce about half a ton of Al, which is about 1000 lb. To be more exact:

$$\text{lb of Al} = 1.00 \text{ ton } Al_2O_3 \times \frac{108 \text{ g Al}}{204 \text{ g } Al_2O_3} \times \frac{1 \text{ lb Al}}{454 \text{ g Al}} \times \frac{454 \text{ g } Al_2O_3}{1 \text{ lb } Al_2O_3} \times \frac{2000 \text{ lb } Al_2O_3}{\text{ton } Al_2O_3}$$

$$= 1060 \text{ lb of Al}$$

By observing this problem carefully, perhaps you can see a shortcut. If the units given for Al_2O_3 and the units sought for Al are the same (ton, for example), the numbers obtained from the chemical equation need no additional factors. (We did not need the two 454 g/lb factors in the preceding numerical solution; they cancel each other.) Thus, the numbers obtained from the chemical equation can have any weight units (tons, pounds, grams, kilograms, and so on) *as long as both weights have the same units.* In other words, the solution to the problem resolves simply into the following:

$$\text{lb of Al} = 1.00 \text{ ton } Al_2O_3 \times \frac{108 \text{ tons Al}}{204 \text{ tons } Al_2O_3} \times \frac{2000 \text{ lb Al}}{\text{ton Al}}$$

$$= 1060 \text{ lb of Al}$$

QUANTITATIVE ENERGY CHANGES IN CHEMICAL REACTIONS

Chemical reactions may produce heat or absorb it. Heat production is an **exothermic** process, and heat absorption is an **endothermic** one. Furthermore, the amount of heat energy involved in a chemical change is just as quantitative as the amounts of chemicals involved. Consider again the reactions given at the beginning of this chapter, but this time add the heat effect:

REACTANTS			*PRODUCTS*	*HEAT EFFECTS** †
CaO Calcium Oxide (Quicklime)	+	H_2O Water	$\longrightarrow$ Ca(OH)$_2$ Calcium Hydroxide (Slaked Lime)	+ 15.6 kcal/mol Ca(OH)$_2$
2 Na Sodium	+	$Cl_2(gas)$ Chlorine	$\longrightarrow$ 2 NaCl Sodium Chloride (Table Salt)	+ 196.4 kcal (98.2 kcal/mol NaCl)
$H_2(gas)$ Hydrogen	+	$I_2(gas)$ Iodine	$\longrightarrow$ 2 HI(gas) Hydrogen Iodide	− 12.4 kcal (−6.20 kcal/mol HI)

* Heat energy is measured in calories (cal). A kilocalorie (kcal) is 1000 cal. A calorie is the amount of heat required to raise the temperature of 1 g of water 1°C.
† + kcal means heat is liberated (exothermic); − kcal means heat is required (endothermic).

In the first reaction, CaO (quicklime) reacts with H_2O to give Ca(OH)$_2$ (slaked lime) with the evolution of heat. In the second reaction, Na reacts with the greenish-yellow gas Cl_2 to give NaCl (table salt). If a piece of hot Na is put into a flask containing Cl_2, the Na burns quickly and liberates a great deal of heat and light. White crystals of NaCl are produced. In the last reaction, gaseous H_2 reacts with gaseous I_2 to produce gaseous HI with the absorption of heat. These facts, along with similar ones, lead to a second generalization about chemical reactions:

> **A given amount of a particular chemical change corresponds to a proportional amount of energy change.**
> For example, the preparation of 1 mol of Ca(OH)$_2$ from CaO and H_2O releases 15.6 kcal. To prepare 2 mol of Ca(OH)$_2$, 2 × 15.6, or 31.2 kcal of heat is released.

Sometimes energy changes in reactions are difficult to observe because of the very slow rate of reaction. An example is the rusting of iron. The reaction involved is complicated, but we can represent it by the simplified equation:

4 Fe + 3 O_2 + 6 H_2O $\longrightarrow$ 4 Fe(OH)$_3$ + 788 kcal [197 kcal/mol Fe(OH)$_3$]
Iron +   Moist Air   Iron   +   Heat
                                      Hydroxide
                                      (Rust)

Ordinarily, the rusting of iron occurs so slowly that the liberation of heat is perceptible only with the aid of special instruments. The total amount of heat evolved in rusting is considerable, but it typically takes place over a long period.

SELF-TEST 6–A

1. In photosynthesis, carbon dioxide is combined with water to form the simple sugar, glucose, and oxygen:
   a. Balance the equation.

   ____ CO$_2$ + ____ H$_2$O $\longrightarrow$ ____ C$_6$H$_{12}$O$_6$ + ____ O$_2$

b. How many molecules of CO_2 are necessary to produce one molecule of sugar?

c. How many moles of CO_2 are necessary to produce 1 mol of sugar?

d. What is the molecular weight for CO_2?
Of $C_6H_{12}O_6$?

e. How many grams of CO_2 are required to make 1 mol of sugar?

2. If 68 kcal of energy is released in the formation of 18 g (1 mol) of water by the combination of H_2 and O_2, how much energy would be released in the formation of 36 g of water?

3. Balance the following equations:

a. _____ Mg + _____ O_2 ⟶ _____ MgO

b. _____ Si + _____ Cl_2 ⟶ _____ $SiCl_4$

c. _____ Al + _____ O_2 ⟶ _____ Al_2O_3

4. An important source of hydrogen, used as a rocket fuel, is the decomposition of water by electrical energy. The reaction is:

$$H_2O \xrightarrow{\text{Electrical Energy}} H_2 + O_2$$

a. Balance the equation.

b. What weight of water is necessary to produce 2.0 g of hydrogen?

c. How many grams of oxygen would be produced as a byproduct?

d. How much water would be necessary to produce 2.0 tons of hydrogen?

THE DRIVING FORCE IN CHEMICAL REACTIONS

Some chemical reactions, such as the burning of sodium metal in air at room temperature, occur **spontaneously** as soon as the two chemicals come into contact. Some spontaneous reactions require no apparent push for them to occur; the contact of the chemicals is all that is necessary. Some spontaneous reactions are so fast and release so much energy that they create explosions. Pouring powdered calcium metal into an atmosphere of fluorine (F_2) would be a very dangerous thing to do if significant amounts of materials are involved. Other spontaneous reactions are very quick or even explosive only if a sufficient amount of energy is added to get the reaction started. Using a match to start a paper fire or the shock wave from the firing pin to set off a gunpowder explosion are common examples. Spontaneous reactions are sometimes so slow that you do not want to wait around for the end, even though the end will surely come if you wait long enough as in the rusting of a shipwreck. Some of the chemical reactions with which we are familiar are **nonspontaneous** in that they will never occur when the chemicals are in contact unless they are pushed from reactants into products. Carbon dioxide and water will never make sugar on contact without having energy applied to facilitate the series of reactions we call photosynthesis. It is natural that you should wonder, "What is the basic reason for chemical reactions and why are some of them spontaneous with varying reaction rates?"

First we observe that spontaneous chemical reactions are usually exothermic, occurring with the evolution of heat energy as in fires and explosions. Chemical systems, like physical systems (balls roll down hill, wound

Figure 6–10 The burning of natural fuels (here methane: CH$_4$) is a spontaneous chemical reaction. (Courtesy of Atlanta Gas Light Company.)

The big bang theory seeks to explain the expansion and scattering of the universe but not the highly energetic origin required.

Figure 6–11 The order of arrangement of colored marbles in a beaker is greater when the marbles are separated by color. Disorder is increased as the marbles are mixed without regard to color. Since entropy is a measure of disorder, the beaker on the left represents a higher level of entropy than the beaker on the right. (Charles Steele.)

A hyperbole, for fun: The shining sun evaporates the water that collects in the clouds, that falls as rain, that is stored in the lake, that falls through the turbine, that generates the electricity, that charges the battery; and, after a finite number of changes, the sun will be cold! But is it really a hyperbole?

springs will unwind, etc.), tend to go from states of high energy to low energy spontaneously. The ashes and carbon dioxide have less energy than the wood and oxygen from which they came. However, even though the tendency to reach a lower energy state is an evident driving force in many chemical systems, some spontaneous reactions are endothermic; that is, they absorb energy. The products eventually contain more energy than that held by the reactants. Here is an interesting classroom demonstration of a spontaneous endothermic chemical reaction: In a well-ventilated area, mix in a beaker with a thermometer two white powders, Ba(OH)$_2 \cdot$8H$_2$O and NH$_4$Cl, and note the dramatic drop in temperature as the thermal energy of the system decreases and the chemical energy increases.

$$Ba(OH)_2 \cdot 8H_2O + 2\ NH_4Cl + energy \longrightarrow BaCl_2 + 2\ NH_3 + 10\ H_2O$$

It becomes evident then that there is a more fundamental consideration than the absorption or the evolution of energy when attempting to understand the basic cause for chemical reactions.

Matter in the universe is going from a state of high orderliness to one of disorder. The natural direction from order to disorder (Fig. 6–11) came to your attention early when you were told to clean your room or to pick up your toys and put them away. It seemed natural and easy to make a mess, but unnatural and difficult to put things in order. Physical scientists measure the disorder in matter in units of **entropy**. The mathematical definition of entropy is beyond the scope of this presentation, but you can think of disorder as being quantitatively measured in entropy units. The entropy of the universe would be zero (total order) if all of the particles of matter were completely separated into kinds, arranged into perfect crystals with absolutely no particle motion. This zero-entropy condition is impossible considering prevailing theories of particle physics! Hence, there is a continuing interest in the interactions of the most minute particles of matter under the highest energetic conditions.

Chemical reactions, like all physical processes, are driven by this unwinding of the material clock—going from order toward disorder. The formation of stars and planets from nebular materials, the formation of continents and oceans, the formation of pure mineral deposits, and, indeed, the formation of a diamond crystal are all ordering processes. Such ordering processes can only occur at the expense of a greater disordering of the universe as a whole. Spontaneous reactions such as the burning of fuels (Fig. 6–10) occur as a part of the natural process of increasing entropy in the universe, whereas nonspontaneous reactions, such as the charging of an automobile battery, increase order in one relatively small volume of space. However, ordering processes occur only when driven by spontaneous and disordering processes.

The net disorder of the universe is increased when a nonspontaneous process is driven by a spontaneous one, the disorder measured by an increase in entropy being greater than the created order measured by a decrease in entropy. The increased order inside the charged battery is more than offset by the disorder resulting from the generation of the required electricity for charging the battery. If the battery was charged with a photovoltaic device powered by the sun, the increasing disorder of the solar material is a part of the whole process. Our sun is winding down very slowly (in our time frame)

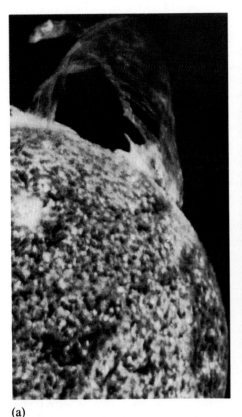

(a)

(b)

Figure 6–12   Some natural processes, such as (a) the burning of the sun, are spontaneous, whereas others, such as (b) photosynthesis, are nonspontaneous. All nonspontaneous processes that occur must be driven by spontaneous processes such that the net of the two processes is a spontaneous one. The result is that the "wound-up" universe in "running down." The sun can power photosynthesis on Earth only for a finite period. (a, Courtesy of NASA; b, Beverly March)

and, in the process, driving many physical and chemical processes on the surface of the earth (Fig. 6–12). For example, photosynthesis, a driven process, produces high-energy fuels which, when oxidized, release the stored chemical energy to run many of the useful chemical events that are so important to our lives.

The unwinding of the universe is phenomenological, and after observing the phenomena of materials formation and separation in nature and the interactions of these materials, which provoke physical and chemical changes, you may wonder how these changes can be explained. We turn to the world of atoms and molecules to understand partially these unfolding events (Fig. 6–13). We explain the extraction of tea into hot water and the resulting homogeneous solution by invoking the concepts of attracting molecules and the scatter of molecules as the result of kinetic energy (the energy of moving molecules that results from previously existing higher energy concentrations). Chemical reactions occur even though the particles of matter are on the move because of attractive interactions between atoms, ions, and molecules. These interactions result from the electric charges within and forces between these fundamental particles. On Earth, essentially all the atoms, except for those of the noble gases, are tied into bonded groups of atoms. The grouping may be elemental, as with H_2, N_2, and O_2, but it is more likely to be in compounds as in H_2O, CO_2, and $C_{12}H_{22}O_{11}$ (sucrose).

Cast a low-entropy deck of cards, ordered by number and suit, across the room to increase their entropy. Pick up the cards and reorder them. Consider the spontaneous and nonspontaneous parts of this process and your ability to repeat the process forever.

The chemical composition of the materials from the moon indicates that atmospheric gases were generated on the moon, but gravitational forces were not strong enough to prevent the natural tendency for scatter.

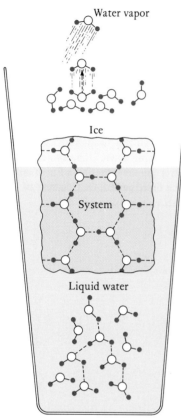

Figure 6–13   Under freezing conditions water molecules "fall" into the water-ice crystalline system. Some but not all of the bonds between the molecules are broken as the ice melts and forms a more disordered system in the liquid state. All of the bonds between molecules are broken as the liquid is vaporized and the scatter (disorder) is maximized. A sample of water has increasingly more entropy as it goes from solid to liquid to gas.

ENERGY, ENTROPY, AND INDUSTRIAL DESIGN

Whether a reaction will go or not depends on the balance of energy and entropy of the reactant and the product. And we can trade off one of these against the other.

In industry, reactions have to work. Both energy and entropy effects determine that. If a reaction does work, if new molecules can be created, then the industrial design is given to the engineers and they focus on energy and materials.

Let's look at an example at the Union Carbide plant in West Virginia. Chemicals are made for 500 different products: detergents, adhesives, plastic wraps and car seats, paints and waxes. Probe the panoply of pipes and towers and you find more than a flow of materials resulting from the scores of chemical reactions. There's a flow of energy, and the engineers must conserve this valuable commodity. Usually a reaction is exothermic, giving off heat. Plant designers want to reuse this heat to drive other reactions, to minimize the waste of energy in the whole plant. So through some of the pipes, they'll transport steam, and out in the plant, steam is piped from point to point, reaction to reaction. The basic raw materials coming into the plant are coal or petroleum, both high in energy. First, ethane gas is produced, and then, by selective addition of oxygen to ethane, we get a variety of industrial chemicals. The plant is constructed so that each product in the chain of reactions has a successively lower level of energy.

The World of Chemistry (Program 13) "The Driving Forces."

It is evident that some atoms are more reactive (Cl for example) and will spontaneously combine to form molecules. Others do not react spontaneously, as is the case with the noble gases. It is also apparent that some molecules, such as those in explosives, are easily disrupted, but that others, such as the molecules in the Teflon on your cookware, are very resistant to chemical reactions. Stability or reactivity in chemical systems then must be tied to order and disorder at the molecular level just as it is in the phenomenological world in which we live (Fig. 6–12). No exception has ever been found in natural or man-made processes — all spontaneous processes occur with an increase in the entropy of the universe.

Spontaneous chemical reactions occur with the increase of disorder (entropy) in the universe. Nonspontaneous reactions increase order but must be driven by a disordering process such that the net result is an increase in disorder in the universe.

Some applications of the entropy law can be found in everyday experiences. Reflection on the realities of birth to death, hot to cold, concentrated to dispersed, available to unavailable, valued materials to waste, order to disorder, and beginning to end tells us that the general idea of entropy is not new. The law of entropy as it describes material order and disorder, like the law of gravity, is a reality of nature. However, ever-increasing entropy in the material universe has nothing to do with your approach to values, purposes, and the meaning of life as these aesthetic concepts are considered and embraced by the human spirit. However, increasing entropy has everything to do with the transient nature of the human experience on planet Earth.

RATES OF CHEMICAL REACTION

How fast do you digest the food you eat? How fast can iron be made from iron ore? How fast does gasoline burn? How fast or slow chemical reactions proceed can be determined quantitatively using reaction rates. The rate of a reaction is always defined in terms of changes in the amounts of chemical substances present per unit of time. Thus, if we consider the burning of sulfur (S) to produce sulfur dioxide (SO_2),

$$S + O_2 \longrightarrow SO_2$$

we can discuss the rate of the reaction in terms of the amount of SO_2 formed per minute or of the amount of S or O_2 consumed per minute. A number of factors affect chemical reaction rates.

Effect of Temperature on Reaction Rate

It is possible to alter the rate of a chemical reaction by changing the temperature. If the temperature is raised, the rates of chemical reactions are increased; if the temperature is reduced, the rates are decreased. We make use of this principle in cooking foods (a roast will cook faster at a higher temperature) and in preserving foods (foods spoil less quickly if refrigerated). Figure 6–14 contrasts the reaction of antimony (Sb) with bromine (Br_2) at 25°C with the same reaction at 75°C.

Raising the temperature speeds up chemical reactions.

For many reactions, a temperature rise of 10°C doubles the rate.

(a)

(b)

Figure 6–14   The elements antimony (Sb) and bromine (Br) react on contact with each other. The reaction at 75°C (b) is much faster than at 25°C (a). Relate this to what you know about the speed of cooking at different temperatures and the retarding of food decay when food is refrigerated. (J. Morgenthaler.)

Effect of Concentration on Reaction Rate

Increasing the concentration of reactants speeds up a reaction.

It is also possible to alter the rate of a reaction by changing the concentrations of the reactants. For example, in the reaction of sulfur and oxygen given earlier, if air replaces oxygen, the reaction will proceed at a slower rate, since air is a mixture of about one part oxygen and four parts nitrogen. The rusting of iron can be retarded by painting or coating the surface of the metal to cut down on the concentration of the oxygen and moisture at the surface. Figure 6–15 illustrates the effect of the concentration of chemicals on the speed of a reaction.

A theoretical (molecular) explanation for the effect of both concentration and temperature on the rate of chemical reactions is illustrated in Figure 6–16. On a molecular basis, the reaction between sulfur and oxygen in air occurs more slowly than in pure oxygen because there are fewer oxygen molecules per volume of air to react (collide) with sulfur molecules.

An interesting and sometimes very dangerous aspect of the concentration effect in reaction rates is the state of subdivision of the chemical reactant. You would find it difficult to impossible to burn a sack of flour in an ordinary fireplace, as the flour would tend to smother the wood and block its contact with oxygen in the air. However, the flour will burn if you can keep it hot enough and keep the air flowing to it. The process is not unlike trying to burn a stack of magazines. Would you be surprised to learn that the same flour as dust in the air of a flour mill forms an explosive mixture so powerful that it can literally blow concrete buildings apart? Indeed, many lives have been lost in such dust explosions. It is simply a matter of having the combustible particles in close contact with the oxygen in the air. Reducing particle size is really a concentration effect in terms of bringing more of the reacting particles together in a given period. The exposed surface of the flour, the

Grain dust has caused explosions destroying entire grain elevators and storage facilities.

Figure 6–15   (a) Lycopodium powder is made up of the ground-up spores of a common moss. If a flame is directed on the powder (left photo), it burns with difficulty, in contrast to the explosive burning (right) when the powder is sprayed into the flame. (Charles D. Winters.) (b) Zinc reacts much faster with a hot sulfuric acid solution (right) than with the same solution at room temperature. (Charles Steele.)

(a)

(b)

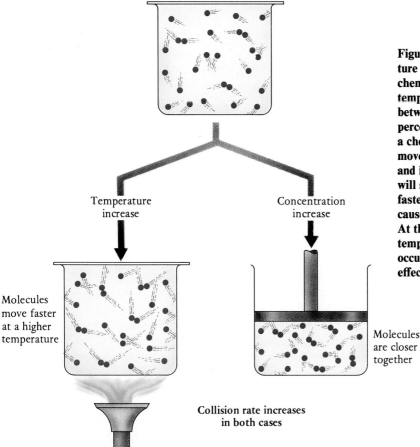

Figure 6–16 **Figure 6–16   Effects of temperature and concentration on rates of chemical reactions. At the higher temperature, more collisons occur between molecules, and a greater percentage of the collisions produces a chemical reaction. (The molecules move faster at high temperatures, and if kept in the same volume, they will strike more often. Moving faster, they will have more energy to cause structural changes to occur.) At the higher concentration (no temperature change), more collisions occur, but the percentage of effective collisions remains the same.**

Temperature increase

Concentration increase

Molecules move faster at a higher temperature

Molecules are closer together

Collision rate increases in both cases

portion of the flour that is in contact with the oxygen, is increased as the particles get smaller and smaller in the same volume of air.

Effect of a Catalyst on Reaction Rate

Slow chemical reactions will often proceed at a much faster rate in the presence of a third chemical or group of chemicals. For example, in the manufacture of sulfuric acid (H_2SO_4), the number one chemical of commerce, it is necessary to convert sulfur dioxide (SO_2) to sulfur trioxide (SO_3):

$$2\ SO_2 + O_2 \longrightarrow 2\ SO_3$$

If the pure chemicals are mixed, the reaction is very slow, much too slow for a profitable industrial process under practical temperature and concentration conditions. However, if some oxides of nitrogen are introduced into the system, the desired reaction proceeds rapidly. Furthermore, the nitrogen compounds are not permanently changed in the process. Such a chemical is referred to as a catalyst. The reaction proceeds faster because a catalyst offers an alternative and easier pathway for the interacting atoms to achieve the new molecular structure.

Catalysts are substances that increase the rate of a chemical reaction without being permanently consumed.

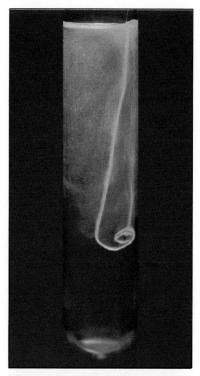

Magnesium metal (Mg) reacts with hydrochloric acid (HCl) to produce bubbles of hydrogen gas (H_2). The reaction can be speeded or slowed by increasing or decreasing the concentration of the acid in the water solution. (Charles D. Winters.)

161

Biological catalysts are called enzymes.

Later, in the study of biochemistry, it will be amazing to note the ability of biological systems to produce catalysts on demand and then destroy them after a particular chemical need is met.

REVERSIBILITY OF CHEMICAL REACTIONS

Chemical reactions are capable of going forward or backward.

Most chemical processes can be reversed under suitable conditions. When a chemical reaction is reversed, some of the products are converted back into reactants. For example, heating calcium hydroxide will drive off water. This process is the reverse of adding water to quicklime (CaO) [See p. 147]:

$$Ca(OH)_2 + heat \longrightarrow CaO + H_2O$$

Calcium
Hydroxide   Quicklime   Water

Other methods can be used to reverse chemical reactions. For example, if we put $Ca(OH)_2$ in a vacuum, there will soon be water vapor in the space around the solid.

It is easier to reverse chemical reactions when they are associated with small heat changes. At slightly elevated temperatures, it is possible to break HI molecules apart into hydrogen and iodine by the application of a relatively small amount of heat:

$$2\ HI + heat \longrightarrow H_2 + I_2$$

On the other hand, water is decomposed into hydrogen and oxygen only by the use of considerable amounts of electrical energy. Although both reactions are reversible, the greater stability of water over hydrogen iodide is associated with the greater amount of heat released per mole in the forma-

Figure 6-17 (a) Electrical energy is required to decompose water into hydrogen (left tube) and oxygen (right tube). Note that two volumes of H_2 are produced for each volume of O_2. (Charles D. Winters.) (b) Hydrogen and oxygen burn to produce water in the gaseous state. The water is condensed on the cooler porcelain dish.

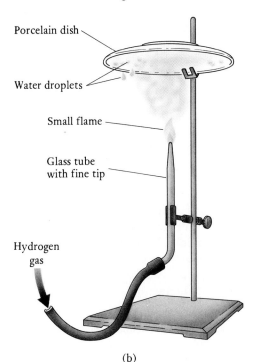

Porcelain dish

Water droplets

Small flame

Glass tube
with fine tip

Hydrogen
gas

(a)   (b)

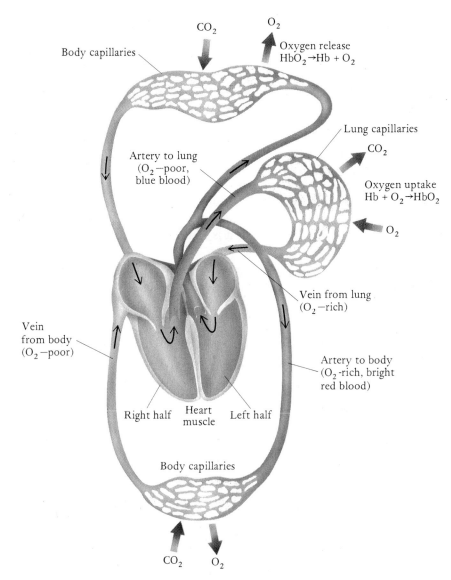

CO₂

O₂
Oxygen release
HbO₂→Hb + O₂

Body capillaries

Lung capillaries

CO₂

Artery to lung
(O₂−poor,
blue blood)

Oxygen uptake
Hb + O₂→HbO₂

O₂

Vein from lung
(O₂−rich)

Vein
from body
(O₂−poor)

Artery to body
(O₂-rich, bright
red blood)

Right half   Heart
muscle   Left half

Body capillaries

CO₂   O₂

Figure 6-18 Chemical reversibility is illustrated in this simplified diagram of human circulation. The heart (shown in the front) is divided into two parallel halves. The right half pumps oxygen-poor blood to the lungs; the left half pumps oxygen-rich blood to the body. Hb = hemoglobin; HbO₂ = oxyhemoglobin. The oxygen is bound chemically in the lungs and, under different conditions, released in the body tissues.

tion of water from its constituent elements. The reaction that is the reverse of the electrolytic decomposition of water is shown in Figure 6–17(b). Hydrogen when burned in air produces water vapor, which can be condensed to liquid water, and much energy.

The fact that many reactions can be controlled to proceed in either direction leads to the conclusion that:

Chemical reactions are generally reversible.

There are many reversible reactions important to human life. One of these involves the transport of atmospheric oxygen from the lungs to the various parts of the body. This task is carried out by hemoglobin, a complex compound found in the blood. This substance takes up oxygen while in the lungs to form oxyhemoglobin.

Hemoglobin + $O_2 \rightleftharpoons$ Oxyhemoglobin

The oxyhemoglobin is then carried by the bloodstream to the various parts of the body where oxygen is released for use in metabolic processes (Fig. 6–18).

The double arrows, $\rightleftharpoons$, indicate a reversible reaction.

163

CHEMICAL EQUILIBRIUM

Chemicals do not always react to form products with the complete extinction of the reactants. In theory, the new arrangement of atoms can rearrange back to the starting point. We may get the idea that all chemical reactions go to completion when we watch a piece of wood "burn up." However, nature quite often displays a reaction in which both reactants and products are present in the reaction medium at constant, but not necessarily the same, concentration levels. When reversible reactions reach the point where the forward reaction is proceeding at the same rate as the reverse reaction, the amount of chemicals present will remain constant because a particular chemical will be produced as fast as it is consumed. At this point, we have **chemical equilibrium**.

> **A chemical change is at equilibrium when products are produced at the same rate the products are consumed in reproducing reactants.**

Consider the equilibrium among limestone ($CaCO_3$), lime (CaO), and carbon dioxide (CO_2). If dry limestone is placed in a vacuum, carbon dioxide gas will soon appear in the container with the limestone. The reaction is:

$$CaCO_3 \rightleftharpoons CaO + CO_2$$

Double arrows ($\rightleftharpoons$) also indicate that the reaction proceeds in both directions at one time and that it has reached equilibrium.

After the CO_2 builds up to a certain concentration level, the system is at equilibrium and the CO_2 level will not rise further. We know that the reaction is proceeding in both directions when the system is at equilibrium because we can introduce radioactive carbon in either the reactant or the product and use a device that detects radiation to trace the radioactive carbon through the reaction in either direction.

For chemical reactions in solution there is an equilibrium constant, which is expressive of the point of equilibrium. The equilibrium constant is a function of the concentrations (at the equilibrium point) of all of the chemicals involved. For a reaction that tends toward more products than reactants at equilibrium, the constant is large. Conversely, a reaction that barely manufactures products in the presence of considerable amounts of reactants has a small equilibrium constant. If the equilibrium constant is too large to be measured (it exists only in theory), the reaction is said to go to completion.

The idea of the equilibrium constant is complicated by the fact that it is a constant only for a particular temperature. Equilibrium constants are often measured at normal laboratory temperature, but there is a quantitative relationship between the constants for a reaction at different temperatures and the amount of heat energy produced or consumed in the reaction. Returning to the limestone and carbon dioxide for a moment, the constant gets quite large as the stone is heated in a kiln because the reaction goes nearer to completion in the conversion of limestone to lime. This is how lime is made in the manufacture of cement.

REACTIONS BY GROUPS OF ATOMS

Under ordinary reaction conditions, chemical reactions proceed with minimal molecular alterations. The evidence is that molecules do not just become unglued completely and break up into individual atoms when entering into a chemical reaction. Rather, a relatively stable group of atoms enter into competition with another stable group for small particles, such as an electron, a proton, an atom, or a group of atoms. For example, H_2SO_4 dissolves in H_2O to form a solution of ions. Evidence indicates that the H_2SO_4 solution contains large amounts of hydronium ions (H_3O^+) and hydrogen sulfate ions (HSO_4^-) along with relatively few sulfate ions (SO_4^{2-}). Apparently, the water removes hydrogen ions (H^+) from the sulfuric acid molecules without completely disrupting the sulfur–oxygen structure in the sulfate ion.

$$H_2O + H_2SO_4 \longrightarrow H_3O^+ + HSO_4^-$$

This is an example of the principle of minimal structural change in chemical reactions:

Ordinary chemical change occurs with a minimum amount of change in the structures of the atoms, molecules, or ions involved.

Consider the following group of changes and note the apparent stability of the sulfate group SO_4^{2-}.

$$\underset{\text{Magnesium}}{\text{Mg}} + \underset{\substack{\text{Sulfuric} \\ \text{Acid}}}{H_2SO_4} \longrightarrow \underset{\substack{\text{Magnesium} \\ \text{Sulfate}}}{MgSO_4} + \underset{\text{Hydrogen}}{H_2}$$

$$\underset{\text{Calcium}}{\text{Ca}} + H_2SO_4 \longrightarrow \underset{\substack{\text{Calcium} \\ \text{Sulfate}}}{CaSO_4} + H_2$$

$$\underset{\text{Strontium}}{\text{Sr}} + H_2SO_4 \longrightarrow \underset{\substack{\text{Strontium} \\ \text{Sulfate}}}{SrSO_4} + H_2$$

$$\underset{\text{Barium}}{\text{Ba}} + H_2SO_4 \longrightarrow \underset{\substack{\text{Barium} \\ \text{Sulfate}}}{BaSO_4} + H_2$$

All alkaline-earth metals react to form similar compounds.

At this point, you may want to glance at some of the formulas in Chapter 21 on medicines and drugs. Our ability to take very complicated molecules and modify them slightly for desired properties both illustrates this important principle and challenges us in our efforts to control the chemical structures for our purposes.

WATER—A VERY SPECIAL COMPOUND ON EARTH

There would be no life on Earth without water with its unique properties. Certainly there are other media on earth and in the universe wherein much chemistry occurs. However, on Earth the chemistry in water solutions and the chemistry of water dominates. Water plays an important role as a reactant, a product, or a coordinating chemical in most of the chemical reactions in our environment. What are water's unique properties, and what are their effects on life as we know it?

Some Physical Properties of Water

1. The density of solid water (ice) is less than that of liquid water. Put another way, water expands when it freezes. If ice were a normal solid, it would be denser than liquid water, and lakes would freeze from the bottom

Figure 6–19   Rock formation in the Strait of Magellan. Water, the universal solvent, covers 72% of the Earth and participates as a reactant, a solvent, or a solvent medium for most of the Earth's natural chemistry.

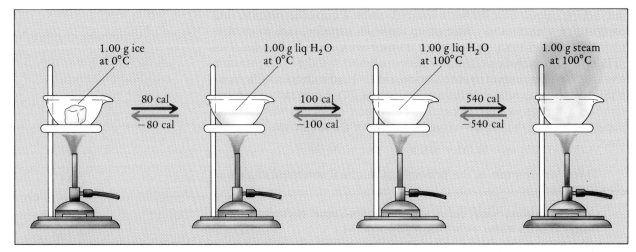

One gram of ice at 0°C requires 80 cal of heat to melt to water at 0°C (no change in temperature). One gram of water at 0°C requires 100 cal of heat to increase its temperature to 100°C (one calorie per gram per degree). One gram of water at 100°C changes to one gram of stream at 100°C on the absorption of 540 cal of heat.

Heat capacity is defined as the amount of heat required to raise the temperature of a sample of matter 1°C.

up. This would have disastrous consequences for marine life, which could not survive in areas with winter seasons.

2. Water is a liquid at room temperature in spite of being composed of very small molecules. In contrast, the hydrogen compounds of the other nonmetals around oxygen in the periodic table are toxic, corrosive gases such as NH_3, H_2S, and HF.

3. Water has a relatively high heat capacity per unit of weight. This means it can absorb large quantities of heat without large changes in temperature. For comparison, the heat capacity of water is about ten times that of copper or iron for equal weights. This property accounts for the moderating influence of lakes and oceans on the climate. Huge bodies of water absorb heat from the Sun and release the heat at night or in cooler seasons. The Earth would have extreme temperature variations if it were not for this property of water. By contrast, the temperatures on the surface of the Moon and the planet Mercury vary by hundreds of degrees through the light and dark cycle.

4. Water has a relatively high heat of vaporization. The heat needed to vaporize 1 g of water at 100°C is 540 cal. A consequence of this is the cooling effect that occurs when water evaporates from moist skin.

5. Water has a large surface tension. The large surface tension of water and its ability to wet surfaces are the bases for capillary action, which carries water to leaves in plants and trees.

Surface molecules of a liquid are pulled inward by the intermolecular interactions with molecules below the surface. Surface tension is a measure of this force.

6. Water is an excellent solvent, often referred to as the universal solvent. As a result, water from natural sources is not pure water but a solution of substances dissolved by contact with water.

The high boiling point, high heat of vaporization, and large heat capacity of water are partially the result of the energy needed to break the hydrogen bonds in liquid water as it is heated or vaporized as described in Chapter 5.

Hydrogen bonding also accounts for the large surface tension of liquid water. The water molecules on the surface are pulled inward by hydrogen bonding to water molecules below the surface. This unbalanced force at the surface causes the surface layer to contract, and energy is required to break

Floating a needle on water illustrates the phenomenon of surface tension. Adding soap to the water will cause the floating needle to sink because the soap lowers the surface tension (see Chapter 22).

this surface. Insects can walk on water because their weight is not sufficient to break through the surface tension.

Interaction of Water with Other Chemicals

The chemical properties of water are also a function of the polar nature of the water molecule—its ability to engage in hydrogen bonding and the strength of the covalent hydrogen–oxygen bond within the molecule. Water will tend to react with any negatively charged species by surrounding it with the positive ends (hydrogen) of the water dipole.

$$\ominus \text{ charge} + x\,H_2O \longrightarrow \ominus \cdot (H_2O)_x \qquad (x = \text{NO. OF WATER MOLECULES})$$

Conversely, water will react with any positively charged species by surrounding it with the negative end of the water dipole.

$$\oplus \text{ charge} + y\,H_2O \longrightarrow \oplus \cdot (H_2O)_y \qquad (y = \text{NO. OF WATER MOLECULES})$$

(See Fig. 9–14 for a visual image of dissolving ions being hydrated by water molecules.)

The number of water molecules involved (x or y) depends on the intensity of the charge and on the size of the charged species.

The ions (charged species) are said to be **hydrated** as they interact with the water molecules, and hydration reactions are generally considered to be physical changes. For example, if NaCl is dissolved in water, the ions separate as they become hydrated by the water molecules.

$$Na^+Cl^-_{(solid)} + (x + y)\,H_2O \longrightarrow Na(H_2O)_y^+ + Cl(H_2O)_x^-$$

On heating, the water can be driven off and the dry salt recovered.

If the intensity of negative charge (or positive charge) is great enough, a hydrogen ion (or hydroxide ion) can be torn from the water molecule and the solution process is not easily reversed. For example, calcium oxide (CaO) will dissolve in (react with) water to give hydrated calcium ions, but essentially no oxide ions exist in solution. The charge on the oxide ion is too concentrated to be stable in the presence of water molecules. The oxide ion removes a hydrogen ion from a water molecule resulting in two hydroxide ions, OH⁻.

$$O^{2-} + H_2O \longrightarrow 2\,OH^-$$

If you evaporate the water from the solution after dissolving CaO in water, the residue will be $Ca(OH)_2$ rather than CaO. Hence, the dissolving of CaO in water is referred to as a chemical change. Although it will not always be easy for you to predict if water reacts physically or chemically with a particular substance, there is a sequence of events that always occur in the interaction of water with other chemicals. First, the water molecules are attracted to any atom displaying a charged atomic surface. Second, the nearby water molecules will be rearranged somewhat due to their interaction with the charged center. Third, if the intensity of the charge is great enough, some of the O—H bonds in the water molecules will be broken resulting in the formation of new chemical species.

An example of a positive ion that is unstable in water is Al^{3+}. If you try to dissolve aluminum chloride ($AlCl_3$) in water, insoluble aluminum hydroxide

The water strider with little weight has an easy time walking on water. (The insect does not provide enough force per unit area to break through the surface tension.) Note that the insect does not walk on the sharp ends of its " toes." (Manfred Danegger/Peter Arnold, Inc.)

The gathering of water molecules around a charged center is somewhat analogous to a magnet picking up a group of paper clips.

Potassium reacts with water to produce hydrogen and a solution of potassium hydroxide. At room temperature enough heat is produced in this reaction to create a fireworks display. (Charles Steele.)

will immediately precipitate as the $+3$ aluminum ion removes hydroxide ions from the water molecules.

$$Al^{3+} + 3\ H_2O \longrightarrow Al(OH)_3 + 3\ H^+$$

We shall see the importance of understanding how water interacts with other chemicals in Chapter 9 as we study the solutions of salts and the reactions of acids and bases. In biochemical reactions (Chapter 15), we shall see that biochemical catalysts facilitate biochemical reactions without the intensity of charge illustrated by the O^{2-} and Al^{3+} reactions cited above. Indeed, the flow of energy through biochemical systems is dependent on the formation and breakup of water molecules while interacting with biochemicals and solution ions.

As we study larger molecules, such as those encountered in organic chemistry (Chapter 12), we shall observe that water will interact with a part of a molecule whenever there is a concentration of either positive or negative charge. For example, complex sugars, like sucrose, which are composed of relatively large molecules will physically dissolve in water because these molecules contain many partially charged $-OH$ groups that hydrogen-bond to water. Then in the presence of appropriate catalysts in solution, sucrose molecules are hydrolyzed (broken down by the reaction with water) into smaller molecules.

Species that will not react with water under normal conditions are the neutral species, with no appreciable concentrations of charge in the molecular or ionic structures. Examples are some of the common gases such as H_2, O_2, N_2, and CH_4. By contrast, the highly electronegative fluorine atoms in F_2 molecules will react with water, releasing oxygen gas.

$$2\ F_2 + 2\ H_2O \longrightarrow 4\ H^+ + 4\ F^- + O_2 + energy$$

The strongly electropositive metals such as Na will release hydrogen from water,

$$2\ Na + 2\ H_2O \longrightarrow H_2 + 2\ Na^+ + 2\ OH^- + energy$$

but some metals will not react with water unless other chemicals are present. As you would expect, the noble gases will not react with water even though they will dissolve somewhat as these small molecules infiltrate the spaces between the groups of water molecules in liquid water. Oils, waxes, grease, plastics, and the like are made up of very large molecules with little to no polarization within the molecules, therefore showing no tendency to interact with water.

The principles presented in this chapter are fundamental and a beginning point for understanding chemical reactions. These particular principles are important to the topics discussed later in this text.

SELF-TEST 6-B

1. Which factor affecting reaction rate (temperature or concentration) is most closely related to freezing foods to prevent spoilage? Why?
2. If water can be produced by burning hydrogen in oxygen, what are the products of the decomposition of water?

3. Name two reversible chemical changes.
4. To what extent is a catalyst (a) used and (b) used up in a chemical reaction?
5. Are equal amounts of reactants and products necessary to achieve chemical equilibrium?
6. In what way is an equilibrium constant not a constant?
7. Which should burn faster: (a) a pound of flour in a sack or (b) the same flour in dust form in the air of a flour mill?
8. What chemical reaction can you name that is constantly reversed in your bloodstream?
9. Name two chemical reactions that are not easily reversed.
10. What is the principle of minimal structural change in chemical reactions?

MATCHING SET

_____	1. Rate of chemical reaction	a. Atom count same on both sides of the arrow
_____	2. Corrosion	b. Speeds up chemical reaction
_____	3. Lower temperature	c. Equal reaction rates
		d. Atomic number ordering
_____	4. Catalyst	e. Slower chemical reaction
_____	5. Equilibrium	f. Formation of rust
_____	6. Balanced equation	g. Amount of matter reacted in a given time
_____	7. 6.02×10^{23}	h. 18 g
_____	8. Gram molecular weight of P_4O_{10}	i. 20
		j. 1 mol
		k. 284 g
_____	9. Weight of one mole of water molecules	l. Mass or weight
		m. 18
_____	10. Conserved during chemical reaction	
_____	11. Number of atoms in the formula $(NH_4)_3PO_4$	

QUESTIONS

1. What is a chemical catalyst?
2. Give an example of a chemical reaction that does not go to completion.
3. Why is it necessary to balance a chemical equation before it can be used in making a calculation?
4. Identify four chemical reactions that we use in our daily lives in which energy plays an important role.
5. List three characteristics of all chemical reactions and illustrate each characteristic with an example.
6. Would you expect all of the chemical bonds in a molecule of aspirin to break as the chemical acts in your body?
7. Is dust a safety factor to be considered in a grain-grinding mill?
8. Fires have been started by water seeping into bags in which quicklime (CaO) was stored. Why would this produce a fire?

9. Write and balance the chemical equation for the reaction between water and hot carbon to form gaseous hydrogen (H_2) and gaseous carbon monoxide (CO).

10. When iron rusts, heat is produced. Why does a rusty piece of iron not feel warm?

11. Balance the following equations:
   a. $SO_2 + O_2 \longrightarrow SO_3$
   b. $K + Br_2 \longrightarrow KBr$
   c. $PbO_2 \longrightarrow PbO + O_2$
   d. $Al + O_2 \longrightarrow Al_2O_3$
   e. $Fe + H_2O \longrightarrow Fe_3O_4 + H_2$
   f. $CH_4 + O_2 \longrightarrow CO_2 + H_2O$
   g. $C_8H_{18} + O_2 \longrightarrow CO_2 + H_2O$
   h. $O_3 \longrightarrow O_2$

12. If the term *endothermic* is used to describe a chemical reaction that absorbs energy as it proceeds, what adjective would be used to describe a reaction that produces energy?

13. In a balanced chemical equation, which is conserved: (a) molecules, or (b) atoms?

14. The kinetic-molecular theory states that matter is made up of molecules and that molecules move faster at higher temperatures. Why should molecules move faster at elevated temperatures?

15. The electrolysis (electrical decomposition) of water is the reverse of what chemical reaction?

16. Firefighters use the methods of controlling the rate of a chemical reaction to combat a fire. Beside each of the following firefighting methods, give the rate-controlling factor that is being applied.
   a. Use of water
   b. Limiting the fuel supply
   c. Use of a fire blanket
   d. Carbon dioxide extinguisher

17. What principle of chemical reactivity applies to the storage of food in a freezer?

18. What can be said about the magnitude of an equilibrium constant for a reaction that appears to go to completion?

19. Add the atomic weights to determine the molecular weights:
   a. H_2O_2 c. $C_2H_4(OH)_2$
   b. H_3BO_3 d. Fe_2O_3

20. What is the weight (in grams) of 1 mol of each of the following?
   a. Xe   c. NH_3
   b. C_2H_5OH d. $Na_2S_2O_3 \cdot 5H_2O$

21. What are some factors that drive reactions toward products?

22. How many moles of KCl can be made using 1 mol of potassium and 1 mol of chlorine?

$$2\,K + Cl_2 \longrightarrow 2\,KCl$$

23. Chlorine can be made by the electrical decomposition of melted sodium chloride:

$$NaCl \xrightarrow{\text{Energy}} Na + Cl_2$$

How many moles of products can be made from 1 mol of sodium chloride? Balance the equation, and be sure to include both products.

24. How many grams of hydrogen are liberated when 75.0 g of sodium metal react with excess water?

$$2\,Na + 2\,H_2O \longrightarrow 2\,NaOH + H_2$$

25. Copper metal can be produced by heating copper sulfide with carbon and air:

$$CuS + O_2 + C \longrightarrow Cu + SO_2 + CO_2$$

   a. Balance the equation.
   b. How many moles of oxygen are required for each mole of CuS?
   c. How many grams of copper can be produced from 100 g of CuS?

26. Silver sulfide (Ag_2S) is the common tarnish on silver objects. What weight of Ag_2S can be made from 1.00 mg of hydrogen sulfide (H_2S) obtained from a rotten egg?

$$4\,Ag + 2\,H_2S + O_2 \longrightarrow 2\,Ag_2S + 2\,H_2O$$

27. Give a reason why some large molecules like sugars will dissolve in water, but others such as oils will not.

28. Which has higher entropy: ice, liquid water, or water vapor? The three samples contain the same mass.

29. What is a spontaneous chemical reaction? Make a general statement about the rate of reaction for spontaneous reactions.

30. Are spontaneous chemical reactions always exothermic? Give an example in your answer.

31. What is the driving force for an endothermic reaction?

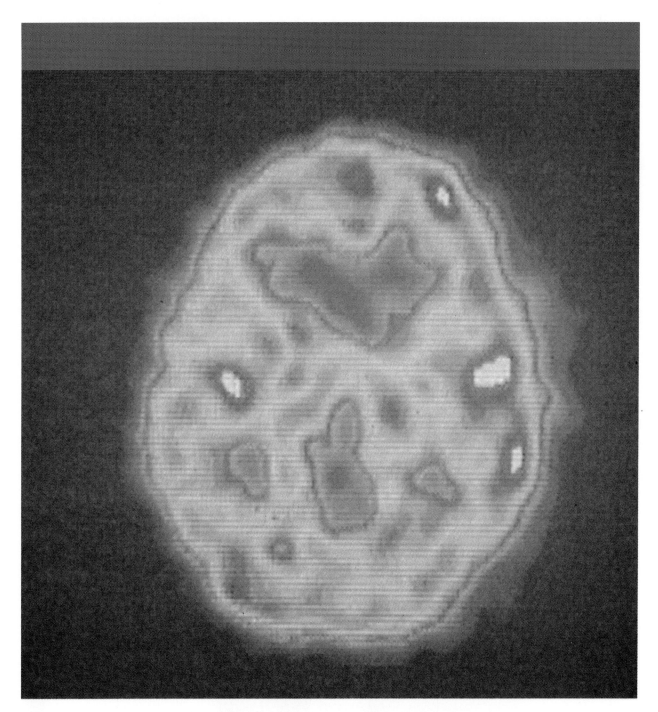

7 Nuclear Reactions

A brain scan using the radioisotope technetium-99m. (*The World of Chemistry,* Program 26, "Futures.")

T he nuclear age is often considered to have started either in the late 1800s and early 1900s with the discovery of radioactive elements by Becquerel and Curie or in 1945 with the first explosion of the atomic bomb. Actually, radioactivity has been a part of the universe since the beginning. The stars are gigantic thermonuclear reactors, and all the planets, moons, and other solid objects in the universe, like the earth and its moon, are thought to contain various radioactive elements. It is true that early atomic theory said nothing about radioactivity, but that was because radiation cannot be directly detected by the five senses. It took the maturing of the sciences—with such diverse discoveries as how to produce a vacuum, photographic film, fluorescent materials, electricity, and magnetic fields—to lead to the knowledge that some atoms spontaneously disintegrate and, in the process, produce radiation.

In February 1896, Henri Becquerel was experimenting in France with materials exposed to the recently discovered X rays. He exposed uranium potassium sulfate to X rays and then positioned the compound on a photographic plate, which was exposed as if by light. Becquerel found by accident that the X-ray exposure was not needed for the uranium compound to expose the photographic plate. In further experiments all uranium compounds, and even the metal itself, were found to expose photographic plates spontaneously. Becquerel showed that uranium (U) and its salts emitted radiation, which was capable of causing ionization in the air. He showed this by charging gold leaves in a vacuum jar (Fig. 7–1) and then discharging them by bringing the uranium samples near the jar. Both positive and negative charges were neutralized this way.

By 1899, Ernest Rutherford had shown that at least two types of radiation were emitted from uranium. One type was alpha (α) radiation. Rutherford found that alpha rays could be stopped by thin pieces of paper and had a range of only about 2.5 cm to 8.5 cm in air before being absorbed. The other form of radiation was beta (β) radiation. Beta rays were capable of penetrating far greater distances in air. Rutherford later found alpha-ray particles to be identical to helium nuclei (He^{2+}) and beta particles to be the same as electrons (e^-).

Figure 7–1   Electroscope used to detect and measure electric charge.

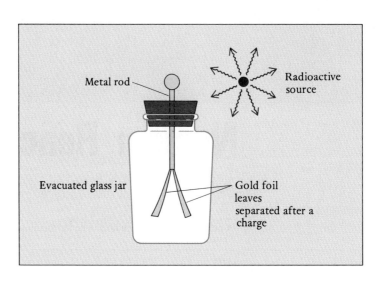

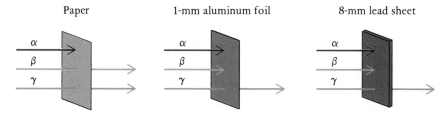

Paper | 1-mm aluminum foil | 8-mm lead sheet

Figure 7–2 Penetrating ability of alpha (α), beta (β), and gamma (γ) radiation. Gamma rays even penetrate an 8-mm lead sheet. Skin will stop alpha rays but not beta rays.

In 1900, Paul Villard characterized a third form of natural radiation, gamma (γ) rays. These, he found, were not streams of particles, but rather had the general characteristics of light or X rays. Gamma rays (a form of electromagnetic radiation) are extremely penetrating; they are capable of passing through over 9 in. of steel and about 1 in. of lead. Figure 7–2 compares the penetrating ability of the three forms of natural radiation.

Gamma rays, like X rays, are part of the electromagnetic spectrum.

Electromagnetic radiation was discussed in Chapter 3.

DETECTING NUCLEAR PARTICLES

In 1911, C. T. R. Wilson invented the **cloud chamber,** which enabled the user to actually "see" a single high-energy, nuclear particle in flight, the collision of such a particle with another nuclear particle, and the path of the products of such a reaction. A single nuclear event could be observed! Furthermore, when the cloud chamber is placed in a magnetic field, the charged particles will follow a curved path and the individual particles can be characterized.

The structure of a simple cloud chamber is illustrated in Figure 7–3. Pressure is exerted on a closed system containing a nuclear particle source (such as an alpha emitter), air, and a layer of water or other liquid, such as ethanol. As pressure is exerted on this system, the concentration of water vapor is increased in the air space around the alpha-emitter. Now, if the pressure is suddenly reduced, the temperature of the air drops, and the air will contain more water vapor than it can normally hold. Consequently,

The cloud track is somewhat analogous to the vapor trails of a high-flying jet airplane; even when the plane is too high to be seen itself, the condensed water from the exhaust clearly marks its pathway.

The Personal Side

Soon after Becquerel's discovery of uranium's radioactivity, Marie Curie, also working in France, studied the radioactivity of thorium (Th: an alpha emitter) and began to search systematically for new radioactive elements. She showed that the radioactivity of uranium was an atomic property—that is, its radioactivity was proportional to the amount of the element present and was not related to the particular compound present. Her experiments indicated that other radioactive elements were present with certain uranium samples. By painstaking technique, she and her husband, Pierre Curie, separated the element radium (Ra) from uranium ore. Radium was found to have an activity over 1 million times greater than that of uranium. Atomic spectra were used to help characterize the new element. In 1903, Marie and Pierre Curie shared the Nobel Prize in physics with Henri Becquerel for their discoveries.

Marie and Pierre Curie.

Figure 7–3   A cloud chamber. This schematic view shows the magnet (a), which produces the field to cause curvature of the particles created and observed inside the chamber (b). The chamber is partly filled with cold vapors of an organic fluid, such as an alcohol. Quickly changing the pressure within the chamber with a plunger (c), allows the tracks of ionizing radiation to be seen as faint trails of condensed liquid.

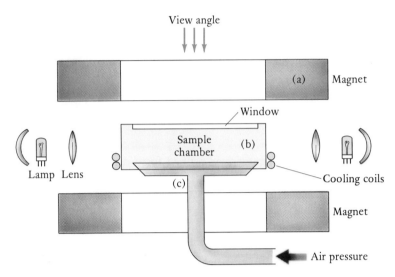

there will be a strong tendency for the water vapor to condense (precipitate). In such a supersaturated system the water molecules readily condense on charged particles. Now, the alpha particle, because of its high energy, ionizes air particles in its path. As a result, there will be a visible path of condensed water (a cloud track, Figure 7–4) tracing the alpha-particle pathway. Any charged particles with sufficient energy to ionize the molecules of the air can

Figure 7–4   A cloud chamber photograph showing an alpha particle striking a nitrogen atom's nucleus. The track going off to the right after the collision is caused by a proton; the track going off to the left is assumed to be that of an oxygen atom.

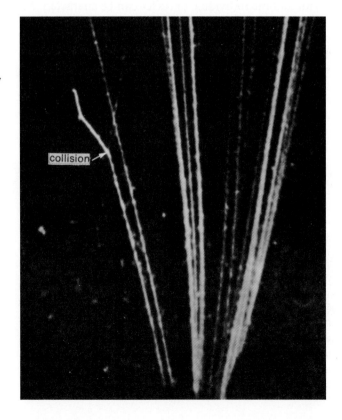

thus be observed. It is a relatively simple matter to photograph such cloud tracks and record nuclear events, such as a collision between an alpha particle and a nitrogen (N_2) molecule, for later study.

NUCLEAR REACTIONS

After the discovery of natural radioactivity in U, Th, and Ra, many other elements were found to have radioactive isotopes. All the elements above bismuth (Bi: atomic number 83) and a few below Bi have naturally occurring radioactive isotopes. Let's now turn our attention to some of the characteristics of nuclear reactions.

Isotopes were discussed in Chapter 3.

Nuclear Reactions	Chemical Reactions
1. New elements are often formed.	1. New elements are never produced.
2. Particles in the nucleus are involved.	2. Usually only outer electrons are involved.
3. Relatively large amounts of energy are involved.	3. Smaller amounts of energy are involved.
4. Rate of reaction is not influenced by temperature, pressure, concentration, or catalysts.	4. Rate of reaction is greatly influenced by temperature, pressure, concentration, or catalysts.

The isotope of U with atomic mass 238 is an alpha-emitter. When the $^{238}_{92}U$ nucleus gives off an alpha particle, made up of two protons and two neutrons, four units of atomic mass and two units of atomic charge are lost. The resulting nucleus has a mass of 234 and a nuclear charge of 90. Atoms containing 90 protons in the nucleus are atoms of Th, not U. This spontaneous nuclear reaction then has changed an atom of one element into an atom of another element and is an example of the **transmutation** of elements.

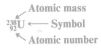

The decomposition of the $^{238}_{92}U$ nucleus is stated briefly by the following nuclear equation:

$$^{238}_{92}U \longrightarrow \, ^4_2He + \,^{234}_{90}Th$$

If the characterized alpha emission were not proof enough that this reaction occurs, additional evidence is supplied by the fact that $^{234}_{90}Th$ is always found with $^{238}_{92}U$ in natural ore deposits and almost always in the concentration predicted by the rates of the U and Th decay.

The nuclear equation for alpha emission of uranium-238 illustrates how nuclear reactions are written. Although atoms and molecules are rearranged in chemical reactions, the number of atoms of a given kind remain the same. However, nuclear reactions involve changes that take place within the nucleus. In a balanced nuclear equation the sum of the mass numbers of the reacting nuclei equals the sum of the mass numbers of the product nuclei. Likewise, the sum of the atomic numbers of the reacting nuclei equal the sum of the atomic numbers of the product nuclei.

One of the dreams of the alchemists (1200–1700 A.D.) was to transmute base metals such as lead and iron into gold. The dream was discarded only after the acceptance of Dalton's indestructible atom.

Consider the decay of radium-226, an alpha-particle-emitter.

$$^{226}_{88}\text{Ra} \longrightarrow ^4_2\text{He} + ^{222}_{86}\text{Rn}$$

Mass Number: $226 = 4 + 222$

Atomic Number: $88 = 2 + 86$

Note how the mass number of the reactant ($^{226}_{88}\text{Ra}$) equals the sums of the mass numbers of the products (^4_2He and $^{222}_{86}\text{Rn}$) and the atomic number of the reactants equals the sum of the atomic numbers of the products. In this nuclear reaction, an atom of one element decays into atoms of two other elements.

Thorium-234 is also radioactive. However, this nucleus is a beta-emitter, which leads to an interesting question: How can a nucleus containing protons and neutrons emit a beta particle, which is an electron? It has been established that an electron and a proton can combine outside the nucleus to form a neutron. Therefore, the reverse process is proposed to occur in the nucleus. A neutron decomposes, giving up an electron and changing itself into a proton:

$$^1_0\text{n} \longrightarrow ^1_1\text{H} + ^{\;\;0}_{-1}\text{e} + \text{energy}$$

Because the mass of the electron is essentially zero compared with that of the proton and neutron, the nucleus would maintain essentially the same mass but would now carry one more positive charge (a proton instead of one of the neutrons). This nucleus is no longer Th, because Th has only 90 protons in the nucleus; it is now a nucleus of element 91, protactinium (Pa). The reaction is:

$$^{234}_{90}\text{Th} \longrightarrow ^{234}_{91}\text{Pa} + ^{\;\;0}_{-1}\text{e} + \text{energy}$$

Gamma radiation may or may not be given off simultaneously with alpha or beta rays, depending on the particular nuclear reaction involved. Gamma rays involve no charge and essentially no mass, so the emission of a gamma photon cannot alone account for a transmutation event.

RATES OF DECAY

The decay of uranium-238 is extremely slow compared with the decay of thorium-234. The rate of decay can be represented by a characteristic **half-life**. A half-life is the period required for half of the radioactive material originally present to undergo transmutation.

The half-life is independent of the amount and chemical form of radioactive material present and is determined only by the type of radioactive nucleus present in the sample. For example, in the preceding reaction, the half-life of $^{234}_{90}\text{Th}$ is 24 days. This means that half of the Th will remain unreacted after 24 days. In another 24 days, half of the half ($\frac{1}{4}$) will remain. This process continues indefinitely, with half of the $^{234}_{90}\text{Th}$ remainder decaying each 24 days.

Figure 7–5 illustrates graphically how the concept of half-life works for a radioactive isotope. Some half-lives are extremely long, and others are

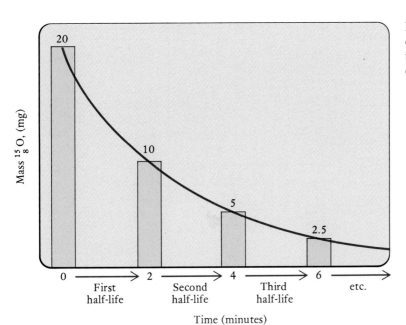

Figure 7–5   The decay of 20 mg of oxygen-15. As each half-life passes, the amount of oxygen remaining is one-half what was present before.

extremely short. The half-life for the $^{238}_{92}U$ alpha decay is 4.5 billion years. As one would expect, relatively large amounts of $^{238}_{92}U$ can be found in nature, whereas only trace amounts of $^{234}_{90}Th$ are present.

The radioactive decay of $^{234}_{90}Th$ into $^{234}_{91}Pa$ is the second step in a series of nuclear decays that starts with $^{238}_{92}U$. After 14 decays the series ends with a stable, nonradioactive isotope of lead, $^{206}_{82}Pb$. This decay series is called the **uranium series** (Figure 7–6.) Table 7–1 gives the half-lives of the isotopes in

TABLE 7–1   Half-Lives of the Naturally Occurring Radioactive Elements in the Uranium-238 ($^{238}_{92}U$) Series

Isotope	Type of Disintegration	Half-Life
^{238}U	α	4.5 billion years
^{234}Th	β	24.1 days
^{234}Pa	β	1.18 min
^{234}U	α	250,000 years
^{230}Th	α	80,000 years
^{226}Ra	α	1620 years
^{222}Rn	α	3.82 days
^{218}Po	α, β	3.05 min
^{214}Pb	β	26.8 min
^{214}Bi	α, β	19.7 min
^{210}Tl	β	1.32 min
^{210}Pb	β	22 years
^{210}Bi	β	5 days
^{210}Po	α	138 days
^{206}Pb	Stable	

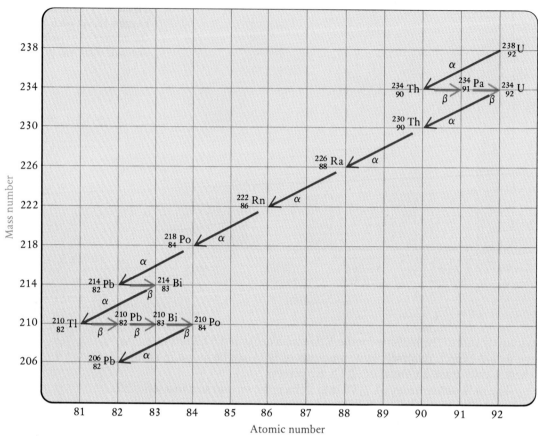

Figure 7–6 The uranium radioactive decay series. An alpha-particle decay results in a decrease of two atomic charge units and four mass units. Beta-particle decay results in no loss of mass and an increase in one atomic charge unit. Gamma emissions are not shown.

the uranium series. Two other natural decay series exist that are similar to the uranium series, but they start out with a different isotope and proceed through a different set of radioactive decay products. The **thorium series** begins with $^{232}_{90}\text{Th}$ (a different isotope from the two Th isotopes that occur in the uranium series) and ends with stable $^{208}_{82}\text{Pb}$. A third series, called the **actinium series,** begins with $^{235}_{92}\text{U}$ and ends with $^{207}_{82}\text{Pb}$. Most of the naturally occurring radioactive isotopes are members of one of these three decay series.

ATOMIC DATING

The concept of radioactive half-life was almost immediately recognized as a useful tool for measuring the age of radioactive materials. As early as 1905, Rutherford suggested during a lecture at Yale University that one indication of age would be the measurement of He formed from alpha particles produced by radioactive decay.

Alpha particles are $^4_2\text{He}^{2+}$ ions.

$$^4_2\text{He}^{2+} \text{ (alpha)} + 2\text{ e}^- \longrightarrow {^4_2\text{He}} \text{ (monatomic gas)}$$

At about the same time, Bertram Boltwood suggested that the $^{238}_{92}\text{U}$/$^{206}_{82}\text{Pb}$ ratio could be measured to date rock samples. To understand how this

the alpha particles is determined by their kinetic energy, which in turn is determined by the parent reaction producing them. Since the alpha particles from a given source all have about the same energy, the alpha tracks in a cloud chamber would be of essentially the same length (Fig. 7–4). In this experiment, Rutherford found some tracks that were much longer than the typical alpha track. Furthermore, these longer tracks did not appear to start at the origin of the alpha tracks but seemed to begin at the termination of an alpha track. When these tracks were studied in a magnetic field, their curvature indicated a particle with a charge-to-mass ratio identical to the value for the proton. Rutherford concluded that the tracks were produced by high-energy protons, which, because of their smaller size and charge, are more penetrating than alpha particles for a given amount of energy. All of the results of the experiment could be explained if one assumed the nuclear reaction to be:

$$\ _{7}^{14}\text{N} + \ _{2}^{4}\text{He} \longrightarrow [\ _{9}^{18}\text{F}] \longrightarrow \ _{8}^{17}\text{O} + \ _{1}^{1}\text{H}$$

where $\ _{9}^{18}\text{F}$ is an unstable compound nucleus. Natural fluorine (F) consists exclusively of the isotope $\ _{9}^{19}\text{F}$. Since both the $\ _{8}^{17}\text{O}$ and the $\ _{1}^{1}\text{H}$ nuclei are stable, the products show no further tendency to undergo nuclear change.

Following Rutherford's original transmutation experiment, there was considerable interest in discovering new nuclear reactions. Many isotopes were subjected to beams of high-energy particles. As you might guess, numerous reactions were found; for example, bombardment of beryllium (Be) with alpha particles produced carbon.

$$\ _{4}^{9}\text{Be} + \ _{2}^{4}\text{He} \longrightarrow \ _{6}^{13}\text{C} \longrightarrow \ _{6}^{12}\text{C} + \underset{\text{Neutron}}{\ _{0}^{1}\text{n}}$$

Chadwick discovered the neutron by this reaction in 1932.

Although the $\ _{6}^{12}\text{C}$ produced in this reaction is stable, the neutron is given off with sufficient energy to provoke additional nuclear reactions in nuclei with which the neutron collides. It was just this nuclear reaction that James Chadwick used in 1932 to prove the existence of the previously postulated neutron.

In 1934, Irène Curie Joliot, daughter of Marie and Pierre Curie (who discovered Ra and polonium [Po]) and her husband Frédéric Joliot, bombarded aluminum (Al) with alpha particles and observed neutrons and a new kind of particle, the **positron.** Positrons are positively charged particles with the same mass as electrons. Being the opposite of electrons, they are a form of **antimatter** — certainly something not seen before. The symbol for a positron is written as $\ _{+1}^{0}\text{p}$. As we shall see later in this chapter, positrons have proven of great practical use in medical diagnosis.

The Joliots discovered that when the alpha particles striking the Al were stopped, the neutrons stopped, but the positron decay continued. They reasoned that the reactions taking place were

$$\ _{13}^{27}\text{Al} + \ _{2}^{4}\text{He} \longrightarrow \ _{15}^{30}\text{P} + \ _{0}^{1}\text{n}$$
$$\ _{15}^{30}\text{P} \longrightarrow \ _{14}^{30}\text{Si} + \ _{+1}^{0}\text{p}$$

The second reaction continued because the phosphorus-30 was decaying at a rate slower than it was being produced. Phosphorus-30 was the first radioac-

Nuclides are the isotopes of the elements.

Figure 7–7   (a) A beryllium nucleus (Be) is struck by an alpha particle, which has sufficient energy to overcome the repulsions of like charges. A nuclear reaction occurs, producing a carbon (C) atom and a neutron. (b) In Rutherford's gold-foil experiment (the experiment that suggested the nuclear atom), the alpha particles were not energetic enough to penetrate the gold nucleus (Au) and were deflected.

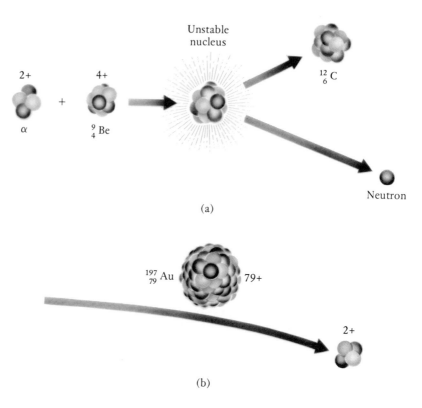

tive isotope to be produced artificially. Today, over 1000 other radioactive nuclides have been produced.

An interesting question arises as to why the alpha particles were scattered by the gold foil in Rutherford's gold-foil experiment (Chapter 3), and yet the same alpha source can produce a nuclear change with a smaller atom such as $^9_4$Be (Fig. 7–7). The answer lies in the fact that the charge on the gold (Au) nucleus is $+79$, whereas the charge on the Be nucleus is $+4$. Most of the alpha particles emitted from natural radioactive decay do not have enough energy to penetrate a heavy, positively charged nucleus such as that of Au. Therefore, if the artificial nuclear reactions are to be studied for the heavier elements, the kinetic energy of the subatomic projectile particles must be increased.

SELF-TEST 7–A

1. A nuclear radiation detection device using a cloud of water vapor is called a _____ .

2. When a $^{87}_{35}$Br nucleus emits a beta particle, the nuclear species that results is _____ .

3. When a $^{216}_{84}$Po nucleus emits an alpha particle, the nuclear species that results is _____ .

4. The half-life of $^{44}_{19}K$ is 22 min. If a 1-g sample of $^{44}_{19}K$ is taken, how much $^{44}_{19}K$ will remain after three half-lives (66 min)?

5. In the following reaction, what is the compound nucleus?

$$^{7}_{3}Li + {}^{1}_{1}H \longrightarrow \underline{\hspace{4cm}} \longrightarrow {}^{7}_{4}Be + {}^{1}_{0}n$$

6. The scientist who discovered the neutron was
_____ . The process was
_____ .

7. In $^{238}_{72}U$-dating of moon rocks, the final decay product measured is
_____ .

8. An art object made of wood was found to contain about 50% of the radioactive carbon concentration found in living wood. Using the concept of carbon-14 dating, approximately how old is the object?
_____ years.

NUCLEAR PARTICLE ACCELERATIONS—NUCLEAR BULLETS

Because charged particles interact with magnetic and electric fields (Chapter 3), these forces can be used to cause nuclear reactions that otherwise would not be observed in nature. Naturally occurring radiation was observed to cause some nuclear reactions. However, many more reactions could be provoked if the nuclear radiation particles could be accelerated, thereby giving them more kinetic energy. The devices used to produce these high-energy nuclear projectiles are called **particle accelerators.** We shall look at a few of these, how they work, and their uses.

Although the construction of particle accelerators is often quite elaborate and different for each apparatus, the basic principles of operation are comparatively simple: (1) opposite charges attract, and (2) the path of a charged particle is curved as it passes through a magnetic field. When these effects are combined with the fact that electric fields do not penetrate to the inside of a charged metal container (the fields are shielded by the "free" electrons in the metal), you have the fundamental facts necessary to explain how particle accelerators work.

Sufficient kinetic energy for a particle to penetrate the electric fields of an atom and to enter a large nucleus cannot be gained readily by accelerating the particle in one step between two electric poles. This would require an impossibly large potential difference of millions of volts. However, if the acceleration is done in several thousand steps, with readily obtainable potential differences of 2000 to 10,000 V being applied to each step, sufficient energy can be imparted to the particle. The **linear accelerator** and the **cyclotron** illustrate the two primary ways this stepwise acceleration is accomplished.

The operation of the linear accelerator does not require a magnetic field and therefore is less complicated than that of the cyclotron. The principle of operation is outlined in Figure 7–8. A source of electrons or protons is provided by ionization (as in a canal-ray rube) at one end of the line of hollow

These accelerators are complicated, but their operating principles are simple.

Figure 7–8   Linear accelerator diagram. Charged particles are produced at the ion source and are attracted toward an oppositely charged electrode, cylinder 1. When the charged particles are passed through cylinder 1, the charge on the cylinder is changed to the same sign as that on the particle being accelerated. At the same instant, the sign of the charge on cylinder 2 becomes opposite to that of the charge on the particle. Thus, the particle is accelerated as it crosses the gap between cylinders. Note that the tubes become successively longer to accommodate increased speeds of the particle.

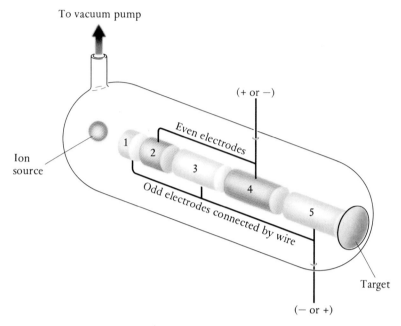

An electron volt is a unit of energy; it is the amount of energy gained by an electron when it is accelerated by an electric field, the electric potential across the field being 1 V.

metal cylinders. A target material is located at the other end. The entire device is enclosed in a near-perfect vacuum so that the accelerated particles may move in a straight path with little possibility of collisions with molecules in the air. Adjacent cylinders of opposite charge cause the traveling charged particles to be repelled by the previous tube and attracted by the upcoming tube. When a given particle enters a cylinder, the sign of the charge on the cylinder changes so the particle will be repelled as it enters the gap between cylinders. Simultaneously, the signs of the charge on all the other cylinders are also changed. As a result, the particle is successively attracted to the upcoming cylinder and repelled by the previous cylinder. At each gap, then, the particles increase their speed and their energy. Often moving near the speed of light, the accelerated particles strike the target atoms and enter their nuclei, producing nuclear changes.

The largest linear accelerator in the world went into operation in 1967 at Stanford, California. It is 2 mi long and accelerates electrons to energies of 20 to 40 billion electron volts (BeV).

The cyclotron was developed in 1931 by Ernest O. Lawrence (element 103, a synthetic element, is named in his honor) and M. S. Livingston. In addition to the electronic circuits and the huge magnets, the instrument consists of two hollow D-shaped metal containers enclosed in a vacuum as shown in Figure 7–9. Charged particles are formed near the center of the gap between the Ds and begin their acceleration to high energy by being attracted into one of the Ds. While the particles are in a D, the influence of the magnetic field causes their path to curve. By the time the particles have completed the semicircle, the signs on the Ds have changed and the repulsion of the previous D and the attraction by the upcoming D accelerate the particles across the gap between the Ds. Each time the particles traverse the gap, they are

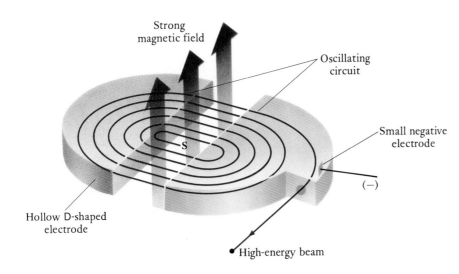

Strong
magnetic field

Oscillating
circuit

Small negative
electrode

(−)

S

Hollow D-shaped
electrode

High-energy beam

Figure 7–9   Schematic drawing of a cyclotron. The source of the particles is shown as S.

accelerated to a greater speed. The increased speed causes the particles to move in a wider arc on each revolution. After many accelerations and revolutions, the arc is sufficiently large for the charged deflector plate to repel the particles through a window and onto the target sample. In some cyclotrons, the particles go too fast to be directed outward by the deflector plate. In this case, the target sample is placed inside one of the Ds.

The first successful model of the cyclotron had Ds with diameters of about 4 in. Later models had diameters of about 2 ft. These models accelerated protons to energies of about 0.5 **million electron volts (MeV)**. Modern cyclotrons have diameters of about 10 ft and accelerate protons to energies of about 250 MeV. Although particle energies this high may seem large, much higher energies have been needed to probe the atom.

A beam of protons (bright stream at right) from a cyclotron. (Argonne National Laboratory photo.)

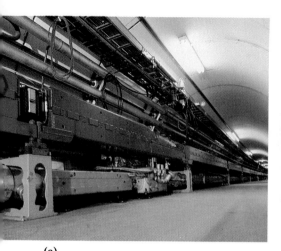

(a)

(b)

Figure 7–10   Particle accelerator at the Fermi National Laboratory, near Batavia, Illinois. (a) Magnets in interior of main accelerator, which is 4 mi in circumference and 1.27 mi in diameter (b) Aerial view of main accelerator. (Fermilab photo.)

In 1978, the world's largest particle accelerator was dedicated in Batavia, Illinois, just southwest of Chicago (Fig. 7–10.) This installation is called the Fermi National Accelerator Laboratory (or Fermilab), after Enrico Fermi, who produced the first nuclear chain reaction (see Chapter 8). This large proton accelerator is 4 mi in circumference, and accelerates protons to energies of 1 **trillion electron volts (TeV),** but it will be dwarfed, when, in about 1995, a vastly larger accelerator will be dedicated in Waxahachie, Texas, near Dallas. This accelerator, called the Superconducting Supercollider (SSC) will be 53 mi in diameter, big enough to fit the District of Columbia inside, and will accelerate two oppositely moving proton beams to energies of 20 TeV. When the two proton beams collide, the particles will have an effective energy of 40 TeV — the most energetic collisions ever to be observed.

Nuclear physicists claim that they are presently at a point in the development of their theories of subatomic matter which is similar to where scientists were when the quantum hypothesis was suggested by Planck (see Chapter 3). The SSC will allow extremely high energy nuclear events to be observed for the first time. The SSC, when it is finished, will have cost over $6 billion, and will be the world's largest machine.

One of the first things scientists will be looking for using the SSC is the so-called Higgs particle, the existence of which is sought to explain mass.

The first synthetic element was prepared in 1940.

TRANSURANIUM ELEMENTS

The heaviest known element before 1940 was U. The invention of the cyclotron and other devices to obtain high-energy particles made it possible for these particles to react with heavy nuclei and to form even more massive nuclei. Thus, **transuranium** elements with atomic numbers greater than 92 were prepared.

In 1940, at the University of California, E. M. McMillan and P. H. Abelson prepared element 93, the synthetic element neptunium (Np). The experiment involved directing a stream of high-energy deuterons ($^2_1$H) onto a

Phillip Abelson (left) and Edwin McMillan (right). (Courtesy of the Lawrence Berkeley Laboratory.)

target of $^{238}_{92}$U. A deuteron is the nucleus of an isotope of hydrogen with one neutron as well as one proton. The initial reaction was the conversion of $^{238}_{92}$U to $^{239}_{92}$U.

$$^{238}_{92}U + {}^{2}_{1}H \longrightarrow {}^{239}_{92}U + {}^{1}_{1}H$$

Uranium-239 has a half-life of 23.5 min and decays spontaneously to the element Np by the emission of beta particles.

$$^{239}_{92}U \longrightarrow {}^{239}_{93}Np + {}_{-1}^{0}e$$

Neptunium is also unstable, with a half-life of 2.33 days; it converts into a second new element, plutonium (Pu).

$$^{239}_{93}Np \longrightarrow {}^{239}_{94}Pu + {}_{-1}^{0}e$$

Plutonium-239, like Np, is radioactive, with a half-life of 24,100 years. Because of the relative values of the half-lives, very little Np could be accumulated, but the Pu could be obtained in larger quantities. The $^{239}_{94}$Pu is important as fissionable material since atomic bombs (see Chapter 8) can be made with it as well as with naturally occurring $^{235}_{92}$U. The names of neptunium and plutonium were taken from the mythological names Neptune and Pluto in the same sequence as the planets Uranus (uranium), Neptune, and Pluto.

Although Neptune and Pluto are the last of the known planets in the solar system, their namesakes are not the last in the list of elements. The rush of transuranium experiments that followed produced additional elements: americium (Am), curium (Cm), berkelium (Bk), californium (C), einsteinium (Es), fermium (Fm), mendelevium (Md), nobelium (No), lawrencium (Lr), and elements 104, 105, 106, 107, and 109.* Until the discovery of element 104, the transuranium elements were named after countries, states, cities, and people. Beginning with element 104, the International Union of Pure and Applied Chemistry (IUPAC) recommended three-letter symbols for the elements. The symbols are based on the numerical roots nil (0), un (1), bi (2), tri (3), quad (4), pent (5), hex (6), sept (7), oct (8), and enn (9). For example, the symbol for element 104 is Unq, a combination of the first letter of each numerical root. The name is obtained by adding "ium" to the

Plutonium is used to make atomic bombs and is also one of the most toxic elements known.

---

* Element 108 is yet to be confirmed.

TABLE 7–3 Nuclear Reactions Used to Produce Some Transuranium Elements

Element	Atomic Number	Reaction
Neptunium (Np)	93	$^{238}_{92}U + ^{1}_{0}n \longrightarrow ^{239}_{93}Np + ^{0}_{-1}e$
Plutonium (Pu)	94	$^{238}_{92}U + ^{2}_{1}H \longrightarrow ^{238}_{93}Np + 2\,^{1}_{0}n$
		$^{238}_{93}Np \longrightarrow ^{238}_{94}Pu + ^{0}_{-1}e$
Americium (Am)	95	$^{239}_{94}Pu + ^{1}_{0}n \longrightarrow ^{240}_{95}Am + ^{0}_{-1}e$
Curium (Cm)	96	$^{239}_{94}Pu + ^{4}_{2}He \longrightarrow ^{242}_{96}Cm + ^{1}_{0}n$
Berkelium (Bk)	97	$^{241}_{95}Am + ^{4}_{2}He \longrightarrow ^{243}_{97}Bk + 2\,^{1}_{0}n$
Californium (Cf)	98	$^{242}_{96}Cm + ^{4}_{2}He \longrightarrow ^{245}_{98}Cf + ^{1}_{0}n$
Einsteinium (Es)	99	$^{238}_{92}U + 15\,^{1}_{0}n \longrightarrow ^{253}_{99}Es + 7\,^{0}_{-1}e$
Fermium (Fm)	100	$^{238}_{92}U + 17\,^{1}_{0}n \longrightarrow ^{255}_{100}Fm + 8\,^{0}_{-1}e$
Mendelevium (Md)	101	$^{253}_{99}Es + ^{4}_{2}He \longrightarrow ^{256}_{101}Md + ^{1}_{0}n$
Nobelium (No)	102	$^{246}_{96}Cm + ^{12}_{6}C \longrightarrow ^{254}_{102}No + 4\,^{1}_{0}n$
Lawrencium (Lr)	103	$^{252}_{98}Cf + ^{10}_{5}B \longrightarrow ^{257}_{103}Lr + 5\,^{1}_{0}n$
Unnilquadium (Unq)	104	$^{242}_{94}Pu + ^{22}_{10}Ne \longrightarrow ^{260}_{104}Unq + 4\,^{1}_{0}n$
Unnilpentium (Unp)	105	$^{249}_{98}Cf + ^{15}_{7}N \longrightarrow ^{260}_{105}Unp + 4\,^{1}_{0}n$
Unnilhexium (Unh)	106	$^{249}_{98}Cf + ^{18}_{8}O \longrightarrow ^{263}_{106}Unh + 4\,^{1}_{0}n$
Unnilseptium (Uns)	107	$^{209}_{83}Bi + ^{54}_{24}Cr \longrightarrow ^{262}_{107}Uns + ^{1}_{0}n$
Unnilennium (Une)	109†	$^{209}_{83}Bi + ^{58}_{26}Fe \longrightarrow ^{266}_{109}Une + ^{1}_{0}n$

† Element 109 was discovered in Germany in August 1982. It was prepared by a technique known as cold fusion. Only a single atom of this element was detected.

numerical roots. Hence, element 104 is Unnilquadium. The other transuranium elements along with the reactions employed to produce them are given in Table 7–3.

RADIATION DAMAGE

We are bombarded constantly by radiation from a number of sources. This radiation includes cosmic rays, medical X rays, radioactive fallout from countries that do nuclear testing, and naturally occurring, widespread radioisotopes. Most radiation damage is too slight to be noticed immediately, although its very presence should be regarded as one of the hazards of everyday life.

As we have seen, a radioisotope either disintegrates into a stable species or becomes part of a decay series. In a sample of radioactive matter large enough to measure, there will be many disintegrations over a given time if the half-life is short or few disintegrations over the same interval if the half-life is long.

Three principal factors render a radioactive substance dangerous: (1) the number of disintegrations per second, (2) the half-life of the isotope, and (3) the type or energy of the radiation produced. In addition, radiation can be very damaging if the radioactive substance is of a chemical nature such that it can be incorporated into a food chain or otherwise enter a living organism.

Radioactive disintegrations are measured in **curies** (Ci); one Ci is 37 billion disintegrations per second. A more suitable unit is the microcurie (μCi), which is 37,000 disintegrations per second. One microcurie of a radioisotope is a potent sample if the energy per disintegration is large enough to

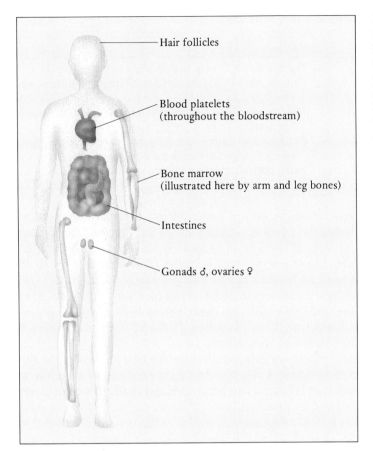

Hair follicles

Blood platelets
(throughout the bloodstream)

Bone marrow
(illustrated here by arm and leg bones)

Intestines

Gonads ♂, ovaries ♀

Figure 7–11   The fast-dividing cells within the body are the ones most harmed by radiation. These include cells in bone marrow, white cells, platelets of the blood, those lining the gastrointestinal tract, hair follicles, and gonads. In addition, the lymphocytes (cells producing the immune responses) are easily killed by radiation.

cause a biochemical change. Normal background radiation to the human body is 2 to 3 disintegrations per second.

The unit **roentgen** (R) is used to measure the intensity of X rays or gamma rays. One roentgen is the quantity of X-ray or gamma-ray radiation delivered to 0.001293 g of air, such that the ions produced in the air carry 3.34×10^{-10} coulomb of charge. A single dental X ray represents about 1 R.

Damage by radiation is due to ionization caused by the fast-moving particles colliding with matter and by the excitation of matter by gamma and X rays, which in turn produce ionization. Neutrons are produced in nuclear explosions, in nuclear reactions, and by background cosmic radiation. A neutron does not produce ionization per se but instead imparts its kinetic energy to atoms, which in turn may ionize or break away from the atom to which they are bonded. Neutrons render many engineering materials, such as plastics and metals, structurally weak over long periods as a result of the decay caused by breaking chemical bonds.

Biological tissue is easily harmed by radiation. A flow of high-energy particles may cause destruction of a vital enzyme, hormone, or chromosome needed for the life of a cell. In general, those cells that divide most rapidly are most easily harmed by radiation (Fig. 7–11).

Whole-body radiation effects are divided into **somatic effects,** which are confined to the population exposed, and **genetic effects,** which are passed on

The coulomb (C) is the standard unit of *quantity* of electrical charge. It is defined as the quantity of electricity transported in one second by a current of one ampere.

Neutrons can damage metals, causing structural failure. This is a severe problem in nuclear reactors.

TABLE 7–4 Effects of Whole Body Radiation

Dose Level (rem)	Effects
0–25	Doses around 25 may reduce white blood cell count.
25–100	Fatigue, blood changes, nausea for about half those exposed.
100–200	Nausea, vomiting, lowered white blood cell count, possibly death.
200–400	Bone marrow, spleen damage. Lethal dose for half of those exposed, especially in absence of treatment.
>600	Fatal, possibly even with treatment.

One *rad* is roughly the energy absorbed by tissue exposed to one roentgen of gamma rays.

1 joule = 10^7 ergs
1 calorie = 4.184×10^7 ergs

to subsequent generations. A unit of measurement of radiation density is helpful in measuring the effect of radiation on tissue. The **rad** is defined as 100 ergs of energy imposed on a gram of tissue. Since alpha and beta radiation vary considerably in energy, that is, some radioisotopes eject these particles with considerably more kinetic energy (velocity) than others, another unit of radiation measurement, the **rem** (for roentgen equivalent man) is often used. One rem is the radiation dosage, regardless of type, that produces the same biological effects in man as 1 rad of X rays. Table 7–4 summarizes whole body radiation effects.

Often somatic effects are delayed. Perhaps the best studied of the delayed effects are the incidences of cancer related to exposure to radiation. It has been estimated that 11% of all leukemia cases and about 10% of all forms of cancer are attributable to background radiations. Certainly an individual who is exposed to a higher than normal level of radiation over considerable time increases the chances of cancer. The alteration of normal cells to cancerous cells caused by radiation is undoubtedly a series of changes, since in almost all cases the onset of cancer lags behind the exposure to radiation by an induction period of 5 to 20 years.

The genetic effects of radiation are the result of radiation damage to the germ cells of the testes (sperm) or the ovaries (egg cells). Ionization caused by radiation passing through a germ cell may break a DNA strand or cause it to be altered in some other way. When this damaged DNA is replicated (i.e., when the DNA structure is copied during cell division; see Chapter 15), the result may be the transmission of a new message to successive generations, a **mutation.** Every type of laboratory animal on which radiation damage experiments have been performed has responded with an increased incidence of mutation. Therefore, the necessity of protecting the population of childbearing age from radiation should be apparent. Theoretically at least, one photon or one high-energy particle can ionize a chromosomal DNA structure and produce a genetic effect that will be carried to the next generation.

SELF-TEST 7–B

1. A particle having an energy of 1 BeV has an energy of _____ electron volts.
2. One type of particle accelerator that moves the charged particles in a circle is called a _____ .

3. The first transuranium element to be "made by man" is
   _____ .

4. The transuranium element of greatest atomic number to be "made by man" is element number _____ .

5. The term *rem* stands for _____ _____
   _____ .

6. What radiation dosage are you receiving each year from background radiation?

7. Near what large city will the proposed Superconducting Supercollider be located?

8. Describe the energies required to produce nuclear particles more fundamental than electrons, protons, and neutrons.

9. Radiation effects passed on to subsequent generations are called _____ effects.

RADON: A DEADLY GAS

Radon (Rn) is the heaviest member of the noble-gas family of elements (He, Ne, Ar, Kr, Xe, Rn). Radon-222, the most common isotope of Rn, is radioactive, with a half-life of 3.82 days. Radon-222 is a product of the uranium decay series (Fig. 7–7) and is a direct result of the decay of radium-226, a naturally occurring isotope that is present at levels of about 1 pCi (picocurie: 1×10^{-12} Ci) per gram in ordinary soil and rocks.

$$^{226}_{88}Ra \longrightarrow {}^{222}_{86}Rn + {}^{4}_{2}He$$

When Rn decays, it produces alpha particles and another short-lived radioisotope, polonium-218.

$$^{222}_{86}Rn \longrightarrow {}^{218}_{84}Po + {}^{4}_{2}He$$

Polonium-218 (half-life 3.05 min) also decays, producing an alpha particle, to lead-214 (half-life 26.8 min).

$$^{218}_{84}Po \longrightarrow {}^{214}_{82}Pb + {}^{4}_{2}He$$

The stable isotope lead-206 is the end result of the decay of all these radioisotopes formed from the decaying radon-222 atom (Table 7–5). Collectively, these radioisotopes are called *radon daughters*.

TABLE 7–5 Radon Daughters from the Decay of Radon-222

Isotope	Decay Particle	Half-Life
$^{218}_{84}Po$	$^{4}_{2}He$ (alpha)	3.05 min
$^{214}_{82}Pb$	$^{0}_{-1}e$ (beta)	26.8 min
$^{214}_{83}Bi$	$^{0}_{-1}e$ (beta)	19.7 min
$^{214}_{84}Po$	$^{4}_{2}He$ (alpha)	1.6×10^{-4} s
$^{210}_{82}Pb$	$^{0}_{-1}e$ (beta)	22 years
$^{210}_{83}Bi$	$^{0}_{-1}e$ (beta)	5.0 days
$^{210}_{84}Po$	$^{4}_{2}He$ (alpha)	46 s
$^{206}_{82}Pb$	stable	

Being a member of the noble-gas family, Rn exists as a gas and is nonreactive chemically. Radon atoms in the air we breathe are inhaled and exhaled without any reaction, although some may dissolve in the fluids found in the lungs. If the Rn atom happens to decay while inside our lungs (recall the half-life of radon-222 is 3.82 days, so half of a sample of these atoms will decay every 3.82 days), atoms of nongaseous and reactive metallic elements such as Po, Bi, and Pb form. These atoms quickly react to form ionic compounds. The ionic compounds, for the most part, remain inside the lung (Fig. 7–12). All of the radon daughters up to lead-210 have short half-lives and emit damaging radiation in close contact to delicate lung tissue. The result is a higher than normal risk of lung cancer.

Figure 7–12   Radon daughters decay inside the lung, leaving long-lived radioactive elements.

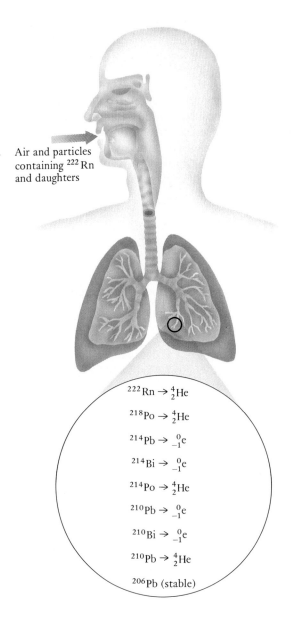

Air and particles containing ^{222}Rn and daughters

^{222}Rn $\rightarrow$ $^{4}_{2}$He

^{218}Po $\rightarrow$ $^{4}_{2}$He

^{214}Pb $\rightarrow$ $^{0}_{-1}$e

^{214}Bi $\rightarrow$ $^{0}_{-1}$e

^{214}Po $\rightarrow$ $^{4}_{2}$He

^{210}Pb $\rightarrow$ $^{0}_{-1}$e

^{210}Bi $\rightarrow$ $^{0}_{-1}$e

^{210}Pb $\rightarrow$ $^{4}_{2}$He

^{206}Pb (stable)

Miners in deep mines are exposed to much higher than normal Rn levels, and as early as 1950, government agencies began monitoring Rn exposures and incidences of lung cancer. Today, it is well known that Rn exposure increases one's chance of developing lung cancer. If you smoke, the chances are even greater, since there seems to be a synergistic effect between smoking and Rn levels in causing lung cancer. Both alpha and beta particles from the decay of radon daughters can cause disruption of chemical bonds in DNA strands in lung cells, which may cause cancerous growth in these cells. Smoking introduces additional chemicals capable of disrupting DNA. It is estimated that about 10,000 of the annual 140,000 lung cancer deaths in the United States are caused by Rn exposure.

When buildings are built over soil containing the heavy radioactive elements that decay to radium-226, some of the Rn gas produced seeps through minute fissures in the soil or rock and migrates into the air in these buildings (Fig. 7–13). The building literally funnels the radon through it by "chimney effects" from such sources as clothes dryers, fireplaces, furnaces and warm air rising and leaving through openings near the roof. If a building is built on a foundation over a crushed stone ballast, holes may be drilled into the ballast. Then a suction pump is attached to produce a negative pressure. For buildings not built on concrete foundations, increasing ventilation both in the basement and inside the building is effective in removing the radon from the building. Of course this means energy losses, but these might be acceptable when the alternatives of the radiation damage from the radon daughters are considered.

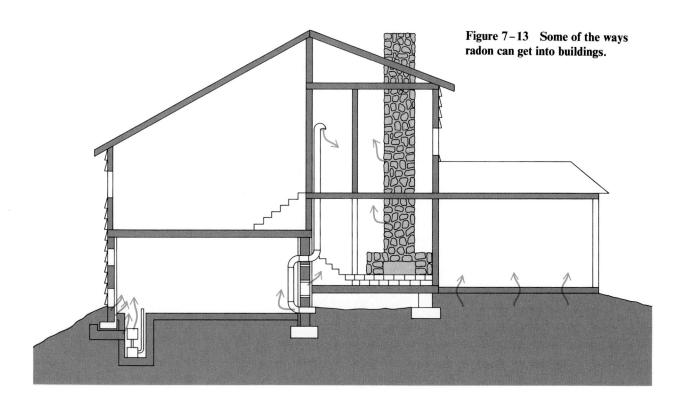

Figure 7–13 Some of the ways radon can get into buildings.

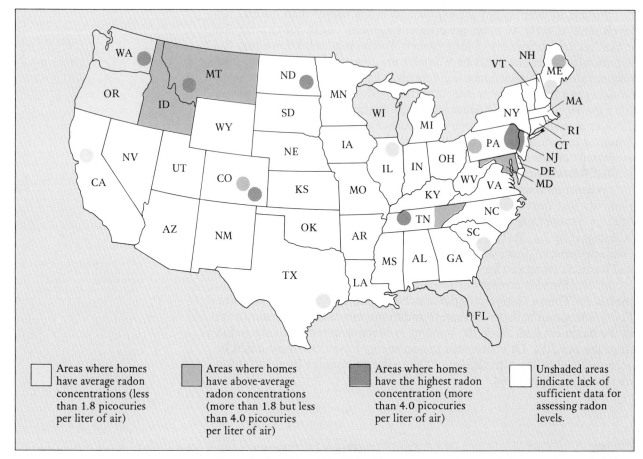

Figure 7–14   **States in which radon concentrations are the highest. Radon has been detected in homes and other buildings in over 30 states.**

It has been estimated that perhaps 8 million homes in the United States are affected by Rn contamination. Radon has been detected in homes in almost every state. Some areas have higher soil Rn than others (see Fig. 7–14.) In 1986, the U.S. Environmental Protection Agency (EPA) surveyed over 11,000 homes for Rn contamination. Almost 39% of the homes in Colorado had levels above the EPA action level of 4 pCi/L. Some homes in the states with the lowest percentage of homes exceeding the action level had the highest measured readings. One home in Alabama had a Rn level of 180 pCi/L although the state as a whole had only 6% of its homes with over 4 pCi/L of Rn.

This EPA action level is important for two reasons. First, levels of Rn above 4 pCi/L should be reduced because it is reasoned that levels above this lead to unacceptable risks of lung cancer. Second, it is difficult to lower the level of Rn in most contaminated homes below 4 pCi/L. This last point is basically an acceptance of the fact that radiation exposure will always be with us.

4 pCi/L is a little less than 1.5 disintegrations every 10 s.

Outdoor Rn concentrations are approximately 0.1 to 0.15 pCi/L worldwide.

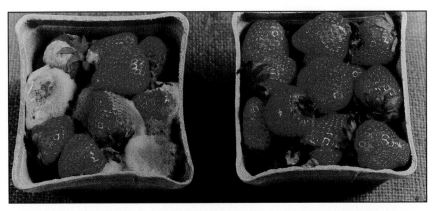

Figure 7–15 **Strawberries irradiated with gamma rays from radioactive isotopes are still fresh after 15 days storage at 4°C (right). Strawberries stored under the same conditions but not irradiated are molded (left). (International Atomic Energy Agency)**

USEFUL APPLICATIONS OF RADIOACTIVITY

Food Irradiation

The damaging aspects of nuclear radiation must always be kept in mind, especially when the possibilities of accidental or unintended exposures are great. However, the harmful radiation from radioisotopes can be put to beneficial use. Consider the important application of killing harmful pests that would destroy our food during storage. In some parts of the world stored-food spoilage may claim up to 50% of the food crop. In our society, refrigeration, canning, and chemical additives lower this figure considerably.

Still, there are problems with food spoilage. Food protection costs amount to a sizable fraction of the final cost of food. Food irradiation using gamma-ray doses from sources such as ^{60}Co and ^{137}Cs is commonly used in European countries, Canada, and Mexico. The U.S. Food and Drug Administration (FDA) has been reluctant to allow this form of food preservation, but changes seem to be coming soon. Foods may be pasteurized by irradiation to retard the growth of organisms such as bacteria, molds, and yeasts. This irradiation prolongs shelf life under refrigeration much in the same way that heat pasteurization protects milk. Normally chicken has a three-day refrigerated shelf life. After irradiation, chicken may have a three-week refrigerated shelf life. The FDA may soon permit irradiation up to 100 kilorads for the pasteurization of foods.

Radiation levels in the 1- to 5-megarad range sterilize; that is, every living organism is killed. Foods irradiated at these levels will keep indefinitely when sealed in plastic or aluminum-foil packages (Fig. 7–15). The FDA is unlikely to approve irradiation sterilization of foods in the near future because of potential problems caused by as yet undiscovered, but possible, "unique radiolytic products." These would-be substances produced by the high-energy irradiation of foods might be harmful in some way.

There are 1000 kilorads in 1 megarad.

TABLE 7–6 Examples of Irradiated Foodstuffs

Food	Purpose	Status
Potatoes	Retardation of sprouts	FDA approved
Wheat	Insect disinfection	FDA approved
Wheat flour	Insect disinfection	FDA approved
Spices	Retardation of microbe growth	FDA approved
Grapefruit	Mold control	For export
Strawberries	Mold control	For export
Fish	Microbe control	For export
Shrimp	Microbe control	For export

For example, irradiation sterilization might produce a chemical substance that is capable of causing genetic damage. To prove or disprove the presence of these substances, animal feeding studies using irradiated foods are presently being conducted.

Presently, over 40 classes of foods are irradiated in 24 countries. In the United States, only a small number of foods may be irradiated (Table 7–6).

Recent findings regarding the potentially harmful health effects of several common agricultural fumigants have indicated that irradiation of fruits and vegetables could be an effective alternative to some chemical fumigants. The agricultural products may be picked, packed, and readied for shipment. After that, the entire shipping container can be passed through a building containing a strong source of radiation (Fig. 7–16). This type of sterilization offers greater worker safety because it lessens chances of exposure to harmful chemicals (see Chapter 16) and protects the environment because it lessens chances of contamination of water supplies with these toxic chemicals (Chapter 17).

Ethylene dibromide (EDB) has been used widely in fumigating fruits. Now EDB is suspected to cause cancer and damage to human reproductive organs. Because of this toxicity, EDB has been banned by the U.S. Environmental Protection Agency.

Figure 7–16 A typical commercial food irradiator. Boxes of food are conveyed into the shielded chamber and around the radiation source (center). When not in use, the source can be lowered into a pool of water below.

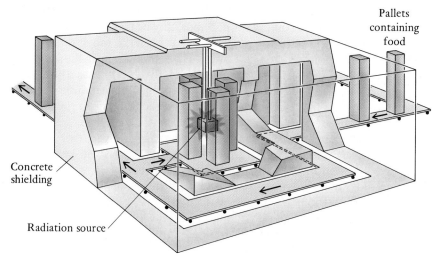

Pallets containing food

Concrete shielding

Radiation source

Materials Testing

The radioisotope ^{60}Co, a gamma-ray emitter, has proved quite useful in the testing of metal castings in industry. Contained within an aluminum thimble, the cobalt radioisotope can be placed inside a casting after a piece of photographic film has been positioned on the outside of the object (Fig. 7–17). The gamma rays penetrate the metal part and make observable any structural flaws in the metal by exposing the photographic film. The intensity of the gamma rays passing through the flawed portion of the casting is different from the intensity passing through the rest of the metal. After development, the photographic film can be examined for the presence of any flaws. Aviation safety has been increased by the use of radiation detection of flaws and weaknesses in structural members of aircraft.

Radioactive Tracers

Because radioisotopes act chemically in a manner almost identical to that of the nonradioactive isotopes of that element, chemists have been using radioactive isotopes as **tracers** in various chemical reactions since their use was discovered in 1945. Several common radiosiotopes used as tracers are listed in Table 7–7. For example, plants are known to take up the element phosphorus (P) from the soil through their roots. The use of the radioactive phosphorus isotope ^{32}P, a beta-emitter, presents a way not only of detecting the uptake of P by a plant but also of measuring the speed of uptake under various conditions. Plant biologists can grow hybrid strains of plants that can absorb P quickly and then test this ability with the radiophosphorus tracer. This type of research leads to faster maturing crops, better yields per acre, and more food or fiber at less expense.

One can measure important characteristics of a pesticide by tagging the pesticide with radioisotopes that have short half-lives and applying it to a test field. Following the tagged pesticide can provide information on its tendency to accumulate in the soil, to be taken up by the plant, and to accumulate in runoff surface water. This is done with a high degree of accuracy by counting the disintegrations of the radioactive tracer. After these tests are completed, the radioisotopes in the tagged pesticides decay to a harmless level in a few days or a few weeks because of the short half-lives of the species used. This type of research leads to safer, more effective pesticides.

Medical Imaging

Radioisotopes are also used in **nuclear medicine** in two distinctly different ways, diagnosis and therapy. In the diagnosis of internal disorders and other maladies, physicians need information on the locations of disorders. This is

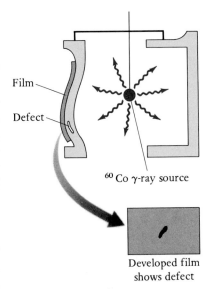

Film

Defect

^{60}Co γ-ray source

Developed film
shows defect

Figure 7–17   A gamma-ray source to detect defects in cast metal parts. The developed photographic film is more strongly exposed where gamma rays passed through a defect.

TABLE 7–7   Radioisotopes Used as Tracers

Isotope	Half-Life	Use
^{14}C	5730 years	CO_2 for photosynthesis research
^{3}H	12.26 years	Tag hydrocarbons
^{35}S	86.7 days	Tag pesticides, air flow
^{32}P	14.3 days	Phosphorus-uptake by plants

Figure 7–18 **Thyroid scan using technetium-99m. The darkened area on the left side is a tumor. (Courtesy of the Department of Radiology, Vanderbilt University.)**

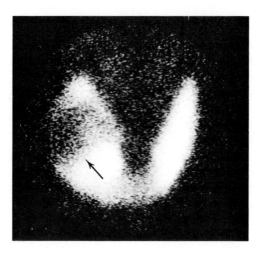

done by **imaging,** a technique in which the radioisotope, either alone or combined with some other chemical, accumulates at the site of the disorder. There, acting like a homing device, the radioisotope disintegrates and emits its characteristic radiation, which is detected. Modern medical diagnostic instruments not only determine where the radioisotope is located in the patient's body but also construct an image of the area within the body where radioisotopes are concentrated (Fig. 7–18).

Four of the most common diagnostic radioisotopes are given in Table 7–8. All of these are made by using a particle accelerator in which heavy charged nuclear particles are made to react with other radioisotopes or stable atoms. Each of these radioisotopes produces gamma radiation, which in low doses is less harmful to the tissue than ionizing radiations such as beta or alpha rays.

By the use of special carriers, these radioisotopes can be made to accumulate in specific areas of the body. For example, the pyrophosphate ion, $P_4O_7^{4-}$, a simple polyatomic ion, can bond to the technetium-99m radioisotope and together they accumulate in the skeletal structure where abnormal bone metabolism is taking place. Such investigations often pinpoint bone tumors.

The *m* in technetium-99m stands for *metastable.* ^{99m}Tc is a gamma-emitter with a half-life of 6 h, which is ideal for medical purposes.

The technetium-99m radioisotope is metastable (denoted by the letter *m*). Metastable isotopes lose energy by disintegrating to a more stable version of the same isotope.

$$^{99m}\text{Tc} \longrightarrow {}^{99}\text{Tc} + \gamma$$

The imaging method is based on the emission of gamma rays from the target organ (Fig. 7–19). The gamma rays strike photosensitive sodium iodide in the imaging device. The photon signal emitted from the sodium iodide is converted to an electric signal and amplified with photomultiplier tubes. The resulting signal is processed by a computer and fed to a video display for construction of the image on the screen.

Positron emission tomography (PET) is a form of nuclear imaging that uses **positron-emitters,** such as carbon-11, fluorine-18, nitrogen-13, or oxygen-15. All of these radioisotopes are neutron deficient, have short half-lives,

Positrons were discovered by Irène Curie Joliot and Frédéric Joliot in 1934.

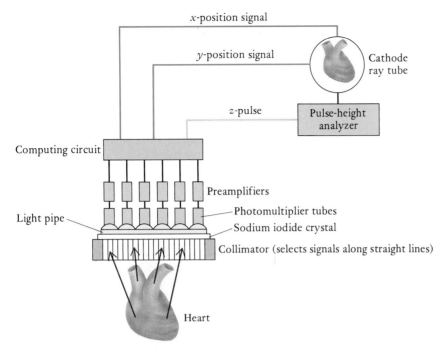

Figure 7–19 The gamma rays emitted from the organ of interest strike a sodium iodide crystal. The photon signal from the sodium iodide crystal is converted to an electric signal, amplified, and then processed by a computer for construction of an image on the cathode-ray tube. (Redrawn from C. Carl Jaffe: Medical imaging. *American Scientist,* Vol. 70, p. 579, 1982, with permission of Sigma Xi)

and therefore must be prepared using a cyclotron immediately before use. When these radioisotopes decay, a proton converts into a neutron, a positron, and a neutrino (v),

$$_1^1p \longrightarrow \ _0^1n + \ _{+1}^0p + v$$

Neutrinos are chargeless and massless "particles" produced in some nuclear reactions.

Since matter is virtually transparent to neutrinos, they escape undetected, but the positron travels less than a few millimeters before it encounters an electron and undergoes antimatter–matter annihilation. The annihilation event produces two gamma rays that radiate

$$\underset{\text{positron}}{_{+1}^0p} + \underset{\text{electron}}{_{-1}^0e} \longrightarrow \underset{\text{gamma ray}}{2\,\gamma}$$

Mass is converted into energy as the particles cease to exist.

in opposite directions and are detected by two scintillation detectors located 180° apart in the PET scanner. By detecting several million annihilation gamma rays within a circular slice about the subject over approximately 10 min, the region of tissue containing the radioisotope can be imaged using computer signal-averaging techniques.

TABLE 7–8 Diagnostic Radioisotopes

Radioisotope	Name	Half-Life (Hours)	Uses
^{99m}Tc*	Technetium-99m	6	As TcO_4^- to the thyroid, brain, kidneys
^{201}Tl	Thallium-201	21.5	To the heart
^{123}I	Iodine-123	13.2	To the thyroid
^{67}Ga	Gallium-67	78.3	To various tumors and abscesses

* The technetium-99m isotope is the one most commonly used for diagnostic purposes. The *m* stands for "metastable," a term explained in the text.

SELF-TEST 7–C

1. Name three uses of radioactive isotopes.
2. Which radioisotope is used for examining metal castings? (a) ^{60}Co, (b) ^{32}P, (c) ^{67}Ga, (d) ^{99m}Tc
3. Name a radioisotope that might be useful as a tracer in agricultural research. (a) ^{32}P, (b) ^{14}C, (c) ^{3}H, (d) all of these
4. The process of concentrating a radioisotope at a particular site of the body in order to locate and measure the extent of a disorder is called: (a) radiotherapy, (b) imaging, (c) sterilization.
5. In the symbol for the radioisotope technetium-99m, the *m* stands for (a) middle, (b) mathematical, (c) metastable.
6. With its half-life of approximately 6 h, how much technetium-99m would remain 18 h after injection into a patient? (a) one-eighth of the original dose, (b) one-half of the original dose, (c) one-sixth of the original dose, (d) one-fourth of the original dose.
7. If two radioisotopes were available for diagnosis, worked equally well, and each decayed by giving off gamma rays, but one had a half-life of 13 h and the other had a half-life of 6 h, which one would you recommend? (a) 13-h half-life isotope or (b) 6-h half-life isotope
8. Radon gas comes from the decay of what naturally occurring radioactive element?
9. What is the name of the element finally produced when radon decays?

MATCHING SET

_____ 1.	Somatic effect	a. Intake stops when organism dies
_____ 2.	1 microcurie	b. Radiation effect on general population
_____ 3.	$^{14}_{6}$C	c. First suggested by Rutherford
_____ 4.	$^{0}_{+1}$p	d. Developed the cyclotron
_____ 5.	Genetic effect	e. Radiation damage to DNA
_____ 6.	$^{238}_{92}$U-dating	f. Tissue easily damaged by radiation
_____ 7.	Element 109	g. Radioisotope used in medical diagnosis
_____ 8.	E. O. Lawrence	h. Time required for half of the nuclei to disintegrate
_____ 9.	$^{218}_{84}$Po	i. 37,000 disintegrations per second
_____ 10.	James Chadwick	j. Used to detect defects in metal castings
_____ 11.	Half-life	k. Opposite of the electron
_____ 12.	C. R. T. Wilson	l. Latest synthetic element
_____ 13.	Bone marrow	m. Developed cloud chamber
_____ 14.	^{99m}Tc	n. Discovered neutron
_____ 15.	^{60}Co	o. Alpha decay product of $^{222}_{86}$Rn
		p. Half-life of 0.5 s

QUESTIONS

1. Describe the operation of a cloud chamber.
2. What does the symbol $^{11}_{5}B$ mean?
3. In general, how have the synthetic transuranium elements been produced?
4. Complete or supply the following nuclear equations:
   a. $^{1}_{1}H + ^{35}_{17}Cl \rightarrow ^{4}_{2}He + ?$
   b. Beta emission of $^{60}_{27}Co$
   c. Alpha emission of $^{238}_{90}Th$
   d. $^{1}_{0}n + ^{60}_{28}Ni \rightarrow ? + ^{1}_{1}H$
   e. $^{238}_{92}U + ^{12}_{6}C \rightarrow ? + 4\,^{1}_{0}n$
   f. $^{2}_{1}H + ^{1}_{1}H \rightarrow ? + ^{0}_{0}\gamma$
   g. $^{238}_{92}U + ^{12}_{6}C \rightarrow ? + 4\,^{1}_{0}n$
5. Look up the origin of the word *mutation* and explain why the word *transmutation* was an apt choice to describe the changing of one element into another.
6. What is the difference between a thermal neutron and a high-energy neutron resulting from cosmic radiation?
7. If a radium atom ($^{226}_{88}Ra$) loses one alpha particle per atom, what element is formed? What is its atomic weight? What is its atomic number?
8. What are the important assumptions made in radiocarbon dating?
9. Name a method by which the age of rocks can be determined. What are the assumptions made in this method?
10. What error would be introduced into the age determination of a tree ring if the amount of cosmic rays had been double their present value at the time the tree grew that ring?
11. What is meant by "delayed somatic effect"? Give an example.
12. The ^{99}Mo (half-life, 67 h) canisters used to prepare solutions of ^{99m}Tc have an effective life of about one week. Can you suggest a reason for this based on half-lives of radioisotopes?
13. Suggest a therapeutic use for the gamma-emitting radioisotope ^{60}Co.
14. Suggest an experiment using a radioisotope tracer in agriculture.
15. Iodine-123 (half-life, 13.2 h) is used to measure iodine-uptake by the thyroid gland. If 1 mg is injected into a patient's bloodstream, how long will it take for the radioisotope to be reduced to less than 1 μg?
16. Ask for an interview with a radiologist at your local hospital. Ask him or her to tell you about the greatest benefits and risks in the use of radioisotopes.
17. Describe Becquerel's discovery of natural radioactivity.
18. What element did Marie Curie discover? What experimental technique did she use?
19. If the two nuclei $^{209}_{83}Bi$ and $^{58}_{26}Fe$ were fused together, what radioisotope would be produced?
20. If 1 Ci of radioactive isotope represents 37 billion disintegrations per second, how many disintegrations per second occur in a sample of 1 μCi?
21. If 1 Ci of a radioisotope that decays to stable products is held in a container, how many curies of radiation would be present after five half-lives?
22. What type of radiation did Irène Curie Joliot and Frédéric Joliot discover? How is this radiation used in medicine?
23. Read a recent magazine or newspaper account of the radon gas problem in your state or a nearby state. Write a short summary.
24. Describe how radon gas is produced and how it gets inside homes.
25. Write a nuclear reaction involving a beta particle and show how charge is conserved.
26. What radiation is produced when a positron and an electron collide?

8

What Every Consumer Should Know About Energy

Energy from the Sun grows trees and powers the Sun Raycer. (Courtesy of GM Hughes Electronics.)

In our industrialized high-tech, appliance-oriented society, the average use of energy per individual is near its highest point in the history of the world (Fig. 8–1). In the United States alone, with only 5% of the world's population, we consume 30% of the daily supply of energy. We are highly dependent on a huge supply of energy. Since 1958, the United States has consumed more energy than it has produced.

Who is using energy?

What is it we use? **Energy**—defined as the ability to do work, which is accomplished only by moving things—is involved every time anything moves or changes. Types of energy involved with matter in motion include heat (molecules in motion), electricity (electrons in motion), sound (compression and expansion of the space between molecules), and mechanical energy (macroscopic objects in motion). All matter in motion involves **kinetic energy,** the energy of motion.

What is energy?

Energy can be stored. Examples include energy stored in chemical bonds (as in wood and food), in the nuclei of atoms (atomic energy), and in gravitational systems (rocks on the top of a hill). Stored energy is **potential energy.**

What are the practical sources of energy? In one way or another, most of our daily energy needs are supplied by the fossil fuels petroleum, coal, and

1. Man without fire (2000 kcal/day)

2. Primitive agriculture (12,000 kcal/day)

3. ca. 1860 (70,000 kcal/day)

4. ca. 1989 (232,000 kcal/day)

Figure 8–1 **Energy use per capita by various types of societies. The energy-per-day use reached an all-time high in 1979 at 243,025 kcal/day per capita in the United States, followed by a low of 207,814 kcal/day in 1983, and a gradual increase to the number given for 1989.**

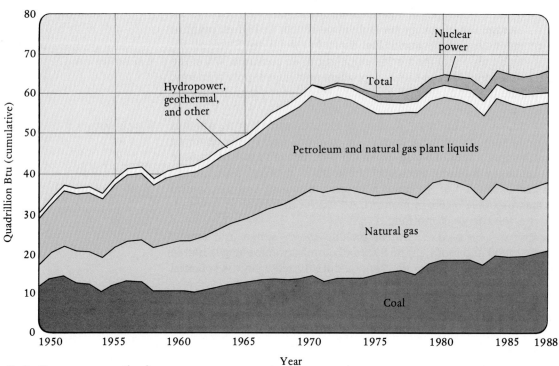

Figure 8-2   Energy consumption in the United States—the immediate view. Note that the burning of fossil fuels (coal, petroleum, and natural gas) furnishes nearly all of our present energy in spite of all the talk about hydroelectric and nuclear energy. *Btu* = British thermal unit (see Table 8-2). (*Annual Energy Review.* Energy Information Administration, 1988.)

Our dependence on the Sun is immense and multifaceted.

natural gas (Fig. 8-2). With the greater amount of coal available, the shift is back toward coal for energy and as a new source for many of the petrochemicals formerly obtained from petroleum. Smaller amounts are supplied by nuclear energy and food, and still smaller amounts are provided in the form of direct solar energy, geothermal energy, wind currents, and ocean currents. These so-called **primary sources** of energy may be converted into electricity (a **secondary** source), the form of energy we may find more useful.

Our dependence on energy from coal, petroleum, and natural gas has been uncomfortably and abruptly realized several times since 1970, but shortages and high prices of energy have been on the horizon for several years. The power brownouts in the eastern United States in the late 1960s were a signal that something was amiss. In the early 1970s, gasoline shortages were caused by inadequate refining capacity, low return on investment capital, and a war in the Middle East. In 1973, the Organization of Petroleum Exporting Countries (OPEC) declared an embargo on oil shipments to all nations that supported Israel, demanding and getting political concessions. This made energy even more a matter of public concern. Since 1973, the price of oil has been raised several times, reaching a high of about $39 a barrel and then decreasing to less than $15 per barrel. Overproduction of oil and an improved efficiency in its use have contributed to the lower prices (Table 8-1).

Our ultimate source of energy is the fusion of nuclei in our Sun. In a major way, the energy of the Sun has been and is being made available to us through photosynthesis. It is thought that photosynthesis of long ago produced the plants that were converted into coal, petroleum, and natural gas,

TABLE 8–1 Excerpts from the 1988 Greenpeace Energy Index

Annual energy bill of the U.S.: $420 billion.

Annual savings attributable to energy efficiency improvements made since 1973: $130+ billion.

Barrels of oil imported to the U.S.: 6.8 million per day.

Barrels of oil saved by energy efficiency improvements made since 1973: 13 million per day.

Amount the U.S. Treasury spends each year in subsidies to the nuclear power industry: $15 billion.

Amount saved in one year if the U.S. converted to the best available lighting technology: $30 billion.

When water falls, potential energy is changed to kinetic energy. (*The World of Chemistry,* Program 5, "A Matter of State.")

though the processes by which this occurred are not completely understood. During every moment of daylight today, yesterday, and tomorrow, photosynthesis produces our food supply either directly or through animal chains. The products of photosynthesis store chemical energy as only a holding form for the almost unlimited amount of nuclear energy in our Sun. In addition to carrying out photosynthesis, the Sun's energy controls wind patterns, precipitation, water power, and seasons. And of course the Sun also provides us with heat, sunsets, suntanning, and the energy for solar batteries. Knowing how enormous, complex, and absolutely necessary are the Sun's contributions to life on Earth, we may be concerned about how long the Sun will last; with a mass of primarily highly concentrated hydrogen (H) 330,000 times greater than the mass of the Earth, the Sun is expected to last an estimated 4 to 5 billion years.

The amount of energy that enters the Earth's atmosphere from the Sun each day is enormous (about 2×10^{15} kcal/min, or 2.0 cal/cm²/min), although it is only about three ten-millionths (0.0000003) of the total energy emitted by the Sun. About half (about 1 cal/cm²/min) of this energy reaches the surface of Earth; the rest goes to reradiation from the atmosphere and to the absorption and scattering of radiant energy by the lower portion of the atmosphere. The actual amount that reaches the surface depends on location, season, and weather conditions. However, even 1 cal/cm²/min is a large amount of energy. For example, at this rate, the roof of an average-sized house receives about 10^8 cal/day, equivalent to the heat energy derived from the burning of about 32 lb of coal a day or to 120 kW-h (kilowatt-hours) of electrical energy a day—more than enough to heat an average American home in the winter.

Chemical potential energy is released when zinc reacts with solid ammonium nitrate.

Large amounts of energy are expressed in *quads.* One quad is one quadrillion (10^{15}) Btu.

Table 8–2 is a chart of energy units based on equivalences of coal, oil, natural gas, and electricity. A **British thermal unit** (Btu) is the amount of heat required to raise the temperature of 1 lb of water 1°F. A smaller unit of energy, the **calorie,** is the amount of heat required to raise the temperature of 1 g of water 1°C. A kilocalorie (kcal) is 1000 cal. From the table you can see that 252 kcal (or 252,000 cal) are the same amount of heat as 1000 Btu. Thus, 252 cal are the same amount of heat as 1 Btu. Equivalent values for other units of energy can also be deduced. For example, 4.18 J **(joules)** are the same as 1 cal, and 860 kcal are the same as 1 kW-h of energy. (The kilowatt-hour will be defined in the later discussion of electricity.)

TABLE 8–2 A Chart of Energy Units*

Cubic Feet of Natural Gas	Barrels of Oil	Tons of Bituminous Coal	British Thermal Units (Btu)	Kilowatt Hours of Electricity	Joules	Kilocalories†
1	0.00018	0.00004	1000	0.293	1.055×10^6	252
1000	0.18	0.04	1×10^6	293	1.055×10^9	0.25×10^6
5556	1	0.22	5.6×10^6	1628	5.9×10^9	1.40×10^6
25,000	4.50	1	25×10^6	7326	26.4×10^9	6.30×10^6
1×10^6	180	40	1×10^9	293,000	1.055×10^{12}	0.25×10^9
3.41×10^6	614	137	3.41×10^9	1×10^6	3.6×10^{12}	0.86×10^9
1×10^9	180,000	40,000	1×10^{12}	293×10^6	1.055×10^{15}	0.25×10^{12}
1×10^{12}	180×10^6	40×10^6	1×10^{15}	293×10^9	1.055×10^{18}	0.25×10^{15}

* Based on normal fuel heating values. $10^6 = 1$ million, $10^9 = 1$ billion, $10^{12} = 1$ trillion, $10^{15} = 1$ quadrillion (quad).
† 1 food calorie = 1000 calories = 1.000 kcal.

The purposes of this chapter are to explain some consumer interest principles about energy, to describe how energy is extracted from chemicals, and to summarize our energy situation.

FUNDAMENTAL PRINCIPLES OF ENERGY

The Distinction Between Energy and Power

Power is the rate at which energy is used.

A joule has fundamental units of kg·m²s⁻².

Energy is the ability to move matter and may be expressed in units of calories, Btu, or joules; in contrast, **power** is the rate at which energy is used. Power is expressed in units of energy used per time, such as calories per second or joules per second (watts). Some consumers confuse energy and power when looking at their electrical bills; consumers pay for the amount of energy they use (kilowatt-hours), not for how fast they use it (kilowatts, or kilojoules per second). Some industries that consume huge amounts of electricity are given restrictions on the rate at which they can use energy.

When Shopping for Energy, Choices Are Limited

By far most substances around us are low-energy compounds. Most mixtures, such as rocks, dirt, earth, and water, are chemically oxygen-saturated. Only a few products of photosynthesis (food, fossil fuels, wood) are in the chemical position of providing energy through oxidation, the principal way in which we obtain energy from chemicals. Our choices of sources of energy from chemical reactions are limited and dwindling.

Energy Extraction from Chemical Change is Net — Not All the Energy in a Chemical

Energy is absorbed (is endothermic) when a chemical bond is broken to yield isolated atoms. The energy required to break 1 mol of a particular kind of bond is the **bond energy** (Table 8–3). The same amount of energy is released (is exothermic) when 1 mol of bonds is formed from isolated atoms.

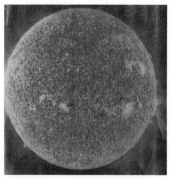

Figure 8–3   The Sun is our ultimate source of energy.

TABLE 8–3 Approximate Bond Energies in kcal/mol

Bond	Energy	Bond	Energy	Bond	Energy
H—H	104	I—I	36	I—Cl	50
C—C	83	H—F	135	O—Cl	49
C=C	146	H—Cl	103	S—Cl	60
C≡C	200	H—Br	88	N—Cl	48
N—N	40	H—I	71	P—Cl	79
N=N	100	H—O	111	C—Cl	79
N≡N	225	H—Se	66	Si—Cl	86
O—O	33	H—S	81	C—O	84
Ȯ—Ȯ in O_2	118	H—N	93	C=O	173
S—S	51	H—P	76	C=O in CO_2	192
F—F	37	H—C	99		
Cl—Cl	58	H—Si	70		
Br—Br	46	Br—Cl	52		

The O_2 molecule has two unpaired electrons, as indicated by the dots in Ȯ—Ȯ.

A given chemical reaction is exothermic if the formation of new bonds liberates more energy than is required to break the bonds in the reaction; a reaction is endothermic if the bonds in the reactants are stronger than those in the products. In summary, we receive energy from a chemical reaction only when more energy is produced by bond-making than is required for bond-breaking.

For example, consider the oxidation of methane (CH_4), the principal component of natural gas:

$$\underset{\underset{H}{|}}{\overset{\overset{H}{|}}{H-C-H}} + 2\,\dot{O}-\dot{O} \longrightarrow O=C=O + 2\,\overset{O}{H^{\diagup\diagdown}H} + 192 \text{ kcal/mol } CH_4$$

A rocket upon takeoff requires much energy in a short time, that is, much power. (NASA.)

A house on fire is normally an unwanted form of reaction of wood with oxygen at temperatures to give off heat and light—a combustion. (*The World of Chemistry,* Program 15, "The Busy Electron.")

TABLE 8–4 Heat Produced by the Combustion of Some Organic Materials

Substance	Heat (kcal/g)
Methane (principal component of natural gas)	13.2
Gasoline, kerosene, crude petroleum, tallow	9.5–11.5
Lipids	9.0–9.5
Carbon (coal)	7.8
Ethyl alcohol	7.1
Proteins	4.4–5.6
Carbohydrates (sugars and starches)	3.6–4.2

These figures correspond to laboratory combustion to yield CO_2, H_2O, and oxides of nitrogen. In the body, proteins are oxidized to CO_2, H_2O, and urea. For the latter process, the heat yield is less than the value indicated above. Thus, proteins and carbohydrates yield (per gram) about the same energy in the body.

Powered iron sprinkled in a flame combines with oxygen to give off heat and light—a combustion.

Thermodynamics is the movement of energy.

First statement of the first law of thermodynamics: "A force [translated: energy] once in existence cannot be annihilated"—Julius Robert Mayer, a ship's doctor, 1840.

The energy released in this exothermic reaction can be used to heat houses, drive gas turbines, and generate electricity. The fact that energy is released means that it takes less energy to break the C—H bonds and O—O bonds than is produced when the C=O bonds and O—H bonds are formed. The bond energies in Table 8–3 bear this out. It takes only 632 kcal to break 2 mol of O—O bonds and 4 mol of C—H bonds [2(118) + 4(99) = 632]. The making of bonds produces 828 kcal generated by the formation of 2 mol of C=O bonds and 4 mol of O—H bonds [2(192) + 4(111) = 828]. This gives a net release of 196 kcal (828 − 632 = 196), which closely agrees with the experimental value of 192 kcal/mol of CH_4 burned.

If the supply of oxygen is sufficient, **combustion** of fossil fuels and wood produces principally carbon dioxide (CO_2) and water (H_2O). The net energy received from the burning of the fuel is the difference between the energy given off in making bonds in CO_2 and H_2O and the energy required to break bonds in O_2 and the fuel. Water and carbon dioxide have their capacities for oxygen satisfied and therefore cannot be burned to extract more energy. The amount of energy derived from the combustion of some of nature's storehouses of energy are given in Table 8–4.

The Law of Conservation of Energy — The First Law of Thermodynamics

Also known as the first law of thermodynamics, the law of conservation of energy asserts that energy is neither lost nor gained in all energy processes. When a beaker of water is heated on a burner, all of the energy given off by the flame can be accounted for in the increased energy of the water and its surroundings; no energy is lost or gained in the transformation of chemical energy to heat energy. Furthermore, as discussed in Chapter 6, the transformation is quantitative, in that a certain amount of gas burned produces a certain amount of energy (see Table 8–4); when one kind of energy is changed into another, the exchange rate is definite, reliable, and reproducible.

The law of conservation of energy also implies that the total amount of energy in the universe is constant. Energy is transformed regularly from one kind to another, but the total remains the same. This means that the Sun and the energy stored in chemicals on the Earth are what we have to use — that is all! There is no creation of new energy.

One last implication of the law applicable to this study concerns perpetual motion. The law recognizes that a machine cannot produce enough energy to run itself, much less create enough energy to be used elsewhere. At a minimum, the machine would have to create enough energy to move its parts and to overcome friction, but this is creation, not transformation, and is therefore impossible.

The law of conservation of energy was extended by Albert Einstein, who showed the interrelationship between mass and energy in the equation $E = mc^2$. This required the inclusion of matter in a more general law:

The total amount of matter and energy in the universe is constant.

Energy Is Conserved in Quantity but Not in Quality — The Second Law of Thermodynamics

What does this second law of thermodynamics mean: that energy is conserved in quantity but not in quality? Perhaps two examples will clarify the concept. Consider first the release of some energy by the burning of coal, petroleum, or wood. Recall from our previous discussion that the main products of these combustion reactions, CO_2 and H_2O, will not burn and release more energy. In the burning process, both matter and energy are conserved, as required by the laws of conservation of matter and energy, respectively. However, the reactants and their stored energy are more useful in energetic terms than the products and their spent energy.

As a second example, consider an electric motor. The electricity that runs the motor is more useful than the heat that comes from the warm motor. Again, energy is conserved in the process of running an electric motor, but the usable energy is not conserved.

In concept, the energy relationships in the second law of thermodynamics can be compared to the relationships among gross income, deductions, and net pay (or realizable income) in a paycheck. A certain amount of energy is available for the process considered; this is analogous to gross income. Some of the energy is not usable owing to frictional losses, electrical shorts and drains, retention of some energy in the chemical products, or some other factor affecting efficiency; this is represented by paycheck deductions. Finally, some of the energy is usable; it is analogous to net pay. The energy, like the money, is accounted for as required by the first law of thermodynamics.

In all processes, then, some energy is wasted — not lost — by conversion into energy that is not usable in doing work. The wasted (or unusable) energy is represented by **entropy,** a measure of the disorder in a physical system (Fig. 8–4). Entropy is not energy per se, but it is a function of energy with units of energy per degree, such as calories/degree.

Another statement of the second law of thermodynamics is based on entropy:

In all natural processes, entropy is increased.

Usable energy is not conserved.

No matter how we try, we can never convert all of the stored energy in a system into usable energy.

Refer to entropy as a driving force in chemical reactions as discussed in Chapter 6.

Entropy means disorder, and measures nonuseful energy.

Figure 8–4 A familiar sight is deteriorating automobiles, which in their assembled splendor are prized possessions. After going through a metal shredder (note the pile on the right), the once valued automobile is now a pile of mixed-up metal junk. The metal shredder created more mixedupiness for the automobile and, hence, created more entropy. Junk yards epitomize entropy . . . a more disordered state.

Where does the heat and light of a flame go? Is all of the emitted energy useful? See the text. (*The World of Chemistry,* Program 13, "The Driving Forces.")

The end of the line for energy is the random motion of molecules.

Taken to its extreme, this means that the entropy of the whole universe is increasing at the expense of stars running down in usable energy at a tremendous rate. This is not a reason for worry, because the universe is so vast that enough usable energy is there for all conceivable purposes for many billions of years. However, sources of usable energy that are not limitless are the so-called fossil fuels (coal, petroleum, and natural gas), which when gone are not easily restored. It would take eons for photosynthesis to regenerate the material for new fossil-fuel deposits.

Why is the energy used to increase entropy instead of usable energy? The derivation of the word *entropy,* meaning "disorder," explains. The ultimate fate of any change in energy is a form of heat energy caused by the random, disordered motion of molecules. Have you ever thought about what happens to the light energy coming from a light bulb, or what happens to electric energy once it is used to run an electric motor, or what happens to the sometimes large amounts of energy that result from an explosion? All forms of energy, including sound, are converted eventually into random molecular motion, in which the molecules move faster and (or) further apart. Molecules moving in all directions are not as useful in bringing about controlled changes as are electrons, photons of light, or molecules when they are moving from one point to another in organized fashion.

Let us summarize this brief discussion of the second law of thermodynamics by describing the energy coming from a burning match. Some of the energy can be used to ignite another object, or to heat an object, or to provide light; this is the organized energy. However, while the usable energy is being used, some of the total energy emanating simply heats molecules in the vicinity and increases the entropy of the molcules. Eventually, all of the heat and light coming from the match becomes increased random motion of the molecules.

What does the second law of thermodynamics mean to the informed citizen? Simply stated, when usable energy-rich chemicals such as coal and petroleum are consumed, the usable energy is less than the total energy produced.

When fossil fuels are gone, then what?

The Efficiency of Energy Use Is Low

In every energy process, the efficiency of energy use is less than 100%, usually far less. Automobiles are about 20% to 25% efficient; that is, about 80% of the useful energy available to do work is lost and not applied to the turning of the wheels. Some fuel cells are about 70% efficient. The human body is about 45% efficient in converting the energy of glucose metabolism to muscle movement. Photosynthesis is about 30% efficient in converting absorbed sunlight into glucose, steam turbines for producing electricity are about 38% efficient, heating homes with electricity is about 38% efficient, and heating homes with natural gas is about 70% efficient. The efficiency is usually greater when a **primary source** of energy is used (gas) than when a **secondary source** is used (electricity). For example, it takes about 10,000 Btu's to produce 1 kW-h of electricity. If this kilowatt-hour is then used for heating, only 3800 Btu's of heat are produced. Natural gas burned on site would be more efficient than natural gas burned in a steam generator plant to produce electricity.

Efficiency = used energy ÷ available energy.

Primary source of energy: one transformation on site (e.g., chemical → heat via combustion). *Secondary source:* usually more than one transformation, plus long-distance transport (e.g., chemical → heat via combustion → steam → mechanical → electricity).

Energy Not Lost Is "Energy Gained"

Energy can be transported through wires (electricity), stored in chemicals (batteries), and carried through space (radio waves). On the other hand, energy can be prevented from moving by means of insulators. The insulation of houses has popularized the **R value** for heat insulators. The R value (the resistance) is inversely proportional to the conductivity of heat through a slab of material; a common unit of R value is (ft^2) (°F)(h/Btu). An R value of 30 is typically recommended for the ceilings of single-family dwellings; an average square foot of such a ceiling would lose heat by conduction at a rate of (1/30) Btu/h for every 1°F difference in temperature. The higher the R value, the fewer Btu's escape per hour per square foot of ceiling. Some R values for 1-in. slabs of material (in units of ft^2°F h/Btu) are air, 5.9; polyurethane foam, 5.9; rock wool, 3.3; fiberglass, 3.0; white pine, 1.3; and window glass, 0.14. Dry, still air has an insulating value (R value) as great as almost any building material. In fact, many commercial materials owe their heat-insulating ability to trapped, isolated pockets of air.

The R factor for heat insulators.

Some Materials Have a Higher Energy Cost Than Others

It costs more energy to produce a ton of some substances than to produce a ton of other substances (Table 8–5). Certain applications now using plastics or metals might more efficiently use ceramics or brick to conserve energy. Of course, other factors such as labor costs also influence the economic decisions involved.

It costs less energy to make a ton of steel than to make a ton of aluminum.

TABLE 8–5 Energy Requirements to Produce Some Common Products*

Product	Millions of Btu/Ton
Titanium	482
Aluminum	244
Copper	112
Polyethylene	100
Polystyrene	64
Polyvinylchloride	49
Plate glass	25
Steel slabs	24
Paper	22
Portland cement	8
Brick	4

* From Peter R. Payne, "Which material uses the least energy," *Chem. Tech.,* September 1980, p. 550.

SELF-TEST 8–A

1. In 1988, what two energy sources provided the most energy for U.S. consumption (see Fig. 8–2)?
2. Which furnishes the most heat energy per gram: coal, petroleum, or natural gas?
3. The typical efficiency of an electric generating plant is about
   ( ) 100%, ( ) 50%, ( ) 33%, or ( ) 10%.
4. Examples of fossil fuels are _____ , _____ , and _____ .
5. Natural gas and petroleum react with _____ to produce CO_2 and _____ .
6. All combustions of fossil fuels give off energy. True ( ) False ( )
7. Energy is the ability to do _____ .
8. One type of energy (for example, light) is always transformed into another type of energy (for example, heat) ( ) quantitatively, ( ) not quantitatively, ( ) sometimes quantitatively, sometimes not quantitatively.
9. The ultimate fate of all types of energy is an increase in _____ .
10. Although the quantity of energy is conserved, the _____ of energy is not conserved.
11. Four units of energy are _____ , _____ , _____ , and _____ .
12. Three units of power are _____ , _____ , and _____ .
13. Which costs more energy to produce, a ton of aluminum or a ton of brick? Which costs more money to buy?

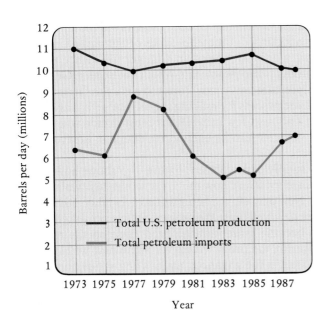

Figure 8-5 The U.S. dependence on imported oil dropped significantly (about 50%) in the early 1980s, but is on the rise now and may soon exceed the 1977 levels.

FOSSIL FUELS

The oxidation of coal, petroleum, and natural gas provides most of the energy for our nation as well as for the world. For each material, the products of complete combustion are CO_2 and H_2O. When coal or petroleum is combusted incompletely or contains certain impurities, major air pollution is possible. For example, the sulfur in coal or petroleum burns to form sulfur dioxide (SO_2), a major cause of acid rain.

Acid rain is discussed in Chapter 18.

No ongoing coal or petroleum production underground has been detected. The fossil fuels that have already been found may be all there is. We must use these limited fossil substances wisely.

Petroleum

Petroleum was first discovered in the United States (in Pennsylvania) in 1859 and in the Middle East (in Iran) in 1908. Today petroleum is pumped from the ground in many parts of the world. As an energy source, petroleum was first used as kerosene for lighting, then as gasoline and aircraft fuel for transportation, and most recently as fuel oil to produce electricity. Of the 80 quads of energy used per year in the United States in recent years, 34 quads (43%) were supplied by petroleum (see Fig. 8-2). Oil produced by the United States and oil imported by the United States over the last several years are shown in Figure 8-5.

Petroleum is a mixture of many hydrocarbons. By refining (separation of the components of the mixture by distillation) and subsequent conversions, much crude oil is turned into gasoline (42.6%, or 6.7 million barrels per day in the United States), and lesser amounts are turned into fuel oil

See Figure 13-2 for a schematic of a distillation tower for petroleum and for a discussion of chains of carbon atoms.

There are 42 gal of oil per barrel.

**A petroleum refinery tower.
(Courtesy of Standard Oil.)**

(26.8%), jet fuel (7.4%), and other miscellaneous fuels. From the small fraction of petroleum that is not burned comes thousands of chemicals known as the petrochemicals; they will be discussed in Chapter 13. Other uses of petroleum include the production of edible fats, which was done in Germany during World War II. Glycerol is now made from petroleum in Germany on a commercial scale, and the process for making sugar from oil has also been developed. The present and next few generations must decide whether to burn our limited supply of petroleum for energy production or to save oil for petrochemical products.

How limited is the supply of petroleum? Figure 8–6 shows how much oil the world has used since 1900 and how much the world has available for about the next 50 years. New oil wells are being drilled every day; other wells come in every day. These new wells raise the projected curve only slightly. As Third World countries become more industrialized, their energy requirements will increase, lowering the projected curve. Some new wells will be required just to keep oil use in the advanced nations at a status quo.

If we are willing to pay the price, there is more oil in the ground that is hard to remove. When an oil well comes in, only the **recoverable oil** is removed. Usually about 30% is obtained, with 70% left in the ground. As supplies decrease and costs go up, it may prove economically feasible to "recover" more oil.

A huge, mostly untapped source of petroleum is **oil shale rock.** Three immense deposits of oil shale rock in the United States are the most-developed deposit in Utah and Colorado, a giant U-shaped formation from Michigan and Pennsylvania to Alabama (believed to hold a trillion barrels of oil), and a vast deposit in north-central Alaska. The Alaskan oil shale yields only a few gallons of oil per ton of rock, whereas saturated ores test at 102 gal of oil per ton of rock.

Most oil shale lies near the surface of the Earth, and in its economic favor is the fact that it's not possible to hit a dry hole. The oil is obtained by

Figure 8–6 World oil use: past and projected. Proved reserves in 1987 were 8.9×10^{11} barrels of petroleum. At the 1987 use rate, all of the known petroleum would be consumed by the year 2025.

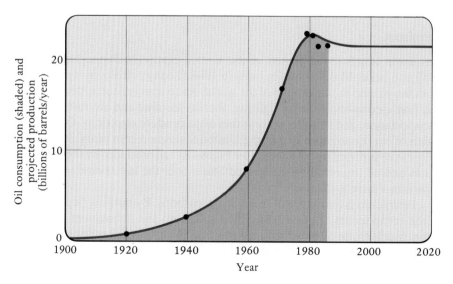

heating of the rock in the absence of air. One technique involves blasting the rock, burning part of the oil underground and using the heat to force the oil out of the rock into underground holding reservoirs, and then pumping the collected oil to the surface.

Since the energy costs of obtaining petroleum from oil shale rock are high and the process of recovery leaves large amounts of waste rock, the production of petroleum from oil shale is not attractive as long as other sources of petroleum provide oil for $30 or less per barrel.

Coal

Coal is a mixture of hydrocarbons with a relatively small amount of sulfur. By way of contrast with petroleum, coal has more rings of carbon atoms (some rings are bonded to each other — fused) and more bonding of chains of carbon atoms to other chains of carbon atoms (Fig. 8–7).

See Chapter 12 for typical chains and rings of carbon atoms.

Figure 8–7 A partial molecular structure of coal. The arrows indicate weak bonds that may be broken easily during heating. The six-sided figures (hexagons) represent six carbon atoms in a ring. When the rings join on a side, those carbon atoms are in two rings. By way of contrast, the structure of petroleum has fewer rings of carbon atoms and less bonding of chains to chains.

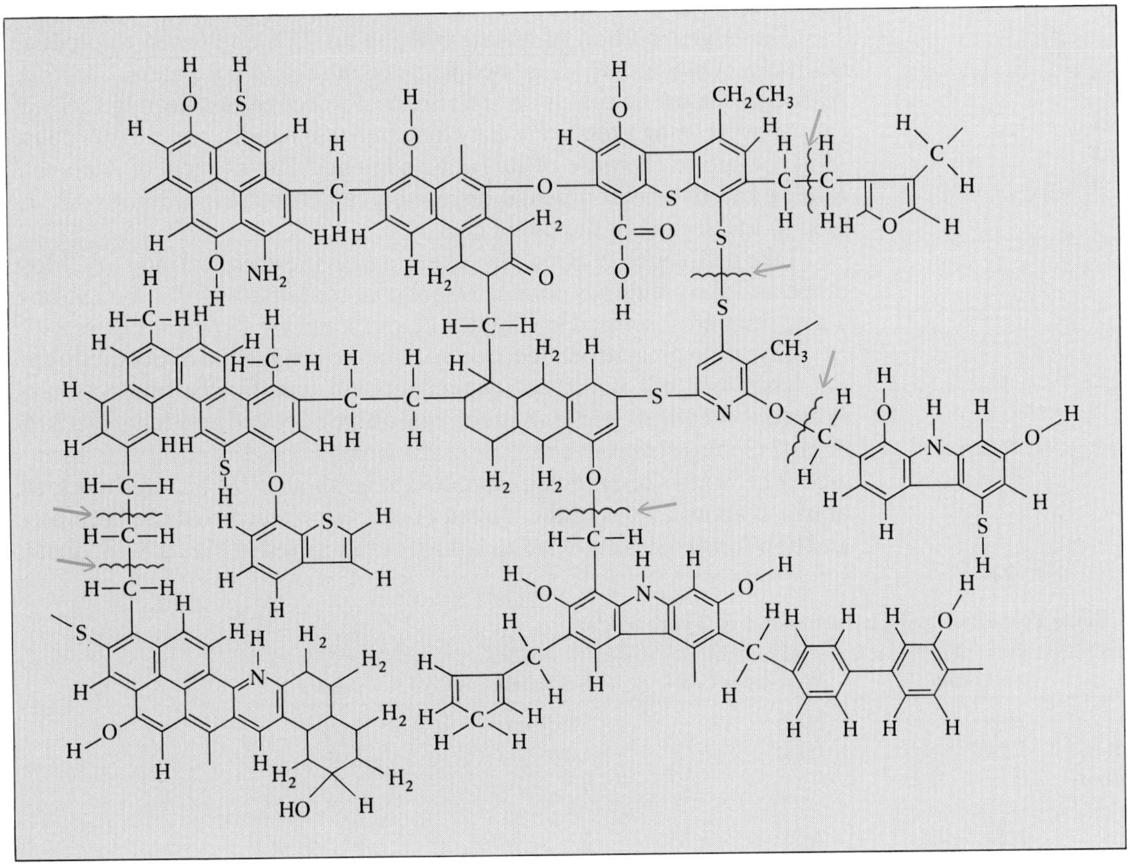

Coal is carried by many forms of transportation to where it is often stockpiled at electrical power plants and various kinds of industries. (Visuals Unlimited/Albert Copley.)

Minable coal is defined as 50% of all coal in a seam at least 12 in. thick and within 4000 ft of the Earth's surface. In the United States the minable coal reserves are divided among anthracite (2%), lignite (8%), subbituminous coal (38%), and bituminous coal (52%). Some properties of the different kinds of coal are listed in Table 8–6.

From the information given in Figure 8–2, it can be seen that in recent years coal provided about 20 quads (25%) of the nation's energy requirements of about 80 quads per year. This required the mining of about 900 million short tons of coal each year.

The largest portion of mined coal (about 75%) is burned to produce electricity. Only about 1% is used for residential and commercial heating. Although the use of coal is on the rise, coal's decline as a heating fuel was caused by its being a relatively dirty fuel, bulky to handle, and a major cause of air pollution (because of its sulfur content). The dangers of deep coal mining and the environmental disruption caused by strip mining contributed to the decline in the use of coal.

Like petroleum, coal supplies chemicals to the chemical industry. Most of the useful compounds obtained from coal contain rings of carbon atoms. These rings are discussed in Chapter 12 and some are shown in Figure 8–7.

Given our great dependence on coal for the production of electricity and our lesser but still significant dependence on coal for the production of industrial chemicals, just how much coal do we have and how long is it likely to last? Geologists believe that all of the world's coal has now been discovered. The world's reserves are estimated to be about 1024 billion short tons, of which about 29% is in the United States. How much coal has been used and how long coal is expected to last are summarized in Figure 8–8. Coal is

TABLE 8–6 Some Properties and Characteristics of Types of Coal

Characteristics	Anthracite	Bituminous Coal	Subbituminous Coal	Lignite
Heat content	High	High	Medium	Low
Sulfur content	Low	High	Low	Low
Hydrogen/carbon mole ratio	0.5	0.6	0.9	1.0
Major deposits	New York, Pennsylvania	Appalachian Mts., Midwest, Utah	Rocky Mts.	Montana

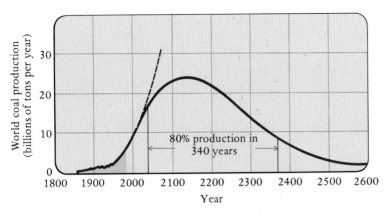

Figure 8-8 The coal mined to date (shaded area) represents only a small fraction of minable coal. The rate of increase in coal consumption (dashed line) is 4% per year. It is obvious that such an exponential rise cannot continue long after the year 2000. At the present usage (held constant at 5 billion short tons per year), the known reserves in 1984 of 1,020 billion short tons of coal would last for another 205 years.

expected to last several hundred years longer than petroleum. New mining techniques that make more of the deposited coal minable can extend the usable life of coal. However, as is the case with petroleum, as more Third World countries advance industrially and technologically, more coal will be used and the total supply will be depleted more quickly.

Some of the problems associated with the use of coal as a fuel can be alleviated by converting coal to combustible gases (coal gasification) or to liquid fuels (coal liquefaction).

Coal Gasification

Coal can be converted into a relatively clean-burning fuel by a process known as **gasification** (Fig. 8-9). In this process, coal is made to react with a limited supply of either hot air or steam. In the reaction of coal with air, the product is a gaseous mixture known as **"power gas,"** and the reaction is exothermic.

Gasification can make coal cleaner to burn and easier to handle from supplier to user.

$$\text{Coal} + \text{Air} \longrightarrow \underset{\text{Power Gas}}{CO(g) + H_2(g) + N_2(g)} + 26.39 \text{ kcal/mol C}$$

Power gas contains up to 50% nitrogen (N_2) by volume and is consequently a relatively poor fuel. In fact, power gas of this composition has only one sixth the heat content of methane.

If the coal is allowed to react with high-temperature steam, a mixture of carbon monoxide and hydrogen known as **synthesis gas** or **coal gas** is obtained. Unlike power gas, this mixture contains no nitrogen.

$$\underset{\text{Coal}}{C} + \underset{\text{Steam}}{H_2O(g)} \longrightarrow \underset{\substack{\text{Synthesis Gas} \\ \text{or Coal Gas}}}{CO(g) + H_2(g)} - 31 \text{ kcal/mol C}$$

This reaction is endothermic.

When air and steam are mixed in the correct proportions, the reaction of the mixture with coal can be self-sustaining, since the production of power gas is exothermic and produces enough energy to drive the endothermic production of coal gas.

In both power gas and coal gas mixtures, the CO and H_2 are burned by oxygen (O_2) in the air to produce heat. The heat produced is about one third that of an equal volume of methane (natural gas).

$$2 \text{ CO} + O_2 \longrightarrow 2 \text{ CO}_2 + 135.3 \text{ kcal} \quad (67.6 \text{ kcal/mol CO})$$
$$2 \text{ H}_2 + O_2 \longrightarrow 2 \text{ H}_2O + 115.6 \text{ kcal} \quad (57.8 \text{ kcal/mol H}_2)$$

These reactions are exothermic.

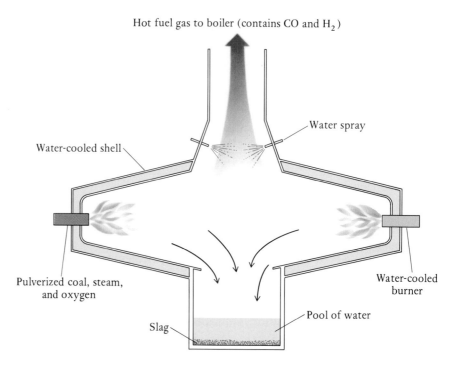

Hot fuel gas to boiler (contains CO and H2)
Water spray
Water-cooled shell
Pulverized coal, steam, and oxygen
Water-cooled burner
Pool of water
Slag

Figure 8–9 Schematic drawing of coal gasifier. A relatively cool combustion of powdered coal in a limited supply of oxygen produces a mixture of carbon monoxide and hydrogen along with other gases. The mineral content in the coal collects in the slag.

In a newer coal gasification process, high-energy methane is the end product. The process uses a catalyst (usually potassium hydroxide or potassium carbonate) and is thermally neutral (neither exothermic nor endothermic) at 700°C, the temperature at which the process is usually run.

In the process, crushed coal is mixed with an aqueous catalyst; the mixture is then dried and sent to a gasifier chamber where CO and H_2 are added. The mixture is then heated to 700°C.

Reactions that occur in the gasifier are (numbers are values at 25°C):

$$2\,C + 2\,H_2O \longrightarrow 2\,CO + 2\,H_2 - 64\ \text{kcal/2 mol C}$$
$$CO + H_2O \longrightarrow CO_2 + H_2 + 8\ \text{kcal}$$
$$CO + 3\,H_2 \longrightarrow CH_4 + H_2O + 54\ \text{kcal}$$

The overall (or net) reaction is:

$$2\,C + 2\,H_2O \longrightarrow CH_4 + CO_2 - 2\ \text{kcal/mol}\ CH_4$$

Any unreacted CO and H_2 are cycled back through the gasifier. Recycled steam is used to help dry the coal before the coal enters the gasifier. The catalyst is recovered and reused.

This process converts solid, messy coal into easily transported, efficiently burned methane, the chief component of natural gas. The energy consumed by the process is small, and, best of all, the combustion of methane is environmentally clean.

Coal Liquefaction

Liquid fuels are made from coal by reacting the coal with H_2 under high pressure in the presence of catalysts (hydrogenating the coal). Some sources of H_2 are power gas, synthesis gas, and the electrolysis of water (see Chapter

10). The process produces more straight chains of C atoms, like those in petroleum. The resulting crude-oil type of material can be fractionally distilled like petroleum into diesel fuel, gasoline, and chemical raw materials for plastics, medicine, and other commodities. About 5.5 barrels of liquid are produced for each ton of coal fed to the plants. The cost is about $35 per barrel, which is about twice the cost of petroleum in 1989.

Although the products of coal gasification and liquefaction burn more cleanly, pollute less, and can be more easily transported than coal, the modification of coal is not expected to contribute significantly to our energy needs until the late 1990s, if at all, but the modifications are expected to contribute significantly to our petrochemical needs.

Natural Gas

Natural gas burns with a high heat output (see Table 8–4), produces little or no residue or pollution from burning, and is transported easily. Practically the only pollution produced by the combustion of natural gas is CO_2, which contributes to the greenhouse effect discussed in Chapter 18. However, the amount of CO_2 produced per energy unit is less for natural gas than for other fossil fuels.

The production of natural gas peaked in the United States in 1973 at 22×10^{12} ft^3/year. The world's reserves of natural gas are estimated to be about 3.8×10^{15} ft^3, most of them in the Middle East, Eastern Europe, and the Soviet Union. Even if significant new deposits of natural gas are discovered in such locations as the outer continental shelves or deep below the Earth's mantle, the North American natural gas deposits are already about 60% depleted, and not much time is available for tapping new sources. Importation is complicated by the danger of transporting and storing concentrated, condensed, volatile natural gas. Yet, liquefied natural gas is produced and stored (Fig. 8–10).

Even with higher prices and an impending depletion of natural gas, more homes are heated by the burning of natural gas than by any other means. About half of the homes in the United States are heated by natural gas, followed by electricity (18.5%), fuel oil (14.9%), wood (4.8%), and lique-

Natural gas is mostly methane, CH_4.

Figure 8–10   Liquefied natural gas storage tanks. (*The World of Chemistry,* Program 5, "A Matter of State.")

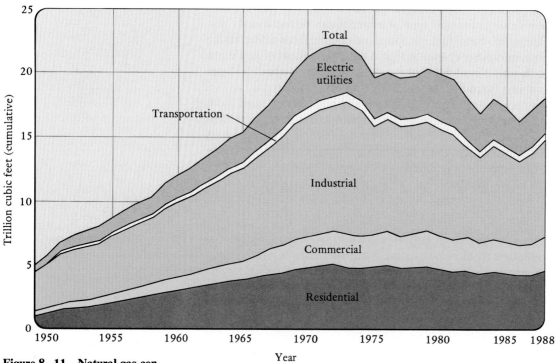

Figure 8–11 Natural gas consumption by various users from 1949 through 1988. Note that residential heating with natural gas has declined in recent years. (*Annual Energy Review,* Energy Information Administration, 1988.)

fied gas such as butane and propane (4.6%). Coal and kerosene come in at a low 0.5%, and solar heating of homes is even lower. Residential heating by the most efficient primary source of energy, natural gas, has gradually decreased since its high point in 1972 to its value of 4.64×10^{12} ft^3 in 1988 (Fig. 8–11).

The Soviet Union produces about 50% more natural gas than is produced in the United States. About 60% of Russia's natural gas is located in northern Siberia where it is extremely cold. Despite production problems, the Russians are pushing development of natural gas-powered transportation. Already 250 automobile gas-filling compressor stations are in operation. By 1990, the number of trucks in Russia burning compressed natural gas will exceed 500,000. This is one of their efforts toward decreasing consumption of gasoline and diesel fuel.

ELECTRICITY: THE MAJOR SECONDARY SOURCE OF ENERGY

A few nuclear reactors, such as some at Oak Ridge, Tennessee, are used to produce radioactive isotopes for medicine and industry, not to produce electricity.

A secondary source of energy is made from a primary source on the way to the end user. Most of the nuclear energy, a primary source, is used to produce electricity. At present other primary sources produce more electricity than nuclear energy (Fig. 8–12). Coal is used to produce more electricity than all of the other primary sources combined.

As is obvious from Figure 8–12, the use of electricity is growing rapidly. Electric motors, the force behind enormous gains in technological innova-

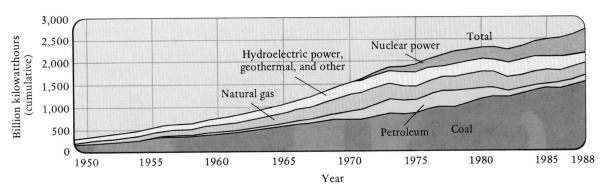

Figure 8–12   The generation of electricity by type of energy source. As shown, coal and nuclear energy are increasing while petroleum is decreasing. Petroleum is mainly used in transportation and to supply petrochemicals. (*Annual Energy Review*, Energy Information Administration, 1988.)

tion and industrial productivity over the last half century, consume two thirds of America's electricity. Electronics, communication devices, elevators, air conditioners, electric irons, microwaves, toothbrushes, flashlights, and typewriters are some of the thousands of electric gadgets that we "cannot do without" in our modern society.

About 36% of all the energy consumed in the United States is used in the production of electricity. In 1988 the 28.6 quads of energy put into the production of electricity yielded about 8.8 quads of electricity in the home or factory using the electricity. The 19.8 quads difference was lost in the production and the transmission of electricity. At least part of this loss is expected because of the second law of thermodynamics (discussed earlier in this chapter), which states that a natural process loses some nonuseful energy as entropy (disorder) is increased.

Electric transmission lines are necessary to carry the electricity, but they incur energy loss to the consumer. (Carolina Biological Supply.)

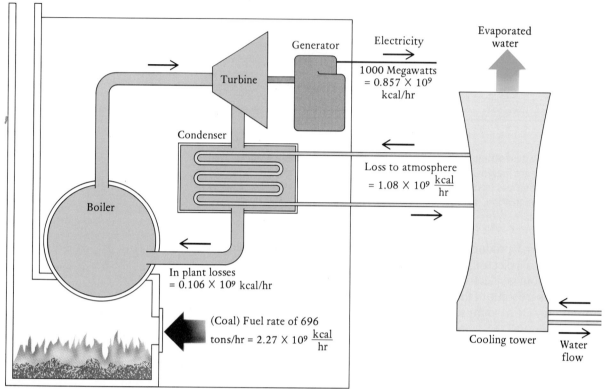
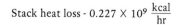

Figure 8–13   The heat balance of a 1000-MW coal-burning electric generating plant. Note that the 696 tons of coal burned per hour furnish 2.27×10^9 kcal of heat energy, but that only 0.857×10^9 kcal of energy, or 38%, is converted to electricity. Note also the large amounts of heat energy lost to the cooling water and atmosphere.

Research on superconductors, described in Chapter 11 may lead to a dramatic reduction in energy loss from transmission of electricity.

When we pay the electricity bill, we pay for energy (kilowatt-hours), not for power (how fast we use the energy, or kilowatts).

A specific example of energy loss in the production of electricity in a coal-burning power plant is shown in Figure 8–13 and summarized below.

For a 1000-MW (megawatt) coal-burning plant, one hour of operation might look like this:

Coal consumed	696 tons producing 2.270 billion kcal
Smokestack heat loss	0.227 billion kcal
Heat loss in plant	0.106 billion kcal
Heat loss in evaporator to cool condenser	1.080 billion kcal
Electric energy delivered to power lines	0.857 billion kcal
Percentage of energy delivered as electricity before transmission losses	$\dfrac{0.857}{2.27} \times 100\% = 37.8\%$

There is a further energy loss in the power lines and the transformers, lowering the useful output of the plant to 30% of the energy consumed. This is the **efficiency** figure for the overall operation. It is important to note that we pay for 300 kcal of heat energy in the form of coal or fuel oil but receive less than 100 kcal of energy in the form of electricity. Obviously, it requires much less fuel to heat homes with the fuel itself than with electricity made from the fuel.

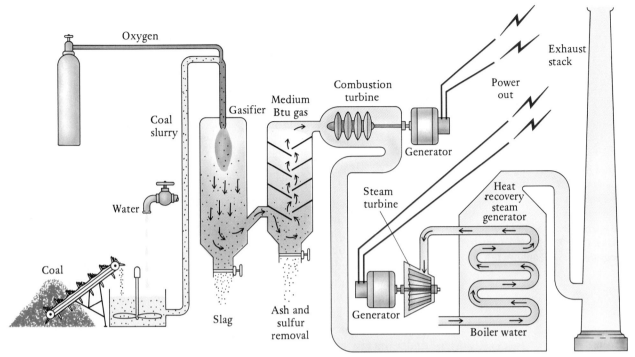

Figure 8-14   A coal gasification process for producing electricity. This process was developed by Texaco.

With more dependence on coal to produce even more electricity, much research is going into improving the efficiency of power plants that burn coal. One such improvement utilizes coal gasification, described earlier. One process involves gasifying the coal and burning the coal in added oxygen (Fig. 8-14). The improvement in efficiency comes from using the generated hot gases twice: once in turning a combustion turbine to turn a generator and utilizing the remaining heat to produce steam to turn another generator.

NOVEL, NONMAINSTREAM SOURCES OF ENERGY

Some energy sources offer more potential than they now supply, though most of these sources are available only in certain areas. For example, in some locales (e.g., Boise, Idaho) it is possible to heat homes and make electricity from geothermal sources, such as hot springs and geysers. Other areas near oceans derive energy from ocean currents and temperature differences between warm surface water and colder, deeper water. For many years wind currents and windmills have provided energy for pumping water and making electricity in some places. In 1987, geothermal sources generated 10.8 billion kW-h of electricity; the combustion of wood and waste generated 1.5 billion kW-h; and wind and other small contributors generated another 14 million kW-h.

One novel source of energy is available in every populated area but is currently tapped in only a few locations. Garbage is everywhere and can be

Everybody has it; nobody wants it. Why not use it to provide energy? Great idea!

Figure 8–15   Burning garbage, ocean currents, and geysers contribute a relatively small but geographically important amount of energy to the energy needs in the United States. (*The World of Chemistry,* Programs 17, 18, and 25.)

used to produce energy by burning or fermenting. Burning the garbage is a solution to an energy problem as well as to the problem of what to do with tons and tons of trash.

Plants for burning garbage to extract energy are in operation in several countries, including France and the United States. The Nashville Thermal Transfer Corporation in downtown Nashville, Tennessee, began operation in February 1974 (Fig. 8–16). The plant supplies steam and (or) cold water to 28 buildings and about 30,000 people in the downtown area. The energy comes from burning 400 tons of residential garbage each day. Pipes laid under the city carry the steam and (or) cold water to the various buildings. The cold water used to cool the buildings is cooled by electricity produced at the plant by steam-powered turbines. The electricity cools the water in a manner similar to the action of a very large refrigerator. The ash from the burning of the garbage has to go to a landfill, but this ash is less than 10% of the volume and 30% of the weight of the original garbage. Oil and gas are available as backup energy sources if needed, but they are rarely used. By burning garbage, the Nashville Thermal Transfer plant alleviates a small part of the enormous solid waste problems that Nashville and most other large cities have.

Figure 8–16   The Nashville, Tennessee, thermal energy plant.

A product of the fermentation of garbage is extracted from the world's largest garbage dump at Fresh Kills on Staten Island, New York. Underneath the huge mounds of garbage, bacteria turn some of the old buried garbage into methane (natural gas). The Brooklyn Union Gas Company has tapped this gas, which provides enough methane to fuel 16,000 homes on Staten Island.

SELF-TEST 8-B

1. The fossil fuel in shortest supply is ——————— .
2. Approximately what percentage of the coal in a mine is left in the ground after "all" of the minable coal is extracted? (25%, 50%, 75%)
3. How many gallons of oil are in one barrel?
4. Is the composition of "power gas," obtained from coal gasification, CO, H_2, N_2 or CO, H_2?
5. Which is more efficient, heating a home by a primary or by a secondary source of energy?
6. What is the major secondary source of energy?
7. Which primary source of energy is used most for producing electricity?
8. Is coal liquefaction expected to contribute significantly to our energy needs in the near future? To our supply of petrochemicals?

NUCLEAR ENERGY

Few issues have captured the awe, imagination, and scrutiny of mankind to quite the extent that nuclear energy has in the past five decades. Nuclear energy has been acclaimed, on the one hand, as the source of all our energy needs, and accused, on the other hand, of being our eventual destroyer.

Part of the interest in nuclear power is the tremendous amount of energy generated by a relatively small amount of fuel. The basics of nuclear reactions are described in Chapter 7, and radioactive wastes are discussed in Chapters 1 and 7. In this section, we focus on the energy that accompanies nuclear reactions.

The vast amounts of energy are released when heavy atomic nuclei split, the **fission** process, and when small atomic nuclei combine to make heavier nuclei, the **fusion** process. Consider the energy contrast between combustion of a fossil fuel and a nuclear fusion reaction. When 1 mol (6.02×10^{23} molecules, or 16 g) of methane is burned, over 200 kcal of heat are liberated:

$$CH_4 + 2\ O_2 \longrightarrow CO_2 + 2\ H_2O + 211\ kcal\ (kcal/mole\ CH_4)$$

In contrast, a lithium (Li) nucleus can be made to react with a H nucleus to form two helium (He) nuclei in a nuclear reaction. The energy released per mole of lithium in this reaction is 23,000,000 kcal. This means that 7 g of Li and 1 g of H produce 100,000 times more energy through fusion of nuclei than 16 g of CH_4 and 64 g of O_2 produce by electron exchange.

$$^7_3Li + {}^1_1H \longrightarrow 2\ {}^4_2He + 23{,}000{,}000\ kcal/mol\ of\ {}^7_3Li$$

Energy changes associated with nuclear events may be many thousands of times larger than those associated with chemical events.

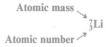

Atomic mass ↘

^7_3Li

Atomic number ↗

Realizing that nuclear changes could involve giant amounts of energy relative to chemical changes for a given amount of matter, Otto Hahn, Fritz Strassman, Lise Meitner, and Otto Frisch discovered in 1938 that $^{235}_{92}U$ is fissionable. Subsequently the dream of controlled nuclear energy became a reality, followed by the bomb and nuclear power plants. In the 1950s it was hoped that nuclear energy would soon relieve the shortage of fossil fuels. To date this has not been accomplished, although the production of nuclear energy has grown in recent years.

Fission Reactions

Fission is the breakup of heavy nuclei.

Fission can occur when a thermal neutron (with a kinetic energy about the same as that of a gaseous molecule at ordinary temperatures) enters certain heavy nuclei with an odd number of neutrons ($^{235}_{92}U$, $^{233}_{92}U$, $^{239}_{94}Pu$). The splitting of the heavy nucleus produces two smaller nuclei, two or more neutrons (an average of 2.5 neutrons for $^{235}_{92}U$), and much energy. Typical nuclear fission reactions may be written:

$$^{235}_{92}U + ^{1}_{0}n \longrightarrow ^{141}_{56}Ba + ^{92}_{36}Kr + 3\ ^{1}_{0}n + energy$$
$$^{235}_{92}U + ^{1}_{0}n \longrightarrow ^{103}_{42}Mo + ^{131}_{50}Sn + 2\ ^{1}_{0}n + energy$$

Recall that in nuclear reactions the sum of the atomic numbers on the left side of the equation equals the sum of the atomic numbers on the right side of the equation. Likewise for the atomic masses.

$^{1}_{0}n$ represents a neutron.

$^{0}_{-1}e$ or $^{0}_{-1}\beta$ represents a beta particle.

A low-energy neutron will disrupt some large nuclei.

Note that the same nucleus may split in more than one way. The fission products, such as $^{141}_{56}Ba$ and $^{92}_{36}Kr$, emit beta particles ($^{0}_{-1}e$) and gamma rays ($^{0}_{0}\gamma$) until stable isotopes are reached.

$$^{141}_{56}Ba \longrightarrow ^{0}_{-1}e + ^{0}_{0}\gamma + ^{141}_{57}La$$
$$^{92}_{36}Kr \longrightarrow ^{0}_{-1}e + ^{0}_{0}\gamma + ^{92}_{37}Rb$$

The products of these reactions emit beta particles, as do their products. After several such steps, stable isotopes are reached: $^{141}_{59}Pr$ and $^{90}_{40}Zr$, respectively.

The neutrons emitted can cause the fission of other heavy atoms if they are slowed down by a moderator, such as graphite. For example, the three neutrons emitted in the preceding uranium reaction could produce fission in three more uranium atoms, the nine neutrons emitted by those nuclei could produce nine more fissions, the 27 neutrons from these fissions could produce 81 neutrons, the 81 neutrons could produce 243, the 243 neutrons could produce 729, and so on. This process is called a **chain reaction** (Fig. 8–17), and it occurs at a maximum rate when the uranium sample is large enough for most of the neutrons emitted to be captured by other nuclei before passing out of the sample. Sufficient sample in a certain volume to sustain a chain reaction is termed the **critical mass.**

In the atomic bomb the critical mass is kept separated into several smaller subcritical masses until detonation, at which time the masses are driven together by an implosive device. It is then that the tremendous energy is liberated and everything in the immediate vicinity is heated to temperatures of 5 to 10 million degrees (Celsius or Fahrenheit). The sudden expansion of hot gases literally explodes everything nearby and scatters the radioactive fission fragments over a wide area. In addition to the movement of

An atomic bomb explosion. (*The World of Chemistry,* Program 6, "The Atom.")

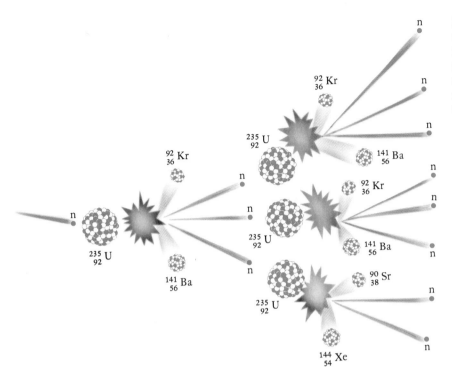

Figure 8–17   A chain reaction. A thermal neutron collides with a fissionable nucleus, and the resulting reaction produces three additional neutrons. If enough fissionable nuclei are present, a chain reaction will be sustained.

gases, there is the tremendous vaporizing heat that makes the atomic bomb so devastating.

There is no danger of an atomic explosion in the uranium mineral deposits in the Earth for two reasons. First, uranium is not found pure in nature—it is found only in compounds, which in turn are mixed with other compounds. Second, less than 1% of the uranium found in nature is fissionable $^{235}_{92}U$. The other 99% is $^{238}_{92}U$, which is not fissionable by thermal neutrons. In order to make nuclear bombs or nuclear fuel for electric generation, a purification enrichment process must be performed on the uranium isotopes, thus increasing the relative proportion of $^{235}_{92}U$ atoms in a sample. Ordinary uranium such as that found in ores is only 0.711% $^{235}_{92}U$.

It is interesting to note that fission products can be found in the Gabon Republic of West Africa, which indicate that a uranium ore deposit "went critical" about 150,000 years ago. At that time the natural uranium-235 content would have been higher than it is now.

Separation of uranium isotopes had to precede the control of atomic energy.

Mass Defect—The Ultimate Nuclear Energy Source

What is the source of the tremendous energy of the fission process? It ultimately comes from the conversion of mass into energy, according to Einstein's famous equation, $E = mc^2$, where E is energy that results from the loss of an amount of mass m, and c^2 is the speed of light squared. If separate neutrons, electrons, and protons are combined to form any particular atom, there is a loss of mass called the **mass defect.** For example, the calculated mass of one $_2^4$He atom from the masses of the constituent particles is 4.032982 amu:

amu = atomic mass unit

$2 \times 1.007826 = 2.015652$ amu, mass of two protons and two electrons
$2 \times 1.008665 = \underline{2.017330}$ amu, mass of two neutrons
total $= 4.032982$ amu, calculated mass of one $_2^4$He atom

Since the measured mass of a $_2^4$He atom is 4.002604 amu, the mass defect is 0.030378 amu:

6.02 × 10²³ amu/gram
or
1 mol amu/gram

$$4.032982 \text{ amu} - 4.002604 \text{ amu} = 0.030378 \text{ amu, mass defect}$$

Because the atom is more stable than the separated neutrons, protons, and electrons, the atom is in a lower energy state. Hence, the 0.030378 amu lost per atom would be released in the form of energy if the $_2^4$He atom were made from separate protons, electrons, and neutrons. The energy equivalent of the mass defect is called the **binding energy.** The binding energy is analogous to the earlier concept of bond energy, in that both are a measure of the energy necessary to separate the package (nucleus or molecule) into its parts.

The mass that is lost leaves in the form of energy: $E = mc^2$.

Atoms with atomic numbers between 30 and 63 have a greater mass defect per nuclear particle than very light elements or very heavy ones, as shown in Figure 8–18. This means the most stable nuclei are the middle-weight ones found in the atomic number range from 30 to 63.

Separated nuclear particles have more mass than when combined in a nucleus.

Intermediate-sized nuclei tend to have the greatest nuclear stability.

Because of the relative stabilities, it is in the intermediate range of atomic numbers that most of the products of nuclear fission are found. Therefore, when fission occurs and smaller, more stable nuclei result, these nuclei will contain less mass per nuclear particle. In the process, mass must be changed into energy. This energy gives the fission process its tremendous energy. It takes only about 1 kg of $_{92}^{235}$U or $_{94}^{239}$Pu undergoing fission to be equivalent to the energy released by 20,000 tons (20 kilotons) of ordinary explosives like TNT. The energy content in matter is further dramatized when it is realized that the atomic fragments from the 1 kg of nuclear fuel

Figure 8–18 Binding energy for different nuclear masses. The most stable nuclei center around $_{26}^{56}$Fe, which has the largest binding energy per nuclear particle. Binding energy is the energy given off when individual nuclear particles form a nucleus. One MeV equals 1.60 × 10⁻¹³ J, which is the same energy as 3.82 × 10⁻¹⁴ cal.

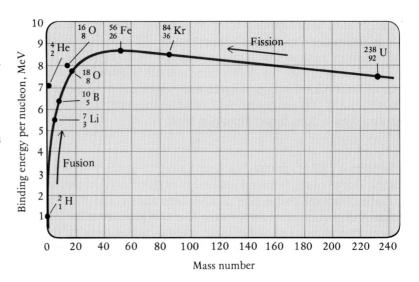

weigh 999 g, so only one tenth of 1% of the mass is actually converted to energy. The fission bombs dropped on Japan during World War II contained approximately this much fissionable material.

Controlled Nuclear Fission

The fission of a $^{235}_{92}U$ nucleus by a slow-moving neutron to produce smaller nuclei, extra neutrons, and large amounts of energy suggested to Enrico Fermi and others that the reaction could proceed at a moderate rate if the number of neutrons could be controlled. If a neutron control could be found, the concentration of neutrons could be maintained at a level sufficient to keep the fission process going but not high enough to allow an uncontrolled explosion. It would then be possible to drain the heat away from such a reactor on a continuing basis to do useful work. In 1942, Fermi, working at the University of Chicago, was successful in building the first atomic reactor, called an **atomic pile.**

An atomic reactor has several essential components. The charge material (fuel) must be fissionable or contain significant concentrations of a fissionable isotope such as $^{235}_{92}U$, $^{239}_{94}Pu$, or $^{233}_{92}U$. Ordinary uranium, which is mostly the nonfissionable $^{238}_{92}U$, cannot be used since it has a small concentration of the $^{235}_{92}U$ isotope. A moderator is required to slow the speed of the neutrons produced in the reactions without absorbing them. Graphite, water, and other substances have been used successfully as moderators. A substance that will absorb neutrons, such as cadmium or boron steel, is present in order to have a fine control over the neutron concentration. Shielding, to protect the workers from dangerous radiation, is an absolute necessity. Shielding tends to make reactors heavy and bulky installations. A heat-transfer fluid provides a large and even flow of heat away from the reaction center.

Once the heat is produced in a nuclear reactor and safety measures are employed to protect against radiation, conventional technology allows this energy to be used to generate electricity, to power ships, or to operate any device that uses heat energy. A system for the nuclear production of electricity is illustrated in Figure 8–19.

What are the fuel requirements in nuclear fission energy production? In a typical fission event such as

$$^{1}_{0}n + ^{235}_{92}U \longrightarrow ^{93}_{37}Rb + ^{141}_{55}Cs + 2\,^{1}_{0}n + 200 \text{ MeV}$$

the energy release, 200 MeV, is equivalent to 7.7×10^{-12} cal per atom of $^{235}_{92}U$, or 4.64×10^9 kcal/mol. Since 1 g of pure $^{235}_{92}U$ contains 2.56×10^{21} atoms, the total energy release for 1 g of uranium-235 undergoing fission would be

$$1\ \text{g} \times 2.56 \times 10^{21}\ \frac{\text{atoms}}{1\ \text{g}} \times 7.7 \times 10^{-12}\ \frac{\text{cal}}{\text{atom}} = 2.0 \times 10^{10}\ \text{cal}$$

This is the amount of energy that would be released if 5.95 tons of coal were burned, or if 13.7 barrels of oil were burned to produce heat to power a boiler. This means that about 3 kg of $^{235}_{92}U$ fuel per day would be required for a 1000-MW (megawatt) electric generator. The fuel used, however, is not pure $^{235}_{92}U$, but **enriched** uranium containing up to 3% $^{235}_{92}U$.

Atomic pile:
1. Carefully diluted fissionable material;
2. Moderator to control fission reaction;
3. Coolant to control heat;
4. Shielding to limit radiation.

The first one was piled together at the University of Chicago in 1942.

1 million electron volts (MeV) = 3.827 × 10^{-14} cal.

When 1 g of ^{235}U undergoes fission, it provides the same energy as burning about 3 tons of coal.

A megawatt is equal to 1 million watts.

Figure 8–19   Schematic illustration of a nuclear power plant.

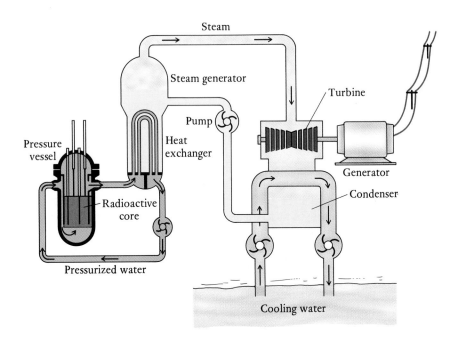

Nuclear energy is mostly used for production of electricity. In 1957, when nuclear energy usage began, and until about 1974, more energy was produced in the United States from burning firewood than from nuclear energy. In 1988, 6.54 quads of energy were produced by nuclear fission, and 0.47 quad was generated by burning firewood. Nuclear energy, then, produced 7.1% of the U.S. energy supply in 1988. Although other countries produce less energy by nuclear means than the United States, nuclear energy is a much larger share of the total energy production in some of these countries. For example, in 1988, France produced 3.13 quads by nuclear means, representing 79.4% of its total energy production; Japan produced 2.16 quads, for 67.2% of its total energy production.

On January 1, 1989, there were 108 operable U.S. nuclear reactors and 16 in various stages of construction but not yet operating (Fig. 8–20). The United States is producing less uranium now, down from a high of 21,850 short tons of U_3O_8 in 1980. Only 6,515 short tons were produced in 1988. Estimates of the reserves are 885,000 short tons of U_3O_8 (reasonably assured) to 4,394,000 short tons (speculated and estimated).

It is possible to convert the nonfissionable $^{238}_{92}U$ and $^{232}_{90}Th$ into fissionable fuels by using a **breeder reactor**. In such a reactor, a blanket of nonfissionable material is placed outside the fissioning $^{235}_{92}U$ fuel (Fig. 8–21), which serves as the source of neutrons in the breeder reactions. The two breeder reaction sequences are

$$^{238}_{92}U + ^1_0n \longrightarrow ^{239}_{92}U \xrightarrow{\beta} ^{239}_{93}Np \xrightarrow{\beta} ^{239}_{94}Pu$$

$$^{232}_{90}Th + ^1_0n \longrightarrow ^{233}_{90}Th \xrightarrow{\beta} ^{233}_{91}Pa \xrightarrow{\beta} ^{233}_{92}U$$

The products of the breeder reactions, $^{233}_{92}U$ and $^{239}_{94}Pu$, are both fissionable with slow neutrons, and neither is found in the Earth's crust.

$^{235}_{92}U$ is in very short supply.

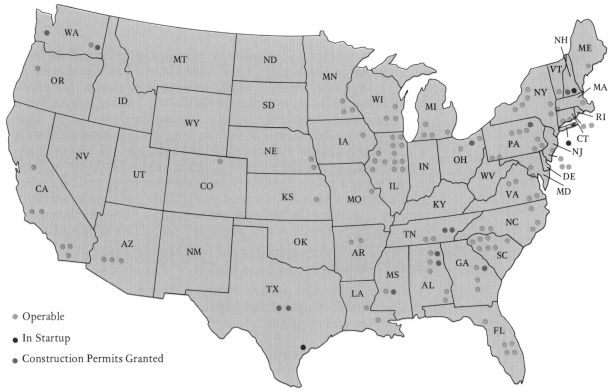

- Operable
- In Startup
- Construction Permits Granted

Figure 8–20 Status of nuclear reactors in the United States on January 1, 1989. No new orders for nuclear reactors have been placed since 1978. (*Annual Energy Review*, Energy Information Administration, 1988.)

The Clinch River Breeder Reactor project died in 1983, but research continues at Argonne National Laboratory outside Chicago. The breeder concept came from Walter Zinn, Argonne's first director. The first nuclear reactor of any kind to produce electricity was a research breeder reactor at Idaho Falls, Idaho. Operation began in 1951. An updated version (EBR-2, experimental breeder reactor) has been in operation for more than 20 years.

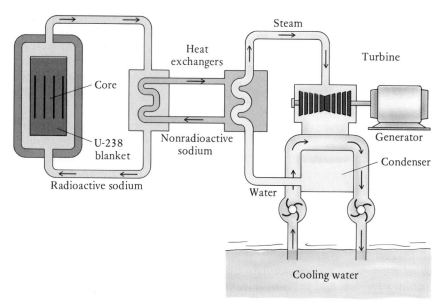

Figure 8–21 Schematic diagram of a fast breeder reactor and steam-turbine power generator.

$^{239}_{94}$Pu is toxic from a radiation as well as a chemical point of view.

Breeder reactors present many technological problems, not the least of which is the potential of a disaster caused by mishandling of the $^{239}_{94}$Pu isotope, which is extremely toxic and can also be fabricated into a fission bomb. Nevertheless, the expected benefit from the breeder program is massive amounts of energy. For example, if all the uranium used for electric generation were used in breeder reactors, instead of running out of uranium fuel in several decades, the breeder fuels would supply U.S. electric requirements for about 2600 years, assuming 1970 electricity-use levels were maintained, something euphemistically called **Zero Energy Growth!** If a breeder program cannot be more widespread soon, then the $^{235}_{92}$U isotope will be used up only to generate electricity and another, as yet undiscovered, source of neutrons will be needed to convert the $^{238}_{92}$U and $^{232}_{90}$Th into fissionable fuels.

Fusion Reactions

Fusion is the combination of very light nuclei.

When very light nuclei, such as H, He, and Li, are combined, or **fused**, to form an element of higher atomic number, energy must be given off consistent with the greater stability of the elements in this intermediate atomic number range (Fig. 8–18). This energy, which comes from a decrease in mass, is the source of the energy released by the Sun and by hydrogen bombs. Typical examples of fusion reactions are:

$^2_1$H = Deuterium
$^3_1$H = Tritium
$^0_{+1}$e = Positron

$$4\,^1_1\text{H} \longrightarrow \,^4_2\text{He} + 2\,^0_{+1}\text{e} + 26.7 \text{ MeV for four } ^1_1\text{H fused}$$
$$^2_1\text{H} + ^2_1\text{H} \longrightarrow \,^3_2\text{He} + ^1_0\text{n} + 3.2 \text{ MeV}$$
$$^2_1\text{H} + ^2_1\text{H} \longrightarrow \,^3_1\text{H} + ^1_1\text{H} + 4.0 \text{ MeV}$$
$$^3_1\text{H} + ^2_1\text{H} \longrightarrow \,^4_2\text{He} + ^1_0\text{n} + 17.6 \text{ MeV}$$

The net reaction for the last three reactions given here is:

$$5\,^2_1\text{H} \longrightarrow \,^4_2\text{He} + ^3_2\text{He} + ^1_1\text{H} + 2\,^1_0\text{n} + 24.8 \text{ MeV for five } ^2_1\text{H fused}$$

Materials for fusion reactions are available in enormous amounts.

The critical mass for ^{235}U tends to limit the size of a fission bomb, but a fusion bomb with more LiH can be made much more powerful.

Deuterium is a relatively abundant isotope—out of 6500 atoms of hydrogen in sea water, for example, one is a deuterium atom. What this means is that the oceans are a potential source of fantastic amounts of deuterium. There are 1.03×10^{22} atoms of deuterium in a single liter of sea water. In a single cubic kilometer of sea water, therefore, there would be enough deuterium atoms with enough potential energy to equal the burning of 1360 billion barrels of crude oil, and this is approximately the total amount of oil originally present on this planet.

Fusion reactions occur rapidly only when the temperature is of the order of 100 million degrees or more. At these high temperatures atoms do not exist as such; instead, there is a **plasma** consisting of unbound nuclei and electrons. In this plasma nuclei merge or combine. In order to achieve the high temperatures required for the fusion reaction of the hydrogen bomb, a fission bomb (atomic bomb) is first set off.

A plasma is a gaseous state composed of ions.

Tritium, $^3_1$H, is radioactive.

One type of hydrogen bomb depends on the production of tritium ($^3_1$H) in the bomb. In this type, lithium deuteride ($^6_3$Li$^2_1$H, a solid salt) is placed around an ordinary $^{235}_{92}$U or $^{239}_{94}$Pu fission bomb. The fission is set off in the usual way. A $^6_3$Li nucleus absorbs one of the neutrons produced and splits into tritium, $^3_1$H, and helium, $^4_2$He.

$$^6_3\text{Li} + ^1_0\text{n} \longrightarrow \,^3_1\text{H} + ^4_2\text{He}$$

The temperature reached by the fission of $^{235}_{92}$U or $^{239}_{94}$Pu is sufficiently high to bring about the fusion of tritium and deuterium:

$$^3_1H + {}^2_1H \longrightarrow {}^4_2He + {}^1_0n + 17.6 \text{ MeV}$$

A 20-megaton bomb usually contains about 300 lb of lithium deuteride, as well as a considerable amount of plutonium and uranium.

Attempts at Controlled Nuclear Fusion

Three critical requirements must be met for controlled fusion. First, the temperature must be high enough for fusion to occur. The fusion of deuterium (^2_1H) and tritium (^3_1H) requires a temperature of 100 million degrees or more.

$$^3_1H + {}^2_1H \longrightarrow {}^4_2He + {}^1_0n + 4.1 \times 10^8 \text{ kcal/mol } {}^2_1H$$

Second, the plasma must be confined long enough to release a net output of energy. Third, the energy must be recoverable in some usable form.

Attractive features that encourage research in controlled nuclear fusion are the rather limited production of dangerous radioactivity and the great abundance of hydrogen fuel (in water), a most abundant resource. Most

Figure 8–22 A magnetic containment device, the Tokomac Fusion Test Reactor at Princeton University, has been able to sustain a fusion reaction for a small fraction of a second. A full second or longer is needed to achieve a net evolution of energy.

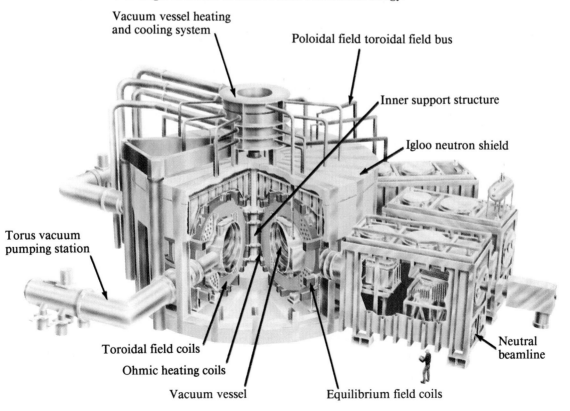

Vacuum vessel heating and cooling system

Poloidal field toroidal field bus

Inner support structure

Igloo neutron shield

Torus vacuum pumping station

Neutral beamline

Toroidal field coils

Ohmic heating coils

Vacuum vessel

Equilibrium field coils

radioisotopes produced by fusion have short half-lives and therefore are a serious hazard for only a short time.

Fusion reactions have not been "controlled." No physical container can contain the plasma without cooling it below the critical fusion temperature. Magnetic "bottles," enclosures in space bounded by a magnetic field, have confined the plasma, but not for long enough periods.

A newer confinement method is based on a laser system that simultaneously strikes tiny hollow glass spheres called **microballoons,** which enclose the fuel, consisting of equal parts of deuterium and tritium gas at high pressures (Fig. 8–23).

Two other attempts to achieve fusion are aneutronic fusion and the particle-beam fusion accelerator (PBFA). Neither approach has delivered enough concentrated energy to sustain fusion.

Aneutronic fusion, also called migma (a Greek word for "mixture"), was presented theoretically by the American scientist, Bogdan Maglich, in 1973. Since this process does not involve neutrons either as products or reactants, a penetrating, hard-to-capture, potentially damaging particle is eliminated. Fusion is achieved by accelerating nuclei (such as deuterons, ^2_1H) in linear accelerators to an energy of 0.7 MeV, which is equivalent to a temperature of $7 \times 10^9 °C$. The high-energy ions are directed on a lithium target. The fusion of a deuteron and a lithium nucleus produces a helium nucleus, two protons, and energy.

In the PBFA process, electric charge is stored in capacitors and discharged in 40-ns (nanosecond) pulses. The energy accelerates lithium ions to a kinetic energy of between 1 and 2 million joules. The lithium ions impinge on a target of deuterium (^2_1H) and tritium (^3_1H). Lithium nuclei fuse with one or the other of the hydrogen isotopes and produce energy.

Containment is one of the biggest problems in developing controlled fusion.

A nanosecond is one billionth of a second.

Figure 8–23   Schematic diagram of an apparatus for laser-induced fusion. Tiny glass pellets (microballoons about 0.1 mm in diameter) filled with frozen deuterium and tritium are subjected to a powerful laser beam, and the contents undergo nuclear fusion.

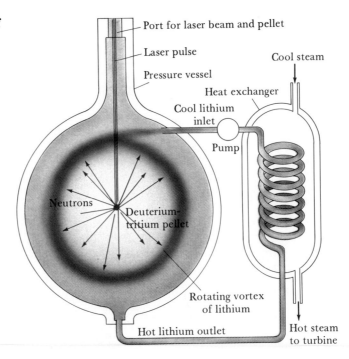

After more than 40 years of intense research, it does not appear that controlled hot fusion is even on the verge of making any contribution to our energy needs in the foreseeable future.

SOLAR ENERGY

Although the Sun is our ultimate source of energy, we use only a small fraction of what the Earth receives from it; efforts are now being made in several directions to use more of the Sun's gift to us. One technique is to use algae to produce hydrogen (Fig. 8–24) which then can be used in fuel cells to produce electricity. Another device is the solar collector, which uses the warmth of sunlight to heat water and air to heat homes. A third device uses photosensitive materials to make a solar electric cell, such as the type that is commonly used to power hand calculators.

Solar energy is transmitted nuclear energy.

Solar devices for heating water or air in buildings are decreasing in sales—from 17 million ft² in 1983 to 5 million ft² in 1986.

Solar Energy and the Solar Cell

Another approach to the direct utilization of solar energy is the *solar cell,* known as a photovoltaic device (Fig. 8–25). A solar cell converts energy from the sun into electron flow. During the 1980s the efficiency of solar cells doubled to the value of 23% routinely and as high as 40% in the laboratory. Routinely, then, solar cells convert sunlight into electric power at the rate of at least 100 W/yd² (watts per square yard) of illuminated surface. Solar cells are now used in calculators, watches, space-flight applications, communication satellites, power for remote water pumps, signals for automobiles and trains, light-weight power supplies for boats and golf carts, and as the source of electricity in utility power plants throughout the world.

A 100-MW utility plant would cover about 600 acres.

Figure 8–24 **Schematic diagram of an electricity-producing photosynthesis process. H_2 and O_2 produced by the blue-green algae, *Anabeana cylindrica,* are separated by palladium metal, which is permeable to H_2 but not to O_2. The H_2 and O_2 are then combined in the fuel cell to produce electricity.**

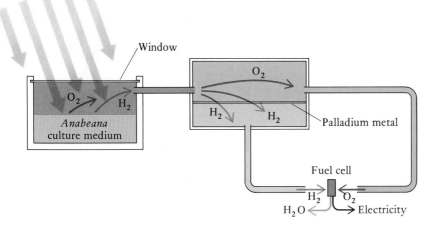

Doped silicon for transistors is discussed in Chapter 11.

Figure 8–25   A bank of photovoltaic, or solar, cells. A photo of an individual silicon solar cell is in Chapter 11. (*The World of Chemistry,* Program 15, "The Busy Electron.")

One type of solar battery consists of two layers of almost pure silicon, doped Si. The lower, thicker layer contains a trace of boron (B), and the upper, thinner layer a trace of arsenic (As). As pointed out in Chapter 11, the As-enriched layer is an *n*-type semiconductor, and the B-enriched layer is a *p*-type semiconductor. Recall that Si has four valence electrons and is covalently bonded to four other Si atoms (Fig. 8–26). Arsenic has five valence

Figure 8–26   Schematic drawing of semiconductor crystal layers derived from silicon.

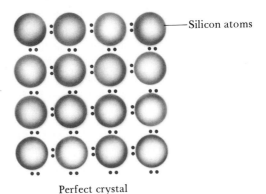

Perfect crystal

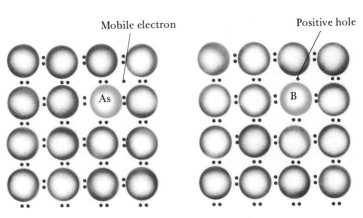

n-type   *p*-type

electrons. When As is included in the Si structure, only four of the five valence electrons of As are used for bonding with four Si atoms; one electron is relatively free to move. B has three valence electrons. When B atoms are included in the Si structure, there is a deficiency of one electron around the B atom; this creates "holes" in the B-enriched layer.

There is a strong tendency for the "free" electrons in the As layer to pair with the unpaired Si electrons in the "holes" in the B layer. If the two layers are connected by an external circuit and quanta of light of sufficient energy strike the surface and are absorbed, excited electrons can leave the As layer and flow through the external circuit to the B layer. As the B layer becomes more negative because of added electrons, electrons are repelled *internally* back into the As layer, which is now positive and attracts the electrons from the B layer. The process can continue indefinitely as long as the cell is exposed to sunlight.

A typical solar cell is constructed on a sheet of plastic or glass (Fig. 8–27). Next to the plastic or glass is a thin sheet of metal that is the electrode giving electrons to the *p*-type semiconductor layer. The topmost *n*-type semiconductor layer, which receives the sun's rays, is nearly transparent. The solar cell is covered with a thin film of indium tin oxide ($InSnO_2$), which acts as an antireflection coating. A metallic grid structure on top of the cell allows as much light as possible to strike the *n*-type layer while functioning as an electrode. The efficiency of the solar cell depends on its ability to absorb photons of light and convert the light to electric energy. Photons that are reflected, pass through, or produce only heat, decrease the efficiency of the cell.

Hyperpure Si is manufactured by zone refining (pages 323–324) in only 3 plants in the world; two in Japan; one in West Germany.

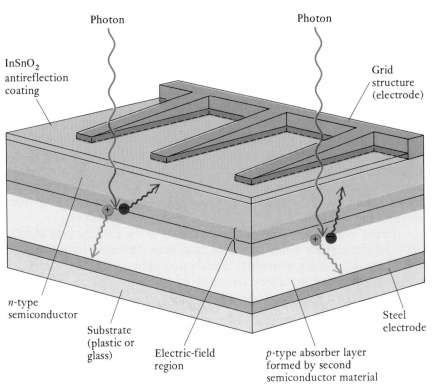

Figure 8–27 Typical photovoltaic cell using crystals of silicon. (Adapted from *Scientific American*.)

Older solar cells were made from single-crystal silicon, but the photovoltaic cells developed in the 1980s use amorphous silicon (a-Si). The amorphous silicon has an irregular array of Si atoms instead of the regular pattern of atoms in a single crystal of Si. The a-Si is as efficient in absorbing light in very thin layers (0.5 μm thick) as single 300-μm-thick crystals were. Other advantages of a-Si include its relatively low production cost (5% of the cost of single-crystal Si) and its production in larger sheets.

Efficiencies are improved, but the expense of producing the cell is increased, by using concentrator cells, which have a lens to concentrate the sunlight onto the solar cell. An added computer turns concentrator cells to track the sun through the sky.

Solar cells are on the threshold of being the next great technological breakthrough, perhaps comparable to the computer chip and other electronic marvels. Although experimental solar-powered automobiles are now available (Fig. 8–28) and many novel applications of solar cells already exist (such as powering an exhaust fan in a luxury automobile to remove hot air accumulated in a car parked in the sun), the breakthrough is the use of banks of solar cells as a utility power plant to produce huge amounts of electricity. One plant already in operation in California uses solar cells to produce 20 MW of electricity, which is enough to supply the electricity needs of a city the size of Tampa, Florida.

Utility plants that presently use coal, oil, or nuclear energy to produce electricity will likely be replaced with solar-powered plants as the older plants wear out or the fuel for the plants becomes too costly or the supply dwindles too far. The North American Electric Reliability Council estimates that 73 1000-MW power plants will be needed by 1997 to supply U.S. energy needs. The Department of Energy reports that dozens of aging coal-, nuclear-, and petroleum-fueled power plants—equivalent to 34 1000-MW plants—must be replaced. To catch up with electricity needs is going to be costly (about $2 billion to build a new coal-fueled power plant) and controversial because of potential pollution generated. With the cost of additional and replaced power plants factored in with the operating costs, new sources such as amorphous-silicon photovoltaic plants could receive very favorable consideration because of their relatively low construction cost, their short construction time (less than a year), absence of pollution, no operating fuel cost, and very low maintenance. Solar utility plants have no moving parts, can be placed anywhere (even in remote places such as in space), can be mass produced, and use plentiful, cheap sand (impure silicon dioxide, SiO_2) as a raw material for the solar cells. At least two U.S. plants are already in operation to mass produce the a-Si cells.

Improvements are being made constantly in the performance of solar cells. Efficiencies are being increased, and manufacturing techniques are being improved. Solar technology is ready now to make a significant contribution to the need for utility power plants. A photovoltaic power plant makes the most economic sense as a new plant or as a replacement for a phased-out old power plant. The cost of electricity from a new photovoltaic plant would be 22¢/kW-h compared with 7¢/kW-h from an *existing* coal-fueled power plant and 9¢/kW-h from an *existing* petroleum-fueled power plant.

Electric cars powered by voltaic batteries are discussed in Chapter 10.

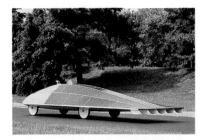

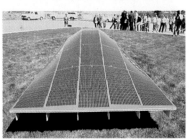

Figure 8–28   The Sun Raycer automobile was built by General Motors. Powered by solar cells, it also has silver-zinc batteries to help on a hill or on a cloudy day. (Courtesy of GM Hughes Electronics.)

Since the first practical use of solar cells in 1955 to power eight rural telephones in Georgia, solar cells have become more efficient and cheaper. Even for a 10% efficient photovoltaic cell, we are effectively utilizing an otherwise neglected energy source that costs nothing. Solar-generated electricity is with us and will steadily increase its contribution to the energy supply. Energy from these and all other solar systems depends on changes in the cloud cover and the seasons. Cloudy and cold days considered, all solar energy devices are expected to meet the official government goal of 20% of our energy needs by the year 2000.

A Hydrogen Economy

When solar cells become cheap enough, the electricity they supply could be used to electrolyze water to yield hydrogen and oxygen. Hydrogen can be transported through pipes far more efficiently than electricity is transported through transmission lines. The hydrogen could be piped to where the energy is needed and burned to heat water to steam, which in turn could generate electricity. Such an arrangement could give rise to a hydrogen economy and further release us from our dependence on fossil fuels and nuclear energy. Liquid hydrogen packs more energy per pound than any other nonnuclear fuel. It is the only fuel that ignites fast enough to boost an aircraft to orbital velocity (about 25,000 ft/s). The principal product is the raw material water.

Hydrogen can be burned in most devices that now burn natural gas.

Some catalysts reported in late 1982 for the decomposition of water in the presence of sunlight are indium phosphide (InP), phosphorus-doped silicon coated with Pt or Ni, and p-type iron oxide semiconductor. All are presently too expensive and (or) inefficient (12%, 12%, and 0.05%, respectively) to compete with hydrogen produced by the reaction of coal with steam.

Some hydrogen-powered buses and cars are now operating on an experimental basis.

$$C + H_2O \longrightarrow CO + H_2$$

ENERGY PROGNOSIS

More than perhaps any other single physical factor, the availability of energy determines our way of life. What, then, can we expect in the way of energy availability (and a way of life) in the year 2000?

We can expect to be less vulnerable to oil-supply disruption than we were in the 1970s. There will be a slower growth of energy demand, a trend toward more efficient use of energy, and a shift from oil to other sources. Hence, a reduced dependence on OPEC is expected. The total oil demand is predicted to be 16.3 million barrels per day, with 10.5 million barrels supplied by the United States and 5.8 million barrels imported. The recent low prices of oil give us a false sense of security about supplies of oil. Oil supplies are expected to be depleted for all practical uses during your lifetime. Energy needs will be shifted more and more to coal, which is in greater abundance.

By the year 2000, use of coal (primarily for electricity generation) is expected to increase to 1700 million short tons, and use of natural gas is expected to decrease to 17.5 trillion ft^3.

Synthetic fuels, such as gasified and liquefied coal, probably will not contribute much until the late 1990s. By the year 2000, the production of coal gas is expected to be 255 billion ft^3/year. Shale oil may contribute 515 thousand barrels per day. Energy from biomass is estimated to be equivalent to 65 thousand barrels of oil per day. This will be energy primarily from alcohol fuels produced from crops, wood, and waste.

Nuclear power may decrease from producing 18% of our electricity, mainly due to fear of the risks associated with the nuclear plants and with the nuclear wastes they produce.

Fuel cells are discussed in Chapter 10.

Electricity use will be expanded. Fuel cells that are 60–80% efficient, will find more commercial use. Electric cars will be seen more. About 70% of all auto trips are within a 100-mile range, and the use of electric cars will cut our oil needs by about half and greatly reduce urban pollution.

The technology and proved workability for solar-powered utility plants for the production of megawatts of electricity will enable them to make a huge contribution to our future energy needs and help in our efforts to control pollution. Look for significant uses of solar-powered products and utility production of electricity by solar-powered photovoltaic cells.

In the next century, we shall undoubtedly be forced to make a transition to an energy economy based on our own natural resources, principally coal and solar energy, followed distantly by energy from geothermal sources and wind, garbage burning, and ocean currents. The transition will be more efficient if done gradually and after careful consideration of the risks and consequences as well as the economics, rather than in reaction to a major energy crisis.

Check these predictions in about 10 to 15 years. We hope the energy situation provides a good lifestyle for us all in the years ahead.

SELF-TEST 8–C

1. The splitting of an unstable nucleus to produce energy is termed
   ( ) fission ( ) fusion.
2. When light nuclei combine to form heavy nuclei and energy, the process is ( ) fission ( ) fusion.
3. The major problem with obtaining fusion energy is ( ) containment of reactants at high temperature, ( ) enough fuel, ( ) costs.
4. A radioactive isotope of hydrogen used as a fuel in experimental fusion reactions is _____ .
5. Progress in the technology of photovoltaic cells has come about from making cheaper, more efficient _____ silicon.
6. When fission is used to produce energy in a breeder reaction, the fuel produced will be _____ and _____ .
7. Flat-plate solar collectors on the roofs of houses use solar energy to heat _____ or _____ in pipes by the _____ effect.
8. The cleanest burning fossil fuel is _____ .
9. The solar battery (cell) contains a small amount of _____ in the *n*-type silicon layer and a small amount of _____ in the *p*-type silicon layer.

MATCHING SET

_____ **1.** User of 35% of the world's energy

_____ **2.** Fossil fuels

_____ **3.** Combustion products of fossil fuels

_____ **4.** Minable coal

_____ **5.** Fissionable isotope

_____ **6.** Product of a fission breeder reactor

_____ **7.** Deuterium

_____ **8.** Date of petroleum discovery in the United States

_____ **9.** Tritium

_____ **10.** Source of deuterium

_____ **11.** Used to confine fusion fuel

_____ **12.** One use of solar radiation

_____ **13.** Approximate efficiency of a solar battery

a. 1859
b. CO_2 and H_2O
c. Uranium-235 ($^{235}_{92}U$)
d. *n*-type transistor
e. Plutonium-239 ($^{239}_{94}Pu$)
f. United States
g. ^2_1H
h. Sea water
i. 90%
j. Within 4000 ft of the Earth's surface
k. Microballoons
l. Coal, petroleum, natural gas
m. Photosynthesis
n. CO and H_2
o. 10–30%
p. ^3_1H
q. China
r. 1740

QUESTIONS

1. What is your attitude toward using up the fossil fuels within a few decades? Do we owe future generations a supply of these resources? Would you agree to give up air-conditioning, private cars, and power tools, to mention a few examples, and to limit heating and cooking if necessary to share these fuels with your grandchildren?

2. Which theoretically yields the greatest energy per mole?
  a. The burning of gasoline
  b. The fission of uranium-235
  c. The burning of methane (natural gas)

3. Which is the more efficient use of energy: burning coal in a house to heat it, or heating the house electrically with energy produced in a coal-burning power plant?

4. Give three examples of systems with stored chemical energy that can be used as a source of heat energy.

5. Is the electric energy where you live produced by burning fossil fuels? If not, what is the energy source? Are there pollution problems associated with the generation of the electric energy?

6. What was the original source of energy that is tied up in fossil fuels?

7. Define *power.* Give a unit (label) for power.

8. What is meant by an insulator R value of 30?

9. Do you pay for electricity as electric power (kilowatts) or electric energy (kilowatt-hours)? What is the difference?

10. What major problem is associated with harnessing the energy from a fusion reaction?

11. Suggest several ways in which solar energy might be harnessed.

12. Name two sources of energy not specifically mentioned in this chapter.

13. Explain how useful energy might be obtained from garbage.

14. Which is more fundamental: a supply of energy or a supply of food? Explain.
15. The energy consumption of the United States in 1970 was 2×10^{13} kW-h. What is this amount of energy expressed in kilocalories? In Btu's?
16. Assume the world population to be 5 billion and calculate the Earth's energy needs if everyone used as much energy as is used in the United States.
17. Which fuel has the greatest energy content per gram of fuel burned: coal, natural gas, or petroleum? Is this factor more important to you, the consumer, than the economics of energy use or pollution caused by fuel use?
18. Define *energy*. Give a unit (label) for energy.
19. List three so-called fossil fuels. Why are they called fossil fuels?
20. Which fuel—coal, petroleum, or natural gas—burns naturally with the least amount of pollution?
21. What are two dangerous properties of plutonium-239?
22. Which is the more efficient transport of energy: gas through pipes, or electricity through wires?
23. If solar energy is so clean, why are we so slow in moving to its use?
24. Do you think the United States should go forward with the use of the breeder reactor? Why?
25. How much has the price of oil changed during your lifetime?
26. Is an energy crisis a crisis of quality or quantity of energy?
27. Is electricity a primary or secondary source of energy? Explain your answer.
28. What are the advantages and disadvantages of using solar-powered banks of photovoltaic cells to produce electricity for cities and industry?
29. How does a photovoltaic cell produce electricity?
30. As a project, update the energy situation in the United States and the world by consulting the most recent edition of the *Annual Energy Review* (published by the Department of Energy, Energy Information Administration), a copy of which is probably in your library.
31. Tritium ($^3_1$H) has a half-life of 12.26 years. Proponents of nuclear disarmament have argued that tritium production for nuclear weapons should cease. Explain why.

STABLE

ESSAY by ROALD HOFFMANN

Words are our friends, and sometimes words are our enemies. We often think in science that words are just a handy way of describing some inner truth that is best expressed by a mathematical equation. Oh, the words matter, we say, but they are not essential for science. We might admit that a poem isn't really translatable, but we argue that the directions for a chemical synthesis could be in Japanese or Arabic or English — if the description is detailed enough, the same molecule would come out of the pot in any laboratory in the world.

Yet words are all we have, and all our precious ideas must be described with them. The meanings of those words are shaped, in ways we may not recognize, by our experience. To people with different backgrounds, the same words often have subtly different shadings.

I was led to reflect on the variation of meanings by the reaction of a friend, a physicist, to my use of the word "stable." Speaking of a form of carbon that has not yet been made, I said that it would be unstable with respect to diamond or graphite. Still, I thought it could be made. My friend said, "Why bother thinking about it at all, if it's unstable?" I said, "Why not?" and there a friendly argument began. Later I pondered why the simple English word "stable" has different meanings for a physicist and a chemist.

First, a little background. Diamond and graphite are the best known forms (allotropes) of pure carbon. A couple of rare allotropes — ones that are related to the diamond and graphite structures — have been prepared, and some others have been predicted but not proved to exist.

One day, I was trying to think up some alternatives to diamond and graphite. Why? Well, it was fun. Also, people have been squeezing nonmetallic elements, trying to make them metallic in order to learn more about the properties of solids. Could there be carbon structures that fill space more densely than in diamond and graphite, yet contain bonds between the atoms in specific directions? If there are, then applying pressure to one of the known allotropes might be a way to make a new form. To sum up a long story, there are *many* possible carbon structures, but we haven't yet found one that is denser than diamond.

Peter Bird and I thought up an allotrope with a density between those of diamond and graphite. It is illustrated in 1. The structure fills space with trigonal carbon atoms (each atom has three equally spaced bonds). Remarkably, according to calculations Tim Hughbanks did, it should be metallic. If only one knew how to make it!

The calculations we were able to make showed that the structure is *unstable* relative to graphite by a whopping 71 kilojoules per mole of carbon atoms. This is what made my physicist friend react when I suggested that this allotrope could exist. He assumed that the structure would collapse as soon as it is made to re-form graphite. But the instability didn't bother me at all.

Why the different reactions? Because the common words "stable" and "unstable" had different meanings for the two of us!

The complete definition of stability is based on a combination of **thermodynamics** (the measurement of energy relationships) and **kinetics** (the study of the speeds of processes by which substances change). In chemistry we distinguish between thermodynamic stability and kinetic stability.

Suppose we have a molecule of substance A that is capable of changing into a molecule of another substance, B. The difference in thermodynamic stability between the two molecules is measured by a quantity called **free energy.** The molecule that has the lower free energy is more stable. The molecule with higher free energy will (theoretically, at least!) spontaneously change into the one with lower free energy.

But life is not so simple. Thermodynamics says what *must* happen, but not how fast it *will* happen. The change from molecule A (say, my metallic carbon structure) to molecule B (graphite) is not straightforward. Many strong bonds in A have to break, and others have to form in the patterns that define B. In almost every case, there are barriers to changes in molecular structure. The energy diagram is typically not that at the left in picture 2, but that at the right.

There's a hill in the way. If the molecule of A has too little energy to climb the left side of the hill, it can't fall into the valley on the right side to become B. This barrier makes A kinetically stable — that is, the change

1

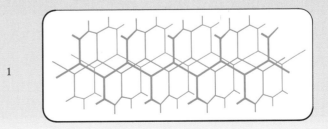

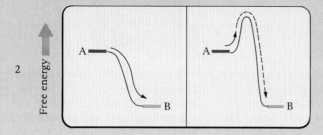

2

occurs slowly, if at all, at ordinary temperatures.

Will the reaction ever occur? Yes, if you wait long enough. How soon, and how many molecules of B form, depends on the size of the hill and on the temperature of the substance. Molecules don't sit still. They bounce around at great speed, colliding randomly with other molecules. It's a crowded dance floor there. The higher the temperature, the faster the molecules move. As a result of collisions, some of the molecules may collect enough energy to pass over the hill. Others don't. If the height of the climb on the left side of the hill is greater than about 30 kilocalories per mole, then at room temperature A will remain A. If you wait a thousand years, you might begin to see a little B.

A chemist would say that A is thermodynamically unstable but kinetically stable with respect to B. These ideas are quite familiar to chemists and physicists. So where is the problem? The difficulty is that our everyday speech is a kind of shorthand that leaves much unsaid. We say "stable" and not "thermodynamically unstable but kinetically stable." Some people would call this sloppy, and say that we should be more precise. I be-

lieve that we wouldn't be human (and therefore have the potential of doing great science) unless we were often imprecise in just this way.

Now we come to the heart of the argument. Into the word "stable" goes the history of who we are and what we have done. When a chemist says "stable," I think 90% of the meaning has to do with kinetic stability and 10% is based on thermodynamic stability. But a physicist, I would guess, applies nearly the opposite extreme: 90% thermodynamic stability and 10% kinetic stability.

Throughout the study of chemistry, kinetic stability is far more important than thermodynamic stability. For example, every organic molecule in the presence of air (the typical condition in the laboratory and in life) is thermodynamically unstable with respect to CO_2 and H_2O. Think of methane (CH_4, natural gas), which is so kinetically stable that it survived beneath the earth for millions of years. Every time you light a gas stove, though, you prove that methane is thermodynamically unstable! It takes the added energy of the match's flame to get the CH_4 and O_2 molecules over the hill so they can combine.

I suspect that thermodynamic stability is more important to the physicist for several reasons. A typical elementary physics course concentrates on motions that occur without such barriers or hindrances as friction. Motion in the presence of barriers is often too complicated to describe exactly, so these problems are rarely mentioned. Also, in thinking about changes in matter, physicists usually begin with motions governed by forces that are the same in all directions. They don't encounter processes whose energy diagrams contain "hills" until much later in their studies.

The subtle differences in emphasis given to various ideas are fixed early in scientific training, and they shape the informal vocabularies that we use. The early experiences matter; this, I think, is why "stable" means different things to chemists and to physicists.

Meanwhile, our metallic carbon allotrope is still waiting to be synthesized. I think it will be pretty stable — sorry, I meant enduring — when it is made. If it is made.

9

Acids and Bases— Chemical Opposites

Many common household items are acids or bases. (Marna G. Clarke.)

When we discover in nature a large group of useful, related compounds, we have simplified categories to study rather than many isolated compounds to characterize. When we learn that the two large groups of compounds react with each other and, as antagonists, ultimately can neutralize (or pacify) the effects each group had originally, we have an additional intriguing reason to find out about these substances. Two such groups of compounds are **acids** and **bases.**

Are acids and bases useful? We cannot live even a minute without them. In the following list are but a few of the routine encounters we have with acids and bases.

Household	Cooking—baking powder (base)
	baking soda (base)
	vinegar (acid)
	Lye and drain cleaners (bases)
	Citrus fruits such as lemons and oranges (acids)
	Acid skin and pimples (acids)
	Toilet bowl cleaners (bases)
Soil	Lime added to "sweeten the soil" (base)
Automobile	Battery acid
	If antifreeze is too acidic, radiators corrode.
Acid rain	
Acid mine drainage	
Streams	Fish die if acidity is too high.
Medicine	Antacids alleviate indigestion (bases).
Body functions	Acidity must be controlled carefully to preserve health and life; this is the every-minute application.
	Some acids and bases are very toxic to the human body.

Beyond our routine encounters with acids and bases, the control of acidity is necessary in many procedures for analyzing chemicals and in many industrial processes. Some of these applications will be discussed later in this textbook. The purpose of this chapter is to provide the fundamentals that will help you understand the world of acids and bases.

First of all, what is an acid? The word *acid* comes from the Latin *acidus,* meaning "sour" or "tart," since in water solutions, acids have a sour or tart taste. Acids in water react with metals such as zinc (Zn) and magnesium (Mg) to liberate hydrogen (H_2), react with bases to produce a salt and water, and change the color of litmus (a vegetable dye) from blue to red. These properties are produced by the release of hydrogen ions, H^+, in water (Fig. 9–1).

A common acidic substance known since antiquity is vinegar, the sour constituent from the fermentation of apple cider. The acid of vinegar is acetic acid, $HC_2H_3O_2$.

Citrus fruit juices turn universal indicator red, which shows citrus fruits to be acidic. Citrus fruits contain citric acid. (Charles Steele.)

Litmus is but one acid–base indicator. Another, phenolphthalein, is colorless in acid and pink in base.

Acid Base

Litmus

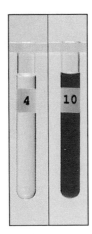

Acid Base

Phenolphthalein

Figure 9–1 Two common acid – base indicators. Litmus is red in acid; blue in base. Phenolphthalein is colorless in acid; pink in base. The number on the test tube is the pH of the solution. (See page 258. Adapted from photograph by Marna G. Clarke.)

Water solutions of bases, which are called alkaline or basic solutions, taste bitter, are slippery or soapy to the touch, change litmus from red to blue, and react with acids to form a salt and water.

Classically, a base is a substance capable of liberating hydroxide ions, OH^-, in water. Some of the most common bases of this type are the hydroxides of the alkali metals—NaOH and KOH—and of the alkaline earth metals—$Ca(OH)_2$ and $Mg(OH)_2$.

In the reaction of an acid with a classic base, the acid supplies hydrogen ions, H^+, which react with hydroxide ions, OH^-, from the base to form water, HOH.

> *A hydrogen ion is a proton. Because of its high charge-to-size ratio, a hydrogen ion in solution is bonded to one or more water molecules.*

$$H^+ \quad + \quad OH^- \quad \longrightarrow \quad HOH$$
(Supplied by Acid)   (Supplied by Base)

Since the common properties of acids and bases and the reactions between acids and bases occur in water solutions, we shall begin with a discussion of the general properties of water, or aqueous, solutions.

LIQUID SOLUTIONS

Recall from Chapter 2 that a solution is a homogeneous mixture. A liquid solution, then, is a uniform distribution of one substance in another, with the mixture having the properties of a liquid.

> *Solution: homogeneous (uniform) mixture of atoms, ions, or molecules.*

How many liquid solutions are familiar to you? How about sugar or salt dissolved in water, or oil paints dissolved in turpentine, or grease dissolved in gasoline? In each of these solutions, the substance present in the greater amount, the liquid, is the **solvent,** and the substance dissolved in the liquid, the one present in a smaller amount, is the **solute**(s). For example, in a glass of tea, water is the solvent, and sugar, lemon juice, and the tea itself are

Figure 9–2 A schematic illustration at the molecular level of sugar solution in water. Large circles represent the sugar molecules and the small circles water. The size of the container and the size of the particles are not to scale.

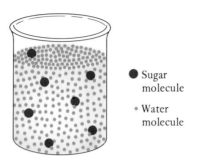

● Sugar
   molecule

• Water
   molecule

solutes. A theoretical concept of a solution of sugar in water pictures a collection of sugar molecules evenly dispersed among the water molecules (Fig. 9–2).

In this presentation of acid–base reactions, most of the chemistry studied will be in water or **aqueous** solutions, where water is the solvent. Generally, one of the species exchanging hydrogen ions is the solute. In addition to being the solvent, water molecules can and do exchange hydrogen ions with solute particles under suitable conditions.

Aqueous solutions are water solutions.

Figure 9–3 A simple test for an electrolytic solution. In order for the light bulb to burn (a), electricity must flow from one pole of the battery and return to the battery via the other pole. To complete the circuit, the solution must conduct electricity. A solution of table salt, sodium chloride, results in a glowing light bulb. Hence, sodium chloride is an electrolyte. In (b), the light bulb does not glow. Hence, table sugar is a nonelectrolyte. In (c), it is evident that the solvent, water, does not qualify as an electrolyte since it does not conduct electricity in this test.

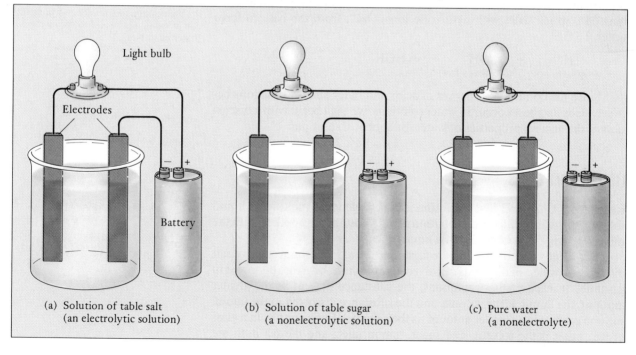

Light bulb

Electrodes

Battery

(a) Solution of table salt
      (an electrolytic solution)

(b) Solution of table sugar
      (a nonelectrolytic solution)

(c) Pure water
      (a nonelectrolyte)

Ionic Solutions (Electrolytes) and Molecular Solutions (Nonelectrolytes)

Solutes in aqueous solutions can be classified by their ability or inability to render the solution electrically conductive. When aqueous solutions are examined to see whether they conduct electricity, we find that solutions fall into one of two categories: **electrolytic** solutions, which conduct electricity, and **nonelectrolytic** solutions, which do not. A simple apparatus such as that shown in Figure 9–3 can be used to determine into which classification a given solution falls.

The conductance of electrolytic solutions is caused by the solute particles in such solutions being ions rather than molecules. Recall that sodium chloride crystals are composed of sodium ions (Na^+), which are positively charged, and chloride ions (Cl^-), which are negatively charged. When sodium chloride dissolves in water, **ionic dissociation** occurs (see Fig. 9–14). The resulting solution (Fig. 9–4a) contains positive sodium ions and negative chloride ions dispersed in water. Each ion is surrounded and insulated by water molecules. This arrangement is represented by Na^+ (*aq*), where *aq* stands for aqueous. Of course, the solution as a whole is neutral, since the total numbers of positive and negative charges are equal, thereby canceling each other.

$$Na^+Cl^- \xrightarrow{\text{water}} Na^+_{(aq)} + Cl^-_{(aq)}$$
$$\text{(Solid)} \qquad \text{(Aqueous)} \quad \text{(Aqueous)}$$
$$\text{Sodium} \quad \text{Chloride}$$
$$\text{Ion} \qquad \text{Ion}$$

The random motions of the sodium and chloride ions are not completely independent. The charges on the particles prevent all of the sodium ions from going spontaneously to one side of the container while all of the chloride ions are going to the other side. However, a net motion of ions

Solute particles may be ions or molecules.

Ionic dissociation is the separation of ions of a solute when the substance is dissolved.

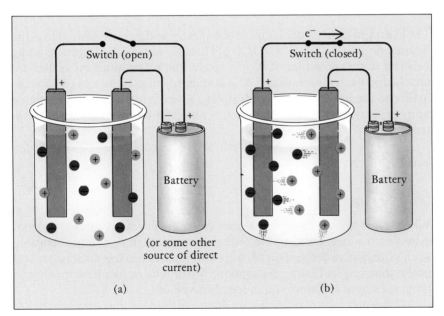

Figure 9–4 Conductance of electricity by ionic solution. (a) The hydrated ions are randomly distributed throughout the salt solution; the net charge is zero. (b) The negative electrode attracts positive ions; the positive electrode attracts negative ions.

Ions migrate toward oppositely charged electrodes in an electric field.

occurs when charged electrodes are placed in an aqueous solution of an electrolyte (Fig. 9–4b). The electric current is carried through the solution by the movement of ions to oppositely charged electrodes. The positive ions move toward the negative electrode; the negative ions move toward the positive electrode. At the electrodes, electrons are interchanged between the ions and the electrodes to complete the circuit.

Nonelectrolytic solutions are composed of solute molecules dispersed throughout solvent molecules, both of which are insensitive to negatively and positively charged electrodes unless the voltage is so great that molecules are changed into ions.

When a molecular solute dissolves in water to produce ions, the process is called ionization.

Sometimes ionic solutions arise when a molecular substance dissolves in water. For example, hydrogen chloride, HCl, is a gas composed of covalent, diatomic molecules. When hydrogen chloride dissolves in water an **ionization** reaction occurs, producing ions from molecules. The resulting solution is composed of hydrogen ions and chloride ions dispersed among the water molecules; consequently the solution conducts electricity and hydrogen chloride in water is properly termed an electrolyte.

$$\underset{\text{Molecule}}{\text{HCl}} \xrightarrow{\text{water}} \underset{\text{Ions}}{H^+_{(aq)} + Cl^-_{(aq)}}$$

The hydrogen ion in aqueous systems does not exist independently. Recall that water molecules are polar. A free proton (isolated hydrogen ion) could not exist in such a medium; the proton becomes attached to the negative end of one of the water dipoles. In fact, the attraction of water dipoles for the polar HCl molecule causes HCl to ionize in the first place.

$$H^+ + H\!:\!\overset{\cdot\cdot}{\underset{H}{O}}\!: \longrightarrow \left[H\!:\!\overset{\cdot\cdot}{\underset{H}{O}}\!:\!H \right]^+$$

Hydronium
Ion

Examples of hydrated ions are shown in Figure 9–14.

H_3O^+ is the hydronium ion.

The hydrogen ion in water is said to be **hydrated** and is often referred to as the **hydronium** ion, H_3O^+ or $H^+(H_2O)$. Through normal hydrogen bonding between water molecules, it is very likely that other water molecules are attached to the hydronium ion. The best representation we can give for the hydrogen ion in water then is $H^+(H_2O)_n$, where n is a constantly changing number, perhaps averaging between 4 and 5 in dilute solutions at room temperature.

Aqueous solutions of many acids and bases conduct electricity readily and are, therefore, electrolytic.

Concentrations of Solutions

When sugar, sodium chloride, alcohol, or any other readily soluble material dissolves in water, we can have either a concentrated or a dilute solution. Such a qualitative description of concentration is much less satisfactory and useful than a quantitative description, which tells us just how much of a given substance is dissolved in a specified volume.

Molar concentration: number of moles of a substance per liter of solution.

Concentrations of solutions are often expressed in the number of *moles* of solute *per liter* of solution. **Molar** and **molarity** are used to denote this

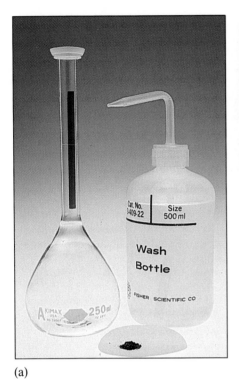

(a)

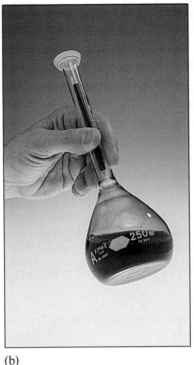

(b)

(c)

Figure 9–5   Preparation of an aqueous solution of known concentration. (a) Have available a volumetric flask with a carefully marked volume, distilled (or deionized) water, and a carefully weighed amount of solid. (b) All of the weighed solid is added to the flask along with some water. The flask is shaken until all of the solid is dissolved. (c) The flask is filled to the mark with distilled (or deionized) water. The flask must be shaken until a homogeneous solution is obtained. (Charles D. Winters.)

concentration unit. For example, 1 L of a 1-molar solution contains 1 mol of solute. If a solution has a molarity of six, the solution has 6 mol of solute dissolved in 1 L of solution. It is convenient to know the molarity of solutions when different solutions are mixed for a reaction and when we want to add a known number of moles of each reactant.

Solutions of known concentration are prepared using volumetric flasks. These are glass vessels with stems marked precisely to indicate specific volumes, such as 1.000 L. The procedure involves the steps shown in Figure 9–5.

To show how concentrations are calculated, let us consider a case where a 25-g sample of NaCl is weighed carefully, then transferred to a 1-L volumetric flask and dissolved in water. The next step is to add water to the flask until the solution has a total volume of 1 L. In order to determine the concentration of such a solution, we need to know the number of moles of NaCl present. The formula weight of NaCl is 23.0 + 35.5, or 58.5. Thus, if we have 58.5 g of NaCl, we have 1 mol. Since we have 25 g, the moles in one liter are:

$$\text{?mole NaCl} = 25 \text{ g NaCl} = \frac{\text{mole NaCL}}{58.5 \text{ g NaCl}}$$

$$= 0.43 \text{ mole NaCl}$$

We usually indicate this as 0.43 M NaCl, where M stands for moles of solute per liter of solution and is read as "molar."

Molarity = M = moles of solute/liter of solution

THE MOLE

The mole concept is the basis for quantitative chemistry. It also provides the foundation for some highly significant industrial processes, like the preparation of intravenous (IV) solutions so important to health.

The Baxter-Travenol Corporation produces very large amounts of IV solutions for hospital use. This company as well as the Food and Drug Administration recognize that the molecular concentrations of sodium chloride and glucose (a simple sugar used as a nutrient) are critical because improper concentrations can result in cell destruction.

The solutions are initially prepared in approximately correct concentrations in very large quantities by mixing materials of known weight with known volumes of distilled water. Once mixed, a key process is then the certification of the concentrations. Here the mole concept enters the picture.

An exact amount of solution is withdrawn from the batch and a titration method for determining chloride is used. A standard solution of silver nitrate, 0.1 M is carefully withdrawn from a buret until all the chloride from the sodium chloride is precipitated as silver chloride. At the end of the reaction, a special indicator that reacts with the first drop of excess silver nitrate produces a salmon-pink color. From the volume of silver nitrate we know how many moles of silver nitrate were consumed because one mole of silver nitrate reacts with one mole of sodium chloride in the reaction. Therefore, we can immediately determine the number of moles of sodium chloride and thus its concentration in the IV solution.
The World of Chemistry (Program 11) "The Mole."

Intravenous, or IV, solutions must be prepared very carefully and exactly to maintain the stability of blood and other body fluids. (Charles Steele.)

Suppose we have 85.5 g of sucrose (table sugar: $C_{12}H_{22}O_{11}$, molecular weight 342) dissolved in a volume of 500 mL. What is the concentration of sugar in this solution?

$$?M = \frac{?\text{moles sucrose}}{\text{liter solution}}$$

$$= \frac{85.5 \text{ g sucrose}}{500 \text{ mL solution}} \times \frac{\text{mole sucrose}}{342 \text{ g sucrose}} \times \frac{1000 \text{ mL}}{L}$$

$$= 0.500 \text{ M}$$

When a more dilute solution is needed, how is the solution prepared? For example, hydrochloric acid is sold commercially as a 12-molar (12-M) solution. If you wanted a 1-M solution of the acid, this solution needs to be one-twelfth as concentrated as the commercial product. One volume of the concentrated acid (a cup or a milliliter or any volume will do) diluted with 11 (not 12) volumes of water will result in the acid being in a volume 12 times larger than the acid was prior to dilution. Hence, the diluted solution is one-twelfth as concentrated as the original solution, and the diluted solution is now 1-M hydrochloric acid.

Commercially available sulfuric acid is 18 M. If you wanted a 4-M solution of the acid, you would take 4 volumes of the concentrated acid and mix this with 14 volumes of water. If you had a 14-M solution and wanted a

5-M solution, you would measure 5 volumes of the concentrated solution and mix it into 9 volumes of water.

The method described for dilution is an approximation—but usually a very good one. Since some ions, such as those with high charge or small size, are highly hydrated, water added for dilution does not increase the volume as much as would be predicted. Hydration restricts the motion of water molecules and thereby diminishes the effective volume the water molecules would have. The more accurate way of diluting is to add enough water to have the total volume correct. Using the last example to illustrate, 5 volumes of concentrated solution would be diluted to 14 volumes of total solution. You may have to add slightly more than 9 volumes of water.

With this knowledge about solutions and their concentrations in mind, let us look at acid–base reactions in solution.

Because of the heat generated when some acids are mixed with water, acids should be added to water to distribute the heat better. This is particularly important when mixing sulfuric acid with water.

AN EXPANDED LOOK AT THE CONCEPTS OF ACIDS AND BASES

In 1923, J. N. Brønsted and T. M. Lowry defined acids and bases as they are generally used by chemists today.

> **Brønsted-Lowry acid: a chemical species that can *donate* hydrogen ions (also called protons of H$^+$ ions) is an acid.**
> **Brønsted-Lowry base: a chemical species that can *accept* hydrogen ions is a base.**

To illustrate these definitions, we again consider the reaction between gaseous hydrogen chloride (HCl) and water:

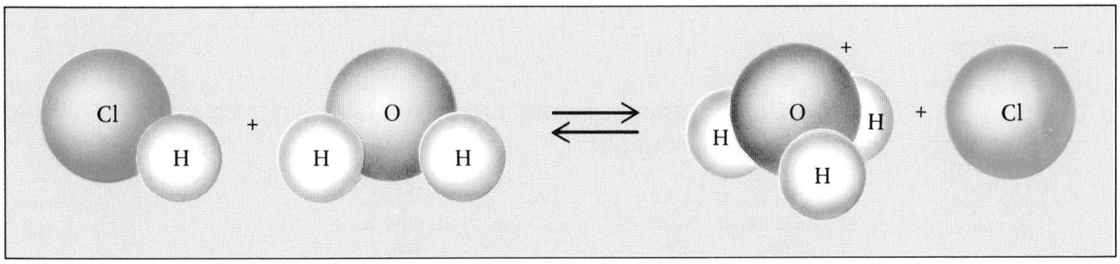

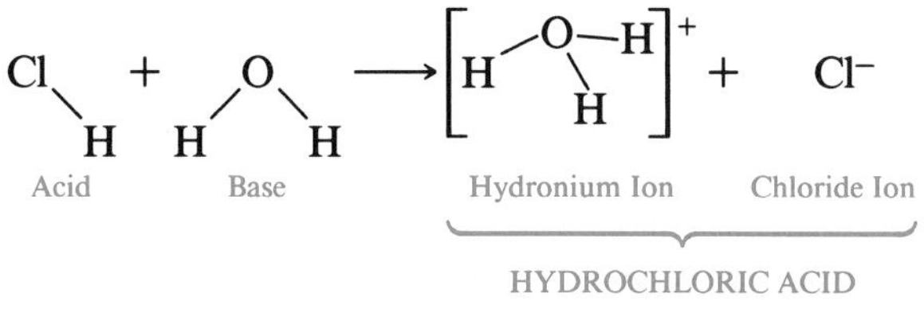

Examination of the preceding reaction shows that the HCl molecule has donated a hydrogen ion (H$^+$) to the water molecule. This transfer of a hydrogen ion is understandable when electronegativity differences between the bonded atoms are considered. First, polar bonds are formed since Cl is more electronegative than H, and O is more electronegative than hydrogen.

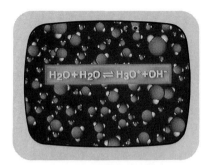

Figure 9–6   Neutral molecules of water are in equilibrium with relatively small amounts of hydronium ions (H_3O^+) and hydroxide (OH^-) ions. (*The World of Chemistry,* Program 16 "The Proton in Chemistry.")

Reaction is:
$Na_2O + H_2O \rightarrow 2\ Na^+ + 2\ OH^-$

The electrical poles formed by the shift of electron pairs toward the more electronegative atom are shown in the following reaction by a delta positive (δ^+) for the positive pole and delta negative (δ^-) for the negative pole.

$$H\!:\!\overset{\delta^-}{\underset{\underset{H}{\delta^+}}{\ddot{O}}}: + \overset{\delta^+}{H}\!:\!\overset{\delta^-}{\ddot{\underset{..}{Cl}}}: \longrightarrow \left[H\!:\!\ddot{O}\!:\!H \atop \ddot{H}\right]^+ + :\!\ddot{\underset{..}{Cl}}\!:^-$$

Base   Acid

Why are hydrogen ions transferred from HCl to water rather than from water to HCl? Experimentally, water is more polar than HCl. Part of water's greater polarity is due to oxygen having greater electronegativity than chlorine. However, polarity is determined by other factors such as the arrangement of atoms in the molecule. Whatever the causes, water does win the battle for the hydrogen ions, and HCl relinquishes its hydrogen ions to the more polar water.

The reaction of HCl with water is practically complete. Almost all of the HCl is converted to H_3O^+ and Cl^-. A concentrated (about 12-M) solution of hydrogen chloride in water is mostly a solution of hydronium (H_3O^+) ions, chloride (Cl^-) ions, and water molecules, with relatively few dissolved HCl molecules, which give the concentrated solution its characteristic odor.

If the ionic solid sodium oxide (Na_2O) is dissolved in water, a vigorous reaction produces a solution containing sodium ions (Na^+) and hydroxide ions (OH^-). In this process, the oxide ion (O^{2-}) reacts with a polar water molecule to form the hydroxide ion. In this, as in other such aqueous reactions, it is understood that the ions are hydrated (i.e., water molecules are bonded to the ions on a transitory basis).

$$:\!\overset{..}{\underset{..}{O}}\!:^{2-} + H\!:\!\overset{..}{\underset{\underset{H}{..}}{O}}: \longrightarrow :\!\overset{..}{\underset{..}{O}}\!:\!H^- + :\!\overset{..}{\underset{..}{O}}\!:\!H^-$$

Base   Acid   Hydroxide Ions

There are many other bases that take a hydrogen ion from a water molecule in this way. For example:

$$\underset{\substack{\text{Sulfide Ion}\\\text{Base}}}{S^{2-}} + \underset{\text{Acid}}{H_2O} \longrightarrow \underset{\substack{\text{Hydrogen}\\\text{Sulfide Ion}}}{HS^-} + \underset{\substack{\text{Hydroxide}\\\text{Ion}}}{OH^-}$$

$$\underset{\substack{\text{Cyanide Ion}\\\text{Base}}}{CN^-} + \underset{\text{Acid}}{H_2O} \longrightarrow \underset{\substack{\text{Hydrogen}\\\text{Cyanide}}}{HCN} + \underset{\substack{\text{Hydroxide}\\\text{Ion}}}{OH^-}$$

Amphiprotic species can be either an acid or a base.

According to the Brønsted-Lowry definition, water acts as an acid in these reactions and donates a hydrogen ion to the other molecule or ion, which acts as a base. A species such as water that can either donate or accept hydrogen ions is called **amphiprotic.** The existence of amphiprotic species implies that acid–base reactions possess a reciprocal nature; an acid and a base react to form another acid (to which a hydrogen ion has just been added) and another base (from which a hydrogen ion has just been removed).

Also, one water molecule can transfer a hydrogen ion to another water molecule (Fig. 9–6). When equilibrium is reached, the reaction has occurred to only a very small extent, as indicated by arrows of unequal length. In neutral water at 25°C, the concentrations of H_3O^+ and OH^- are the same,

0.0000001 M (or 10^{-7} M). Since H_3O^+ and OH^- are produced in equal amounts when only water is present, pure water is neither acidic nor basic, but is described as *neutral*.

Because water is the most commonly used solvent, it is also the most usual reference compound for acid–base reactions. A chemical species in water solution is commonly spoken of as an acid if it donates hydrogen ions to water and increases the concentration of H_3O^+ or $H^+_{(aq)}$. Similarly, a base in water solution is commonly described as a compound whose addition to water increases the concentration of OH^-. Since water is not the only possible solvent, these concepts are too narrow for general scientific use; they have been extended by the Brønsted and Lowry definitions, which focus on the essential feature of such acid–base behavior—that is, the donation or acceptance of a hydrogen ion (H^+) in a reaction.

The Happy Medium Between Acids and Bases — Neutralization

When acids react with bases, the properties of both species disappear. The process involved is called **neutralization.** To get a more precise picture of acid–base neutralization reactions, we shall consider what happens when a solution of hydrochloric acid is mixed with a solution of sodium hydroxide. The hydrochloric acid contains H_3O^+ and Cl^- ions; the sodium hydroxide solution contains Na^+ and OH^- ions. When these two solutions are mixed, a reaction occurs between H_3O^+ and OH^-.

$$Na^+ + Cl^- + \underset{\text{Acid}}{H_3O^+} + \underset{\text{Base}}{OH^-} \longrightarrow H_2O + H_2O + Na^+ + Cl^-$$

If we have an equal number of H_3O^+ and OH^- ions, they will react to produce a neutral solution, with the hydronium ions (H_3O^+) donating their hydrogen ions to the hydroxide ions (OH^-), forming molecules of water. Such reactions are called neutralization reactions because the acids and bases neutralize each other's properties. Often, the products are water and a salt. Sometimes only a salt is made, as when NH_3 reacts with HCl to form NH_4Cl (Fig. 9–7). If we have more H_3O^+ ions than OH^- ions, the extra H_3O^+ will make the resulting solution **acidic.** If we have more OH^- ions than H_3O^+ ions, only a fraction of the OH^- ions will be neutralized, and the extra OH^- ions will make the resulting solution **basic.**

Figure 9–7   In a neutralization reaction, gaseous HCl and gaseous NH_3 form a white cloud, the salt NH_4Cl. (*The World of Chemistry*, Program 16, "The Proton in Chemistry.")

KINSHIP OF SOME ACIDS AND BASES

When an acid ionizes, a hydronium ion is produced along with a species called the **conjugate base** of that acid. For example:

$$\underset{\substack{\text{Nitric} \\ \text{Acid}}}{HNO_3} + \underset{\text{Water}}{H_2O} \longrightarrow \underset{\substack{\text{Hydronium} \\ \text{Ion}}}{H_3O^+} + \underset{\substack{\text{Nitrate Ion, the} \\ \text{Conjugate Base of} \\ \text{the Acid } HNO_3}}{NO_3^-}$$

In the same manner we speak of nitric acid, HNO_3, as being the **conjugate acid** of the nitrate ion, NO_3^-, a base. Conjugate acids and bases differ by one hydrogen ion. The conjugate acid has one more hydrogen ion than its conjugate base.

SELF-TEST 9-A

1. When sugar dissolves in water, the resulting solution does not conduct electricity. A sugar solution is therefore a(n) _____ .
2. A chemical species that can accept a hydrogen ion is a(n) _____ . A chemical species that can donate a hydrogen ion is a(n) _____ .
3. A compound HA is found to undergo a reaction forming a product H_2A^+. Therefore HA is a(n) ( ) acid ( ) base. If compound HA reacted to form A^-, then HA would be a(n) ( ) acid ( ) base.
4. If the aforementioned compound HA undergoes both reactions described, then HA is termed _____ .
5. A solution that contains equal concentrations of OH^- and H^+ ions is termed _____ .
6. The word *aqueous* means _____ .
7. A 1-M solution contains how many moles in 2 L of the solution? _____
8. The molecular weight of sulfuric acid (H_2SO_4) is 98. If 49 g of this acid are dissolved in 1 L of water, the molarity is _____ .
9. In a neutralization reaction, a(n) _____ reacts with a(n) _____ .
10. In order to make a 5-M solution from a 17-M solution, you would mix five volumes of the concentrated solution with _____ volumes of water.

THE STRENGTHS OF ACIDS AND BASES

Strong acids relinquish practically all of their hydrogen ions to water. For example, strong acids such as sulfuric acid (H_2SO_4), nitric acid (HNO_3), and hydrochloric acid (HCl) release their hydrogen ions to form H_3O^+ and negative ions such as HSO_4^-, SO_4^{2-}, NO_3^-, and Cl^-, respectively. Consequently, the following reactions would be converted almost completely to products.

$$
\begin{array}{ccccc}
\textit{REACTANTS} & & & \textit{PRODUCTS} & \\
HCl & + H_2O & \longrightarrow & H_3O^+ + & Cl^- \\
HNO_3 & + H_2O & \longrightarrow & H_3O^+ + & NO_3^- \\
H_2SO_4 & + H_2O & \longrightarrow & H_3O^+ + & HSO_4^- \\
& & & \text{Conjugate} & \\
\text{Acid} & \text{Base} & & \text{Acid} & \text{Base}
\end{array}
$$

Since many ions are present, the solutions conduct electricity well; HCl, HNO_3, and H_2SO_4 are **strong electrolytes** (Fig. 9-8).

Not all acids lose hydrogen ions as readily to water as do nitric acid and hydrochloric acid. Some negative ions are capable of competing with water for the hydrogen ion being exchanged. The result of this competition is the establishment of an **equilibrium** (or balance) between neutral acid molecules and hydronium ions in water solution.

Strong electrolytes dissociate completely in water.

Under equilibrium conditions, the concentrations of the species in solution remain unchanged even though reactions in the forward and reverse directions continue.

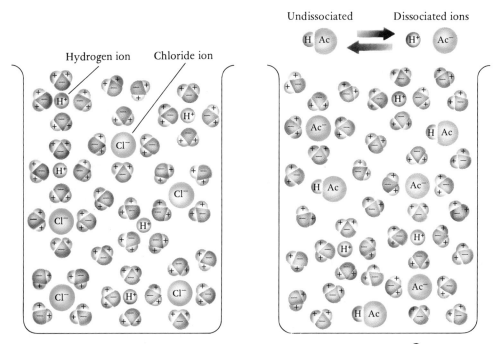

Figure 9–8 An illustration of what it may be like in an aqueous (water = ⊕) solution of a strong acid (HCl) and an aqueous solution of a weak acid (acetic acid, HAc). The strong acid is practically all ions; very few molecules. The weak acid is mostly molecules with only a few ions.

Acetic acid, found in vinegar, is a weak acid; that is, its conjugate base, acetate ion, competes well with water for a hydrogen ion. The molecular structure of acetic acid ($HC_2H_3O_2$) is

$$
\begin{array}{c}
\text{H} \qquad \text{O} \\
| \qquad\quad \parallel \\
\text{H}-\text{C}-\text{C} \\
| \qquad\quad \diagdown \\
\text{H} \qquad \text{O}-\text{H} \\
\end{array}
\quad \text{Acidic Hydrogen Atom}
$$

The hydrogen atom bonded to a highly electronegative oxygen is the only hydrogen positive enough to be taken by a base in water solution; for that reason, the most positive hydrogen is designated an acidic hydrogen in the preceding formula. The other hydrogen atoms in the acetic acid molecule are not acidic because the C—H bonds are almost nonpolar since the electronegativities of carbon and hydrogen are nearly the same value.

When acetic acid is dissolved in water, some ions are produced. However, most of the acetic acid molecules do not relinquish hydrogen ions to water molecules. The result is a mixture containing a few H_3O^+ and $C_2H_2O_2^-$ ions and many $HC_2H_3O_2$ molecules. The reaction is:

$$HC_2H_3O_2 + H_2O \rightleftharpoons H_3O^+ + C_2H_3O_2^-$$

Acetic Acid   Water   Hydronium   Acetate
Ion   Ion

TABLE 9–1 Relative Strengths of Conjugate Acid–Base Pairs

Acid		Base

The relatively few ions in an acetic acid solution do not conduct electricity very effectively; consequently, acetic acid is a **weak electrolyte** (Fig. 9–8). (A dilute solution of acetic acid would barely conduct in the apparatus shown in Fig. 9–3.)

Another way of looking at this reaction is to realize that two bases are in the mixture: the water molecule, H_2O, and the acetate ion, $C_2H_3O_2^-$. Since the reverse reaction dominates, the acetate ion must be a stronger base than the water molecule.

The same kind of considerations can be made for other bases. Ammonia dissolved in water is a weak base. The reaction between ammonia and water produces relatively few ions; ammonia remains mostly in the molecular form:

$$NH_3 + H_2O \rightleftharpoons NH_4^+ + OH^-$$

Ammonia   Water   Ammonium   Hydroxide
Ion   Ion

Consequently, the relatively few ions present do not conduct electricity well, and ammonia may thus be called a weak electrolyte.

Table 9–1 gives some common acids and bases ranked according to their relative strengths.

The pH Scale of Acidity and Alkalinity

Pure water is a very weak conductor of electricity. Distilled water will not conduct electricity in the conductivity apparatus shown in Figure 9–3; however, very sensitive equipment shows that water does conduct electricity slightly. Thus, water must be slightly ionized.

$$H_2O + H_2O \rightleftharpoons H_3O^+ + OH^-$$

If the forward and reverse arrows were drawn to scale in length, the reverse arrow would have to be 550,000,000 times longer than the forward arrow, since water is only 0.00000018% ionized at room temperature.

Ammonia is the number five commercial chemical in quantity produced, sodium hydroxide is number seven and nitric acid is number twelve. (See inside front cover.)

Purple cabbage boiled in water yields dyes that show a variety of colors over the pH range of 1 through 14. Shown (left to right) is purple cabbage juice in solutions of pH 1, 4, 7, 10, and 13. (Charles Steele.)

In pure water at room temperature, the concentration of the H_3O^+ ion is 0.0000001 mol/L (10^{-7} mol/L). The concentration of the OH^- ion is the same as the concentration of the H_3O^+ ion, since the two are produced in equal amounts as the water ionizes. When the concentrations of H_3O^+ and OH^- are equal, the solution is *neutral*. Chemical equilibrium is established when the product of the concentration of the hydronium ion and the concentration of the hydroxide ion is 1.00×10^{-14}.

$$[H_3O^+][OH^-] = (1.00 \times 10^{-7} \text{ mol/L})(1.00 \times 10^{-7} \text{ mol/L})$$
$$= 1.00 \times 10^{-14} \text{ mol}^2/\text{L}^2$$

If we add acid to water, the concentration of H_3O^+ will become greater than 1.00×10^{-7} mol/L; the concentration of the OH^- ion will have to go down to a smaller number so that the product of the two concentrations will still be 1.00×10^{-14}. One variable goes down as the other goes up such that their product is always the same constant.

Now look at two extreme cases:

1. What are the $[H_3O^+]$ and $[OH^-]$ concentrations in 0.1-M hydrochloric acid?

   Answer: The strong acid is 100% ionized, so the H_3O^+ concentration is the same as the acid, 0.1, or 10^{-1} M. The OH^- concentration is then the constant divided by the H_3O^+ concentration, or

$$\frac{1.00 \times 10^{-14}}{1.00 \times 10^{-1}} = 1.00 \times 10^{-13} \text{ M}$$

2. What are the $[H_3O^+]$ and $[OH^-]$ concentrations in 0.1-M NaOH?

   Answer: The OH^- concentration from the strong base is 0.1, or 10^{-1} mol/L. The H_3O^+ concentration is then the constant divided by the OH^- concentration, or

$$\frac{1.00 \times 10^{-14}}{1.00 \times 10^{-1}} = 1.00 \times 10^{-13} \text{ M}$$

In summary, the concentrations of the H_3O^+ ion in the three cases studied are:

0.1-M hydrochloric acid	1.00×10^{-1}
Pure water	1.00×10^{-7}
0.1-M sodium hydroxide	1.00×10^{-13}

Chemists, like others, look for concise expressions. Note that the exponent goes from -1 in a strong acid solution to -7 in a neutral solution and

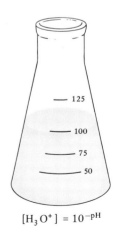

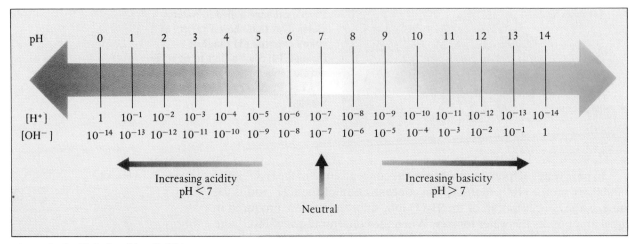

Figure 9–9   Relationship of pH to the concentration of hydrogen ions (H⁺ or H₃O⁺) and hydroxide ions (OH⁻) in water at 25° C.

then to -13 in a strong basic solution. The exponent tells the strength of the acidity or alkalinity. Since there is no need to distinguish between negative and positive (all the exponents are negative), we might as well communicate in the positive. Thus, the definition of pH was obtained:

$$\mathbf{pH} = -\mathbf{log}\ [\mathbf{H^+}]$$

The pH is the negative log (exponent of ten) of the hydrogen ion concentration.

Pure water has a pH of 7 and is a neutral solution. If the pH is below 7, the solution is acidic, and each drop of one pH unit represents a tenfold increase in acidity, or hydronium ion concentration. A pH above 7 is alkaline, with each unit of increase on the exponent scale representing a decrease

Figure 9–10   A universal indicator composed of several dyes at pHs 1 through 12.

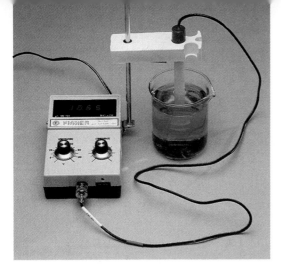

Figure 9–11 An electric pH meter detects the hydrogen ion concentration and expresses it as the negative exponent of 10. The positive exponent is the pH. The solution shown has a pH of 10.86, which means the solution is basic. (Marna G. Clarke).

in the hydronium ion concentration by a factor of one-tenth (Fig. 9–9). In summary:

if pH < 7.0,   solution is acidic
if pH = 7.0,   solution is neutral
if pH > 7.0,   solution is basic

The pH can be approximated by acid–base indicators (Fig. 9–10) or given more exactly by electronic pH meters (Fig. 9–11).

Consumers are frequently asked to deal with pH. Figure 9–12 gives the pH of some common materials with which you are probably familiar.

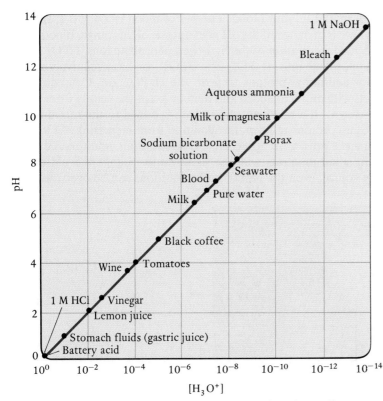

Hydrogen or hydronium ion concentration in moles per liter

Figure 9–12 A plot of pH versus hydrogen ion concentration $[H_3O^+]$. Note that the pH increases as the $[H_3O^+]$ decreases. The pH values of some common solutions are given for reference. (A solution in which $[H_3O^+] = 1$ M has a pH of 0, since $1 = 10^0$.)

Figure 9–13 Acid–base buffers resist a change in hydrogen ion concentration when either acid or base is added to a system. The three tubes on the left originally contained only water. Nothing was added to tube A, a few drops of a strong acid were added to tube B, and a few drops of strong base were added to tube C. Compare the colors of the solutions to the colors in Figure 9–10. The three tubes on the right contained originally a buffer solution at pH 7. To tube B was added a few drops of strong acid, and to tube C was added a few drops of strong base. The buffer solution is essentially unaffected by the addition of acid or base. (Marna G. Clarke.)

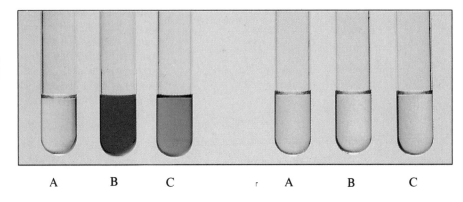

A   B   C   A   B   C

ACID–BASE BUFFERS

The general idea of a buffer is that of a shock absorber—something to absorb a disturbance while retaining the original conditions or structure. The control of pH involves maintaining a steady level of acidity even when sudden "shocks" of acid or base are added.

> **Buffer solutions have chemical species that can react with added acid or base and maintain a pH very close to the original value (Fig. 9–13).**

Buffers control pH.

The pH of a buffer is dependent on the chemicals used to make the buffer.

The control of pH is necessary in many industrial and natural processes. It is a critical matter in your blood. The pH of blood is 7.4 ± 0.1, slightly alkaline, and life is in danger if the pH goes outside of this range. Blood has a relatively high concentration of the hydrogen phosphate ions, HPO_4^{2-}, and dihydrogen phosphate ions, $H_2PO_4^-$. The HPO_4^{2-} ion has a relatively strong attraction for additional protons to form the $H_2PO_4^-$ ion. Consequently, HPO_4^{2-} reacts with any acid added and thereby keeps the acidity from going up (the pH down). If a relatively strong base is added to the mixture, the $H_2PO_4^-$ gives up protons to keep the alkalinity from increasing. As long as this buffer pair is present in appreciable concentration, additions of acids and bases in small amounts will not significantly change the pH of the solution. Of course, even the best of buffer solutions can be overwhelmed.

A buffer is made of a conjugate (kin to) pair of species. The species differ by only one proton.

Many consumer products are buffered. Aspirin and blood plasma are two examples. Capsules shown on the right are available for making a buffered solution of desired pH. (Marna G. Clarke.)

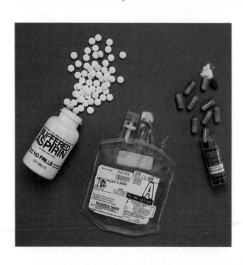

STOMACH ACIDITY

The pH of stomach fluids, even in the normal stomach is 1. The acid is hydrochloric acid. The stomach produces a small amount of acid all the time, but then that amount of acid may be stimulated by food. Even the sight and smell of appetizing food is enough to make the stomach produce more acid.

What actually happens when you eat a meal? Both hydrogen ions and chloride ions, maintaining an electrochemical balance, move through the stomach lining from the surrounding blood plasma. In the stomach the result is a highly acidic medium. That's what it takes to activate certain enzymes for the process of digestion. If the acidity of the stomach becomes excessive, problems can occur, problems that often need antacid solutions.

An expert on stomach acid is Dr. Paul Maton of the National Institutes of Health; he says that:

There are a variety of different compounds, basically bases, that can function as antacids. For example, sodium bicarbonate could be used as an antacid, or magnesium hydroxide or calcium carbonate.

All antacids neutralize acid. And if they're given in sufficient amounts, any antacid is as good as another at neutralizing acid.

One hears a lot of talk about hyperacidity—too much acid. In fact, there's very little or no evidence that people with ulcers actually produce more acid than many of the rest of us.

For most of those acid stomach discomforts, then, the tried and tested over-the-counter remedies do their job of neutralization well enough.

The World of Chemistry (Program 16) "The Proton in Chemistry."

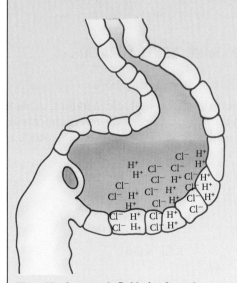

The pH of stomach fluids is about 1.

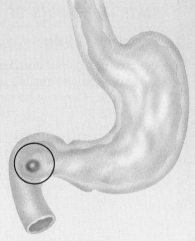

A duodenal ulcer.

TABLE 9–2 The Chemistry of Some Antacids

Compound	Reaction in Stomach	Examples of Commercial Products
Milk of magnesia: $Mg(OH)_2$ in water	$Mg(OH)_2 + 2\,H^+$ $\longrightarrow Mg^2 + 2\,H_2O$	Phillips Milk of Magnesia
Calcium carbonate: $CaCO_3$	$CaCO_3 + 2\,H^+$ $\longrightarrow Ca^{2+} + H_2O + CO_2$	Tums, Di-Gel
Sodium bicarbonate: $NaHCO_3$	$NaHCO_3 + H^+$ $\longrightarrow Na^+ + H_2O + CO_2$	Baking soda, Alka-Seltzer
Aluminum hydroxide: $Al(OH)_3$	$Al(OH)_3 + 3\,H^+$ $\longrightarrow Al^{3+} + 3\,H_2O$	Amphojel
Dihydroxyaluminum sodium carbonate: $NaAl(OH)_2CO_3$	$NaAl(OH)_2CO_3 + 4\,H^+$ $\longrightarrow Na^+ + Al^{3+} + 3\,H_2O + CO_2$	Rolaids

INDIGESTION: WHY REACH FOR AN ANTACID?

The contents of the stomach are highly acidic.

The walls of a human stomach contain thousands of cells that secrete hydrochloric acid, the main purposes of which are to suppress the growth of bacteria and to aid in the hydrolysis (digestion) of certain foodstuffs. Normally the stomach's inner lining is not harmed by the presence of hydrochloric acid, since the mucosa, the inner lining of the stomach, is replaced at the rate of about a half million cells per minute. However, when too much food is eaten, the stomach often responds with an outpouring of acid, which lowers the pH to a point at which discomfort is felt.

If the reduction of acidity is too great, the stomach responds by secreting an excess of acid. This is "acid rebound."

Antacids are basic compounds used to decrease the amount of hydrochloric acid in the stomach. The normal pH of the stomach ranges from 0.9 to 1.5. Some alkaline compounds used for antacid purposes and their modes of action are given in Table 9–2.

The fizz of antacids when expelled from the stomach relieves the pressure in the stomach. (Charles D. Winters.)

SALTS—PRODUCTS OF ACID–BASE NEUTRALIZATION

Preparation of Salts

The chemical compounds known as **salts** play a vital role in nature, in plant and animal growth and life, and in the manufacture of various chemicals for human use. They can be formed as the products of acid–base neutralizations, as in the following example:

$$(K^+ + OH^-) + (H_3O^+ + Cl^-) \longrightarrow \underbrace{K^+ + Cl^-}_{} + 2\,H_2O$$

Potassium Hydroxide in Water BASE   Hydrochloric Acid in Water ACID

Crystalize the Salt by Removal of Solvent, Water

$$KCl\ (solid)$$
SALT

Most salts contain ions held together by **ionic bonding** (see Chapter 5). Solid potassium chloride, for example, is composed of an equal number of K^+ ions and Cl^- ions arranged in definite positions with respect to one another in an ionic structure or lattice (see Fig. 5–6). Since the salt crystal Salts are ionic compounds. must be electrically neutral, it can have neither an excess nor a deficiency of positive or negative charge.

Let us imagine that we have at our disposal the ions in the following list, and let us see what salts could result.

Negative Ions	Positive Ions	
	Na^+ sodium	Ca^{2+} calcium
Cl^- chloride	NaCl sodium chloride	$CaCl_2$ calcium chloride
NO_3^- nitrate	$NaNO_3$ sodium nitrate	$Ca(NO_3)_2$ calcium nitrate
SO_4^{2-} sulfate	Na_2SO_4 sodium sulfate	$CaSO_4$ calcium sulfate
$C_2H_3O_2^-$ acetate	$NaC_2H_3O_2$ sodium acetate	$Ca(C_2H_3O_2)_2$ calcium acetate

In the examples just given, notice that in order to attain an electrically neutral lattice, it is necessary to balance the charges of the ions. A sodium (Na^+) ion requires just one chloride ion (Cl^-), and the NaCl lattice contains an equal number of Na^+ and Cl^- ions. A sulfate ion (SO_4^{2-}) with two negative charges must have its negative charge balanced by two positive charges. This may be done by using two Na^+ ions

$$2\ Na^+ + SO_4^{2-} \longrightarrow Na_2SO_4$$

In the formula of a salt, the positive and negative charges are equal.

or one Ca^{2+} ion

$$Ca^{2+} + SO_4^{2-} \longrightarrow CaSO_4$$

It is possible to form many solid salts by mixing water solutions of different soluble salts with each other. For example, both lead acetate and sodium chloride are soluble in water. If we prepare solutions of these salts and then mix the solutions, we find the insoluble salt, lead(II) chloride, precipitates from the mixture and may be removed by filtration.

$$\underbrace{2\ Na^+ + 2\ Cl^-}_{\substack{\text{Sodium Chloride} \\ \text{in Solution}}} + \underbrace{Pb^{2+} + 2\ C_2H_3O_2^-}_{\substack{\text{Lead(II) Acetate in} \\ \text{Solution}}} \longrightarrow \underset{\substack{\text{Solid} \\ \text{Lead(II)} \\ \text{Chloride}}}{PbCl_2} + \underbrace{2\ Na^+ + 2\ C_2H_3O_2^-}_{\substack{\text{Sodium Acetate} \\ \text{in Solution}}}$$

Sodium acetate may be recovered by evaporating the water.

This reaction illustrates an important principle in solubility. If the component ions of a compound of low solubility are mixed in solution in great enough concentrations, the compound containing those ions will precipitate from solution.

Salts in Solution

An important property of many salts is their solubility in suitable solvents. The amount of a salt that will dissolve in a given quantity of solvent tells us the salt's solubility in that solvent. The preparation of lead(II) chloride just Although some salts are very soluble in water, others are quite insoluble. Salts are found with a wide range of water solubilities.

Figure 9–14 Dissolution of sodium chloride in water. (a) Geometry of the polar water molecule. (b) Solvation of sodium and chloride ions due to interaction (bonding) between these ions and water molecules. (c) Dissolution occurs as collisions between water molecules and crystal ions result in the removal of the crystal ion. In the process the ion becomes completely solvated.

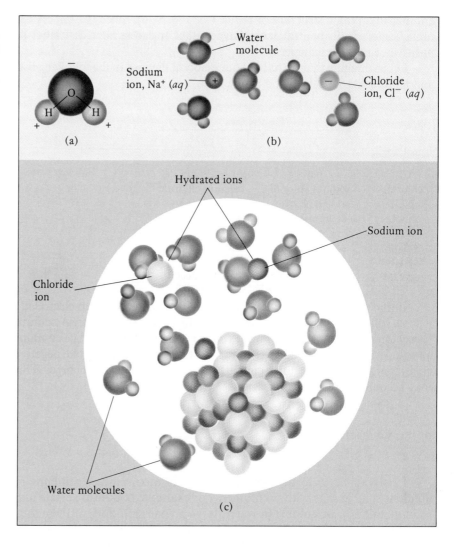

The two most abundant ions in ocean water are Cl^- and Na^+ ions, sufficient to recover about 27 g of NaCl per kilogram of sea water. To put it another way, there are about 128 million tons of NaCl per cubic mile of sea water.

shown was made possible by the differences in solubilities of different salts in the same solvent. As a result of these differences in solubilities, we can make roads out of calcium carbonate (limestone), which is insoluble in water, but not calcium chloride, which is water-soluble.

Consider what happens at the ionic level when a sodium chloride crystal is placed in contact with water. We know that sodium chloride is soluble in water. This means that most, if not all, of the attractive forces between the ions in the crystal lattice are somehow overcome in the solution process.

The surface of the salt crystal appears calm when the crystal is placed in water, but on the ionic level there is a great deal of agitation. Water molecules have sufficient polarity to interact strongly with the ions and bond with them. Once this occurs, the ion is less strongly bound in the lattice and so can be removed from the crystal. The crystal lattice now has a gap in it where the ion was removed. Ions that are bonded by solvent molecules are termed solvated ions, and the process of ion–solvent interaction is called **solvation.** But just how are these ions solvated? What causes this solvation?

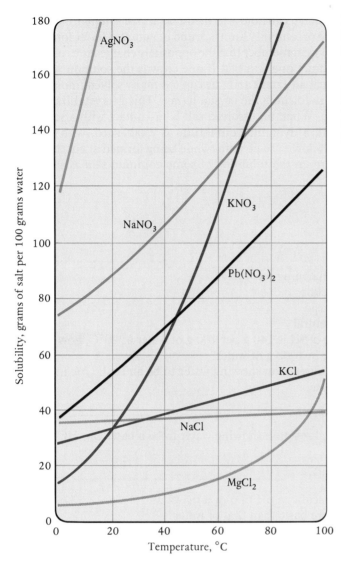

Figure 9–15 The effect of temperature on the solubility of some common salts in water. $AgNO_3$ is silver nitrate; $NaNO_3$ is sodium nitrate; KNO_3 is potassium nitrate; $Pb(NO_3)_2$ is lead nitrate; KCl is potassium chloride; NaCl is sodium chloride; and $MgCl_2$ is magnesium chloride.

The answers lie in the structure of the water molecule and its ability to interact with ions. As we saw in Chapter 5, water is a **polar** compound. The negative ends of its molecules are attracted to positive ions, and the positive ends are attracted to negative ions. As a result, several water molecules will interact with each ion. Figure 9–14 shows several water molecules solvating a Na^+ ion. The positive end of the water molecule will tend to interact with a negative ion; this is shown for the Cl^- ion in Figure 9–14.

Every salt has what may be termed a solubility limit for a given solvent: at a given temperature, a certain number of grams of salt, and *no more,** will dissolve in a certain quantity of solvent (Fig. 9–15). A solution that contains

Because of its polar nature, the water molecule is ideally suited to interact with ions.

---

* Supersaturated solutions can be formed under special conditions but are not stable in the presence of the solid solute. The presence of the solid solute causes the excess dissolved solute to crystallize, and a saturated solution is formed.

all the dissolved salt that it can hold is termed a **saturated solution.** One might ask just why this type of solubility limit is found in nature. The reason for this can be understood if we remember that the oppositely charged ions of the salt solution actually attract one another. If one crowds the solution with solvated ions to too great an extent and "ties up too many solvent molecules," the ions will begin re-forming the crystal lattice. This is **crystallization,** or solvation in reverse. When undissolved salt is in contact with a saturated solution of that salt, a dynamic equilibrium is established, with the salt crystal being broken down at one point while being formed at another. The effect of temperature on the solubility of some common salts is shown in Figure 9–15.

Generally, for a given solvent, a salt will dissolve to a greater extent in the hot solvent than in the cold solvent.

SELF-TEST 9–B

1. Complete the matching set.
   pH 10   **a.** acidic
   pH 7   **b.** basic
   pH 3   **c.** neutral
2. If the solubility of KI is 140 g per 100 g of water at 20°C, how much KI will dissolve in 1000 g of water?
3. Write a chemical reaction showing water to be an acid. Use ammonia (NH_3) as the base.

$$H_2O + NH_3 \longrightarrow \text{_____} + \text{_____}$$

4. Write a chemical reaction showing water to be a base. Use HCl as the acid.

$$H_2O + HCl \longrightarrow \text{_____} + \text{_____}$$

5. Which is more acidic, a pH of 6 or a pH of 2?
6. Is a pH of 8 more basic than that of water?
7. The pH of the blood is maintained at 7.4 by agents generally known as _____ .
8. Give examples of the following:
   a. A strong acid
   b. A weak acid
   c. A strong base
   d. A weak base
9. The pH of pure water at room temperature is _____ .
10. The hydronium ion concentration at a pH of 5 is how many times its value at a pH of 3?
11. In water, if the hydronium ion concentration goes up, the _____ ion concentration must go down such that the product of these two concentrations is a _____ value.
12. Two of the ions that are used to buffer the pH of your blood are: _____ and _____ .
13. High pH means (   ) high (   ) low hydrogen ion concentration.
14. Low pH means (   ) high (   ) low hydrogen ion concentration.

MATCHING SET

_____ 1. Conjugate base of H_2A
_____ 2. Conjugate acid of A^{2-}
_____ 3. M
_____ 4. pH of pure water
_____ 5. Solution
_____ 6. Strong acid
_____ 7. Alkaline solution
_____ 8. Weak conjugate acid
_____ 9. Acid definition
_____ 10. Base definition
_____ 11. Weak acid
_____ 12. Electrolyte
_____ 13. Buffer

a. 7
b. Molarity; moles of solute per liter of solution
c. Hydrogen ion donor
d. Maintains pH
e. Hydrogen ion acceptor
f. Homogeneous mixture of atoms, molecules, and/or ions
g. $HC_2H_3O_2$, acetic acid
h. Strong conjugate base
i. Causes solution to conduct electricity
j. NaOH in HOH
k. HA^-
l. H_2SO_4
m. A^{2-}
n. 3

QUESTIONS

1. Define acid–base reactions in terms of hydrogen ions.
2. Indicate the solute and solvent in (a) a cup of coffee, (b) a 5% solution of alcohol in water, (c) a 5% solution of water in alcohol, and (d) a solution of 50% alcohol and 50% water.
3. Describe a test to determine whether a solution is a weak acid or a strong acid.
4. Describe a test to determine whether a solution is acidic or basic.
5. Why can boric acid (H_3BO_3) be used in eyewashes, but hydrochloric acid (HCl) is not safe to use?
6. Write a neutralization reaction between lye (NaOH) and muriatic acid (HCl).
7. What is the main distinction between water solutions of strong and weak electrolytes?
8. **a.** What is the pH of a neutral solution?
   b. Which pH is more acidic, a pH of 5 or a pH of 2?
   c. Which pH is more basic, a pH of 5 or a pH of 10?
9. Classify each of the following as acids or bases, using the Brønsted-Lowry definitions: H_2SO_4, CO_3^{2-}, Cl^-, HCO_3^-, O^{2-}, H_2O.
10. Would liquid Ajax be more likely to have a pH greater or less than 7? (It has a strong smell of ammonia.)
11. Two solutions contain 1% acid. Solution A has a pH of 4.6, and solution B has a pH of 1.1. Which solution contains the stronger acid?
12. Predict the formulas of salts formed with the following pairs of ions:

Na^+ and SO_4^{2-}
Ca^{2+} and I^-
Mg^{2+} and NO_3^-
Ca^{2+} and PO_4^{3-}
K^+ and Br^-

13. Moist baking soda is often put on acid burns. Why? Write an equation for the reaction assuming the acid to be hydrochloric (HCl).
14. Hydrochloric acid is the acid present in the human stomach. Is this a strong or a weak acid? Why does the stomach not dissolve itself?
15. From a practical standpoint and for safety reasons, why should you know if the acid you are using is strong or weak?
16. If a drug consumes 37 times its weight in excess stomach acid, the reaction is called _____ and the drug must be a(n) _____.
17. Describe what happens when an ionic solid dissolves in water.
18. Describe vividly the scenario of HCl molecules being added to water. Compare and contrast this scenario with what happens when acetic acid (HAc) molecules are added to water.
19. What ions are present in water solutions of the following salts: Na_2SO_4, $CaBr_2$, $Mg(NO_3)_2$?
20. In terms of the hydrogen ion (H^+) concentration, when is a solution acidic and when is it basic?
21. What is meant by an alkaline solution?

22. Identify the conjugate acid–base pairs in the following equations:

  a. $HCl + H_2O \rightarrow H_3O^+ + Cl^-$
  b. $NH_3 + H_3O^+ \rightarrow NH_4^+ + H_2O$
  c. $HC_2H_3O_2 + H_2O \rightleftarrows H_3O^+ + C_2H_3O_2^-$
  d. $HC_2H_3O_2 + OH^- \rightarrow H_2O + C_2H_3O_2^-$
  e. $CN^- + HC_2H_3O_2 \rightleftarrows HCN + C_2H_3O_2^-$
  f. Ionization of sulfuric acid (2 steps)
    $H_2SO_4 + H_2O \rightarrow H_3O^+ + HSO_4^-$
    $HSO_4^- + H_2O \rightleftarrows H_3O^+ + SO_4^{2-}$

23. Why does molten sodium chloride conduct electricity when solid NaCl, though ionic, does not?

24. Explain how a buffer such as the $H_2PO_4^- - HPO_4^{2-}$ system consumes hydrogen ions released into the bloodstream.

25. What is the function of a buffer?

26. On a simplified basis, how much water will have to be mixed with the volume of concentrated solution given in order to prepare the desired dilute solution? Fill in the blanks in the table below.

27. A mole of ethyl alcohol weighs 46 g. How many grams of ethyl alcohol would be required to make 1 L of 1.5-M solution?

28. Which has more grams of solute dissolved per liter of solution, a 0.50-M solution of sucrose (molecular weight 342) or a 0.50-M solution of sodium chloride (molecular weight 58.5)?

29. What is the molar (M) concentration of a solution containing 12.0 g of NaCl dissolved in 500 mL of solution?

30. If a hydrochloric acid solution is 0.1 M, how many grams of HCl are dissolved in 1 L of this solution?

31. What is the purpose of antacids? Give two chemical examples of antacids in commercial products.

Concentrated Solution		Dilute Solution		
Concentration	Volume to Be Used (mL)	Concentration	Volume of Water to Be Used (mL)	Total Volume (mL)
a. 18 M	5 mL	5 M	——	——
b. 18 M	——	7 M	——	36
c. 15 M	12 mL	6 M	——	——
d. 14 M	——	4 M	——	28
e. 12 M	——	3 M	——	48

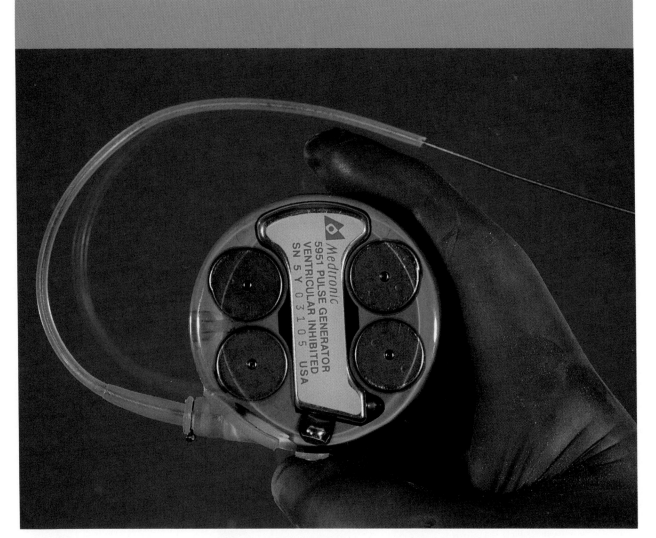

10

Oxidation – Reduction: Electron Transfer Chemistry

Pacemaker. (Martin M. Rotker.)

E qual in importance to the hydrogen ion transfer in acid–base chemistry are the processes called oxidation and reduction. As we shall find out later, oxidation and reduction reactions always occur together, so they are often named together as **oxidation–reduction.** In this chapter we shall look at oxidation–reduction from both the theory, which seeks to explain what is going on at the molecular level, and the applications, which are of vast importance in our lives.

Oxidation got its name from the chemical changes associated with the element oxygen combining with other elements. In fact, oxygen combines with every element except helium, neon, and argon. Prior to the discovery of the electron, oxidation was considered a simple combination of two elements that produced a compound called an **oxide.** Recall from Chapter 3 that Antoine Lavoisier used the decomposition of mercuric oxide (HgO), along with other reactions, to develop the law of conservation of matter.

When oxygen combines with another element, heat is almost always produced. If this energy (as heat) is given off rapidly enough, the oxidation is called *combustion,* or *burning.* An example of rapid combustion is shown in Figure 10–1.

Neither oxidation nor combustion is limited to oxygen combining with just elements. Compounds may be oxidized as well. Automobile engines burn hydrocarbon fuels (Chapter 13) and produce the oxides of hydrogen (water) and carbon (carbon monoxide and carbon dioxide). Oxides of nitrogen are produced as well. These nitrogen oxides come from the oxidation of some of the nitrogen in the air that is mixed with the fuel and ignited in the combustion chamber. Most oxidation is controlled, such as the combustion of fuels in engines, furnaces, fireplaces, and stoves, but some oxidation, such as rusting and forest and house fires, is not easily controlled, is unwanted, and may be life-threatening (Fig. 10–2).

Elemental oxygen makes up 21% by volume of our atmosphere. Most of the remainder of our atmosphere is nitrogen. Because of its many commercial uses, oxygen is extracted from the air in large quantities by liquefaction followed by distillation (see Chapter 11).

Combustion is always accompanied by heat and light.

If the atmosphere were composed of a greater concentration of oxygen, then fires could more readily get out of control; rates of chemical reactions are related to the concentrations of the reactants.

Figure 10–1   The hydrogen-filled dirigible *Hindenburg,* May 1939 at Lakehurst, New Jersey. (United Press International.)

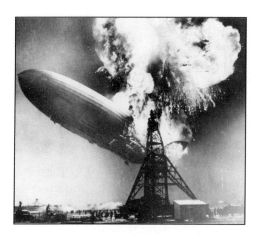

Figure 10–2 Forest fire. (Cunningham/Visuals Unlimited.)

Here are some properties of oxygen:

Formula	O_2
Molecular weight	32.00
Melting point	$-218.4°C$
Boiling point	$-183.0°C$
Description	Colorless and odorless gas
Solubility in water	48.9 mL per liter of water at 0°C

WHAT IS OXIDATION?

Whenever oxygen combines with another element or compound, the chemical reaction is one form of **oxidation**. The products of the reaction are called **oxidation products.** Most metals react readily with oxygen to form oxides. A few metals like gold and platinum do not readily oxidize, but can form oxides using indirect means. When iron, an easily oxidized metal, reacts with oxygen, a red-brown oxide forms.

$$4\ Fe + 3\ O_2 \longrightarrow 2\ Fe_2O_3$$

In the presence of moisture, usually found in the air, a *hydrate* of iron oxide forms. This iron oxide hydrate is known as **rust** (Fig. 10–3).

$$4\ Fe + 3\ O_2 + x H_2O \longrightarrow 2\ Fe_2O_3 \cdot x H_2O$$

In the formula for rust, the x represents a varying number of water molecules.

Oxygen also combines with nonmetals to form oxides. Carbon burns to form carbon monoxide and carbon dioxide.

$$2\ C + O_2 \longrightarrow 2\ CO$$
$$C + O_2 \longrightarrow CO_2$$

The formation of carbon monoxide when carbon dioxide could be formed is called **incomplete combustion.** In a limited supply of oxygen, carbon monoxide is the likely product. The carbon monoxide can be further oxidized to carbon dioxide.

Figure 10–3 Rusting of iron and steel objects costs billions of dollars each year. (*The World of Chemistry,* Program 15, "The Busy Electron.")

A hydrate is a stable molecular or ionic substance associated with water.

Carbon monoxide is highly toxic; see Chapter 16.

273

$$2\,CO + O_2 \longrightarrow 2\,CO_2$$

Carbon monoxide formed by the incomplete combustion of hydrocarbon fuels is a major component of urban air pollution.

While providing about 90% of all the energy needs for our society through the combustion of fuels, oxygen combines with other elements either in the air or in the fuels themselves to produce air pollutants (see Chapter 18).

OXIDATION DEFINED FROM THE ELECTRON POINT OF VIEW

Oxidation can be defined in three ways, each of which explains what oxidation actually is. The theorist can even explain the first two definitions in terms of the third.

Oxidation is the gain of oxygen.

The first definition of oxidation, already given, is in terms of oxygen combining with some element or compound. Oxygen is said to be the **oxidizer,** and the other reactant gets **oxidized** to form oxidation products.

Glucose ($C_6H_{12}O_6$) is oxidized by oxygen in living cells to the oxidation products carbon dioxide and water.

$$C_6H_{12}O_6 + 6\,O_2 \longrightarrow 6\,CO_2 + 6\,H_2O + 669.5 \text{ kcal}$$

This reaction provides the energy of life for almost all living cells.

The carbon in carbon dioxide is more oxidized than the carbon in carbon monoxide (CO). In general, when elements form several different compounds with oxygen, there is a **degree of oxidation.** Elements that are highly oxidized are often themselves capable of causing oxidation to occur. One name used for the compounds of these highly oxidized elements is **oxidizing agent.** Table 10–1 shows several of these oxidizing agents. Note the oxygen in their formulas.

Oxidation is the loss of hydrogen atoms.

A second definition of oxidation involves the loss of hydrogen atoms in organic molecules. The loss of hydrogen is not the cause of the oxidation but merely one way to recognize when oxidation has occurred in organic molecules. For example, ethanol is oxidized to the compound acetaldehyde in the liver with the aid of enzymes. In the process, two hydrogen atoms are removed from each molecule of ethanol.

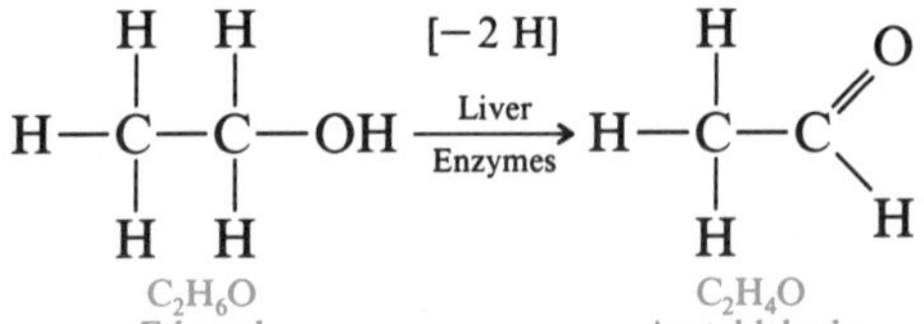

The acetaldehyde is more highly oxidized than ethanol because of the loss of the two hydrogen atoms. Other examples of reactions of this type will be discussed in Chapters 12 and 13 on the chemistry of organic compounds and in Chapter 15 on biochemistry.

Oxidation is the loss of electrons.

The third and most general definition of oxidation involves **electron loss.** An element is said to be oxidized when it loses electrons. When a

TABLE 10-1 Some Oxidizing Agents and Their Uses

Name	Formula	Uses as Oxidizing Agent
Potassium dichromate	$K_2Cr_2O_7$	Tests for alcohol in breath
Potassium nitrate	KNO_3	Gunpowder
Calcium hypochlorite	$Ca(OCl)_2$	Bleach, swimming pool disinfectant
Lead dioxide	PbO_2	Lead storage batteries
Manganese dioxide	MnO_2	Alkaline and lithium batteries
Hydrogen peroxide	H_2O_2	Disinfectant, antiseptic
Potassium peroxydisulfate	$K_2S_2O_8$	Denture cleansers

Figure 10-4   The burning of metallic sodium in bromine vapors. The product is white sodium bromide. (*The World of Chemistry,* Program 8, "Chemical Bonds.")

neutral atom becomes a positive ion, it has lost electrons and has been oxidized. Sodium is oxidized by bromine to produce sodium ions and bromide ions (Fig. 10-4).

$$2 \, Na + Br_2 \longrightarrow 2 \, Na^+ + 2 \, Br^-$$

These ions, in equal numbers, form sodium bromide (NaBr), a white solid.

In addition to oxygen and bromine, the elements chlorine (Cl) and fluorine (F) combine with elements and compounds in ways that can be called oxidation. Fluorine is rather exotic in its applications, but Cl_2 is commonly used in oxidation applications such as disinfecting water supplies and in bleaches and cleaning compounds.

SELF-TEST 10-A

1. A type of oxidation that produces heat and light is _____ .
2. When gasoline burns in plentiful air, the principal oxidation products are _____ and _____ .
3. When oxygen combines with an element to form a compound, that compound is often known as a(n) _____ .
4. When iron oxidizes in the presence of moisture, the product of the reaction is called _____ .
5. The products of the complete combustion of methane (CH_4) are _____ and _____ .
6. The incomplete combustion of carbon produces _____ .
7. When oxygen reacts with glucose to form carbon dioxide and water, the reactant oxygen is called the _____ .
8. Which is more highly oxidized? (a) ethanol or (b) acetaldehyde.
9. Which is the oxidized form of sodium? (a) Na^+ ion, (b) Na atom.

REDUCTION—THE OPPOSITE OF OXIDATION

Reduction always accompanies oxidation. When something is oxidized, something else is reduced. When something gains oxygen and gets oxidized, something else loses oxygen and gets reduced. When something loses hydro-

Oxidation is always accompanied by reduction and vice versa. One cannot occur without the other.

gen and gets oxidized, something else gains hydrogen and gets reduced. When something loses electrons and gets oxidized, something else gains electrons and gets reduced. As an oxidizing agent causes oxidation, a **reducing agent** *causes* reduction.

	Oxidation	**Reduction**
In terms of oxygen:	Gain of oxygen	Loss of oxygen
In terms of hydrogen:	Loss of hydrogen	Gain of hydrogen
In terms of electrons:	Loss of electrons	Gain of electrons

By observing the hydrogen and oxygen exchanges, one is limited to the compounds that contain H and O. The electron exchange is the broader concept. Lavoisier's experiments with mercuric oxide illustrate the reduction of mercury. When mercuric oxide is heated, oxygen is oxidized and becomes an oxygen molecule.

$$2\ HgO \longrightarrow 2\ Hg + O_2$$

The mercury loses oxygen and is reduced.

The chemistry in the blast furnace is discussed in Chapter 11.

Many metal ores consist of metal oxides. Iron ore contains the oxide hematite ($Fe_2O_3 \cdot xH_2O$). When iron ore is fed into a blast furnace, the iron in iron oxide is reduced, and metallic iron is one of the products. This reduction reaction occurs with the aid of the reducing agent coke, a form of carbon produced from coal. The overall reaction is

$$2\ Fe_2O_3 \cdot xH_2O + 3\ C \longrightarrow 4\ Fe + 3\ CO_2 + 2x\ H_2O$$

Looking at the equation for iron oxide reduction in the blast furnace, we see the iron in iron oxide loses oxygen (gets reduced). The reducing agent is carbon, which has no oxygen associated with it at the start of the reaction. At the end of the reaction, carbon is combined with oxygen as carbon dioxide. The carbon gets oxidized.

Another very important reaction involving oxidation and reduction is the reduction of carbon monoxide, which can be reduced to organic compounds in the presence of catalysts, with hydrogen as the reducing agent.

Methanol as an automotive fuel is discussed in Chapter 13.

$$CO + 2\ H_2 \longrightarrow CH_3OH$$
$$\text{Methanol}$$

This reaction is currently used for making alcohol fuels.

Substances called **fuels** are reducing agents. Common fuels like gasoline and diesel fuel contain compounds of carbon and hydrogen and are reducing agents. Table 10–2 looks at some common and some uncommon fuels.

Often oxidizers and fuel react out of control. Forest and building fires are examples. Grain dust explosions are another. Grain dust, a carbohydrate similar in composition to the major component found in wood, is the fuel, and the oxygen in the air mixed with the dust is the oxidant.

The tragic explosion of the space shuttle *Challenger* in January, 1986, in which seven astronauts perished was caused by hot exhaust gases rupturing the hydrogen fuel tanks. The released hydrogen burned in the atmosphere with explosive force. The pure oxygen in the adjacent oxygen tank accelerated the reaction even more when the tank ruptured.

Prototype car running on hydrogen gas stored in a metal hydride. (Courtesy of Mercedes-Benz.)

TABLE 10–2 Some Fuels and Their Reactions

Fuel	Application	Oxidizer	Products
Gasoline	Passenger cars	Oxygen in air	$CO_2 + H_2O$ + some CO
Ethanol	Race cars	Oxygen in air	$CO_2 + H_2O$
Hydrogen	Space shuttle main engine	Pure Oxygen	H_2O
Hydrazine	Rocket motors	Nitrogen Tetroxide	$N_2 + H_2O$
Propane	Home heating	Oxygen in air	$CO_2 + H_2O$
Coal	Home heating	Oxygen in air	$CO_2 + H_2O$ + some CO

REDUCTION OF METALS FROM THEIR ORES

One of the most practical applications of the oxidation–reduction principle is the separation of metals from their ores. The majority of metals are found in nature as compounds; that is, the metals are in an oxidized state. To be obtained, the metal must be reduced from the oxidized to the neutral elemental state. This requires a gain of electrons. In addition to electricity, a variety of chemicals can supply the electrons for the reduction process. Some ways of reducing metals from their ores are given in Table 10–3.

Hydrogen is the fuel for the space shuttle main engine. (NASA.)

ELECTROLYSIS

Several metals either are separated from their ores or are purified afterward by electrolysis. **Electrolysis** is a type of chemical reaction caused by the application of electrical energy.

The suffix -*lysis* means "splitting" or "decomposition"; electrolysis is decomposition by electricity.

TABLE 10–3 Some Methods Used to Separate Metals from Their Ores by Reduction*

Metal	Occurrence	Reduction Process	Uses of Metal
Cu	Cu_2S, chalcocite	Air blown through melted ore $Cu_2S + O_2 \longrightarrow 2\ Cu + SO_2(g)$ $(Cu^+ + e^- \longrightarrow Cu)$	Electrical wiring, boilers, pipes, brass (Cu 85%, Zn), bronze (Cu 90%, Sn, Zn), other alloys
Mg	Mg^{2+}, sea water	Electrolysis of fused chloride $MgCl_2 \longrightarrow Mg + Cl_2$ $(Mg^{2+} + 2\ e^- \longrightarrow Mg)$	Light alloys such as duralumin (0.5% Mg, Al), Dowmetal H (90.7% Mg, Al, Zn, Mn), flares, some flash bulbs
Al	$Al_2O_3 \cdot H_2O$, bauxite	Electrolysis in fused cryolite, Na_3AlF_6, at 800–900°C $2\ Al_2O_3 \longrightarrow 4\ Al + 3\ O_2$ $(Al^{3+} + 3\ e^- \longrightarrow Al)$	Packaging, airplane and automobile parts, alloys, roofing, siding
Hg	HgS, cinnabar	Roasting in air $HgS + O_2 \xrightarrow{heat} Hg + SO_2(g)$ $(Hg^{2+} + 2\ e^- \longrightarrow Hg)$	Some thermometers, electrical switches, blue-green street lights, amalgams in dentistry, fluorescent lighting

* Other methods are discussed in Chapter 11.

Figure 10–5   Electroplating from a copper sulfate solution.

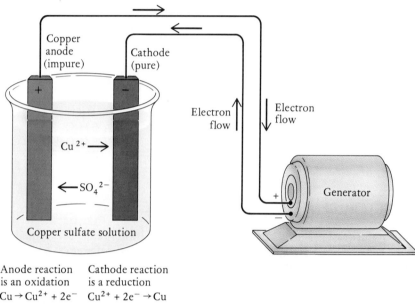

Anode reaction   Cathode reaction
is an oxidation   is a reduction
$Cu \rightarrow Cu^{2+} + 2e^-$ $Cu^{2+} + 2e^- \rightarrow Cu$

Reduction at cathode
Oxidation at anode

The principal parts of an electrolysis apparatus are shown in Figure 10–5. Electrical contact between the external circuit and the solution is obtained by means of electrodes, which are often made of graphite or metal. The electrode at which electrons enter an electrolysis cell is termed the **cathode,** and this is the electrode at which reduction takes place. The electrode at which the electrons leave the cell is the **anode.** At the anode, oxidation takes place.

The battery or generator produces a current of electrons, which flow toward one electrode and make it negatively charged, and away from the other electrode and make it positively charged. When the switch is closed, the positive ions in solution migrate toward the cathode. Soon, a chemical reaction is evidenced at the electrodes. Depending on the substances present in the solution, gases may be evolved, metals deposited, or ionic species changed at the electrodes. The ions that migrate to the electrodes are not necessarily the species undergoing reaction at the electrodes, because sometimes the solvent undergoes reaction more easily. Whatever happens, the chemical reactions occurring at the cathode and anode are due to electrons going into and coming out of the solution. The chemical reaction at the cathode furnishes electrons to solution species (reduction). At the anode, electrons are taken from species in solution, so the chemical reaction at the anode gives up electrons (oxidation).

The electroplating of copper is illustrated in Figure 10–6. Such an electrolysis can be used either to plate an object with a layer of pure copper or to purify an impure sample of copper metal; copper is transferred from the positive electrode into the solution and eventually to the negative electrode. If the positive electrode is impure copper to be purified, electrolysis deposits the copper as very pure copper on the negative electrode.

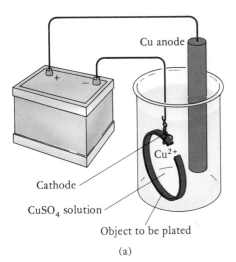

Figure 10-6 **Electroplating with copper. (a) The copper anode dissolves into a copper solution and plates out on the cathode. (b) Baby shoes electroplated with copper (C. Steele.)**

Cu anode

Cathode

CuSO₄ solution

Object to be plated

(a)

(b)

Now let us examine how the electrolysis transfers the copper from the positive electrode to the negative electrode. Electrons flow out of the negative terminal of the generator through the wire and into the negative electrode. Somehow this negative charge must be used up at the surface of the electrode.

Consider what happens when the electrons build up on the negative electrode. The positive copper ions nearby will be attracted to the surface and will take the electrons. Thus, the Cu^{2+} ions are reduced:

$$Cu^{2+} + 2\ e^- \longrightarrow Cu \qquad \text{(CATHODE REACTION, REDUCTION)}$$

In a similar way the negative sulfate ions migrate to the positive electrode (anode). However, it is easier to get electrons from the copper metal of the electrode than from the sulfate ions. As each copper atom gives up two electrons, the copper ion passes into solution:

$$Cu \longrightarrow Cu^{2+} + 2\ e^- \qquad \text{(ANODE REACTION, OXIDATION)}$$

In effect, then, the copper of the positive electrode is oxidized (the anode reaction) and passes into solution; the copper ions in solution migrate to the negative electrode, are reduced (the cathode reaction), and plate out as copper metal. Large amounts of copper are purified in this way each year. Silver and gold can be purified similarly.

If we desire to plate an object with copper, we have only to render the surface conducting and make the object the negative electrode in a solution of copper sulfate. The object will become coated with copper, with the copper coating growing thicker as the electrolysis is continued. If the object is a metal, it will conduct electricity by itself. If the object is a nonmetal, its surface can be lightly dusted with graphite powder to render it conducting.

A potentially very important electrolysis reaction is the electrolysis of water (Fig. 10-7). When electricity is passed into graphite electrodes im-

Figure 10-7 **Electrolysis of water. The electrodes are immersed in a dilute salt solution (the salt causes the solution to conduct electricity). (Charles D. Winters.)**

Copper can be plated onto an object by making that object the negative electrode in a cell containing dissolved copper salts.

COLD FUSION — ENERGY *FROM* ELECTROLYSIS

Electrolysis of water *requires* energy. Hydrogen gas is produced at the cathode and oxygen gas at the anode. This is a reaction you can do at the kitchen sink with a battery, some dilute salt water, and two strips of uninsulated copper wire. If the source of electricity is cut off by breaking the circuit, the reaction stops. This is evident by the stopping of the production of the two gases at the electrodes. If the electrodes are made of the metal palladium — a very expensive metal — some of the hydrogen gas produced at the cathode actually dissolves in the palladium. Palladium is not the only metal with this property, but it does seem to dissolve hydrogen the best of the elements.

If heavy water, D_2O*, is electrolyzed, D_2 is produced at the cathode.

$$2 \, D_2O \longrightarrow 2 \, D_2 + O_2$$

Like ordinary hydrogen, some of the deuterium dissolves in the palladium, and if the electrolysis is run long enough, the amount of dissolved deuterium reaches a saturation point. When this happens, the deuterium atoms are crowded quite close together in the lattice of palladium atoms.

In March of 1989, two electrochemists, Stanley Pons of the University of Utah, and Martin Fleischmann of the University of Southampton in England, claimed that their electrolysis experiments with deuterium oxide (heavy water) using palladium electrodes had actually *produced* more energy than was used for electrolysis. Furthermore, they claimed that nuclear **fusion**† of the deuterium atoms was taking place in the closeness of the palladium metal. If fusion did occur, it was a form of **cold fusion.** Cold fusion would be a drastic departure from the normal means of producing fusion, which requires temperatures reached in an atomic explosion (see Chapter 8).

* Hydrogen has three isotopes, hydrogen-1, deuterium (hydrogen-2), and tritium (hydrogen-3). Deuterium is very abundant in the world's oceans (about one atom of deuterium for every 10,000 atoms of hydrogen-1). Tritium is very rare and naturally radioactive.

† The nuclear reaction for the fusion of two deuterium atoms can be written as

$$^2_1D + {}^2_1D \longrightarrow {}^3_2He + {}^1_0n + energy$$

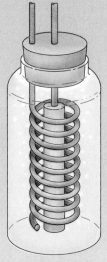

Schematic of the cold fusion apparatus.

Gold plating. (*The World of Chemistry,* Program 15, "The Busy Electron.")

mersed in a dilute salt solution, water is reduced to hydrogen and hydroxide ions at the cathode:

$$2 \, H_2O + 2 \, e^- \longrightarrow H_2 \, (g) + 2 \, OH^- \quad \text{(CATHODE REACTION, REDUCTION)}$$

At the anode, water is oxidized to oxygen and hydrogen ions:

$$2 \, H_2O \longrightarrow O_2 + 4 \, H^+ + 4 \, e^- \quad \text{(ANODE REACTION, OXIDATION)}$$

The OH^- and H^+ ions combine to re-form water. The overall, or net, cell reaction is:

$$2 \, H_2O \xrightarrow{\text{electricity}} 2 \, H_2 \, (g) + O_2 \, (g)$$

The hydrogen produced by the reduction of water can be stored and used as a fuel — for example, to power rockets into space. Someday, if electricity becomes inexpensive enough (see Chapter 8), water may be electrolyzed to produce hydrogen, which can then be piped to the point of use, just as natural gas is today.

SELF-TEST 10-B

1. Conversion of carbon monoxide to methanol by the addition of hydrogen is an example of carbon monoxide being (   ) oxidized, (   ) reduced.
2. When coke (a form of carbon) reacts with iron ore in a blast furnace, the coke is the (   ) oxidation product, (   ) reducing agent, (   ) oxidizing agent.
3. When a metal ion in solution is plated on an object, the metal ion is (   ) oxidized, (   ) reduced.
4. A substance called a fuel is (   ) an oxidizing agent, (   ) a reducing agent.
5. An oxidized form of sodium is in the compound NaCl. The reduced form of sodium is _____ .
6. Magnesium in sea water is in the oxidized form, the Mg^{2+} ion. The reduced form of magnesium is _____ .
7. A chemical reaction caused by the application of electricity is called _____ .
8. An electrode that supplies electrons will have (   ) oxidation, (   ) reduction taking place on its surface.
9. An electrode that takes away electrons will have (   ) oxidation, (   ) reduction taking place on its surface.
10. When electricity is supplied to a dilute salt solution, water is electrolyzed. The products of this reaction are _____ and _____ .

Figure 10-8 Metallic copper plates onto a zinc rod from a blue copper solution. After some time the blue color of the solution disappears, indicating that no more copper ions are in solution. (Marna G. Clarke.)

RELATIVE STRENGTHS OF OXIDIZING AND REDUCING AGENTS

When a piece of metallic zinc is placed in a solution containing hydrated copper ions (Cu^{2+}), an oxidation–reduction reaction occurs:

$$Zn + Cu^{2+} \longrightarrow Zn^{2+} + Cu$$

Evidence for this reaction is the deposit of copper on the zinc (Fig. 10-8). The gradual decrease in the intensity of the blue color of the solution indicates removal of the Cu^{2+} ions.

The oxidation of zinc by copper ions can be thought of as a competition between zinc ions (Zn^{2+}) and copper ions (Cu^{2+}) for the two electrons. Since the reaction proceeds almost to completion, the Cu^{2+} ions obviously win out in the competition. Other metals can compete similarly for electrons.

The **activity** of a metal is a measure of its tendency to lose electrons. Zinc is a more active metal than copper on the basis of the experiment just described. This means that given an equal opportunity, the first reaction will take place to a greater extent:

A copper ion has a greater attraction for electrons than does a zinc ion.

The active metals lose electrons more easily; hence, these free metals are not found in nature.

$$Zn \longrightarrow Zn^{2+} + 2\ e^- \quad \text{(MORE LIKELY TO OCCUR)}$$
$$Cu \longrightarrow Cu^{2+} + 2\ e^- \quad \text{(LESS LIKELY TO OCCUR)}$$

Experiments of this type with various pairs of metals and other reducing agents yield an **activity series** of the elements, which ranks each oxidizing and reducing agent according to its *strength* or *tendency* for the electron transfer to take place. An iron nail will be partly dissolved in a solution of a copper salt containing Cu^{2+} ions, with copper being deposited on the nail that remains. From this, it is determined that iron, like zinc, is more active than copper. The reaction that occurs is

$$Fe + Cu^{2+} \longrightarrow Fe^{2+} + Cu$$

Now, which is more active, zinc or iron? This question can be answered by placing an iron nail in a solution containing Zn^{2+} ions and, in a separate container, a strip of zinc in a solution containing Fe^{2+} ions. The zinc strip is found to be eaten away in the solution containing Fe^{2+} ions. The reaction, then, is

$$Zn + Fe^{2+} \longrightarrow Fe + Zn^{2+}$$

Nothing happens to the iron nail in the solution of Zn^{2+} ions. We deduce that Zn loses electrons more readily than Fe.

Such an activity series can be extended to include other metals and even nonmetals. The concentrations of the ions in solution and other factors often must be considered for accurate work, but for our purposes these will be ignored. Table 10–4 is an activity series of some oxidizing and reducing agents.

Activity: Zn > Fe > Cu

The activity series can be used to predict whether a reaction will occur. Thus, a reducing agent in Table 10–4 (right column) is able to reduce the oxidized form of any species below it. For example, magnesium can reduce Cu^{2+} to Cu:

$$Mg + Cu^{2+} \longrightarrow Cu + Mg^{2+}$$

Magnesium can also reduce Ag^+ to Ag:

$$Mg + 2\ Ag^+ \longrightarrow Mg^{2+} + 2\ Ag$$

Zinc can also reduce silver ions:

$$Zn + 2\ Ag^+ \longrightarrow 2\ Ag + Zn^{2+}$$

Zinc cannot reduce calcium ions, since calcium is above zinc in the series:

$$Zn + Ca^{2+} \longrightarrow\!\!\!| \quad No\ reaction$$

The series also arranges oxidizing agents in the order of their effectiveness. Fluorine (F_2) can oxidize water, silver, iron (II) ion, or any species above it in the series; copper (II) ion (Cu^{2+}) can oxidize H_2, Fe, Zn, and the metals above Cu in the table, because the Cu^{2+} ion has a greater tendency to take on electrons than do the ions that are formed. Thus:

$$Cu^{2+} + Fe \longrightarrow Fe^{2+} + Cu$$
$$Cu^{2+} + Mg \longrightarrow Mg^{2+} + Cu$$
$$Cu^{2+} + Ag \longrightarrow\!\!\!| \quad No\ reaction$$

TABLE 10-4 Relative Strengths of Some Oxidizing and Reducing Agents: The Activity Series

	Oxidizing Agents		Reducing Agents	
Increasing Strength of Oxidizing Agent	$Li^+ + e^-$	$\rightleftharpoons$	Li	*Increasing Strength of Reducing Agent*
	$Ca^{2+} + 2\,e^-$	$\rightleftharpoons$	Ca	
	$Na^+ + e^-$	$\rightleftharpoons$	Na	
	$Mg^{2+} + 2\,e^-$	$\rightleftharpoons$	Mg	
	$Zn^{2+} + 2\,e^-$	$\rightleftharpoons$	Zn	
	$Fe^{2+} + 2\,e^-$	$\rightleftharpoons$	Fe	
	$2\,H^+ + 2\,e^-$	$\rightleftharpoons$	H_2	
	$Cu^{2+} + 2\,e^-$	$\rightleftharpoons$	Cu	
	$Fe^{3+} + e^-$	$\rightleftharpoons$	Fe^{2+}	
	$Ag^+ + e^-$	$\rightleftharpoons$	Ag	
	$O_2 + 4\,e^- + 4\,H^+$	$\rightleftharpoons$	$2\,H_2O$	
	$F_2 + 2\,e^-$	$\rightleftharpoons$	$2\,F^-$	

CORROSION—UNWANTED OXIDATION-REDUCTION

In the United States alone, more than $10 billion is lost each year to corrosion. Much of this corrosion is the rusting of iron and steel, although other metals may oxidize as well. The problem with iron is that its oxide, rust, does not adhere strongly to the metal's surface once the rust is formed. Because the rust flakes off or is rubbed off easily, the metal surface becomes pitted. The continuing loss of surface iron by rust formation eventually causes structural weakness.

The corrosion of metals involves oxidation and reduction. The driving forces behind corrosion are the activity of the metal as a reducing agent and the strength of the oxidizing agent. Whenever a strong reducing agent (the metal) and a strong oxidizing agent (like oxygen) are together, a reaction between the two substances is likely. Factors governing the rates of chemical reaction such as temperature and concentration will affect the rate of corrosion as well. Consider the corrosion of an iron spike (Fig. 10-9). The surface of the iron is far from perfect. There are tiny microcrystals composed of

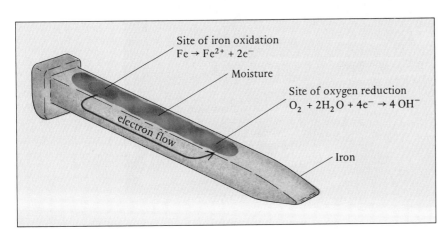

Site of iron oxidation
$Fe \rightarrow Fe^{2+} + 2e^-$

Moisture

Site of oxygen reduction
$O_2 + 2H_2O + 4e^- \rightarrow 4\,OH^-$

electron flow

Iron

Figure 10-9 The site of iron oxidation may be different from the point of oxygen reduction owing to the ability of the electrons to flow through the iron. The point of oxygen reduction can be located with an acid–base indicator because of the OH^- ions produced.

loosely bound iron atoms on the surface of the metal. The iron can readily ionize into any water present on the surface of the metal.

$$Fe \longrightarrow Fe^{2+} + 2\ e^- \qquad \text{(OXIDATION)}$$

The ionization of iron atoms into Fe^{2+} ions is an oxidation process and a result of the position of iron in the activity series (Table 10–4). Iron is a fairly active metal, that is, it tends to give its electrons up rather easily. Since iron is a good conductor of electricity, the electrons produced at this site can migrate to some point where they can reduce something. If these electrons did not migrate, the corrosion of iron would come to an abrupt halt as a result of a buildup of excessive negative charge. One location on the surface of the iron where electrons can be used would be any tiny drop of water containing dissolved oxygen. Here, the oxygen gains the electrons, forming hydroxide ions.

Oxidation cannot occur without reduction.

$$O_2 + 2\ H_2O + 4\ e^- \longrightarrow 4\ OH^- \qquad \text{(REDUCTION)}$$

This reduction of oxygen occurs so readily that when Fe^{2+} ions are encountered, they are further oxidized to Fe^{3+} ions. This happens to the dissolved Fe^{2+} ions in the water on the surface of the metal. The reaction is

$$4\ Fe^{2+} + O_2 + 2\ H_2O \longrightarrow 4\ Fe^{3+} + 4\ OH^-$$

Finally, the Fe^{+3} ions combine with hydroxide ions to form the iron oxide we call rust.

$$2\ Fe^{3+} + 6\ OH^- \longrightarrow Fe_2O_3 \cdot 3H_2O$$

The rate of rusting is enhanced by salts, which dissolve in the water on the surface of the iron and act like tiny salt bridges of an electrochemical cell (discussed in the next section). The hydroxide ions and Fe^{2+} and Fe^{3+} ions migrate more easily in the ionic solutions produced by the presence of the dissolved salts. Automobiles rust out more quickly when exposed to road salts in wintery climates. If road salts are used in your driving area, it's a good idea after snowy seasons to wash the undersides of automobiles to remove the accumulated salts.

Chrome plating for beauty and corrosion protection. (Tom Stack and Associates.)

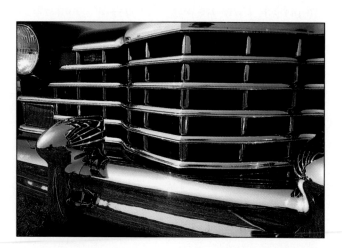

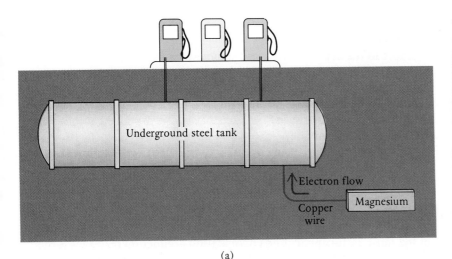

(a)

(b)

Figure 10–10   (a) A schematic of how cathodic protection works for an underground storage tank. The more active magnesium metal protects the steel (iron). (b) A ship's hull is cathodically protected by strips of titanium (shown as four strips along the propeller axis).

Rusting can be prevented by protective coatings such as paint, grease, oil, enamel, or some corrosion-resistant metal like chromium. Some metals are more active than iron, but when these metals corrode, they form adherent oxide coatings. Coatings with these metals provide corrosion protection. One of these metals is zinc. Zinc coating of iron and steel is called **galvanizing** and may be done by dipping the object into a molten bath of zinc metal or by electroplating zinc onto the surface of an iron or steel object. In galvanized objects in which the zinc coating is exposed to air and water, a thin film of zinc oxide forms that protects the zinc from further oxidation. Galvanizing is a type of **cathodic protection.** As the name implies, a cathode is protected by using a more active metal in good electrical contact with the metal to be protected. The electrons for the reduction of oxygen,

$$O_2 + 2\,H_2O + 4\,e^- \longrightarrow 4\,OH^-$$

are supplied by the more active metal. Thus, a more active metal, electrically connected to a piece of iron, would be oxidized before the iron is oxidized.

Some cathodic protection relies on the cathode being sacrificed. An important application is the cathodic protection of underground steel storage tanks (Fig. 10–10) that hold gasoline and other hazardous liquids. These tanks must be protected since leakage would contaminate groundwater supplies. Beginning in 1986, these tanks must be cathodically protected under new federal regulations designed to protect groundwater.

The importance of groundwater purity is discussed in Chapter 17.

SELF-TEST 10–C

1. When electrons enter an electrode and cause a chemical reaction, that reaction is called ———————— .
2. When electrons are removed from a chemical at an electrode, that reaction is called ———————— .
3. What three reactants are necessary for rusting?

4. Which metal would likely be used to protect iron cathodically?
   ( ) copper, ( ) magnesium, or ( ) silver.
5. According to the activity series, can magnesium reduce copper ions?
6. According to the activity series, which is the stronger oxidizing agent,
   ( ) fluorine, ( ) oxygen?
7. When water is electrolyzed, at the reduction electrode (cathode),
   _____ is produced. _____ is produced at the
   oxidation electrode (anode).
8. Which is the stronger reducing agent? ( ) iron (Fe), ( ) lithium (Li).

BATTERIES

One of the most useful applications of oxidation–reduction reactions is the production of electrical energy. A device that produces an electron flow (current) is called an **electrochemical cell.** Although a series of such cells is a **battery,** the term *battery* is commonly used even for single cells such as those we shall describe.

Consider the reaction between zinc atoms and copper ions that was discussed previously. If zinc is placed in a solution containing Cu^{2+} ions, the electron transfer takes place between the zinc metal and the copper ions, and the energy liberated simply causes a slight heating of the solution and the zinc strip. If the zinc could be separated from the copper solution and the two connected in such a way to allow current flow, the reaction can proceed, but

Figure 10–11   A simple battery involving the oxidation of zinc metal and the reduction of Cu^{2+} ions.

Historic battery used in the early telegraph

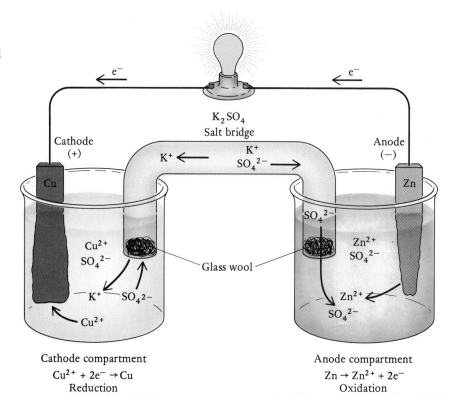

Cathode compartment
$Cu^{2+} + 2e^- \rightarrow Cu$
Reduction

Anode compartment
$Zn \rightarrow Zn^{2+} + 2e^-$
Oxidation

now the electrons are transferred through the connecting wires. Figure 10–11 shows a battery that can be constructed to make use of the oxidation–reduction involved in the reaction of Zn with Cu^{2+}.

The anode reaction is the oxidation of zinc to Zn^{2+} ions.

$$Zn \longrightarrow Zn^{2+} + 2\ e^-$$

The electrons flow from the Zn electrode through the connecting wire, light the lamp in the circuit, and then flow into the copper cathode where reduction of Cu^{2+} ions occurs:

$$Cu^{2+} + 2\ e^- \longrightarrow Cu$$

The copper is deposited on the copper cathode.

This flow of electrons (negative charge) from the anode to the cathode compartment in the battery must be neutralized electrically. This is done by using a "salt bridge" provided to connect the two compartments. The salt bridge contains a solution of a salt such as K_2SO_4. Its purpose is to keep the two solutions neutral. Around the cathode the deposition of positive copper ions (Cu^{2+}) would tend to cause the solution to become negative owing to the presence of excess negative sulfate ions (SO_4^{2-}). Two actions can keep the solution around the cathode neutral: either positive potassium ions (K^+) pass into the solution, or negative sulfate ions pass out of the solution and into the salt bridge. Actually, both processes occur. Similarly, around the anode, the solution would tend to become positive because positive zinc ions (Zn^{2+}) are put into the solution. Two actions can keep the solution around the anode neutral: either negative sulfate ions pass into the solution or positive zinc ions pass out of the solution and into the salt bridge. In all of this exchange, it is necessary for the solution to maintain the same number of positive charges as negative charges. The reaction between zinc atoms and copper ions continues until one or the other is consumed.

The most common battery in use is the "dry cell," invented by Georges Leclanché in 1866. The container of the cell is made of zinc, which serves as the anode for the cell. The zinc is separated from the other chemicals (Fig. 10–12) by a porous paper. The center electrode, the cathode, of the dry cell is

In commercial batteries, the salt bridge is often replaced by a porous membrane.

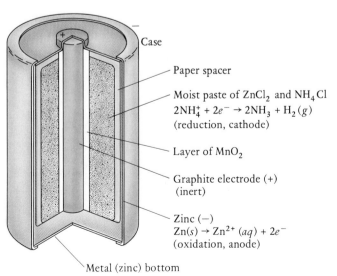

Figure 10–12 A common dry cell battery found in such items as flashlights.

Case

Paper spacer

Moist paste of $ZnCl_2$ and NH_4Cl
$2NH_4^+ + 2e^- \rightarrow 2NH_3 + H_2\,(g)$
(reduction, cathode)

Layer of MnO_2

Graphite electrode (+)
(inert)

Zinc (−)
$Zn(s) \rightarrow Zn^{2+}\,(aq) + 2e^-$
(oxidation, anode)

Metal (zinc) bottom

The dry cell really isn't dry.

graphite, a form of the element carbon, and is inserted into the surrounding moist mixture of ammonium chloride, zinc chloride, and manganese dioxide. As electrons flow from the cell through a flashlight bulb, for example, the zinc is oxidized

$$Zn \longrightarrow Zn^{2+} + 2\,e^-$$

and the ammonium ions are reduced.

$$2\,NH_4^+ + 2\,e^- \longrightarrow 2\,NH_3 + H_2$$
Ammonium   Ammonia
Ion

The ammonia formed reacts with zinc ions to form a zinc-ammonia complex ion. This prevents a buildup of gaseous ammonia.

$$Zn^{2+} + 4\,NH_3 \longrightarrow [Zn(NH_3)_4]^{2+}$$
Zinc-Ammonia
Complex Ion

The hydrogen produced is oxidized by the MnO_2 in the cell. In this way, hydrogen gas does not accumulate.

$$H_2 + 2\,MnO_2 \longrightarrow 2\,MnO(OH)$$

Alkaline dry cells (alkaline batteries) are similar to Leclanché dry cells, except that the electrolyte mixture contains potassium hydroxide (KOH), which is strongly alkaline, and the surface area of the zinc electrode is increased. The reactions are

$$Zn + 2\,OH^- \longrightarrow Zn(OH)_2 + 2\,e^- \quad \text{(ANODE)}$$
$$2\,MnO_2 + 2\,H_2O + 2\,e^- \longrightarrow 2\,MnO(OH) + 2\,OH^- \quad \text{(CATHODE)}$$

The mercury cell (Fig. 10–13) is similar to the alkaline battery in that the anode reactions are identical. The cathode reaction is the reduction of mercuric oxide.

$$HgO + H_2O + 2\,e^- \longrightarrow Hg + 2\,OH^- \quad \text{(CATHODE)}$$

Mercury cells should never be disposed of in fire because the mercury will vaporize and rupture the sealed container. Mercury's toxicity means that these batteries represent an environmental hazard if disposed of improperly (see Chapter 17).

Lithium battery. (Custom Medical Stock Photo.)

Figure 10–13   The mercury battery. (Photo courtesy Eveready Batteries.)

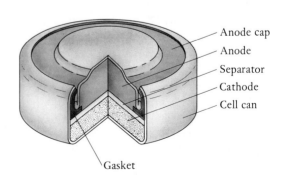

Anode cap
Anode
Separator
Cathode
Cell can

Gasket

THE PACEMAKER STORY

Sometimes an advance in science can come from an unlikely source. Several years ago, the inventor Wilson Greatbatch had an outrageous dream to prolong human life. His story is fascinating:

I quit all my jobs and with two thousand dollars, I went out in the barn in the back of my house and built 50 Pacemakers in two years.
I started making the rounds of all the doctors in Buffalo who were working in this field, and I got consistently negative results. The answer I got was, well, these people all die in a year, you can't do much for them, why don't you work on my project, you know.
When I first approached Dr. Shardack with the idea of the Pacemaker, he, alone, thought that it really had a future. He looked at me sort of funny, and he walked up and down the room a couple of times. He said, "you know, . . . if you can do that, . . . you can save a thousand lives a year."

In 1958, a medical team implanted the first heart pacemaker, but for the next few years there was one major problem.

After the first ten years, we were still only getting one or two years out of pacemakers, two years on average, and the failure mechanism was always the battery. It didn't just run down, it failed. The human body is a very hostile environment, it's worse than space, it's worse than the bottom of the sea. You're trying to run things in a warm salt water environment. The first pacemakers could not be hermetically sealed, and the battery just didn't do the job. Well, after ten years, the battery emerged as the primary mode of failure, and so we started looking around for new power sources. We looked at nuclear sources, we looked at biological sources, of letting the body make its own electricity, we looked at rechargeable batteries, and we looked at improved mercury batteries. And we finally wound up with this lithium battery. It really revolutionized the pacemaker business. The doctors have told me that the introduction of the lithium battery was more significant than the invention of the Pacemaker in the first place.
The World of Chemistry (Program 15) "The Busy Electron."

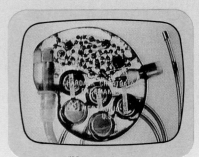

Pacemaker. (*The World of Chemistry,* Program 15, "The Busy Electron.")

Lithium batteries, which are so popular because of their light weight and high energy content per pound, use lithium metal as the anode. Some lithium batteries use MnO_2 as the oxidizer, and others, like some pacemaker batteries, use exotic chemicals like sulfuryl chloride ($SOCl_2$). The reactions are

$$4\ Li \longrightarrow 4\ Li^+ + 4\ e^- \quad \text{(ANODE)}$$
$$2\ SOCl_2 + 4\ e^- \longrightarrow 4\ Cl^- + S + SO_2 \quad \text{(CATHODE)}$$

Lithium should react with any oxidizing media right away because of its strong reducing tendencies (see Table 10–4), but lithium forms a protective chloride coating on its surface that prevents rapid oxidation. Virtually every lithium battery contains some reactant capable of forming a protective chlo-

Lithium metal has the lowest density of any nongaseous element: 0.534 g/cm².

Aluminum, another very reactive metal, also forms a strong protective oxide coating in air that prevents further oxidation.

TABLE 10–5 Characteristics of Some Batteries

System	Anode (oxidation)	Cathode (reduction)	Electrolyte	Typical Operating Voltage Per Cell
Dry cell	Zn	MnO_2	$NH_4Cl–ZnCl_2$	0.9–1.4
Edison storage	Fe	Ni oxides	KOH	1.2–1.4
Nickel-cadmium (NiCd)	Cd	Ni oxides	KOH	1.1–1.3
Silver cell	Cd	Ag_2O	KOH	1.0–1.1
Lead storage	Pb	PbO_2	H_2SO_4	1.95–2.05
Mercury cell	Zn(Hg)	HgO	KOH–ZnO	1.30
Alkaline cell	Zn(Hg)	MnO_2	KOH	0.9–1.2
Lithium battery	Li	MnO_2	KOH	3.4

The mercury batteries that are so popular for small electronic devices must be discarded carefully. If heated, these hermetically sealed batteries will rupture explosively due to expanding vapors within the package.

ride coating on the lithium surface. One of the most common is lithium perchlorate.

Many different oxidation–reduction combinations are used in commercial batteries to produce a flow of electrons. A few of the more popular ones are listed in Table 10–5.

Batteries in which the stored chemical energy is simply used up are called **primary** batteries. In such batteries, the oxidation products produced at the anode are allowed to mingle with the reduction products formed at the cathode. Because of this mixing, the battery may be used only once and then discarded (or recycled). Many of the less expensive batteries used to power flashlights, toys, radios, watches, cameras, and hand-held calculators are primary batteries.

Some batteries can be recharged. These are called **secondary** batteries. In these batteries, the oxidation products stay at the anode, and the reduction products remain at the cathode. Under favorable conditions, these secondary batteries may be discharged and recharged many times over.

One of the most widely used secondary batteries is the lead storage battery (Fig. 10–14). As this battery is discharged, metallic lead is oxidized to lead sulfate at the anode, and lead dioxide is reduced at the cathode.

A formula underlined like $\underline{PbSO_4}$ indicates that the substance is insoluble.

$$Pb + SO_4^{2-} \rightleftharpoons \underline{PbSO_4} + 2\ e^- \qquad \text{(ANODE)}$$
$$2\ e^- + PbO_2 + 4\ H^+ + SO_4^{2-} \rightleftharpoons \underline{PbSO_4} + 2\ H_2O \qquad \text{(CATHODE)}$$

The lead sulfate formed at both electrodes is an insoluble compound so it stays on the electrode surface. Since sulfuric acid is used in both the anode and the cathode reactions, the concentration of the sulfuric acid electrolyte decreases as the battery discharges. A measurement of the density of this battery acid gives a measure of the state of charge of the battery. The lower the density of the battery acid, the lower the state of charge.

Recharging a secondary battery requires reversing the electric current flow through the battery. When this occurs, the anode and cathode reactions are reversed.

AT THE NEGATIVE ELECTRODE:

$$Pb + SO_4^{2-} \underset{\text{Charge}}{\overset{\text{Discharge}}{\rightleftharpoons}} \underline{PbSO_4} + 2\ e^-$$

AT THE POSITIVE ELECTRODE:

$$PbO_2 + SO_4^{2-} + 4\ H^+ + 2\ e^- \underset{\text{Charge}}{\overset{\text{Discharge}}{\rightleftharpoons}} \underline{PbSO_4} + 2\ H_2O$$

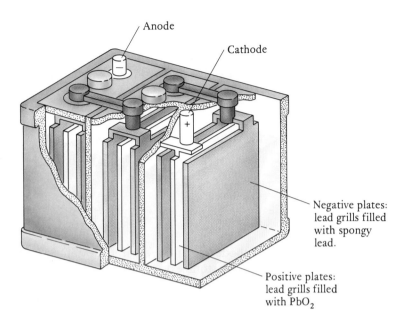

Anode

Cathode

Negative plates:
lead grills filled
with spongy
lead.

Positive plates:
lead grills filled
with PbO_2

Normal charging of an automobile lead storage battery occurs during driving. The voltage regulator senses the output from the alternator, and when the alternator voltage exceeds that of the battery, the battery is charged. During the charging cycle in most batteries, some water is reduced at the cathode, while water is oxidized at the anode.

The lead–acid battery was first presented to the French Academy of Science in 1860 by Gaston Planté.

OXIDATION OF WATER: $\qquad 2\,H_2O \longrightarrow O_2 + 4\,H^+ + 4\,e^-$ (ANODE)

REDUCTION OF WATER: $\quad 4\,H_2O + 4\,e^- \longrightarrow 2\,H_2 + 4\,OH^-$ (CATHODE)

These reactions produce a mixture of hydrogen and oxygen in the atmosphere in the top of the battery. If this mixture is accidentally sparked, an explosion results. For this reason it is a good idea always to open a battery carefully and not introduce any sparks or open flames near a lead storage battery.

During starts, especially during extremely cold weather, the battery works very hard. This means that the battery must be recharged. Recharging often causes elongated crystals to grow on the electrode surfaces as the lead and lead oxide are redeposited on the negative and positive electrodes. Often these crystals of lead and lead oxide grow between the electrodes, causing internal short circuits. Usually, when this happens, the battery is "dead" and must be replaced. If electrolyte fluid runs low, the electrode surfaces dry, tending to make the surfaces not recharge properly.

All in all, the lead storage battery is relatively inexpensive, reliable, and relatively simple and has an adequate life. Its high weight is its major fault. Newer secondary batteries have found use in some applications such as electronics, but none of these newer batteries can perform like the lead storage battery does for its cost.

Nickel-cadmium batteries ("Ni-Cad") are another popular secondary battery. Being lightweight and producing a constant voltage until discharge, these batteries are popular in cordless appliances, video camcorders, portable radios, and other applications. They suffer somewhat by having a dis-

Edison invented the NiCd battery in 1900.

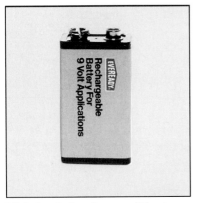

Nickel-cadmium battery. (Courtesy Eveready Battery Company.)

Fuel cells are energy conversion devices. Batteries are energy storage devices.

Figure 10–15   How an alkaline fuel cell works. The net reaction is $2 H_2 + O_2 \rightarrow 2 H_2O$. Hydrogen is oxidized at the anode, and oxygen is reduced at the cathode. The water molecules produced are discharged into the alkaline electrolyte solution, which consists of potassium hydroxide (KOH) and water.

charge "memory," so the user should carefully follow the manufacturer's charging suggestions for maximum battery life.

NiCd batteries can be recharged because the reaction products are insoluble hydroxides that remain at the electrode surface. The anode reaction involves cadmium.

$$Cd + 2\,OH^- \longrightarrow Cd(OH)_2 + 2\,e^- \quad \text{(ANODE)}$$

The cathode reaction involves the reduction of the nickel ion.

$$NiO(OH) + H_2O + e^- \longrightarrow Ni(OH)_2 + OH^- \quad \text{(CATHODE)}$$

FUEL CELLS

Fuel cells, like batteries, have a cathode and anode separated by an electrolyte. Unlike batteries, which are energy storage devices, fuel cells are energy conversion devices. Most fuel cells convert the energy of oxidation–reduction reactions of gaseous reactants directly into electricity. They are a special application of oxidation–reduction chemistry. The most popularized application of fuel cells has been in the space program on board the Gemini, Apollo, and Space Shuttle missions.

Consider the reaction between hydrogen and oxygen to produce water and energy.

$$2 H_2 + O_2 \longrightarrow 2 H_2O + \text{energy}$$

As mentioned earlier in this chapter, if a mixture of hydrogen and oxygen is sparked, the energy is released suddenly in the form of a violent explosion. In the presence of a platinum gauze, these gases will react at room temperature, slowly heating the catalytic surface to incandescence. In a fuel cell (Fig. 10–15), the oxidation of hydrogen by oxygen takes place in a controlled manner, with the electrons lost by the hydrogen molecules flowing out of the

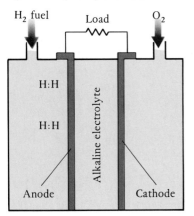

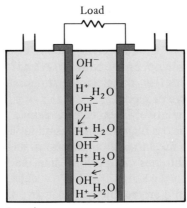

Reaction:
$$2 H_2 + 4\,OH^- \rightarrow 4 H_2O + 4\,e^-$$

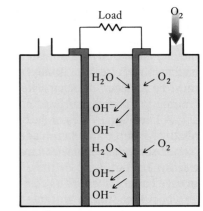

Reaction:
$$2 H_2O + O_2 + 4\,e^- \rightarrow 4\,OH^-$$

Figure 10-16   A hydrogen-oxygen fuel cell like the kind used on spacecraft. (United Technologies.)

fuel cell and back in again at the electrode where oxygen is reduced. This electron flow powers the electrical needs of the spacecraft, or whatever else is connected to the fuel cell. The water produced in the fuel cell can be purified for drinking purposes.

Because of their light weight and their high efficiencies compared to batteries, fuel cells like the one shown in Figure 10-16 have proved valuable in the space program. Beginning with Gemini 5, alkaline fuel cells have logged over 10,000 hours of operation in space. The fuel cells used aboard the Space Shuttle deliver the same power that batteries weighing ten times as much would provide. On a typical 7-day mission, the Shuttle fuel cells consume 1500 lb of hydrogen and generate 190 gal of potable water.

Other types of fuel cells that have been developed use air as the oxidizer and less-pure hydrogen or carbon monoxide as the fuel. It is hoped that fuel cells capable of direct air oxidation of cheap gaseous fuels such as natural gas will eventually be developed. Recent advances in ceramics (Chapter 11) have led to fuel cells that may be able to burn much cheaper fuels.

Fuel cells are about 60% efficient in converting chemical energy to electricity.

Potable means drinkable.

ELECTRIC AUTOMOBILES

Automobiles that derive their power from chemical energy stored in batteries have been around since the late 1800s. The earliest electric cars were very elegant and expensive for their time. They derived their electrical power from lead-acid batteries (see Fig. 10-14). Compared with the problems of the early gasoline-powered cars, such as hard starting and temperamental engines that sometimes didn't run too well on the gasoline of the day, electric cars seemed liked the better choice—just get in, push the accelerator, and go. By about 1920, however, electric cars had all but disappeared from the roads as gasoline engines became more powerful and offered far greater speed and range. By then the electric starter had become standard equipment.

Electric cars can be manufactured now. In fact large fleets of electric trams, postal vehicles, delivery vans, and recreational vehicles are currently in use. The problem with these vehicles is the same as when the electric car lost out to the gasoline car—ease of use, speed, range, and now, three generations of drivers who have grown up with these features and who don't seem to want to give them up.

The electric-powered G van. (Courtesy of Electric Power Research Institute.)

SELF-TEST 10–D

1. What is the strongest oxidizing agent?
2. What is the strongest reducing agent?
3. What happens when a zinc rod is placed in a copper salt solution?
4. What acid is found in the lead acid battery?
5. When magnesium is connected to a piece of iron, this is an example of _____ protection.
6. Batteries that can be recharged are called _____ cells.
7. Batteries that use up their chemical energy and cannot be recharged are called _____ cells.
8. A cell in which hydrogen and oxygen react on a catalytic surface is called a _____ cell.

MATCHING SET

_____ 1.	Chemicals in automobile lead storage battery	a. Fluorine
_____ 2.	Most active metal in Table 10–4	b. $2 H_2 + O_2 \rightarrow 2 H_2O$
_____ 3.	Ore of mercury	c. Nitric oxide
_____ 4.	Oxidizing agent used to purify drinking water	d. Nickel-cadmium
_____ 5.	Product of incomplete combustion of carbon	e. Reduction
_____ 6.	Most oxidized form of carbon	f. Carbon monoxide
_____ 7.	Oxidation	g. Combustion
_____ 8.	Reduction	h. Pb, PbO_2, H_2SO_4
_____ 9.	Gain of hydrogen	i. Causes solutions to conduct electricity
_____ 10.	Pollutant from oxidation of sulfur in fuels	j. Loss of electrons
_____ 11.	Pollutant from oxidation of nitrogen in air	k. Cinnabar
_____ 12.	Electrolyte	l. Carbon dioxide
_____ 13.	Strongest oxidizing agent	m. Acetaldehyde
_____ 14.	Secondary battery	n. Gain of electrons
_____ 15.	A definitive name for oxidation	o. Li
_____ 16.	Fuel-cell reaction	p. SO_2
_____ 17.	Product of ethanol oxidation in the liver	q. Chlorine
_____ 18.	Metal used as the anode in a lightweight commercial battery.	r. Methane
		s. $Fe_2O_3 \cdot x H_2O$
		t. Reduction

QUESTIONS

1. Write equations for the following reactions:
   a. The oxidation of methanol to formaldehyde
   b. The reduction of copper oxide (CuO) with hydrogen
   c. The reduction of copper ions (Cu^{2+}) to metallic copper at a cathode
   d. The burning of ethyl alcohol (C_2H_5OH) to form carbon dioxide and water
2. Give three examples of undesired oxidation.
3. Write formulas for four metal oxides and four non-metal oxides.
4. Why would rusting of automobiles be less of a problem in Arizona than in Chicago?
5. Which do you think is more highly oxidized?
   a. MnO_2 or MnO
   b. PbO or PbO_2
   c. H_2O or H_2O_2
   d. K_2SO_4 or $K_2S_2O_8$
6. Give three uses for common oxidizing agents.
7. Which is the oxidized form of carbon, CO or C?
8. When iron is converted to rust, what is the iron said to be?
9. When iron ore is converted to iron, what is the iron said to be?
10. In electrolysis, reduction takes place at the negative electrode. What is this electrode called?
11. If an electrode is negatively charged, it will tend to attract ions of what charge?
12. When water is electrolyzed, what are the products of oxidation and reduction?
13. Using Table 10–4, which gives relative strengths of oxidizing and reducing agents, predict whether the following reactions would be expected to occur.

   a. $2\ Na + Fe^{2+} \rightarrow 2\ Na^+ + Fe$
   b. $Ca + F_2 \rightarrow Ca^{2+} + 2\ F^-$
   c. $2\ H^+ + Cu \rightarrow Cu^{2+} + H_2$
   d. $Zn^{2+} + Fe \rightarrow Fe^{2+} + Zn$
14. What is the purpose of a salt bridge in a battery?
15. Name two other examples of metallic corrosion besides rusting.
16. What is the effect of salts in the rusting process?
17. Would you expect an iron object to rust on the surface of the moon? Explain.
18. Describe two kinds of cathodic protection—one in which the cathode is sacrificed and one in which it is not.
19. What is the name given to coating steel with zinc?
20. State a general difference between a primary and a secondary battery.
21. What does the density of the electrolyte solution in a lead storage battery tell about the state of charge of the battery?
22. What is one cause of short circuits in lead storage batteries?
23. How is a fuel cell similar to most batteries?
24. How is a fuel cell different from most batteries?
25. What is the fuel in the type of fuel cells used on board spacecraft? What is the oxidant?
26. Explain why the spacecraft fuel cells are expensive to operate.
27. In many fuel cells the oxidant is pure oxygen. What would be a less expensive oxidant to use in fuel cells?
28. Besides electricity, what is produced in the NASA fuel cells?

11

Chemical Raw Materials and Products from the Earth, Sea, and Air

A basic oxygen furnace in operation. Molten pig iron is poured into the furnace in which a blast of oxygen removes excess phosphorus, carbon, and sulfur; the result is a steel, the first metal in the operation of our technological society. (Brownie Harris.)

The long view from space has dramatized what we already knew — the crust of the Earth is a very unusual environment, uniquely suited, at least in this solar system, for the production and support of life forms. Note again from Figure 2–12 that the chemical composition of the Earth's crust differs dramatically from that of the universe or even the composition of the Earth as a whole. The element oxygen (O_2) dominates at about 50%, being in the air, water, and rocks. Silicon (Si), at about 25%, is a major part of silicate rocks, clays, and sand. Then come the major metals, iron (Fe), calcium (Ca), sodium (Na), potassium (K), and magnesium (Mg), found mostly in mineral deposits and in the sea. Hydrogen (H_2), at less than 1% by weight, is ninth in abundance, and carbon (C), the central element in all life forms, is present in little more than trace amounts (Fig. 11–1). It is apparent, then, that life forms are a very small part of the whole and are clinging to the edge of the spaceship Earth.

Our environment is also quite heterogeneous in nature. Mixtures abound; everywhere we look the elements and compounds are almost lost in the complicated array of mixtures that have resulted from natural forces over a very long time.

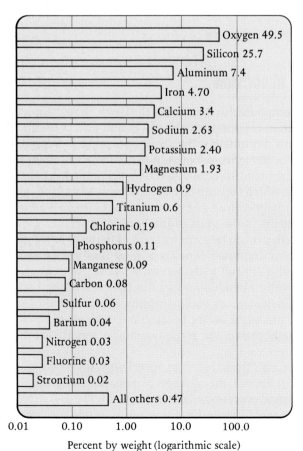

Figure 11–1   Abundance of the elements in the crust of the Earth (percentage by weight).

Oxygen 49.5
Silicon 25.7
Aluminum 7.4
Iron 4.70
Calcium 3.4
Sodium 2.63
Potassium 2.40
Magnesium 1.93
Hydrogen 0.9
Titanium 0.6
Chlorine 0.19
Phosphorus 0.11
Manganese 0.09
Carbon 0.08
Sulfur 0.06
Barium 0.04
Nitrogen 0.03
Fluorine 0.03
Strontium 0.02
All others 0.47

0.01   0.10   1.00   10.0   100.0

Percent by weight (logarithmic scale)

Throughout most of history, we had not developed the power to alter our environment significantly. Most of the materials used, such as the wooden hammer or plow, were only physically changed from the natural material. Then came the chemical reduction of copper from its ores, followed by iron, and now a flood of new chemicals are produced each year. We have now developed, beyond question, the power to change the chemical mixtures that are immediately around us. It is quite possible that we can control human-produced changes for the advantage of living things or, in contrast, be capable of destroying all life on Earth, and our present record is somewhere in between these extremes. Good solutions to the problems related to the control and use of these human-made changes in nature will require the common sense of all who will participate. Certainly information and advice will be required from social, biological, and physical scientists if we are to achieve solutions that are economically, environmentally, and ethically sound.

In this chapter we shall look at the chemistry of some major industrial operations on which our culture depends. However, the story only begins here, for the remainder of this text will deal with an ever-increasing number of chemical changes that utilize the natural mixtures of the Earth, sea, and air around us.

We shall begin with the study of the chemistry of the major metals used in our society.

METALS AND THEIR REDUCTION

Metals occur mostly as compounds in the crust of the Earth, though some of the less active metals such as copper (Cu), silver (Ag), and gold (Au) can be found also as free elements. Fortunately, the distribution of elements in the crust is not uniform. Some elements that are not particularly abundant are familiar to us because they tend to occur in very concentrated, localized deposits, called **ores,** from which they can be extracted economically. Examples of these are lead (Pb), copper (Cu), and tin (Sn), none of which is among the more abundant elements in the crust of the Earth (Fig. 11–1). Other elements that actually form a much larger percentage of the crust are almost unknown to us because concentrated deposits of their ores are less commonly found or the metal is difficult to extract from its ore. An example is titanium (Ti), the tenth most abundant element in the crust of the Earth. Although the ores of Ti, known as rutile (mostly TiO_2), and ilmenite ($FeTiO_3$), are common, the metal is rare because it is difficult to reclaim from the ores. (Some useful metals and related common ores are listed in Table 11–1.)

The preparation of metals from their ores, **metallurgy,** involves chemical reduction (Chapter 10). Indeed, the concepts of oxidation and reduction developed from metallurgical operations. Iron in iron oxide (Fe_2O_3) is in the form of Fe^{3+}. If we *reduce* Fe^{3+} ions to Fe atoms, we must find a source of electrons. (Recall that oxidation is the loss of electrons and that reduction is the gain of electrons.) Sometimes the desired metal is in solution (e.g.,

A continual search is under way for new ore deposits.

TABLE 11–1 Some Common Metals and Their Minerals

Metal	Chemical Formula of Compound of the Element	Name of Mineral
Aluminum	$Al_2O_3 \cdot xH_2O$	Bauxite
Calcium	$CaCO_3$	Limestone
Chromium	$FeO \cdot Cr_2O_3$	Chromite
Copper	Cu_2S	Chalcocite
Iron	Fe_2O_3	Hematite
	Fe_3O_4	Magnetite
Lead	PbS	Galena
Manganese	MnO_2	Pyrolusite
Tin	SnO_2	Cassiterite
Zinc	ZnS	Sphalerite
	$ZnCO_3$	Smithsonite

Iron ore samples. (Courtesy of Bethlehem Steel Company.)

magnesium in the sea, where Mg exists in the oxidized form as Mg^{2+} ions). To obtain free metallic magnesium, we must add electrons to Mg^{2+} ions (reduction) to produce neutral atoms.

Reduction of magnesium: $Mg^{2+} + 2\,e^- \rightarrow Mg$

Iron and Steel

The sources of most of the world's iron are large deposits of the iron oxides in Minnesota, Sweden, France, Venezuela, Russia, Australia, and England (Fig. 11–2). In nature these oxides are frequently mixed with impurities, so the production of iron usually incorporates steps to remove such impurities. Iron ores are then reduced to the metal by using carbon, in the form of coke, as the reducing agent.

Iron ore is reduced in a blast furnace (Fig. 11–3). The solid material fed into the top of the blast furnace consists of a mixture of an oxide of iron (Fe_2O_3), coke (C), and limestone ($CaCO_3$). A blast of heated air is forced into the furnace near the bottom. Much heat is liberated as the coke burns, and

Iron ores are mixtures containing iron compounds. To get iron from the ores, the iron in the compounds must be reduced.

Figure 11–2   Open-pit mining of iron ore. (Courtesy of Bethlehem Steel Company.)

Figure 11–3   Diagram of a blast furnace used for the reduction of iron from iron ore.

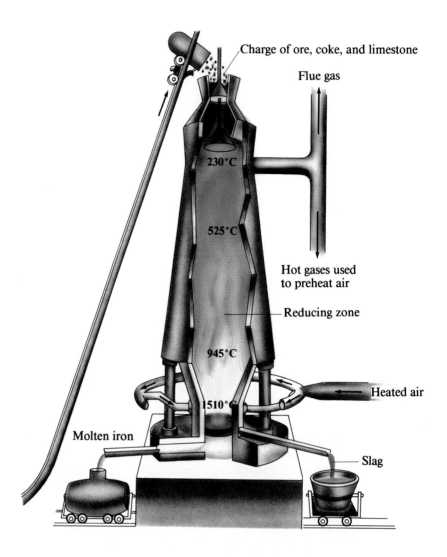

the heat speeds up the reaction, the speed being important in making the process economical. The reactions that occur within the blast furnace are

$$2\,C + O_2 \longrightarrow 2\,CO + heat$$

Carbon   Oxygen   Carbon Monoxide

$$Fe_2O_3 + 3\,CO \longrightarrow 2\,Fe + 3\,CO_2 + heat$$

Iron Oxide   Carbon Monoxide   Iron   Carbon Dioxide

Limestone (calcium carbonate) is added to remove the silica (SiO_2) impurity.

$$CaCO_3 \xrightarrow{\text{Heat}} CaO + CO_2$$

Calcium Carbonate   Calcium Oxide   Carbon Dioxide

$$CaO + SiO_2 \longrightarrow CaSiO_3$$

Calcium Oxide   Silicon Dioxide   Calcium silicate

The calcium silicate, or **slag,** exists as a liquid in the furnace. Consequently, as the blast furnace operates, two molten layers collect in the bot-

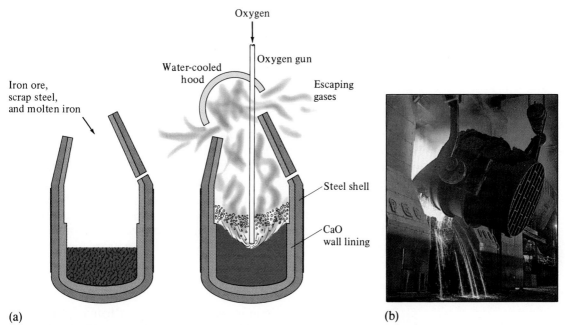

(a)   (b)

**Figure 11–4   (a) The basic oxygen process furnace. Much of the steel manufac-
tured today is refined by blowing oxygen through a furnace charged with scrap and
molten iron from a blast furnace. Measured amounts of alloying elements determine
the particular steel produced. (b) Molten steel poured from a basic oxygen furnace.
(Courtesy of Bethlehem Steel.)**

tom. The lower, denser layer is mostly liquid iron that contains a fair amount
of dissolved carbon and often smaller amounts of other impurities. The
upper, lighter layer is primarily molten calcium silicate with some impurities. From time to time the furnace is tapped at the bottom, and the molten
iron is drawn off. Another outlet somewhat higher in the blast furnace can be
opened to remove the liquid slag.

As it comes from the blast furnace, the iron contains too much carbon
for most uses. If some of the carbon is removed, the mixture becomes
structurally stronger and is known as **steel.** Steel is an alloy of iron with a
relatively small amount of carbon (less than 1.5%); it may also contain small
percentages of other metals. In order to convert iron into steel, the excess
carbon is burned out with oxygen.

*An alloy is a metal mixture consisting of
two or more elements.*

There are several techniques for burning the excess carbon. A recent
development that has been very widely adopted is the basic oxygen process
(Fig. 11–4). In this process pure oxygen is blown into molten iron through a
refractory tube (oxygen gun) that is pushed below the surface of the iron. At
elevated temperatures, the dissolved carbon reacts very rapidly with the
oxygen to give gaseous carbon monoxide and carbon dioxide, which then
escape.

After the carbon content has been reduced to a suitable level, the molten
steel is formed into desired shapes. During processing, the steel is subjected
to carefully regulated heat treatment to ensure that it has a uniform crystal-
linity, which in turn determines its malleability, toughness, and other useful
mechanical properties.

All of the processes in steelmaking, from the blast furnace to the final heat treatment, use tremendous quantities of energy, mostly in the form of heat. In the production of a ton of steel, approximately one ton of coal or its energy equivalent is consumed.

Aluminum

Aluminum, in the form of Al^{3+} ions, constitutes 7.4% of the Earth's crust. However, because of the difficulty of reducing Al^{3+} to Al, only recently have we learned to isolate and use this abundant element. Aluminum metal is soft and has a low density. Many of its alloys, however, are very strong. Hence, aluminum is an excellent choice when a lightweight, strong metal is required. In structural aluminum, the high chemical reactivity of the element is offset by the formation of a transparent, hard film of aluminum oxide, Al_2O_3, over the surface, which protects the aluminum from further oxidation:

$$4\,Al + 3\,O_2 \longrightarrow 2\,Al_2O_3$$

The principal ore of aluminum contains the mineral bauxite, a hydrated aluminum oxide, $Al_2O_3 \cdot xH_2O$. Because impurities such as iron oxides in the ore have undesirable effects on the properties of aluminum, the ore is purified by the Bayer process. In the Bayer process the mixture of oxides is treated with a sodium hydroxide solution that dissolves aluminum oxide and leaves iron oxide, which is insoluble in the solution:

$$\underset{\text{A Solid Mixture}}{Al_2O_3 \cdot xH_2O \text{ and } Fe_2O_3} \xrightarrow{\text{NaOH}} \underset{\text{Solution}}{Al(OH)_4^- + Na^+ + Fe_2O_3}$$

The mixture is filtered, and $Al(OH)_3$ is then carefully precipitated out of the clear solution by the addition of carbon dioxide, an acid.

$$CO_2 + Al(OH)_4^- \longrightarrow Al(OH)_3 + HCO_3^-$$

The aluminum hydroxide is heated to transform it into pure anhydrous aluminum oxide:

$$2\,Al(OH)_3 \xrightarrow{\text{Heat}} Al_2O_3 + 3\,H_2O$$

Metallic aluminum is obtained from the purified oxide by the Hall-Heroult process, an electrolytic process that uses molten cryolite, Na_3AlF_6 (melting point 1000°C) [Fig. 11–5]. Cryolite dissolves considerable amounts of aluminum oxide, which in turn lowers the melting point of the cryolite solution. This mixture of cryolite and aluminum oxide is electrolyzed in a cell with carbon anodes and a carbon cell lining that serves as the cathode on which aluminum is deposited. As the operation of the cell proceeds, molten aluminum sinks to the bottom of the cell. From time to time the cell is tapped and molten aluminum is allowed to run off into molds.

Aluminum is used both as a structural and decorative metal and as an electrical conductor in high-voltage transmission lines. Aluminum competes with copper as an electric conductor because of the lower cost of

As a college chemistry student at Oberlin College, Charles Martin Hall was intrigued by the potential uses of aluminum and the difficulties involved in reducing this chemically active metal from its oxide. In electrolysis experiments in his family woodshed, Hall used batteries and a blacksmith's fire in 1886 to reduce the Al_2O_3 dissolved in a high-melting salt, cryolite, to metallic aluminum. Hall was 22 years old when he made this great discovery. Later he founded the Aluminum Corporation of America and died a multimillionaire in 1914. Independently, Paul Heroult, a Frenchman, made the same discovery at approximately the same time.

The top of the Washington Monument is a casting of aluminum made in 1884.

A BETTER ALUMINUM FOIL

The forte of the materials scientist is to try to control, using changes in chemistry and processing, the production of a material that possesses the unique properties required for a particular application. It is relatively easy to have ideas and dreams for new materials for new technological applications, but these applications must be facilitated or limited by our ability to transform the raw materials into the new material required.

One example is the story of a new kind of aluminum foil developed by researchers at Allied Signal Corporation. Chemically and physically, it's very different from household aluminum foil. The new kind of aluminum foil is an alloy containing iron, silicon, and other elements. In ordinary aluminum casting, the metal cools slowly and any added elements can separate partially and form different kinds of crystals within the foil. The interfaces between the different crystalline forms are weak points along which cracks could easily develop. Researchers at Allied Signal Corporation found that if they cooled the molten aluminum alloy quickly, up to a hundred degrees in a fraction of a second, the aluminum solidified before the different crystals could form. This quick cooling of the metal casting created a new aluminum alloy. Its enhanced strength comes from the iron and its heat resistance comes from the silicon.

The World of Chemistry (Program 19) "Metals."

aluminum. Larger diameter aluminum wires must be used to offset the lower electric conductivity of aluminum compared with copper.

About ten times more energy is needed to produce a ton of aluminum than to produce a ton of steel. List some of the ways energy is used in its various forms to prepare aluminum from its ore. Recycled aluminum stock from aluminum cans requires approximately one-half the energy needed to produce the same amount of the metal from the ore.

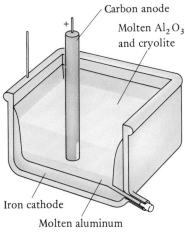

Figure 11–5 **(a) Schematic drawing of a furnace for producing aluminum by electrolysis of a melt of Al_2O_3 in Na_3AlF_6. The molten aluminum collects in the cathode container. (b) Molten aluminum flows from the electrolytic furnaces in the background into molds for casting the commercial product. (John Shaw/Tom Stack and Associates.)**

(a)   (b)

Figure 11–6   A mixture of two minerals containing copper. Azurite [2 CuCO₃ · Cu(OH)₂] is blue and malachite [CuCO₃ · Cu(OH)₂] is green. Based on the colors you may have observed, which mineral is likely to be found on copper statues? (Bryan Parker/Tom Stack and Associates.)

Gangue is composed of unwanted substances mixed with the desired mineral.

Copper

Although copper metal occurs in the free state in some parts of the world, the supply available from such sources is quite insufficient for the world's need. The majority of the copper obtained today is from various copper sulfide ores, such as $CuFeS_2$ (chalcopyrite), Cu_2S (chalcocite), and CuS (covellite). Because the copper content of these ores is around 1% to 2%, the powdered ore is first concentrated by the flotation process (Fig. 11–6).

In the flotation process, the powdered ore is mixed with water and a frothing agent such as pine oil. A stream of air is blown through the mixture to produce froth (Fig. 11–7). The **gangue** in the ore, which is composed of

Figure 11–7   (a) The flotation process. The lighter copper sulfide particles, containing the copper, are trapped in the bubbles of the foam and are passed over the spillway while the heavier waste is removed from the bottom of the mixing container. (b) The copper-containing foam flows out of the top of the flotation apparatus.

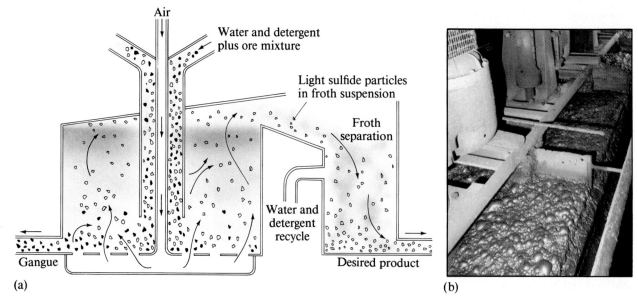

(a)

(b)

Figure 11–8 Near Bagdad, Arizona, copper ore is mined in an open-pit operation. (James Cowlin.)

sand, rock, and clay, is easily wet by the water and sinks to the bottom of the container. In contrast, a copper sulfide particle is hydrophobic — it is not wet by the water. The copper sulfide particle becomes coated with oil and is carried to the top of the container in the froth. The froth is removed continuously, and the floating copper sulfide minerals are recovered.

The preparation of copper metal from copper sulfide ore involves roasting in air to convert some of the copper sulfide and any iron sulfide present to the oxides:

$$2 \, Cu_2S + 3 \, O_2 \longrightarrow 2 \, Cu_2O + 2 \, SO_2$$
$$2 \, FeS + 3 \, O_2 \xrightarrow{\text{Roasting}} 2 \, FeO + 2 \, SO_2$$

Subsequently the mixture is heated to a higher temperature, and some copper is produced by the reaction:

$$Cu_2S + 2 \, Cu_2O \xrightarrow{\text{Heat}} 6 \, Cu + SO_2$$

The product of this operation is a mixture of copper metal and sulfides of copper, iron, other ore constituents, and slag. The molten mixture is heated in a converter with silica materials. When air is blown through the molten material in the converter, two reaction sequences occur. First, the iron is converted to a slag:

$$2 \, FeS + 3 \, O_2 \longrightarrow 2 \, FeO + 2 \, SO_2$$
$$FeO + SiO_2 \xrightarrow{\text{Molten Slag}} FeSiO_3$$

The remaining copper sulfide is then converted to copper metal. The reactions are:

$$2 \, Cu_2S + 3 \, O_2 \longrightarrow 2 \, Cu_2O + 2 \, SO_2$$
$$Cu_2S + 2 \, Cu_2O \longrightarrow 6 \, Cu + SO_2$$

The copper produced in this manner is crude or "blister" copper, the blistered surface resulting from the escaping gas. The blister copper is later purified electrolytically.

Review the electrolysis of metal solutions in Chapter 10.

305

Figure 11–9 **(a)** Copper from a smelter is purified in an electrolysis cell. The impure copper ore is oxidized at the anode and passes into solution. The oxidized copper is reduced from the solution onto the cathode as 99.95% pure copper. **(b)** Electrolysis cells for purifying copper.

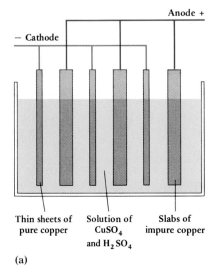

Thin sheets of pure copper   Solution of $CuSO_4$ and H_2SO_4 Slabs of impure copper

(a)

(b)

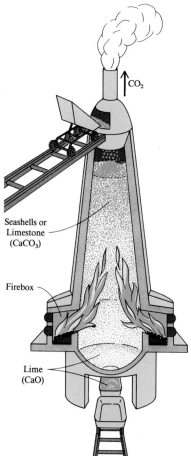

Commercial quantities of lime are prepared from sea shells or limestone by using heat to decompose the calcium carbonate into carbon dioxide and lime.

In the electrolytic purification of copper, the crude copper is first melted and cast into bars (anodes) that are placed in a solution of copper sulfate and sulfuric acid in water. The cathodes are made of pure copper. As electrolysis proceeds, copper is oxidized at the anode, moves through the solution as Cu^{2+} ions, and is deposited on the cathode. The voltage of the cell is regulated so that more active impurities (such as iron) are left in the solution, and less active ones are not oxidized at all. The less active impurities include gold and silver, which collect as "anode slime," an insoluble residue beneath the anode. The anode slime is subsequently treated to recover the valuable metals.

The copper produced by the electrolytic cell is 99.95% pure and is suitable for use as an electric conductor (Fig. 11–9). Copper for this purpose must be pure because very small amounts of impurities, such as arsenic, considerably reduce copper's electric conductivity.

Magnesium from the Sea

Magnesium, with a density of 1.74 g/mL, is the lightest structural metal in common use. For this reason magnesium is most often used in alloys designed for light weight and great strength. Magnesium is a relatively active metal chemically because it loses electrons easily. Magnesium "ores" include sea water, which has a magnesium concentration of 0.13%, and dolomite, a mineral with the composition $CaCO_3 \cdot MgCO_3$. Because there are 6 million tons of magnesium present as Mg^{2+} salts in every cubic mile of sea water, the sea can furnish an almost limitless amount of this element.

The recovery of magnesium from sea water (Fig. 11–10) begins with the precipitation of magnesium hydroxide by the addition of lime to sea water:

$$CaO + H_2O \longrightarrow Ca^{2+} + 2\ OH^-$$

$$Mg^{2+} + 2\ OH^- \longrightarrow Mg(OH)_2$$

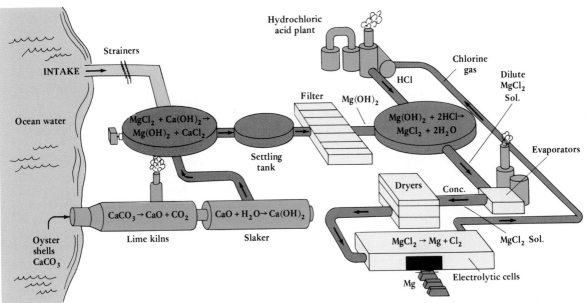

Figure 11–10   Diagram of an industrial plant showing steps necessary for recovering magnesium metal from seawater.

The magnesium hydroxide is removed by filtration and then neutralized with hydrochloric acid to form the chloride:

$$Mg(OH)_2 + 2\,H^+ + 2\,Cl^- \longrightarrow Mg^{2+} + 2\,Cl^- + 2\,H_2O$$

The water is then evaporated. This is followed by the electrolysis of molten magnesium chloride in a huge steel pot that serves as the negative electrode, or cathode. Graphite bars serve as the positive electrodes, or anodes.

$$\underset{\text{Melted}}{Mg^{2+} + 2\,Cl^-} \longrightarrow Mg + Cl_2$$

REDUCTION AT THE CATHODE: $Mg^{2+} + 2\,e^- \longrightarrow Mg$

OXIDATION AT THE ANODE: $2\,Cl^- \longrightarrow Cl_2 + 2\,e^-$

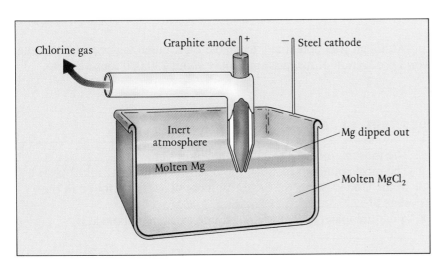

A cell for electrolyzing molten $MgCl_2$. Liquid magnesium metal is formed on the steel cathode and rises to the top where it is dipped off periodically. Chlorine gas is formed on the graphite anode and is piped off.

MATERIALS FROM THE EARTH

In the African bush veldt, an area the size of New England, located in southern Africa, interesting and valuable natural chemical separations have been made. Just beneath the surface lie some of the world's richest deposits of platinum, rhodium, and chromium. A mineral is a naturally occurring substance with a characteristic chemical composition. When minerals are concentrated and have economic value, they are called ores. Because some places are rich with valuable ores, like southern Africa, and others are not, the fortunes of whole countries can rise and fall based on the wealth found in the natural treasuries. Since prehistoric times, we have recovered and used a variety of minerals. How did our ancestors obtain the minerals and elements they needed from the Earth? One of the early techniques used in an ancient tin mine in Cornwall, England, was to build large fires at the base of the rock cliffs. The intense heat cracked the rocks, exposing the tin ore, which was then removed. Today we use dynamite and ammonium nitrate to blast rock away from ore-rich veins. But even though we have greatly increased our power to remove useful ores from the earth, we are still limited by effective and economical mining techniques and, to an ever increasing degree, the availability of exploitable ore deposits.

The World of Chemistry (Program 18) "The Chemistry of Earth."

Ancient tin mines at Cornwall, England. (*The World of Chemistry,* Program 18, "The Chemistry of Earth.")

Lime from oyster shells, methane from natural gas, and electricity are used to produce magnesium from sea water.

As the magnesium melts, it floats to the surface and is removed. The chlorine is recovered and reacts with air and natural gas (methane, CH_4) to form hydrochloric acid, which in turn is used to neutralize and dissolve the magnesium hydroxide:

$$4 \, Cl_2 + 2 \, CH_4 + O_2 \longrightarrow 2 \, CO + 8 \, HCl$$
$$\text{Methane}$$

The lime used to precipitate the magnesium as the hydroxide is obtained by heating limestone or oyster shells:

$$CaCO_3 \xrightarrow{\text{Heat}} CaO + CO_2$$
$$\text{Lime}$$

Although it is potentially available on a much larger scale, the total world production of magnesium is only about 250,000 tons per year.

SELF-TEST 11–A

1. The most abundant element in the Earth's crust is ＿＿＿＿＿＿ .
2. The most abundant metal in the Earth's crust is ＿＿＿＿＿＿ .
3. In the United States the largest iron ore deposits are found in the state of ＿＿＿＿＿＿＿＿＿ .
4. A natural material that is almost pure calcium carbonate is ＿＿＿＿＿＿ .

5. Which of the following metals may occur in the free or metallic state in mineral deposits: iron, copper, aluminum, and/or magnesium?

6. Which of these metals (iron, copper, aluminum, and/or magnesium) are either produced or purified using electricity?

7. Another name for calcium silicate as it applies to production of iron is _____ .

8. For most metals to be prepared from their ores, they must be ( ) oxidized ( ) reduced.

9. In an electrical refining process for metals, the purest metal will always be found at the ( ) anode ( ) cathode.

10. Which metal is sufficiently concentrated in the oceans to be extracted commercially: magnesium, aluminum, copper, or iron?

11. Most metals are found in the Earth as ( ) neutral atoms, ( ) positive ions, ( ) negative ions.

THE FRACTIONATION OF AIR

The atmosphere of the Earth is a fantastically large source of the elements nitrogen (N_2) and oxygen (O_2) and much smaller amounts of certain of the noble gases, including argon (Ar), neon (Ne), and xenon (Xe) (see Table 11–2). Figure 11–11 presents some of the basic facts about our atmosphere, including the naming of the stratified layers, the chemical species present in the layers, atmospheric pressures, and human penetrations.

Before pure O_2 and N_2 can be obtained from the air, water vapor (H_2O) and carbon dioxide (CO_2) must be removed. This is usually done by precooling the air through refrigeration or by using silica gel to absorb H_2O and lime to absorb CO_2. Afterward, the air is compressed to a pressure exceeding 100 times normal atmospheric pressure, cooled to room temperature, and allowed to expand into a chamber. This expansion produces a cooling effect (the **Joule-Thompson effect**) due to breaking the weak attractive London forces between the gaseous molecules. Recall that breaking bonds requires energy, so the expanding gas absorbs energy from its own moving molecules, thus slowing the molecules and thereby cooling the gas. If this expansion is repeated and controlled properly, the expanding air actually cools to the

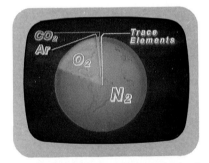

Atmospheric gases. (*The World of Chemistry,* Program 17, "The Precious Envelope.")

TABLE 11–2 The Dry Atmospheric Composition of the Earth at Sea Level

Gas	Percentage by volume
Nitrogen	78.084
Oxygen	20.948
Argon	0.934
Carbon dioxide	0.033
Neon	0.00182
Hydrogen	0.0010
Helium	0.00052
Methane	0.0002
Krypton	0.0001
Xenon	0.000008

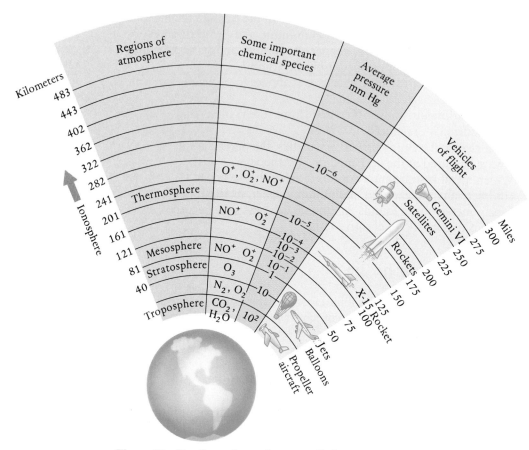

Figure 11–11   Some facts about our limited atmosphere. The troposphere was named by British meteorologist Sir Napier Shaw from the Greek word *tropos,* meaning "turning." The stratosphere was discovered by the French meteorologist Léon Philippe Teisserenc de Bort, who believed that this region consisted of an orderly arrangement of layers with no turbulence or mixing. The word *stratosphere* comes from the Latin word *stratum,* meaning "layer."

point of liquefaction (Fig. 11–12). The temperature of the liquid air is usually well below the boiling points of N_2 ($-195.8\,°C$), O_2 ($-183\,°C$), and Ar ($-189\,°C$). This liquid air is then allowed to partially vaporize again, and since N_2 is more volatile than O_2 or Ar (N_2 has a lower boiling point), the liquid becomes more concentrated in O_2 and Ar. This process, known as the **Linde process,** produces high-purity N_2 ($99.5+\%$) and O_2 with a purity of 99.5% (Fig. 11–13). Further processing produces pure Ar, Ne (boiling point $-246\,°C$), and even helium (He; boiling point $-268.9\,°C$), but most He used in the United States is separated from natural gas.

Oxygen

A small but vital use of oxygen is in breathing therapy.

Most oxygen produced by the fractionation of liquid air is used in steel-making, although some is used in rocket propulsion (to oxidize hydrogen) and in controlled oxidation reactions of other types. Liquid oxygen (LOX)

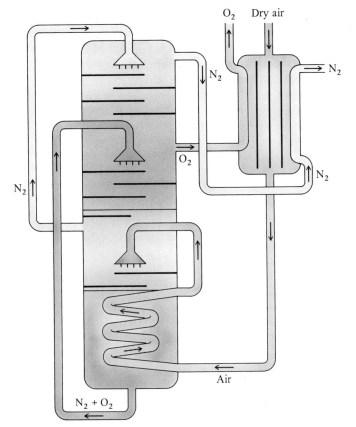

O₂   Dry air

N₂

N₂

O₂

N₂

N₂

N₂

Air

N₂ + O₂

Figure 11–12   Diagram of a fractionating column for separating oxygen and nitrogen from an air supply. Air cools when expanded through valves (three sets shown). In each expansion, the more volatile nitrogen rises toward the top of the column, where it is removed. Oxygen is removed below.

Figure 11–13   Liquid nitrogen. Note that the nitrogen is boiling at room temperature. Relate this to the boiling of water on a cook stove hot plate. (Charles D. Winters.)

can be shipped and stored at its boiling temperature of −183°C under atmospheric pressure. Substances this cold are called **cryogens** (from the Greek *kryos,* meaning "icy cold"). Cryogens represent special hazards since contact produces instantaneous frostbite, and structural materials such as plastics, rubber gaskets, and some metals become brittle and fracture easily at these low temperatures. Liquid oxygen can accelerate oxidation reactions to the point of explosion because of the high oxygen concentration. For this reason, contact between liquid oxygen and substances that will ignite and burn in air must be prevented.

Special cryogenic containers holding liquid oxygen are actually huge vacuum-walled bottles much like those used to carry hot soup or coffee. These containers can be seen outside hospitals or industrial complexes, on highways and railroads, and even aboard ocean-going vessels.

Nitrogen

Liquid nitrogen is also a cryogen. It has uses in medicine (cryosurgery), for example in cooling a localized area of skin prior to removal of a wart or other unwanted or pathogenic tissue. Since nitrogen is so chemically unreactive, it is used as an inert atmosphere for certain applications such as welding, and liquid nitrogen is a convenient source of high volumes of the gas. Because of its low temperature and inertness, liquid nitrogen has found wide use in

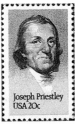

Joseph Priestley
USA 20c

Figure 11–14   A U.S. postage stamp honoring Joseph Priestley (1733–1804), who discovered elemental oxygen in 1774. He used sunlight and a magnifying glass to heat an oxide of mercury that released oxygen gas. Priestley left his native England for religious reasons to spend the last ten years of his life in Northumberland, Pennsylvania.

frozen food preparation and preservation during transit. Containers with nitrogen atmospheres, such as railroad boxcars or truck vans, present health hazards since they contain little (if any) oxygen to support life, and workers have died when they entered such areas without special breathing apparatus.

Large industrial amounts of nitrogen from air reduction are used in the Haber process (Chapters 1 and 19), wherein nitrogen is combined with hydrogen to produce ammonia (Fig. 11–15). The cost of the nitrogen reflects only the industrial cryogenic and distillation operations, because the raw material is free.

The first commercial U.S. ammonia plant began production in 1921.

$$N_2 + 3\ H_2 \rightleftharpoons \underset{\text{Ammonia}}{2NH_3}$$

Hydrogen for the Haber process is more difficult to obtain. At present, the principal source is the reaction of petroleum products such as propane $(CH_3CH_2CH_3)$ with steam in the presence of catalysts to produce hydrogen:

$$CH_3CH_2CH_3 + \underset{\text{Steam}}{6\ H_2O} \xrightarrow{\text{Catalyst}} 3\ CO_2 + 10\ H_2$$

Hydrogen can also be prepared by the electrolysis of water:

$$2\ H_2O \xrightarrow[\text{KOH}]{\text{Electricity}} 2\ H_2 + O_2$$

Note that the hydrogen is not free since industrial quantities of thermal or electrical energy are required.

Nitrogen in compounds is referred to as **fixed nitrogen,** which is readily available for both biological transformations and industrial manipulations, in contrast to nitrogen gas, which is relatively inert.

Noble Gases

Approximately 250,000 tons of argon, the most abundant noble gas in the air, are recovered each year in the United States. Most of the Ar is used to provide inert atmospheres for high-temperature metallurgical processes. If air is not excluded from these processes, unwanted oxides of the hot metals will form. Argon is also used as a filler gas in incandescent light bulbs in order to prolong the life of the hot filament.

Some background material on the noble gases appears in Chapter 4.

Relatively small amounts of Ne, krypton (Kr), and Xe are recovered from air for commercial purposes. Although Ne is used in the greatest amounts, all three are used in "neon" advertising signs (Fig. 11–16). There is an interesting potential medical use for Xe. Xenon, which is readily soluble in blood, acts as an inhalation anesthetic in much the same way as laughing gas, dinitrogen oxide (N_2O).

Helium (He) is not commercially recovered from air because it is cheaper to isolate it from natural gas where it is sometimes as much as 7% by volume. Helium is used for inert atmospheres, especially in welding. Deep-sea diving gases are mixtures of He and O_2 to avoid the dissolution of excess N_2 in the blood if air is used. Liquid He, boiling point $-268.9\,°C$, is used as a refrigerant when extremely cold temperatures are required.

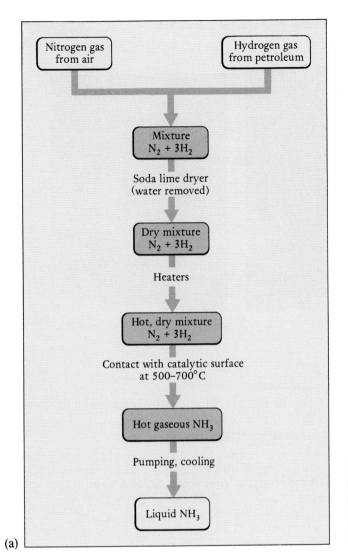

(a)

Figure 11–15 (a) A block diagram for the Haber process for synthesizing ammonia. (b) The mixed gases are pumped over the catalytic surface to produce the ammonia. Note that the ammonia is collected as a liquid and the uncombined nitrogen and hydrogen are recycled to the catalytic chamber.

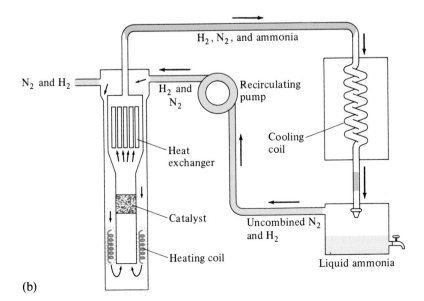

(b)

Figure 11 – 16   (a) The familiar neon light results from a high-voltage electric discharge through a tube of low-pressure neon gas. (b) The color of argon in a discharging tube is not as attention-getting. However, the more abundant argon is used to surround the filament in a low-voltage incandescent light bulb to protect the filament from oxidation and to reduce its evaporation due to heat. (Charles D. Winters.)

(a)

(b)

SELF-TEST 11 – B

1. Name the two elements that are commercially prepared by the fractionation of the air.
2. LOX is the industrial abbreviation for

   _____ .

3. What does the word *cryogen* mean?
4. Give four uses of the products from liquid air, and identify each with a particular element.

SILICON MATERIALS — OLD AND NEW

Silicon and oxygen make up 75% of the crust of the Earth; the bonding between these two elements in clays and rocks literally holds together the Earth's skin. The chemical structures involved are many and complex. However, a few typical molecular structures are common to some of the most important materials in our society, from glass to the computer chip.

Glass

Silicon dioxide, SiO_2, (also called silica), occurs naturally in large amounts in rocks and sand, or more rarely in much larger crystals called quartz (Fig. 11 – 17a). Silicon dioxide has a melting point of 1710°C. If the melted material is cooled rapidly, a noncrystalline solid is obtained. By way of contrast, crystalline quartz consists of an extended structure in which each Si atom is bonded tetrahedrally to four O atoms (Fig. 11 – 18). Each O atom is bonded to two Si atoms, and the bonding extends throughout the crystal. When silica is melted, some of the bonds are broken and the units move with respect to each other. When the liquid is cooled, the re-formation of the original solid requires a reorganization that is hard to achieve because of the difficulty the

(a)   (b)

Figure 11–17   Quartz is a crystal-line form of silicon dioxide. (a) In the pure form the crystals are clear. (Ward's Natural Science Establishment.) (b) Impurities in the quartz crystals can add color as in amethystine quartz. Iron produces the color in amethyst. (Allen B. Smith/Tom Stack and Associates.)

groups experience in moving. The very viscous liquid structure is thus partially preserved on cooling to give the characteristic feature of a glass, which is an apparently solid material (pseudo-solid) with some of the randomness in structure characteristic of a liquid. This random structure accounts for one of the typical properties of a glass: it breaks irregularly rather than splitting along a plane like a crystal.

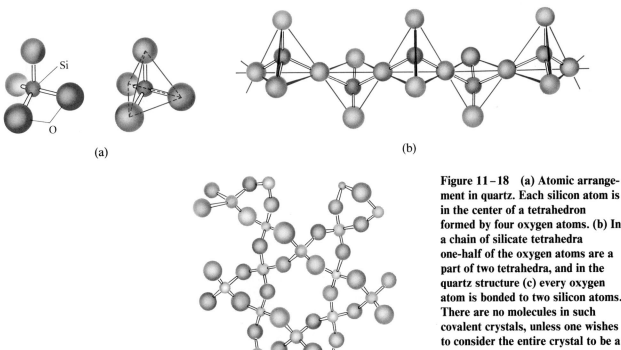

(a)   (b)

(c)

Figure 11–18   (a) Atomic arrangement in quartz. Each silicon atom is in the center of a tetrahedron formed by four oxygen atoms. (b) In a chain of silicate tetrahedra one-half of the oxygen atoms are a part of two tetrahedra, and in the quartz structure (c) every oxygen atom is bonded to two silicon atoms. There are no molecules in such covalent crystals, unless one wishes to consider the entire crystal to be a macromolecule.

Tetrahedral structure of silicon and oxygen in silicates. Silicate tetrahedra can be linked to form single, double, or multiple-width chains (a). Sheets (b) result when chain formation is extensive in two directions. (*The World of Chemistry,* Program 18, "The Chemistry of Earth.")

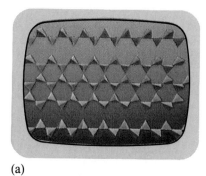

(a)

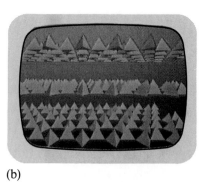

(b)

Sodium carbonate, or soda ash, (Na_2CO_3) is the number 11 commercial chemical. Calcium oxide is number 6. See inside the front cover.

$$Na_2CO_3 \xrightarrow{\text{Heat}} Na_2O + CO_2;$$
$$CaCO_3 \xrightarrow{\text{Heat}} CaO + CO_2$$

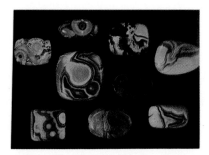

Figure 11–19 The fibrous crystals formed from chains in quartz make possible a mineral known as chalcedony. A special case of chalcedony, agate (shown above) has bands of color due to the layers of fibrous crystals colored with different impurities. (Gemological Institute of America.)

Cut and natural quartz. (Photo copyright Gemological Institute of America.)

By the addition of metal oxides to silica, the melting temperature of the mixture can be reduced from 1710°C to about 700°C. The oxides most often added are sodium oxide (added as Na_2CO_3, soda ash) and calcium oxide (added as $CaCO_3$). The addition of a metal oxide to molten silica supplies additional oxygen atoms to occupy the tetrahedral points around each silicon atom. The result is a **silicate** in which some of the oxygen atoms are bonded to only one silicon atom. This allows for chains of silicate tetrahedra (as in pyroxenes), linked chains (as in amphiboles), or sheets (as in clays and mica). Each added oxygen from the metal oxide carries a −2 charge, so the metal ions are ionically bonded to the negatively charged silicate chains or sheets. Like quartz, pyroxenes and amphiboles are crystalline minerals that readily break along crystalline lines. Mica is more familiar as it has been used widely as an electric and thermal insulator. Sheets, or plates, of mica are easily separated from a crystal with a pen knife or even your thumb nail. In glass-making, the crystalline order of natural silicates is not achieved. Instead glass has a pseudo-solid form with a random and snarled arrangement of the chain-linked silicate tetrahedra (Fig. 11–18a). As you know, glass breaks along random lines reflecting no crystalline structure. Sodium glass, so-called soda-lime glass, has a relatively low melting temperature and is widely used because the furnace used to melt the glass can be operated at a lower temperature; energy costs are thus reduced.

It is thought that Phoenician sailors were the first humans to make glass, around 5000 B.C. Blocks of sodium carbonate, from the ship's cargo, and sand were used to make a fire bed insulated from the wooded deck. The glass that flowed from the charcoal fire resembled obsidian, a natural glassy mate-

TABLE 11–3 Substances Used in Colored Glasses

Substance	Color
Copper (I) oxide	Red, green, or blue
Tin (IV) oxide	Opaque
Calcium fluoride	Milky white
Manganese (IV) oxide	Violet
Cobalt (II) oxide	Blue
Finely divided gold	Red, purple, or blue
Uranium compounds	Yellow, green
Iron (II) compounds	Green
Iron (III) compounds	Yellow

rial that has been valued since antiquity. Blowpipe control in making glass vessels came with the iron age some 4700 years later.

Soda-lime glass will be clear and colorless only if the purity of the ingredients has been controlled carefully. If, for example, too much iron oxide is present, the glass will be green. Other metal oxides produce other colors (see Table 11–3). To some extent, one color can counteract another.

The substances are melted together in a gas- or oil-fired furnace. As they react, bubbles of CO_2 gas are evolved.

$$CaCO_3 + SiO_2 \longrightarrow CaSiO_3 + CO_2$$

$$Na_2CO_3 + SiO_2 \longrightarrow Na_2SiO_3 + CO_2$$

The mixture is heated to about 1500°C to remove the bubbles of CO_2. At this temperature the viscosity is low, and the bubbles of entrapped gas escape easily. The mixture is cooled somewhat and then is blown into bottles by

Figure 11–20 Commercial varieties of asbestos are silicates and are valued because of their high tensile strength and resistance to heat. Asbestos has attracted much attention in commercial use since it has been learned that particulates of this material when inhaled cause lung cancer. (Particulate Mineralogy Unit, Avondale Research Center, U.S. Bureau of Mines.)

(a)

(b)

Figure 11–21   (a) The color of glass depends on the metallic content. The sodium and calcium glasses are clear and colorless like pure quartz. (b) Selenium glass is red due to the presence of colloidal particles of this element. (Photographs by J. Morgenthaler.)

TABLE 11–4 Special Glasses

Special Addition or Composition	Desired Property
Large amounts of PbO with SiO_2 and Na_2CO_3	Brilliance, clarity, suitable for optical structures: crystals or flint glass
SiO_2, B_2O_3, and small amounts of Al_2O_3	Small coefficient of thermal expansion: borosilicate glass. "Pyrex," "Kimax," and others
One part SiO_2 and four parts PbO	Ability to stop (absorb) large amounts of X rays and gamma rays: lead glass
Large concentrations of CdO	Ability to absorb neutrons
Large concentrations of As_2O_3	Transparency to infrared radiation
Suspended Se particles	Red color

machines, drawn into sheets, or molded into other forms (Fig. 11–21). It is possible to incorporate a wide variety of materials into glass for special purposes. Some examples are given in Table 11–4.

Ceramics

Ceramic materials have been made since well before the dawn of recorded history. They are generally fashioned from clay or other natural earths at room temperature and then permanently hardened by heat. Clays have a wide variety of properties that are found in a considerable range of ceramic materials, from bricks to table china. The techniques developed with natural clay have been applied to numerous other inorganic materials in recent years. The result has been a considerable increase in the kinds of ceramic materials available. One can now obtain magnetic ceramics as well as ceramics suitable for rocket nozzles—both were developed from mixtures of inorganic oxides by the use of ceramic technology. Basic to the process is the heating of the materials to make them hard and resistant to wear.

The three major ingredients of common pottery are silicate minerals: clay, sand, and feldspar. The term **clay** includes materials with a wide range of chemical compositions that are produced from the weathering of granite and gneiss rocks. Sand is mainly silicon dioxide, SiO_2. Feldspars are aluminosilicates containing potassium, sodium, and other ions in addition to silicon and oxygen. An approximation of a part of the weathering process of granite and gneiss rocks can be made if we write feldspar as a mixture of oxides: $K_2O \cdot Al_2O_3 \cdot 6SiO_2$; weathering includes, among other reactions, the reaction of the mineral with water containing dissolved carbon dioxide to form clay.

$$K_2O \cdot Al_2O_3 \cdot 6SiO_2 + 2\ H_2O + CO_2 \longrightarrow$$
$$\underbrace{Al_2O_3 \cdot 2SiO_2 \cdot 2H_2O}_{\text{A Clay}} + 4\ SiO_2 + 2K^+ + CO_3^{2-}$$

The essential feature of the clay mineral is that it occurs in the form of extremely minute platelets, which, when wet, are malleable, meaning that they can be shaped easily. When dry, the clay platelets are locked without the lubrication provided by water between the platelets; if heated to an elevated temperature, the mass of platelets becomes permanently rigid due to extensive formation of silicate chains, and is no longer subject to dispersion in water. When natural clays are mixed with feldspars (the parent material for clay formation) and silica and heated, the mixture of crystals produced are held together by the matrix of glasslike material. The clays can be used by themselves to make bricks, flowerpots, and clay pipe, but finer quality ceramic materials contain purified clays and other ingredients in carefully controlled proportions. The clay is mixed with potter's flint (a form of silica) and feldspar in various proportions. The clay makes the mixture pliable; the silica decreases the amount of shrinkage that occurs after drying and firing in a kiln; the feldspar lowers the temperature needed for adequate firing. Clay must usually be dried before being fired; otherwise the rapid loss of water from the surface and the slower loss from the interior of the object will cause cracks in the finished product.

In general, natural clays are extremely complex mixtures. If these are used in ceramics without treatment, the finished materials have a color and physical properties characteristic of the impurities present. The first pieces of fine Oriental chinaware arrived in Europe during the late Middle Ages, and European potters envied and admired the obviously superior product. This led to the beginning of systematic studies on the effect of composition on the nature of the ceramic produced and to a keen appreciation of the role of the purity of the clay in determining the color and potential quality and value of the piece (Fig. 11–22).

Alchemists made notable achievements in the production of fine china. One of these, Johann Friedrich Bottger, worked from about 1705 to 1719 for King Augustus of Saxony, who kept him almost as a prisoner. The king hoped to gain power from the alchemist's discoveries. Bottger succeeded in developing several novel ceramic materials, of which the most important was the first white glazed porcelain made in Europe (in 1709). Bottger devoted the rest of his life to the perfection of the manufacture and decoration of this material, in which he enjoyed considerable success. His china was made in the town of Meissen and was both glazed and vitrified. Glazing is the coating of ceramic pieces with a material that melts during the firing, producing an impermeable layer on the surface. Vitrification is a process whereby the clay is fired at a temperature sufficient to melt a portion of the material and, in effect, produce an impermeable glass that holds the remaining particles together.

With a market at $10 billion per year for advanced ceramics, industrial need for new materials as well as fundamental research are causing an explosion in new ceramic materials being produced. New ceramic materials are attractive for several reasons. The starting materials for making them are readily available and cheap. Ceramics are lightweight in comparison to metals and retain their strength at high temperatures (above 1000°C), where metal parts tend to fail. Ceramic materials are now being used in cutting tools, and in electrical, optical, and magnetic applications in the computer and electronic industries. The potential uses of ceramics in transportation appear dramatic.

However, ceramics deform very little before they fail catastrophically, the failure resulting from a weak point in the bonding within the ceramic matrix. Such weak points are not consistent from object to object so that the

Figure 11–22   Fine china. The qualities of the china are dependent on the additives in the clay mix. This cup is bone china, the white color resulting from pyrophosphate (from bone meal)-iron complexes.

Vitrification is the formation of glass by the heating of silicate materials.

predictability of failure is poor. The one severely limiting problem in utilizing ceramics is their brittle nature.

The single greatest use for advanced ceramics is in the electrical insulation of integrated-circuit packages and substrates to support electrical elements in all sorts of electrical instrumentation. Kyoto Ceramics (Kyocera in the U.S.) has two-thirds of the world market for the "new stone age" scissors and knives that are widely available. Nissan has equipped a deluxe series of its cars with a turbocharger containing ceramic blades that operate more efficiently. Garrett Corporation and Ford Motor Company are jointly testing a 100-hp (horsepower) ceramic turbine engine. Each new application calls for experimentation with new human-made ceramics. In the 1950s and 1960s, there was an explosion in the production of new high-molecular-weight organic compounds, the synthetic polymers (Chapter 14), which literally changed the way we live. Quite likely we are now poised on the verge of another chemical revolution, as we copy nature again, in making what appears to be a limitless number of high-molecular-weight inorganics, the applied uses of which will again change the way we live.

Since the stress failure of ceramic materials is due to molecular abnormalities resulting from impurities or disorder in the basic atomic arrangements, much attention is now being given to purer starting materials and the control of the processing steps. If one mixes the oxide powders and heats to 1500°C, anomalies in structure can be up to 50 μm in size, the size of the powder grain. The **solution-sol-gel technique,** developed after World War II, allows for intimate mixing of the ingredients down to the level of 0.5 nm. The oxides are dissolved, usually in an aqueous medium, so that solution species are homogeneously mixed. Solution concentrations are gradually changed, usually by changing pH, so some of the solution species begin to precipitate. The sol stage, colloidal solid particles dispersed in a liquid, goes to a gel stage where the solid particles link into a network trapping the liquid (just like making Jello). The gel can be formed in the desired shape, dried, and then fired to produce the ceramic object.

The possibilities for placing chemically active structures or catalytically active surfaces on ceramic materials appear unlimited in what has come to be known as **ultrastructure processing.** For example, imagine a ceramic structure put together like a stack of corrugated pieces of cardboard. At 1000°C pass a fuel such as gasoline and air through adjacent channels. If the ceramic material between the air and fuel is chemically modified to pass ions but not electrons, if the walls of the fuel channels are catalytically active to facilitate oxidation (loss of electrons from fuel molecules), if the air channels are active to reduce the oxygen to oxide ions, and if separate strips of electron-conducting ceramics connect together all of the fuel-channel walls on one side and also all of the air-channel walls on the other side, we have a fuel cell (Chapter 10, Figure 10–15). At the present time, fuel cells used in space travel are too heavy to power automobiles efficiently. But researchers at Argonne National Laboratory are working on the kind of fuel cell described above, which at 600 lb will develop 231 hp of electric energy. Test models show that fuel efficiency will go from 30% in the internal combustion engine to 50% in the fuel-cell electric car. At operating temperatures for the fuel cell, temperatures where metals would fail, the electric motor is controlled by

An old demonstration: Water glass is a solution of silica in strong base.
$SiO_2 + 2\ NaOH \rightarrow Na_2SiO_3 + H_2O$.
Add water to a solution of water glass to lower the pH and a gel forms. Reach in and grab a handful of the gel and press into a ball. The ball of sodium silicate-water gel will bounce nicely off the floor—until it begins to dry.

current flow which, in turn, is controlled by fuel flow. The products from such a propulsion system would be carbon dioxide and water. There are no polluting wastes such as ozone or sulfur dioxide.

Recently, a new class of materials, the glass ceramics, has been commercialized. These have unusual but very valuable properties. As noted, glass breaks because once a crack starts, there is nothing to stop it from spreading. It was discovered that if glass is treated by heating until many tiny crystals have developed throughout the sample, the resulting material, when cooled, is much more resistant to breaking than normal glass. In molecular terms, the randomness of the glass structure has been partially replaced by the order in a crystallizing silicate. The process must be controlled carefully to obtain the desired properties. The materials produced in this way are generally opaque and are used for cooking utensils and kitchen ware. An example is products marketed under the name Pyroceram®. The initial manufacturing process is similar to that of other glass objects, but once the materials have been formed into their final shapes, they are heat treated to develop their special properties.

Portland Cement and Concrete

Cements bind other materials together. Portland cement contains calcium, iron, aluminum, silicon, and oxygen in varying proportions. Portland cement has a structure somewhat similar to that described earlier for glass, except that in cement some of the silicon atoms have been replaced by aluminum atoms. Cement reacts in the presence of water to form a hydrated colloid of large surface area that subsequently undergoes recrystallization and reaction to bond to itself and to bricks, stone, or other silicate materials. The cement is made by roasting a powdered mixture of calcium carbonate (limestone or chalk), silica (sand), aluminosilicate mineral (kaolin, clay, or shale), and iron oxide at a temperature of up to 870°C in a rotating kiln. As the materials pass through the kiln, they lose water and carbon dioxide and ultimately form a "clinker," in which the materials are partially fused. The clinker is then ground to a very fine powder after the addition of a small amount of calcium sulfate (gypsum). A typical composition of a Portland cement can be expressed as follows: 60–67% CaO; 17–25% SiO_2; 3–8% Al_2O_3; up to 6% Fe_2O_3; and small amounts of magnesium oxide, magnesium sulfate, and potassium and sodium oxides. As in the structure of glass, the oxides are not isolated into molecules or ionic crystals, and the submicroscopic structure, which tends to be composed of very large molecular species, is quite complex.

Many different reactions occur during the setting of cement. The various constituents react with water and subsequently at the surface with the carbon dioxide in air. The initial reaction of cement with water gives a sticky gel that results from the hydrolysis of the calcium silicates. This sticks to itself and to the other particles (sand, crushed stone, or gravel). The gel has a very large surface area and is responsible for the strength of concrete. The setting process also involves the formation of small, densely interlocked crystals after the initial solidification of the wet mass. This continues for a long time after the initial setting and increases the compressive strength of the cement.

Water is required for the setting process since the reactions involve hydration. For this reason, freshly poured concrete is kept moist for several days. Over 400 million tons of cement are manufactured each year, most of which is used to make concrete. Concrete, like many other materials containing Si—O bonds, is highly noncompressible but lacks tensile strength. If concrete is to be used where it is subject to tension, it must be reinforced with steel.

"Pure" Silicon and "The Chip"

Silicon of about 98% purity can be obtained by heating silica and coke at 3000°C in an electric arc furnace.

$$SiO_2 + 2\,C \longrightarrow Si + 2\,CO$$

Silicon of this purity is alloyed with aluminum and magnesium to increase the hardness and durability of the metals and is used in making silicone polymers.

High-purity silicon can be prepared by reducing $SiCl_4$ with magnesium.

$$SiCl_4 + 2\,Mg \longrightarrow Si + 2\,MgCl_2$$

The magnesium chloride, being water soluble, is then washed from the silicon. The final purification of the silicon takes place by a melting process called zone-refining (Fig. 11–23), which produces silicon containing less than one part per billion of impurities such as boron, aluminum, and arsenic.

One outstanding property of silicon in a high state of purity is its electric conductivity. Unlike a metal, which easily conducts electricity, and unlike a nonmetal, which does not, silicon is a **semiconductor.** That is, it does not conduct until a certain electric voltage is applied, but beyond that it conducts moderately. By placing other atoms in a crystal of pure silicon, a process known as **doping,** experimenters have found that its conductivity properties can be changed. Recall from Chapter 8, Figure 8–26 that *n*-type silicon contains extra electrons and *p*-type silicon has a deficiency of electrons in comparison to pure silicon. The excesses and deficiencies of electrons result from inserting other atoms in the place of silicon atoms in the crystalline structure.

In 1947, an electrical device called the **transistor** was invented. The simplest device used layers of *n-p-n-* or *p-n-p*-doped silicon. Germanium (Ge), a group IV element just below Si in the periodic table, was also used in place of Si. Later, scientists used electric fields to control conductivity in silicon transistors. These **field-effect transistors** have been put to good use by engineers, who have designed low-noise amplifiers, receivers, and other forms of electronic equipment.

The most revolutionary application of silicon's semiconductor properties has been the design of **integrated electric circuits,** computer memories, and even whole computers called **microprocessors** on tiny chips of silicon scarcely larger than a millimeter or so in diameter. As mentioned in Chapter 1, these devices have begun to permeate our whole society. You will find

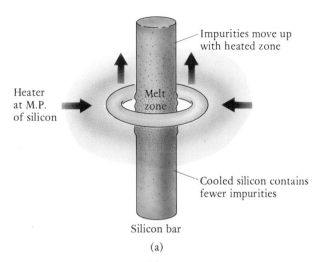

Impurities move up
with heated zone

Heater
at M.P.
of silicon

Melt
zone

Cooled silicon contains
fewer impurities

Silicon bar

(a)

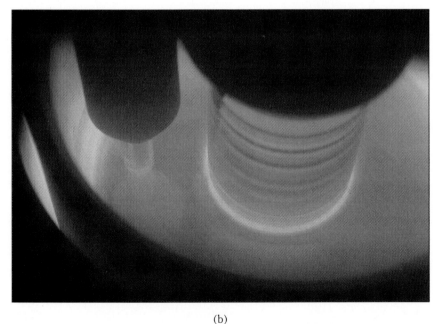

(b)

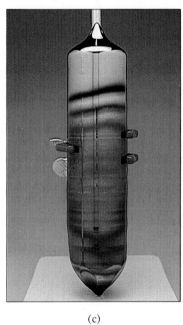

(c)

Figure 11–23 Zone refining. (a) The hot zone moves upward on the silicon bar. As the silicon melts, impurities become mobile and move with the hot molten zone. Repeated passes of the heater produce a crystalline silicon bar with fewer than one part impurity per billion parts of silicon (1 ppb). (b) The zone heater ring surrounds the bar in the purification process. (Courtesy of Great Western Silicon Company.) (c) A 10-in. bar of "pure" silicon. (Charles D. Winters.)

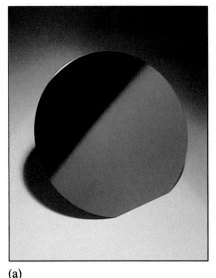

(a)   (b)

Figure 11–24   (a) A five-inch wafer of ultrapure silicon that has been cut from a rod purified by zone-refining. (Charles Steele.) (b) A use of pure silicon in the structure of photo cells for the conversion of sunlight to electric energy. [Courtesy of Standard Oil Company (Ohio).]

them in calculators, cameras, watches, toys, coin changers, cardiac pacemakers, and many other products. Truly, silicon is both the world we walk on and at the same time our constant companion in communications and electronic controls (Figs. 11–24 and 11–25).

GEMSTONES—CRYSTALS COLORED BY IMPURITIES

The pure color of a gemstone comes from impurities in natural crystals that are clear in the pure state. Corundum, a crystalline form of aluminum oxide, Al_2O_3, is a clear crystal with an ionic structure as shown in the margin. Note that each Al^{3+} ion is enclosed in a cage of six O^{2-} ions. If about 1% of the aluminum ions are replaced by Cr^{3+} ions, the crystal is a ruby. A ruby is red because the presence of the chromium ion causes light in the blue end of the visible spectrum to be absorbed; the ruby-red light is passed through the crystal.

Figure 11–25   A tiny microcomputer can be fabricated from a single piece of highly purified silicon. Such computers are capable of many millions of computations per second. Their speed and small size have revolutionized computers and their applications. (Courtesy of AT & T Bell Laboratories.)

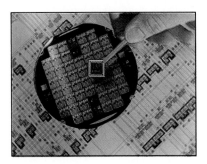

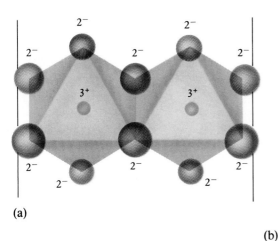

(a)

(b)

(a) A crystalline form of aluminum oxide, corundum, contains aluminum ions in octahedral cages of six oxide ions. Corundum is a clear crystal but takes on different colors if a small percentage of the aluminum ion is replaced by other metal ions. (b) Synthetic ruby. (Kurt Nassau.)

Another example of gem color due to impurities is the emerald. Beryl, a clear crystal, is beryllium aluminum silicate ($3BeO \cdot Al_2O_3 \cdot 6SiO_2$). An emerald is a crystal of beryl having a small fraction of the aluminum replaced by chromium. Table 11–5 lists traditional birthstones with their color-causing impurities, the chromophores.

SULFURIC ACID

Sulfuric acid is the number one chemical in commerce. More than 80 billion lb are produced and sold each year in the United States alone. This acid is used in huge quantities in the manufacture of fertilizers, in the petroleum industry, and in the production of steel. Also, sulfuric acid plays an important role in the manufacture of organic dyes, plastics, drugs, and many other products. Since sulfuric acid is the cheapest acid, it is always the first acid of

TABLE 11–5 Birthstones and Color

Month	Gem	Formula	Color	Chromophore
January	Garnet	$(Mg,Fe^{3+})_2Al_2(SiO_4)_3$	Red	Fe^{2+}
February	Amethyst	SiO_2	Purple	Fe^{3+}/Fe^{4+}
March	Aquamarine	$Be_3Al_2Si_6O_{18}$	Blue	Fe^{2+}/Fe^{3+}
April	Diamond	C	Clear	
May	Emerald	$Be_3Al_2Si_6O_{18}$	Green	Cr^{3+}
June	Alexandrite	Al_2BeO_4	Red/green	Cr^{3+}
July	Ruby	Al_2O_3	Red	Cr^{3+}
August	Peridot	$(Mg,Fe^{2+})_2SiO_4$	Yellow-green	Fe^{2+}
September	Sapphire	Al_2O_3	Blue	Fe^{2+}/Ti^{4+}
October	Opal	$SiO_2 \cdot H_2O$	Opalescent	
November	Topaz	$Al_2SiO_4(OH,F)$	Colorless	
December	Turquoise	$CuAl_6(PO_4)_4(OH)_8 \cdot 4H_2O$	Blue	Cu^{2+}

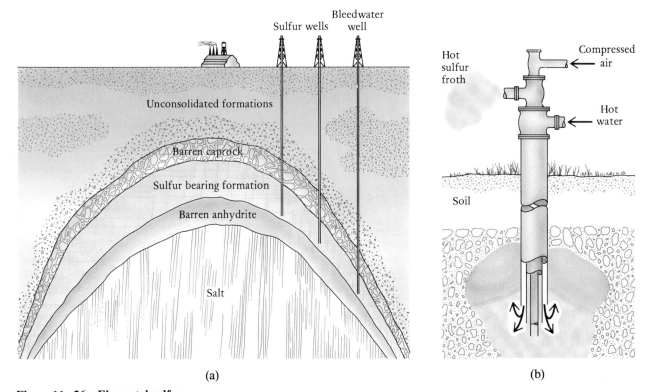

(a)

(b)

Figure 11 – 26   Elemental sulfur recovery and storage. (a) Deposits of sulfur are associated with the cap rock formation over salt domes. Theory suggests that the sulfur was released from gypsum, CaSO$_4$ · 2H$_2$O, by bacterial action. (b) The Frasch process for mining sulfur uses water at 170°C under a pressure of 100 lb/in^2. The hot water melts the sulfur, and the compressed air forces the water-sulfur mixture to the surface. (c) A "mountain" of sulfur at Pennzoil's terminal in Galveston awaits shipment. (Courtesy of Pennzoil Company.)

(c)

choice. The cost of sulfuric acid, about 1 cent per pound, has not changed much in 300 years, a tribute to improving technology in its production from natural sources and pollution wastes.

Sulfur, in very pure form, is found in underground deposits and is brought to the surface by the **Frasch process,** which utilizes the fact that sulfur can be melted by superheated steam. The molten sulfur is raised to the surface of the Earth by means of compressed air and is then allowed to cool in large vats (Fig. 11–26). Large amounts of sulfur are also produced from petroleum, which often contains sulfur compounds.

Sulfur is converted to sulfuric acid (H_2SO_4) in four steps, called the **contact process.** In the first step the sulfur is burned in air to give mostly sulfur dioxide:

$$S + O_2 \longrightarrow SO_2(g)$$

The gaseous SO_2 is then converted to SO_3 by passing SO_2 over a hot catalytically active surface, such as platinum or vanadium pentoxide:

$$2\,SO_2(g) + O_2 \xrightarrow{\text{Catalyst}} 2\,SO_3(g)$$

Although SO_3 can be converted directly into H_2SO_4 by passing SO_3 into water, the enormous amount of heat released in the reaction causes the formation of a stable fog of H_2SO_4. This is avoided by passing the SO_3 into H_2SO_4:

$$\underset{\text{Sulfuric Acid}}{SO_3 + H_2SO_4} \longrightarrow \underset{\text{Pyrosulfuric Acid}}{H_2S_2O_7}$$

and then diluting the $H_2S_2O_7$ with water:

$$H_2S_2O_7 + H_2O \longrightarrow 2\,H_2SO_4$$

Native sulfur surrounding Emerald Lake in Yosemite National Park. Such deposits are often found at hot springs or geysers. (Don and Pat Valenti/Tom Stack and Associates.)

Two sulfur-containing minerals. (a) Iron pyrite, commonly called fool's gold, is iron disulfide, FeS_2. (b) Gypsum deposits, $CaSO_4 \cdot 2H_2O$ are often observed in caves. (Ward's Natural Science Establishment.)

(a)   (b)

Figure 11–27　Cross section of cell for the electrolysis of brine to yield chlorine, hydrogen, and caustic soda (NaOH).

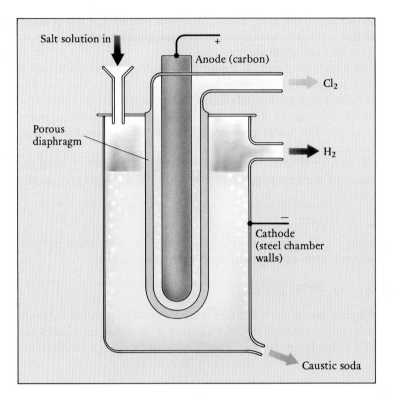

SODIUM HYDROXIDE, CHLORINE, AND HYDROGEN CHLORIDE

Sodium hydroxide (NaOH), hydrogen (H_2), and chlorine (Cl_2) are prepared simultaneously by the electrolysis of a concentrated solution of sodium chloride in water (Fig. 11–27). There are several variations in the basic process that improve its efficiency and the purity of the products. The basic reactions, which occur at nonreactive solid electrodes, are:

ANODE:　$2\ Cl^- \longrightarrow Cl_2(g) + 2\ e^-$　　OXIDATION

CATHODE:　$2\ H_2O + 2e^- \longrightarrow 2\ OH^- + H_2$　　REDUCTION

The reaction replaces the Cl^- of NaCl with OH^-.

Figure 11–28　Mined chlorides. Chlorides occur in salt beds. The annual production of chlorine from salt in the United States is about 20 billion lb per year. (Kevin Schafer/ Tom Stack and Associates.)

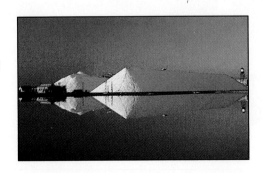

Purer sodium hydroxide is produced when the process is carried out using a mercury cathode. In this case, Na^+ is reduced to Na metal, which dissolves in the mercury electrode:

CATHODE: $Na^+ + e^- \xrightarrow{Hg} Na(Hg)$
          Amalgam

The sodium-mercury amalgam* is a liquid. The amalgam is removed from the cell continuously and reacted with pure water:

$$2\,Na(Hg) + 2\,H_2O \longrightarrow 2\,Na^+ + 2\,OH^- + H_2(g)$$

Chlorine gas is collected from the anode compartment and compressed into tanks. The hydrogen gas is also collected and sold. An important reaction between hydrogen and chlorine is the production of hydrogen chloride (HCl) gas and hydrochloric acid:

$$H_2(g) + Cl_2(g) \longrightarrow 2\,HCl(g)$$

This process can be used to prepare hydrogen chloride, which is quite pure. Hydrochloric acid is a solution of hydrogen chloride in water. The use of mercury in the electrolysis of brine is a potential hazard, because mercury contamination of streams results when the waste solutions are discarded (see Chapters 1 and 17).

Because electric energy is expensive, an electrochemical process is more likely to be economical if all the products can be sold profitably. In recent years the demand for chlorine has grown very rapidly, but the demand for sodium hydroxide has not. Consequently it became necessary to devise some way of disposing of relatively large amounts of sodium hydroxide. This has been accomplished, in part, by transforming NaOH into sodium hydrogen carbonate:

$$Na^+ + OH^- + CO_2(g) \longrightarrow Na^+ + HCO_3^- \longrightarrow NaHCO_3$$

Sodium hydrogen carbonate, or sodium bicarbonate, $NaHCO_3$, is often referred to in the kitchen as baking soda or bicarbonate of soda.

PHOSPHORIC ACID

Phosphoric acid (H_3PO_4) and its salts are used in the manufacture of many materials encountered in our daily lives. These products include baking powder, carbonated beverages, detergents, fertilizers, and fire-resistant textiles. Several processes are available for the manufacture of phosphoric acid, but the major raw material is usually the calcium phosphate that occurs naturally in apatite minerals with the general formula $Ca_5X(PO_4)_3$, where X may be F^-, Cl^-, or OH^-.

An electric furnace is used to heat a mixture of the phosphate ore ($Ca_5F[PO_4]_3$), silica (SiO_2), and coke (C) [Fig. 11-29]. At an elevated temperature a reaction produces elemental phosphorus vapor (P_4):

$$4\,Ca_5F[PO_4]_3 + 18\,SiO_2 + 15\,C \longrightarrow 18\,CaSiO_3 + 2\,CaF_2 + 15\,CO_2 + 3\,P_4$$

---

* An amalgam is an alloy of a metal with mercury; the alloy can either be a liquid or solid.

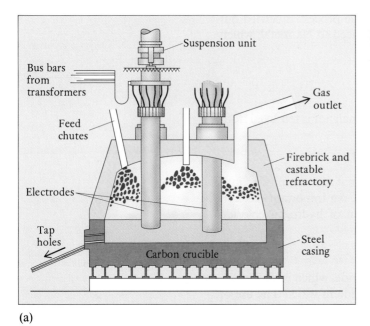

(a)

(b)

Figure 11–29 (a) A phosphorus production furnace is fed with a mixture of phosphate rock [$Ca_5F(PO_4)_3$], sand (SiO_2), and carbon (C). Heat is produced with electric energy, and elemental phosphorus is condensed from the exiting vapors. (b) Two allotropic forms of phosphorus are red and white phosphorus. The white phosphorus is more reactive with the air and has to be kept under water to prevent oxidation. (Charles D. Winters.)

Substances that spontaneously ignite on contact with air are described as pyrophoric.

The phosphorus is then condensed from the gas state and purified. The low melting point of white phosphorus (44.1°C) allows storage and handling as a liquid if protected from air. When exposed to the air, white phosphorus ignites spontaneously. Red phosphorus is stable in dry air (Fig. 11–30).

Elemental phosphorus is transformed into phosphoric acid by oxidation with air to give P_4O_{10},

$$P_4\,(g) + 5\,O_2 \longrightarrow P_4O_{10}\,(g)$$

which is then hydrated by absorption into hot phosphoric acid containing about 10% water. The reaction of P_4O_{10} with water produces more phosphoric acid, H_3PO_4,

$$P_4O_{10} + 6\,H_2O \longrightarrow 4\,H_3PO_4$$

Arsenic, when present in the original ore, is carried through to the phosphoric acid produced, at which point the arsenic can be precipitated by treatment with H_2S. The removal of arsenic is necessary if the phosphoric acid is to be used in the manufacture of food products.

SODIUM CARBONATE

Figure 11–30 The red phosphorus is apparent in strike-anywhere matches. (Charles Steele.)

As noted in our discussion of glass-making, sodium carbonate (Na_2CO_3) is a major ingredient of glass. Sodium carbonate is also important as a cheap alkaline material in the production of numerous chemicals and in the paper

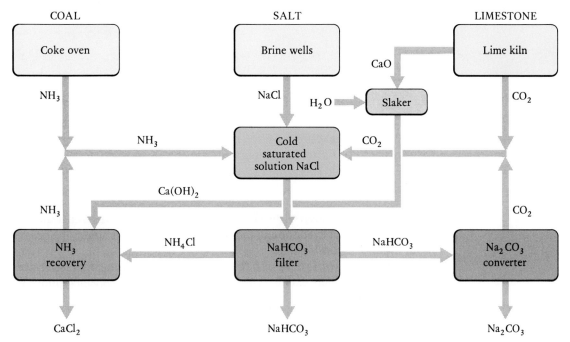

Figure 11–31 Although Trona ore [Impure $Na_2CO_3 \cdot NaHCO_3 \cdot H_2O$] supplies most of the sodium carbonate and sodium bicarbonate used in the United States, it is not an adequate supply on a world-wide basis. The Solvay process for the commercial synthesis of these important chemicals was named after Ernest Solvay, an Englishman who first succeeded in controlling the process on an economically sound basis in 1869.

industry. It is especially useful to the homeowner as a water softener because carbonate ions will precipitate the metal ions of iron, calcium, and magnesium, which cause water hardness (Chapter 17).

Trona ore, an impure form of sodium carbonate, is insufficient as a source of this important chemical. Several chemical methods have been employed in the production of sodium carbonate; the **Solvay Process** is the dominant method (Fig. 11–31). In this process, ammonia from the Haber process (Chapters 1 and 19) and carbon dioxide from the lime kiln are dissolved in a concentrated brine (NaCl) solution. Sodium hydrogen carbonate ($NaHCO_3$) precipitates from this solution:

$$NH_3 + CO_2 + H_2O \rightleftharpoons NH_4^+ + HCO_3^-$$

$$NH_4^+HCO_3^- + Na^+Cl^- \rightleftharpoons NaHCO_3(s) + NH_4^+Cl^-$$

On gentle heating, the dry sodium hydrogen carbonate yields the anhydrous sodium carbonate:

$$2\, NaHCO_3 \xrightarrow{\text{Heat}} Na_2CO_3 + H_2O(g) + CO_2(g)$$

Anhydrous sodium carbonate is known industrially as soda ash, and washing soda (a common ingredient of cleansers) is the decahydrate,

$Na_2CO_3 \cdot 10H_2O$, which is prepared by crystallization directly from water solution.

SEABED MINERALS — A VAST POTENTIAL RESOURCE

Minerals and ores result from the natural separation processes occurring in the crust of the Earth. The separation processes under the ocean waters, where hot and cold solids interact with hot and cold liquids, is very different from those on the continents where the air and weathering processes greatly modify the materials thrust up from the magma under the crust. Since the world is mostly covered with water, the amounts of chemicals under the seas are relatively great even though the Earth's crust is thinner under the oceans than under the continents.

In the deep sea, it is around the vents in the developing undersea ridges and mountains where extensive high-quality minerals have been found. The next island of Hawaii, Loihi (currently submerged), is now within 980 m of the surface. Low-temperature vents (30°C), at depths of 980–1360 m, are surrounded with volcanic cones of iron oxide. The vents are often called black smokers since the hot, dark, mineral-laden water coming from the vent produces a dark convection stream. Solid material carried along in the stream tends to settle around the vent as well as solids from the crystallization of dissolved minerals as the hot water cools. A second class of vents along the faults and fissures of the steep slopes of the mountain have developed iron oxide deposits as thick as 30 m. A third class of vents inside the developing craters produce rich deposits of iron, manganese (Mn), cobalt (Co), copper, chromium (Cr), zinc (Zn), nickel (Ni), and gold minerals. Also, large hydrothermal fields (areas of warmer water) occur in the deep sea where the tectonic plates of the crust are separating and allowing seawater to come in contact with the underlying volcanic rocks. Water temperatures as high as 500°C have been measured in hydrothermal fields containing mineral deposits of copper, iron, lead, and zinc. Theoretical computer models of the separation of these minerals involve conduits as large as 200 m in diameter of superheated water (on average at about 350°C) moving up from just above the liquid magma to the ocean floor. Such models can account for as much as 50% of the zinc and 90% of the copper being extracted from the crust material.

Sedimentary deposits of minerals appear not to be extensive over the deep sea floor, which is probably related to the fluid nature of the medium and the extensive solvent property of water. However, nodules of minerals are found on areas of the ocean floor; an example is the extensive field of manganese nodules in the Pacific Ocean.

In the continental coastal shelf regions, mineral deposits are often found as extensions of deposits in the nearby land areas. For example, as the phosphate ores in North Carolina and Florida are exhausted in the near future, further consideration of the mining of the nearby underwater phosphate beds may be made.

A tectonic plate in the surface of the Earth is a moving solid mass relative to an adjacent mass. Interfaces between plates are the earthquake fault lines.

Except for oil and gas, it is still an open question as to whether there are enough mineable ores of sufficient quality to justify the development of the seabed as a source of needed materials until the shortages of the land sources are much more acute than at present.

SOURCES OF NEW MATERIALS

For most of history, we have been limited to the materials found on the surface of the Earth and to relatively few chemical transformations. Technological advances in mining techniques and the control of chemical change through basic research have radically changed the materials available for human use. "The era of discovery has come to an end, and we are facing the dawn of the era of creation," said Jiro Kondo, president of the Science Council of Japan, to a recent international CHEMRAWN (Chemical Research Applied to World Needs) conference sponsored by the International Union of Pure and Applied Chemistry. Whether in energy, transportation, or communication, the call is for new materials that are stronger, lighter, stiffer, tougher, more durable, more heat-resistant, cheaper, easier to fabricate, having better electronic properties, nonpolluting, and with no disposal problems. Remarkable progress has been made in developing new materials for specific needs in each of the major areas in materials science: ceramics, glasses, metals, semiconductors, organics and polymers, and composites.

Five sources of new materials are in use or are promised: (1) Thousands of new compounds will be produced, described, and catalogued each year as a continuation of ongoing basic and applied research efforts. (2) New processing techniques, such as ultrarapid heating and cooling, and vapor deposition allowing atoms, ions, or molecules to be laid down like bricks in controlled order, result in new and very different materials from old resources. (3) New stable elements appear possible. If elements are made with atomic numbers up to 150 as predicted, some of the elements should be stable and result in a multitude of new pure substances and mixtures. (4) Our reach-through-space programs come ever closer to the exploitation of extraterrestrial materials from the moon, meteors, and nearby planets. Indications now are that the elements are the same, but the mineral configurations are different for space materials. (5) Recent experiments make it clear that chemicals formed and purified in a gravity-free environment are significantly different from the "same" chemicals formed and purified in the presence of Earth's gravity.

In addition to the enhanced effects of purified copper in electric conduction and ultrapure silicon in the control of electric circuits and in the processing of information, there are excellent prospects for electrically conducting polymers, superconductive ceramics at higher and higher temperatures, organic semiconductors, and polymer fibers that are stronger per unit weight than steel or carbon fibers. One has to keep reading the current literature to keep up with the new materials, and new applications are in the offing. For example, consider again the making of glass. One of the biggest problems in controlling the purity of a particular glass is contamination from the container in which it is made. It is just a matter of time until we make many new glasses in space where no container is needed! Presently used optical fibers have reduced absorbing impurities to levels below one part per billion and have only 0.1% of the absorption loss characteristic of the fibers first used about 20 years ago. Light fibers have successfully carried 20 billion laser pulses per second, equivalent to 300,000 simultaneous conversations over a 60-mile distance.

Theory and early laboratory experiments indicate other optical materials, such as the metal fluorides, to be much superior to glass for making optical fibers. However, technologies are not now available for making the fibers commercially.

Perhaps the biggest breakthrough in optical communication in the next few years will be the fabrication of the silicon chip that is able to detect, process, and emit light energy at the same time it is handling the electrical signals. Problems related to junction limitations, channel switching, and circuit controls will be dramatically reduced when such a chip is made. In theory, computers based on light (photon) processing will operate much faster than the current semiconductor devices, which are limited by the rates at which charges can be moved in the semiconductors.

SUPERCONDUCTORS

A recent discovery of materials that exhibit superconductivity at temperatures as high as $-175°C$ deserve special attention because of the potentially practical uses that can be made of them in the transmission of electric energy and in other applications. A superconductor is a material that has the ability to conduct electric current with no apparent resistance, in contrast to all previous transmissions of electric energy in which an appreciable amount of the energy is lost to wasted heat (heat energy that cannot be recovered for useful work, which increases the entropy of the universe).

K is Kelvin temperature: $K = 273 + °C$

Early superconductors made of niobium (Nb) alloys required cooling to temperatures at or below 23 K ($-250°C$) in order to exhibit this unusual property in electric conductance. To maintain such cold temperatures required liquid helium at $4 per liter — an expensive operation, to say the least. Recent discoveries of superconductors composed of mixed metal oxides (Fig. 11–32) superconduct at temperatures as high as 98 K. Temperatures in this range can be maintained with liquid nitrogen at a cost of 22¢ per gallon. The effort goes on to find combinations of atoms that will superconduct at higher and higher temperatures.

Why the great excitement over the potentialities of superconductivity? Superconducting materials are used to build more powerful electromagnets, such as those used in nuclear particle accelerators (Chapter 7) or in magnetic resonance imaging (MRI) machines, which are used in medical diagnosis. The main drawback from wider application is the cost in cooling the magnet. Many scientists are saying that this discovery is more important than the discovery of the transistor because of its potential effect on electrical and electronic technology. For example, the use of superconducting materials for

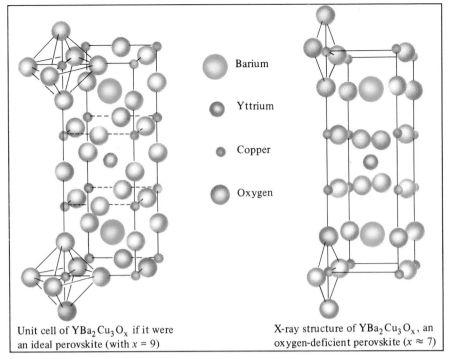

Barium

Yttrium

Copper

Oxygen

Unit cell of $YBa_2Cu_3O_x$ if it were an ideal perovskite (with $x = 9$)

X-ray structure of $YBa_2Cu_3O_x$, an oxygen-deficient perovskite ($x \approx 7$)

(a)

(b)

Figure 11–32 (a) Some substances lose all resistance to the flow of electric energy at low temperatures. The most promising to date are ceramic materials with the formula $YBa_2Cu_3O_x$ and with a structure similar to the mineral perovskite. The structure allows for the x to have a value as high as 9. In most working superconducting samples the value of x is between 6.5 and 7.2. (b) Note the levitating weight (representing a train moving on air) in the beaker illustrates one of the remarkable physical possibilities using superconductivity. (Courtesy of Argonne National Laboratory.)

transmission of electric power could save as much as 30% of the energy now lost because of the resistance of the wire. Superchips for computers could be up to 1000 times faster than existing conventional silicon chips. Electromagnets could be both more powerful and smaller, which could hasten the day of a practical nuclear fusion reactor. Since there is a magnetic field around the superconducting material, it is conceivable that cars and trains can be moved along magnetic tracks so that no parts touch, thereby eliminating friction.

Although the discovery of superconductivity is a significant one, translating the research into practical applications such as those described above will take some time. The technology is just beginning to be developed, but there is hope that problems related to applying the discovery to a variety of applications will be overcome. In addition, there is reasonable hope that superconductors at room temperatures will be discovered.

A fundamental reversal in the relationship between humans and materials is taking place. Its economic consequences are likely to be profound. Historically, humans have adapted such natural materials as stone, wood, clay, vegetable fiber, and animal tissue to economic uses. The smelting of metals and the production of glass represented a refinement of this relationship. Yet it is only recently that advances in the theoretical understanding of structural and biological material, in experimental technique, and in processing technology have made it possible to start with a need and then develop a material to meet it, atom by atom.

Joel Clark and Merton Flemings,
Professors of Materials Science and Engineering
Massachusetts Institute of Technology.

ARE THE RAW MATERIALS EXHAUSTIBLE?

The resource materials of the Earth are not exhausted in the sense that they are placed outside the reach of the human reclamation processes; it takes a spaceship to do that. Such materials are becoming exhausted because it is easier and less costly to start with new raw materials than to recycle used ones. Except for radioactive wastes, we know how to recycle all of the materials we use. We can cycle them back into a useful product, such as steel into a new automobile, or back into a natural material as when neon from a light bulb is released into the atmosphere.

The problem of waste disposal then becomes a question of human values and related costs. The record has been dismal. Nature cannot clean up after 5 billion people as they employ more and more technology in the manipulation of the Earth's materials. **Open dumps** could only survive in unoccupied regions. **Sanitary landfills** have offered considerable hope in theory, but the management and containment of such operations has been sorely deficient. **Chemical dumps** have become a national crisis. **Incineration,** which allows for energy recovery, is a clean operation only if elaborate procedures allow for the removal of the harmful combustion products, products which will vary depending on the particular wastes being burned.

Consider the following possible solution to the sensible use of the Earth's natural resources:

> **The consumer cost of any material or product must include reuse and/or recycling costs so that the resources of the Earth are not depleted.**

Such an ideal is not likely to be achieved soon. As a species we seem more dedicated to depleting fossil fuels, removing the Earth's forests, and exhausting the mineable ores of essential minerals. However, thoughtful citizens will surely accept possible advances in waste management as a necessary part of the maintenance and improvement of our civilization.

It is very important for the student to understand that the human knowledge is presently at hand to protect our natural environment. With the exception of radioactive nuclear wastes, we can, if we wish to pay the cost, cycle and recycle our natural resources through countless uses. We are learning that we do not have to use our iron, aluminum, glass, and paper just once and then haul them to the dump. In contrast, perhaps the most striking aspect of natural chemistry is its cycling of elements and compounds through countless numbers of like uses.

SELF-TEST 11-C

1. The two principal nonmetals in glass are _____ and _____ .

2. Flint glass, or crystal, contains a large amount of a compound of what metal?

3. What element is associated with the miniaturization of electronic components?

4. What three silicate minerals are used to make common pottery materials?
5. Roasting a mixture of powdered limestone, clay, sand, and iron oxide produces what important commercial building material?
6. Which element is pumped from underground deposits as a molten material mixed with hot water?
7. The electrolysis of brine produces what element at the anode where oxidation occurs?
8. Phosphorus can be moved about as a liquid because of its low melting point provided it is kept out of contact with the _____ .
9. What is the chemical name of washing soda? What related chemical is likely to be on the shelf in the kitchen?

MATCHING SET

_____ 1. Copper
_____ 2. Aluminum
_____ 3. Milk glass
_____ 4. Magnesium
_____ 5. Oyster shells
_____ 6. Sodium bicarbonate
_____ 7. Sulfur
_____ 8. Iron
_____ 9. Phosphate
_____ 10. Silicon

a. Reduced in a blast furnace
b. Calcium fluoride added to glass
c. Contains phosphorus and oxygen
d. Used in making transistors
e. $NaHCO_3$
f. Mined with superheated water
g. A limitless supply in sea water
h. Supply calcium hydroxide for magnesium production
i. Purified electrolytically
j. The most abundant metal in the Earth's crust
k. Sodium carbonate
l. An element distilled from air

QUESTIONS

1. Name three metals that you would expect to find free in nature. Name three that you would not.
2. What is the primary reducing agent in the production of iron from its ore?
3. Why is CaO necessary for the production of iron in a blast furnace?
4. What is the chemical difference between iron and steel?
5. Are natural materials most likely to be elements, compounds, or mixtures?
6. What metal is most used in industry?
7. What metal is recovered from sea water in industrial quantities?
8. Describe the solution used in a commercial cell for the electrolytic reduction of aluminum.
9. Why is it so important to purify industrial quantities of copper electrolytically to a level above 99.9% pure?
10. What chemical is obtained from oyster shells in the production of magnesium from sea water? What is the role of this chemical in the process?
11. Explain how the structure of glass and a liquid are similar.
12. What oxide is the main ingredient in glass?
13. Give reactions involved in the preparation of:
    a. $Ca(OH)_2$ from $CaCO_3$
    b. NH_3 from N_2 and H_2
    c. Iron from iron oxide
    d. Sulfuric acid from sulfur
    e. Phosphoric acid from phosphorus
14. A typical soda-lime glass has a composition reported as 70% SiO_2, 15% Na_2O, and 10% CaO. What ratio of weights of silica (SiO_2), sodium carbonate (Na_2CO_3), and calcium carbonate ($CaCO_3$) must be melted to-

gether to make this glass? The carbonates are decomposed by heat to evolve carbon dioxide.

15. What is the maximum weight (in pounds) of magnesium that can be obtained from 100 lb of sea water? (See p. 306)

16. What is the purpose of using sodium carbonate in the family wash?

17. What is the purpose of oxidizing phosphorus in air after it has been reduced from phosphate rock in a furnace?

18. Two elements and a compound are produced in the electrolysis of brine. Name them and write the reactions involved in their production.

19. Why is it necessary to have two oxidation steps in the production of sulfuric acid from elemental sulfur?

20. Some argue that atomic energy is the most significant scientific and technical event of the 20th century. Others say the silicon chip has the greatest impact for change in our society. Should genetic engineering also be mentioned in this context? What do you think? Give reasons for your answer.

21. What chemicals are in Portland cement? Give the chemical source of each.

22. When clay is fired, a rigid glasslike framework is established that is not attacked by water. Where do you think the "glass" comes from in the fired clay?

23. Is glass a mixture or a compound? Explain.

24. Is the recovery of oxygen from the air a chemical or a physical process? Give a reason for your answer.

25. Name two commercial sources for lime (CaO), and give the equations involved. Also, name two uses for this important chemical.

26. Two elements discussed in this chapter have special electrical properties only when they are in a high state of purity. What are they, and what are the applications for the pure materials?

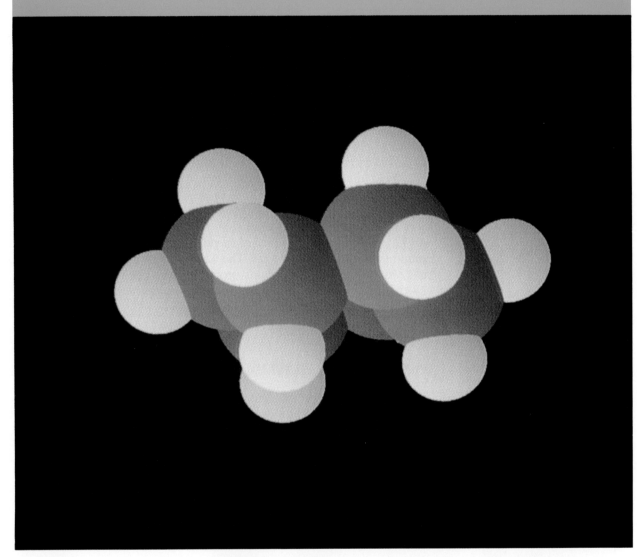

12

The Ubiquitous Carbon Atom—An Introduction to Organic Chemistry

Space-filling model of cyclohexane, chair form. (Courtesy of Phillips Petroleum.)

339

Carbon and its compounds are necessary for life on this planet.

Carbon compounds hold the key to life on Earth. Consider what the world would be like if all carbon compounds were removed; the result would be much like the barren surface of the moon. If carbon compounds were removed from the human body, there would be nothing left except water and a residue of minerals. The same would be true for all living things. Carbon compounds are also an integral part of our lifestyle. Fossil fuels, foods, and most drugs are made of carbon compounds. Since we live in an age of plastics and synthetic fibers, our clothes, appliances, and most other consumer goods contain a significant portion of carbon compounds.

About nine million of the over ten million known compounds are carbon compounds, and thousands of new ones are reported every year. Although there are 88 other naturally occurring elements, there are many more known carbon compounds than those that contain no carbon. The very large and important branch of chemistry devoted to the study of carbon compounds is **organic chemistry.** The name *organic* is actually a relic of the past, when chemical compounds produced from once-living matter were called "organic" and all other compounds were called "inorganic."

Organic chemistry is the study of the nonmineral compounds of carbon.

WHY ARE THERE SO MANY ORGANIC COMPOUNDS?

The enormous number of organic compounds has intrigued chemists for over a hundred years. The atomic theory, as developed earlier for all atoms, describes a structure for the carbon atom that explains this multiplicity of carbon compounds. Carbon has four valence electrons and follows the octet rule by having four covalent bonds with itself or with other nonmetal atoms. Table 12–1 summarizes the possibilities and gives a few examples for each type.

Review covalent bonding in Chapter 5.

A single organic molecule may contain a single carbon-carbon bond,

$$-\overset{|}{\underset{|}{C}}-\overset{|}{\underset{|}{C}}-$$

or thousands of such bonds.

$$-\overset{|}{\underset{|}{C}}-\overset{|}{\underset{|}{C}}-\overset{|}{\underset{|}{C}}-\overset{|}{\underset{|}{C}}-\overset{|}{\underset{|}{C}}-\overset{|}{\underset{|}{C}}-\overset{|}{\underset{|}{C}}-\overset{|}{\underset{|}{C}}-\overset{|}{\underset{|}{C}}-\overset{|}{\underset{|}{C}}-\overset{|}{\underset{|}{C}}-\overset{|}{\underset{|}{C}}-\overset{|}{\underset{|}{C}}-\overset{|}{\underset{|}{C}}-\overset{|}{\underset{|}{C}}-\overset{|}{\underset{|}{C}}-\overset{|}{\underset{|}{C}}-\overset{|}{\underset{|}{C}}-\overset{|}{\underset{|}{C}}-\overset{|}{\underset{|}{C}}-$$

A few other elements are capable of forming stable bonds between like atoms. These include such elements as nitrogen (N_2), oxygen (O_2), and sulfur (S_8), to name a few. But of these, only S, tin (Sn), silicon (Si), and phosphorus (P) form molecules with more than three like atoms in a chain, and the chains of these elements are unstable and undergo reaction with either water or oxygen, or both, much more easily than do carbon chains.

Another reason for the large number of organic compounds is the ability of a given number of atoms to combine in more than one molecular pattern

TABLE 12–1 Different Arrangements of Covalent Bonds for the Carbon Atom

Bond Arrangement	Number and Type of Bond	Example
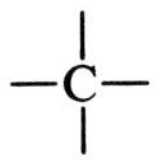 $-\overset{\textstyle\mid}{\underset{\textstyle\mid}{C}}-$	four single bonds	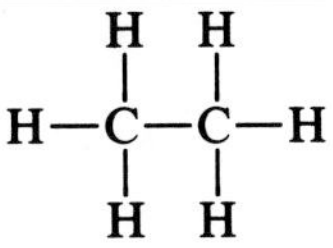 $H-\overset{\textstyle H}{\underset{\textstyle H}{C}}-\overset{\textstyle H}{\underset{\textstyle H}{C}}-H$ ethane
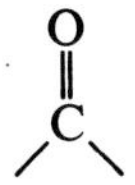 $\overset{\textstyle O}{\underset{\diagup\ \diagdown}{C}}$ or $\diagdown\!\!\diagup_{C=C}\diagup\!\!\diagdown$ 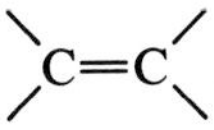	two single bonds and one double bond	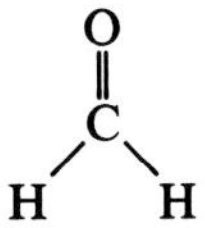 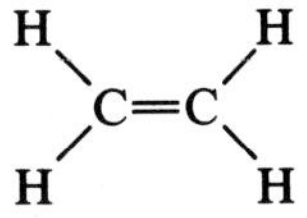 $\overset{\textstyle O}{\underset{H\diagup\ \diagdown H}{C}}$ formaldehyde $\underset{H\diagup\ \ \diagdown H}{\overset{H\diagdown\ \ \diagup H}{C=C}}$ ethene (ethylene)
$=C=$	two double bonds	$O=C=O$ carbon dioxide
$-C\equiv$ or $-C\equiv C-$	one triple bond and one single bond	$H-C\equiv N$ hydrogen cyanide $H-C\equiv C-H$ ethyne (acetylene)

and, hence, produce more than one compound. For example, a compound with a formula of C_4H_{10} could have a straight-chain pattern

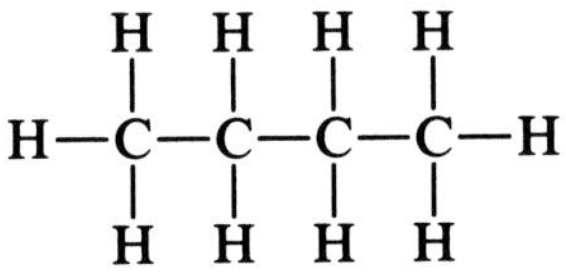

$H-\overset{\textstyle H}{\underset{\textstyle H}{C}}-\overset{\textstyle H}{\underset{\textstyle H}{C}}-\overset{\textstyle H}{\underset{\textstyle H}{C}}-\overset{\textstyle H}{\underset{\textstyle H}{C}}-H$ *n*-BUTANE

or a branched-chain pattern.

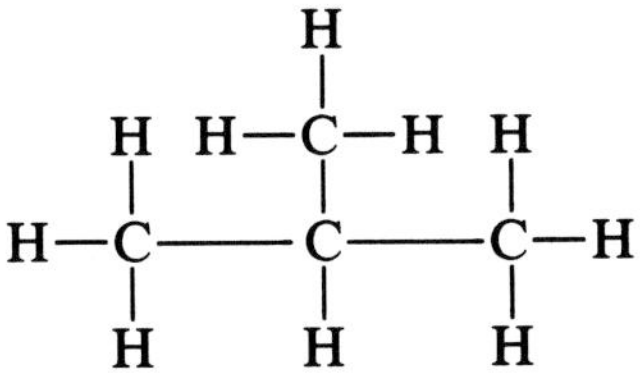

$H-\overset{\textstyle H}{\underset{\textstyle H}{C}}——\overset{\textstyle H-\overset{\textstyle H}{\underset{}{C}}-H}{\underset{\textstyle H}{C}}——\overset{\textstyle H}{\underset{\textstyle H}{C}}-H$ ISOBUTANE

These molecules have different properties even though they have the same number of atoms per molecule. Such compounds, each of which has molecules containing the same number and kinds of atoms, but arranged differently relative to each other, are called **isomers**. Another example of how a difference in molecular pattern can cause a large difference in properties can be seen by considering two molecules that have the formula C_2H_6O. Possible arrangements are

Isomers are two or more different compounds with the same number of each kind of atom per molecule.

$$H-\overset{\displaystyle H}{\underset{\displaystyle H}{\overset{|}{\underset{|}{C}}}}-\overset{\displaystyle H}{\underset{\displaystyle H}{\overset{|}{\underset{|}{C}}}}-O-H \qquad H-\overset{\displaystyle H}{\underset{\displaystyle H}{\overset{|}{\underset{|}{C}}}}-O-\overset{\displaystyle H}{\underset{\displaystyle H}{\overset{|}{\underset{|}{C}}}}-H$$

The molecular structure on the left represents a molecule of ethyl alcohol, and the molecular structure on the right represents a molecule of dimethyl ether—two completely different compounds.

A final factor explaining the large number of organic compounds is the ability of the carbon atom to form strong covalent bonds with atoms of numerous other elements, such as N, O, S, chlorine (Cl), fluorine (F), bromine (Br), iodine (I), Si, boron (B), and even many metals. As a result, there are many classes of organic compounds. A distinguishing **functional group,** a particular combination of atoms, appears in each member of a class. For example, ethyl alcohol and dimethyl ether are members of the alcohol and ether classes, respectively.

In summary, the large number of carbon compounds is due to

1. the stability of chains of carbon atoms
2. the occurrence of isomers
3. the variety of functional groups that bond to carbon atoms.

The largest class of organic compounds are the hydrocarbons, compounds of hydrogen and carbon. Complex mixtures of hydrocarbons occur in enormous quantities as natural gas, petroleum, and coal. It is generally believed that these materials were formed from organisms that lived millions of years ago. After their death, they became covered with layers of sediment and ultimately were subjected to high temperatures and pressures in the depths of the Earth's crust. In the absence of oxygen, these conditions converted once-living tissue into natural gas, petroleum, and coal. Hence, they are known as **fossil fuels.**

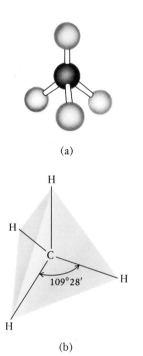

(a)

(b)

(c)

Figure 12–1   Methane. (a) Ball-and-stick model showing tetrahedral structure. (b) Geometry of regular tetrahedron. (c) Space-filling model showing relative size of atoms in relationship to interatomic distances. (Charles Steele.)

HYDROCARBONS

There are four classes of hydrocarbons: the **alkanes,** which contain C—C bonds; the **alkenes,** which contain one or more C=C bonds; the **alkynes,** which contain one or more C≡C bonds; and the **aromatics,** which consist of benzene, benzene derivatives, and fused benzene rings.

Alkanes

Alkanes are hydrocarbons that contain only single bonds. They are often referred to as **saturated hydrocarbons** because they contain the highest ratio of hydrogen to carbon possible. The simplest alkane is **methane,** the principal component of natural gas. Four ways of representing the methane molecule are

$$CH_4 \quad H:\overset{\overset{\displaystyle H}{..}}{\underset{\overset{..}{H}}{C}}:H \quad H-\overset{\overset{\displaystyle H}{|}}{\underset{\overset{|}{H}}{C}}-H \quad \overset{\displaystyle H}{\underset{H \ H \ H}{C}}$$

(a)　　(b)　　　(c)　　　　(d)

The condensed formula (a) conveys the elemental composition of the substance. Structure (b) shows the presence of four bonding pairs of electrons, and the structural formula (c) uses a dash for each bonding electron pair. The last representation (d) depicts in a perspective drawing the geometrical shape of CH_4; the solid wedges represent bonds extending above the page, and the dashed line represents a bond below the page. Recall from Chapter 5 that a tetrahedral structure is predicted when the central atom has four bonding pairs. Methane molecules are tetrahedral and have four C—H bonds, as shown in Figure 12–1.

The next member of the alkane family is **ethane,** C_2H_6, a hydrocarbon gas with two carbon atoms. The bonding and atomic arrangements in ethane are illustrated by the following formulas:

$$C_2H_6 \quad H:\overset{\overset{\displaystyle H}{..}}{\underset{\overset{..}{H}}{C}}:\overset{\overset{\displaystyle H}{..}}{\underset{\overset{..}{H}}{C}}:H \quad H-\overset{\overset{\displaystyle H}{|}}{\underset{\overset{|}{H}}{C}}-\overset{\overset{\displaystyle H}{|}}{\underset{\overset{|}{H}}{C}}-H \quad \overset{H \qquad H}{\underset{H \quad H}{H-C-C-H}}$$

(a)　　　(b)　　　　(c)　　　　　(d)

The condensed structural formula in (a) is used by chemists to represent hydrocarbons without drawing all the bonds. In order to save time and space, chains of carbon atoms are usually represented with straight lines as in structure (c) rather than the tetrahedral structure represented in (d). However, keep in mind that the "straight-chain" representation is not an accurate representation of the tetrahedral bond angles, which are 109.5°.

Figure 12–2 illustrates two possible rotational forms of the ethane molecule. These are referred to as **conformations** of the molecule and are not regarded as different molecules because rotational motion about the single carbon-carbon bond occurs readily at ordinary temperatures. As a result, a sample of a pure hydrocarbon with two or more carbon-carbon single bonds

Rotational forms resulting from the twisting around C—C single bonds are called conformations; they are not isomers.

Staggered conformation

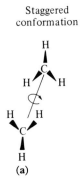

Eclipsed conformation

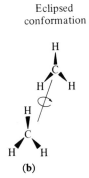

Figure 12–2 Two possible rotational forms (conformations) of the ethane molecule. (a) Staggered. (b) Eclipsed. In ethane the two methyl groups can rotate easily about the carbon-carbon bond to convert one conformation to another, as shown by the arrows. (Photographs by Charles Steele.)

(a)　　　　　　　(b)

Figure 12–3   Ball-and-stick and space-filling models of propane, C_3H_8. (Charles Steele.)

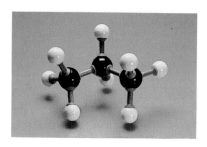

contains a mixture of molecules in all of the possible conformations; that is, molecules rotated into all of their possible shapes.

By applying what we have learned, it is a simple matter to extend the concept of carbon-carbon bonding to a three-carbon molecule such as that of **propane** (C_3H_8), one of the principal components of bottled gas. In Figure 12–3, note that the three carbon atoms in propane do not lie in a straight line because of the tetrahedral bonding about each carbon atom. Also, because of the rotation about the two C—C single bonds, the molecule is "flexible."

It is apparent that these bonding concepts can be extended to a four-carbon molecule and to a "limitless" number of larger hydrocarbon molecules. Actually, many such compounds are known; some, such as natural rubber, are known to contain over a thousand carbon atoms in a chain.

The first ten straight-chain alkanes are listed in Table 12–2. Remember that straight-chain means that the carbon atoms are bonded together in succession, not in a straight line. Notice that each succeeding formula in Table 12–2 is obtained by adding CH_2 to the previous formula. Alkanes are an example of a **homologous series**—a series of compounds of the same chemical type that differ only by a fixed increment. In this case the fixed increment is CH_2. The alkane homologous series can be represented by the general formula C_nH_{2n+2}, where n is the number of carbon atoms for a member of the series.

The names of the first ten straight-chain alkanes are given in Table 12–2. Much attention has been given to naming organic compounds in a consistent way, and several international conventions have been held to work out a satisfactory system that can be used throughout the world. The International Union of Pure and Applied Chemistry has given its approval to a systematic nomenclature system (IUPAC system), which is now in general use. The name of each of the members of the hydrocarbon classes has two parts. The first part—*meth-, eth-, prop-, but-*, and so on—reflects the number of carbon atoms. When more than four carbons are present, the Greek number prefixes are used: *pent-, hex-, hept-, oct-, non-*, and *dec-*. The second part of the name, or the suffix, tells the class of hydrocarbon. Alkanes have carbon-carbon single bonds, alkenes have carbon-carbon double bonds, and alkynes have carbon-carbon triple bonds.

Chemists from all over the world belong to IUPAC. Nomenclature recommendations from this organization are accepted worldwide.

Structural Isomers: Straight- or Branched-Chain Variations

The two possible structural arrangements for **butane,** C_4H_{10}, were given earlier in this chapter. Four carbon atoms can be bonded in either a straight chain or a branched chain. These are different compounds with different

TABLE 12–2 The First Ten Straight-Chain Saturated Hydrocarbons

Name	Formula	Boiling Point, °C	Structural Formula	Use
Methane	CH_4	−162		Principal component in natural gas
Ethane	C_2H_6	−88.5		Minor component in natural gas
Propane	C_3H_8	−42		Bottled gas for fuel
n-Butane	C_4H_{10}	0		
n-Pentane	C_5H_{12}	36		Some of the components of gasoline
n-Hexane	C_6H_{14}	69		
n-Heptane	C_7H_{16}	98		
n-Octane	C_8H_{18}	126		
n-Nonane	C_9H_{20}	151		
n-Decane	$C_{10}H_{22}$	174		Found in kerosene

(a)

(b)

Figure 12–4   Ball-and-stick models of the two isomeric butanes C_4H_{10}. (a) Normal butane, usually written *n*-butane. (b) Methylpropane (isobutane). (Charles D. Winters.)

properties, but because they have the same formula, they are examples of structural isomers. Ball-and-stick models of these two isomers are shown in Figure 12–4. Structural isomerism can be compared to the results you might expect from a child building many different structures with the same collection of building blocks, and using all of the blocks in each structure.

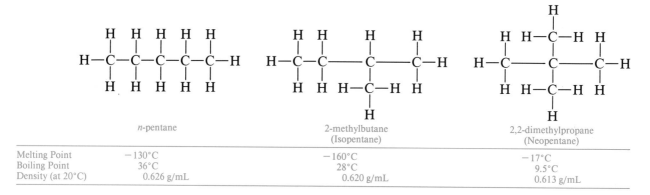

	n-butane	Methylpropane (Isobutane)
Melting Point	−138.3°C	−160°C
Boiling Point	−0.5°C	−12°C
Density (at 20°C)	0.579 g/mL	0.557 g/mL

The two butanes (and all hydrocarbon molecules) are essentially non-polar, since the C—C bonds are nonpolar, and the slightly polar C—H bonds are symmetrically arranged to cancel each other out. The forces holding these molecules together in the liquid, therefore, are London forces, which depend on the surface area of a molecule and the closeness with which the molecules can approach each other. In general, a branched-chain isomer has a lower boiling point than a straight-chain isomer, since the branched-chain isomer does not permit intermolecular distances as short as those of a straight chain; also the branched-chain isomer has less surface area. The result is less intermolecular attraction. Melting points of isomers generally do not follow the same pattern, since they also depend on the ease with which the molecules fit into a crystalline array.

Consider the isomeric pentanes, C_5H_{12}. There are three of these (Fig. 12–5).

	n-pentane	2-methylbutane (Isopentane)	2,2-dimethylpropane (Neopentane)
Melting Point	−130°C	−160°C	−17°C
Boiling Point	36°C	28°C	9.5°C
Density (at 20°C)	0.626 g/mL	0.620 g/mL	0.613 g/mL

The IUPAC names are given under the structures with the common names in parentheses. There are no other ways to unite the 17 atoms in the isomeric pentanes and still follow bonding rules for carbon (eight valence-shell electrons) and hydrogen (two valence-shell electrons). These three isomers of pentane are well known, and no others have been found.

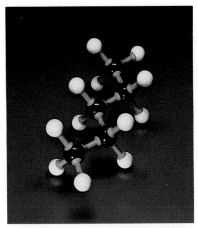  

Figure 12–5   Ball-and-stick models of the three isomeric pentanes C₅H₁₂. (Charles Steele.)

The IUPAC rules for naming structural isomers are illustrated with the names for the pentane isomers. The italic prefix *n-* stands for *normal* and identifies the straight-chain isomer. For branched-chain isomers, it becomes necessary to name submolecular groups. The —CH₃ group is called the methyl group; this name is derived from methane by the deletion of the *-ane* and the addition of *-yl*. Any of the alkanes can give rise to a similar subgroup, referred to generally as an alkyl group, by using the first part of the alkane name and the substitution of *-yl* for *-ane*. Some examples are given in Table 12–3.

The rules for naming branched-chain isomers are:

1. Find the longest continuous chain of carbon atoms: this chain determines the parent name for the compound.

The two branched-chain isomers of pentane have a *butane* parent and a *propane* parent. The longest continuous chain may not be obvious from the

TABLE 12–3 Some Alkyl Groups

Name	Condensed Structural Representation*
Methyl	CH₃—
Ethyl	CH₃CH₂— or C₂H₅—
n-propyl	CH₃CH₂CH₂— or C₃H₇—
Isopropyl	CH₃CH— | CH₃

* CH₃— represents H—C—. Although a more accurate representation would be —CH₃ to show that the available bond is to carbon, not hydrogen, the conventional representation is CH₃—.

way the formula is written, especially for the straight-line format that is commonly used. For example, the longest continuous chain of carbon atoms in the following chain is *eight, not four or six.*

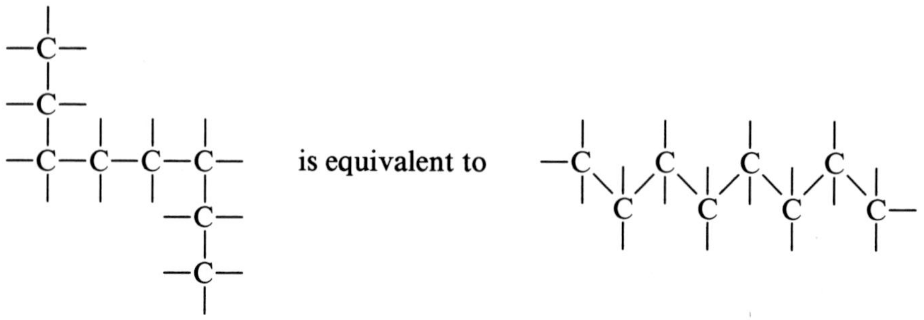

When these are drawn in zig-zag fashion, the number in the longest continuous chain is obvious. However, the present convention is to use straight lines with the understanding that the geometry of each carbon is tetrahedral. This can cause confusion if the approximate character of the straight-chain drawings isn't kept in mind.

2. Number the longest chain beginning with the end of the chain nearest the branching. Use these numbers to designate the location of the attached group.

   The name of the compound below is 3-methylheptane, *not* 5-methylheptane or 2-ethylhexane.

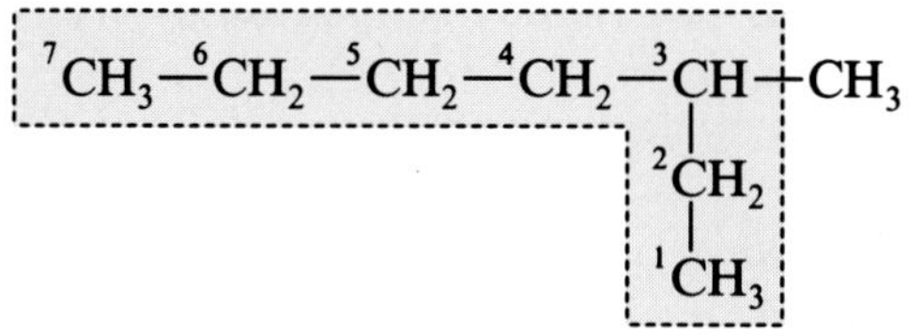

3-methylheptane

3. When two or more substituents are identical, indicate this by the use of the prefixes di-, tri-, tetra-, and so on. Positional numbers of the groups should have the smallest possible sum.

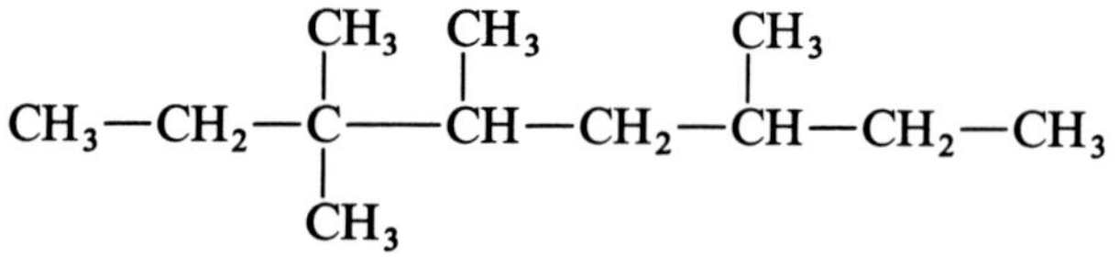

The correct name of the above compound is 3,3,4,6-tetramethyloctane.

4. If there are two or more different groups, the groups are listed alphabetically.

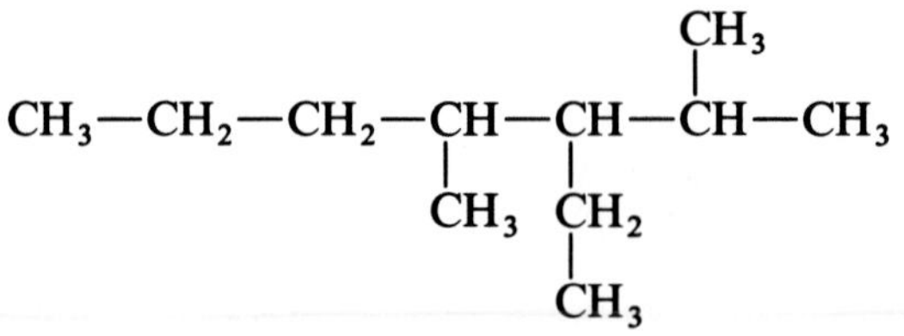

The correct name of the above compound is 3-ethyl-2,4-dimethylheptane. Note that the prefix "di" is ignored in determining alphabetical order.

Table 12–4 gives the number of structural isomers predicted for some larger alkane molecules, starting with C_6H_{14}. Every predicted isomer, *and no more,* has been isolated and identified for the C_6, C_7, and C_8 groups. Although not all of the C_{15} molecules have been isolated, there is reason to believe that with enough time and effort they could be, so structural isomerism certainly helps to explain the vast number of carbon compounds. However, as the number of carbon atoms gets larger than 17, the number of isomers that could be expected to be stable enough for isolation is much smaller than the predicted number. The reason for this difference is that the calculation of predicted isomers does not include a consideration of space requirements of atoms within the molecules for the various isomers. As a result, many of the predicted isomers for these larger molecules would require such overcrowding of atoms within the molecules that they would not be stable. (See the reference listed in Table 12–4 for further discussion of this point.)

Alkenes

Molecules of alkenes have one or more carbon-carbon double bonds (C=C). The general formula for alkenes with one double bond is C_nH_{2n}. The first two members of the homologous alkene series are ethene (C_2H_4), and propene (C_3H_6) and their structural formulas are:

$$
\begin{array}{cc}
\text{H} \qquad \text{H} & \text{H} \qquad \text{H} \\
\diagdown \quad \diagup & \diagdown \quad \diagup \\
\text{C=C} & \text{C=C} \\
\diagup \quad \diagdown & \diagup \quad \diagdown \\
\text{H} \qquad \text{H} & \text{H}_3\text{C} \qquad \text{H} \\
\text{Ethene} & \text{Propene} \\
\text{(Ethylene)} & \text{(Propylene)}
\end{array}
$$

Ball-and-stick models for these formulas are shown in Figure 12–6. The common names, ethylene and propylene, are often used, particularly when

Figure 12–6   Ball-and-stick models of ethene (left) and propene (right). (Charles D. Winters.)

TABLE 12–4 Structural Isomers of Some Hydrocarbons*

Formula	Isomers Predicted	Found
C_6H_{14}	5	5
C_7H_{16}	9	9
C_8H_{18}	18	18
$C_{15}H_{32}$	4,347	—
$C_{20}H_{42}$	366,319	—
$C_{30}H_{62}$	4,111,846,763	—
$C_{40}H_{82}$	62,491,178,805,831	—

* R. E. Davies and P. J. Freyd, "$C_{167}H_{336}$ Is the Smallest Alkane with More Realizable Isomers than the Observed Universe Has 'Particles'," *Journal of Chemical Education,* Vol. 66, pp. 278–281, 1989.

Plastics are discussed in Chapter 14.

referring to polyethylene and polypropylene, plastics prepared from ethylene and propylene, respectively.

The structural formulas illustrate why alkenes are said to be **unsaturated hydrocarbons.** They contain fewer hydrogen atoms than the corresponding alkanes and can be made to react with hydrogen to form alkanes.

$$\underset{\substack{\text{Ethene} \\ \text{(Unsaturated)}}}{\overset{\substack{H \qquad\quad H \\ }}{\text{C=C}}} + H_2 \xrightarrow{\text{Platinum}} \underset{\substack{\text{Ethane} \\ \text{(Saturated)}}}{H-\overset{H}{\underset{H}{C}}-\overset{H}{\underset{H}{C}}-H}$$

Alkenes are named by using the prefix to indicate the number of carbons and the suffix *-ene* to indicate one or more double bonds. The first member, ethene, is the most important raw material used in the organic chemical industry. It ranks fourth in the top 50 chemicals (see Table 13–1) and is the number-one organic chemical. Over 36 billion pounds were produced in 1988 for use in making polyethylene, antifreeze (ethylene glycol), ethyl alcohol, and other chemicals.

The next alkene in the series, butene, has two possible locations for the double bond.

$$\underset{\text{1-butene}}{H-\overset{H}{C}=\overset{H}{C}-\overset{H}{\underset{H}{C}}-\overset{H}{\underset{H}{C}}-H} \qquad\qquad \underset{\text{2-butene}}{H-\overset{H}{\underset{H}{C}}-\overset{H}{C}=\overset{H}{C}-\overset{H}{\underset{H}{C}}-H}$$

The possibilities for locating the double bond between different carbon atoms has to be considered when thinking about the number of structural isomers of alkenes. Substituent groups are named as in the rules for alkanes, but the longest hydrocarbon chain is numbered from the end that will give the double bond the lowest number.

$$H-\overset{H}{\underset{H}{C}}-\overset{H}{C}=\overset{CH_3}{C}-\overset{CH_3}{\underset{H}{C}}-\overset{H}{\underset{H}{C}}-\overset{CH_3}{\underset{H}{C}}-\overset{H}{\underset{H}{C}}-H$$

The above compound is correctly named 3, 4, 6-trimethyl-2-heptene.

Stereoisomerism—Geometric Isomers in Alkenes

Think of stereoisomers as "space" isomers.

Organic molecules can exhibit a property called **stereoisomerism,** in which the same bonds in the molecule are oriented differently in space. This differs from structural isomerism, in which the atoms are bonded together in a different order. There are two types of stereoisomers—**geometric** and **optical.** Geometric isomers in alkenes will be discussed here, and examples of optical isomers will be given later in this chapter.

Geometric isomers can exist in alkene compounds because the carbon-carbon double bond does not allow free rotation about the bond. Consider the compound ethene, C_2H_4. Its six atoms lie in the same plane, with bond angles of approximately 120°.

$$\begin{array}{c} \text{H} \\ \diagdown \\ \text{H} \end{array} \text{C} = \text{C} \begin{array}{c} \text{H} \\ \diagup \\ \text{H} \end{array}$$

If two chlorine atoms replace two hydrogen atoms, one on each carbon atom of ethene ($H_2C{=}CH_2$), the result is ClHC$=$CHCl. Experimental evidence confirms the existence of *two* compounds with the same set of bonds. If the two chlorine atoms are closer together, this is characteristic of one isomer (the **cis** isomer), and if they are farther apart, another isomer (the **trans** isomer) is indicated. The ball-and-stick models shown in Figure 12–7 help to visualize the difference between these two isomers. Both compounds are called 1,2-dichloroethene (the 1 and 2 indicate that the two chlorine atoms are attached to different carbon atoms). The two arrangements are distinguished from each other by the prefixes *cis* and *trans*. Note that the two isomeric compounds have significant differences in their properties.

If free rotation occurred around a carbon-carbon double bond, these two compounds would be one.

	Cis-1,2-dichloroethene	$Trans$-1,2-dichloroethene
Melting Point	−80.5°C	−50°C
Boiling Point	60.3°C	47.5°C
Density (at 20°C)	1.284 g/mL	1.265 g/mL

The third possible isomer, 1,1-dichloroethene (a structural isomer of the *cis* and *trans* isomers), does not have *cis* and *trans* structures since each carbon atom has two identical groups.

1,1-dichloroethene

Melting Point	−122.1°C
Boiling Point	37°C
Density (at 20°C)	1.218 g/mL

As a general rule, *trans* isomers have higher melting points than *cis* isomers because of the greater ease with which the *trans* molecules can fit into a solid structure and form strong intermolecular bonds.

Figure 12–7   Geometric isomers of 1,2-dichloroethene. (Charles Steele.)

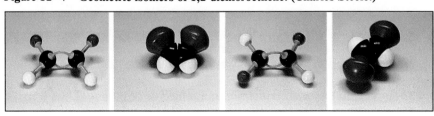

Cutting steel with oxyacetylene torch. (© Joseph Nettis, Photo Researchers, Inc.)

When there are four or more carbon atoms in an alkene, the possibility exists for *cis* and *trans* isomers even when only carbon and hydrogen are present. For example, 2-butene has both *cis* and *trans* isomers.

	Cis-2-butene	*Trans*-2-butene
Melting Point	−138.9°C	−105.5°C
Boiling Point	3.7°C	0.9°C
Density (at 20°C)	0.621 g/mL	0.604 g/mL

Alkynes

The alkynes have one or more triple bonds ($-C\equiv C-$) per molecule and have the general formula C_nH_{2n-2}. The simplest one is ethyne, commonly called acetylene (C_2H_2).

$$H-C\equiv C-H$$

Alkynes do not allow for geometric isomers because all of the bond lines attached to the triple-bond carbon atoms are in a straight line.

A mixture of acetylene and oxygen burns with a flame hot enough to cut steel (3000°C). The naming of alkynes is similar to that of alkenes with the lowest number possible being used for locating the triple bond.

The name of the above compound is 4-methyl-2-pentyne.

Figure 12-8 Ball-and-stick model of acetylene. (Charles Steele.)

SELF-TEST 12-A

1. The branch of chemistry that deals with compounds of carbon is known as _____ chemistry.
2. Each carbon in a saturated hydrocarbon has _____ geometry.
3. How many covalent bonds are in a molecule of butane? Does it make a difference which butane is considered?

4. A straight-chain hydrocarbon, such as pentane, actually has all of its carbon atoms in a straight line ( ) true ( ) false.

5. How many different isomers of C_5H_{12} are shown below?

a.

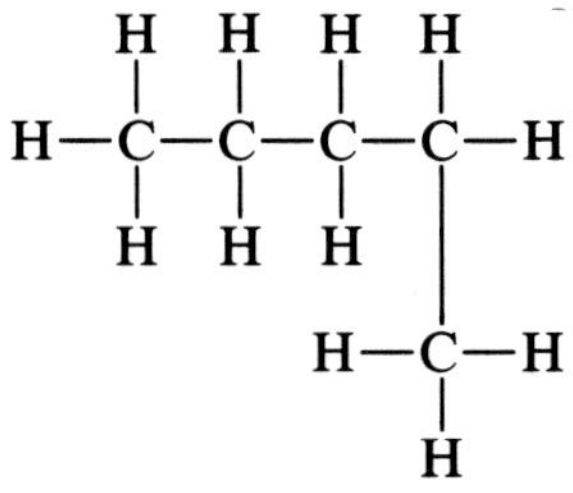

$$H-\underset{\underset{H}{|}}{\overset{\overset{H}{|}}{C}}-\underset{\underset{H}{|}}{\overset{\overset{H}{|}}{C}}-\underset{\underset{H}{|}}{\overset{\overset{H}{|}}{C}}-\underset{\underset{\underset{H}{|}}{\overset{\overset{|}{}}{C}-H}}{\overset{\overset{H}{|}}{C}}-H$$

b.

$$H-\underset{H}{\overset{H}{\overset{|}{C}}}-H$$

$$H-\overset{H}{\underset{H}{\overset{|}{C}}}-\overset{H}{\underset{\underset{\underset{H}{|}}{C}-H}{\overset{|}{C}}}-\overset{H}{\underset{H}{\overset{|}{C}}}-H$$

c.

$$H-\overset{H}{\overset{|}{C}}=\overset{H}{\underset{\underset{\underset{\underset{H}{|}}{C}-H}{\overset{|}{C}}}{\overset{|}{C}}}-\overset{H}{\underset{H}{\overset{|}{C}}}-\overset{H}{\underset{H}{\overset{|}{C}}}-H$$

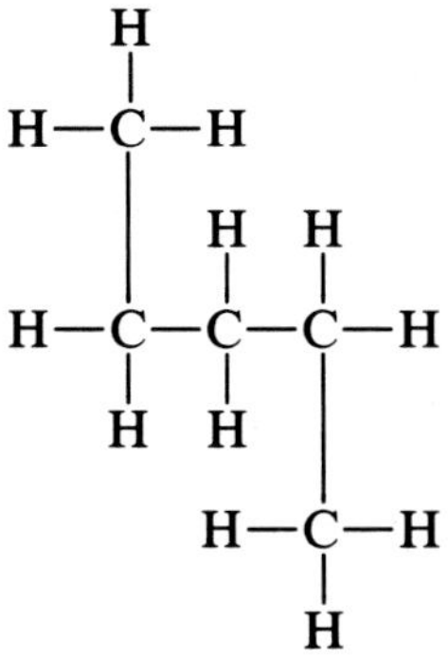

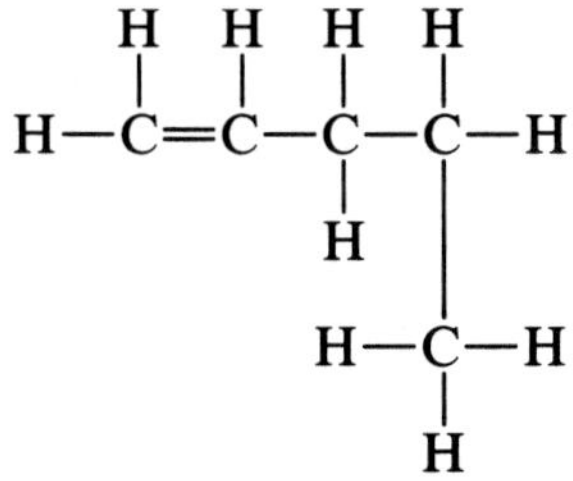

d.

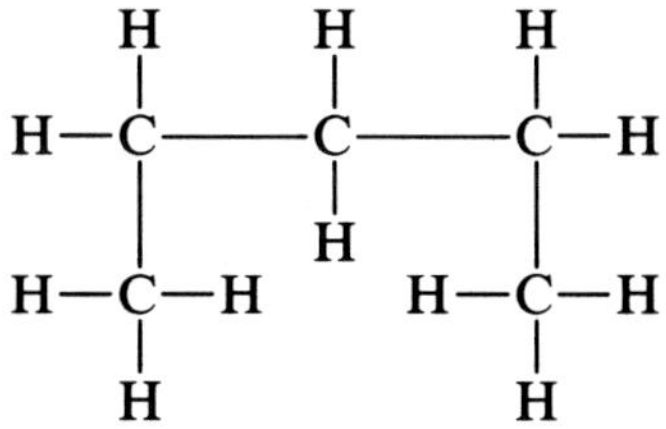

$$H-\overset{H}{\overset{|}{C}}-\overset{H}{\underset{\underset{\underset{H}{|}}{C}-H}{\overset{|}{C}}}-\overset{H}{\underset{\underset{\underset{H}{|}}{C}-H}{\overset{|}{C}}}-H$$

e.

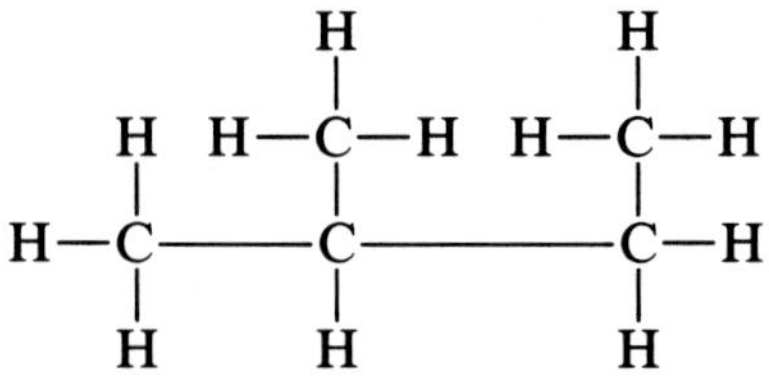

$$H-\underset{H}{\overset{H}{\overset{|}{C}}}-\overset{H-\overset{H}{\overset{|}{C}}-H}{\underset{H}{\overset{|}{C}}}\ \ \overset{H-\overset{H}{\overset{|}{C}}-H}{\underset{H}{\overset{|}{C}}}-H$$

f.

$$H-\overset{H}{\overset{|}{C}}-\overset{H}{\overset{|}{C}}-\overset{H}{\underset{\underset{\underset{H}{|}}{C}-H}{\overset{|}{C}}}-\overset{H}{\overset{|}{C}}-H$$

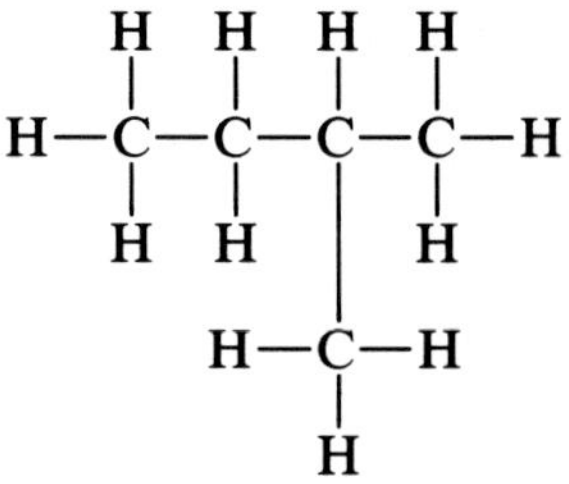

6. Butane and isobutane are examples of _____ isomers.

7. The number-one organic chemical produced in the United States is

_____ .

8. _____ is the first member of the alkyne series of hydrocarbons.

9. When the name of a compound ends in *-ene* (for example, *butene*), what structural feature is indicated?

10. Name this compound:

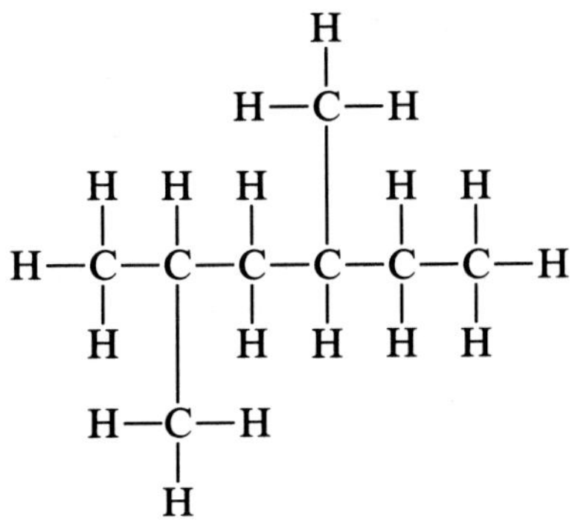

$$\overset{H}{\underset{\underset{\underset{\underset{H}{|}}{C}-H}{\overset{|}{H-C-H}}}{}}$$

$$H-\overset{H}{\underset{H}{\overset{|}{C}}}-\overset{H}{\underset{\underset{\underset{H}{|}}{C}-H}{\overset{|}{C}}}-\overset{H}{\underset{H}{\overset{|}{C}}}-\overset{H}{\underset{H}{\overset{|}{C}}}-\overset{H}{\underset{H}{\overset{|}{C}}}-H$$

11. The formula for the ethyl group is _____ .
12. The rigidity of the double carbon-carbon bond allows for the possibility of _____ isomers.
13. Draw the structure of 3-ethyl-2-pentene.

THE CYCLIC HYDROCARBONS

Hydrocarbons can form rings as well as straight chains and branched chains. The cyclic hydrocarbons include **cycloalkanes** (all single bonds), **cycloalkenes** (one carbon-carbon double bond), **cycloalkynes** (one carbon-carbon triple bond), and the **aromatics** (a unique merging of single bonds and double bonds resulting in electron delocalization around the ring).

Cycloalkanes

The simplest cycloalkane is cyclopropane, a highly strained ring compound:

Cycloalkanes have the same general formula as alkenes: C_nH_{2n}.

Symbols like △ are just more chemical shorthand.

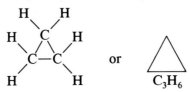

The bonds are strained because of the 60° angles in the ring; angles above 90° show a much greater stability. Cyclopropane, a volatile, flammable gas (b.p. is −32.7°C), is a rapidly acting anesthetic. A cyclopropane-oxygen mixture is useful in surgery on babies, small children, and "bad risk" patients because of its rapid action and the rapid recovery of the patient. Helium gas is added to the cyclopropane-oxygen mixture to reduce the danger of explosion in the operating room. Cyclopropane is an isomer of propene, C_3H_6, the second member of the alkene series (see earlier discussion of alkenes). Although cycloalkanes are isomers of alkenes, they are called cycloalkanes because all the bonds in the molecules are single bonds.

The cycloalkanes are commonly represented by a polygon. Each corner represents a carbon atom and two hydrogen atoms, and the lines represent C—C bonds. The C—H bonds are not shown, but are understood. Other common homologous cycloalkanes include cyclobutane, cyclopentane, and cyclohexane. These are represented as:

Cyclohexane exists in two conformations referred to as the "boat" and the "chair" forms (Fig. 12–9). These are conformations, not isomers, because they result from twisting around the C—C single bonds rather than breaking bonds to put atoms in different positions, such as in *cis* and *trans* 2-butene.

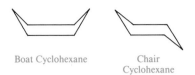

Boat Cyclohexane   Chair
                        Cyclohexane

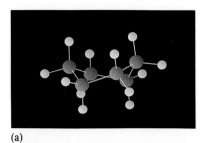

(a)

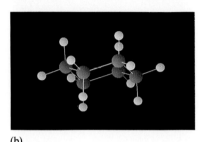

(b)

Since the end groups are farther apart in the chair form, steric (spatial) repulsions between the end groups will be less than in the boat form. As a result, the chair form is more stable. The chair form of cyclohexane is the prototype of the six-membered rings found in glucose and other sugars (Chapter 15).

Cycloalkenes

Examples of the unsaturated cyclic hydrocarbons include cyclohexene, which is used as a stabilizer in high-octane gasoline.

Cyclohexene

Figure 12–9   Ball-and-stick models of the (a) boat and (b) chair forms of cyclohexane. (Courtesy of Phillips Petroleum.)

Aromatic Compounds

Hydrocarbons containing one or more benzene rings (Fig. 12–10) are called **aromatic compounds.** The word *aromatic* was derived from *aroma,* which describes the rather strong and often pleasant odor of these compounds. However, benzene and some other aromatic compounds, such as benzo(α)pyrene, are both toxic and carcinogenic. The main structural feature, which is responsible for the distinctive chemical properties of the aromatic compounds, is the benzene ring.

Carcinogenic means cancer-causing. The type of cancer caused may vary from one carcinogen to another. Benzene causes a form of leukemia.

Figure 12–10   (a) Bonding in benzene. (b) Ball-and-stick model of benzene. (Charles D. Winters.)

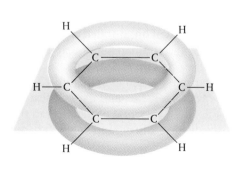

(a)

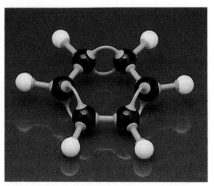

(b)

The molecular structure of benzene is a ring of carbon atoms in a plane, with one hydrogen atom bonded to each carbon atom (Fig. 12–10a). The bonds between these carbon atoms are shorter than single bonds but longer than double bonds. The measured bond angles are 120°. The benzene structure is sometimes written with alternating double bonds, though these structures do not conform to experimental facts about benzene.

The six electrons in the three double bonds are not localized between atoms like they are in the double bonds in alkenes, but are spread over the ring. Figure 12–10a illustrates the delocalization of six electrons above and below the plane of the ring. In other words, all six carbon–carbon bonds are equivalent, so that a better representation of benzene is:

where the circle represents the evenly distributed, delocalized electrons.

When hydrogen and carbon atoms are not shown, benzene is represented by a circle in a hexagon. Each corner in the hexagon represents one carbon atom and one hydrogen atom. Remember that this diagram of benzene stands for C_6H_6.

The delocalization of electrons around the ring accounts for the greater stability of aromatic compounds relative to that of unsaturated compounds, which contain double or triple bonds. (See the comparison of reactivity in the next section.)

Examples of some aromatic hydrocarbons are shown in Figure 12–11. Many aromatic compounds, such as benzene, toluene, and the xylenes, are on the list of top 50 chemicals (see Table 13–1) because of their use in the manufacture of plastics, detergents, pesticides, drugs, and other organic chemicals.

Xylenes rank numbers 25 and 27 among commercial chemicals.

The three xylenes shown in Figure 12–11 are structural isomers— different compounds with the same formula, C_8H_{10}. Each of these isomers has two methyl groups substituted for hydrogen atoms on the benzene ring. For two groups, either the prefixes *para-, meta-, ortho-* or numbers are used.

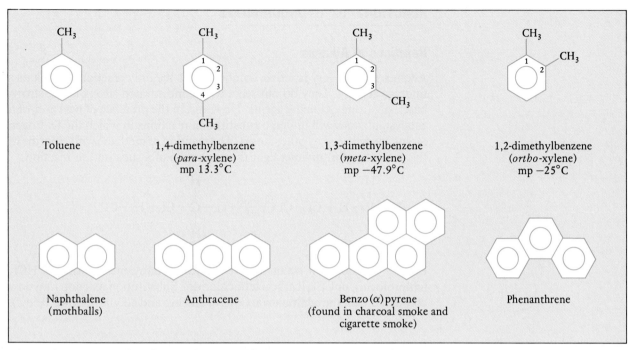

Figure 12–11 Examples of aromatic hydrocarbons.

If more than two groups are attached to the benzene ring, numbers must be used. Consider the following compounds:

1,2,3-trichlorobenzene   1,2,4-trichlorobenzene   1,3,5-trichlorobenzene

There is no other way to attach three atoms of chlorine to a benzene ring, and only three trichlorobenzenes have been isolated in the laboratory.

Some rings have nitrogen, oxygen, or sulfur atoms in place of carbon atoms in the ring. Of these possibilities only aromatic ring structures with nitrogen will be used later in the text. Examples are pyridine and pyrimidine. If a nitrogen atom replaces a carbon atom in an aromatic molecule, there is one less hydrogen atom in the molecular structure. The corner at the N does not have an H attached, but there is a pair of unshared electrons on the nitrogen atom. More will be made of this point in the discussion of biochemistry in Chapter 15.

Note that 1,2,3- is the same as 4,5,6- if the molecule is flipped over and turned around. Molecules don't know external directions.

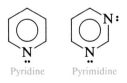

Pyridine   Pyrimidine

REACTIONS OF HYDROCARBONS

Reactions of Alkanes

Alkanes are not very reactive; combustion is the only reaction that alkanes undergo readily. They do not react with concentrated strong acids, strong bases, or strong oxidizing agents. However, in the presence of heat and/or a catalyst, alkanes will undergo **substitution** reactions in which the hydrogen atoms are replaced by other atoms. The reaction of methane with chlorine in the presence of ultraviolet light is an example of a substitution reaction.

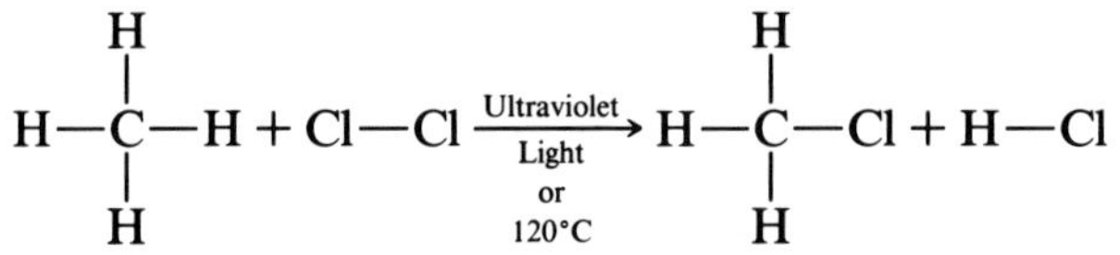

Methane Chloromethane

Further reaction can occur to give CH_2Cl_2 (methylene chloride), $CHCl_3$ (chloroform), or CCl_4 (carbon tetrachloride). Substitution reactions are slow compared with the **addition** reactions of alkenes and alkynes (shown below).

Reactions of Alkenes and Alkynes

Alkenes and alkynes are more reactive than alkanes because of the presence of double or triple bonds that can add atoms of the halogens, hydrogen, or even groups of atoms. For comparison with alkanes, the reactions of alkenes and alkynes with chlorine take place at room temperature in the dark.

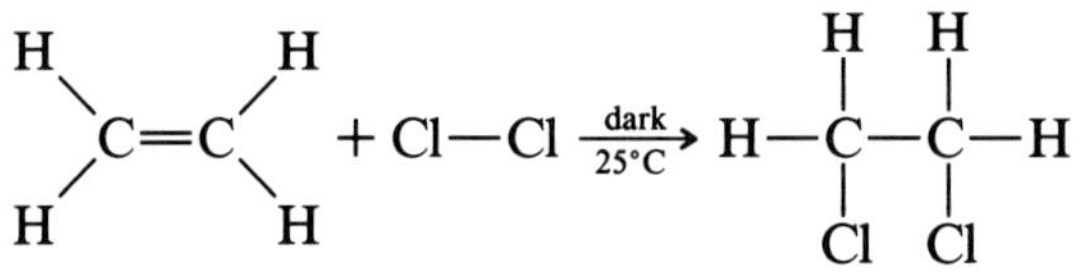

Ethene 1,2-dichloroethane (Methylene Chloride)

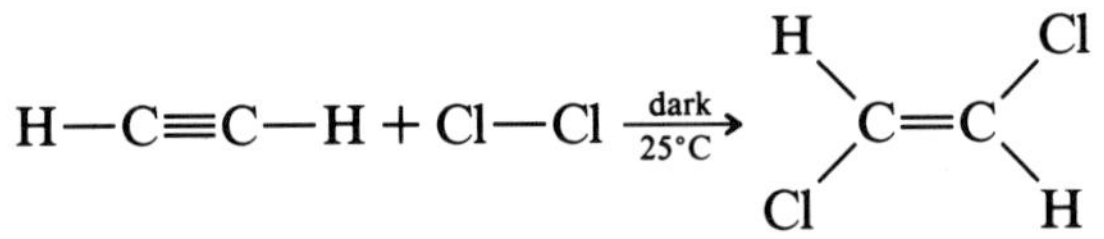

Ethyne Dichloroethene

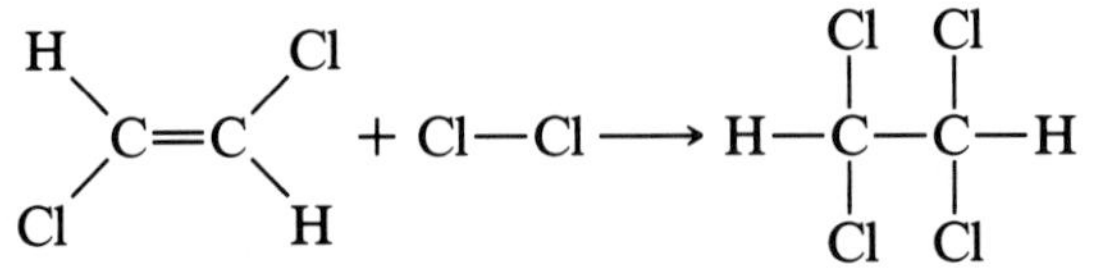

Dichloroethene Tetrachloroethane

Examples of other addition reactions that occur readily are given below. **Hydrogenation** is the addition reaction of hydrogen to an alkene or alkyne to form an alkane. The reaction requires a platinum, nickel, or palladium catalyst.

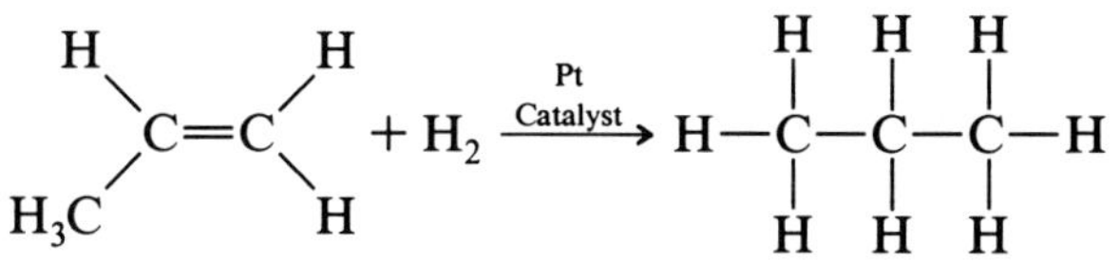

Propene + H$_2$ $\xrightarrow{\text{Pt Catalyst}}$ Propane

Hydrogen chloride gas will react to give chlorinated hydrocarbons.

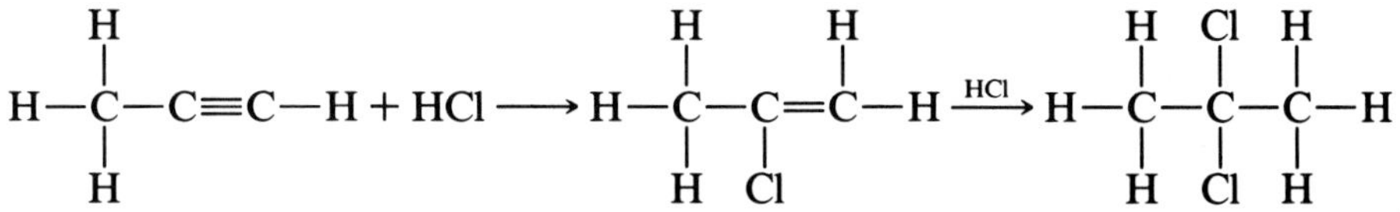

Propyne + HCl ⟶ 2-chloropropene $\xrightarrow{\text{HCl}}$ 2,2-dichloropropane

An acidic solution will react to give an alcohol.

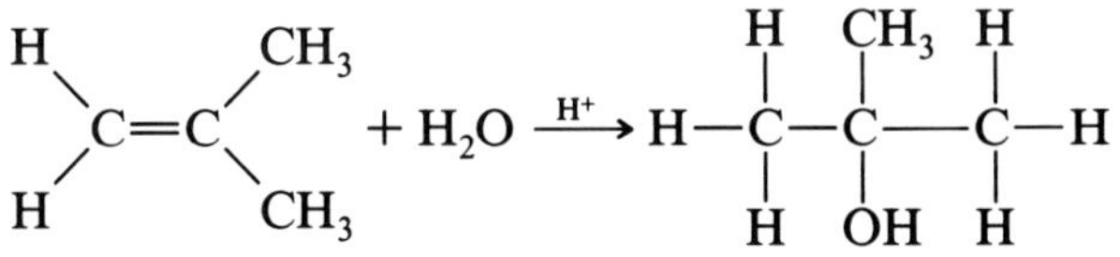

2-methylpropene + H$_2$O $\xrightarrow{\text{H}^+}$ 2-methyl-2-propanol

Addition reactions are used to prepare some polymers, an important class of compounds discussed in Chapter 14.

Aromatic compounds are less reactive than alkenes and alkynes and undergo substitution reactions similar to the alkanes in the presence of a suitable catalyst such as iron (III) chloride.

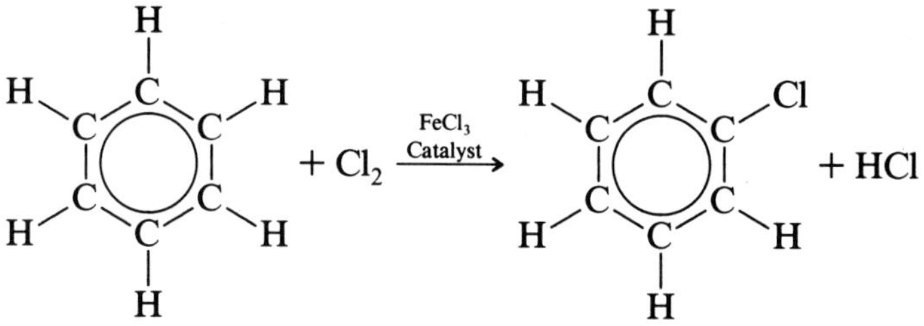

Benzene + Cl$_2$ $\xrightarrow{\text{FeCl}_3 \text{ Catalyst}}$ Chlorobenzene + HCl

or

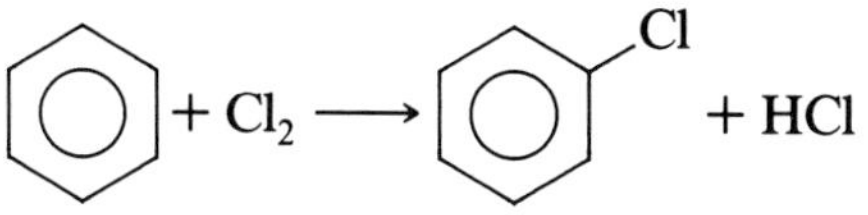

It may seem inconsistent that the hydrocarbon classes with the strongest bonds, the alkenes and the alkynes, are also more reactive than alkanes and aromatics. Their reactivity relates to the degree of unsaturation since the unsaturated compounds can *add* atoms rather than having to *substitute* atoms for hydrogen.

Combustion: A Common Reaction for all Hydrocarbons

Saturated and unsaturated hydrocarbons do undergo some similar reactions. One type shared by all hydrocarbons is **combustion** reactions, in which the hydrocarbon reacts with oxygen, that is, burns in air. This is the most common reaction since over 95% of the use of fossil fuels is for energy. If the

TABLE 12–5 Classes of Organic Compounds Based on Functional Groups*

General Formulas of Class Members	Class Name	Typical Compound	Compound Name	Common Use of Sample Compound
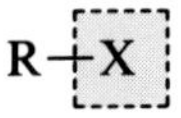 R—X	Halide	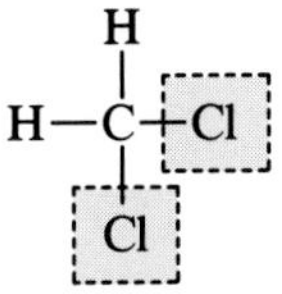 H—C(—Cl)(Cl)(H) Dichloromethane structure	Dichloromethane (methylene chloride)	Solvent
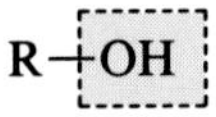 R—OH	Alcohol	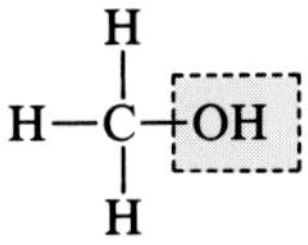 H—C(—OH)(H)(H)	Methanol (wood alcohol)	Solvent
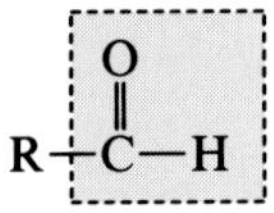 R—C(=O)—H	Aldehyde	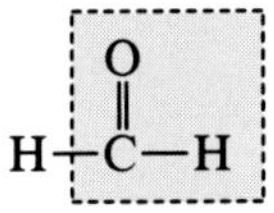 H—C(=O)—H	Methanal (formaldehyde)	Preservative
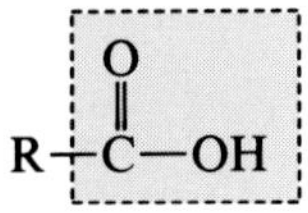 R—C(=O)—OH	Carboxylic acid	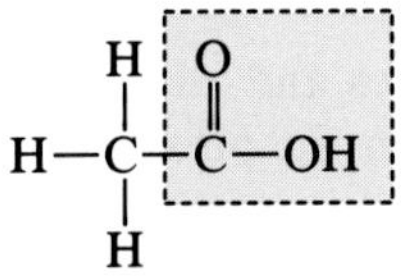 H—C(H)(H)—C(=O)—OH	Ethanoic acid (acetic acid)	Vinegar
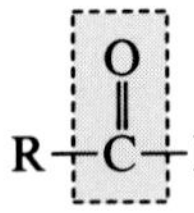 R—C(=O)—R′	Ketone	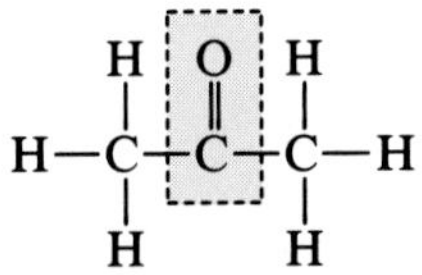 H—C(H)(H)—C(=O)—C(H)(H)—H	Propanone (acetone)	Solvent
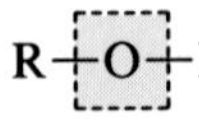 R—O—R′	Ether	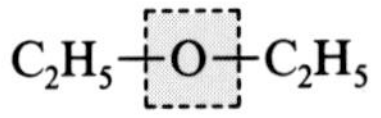 C_2H_5—O—C_2H_5	Diethyl ether (ethyl ether)	Anesthetic
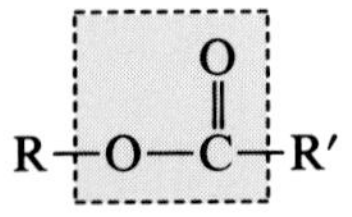 R—O—C(=O)—R′	Ester	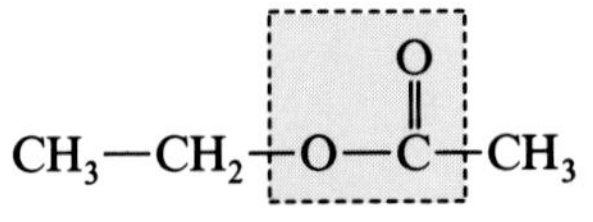 CH_3—CH_2—O—C(=O)—CH_3	Ethyl ethanoate (ethyl acetate)	Solvent in fingernail polish
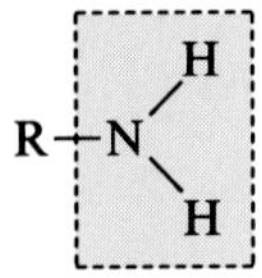 R—N(H)(H)	Amine	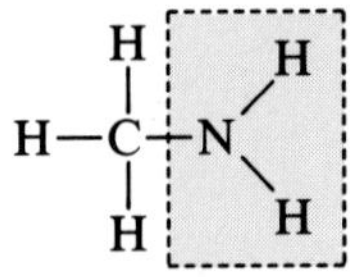 H—C(H)(H)—N(H)(H)	Methylamine	Tanning (foul odor)
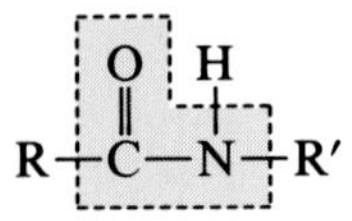 R—C(=O)—N(H)—R′	Amide	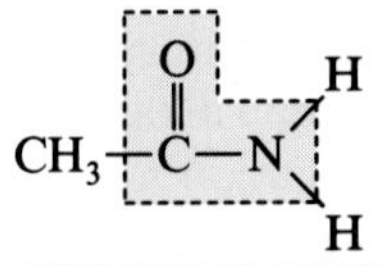 CH_3—C(=O)—N(H)(H)	Acetamide	Plasticizer

* R stands for an H or a hydrocarbon group such as —CH_3 or —C_2H_5. R′ could be a different group from R.

amount of air is sufficient, the products are carbon dioxide (CO_2) and water (H_2O).

$$2\ CH_3-CH_3 + 7\ O_2 \longrightarrow 4\ CO_2 + 6\ H_2O + 373\ \text{kcal/mole } C_2H_6$$

$$CH_2=CH_2 + 3\ O_2 \longrightarrow 2\ CO_2 + 2\ H_2O + 337\ \text{kcal/mole } C_2H_4$$

$$2\ HC\equiv CH + 5\ O_2 \longrightarrow 4\ CO_2 + 2\ H_2O + 311\ \text{kcal/mole } C_2H_2$$

Note that the energy released is directly related to the number of moles of oxygen per mole of hydrocarbon: $3\frac{1}{2}$, 3, and $2\frac{1}{2}$, respectively.

FUNCTIONAL GROUPS

The millions of organic compounds include classes of compounds that are obtained by replacing hydrogen atoms of hydrocarbons with atoms or groups of atoms known as **functional groups.** The important classes of compounds that result from attaching functional groups to a hydrocarbon framework are shown in Table 12–5. The "R" attached to the functional group represents the nonfunctional, hydrocarbon framework with one hydrogen atom removed for each functional group added. Refer back to Table 12–3 for some examples of alkyl groups that are represented by R.

Both common names and the IUPAC name for each class of functional group are illustrated in Table 12–5. The chemist needs to know both the common names and the IUPAC names since both are widely used. The IUPAC system provides a systematic method for naming all members of a given class. For example, alcohols end in -*ol* (methan*ol*); aldehydes end in -*al* (methan*al*); carboxylic acids end in -*oic* (ethan*oic* acid); and ketones end in -*one* (propan*one*). Note that in each of these cases the prefix represents the total number of carbon atoms in the molecule.

Isomers are also possible for molecules containing functional groups. For example, a single hydrocarbon molecule can give rise to several alcohols if there are different isomeric positions for the —OH group. Three different alcohols result when a hydrogen atom is replaced by an —OH group in *n*-pentane, depending on which hydrogen atom is replaced (Table 12–6).

Alcohols are classified as primary, secondary, and tertiary based on the number of other carbons bonded to the —C—OH carbon.

Primary

H
|
R—C—OH
|
H

CH₃CH₂OH
Ethanol
(Grain Alcohol)

Secondary

R
|
R′—C—OH
|
H

CH₃
|
CH₃—C—OH
|
H

2-propanol
(Rubbing Alcohol)

Tertiary

R
|
R′—C—OH
|
R″

CH₃
|
CH₃—C—OH
|
CH₃

2-methyl-2-propanol
(Gasoline Additive)

The use of R, R′, and R″ indicates all R groups can be different.

The common name of 2-propanol is isopropyl alcohol.

The common name of 2-methyl-2-propanol is tertiary-butyl alcohol.

When one or more functional groups appear in a molecule, the IUPAC name reveals the functional group name and position. For example, the

TABLE 12–6 Alcohols Derived from Pentane (C_5H_{12})

$$\begin{array}{c}
\text{H} \quad \text{H} \quad \text{H} \quad \text{H} \quad \text{H} \\
| \quad\;\; | \quad\;\; | \quad\;\; | \quad\;\; | \\
\text{H}-\text{C}-\text{C}-\text{C}-\text{C}-\text{C}\!-\!\boxed{\text{H}}\!\leftarrow \\
| \quad\;\; | \quad\;\; | \quad\;\; | \quad\;\; | \\
\text{H} \quad \text{H} \quad \text{H} \quad \text{H} \quad \text{H}
\end{array}$$

gives

$$\begin{array}{c}
\text{H} \quad \text{H} \quad \text{H} \quad \text{H} \quad \text{H} \\
| \quad\;\; | \quad\;\; | \quad\;\; | \quad\;\; | \\
\text{H}-\text{C}-\text{C}-\text{C}-\text{C}-\text{C}\!-\!\boxed{\text{OH}} \\
| \quad\;\; | \quad\;\; | \quad\;\; | \quad\;\; | \\
\text{H} \quad \text{H} \quad \text{H} \quad \text{H} \quad \text{H} \\
\text{1-pentanol}
\end{array}$$

Substitution of an —OH for an end hydrogen

$$\begin{array}{c}
\text{H} \quad \text{H} \quad \text{H} \quad \text{H} \quad \text{H} \\
| \quad\;\; | \quad\;\; | \quad\;\; | \quad\;\; | \\
\text{H}-\text{C}-\text{C}-\text{C}-\text{C}-\text{C}-\text{H} \\
| \quad\;\; | \quad\;\; | \quad\;\; | \quad\;\; | \\
\text{H} \quad \text{H} \quad \text{H} \quad \boxed{\text{H}} \quad \text{H} \\
\uparrow
\end{array}$$

gives

$$\begin{array}{c}
\text{H} \quad \text{H} \quad \text{H} \quad \text{H} \quad \text{H} \\
| \quad\;\; | \quad\;\; | \quad\;\; | \quad\;\; | \\
\text{H}-\text{C}-\text{C}-\text{C}-\text{C}-\text{C}-\text{H} \\
| \quad\;\; | \quad\;\; | \quad\;\; | \quad\;\; | \\
\text{H} \quad \text{H} \quad \text{H} \quad \boxed{\begin{array}{c}\text{O}\\\text{H}\end{array}} \quad \text{H} \\
\text{2-pentanol}
\end{array}$$

Substitution of an —OH for a 2-carbon hydrogen

$$\begin{array}{c}
\text{H} \quad \text{H} \quad \text{H} \quad \text{H} \quad \text{H} \\
| \quad\;\; | \quad\;\; | \quad\;\; | \quad\;\; | \\
\text{H}-\text{C}-\text{C}-\text{C}-\text{C}-\text{C}-\text{H} \\
| \quad\;\; | \quad\;\; | \quad\;\; | \quad\;\; | \\
\text{H} \quad \text{H} \quad \boxed{\text{H}} \quad \text{H} \quad \text{H} \\
\uparrow
\end{array}$$

gives

$$\begin{array}{c}
\text{H} \quad \text{H} \quad \text{H} \quad \text{H} \quad \text{H} \\
| \quad\;\; | \quad\;\; | \quad\;\; | \quad\;\; | \\
\text{H}-\text{C}-\text{C}-\text{C}-\text{C}-\text{C}-\text{H} \\
| \quad\;\; | \quad\;\; | \quad\;\; | \quad\;\; | \\
\text{H} \quad \text{H} \quad \boxed{\begin{array}{c}\text{O}\\\text{H}\end{array}} \quad \text{H} \quad \text{H} \\
\text{3-pentanol}
\end{array}$$

Substitution of an —OH for a 3-carbon hydrogen

name of an alcohol includes the name of the hydrocarbon to which it corresponds and indicates the number of its carbon atoms, and the suffix *-ol* denotes an alcohol. As before, a number is used to indicate the position of the alcohol group. The IUPAC name for the alcohol used as permanent antifreeze is 1,2-ethanediol because there are two alcohol groups replacing two hydrogen atoms in ethane.

$$\begin{array}{c}
\text{H} \quad \text{H} \\
| \quad\;\; | \\
\text{H}-\text{C}-\text{C}-\text{H} \\
| \quad\;\; | \\
\text{HO} \quad \text{OH}
\end{array}$$

1,2-ethanediol
(Ethylene Glycol)

Characteristic chemistry for important members of each functional group class is given in Chapter 13.

STEREOISOMERISM—OPTICAL ISOMERS

Naming Optical Isomers

Review earlier discussion in this chapter on geometric isomers.

Chiral is pronounced ki-ral.

Optical isomers are possible when a molecular structure is **asymmetric** (without symmetry). One common example of an asymmetric molecule is one containing a tetrahedral carbon atom bonded to four *different* atoms or groups of atoms. Such a carbon atom is called a **chiral** (or asymmetric) carbon atom; an example is the carbon atom in the molecule CHIBrCl.

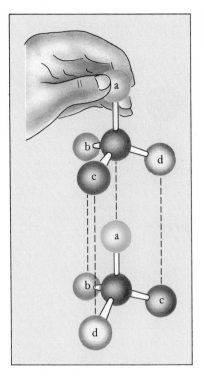
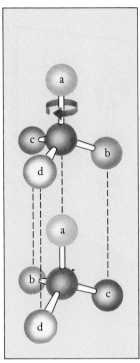

Figure 12–12   Four different atoms, or groups of atoms, are bonded to tetrahedral center atoms so that the upper isomeric form cannot be turned in any way to match exactly the lower structure. The upper structure and the lower structure are nonsuperimposable mirror images. (See also Fig. 12–14.)

Figure 12–12 shows the two ways to arrange four different atoms in the tetrahedral positions about the central carbon atom. These result in two nonsuperimposable, mirror-image molecules that are called **optical isomers** or **enantiomers.**

There are many examples of nonsuperimposable mirror images in the macroscopic world. Consider your hands or right- and left-hand gloves, for instance. They are mirror images of one another and are nonsuperimposable (Fig. 12–13).

All amino acids except glycine can exist as one of two optical isomers. In Figure 12–14, the mirror-image relationship is shown for optical isomers of alanine, an amino acid with a tetrahedral carbon atom surrounded by an

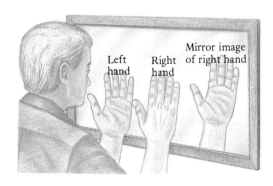

Figure 12–13   Mirror images. Your left hand is a nonsuperimposable mirror image of your right hand. For example, the mirror image of your right hand looks like your left hand. If you place one hand directly over the other, they are not identical; hence, they are nonsuperimposable mirror images.

Figure 12–14 Optical isomers of the amino acid alanine, 2-aminopropionic acid. The D-form is the nonsuperimposable mirror image of the L-form.

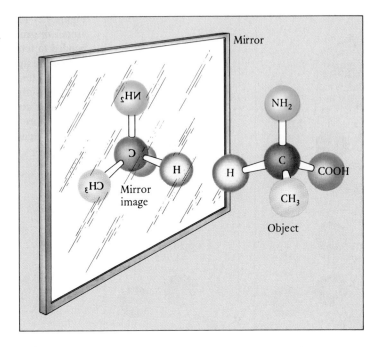

All amino acids have an amine group (—NH₂) and an acid group (—COOH).

The formula of glycine is H₂NCH₂COOH. Why doesn't glycine have optical isomers?

amine group (—NH₂), a methyl group (—CH₃), an acid group (—COOH), and a hydrogen atom. Note that the carbon atoms in the methyl and acid groups are not asymmetric since these carbon atoms are not bonded to four different groups.

The "handedness" of optical isomers is represented by D for right-handed (D stands for dextro from the Latin *dexter* meaning "right") and L for left-handed (L stands for levo from the Latin *laevus* meaning "left"). The orientation of the L-form of amino acids is pictured in Figure 12–15.

The properties of optical isomers of a compound are almost identical — they have the same melting point, the same boiling point, the same density, and many other identical physical and chemical properties. However, they always differ with respect to one physical property: they rotate the plane of **polarized** light in opposite directions. According to the wave theory of light, a light wave traveling through space vibrates at right angles to its path but in

Figure 12–15 Space orientation of L-form of amino acids. When walking along the C — C — N bridge from the C — O end to the NH₂ end, the acid is in the L-form if the R group is on the left. If R is on the right, it is a D-acid. Only L-isomers of amino acids are found in nature.

L-amino acid side chain

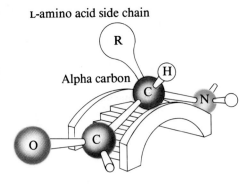

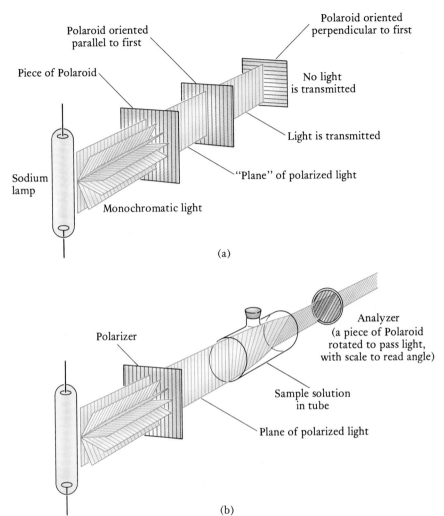

Polaroid oriented
parallel to first

Piece of Polaroid

Polaroid oriented
perpendicular to first

No light
is transmitted

Light is transmitted

Sodium
lamp

"Plane" of polarized light

Monochromatic light

(a)

Polarizer

Analyzer
(a piece of Polaroid
rotated to pass light,
with scale to read angle)

Sample solution
in tube

Plane of polarized light

(b)

Figure 12–16 Rotation of plane-polarized light by an optical isomer. (a) A sodium lamp provides a monochromatic yellow light. The original beam is nonpolarized; it vibrates in all directions at right angles to its path. After passing through a Polaroid filter, the light is vibrating in only one direction. This polarized light will pass through another Polaroid filter if the filter is lined up properly, but will not pass through the third Polaroid filter if it is at right angles to the other two. The direction of the Polaroid filters determines the direction of the polarization. (b) The plane of polarized light is rotated by a solution of an optically active isomer. The analyzer can be a second Polaroid filter that can be rotated to find the angle for maximum transmission of light. If the solution rotates the plane of polarized light, the analyzer will not be at the same angle as the polarizer for maximum transmission.

many planes (Figure 12–16). If a group of waves is passed though a polarizing crystal, such as Iceland spar (a form of $CaCO_3$), or through a sheet of Polaroid material, the light is split into two rays, and the waves emerging along the incoming axis vibrate in only one plane perpendicular to the light path. Such light is said to be **plane-polarized**. An **optically active** substance rotates the plane-polarized light either clockwise or counterclockwise.

Optical Isomers and Life

Optical isomers can also differ with respect to biological properties. An example is the hormone adrenalin (or epinephrine). Adrenalin is the L-form of a pair of optical isomers.

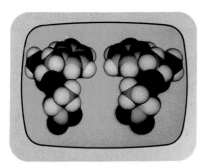

Models of the optical isomers of aspartame. (*The World of Chemistry,* Program 9, "Molecular Architecture.")

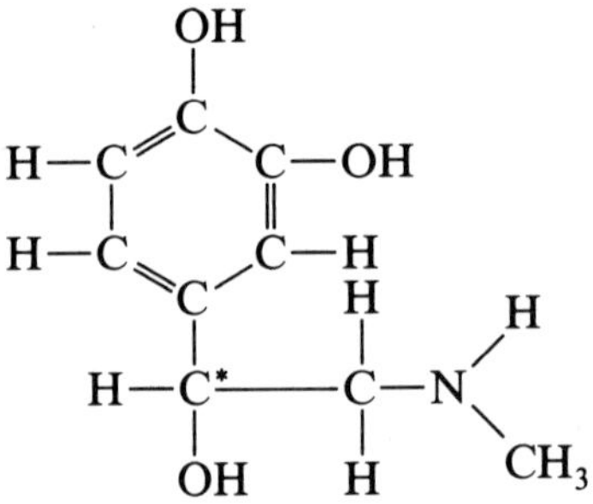

Adrenalin
(Epinephrine)

Only the L-isomer is effective in starting a heart that has stopped beating momentarily, or in giving a person unusual strength during times of great emotional stress. The other isomer is inactive.

The optical isomers of aspartame have different properties. The L-isomer is used as an artificial sweetener (NutraSweet®), but the D-isomer is bitter.

Only L-isomers of optically active amino acids are found in proteins. Nature's preference for L-amino acids has provoked much discussion and speculation among scientists since Pasteur's discovery of optical activity in 1848 from studies of crystals of tartaric acid salts.

Enzymes, the catalysts for biochemical reactions, also have a handedness and, as a left-hand glove only fits a left hand, bind to only one of the

---

C* designates the chiral carbon atom in the structure above.

MOLECULAR ARCHITECTURE

There are various types of isomerism common to organic chemistry. Of these, optical isomers are among the most fascinating. This is because they play such important roles in life processes. In living things, chiral molecules exist in only one form or the other but not both. How did the selection of one optical isomer over the other occur in nature? This question is addressed by Nobel laureate Christian Anfinsen who says

How this selection began in nature is anybody's guess. One assumption is that some naturally occurring minerals, for example, might have been involved in binding one form and not the other. In the process a concentration of the form we have now was built up so that when life started, it was stuck with that form. In nature we're stuck pretty much with one isomer. The world has become so evolved that living things are in general composed of one of the two possible mirror images of the basic compounds.

The World of Chemistry (Program 9) "Molecular Architecture."

Christian Anfinsen. (*The World of Chemistry,* Program 9, "Molecular Architecture.")

optical isomers. For example, during contraction of muscles the body produces only the L-form of lactic acid and not the D-form.

Large organic molecules may have a number of chiral carbon atoms within the same molecule. At each such carbon atom there exists the possibility of *two* arrangements of the molecule. The total number of possible molecules, then, increases exponentially with the number of chiral centers. With two asymmetric carbon atoms there are 2^2, or four, possible structures. It should be emphasized that each of the four isomers can be made from the *same* set of atoms with the *same* set of chemical bonds. Glucose, a simple blood sugar also known as dextrose, contains four asymmetric carbon atoms per molecule. Thus, there are 2^4 (16) isomers in the family of stereoisomers to which D-glucose belongs. However of the 16 possible isomers, only 3 are important. These are D-glucose, D-mannose, and D-galactose. Of these, D-glucose is by far the most common.

The concentration of lactic acid in the blood is associated with the feeling of tiredness, and a period of rest is necessary to reduce the concentration of this chemical by oxidation.

$$\begin{array}{c} COOH \\ | \\ C \\ H \diagup \vdots \diagdown OH \\ CH_3 \end{array}$$
D-lactic acid

$$\begin{array}{c} COOH \\ | \\ C \\ HO \diagup \vdots \diagdown H \\ CH_3 \end{array}$$
L-lactic acid

$$\begin{array}{ccccc}
& O & & O & & O \\
& \parallel & & \parallel & & \parallel \\
& C-H & & C-H & & C-H \\
& | & & | & & | \\
H-&C^*-OH & HO-&C^*-H & H-&C^*-OH \\
& | & & | & & | \\
HO-&C^*-H & HO-&C^*-H & HO-&C^*-H \\
& | & & | & & | \\
H-&C^*-OH & H-&C^*-OH & HO-&C^*-H \\
& | & & | & & | \\
H-&C^*-OH & H-&C^*-OH & H-&C^*-OH \\
& | & & | & & | \\
HO-&C-H & HO-&C-H & HO-&C-H \\
& | & & | & & | \\
& H & & H & & H \\
& \text{D-glucose} & & \text{D-mannose} & & \text{D-galactose}
\end{array}$$

(C* = Asymmetric Carbon Atom)

SELF-TEST 12-B

1. The simplest cycloalkane is _____ .
2. The difference between cyclohexane and benzene is the number of _____ atoms.
3. The most reactive hydrocarbon class is the _____ .
4. The benzene ring has both localized electrons and _____ electrons.
5. All ring structures contain only carbon atoms in the ring and have delocalized electrons. ( ) True ( ) False.
6. To have optical isomers in carbon compounds, a carbon atom must have _____ different groups attached.
7. In what physical property do optical isomers differ?
8. Identify the functional groups present in each of the following molecules:

a. R—OH   c. $\overset{\overset{O}{\|}}{R—C—H}$

b. $\overset{\overset{O}{\|}}{R—C—OH}$ d. $\overset{\overset{O}{\|}}{R—C—R'}$

9. **a.** Write the structural formulas for ethane, ethanol, ethanal, ethanoic acid, diethyl ether, and ethyl amine.
   b. Give common names where possible.
   c. What R group is present in these compounds?

10. Give examples of the following:
    a. An ether
    b. An alcohol
    c. An organic acid
    d. A ketone

11. How many atoms does the symbol 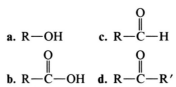 represent?

12. Name the following compound.

$$\overset{CH_3}{\underset{CH_3}{\bigcirc}}—CH_3$$

_____ 1.	Organic chemistry	**a.** Compounds of carbon and hydrogen
_____ 2.	Isomers	**b.** Found in benzene
_____ 3.	Hydrocarbon	**c.** Chemistry of nonmineral carbon compounds
_____ 4.	Methyl group	
_____ 5.	Delocalized electrons	**d.** Alcohol
		e. Aldehyde
		f. Carboxylic acid
		g. Ester
_____ 6.	$\overset{\overset{O}{\|}}{R—C—OH}$	**h.** Ether
		i. Same number and kinds of atoms arranged differently
_____ 7.	$\overset{\overset{O}{\|}}{R—C—H}$	
_____ 8.	R—O—R'	**j.** —CH_3
		k. Ketone
_____ 9.	$\overset{\overset{O}{\|}}{R—O—C—R'}$	**l.** Found in cyclohexane
_____ 10.	R—OH	

QUESTIONS

1. Saturated hydrocarbons are so named because they have the maximum amount of hydrogen present for a given amount of carbon. The saturated hydrocarbons have the general formula C_nH_{2n+2}, where n is a whole number. What are the names and formulas of the first four members of this series of compounds?
2. What is the simplest aromatic compound?
3. Draw the structural formula for each of the five isomeric hexanes, C_6H_{14}.
4. What is the difference between benzene and cyclohexane?
5. How many trichloroethene structures are possible?
6. Draw the *cis* and *trans* isomers for:
   a. 1,2-dibromoethene
   b. 1-bromo-2-chloroethene
7. What unique bond is present in an alkyne hydrocarbon?
8. How do primary, secondary, and tertiary alcohols differ?
9. Draw the structure of 1,1,1-trichlorethane.
10. Describe the bonding in benzene.
11. Give three reasons why there are over nine million known organic compounds.
12. Draw the three trichlorobenzene structures and name them.
13. Define *ubiquitous,* and explain how this word is descriptive of the carbon atom.
14. Are more organic or inorganic compounds known?
15. Why is carbon the "central" element in organic compounds?
16. a. Carbon and hydrogen have almost the same electronegativities. On the basis of this information and the tetrahedral structure around each carbon atom in a hydrocarbon such as octane (C_8H_{18}), would octane be polar or nonpolar?
    b. Would octane likely dissolve appreciably in water?
17. What is the difference between an addition reaction and a substitution reaction?
18. Compare the relative reactivities of alkanes, alkenes, alkynes, and aromatic compounds.
19. Give the IUPAC name for the following compounds.
    a. CH_3OH
    b. CH_3CH_2OH
    c. $HCOOH$
    d. $CH_3CH_2CHCH_3$ with OH below
    e. CH_3COOH
    f. CH_2-CH_2 with OH and OH below
    g. CH_3CHO
    h. CH_3CHCH_3 with Br below
20. What is the structural formula for 1-pentene?

21. An alkene always contains at least one _____ bond.
22. Write the structural formulas for butanoic acid, aminomethane, 2-butanol, and 3-aminopentane.
23. Write the structural formulas for two compounds that can have each of the molecular formulas listed.
    a. $C_5H_{12}O$
    b. C_3H_6O
    c. $C_5H_{10}O_2$
24. Write the structural formulas for:
    a. 2-methylbutane
    b. 4,4-dimethyl-5-ethyloctane
    c. methylbutane
    d. 2-methyl-2-hexene
25. Which propanol is used as rubbing alcohol?
26. What functional groups are found in glucose?
27. Give an example of
    a. an alkane   b. an amine
    c. a carboxylic acid d. an ether
    e. an ester   f. an alkene
    g. an alkyne   h. an alcohol
    i. a ketone   j. an aldehyde
28. Indicate the functional groups present in the following molecules.
    a. $CH_3CH_2CH_2COOH$
    b. $CH_3CH_2NH_2$
    c. $CH_3CHCH_2CH_2COOH$ with NH_2 below
    d. CH_3CHCH_2COOH with OH below
    e. $CH_3CCH_2CH_2COOH$ with O (double bond) below
    f. CH_3CHCH_2OH with NH_2 below
29. Give IUPAC names for the following compounds.

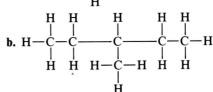

c.

$$H-\underset{\underset{H}{|}}{\overset{\overset{H}{|}}{C}}\underset{}{\overset{}{\quad}}\underset{\underset{H}{|}}{\overset{\overset{H-C-H}{|}}{C}}=\underset{\underset{H}{|}}{\overset{\overset{H-C-H}{|}}{C}}\underset{}{\quad}\underset{\underset{H}{|}}{\overset{\overset{H}{|}}{C}}-H$$

d.

$$H-\underset{\underset{H}{|}}{\overset{\overset{H}{|}}{C}}-C\equiv C-\underset{\underset{H}{|}}{\overset{\overset{H}{|}}{C}}-H$$

30. a. How can optical isomers be distinguished from each other experimentally?
   b. Which arrangement has a mirror image that is non-superimposable?

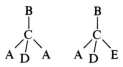

   c. Use the appropriate structure in (b) to explain the term *asymmetric carbon atom.*

MOLECULAR BEAUTY II: FROGS ABOUT TO BE KISSED

ESSAY by ROALD HOFFMANN

Let's look at a second source of molecular beauty. At first glance, structure 1 doesn't appear to be beautiful at all. The dangling $(CH_2)_{12}$-Cl chains at left and the asymmetrical $(CH_2)_{25}$ loop at right are rather odd. This molecule is, if not ugly (I maintain that there are no ugly molecules), at best plain. It isn't an essential component of life, nor is it produced commercially in megakilogram lots. Its purpose is not obvious.

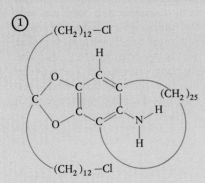

In fact, structure 1 resembles the fairy tale frog that, when kissed, becomes a handsome prince. Chemistry is about molecules, but it is also about chemical change—the transformations of molecules. The beauty of a molecule may lie in its structure (as we saw for $NaNb_3O_6$), or it may be hidden in the molecule's potential to undergo an interesting change. This is the case for structure 1, which is an **intermediate** in a fascinating sequence of reactions. Where are we going? To a **catenane** (structure 2)— two interlocking rings of carbon atoms, not chemically combined but held together like the links of a chain. Why would anyone want to make a catenane? The best reason is that one had never been made before.

It is not at all obvious how to go about making such a molecule. Here's one strategy, which I will call a "statistical" one. Scheme 3 shows a typical chemical reaction of carbon chains, called a **cyclization**. Here X, Y, and R represent some unspecified atoms or groups of atoms that react to form one or more molecules (shown as {X,Y,R}). They leave behind a hydrocarbon ring, the structure in which we are interested. If the starting material is a very long carbon chain, then some fraction of the time— purely by chance—two chains will be intertwined or a chain will thread itself through an existing ring, in such a way that cyclization forms a caten-

ane. There are about 10^{20} molecules in a typical laboratory flask, and in this multitude statistical probabilities have a chance to work. E. Wasserman peformed this experiment in 1960, synthesizing a catenane for the first time.

The statistical method works, but it forms relatively few catenane molecules. There are other, more efficient ways to craft the catenane structure. One elegant synthesis was devised by G. Schill and A. Lüttringhaus. Their scheme is summarized in 4. The starting point is a molecule with lots of carefully planned functional groups. In chemistry, a **functional group** is a set of bonded atoms whose properties are more or less constant from one kind of molecule to another. The most important of these properties is the chemical reactivity, or the "function," of the group. To

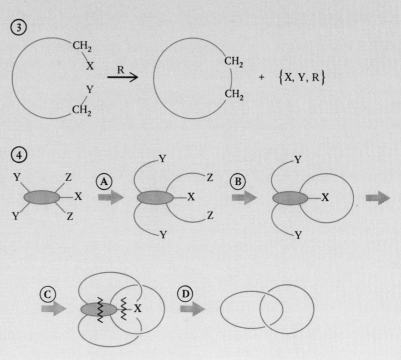

put it another way, functional groups are the "handles" of a molecule. The predictable reactions of functional groups are essential in designing syntheses in organic chemistry. Some common functional groups are —OH (alcohols), —COOH (organic acids), —COH (aldehydes), and the halides (—F, —Cl, —Br, and —I). In scheme 4, the letters X, Y, and Z represent functional groups.

Step A in 4 places the Y and Z functional group handles at the ends of long carbon chains. Step B closes up, or "cyclizes," the pair of chains containing Z. Step C is a different kind of cyclization, linking the ends of the other pair of chains to the remaining functional group X. Step D, which is perhaps several reactions, is a fragmentation in which the core — from which the sequence began, on which all was built — is torn apart to reveal the catenane.

Several points about this process are important. (1) The synthesis is a building process, and (2) like all construction, it requires work. It's easy to write down the logical sequence of steps as I have done. However, each step may be several chemical reactions, and each reaction a more or less elaborate set of physical processes. These take time and money! (3) The nature of the synthesis requires that the molecules in the middle of the construction be more complex than those at the beginning and at the end.

To return to specific molecules, it becomes clear what molecule 1 is when you compare it to scheme 4. It's the crucial point in the middle, after step B and before step C. It is poised to cyclize, the chlorine atoms at the ends of the $(CH_2)_{12}$ chains set to react with the NH_2 group. The synthesis by Schill and Lüttringhaus begins with molecule 5 and ends with catenane 6, in which a ring of 28 carbons interlocks with a ring of 25 carbons and one nitrogen. While getting to structure 6 is sweet, it should be clear that

what is important is the process. That process is reasonably straightforward (although the lack of detail in scheme 4 makes it appear far simpler than it is). You might suppose that any point in such a sequence of changes could be claimed as the "most significant" one. The steps in any synthesis vary greatly in difficulty, and therefore in the cleverness and effort required to accomplish them. The unpublished lore of chemistry has many tales of elaborate syntheses in which only the last step fails, although it was thought to be trivial.

I claim the honor of "most significant" for the molecule somewhere in the middle of the scheme, the one that is most complicated compared with the starting materials and the goal. It is the molecule that is most disguised, yet the one bearing in it the surprise, the essence of what is to come. On the way to a catenane, that's what molecule 1 is. This frog, warts and all, is about to become a prince.

372

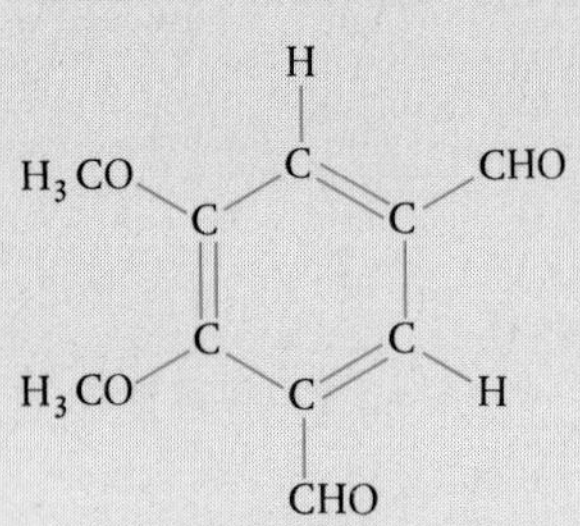

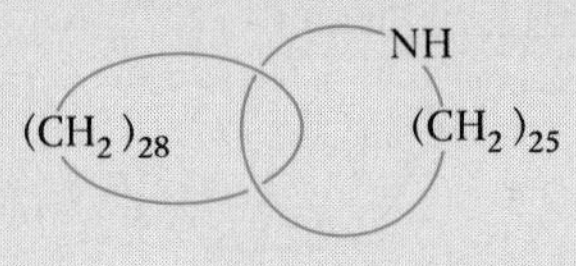

13 Organic Chemicals: Energy and Materials for Society

Eastman Kodak "chemicals-from-coal" plant in Kingsport, TN. (Courtesy of Tennessee Eastman.)

The millions of organic compounds are either hydrocarbons or derivatives of hydrocarbons. In Chapter 12 the hydrocarbon derivatives were organized into classes based on functional groups (Table 12–5), which replace one or more hydrogen atoms in the parent hydrocarbon. Fossil fuels are the major source of hydrocarbons, and among these, about 10% of the petroleum refined today is the source of most of the organic chemicals used to make plastics, synthetic rubber, synthetic fibers, fertilizers, and thousands of other consumer products. Figure 13–1 summarizes the organic chemicals obtained from fossil fuels and their uses in the synthesis of a wide range of commercial products. For this reason the organic chemical industry is often referred to as the **petrochemical industry.**

Another way of examining the importance of organic chemistry to society is to look at the list of the top 50 chemicals produced in the United States. Of the top 50 listed in Table 13–1, 29 are organic chemicals. The chemistry of many of these will be described later in this chapter and in Chapter 14. However, since over 90% of the fossil fuels are burned to obtain energy for transportation, heating, and the production of electricity, the use of fossil fuels as an energy source will be examined first.

PETROLEUM

Petroleum is a complex mixture of alkanes, cycloalkanes, alkenes, and aromatic hydrocarbons. Thousands of compounds are present in crude petroleum, and the actual composition of petroleum varies with location. For example, Pennsylvania crude oils are primarily chain hydrocarbons, whereas California crude is composed of a larger portion of aromatics. The composition of petroleum can be classified according to boiling point ranges (Table 13–2). Each of these ranges has important uses. The **refining of petroleum** is the separation of fractions with a certain boiling point range by a process called **fractional distillation.**

Refining Petroleum

Figure 13–2 is a schematic drawing of a fractional distillation tower used in the petroleum refining process; a picture of a fractionating tower is shown in Figure 13–3a. The crude oil is heated to about 400°C to produce a hot vapor and liquid mixture that enters the fractionating tower. The vapor rises and condenses at various points along the tower. The lower boiling fractions (those that are more volatile) will remain in the vapor stage longer than the higher boiling fractions. These differences in boiling point ranges allow the separation of fractions in the same way that simple distillation allows the partial separation of water and ethanol. Some of the gases do not condense and are drawn off the top of the tower. The unvaporized residual oil is collected at the bottom of the tower. Typical products of the fractionation of petroleum are listed in Table 13–2. From the time petroleum was discov-

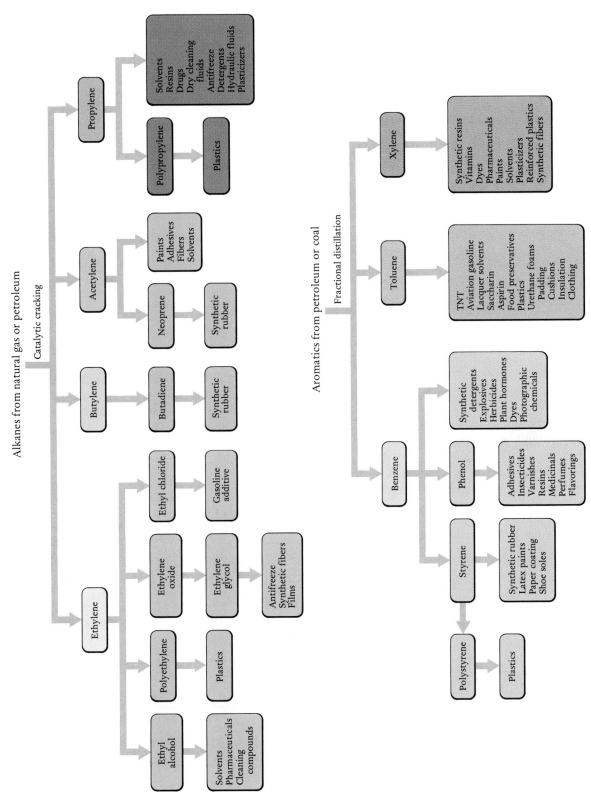

Alkanes from natural gas or petroleum

Catalytic cracking

Ethylene

Ethyl alcohol → Solvents, Pharmaceuticals, Cleaning compounds

Polyethylene → Plastics

Ethylene oxide → Ethylene glycol → Antifreeze, Synthetic fibers, Films

Ethyl chloride → Gasoline additive

Butylene → Butadiene → Synthetic rubber

Acetylene → Neoprene → Synthetic rubber

Acetylene → Paints, Adhesives, Fibers, Solvents

Propylene → Polypropylene → Plastics

Propylene → Solvents, Resins, Drugs, Dry cleaning fluids, Antifreeze, Detergents, Hydraulic fluids, Plasticizers

Aromatics from petroleum or coal

Fractional distillation

Benzene

Styrene → Polystyrene → Plastics

Styrene → Synthetic rubber, Latex paints, Paper coating, Shoe soles

Phenol → Adhesives, Insecticides, Varnishes, Resins, Medicinals, Perfumes, Flavorings

Benzene → Synthetic detergents, Explosives, Herbicides, Plant hormones, Dyes, Photographic chemicals

Toluene → TNT, Aviation gasoline, Lacquer solvents, Saccharin, Aspirin, Food preservatives, Plastics, Urethane foams, Padding, Cushions, Insulation, Clothing

Xylene → Synthetic resins, Vitamins, Dyes, Pharmaceuticals, Paints, Solvents, Plasticizers, Reinforced plastics, Synthetic fibers

Figure 13–1   Some of the organic chemicals obtained from fossil fuels and their uses as raw materials.

TABLE 13–1 Top 50 Chemicals Produced in the United States in 1988*

1. Sulfuric acid	18. Terephthalic acid	35. Butadiene
2. Nitrogen	19. Carbon dioxide	36. Acetic acid
3. Oxygen	20. Vinyl chloride	37. Propylene oxide
4. Ethylene	21. Styrene	38. Carbon black
5. Ammonia	22. Methanol	39. Aluminum sulfate
6. Lime	23. Formaldehyde	40. Acrylonitrile
7. Sodium hydroxide	24. Toluene	41. Vinyl acetate
8. Phosphoric acid	25. Xylene	42. Cyclohexane
9. Chlorine	26. Hydrochloric acid	43. Acetone
10. Propylene	27. *p*-Xylene	44. Titanium dioxide
11. Sodium carbonate	28. Ethylene oxide	45. Sodium sulfate
12. Nitric acid	29. Ethylene glycol	46. Sodium silicate
13. Urea	30. Cumene	47. Adipic acid
14. Ammonium nitrate	31. Methyl *t*-butyl ether	48. Isopropyl alcohol
15. Ethylene dichloride	32. Ammonium sulfate	49. Calcium chloride
16. Benzene	33. Phenol	50. Caprolactam
17. Ethylbenzene	34. Potash	

* Data from *Chemical and Engineering News,* April 10, 1989, p. 12.

ered in the United States in 1859 until the automobile became popular in 1900, most oil was refined to yield kerosene, a mixture of hydrocarbons in the C_{12} to C_{16} range. This liquid was used principally as a fuel for heating and lighting.

The internal combustion engines used in early automobiles were designed to burn a more volatile mixture of hydrocarbons, the C_6 to C_{10}

Figure 13–2   Schematic diagram of a fractionating column for distilling petroleum.

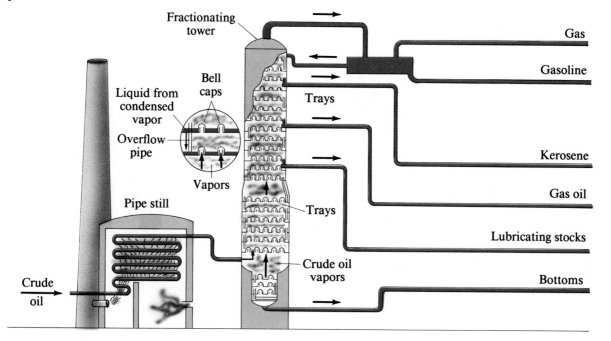

TABLE 13–2 Hydrocarbon Fractions from Petroleum

Fraction	Size Range of Molecules	Boiling Point Range (°C)	Uses
Gas	C_1–C_4	−160 to 30	Gaseous fuel, production of H_2
Straight-run gasoline	C_5–C_{12}	30 to 200	Motor fuel
Kerosene, fuel oil	C_{12}–C_{18}	180 to 400	Diesel fuel, furnace fuel, cracking
Lubricants	C_{17} and up	350 and up	Lubricants
Paraffins	C_{20} and up	Low-melting solids	Candles, matches
Asphalt	C_{36} and up	Gummy residues	Surfacing roads, fuel

fraction, which became known as **gasoline.** The vapors of these lower molecular weight hydrocarbons mix readily with air in simple carburetors and burn fairly completely. With the increasing popularity of the automobile, petroleum refiners had to shift the output of a barrel of crude oil from a reasonably large fraction of kerosene to almost no kerosene and a much greater fraction of gasoline (Table 13–3). This dramatic increase in the amount of gasoline from a barrel of crude oil was accomplished by the discovery of chemical processes that convert nongasoline molecules into ones that burn well in an automobile engine.

A barrel of crude oil is 42 gal.

    Thermal cracking involves heating saturated hydrocarbons under pressure in the absence of air. The hydrocarbons break into shorter-chain

Cracking breaks larger molecules into smaller ones.

Figure 13–3 (a) A view of the towers at an oil refinery used for the fractional distillation of petroleum. (© Four By Five.) (b) A view of the catalytic cracking unit. (*The World of Chemistry,* Program 22, "The Age of Polymers.")

(a)

(b)

TABLE 13–3 Division of a Barrel of Crude Oil

	1920s %	1960s %	1980s %
Gasoline	26.1	44.8	42.7
Kerosene	12.7	2.8	0.9
Jet fuel	—	7.6	7.5
Heavy distillates	48.6	22.2	19.1
Other (asphalt, road oil, residual fuel oil, etc.)	12.6	22.6	29.8

hydrocarbons—both alkanes and alkenes, some of which will be in the gasoline range.

$$C_{16}H_{34} \xrightarrow[\text{Heat}]{\text{Pressure}} C_8H_{18} + C_8H_{16}$$

An Alkane   An Alkane   An Alkene
in the Gasoline Range

Review the discussion of catalysis in Chapter 6.

Later, **catalytic cracking** was carried out in units such as the one shown in Figure 13–3b. These units use catalysts, which allow the processes to proceed at lower pressures and result in even higher yields of gasoline. Today, these catalysts include specially processed clays, known as **zeolites.** Refiners of petroleum use their own special methods, which offer different advantages in cost and type of crude oil handled. The hydrocarbon molecules produced are much the same regardless of the methods used.

The "straight-run" gasoline fraction obtained from the fractional distillation of petroleum contains primarily straight-chain hydrocarbons that burn too rapidly to be suitable for use as a fuel in internal combustion engines. Rapid ignition causes a "knocking" or "pinging" sound in the engine that reduces engine power and may damage the engine. Cracking of petroleum fractions brought about not only an increase in the quantity of gasoline available from a barrel of crude oil, but also an increase in *quality.* That is, gasoline from a cracking process can be used at higher efficiency (in a high-compression engine) than can straight-run gasoline, because the molecular structures of the hydrocarbons in the cracked gasoline allow them to oxidize more smoothly at high pressure without knocking. Knocking can spell trouble to the owner of an automobile because it will eventually lead to the breakdown of the internal parts of the automobile's engine.

Knocking in gasoline engines is a sign of improper combustion.

Octane Rating

An arbitrary scale for rating the relative knocking properties of gasolines has been developed based on the operation of a standard test engine. Normal heptane, typical of straight-run gasoline, knocks considerably and is assigned an octane rating of 0:

$$CH_3CH_2CH_2CH_2CH_2CH_2CH_3 \qquad (\text{OCTANE RATING} = 0)$$

n-heptane

whereas 2,2,4-trimethylpentane (isooctane) is far superior in this respect and is assigned an octane rating of 100:

$$CH_3-\overset{\overset{\displaystyle CH_3}{|}}{\underset{\underset{\displaystyle CH_3}{|}}{C}}-CH_2-\overset{\overset{\displaystyle CH_3}{|}}{\underset{\underset{\displaystyle H}{|}}{C}}-CH_3 \qquad \text{(OCTANE RATING = 100)}$$

2,2,4-trimethylpentane

The octane rating of a gasoline is determined by first using the gasoline in a standard engine and recording its knocking properties. This is compared to the behavior of mixtures of *n*-heptane and isooctane, and the percentage of isooctane in the mixture with identical knocking properties is called the octane rating of the gasoline. Thus, if a gasoline has the same knocking characteristics as a mixture of 9% *n*-heptane and 91% isooctane, it is assigned an octane rating of 91. This corresponds to a regular grade of gasoline. Since the octane rating scale was established, fuels superior to isooctane have been developed, so the scale has been extended well above 100.

Table 13–4 lists octane ratings for some hydrocarbons and octane enhancers. The straight-run gasoline fraction obtained from the fractional distillation of petroleum has an octane rating of 50 to 55, which is too low for use as a fuel in vehicles. From Table 13–4 we can see that the octane rating of a gasoline can be increased either by increasing the percentage of branched-chain and aromatic hydrocarbon fractions or by adding octane enhancers (or a combination of both).

The **catalytic re-forming** process is used to produce branched-chain and aromatic hydrocarbons. Under the influence of certain catalysts, such as finely divided platinum, straight-chain hydrocarbons with low octane numbers can be re-formed into their branched-chain isomers, which have higher octane numbers.

Catalytic re-forming is also used to produce aromatic hydrocarbons such as benzene, toluene, and xylenes by using different catalysts and petroleum mixtures. For example, when the vapors of straight-run gasoline, kero-

The octane scale measures the ability of a mixture to burn without knocking in a gasoline engine.

The hydrogen produced here can be used in the synthesis of ammonia by the Haber process. (See Chapter 19)

TABLE 13–4 Octane Numbers of Some Hydrocarbons and Gasoline Additives

Name	Octane Number
n-heptane	0
n-hexane	25
n-pentane	62
1-pentene	91
2,2,4-trimethylpentane (isooctane)	100
benzene	106
o-xylene	107
methanol	107
ethanol	108
t-butyl alcohol	113
methyl *t*-butyl ether	116
p-xylene	116
toluene	118
ethyl *t*-butyl ether	118

sene, and light oil fractions are passed over a copper catalyst at 650°C, a high percentage of the original material is converted into a mixture of aromatic hydrocarbons, from which benzene, toluene, xylenes, and similar compounds may be separated by fractional distillation. For example, *n*-hexane is converted into benzene

$$CH_3CH_2CH_2CH_2CH_2CH_3 \longrightarrow \bigcirc + 4\ H_2$$

n-hexane   Benzene

and *n*-heptane is changed into toluene.

$$CH_3CH_2CH_2CH_2CH_2CH_2CH_3 \longrightarrow \overset{CH_3}{\bigcirc} + 4\ H_2$$

n-heptane   Toluene

This process is one example of the use of expensive noble metals [platinum (Pt), palladium (Pd), rhodium (Rh), iridium (Ir), gold (Au), and silver (Ag)] as catalysts in many industrial processes. The surfaces of these metals are often used to catalyze gas-phase reactions. For example, the platinum catalyst in the catalytic re-forming process adsorbs low-octane alkanes such as *n*-hexane and converts them to compounds with higher octane numbers, such as benzene and branched or cyclic alkanes. The product depends on the pattern of platinum atoms (crystal face) at the surface where adsorption takes place. Figure 13–4 illustrates how products depend on the particular characteristics of the exposed surface. Note in Figure 13–4 that undesirable alkanes are produced at the "kinked" surface. The catalytic converter found in American automobiles produced since 1975 is also a good example of a catalytic process that uses platinum as a surface catalyst.

The octane number of a given blend of gasoline can also be increased by adding "antiknock" agents or octane enhancers. Prior to 1975, the most widely used antiknock agent was tetraethyllead, $(C_2H_5)_4Pb$. The addition of 3 g of $(C_2H_5)_4Pb$ per gallon increases the octane rating by 10 to 15, and before the Environmental Protection Agency (EPA) required reductions in lead content, both regular and premium gasoline contained an average of 3 g of $(C_2H_5)_4Pb$ or $(CH_3)_4Pb$ per gallon. Scheduled reductions in the lead content of gasoline have led to as little as 0.1 g of lead per gallon, and the goal is lead-free gasoline.

With the decreased use of tetraethyllead, other octane enhancers are being added to gasoline to increase the octane rating. These include toluene, 2-methyl-2-propanol (also called **tertiary**, or *t*-butyl alcohol), methyl-*t*-butyl ether (MTBE), methanol, and ethanol. In 1988, the most popular octane enhancer was MTBE, which joined the top-50 chemical list for the first time in 1984 and was number 31 in 1988. Another ether, ethyl *t*-butyl ether (ETBE), is receiving increased interest as an octane enhancer because it is less volatile than MTBE. ETBE also has a slightly higher octane rating than MTBE.

t-butyl Alcohol

MTBE

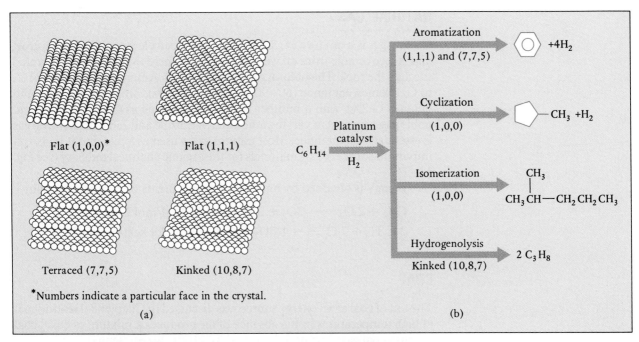

Figure 13-4 (a) Chemistry on a platinum surface depends on which surface is exposed. (b) Different surfaces favor different products. (Redrawn from George C. Pimentel: Surfaces and Condensed Phases. *Chem Tech*, September 1986, p. 537).

Gasoline blends that contain methanol and ethanol are also being used as fuels. The EPA and all U.S. car manufacturers have approved the use of ethanol–gasoline blends up to 10% ethanol (known as **gasohol** when introduced in the 1970s). However, methanol is receiving much attention because it offers several advantages as an octane enhancer. When properly blended, methanol is more economical, has a higher octane rating, and can reduce emission levels of particulates, hydrocarbons, carbon monoxide, and nitrogen oxides. However, the biggest disadvantage of methanol relates to moisture. Small amounts of moisture destabilize the methanol–gasoline mixture, and metal corrosion of the engine becomes a serious problem.

The methanol moisture problem is solved by using another alcohol (ethanol, propanols, butanols) as a cosolvent in methanol blends. The EPA has approved several methanol blends that meet the vehicle emission standards and provide a high-octane gasoline. Most methanol blends contain about 2.5% methanol, 2.5% *t*-butyl alcohol, 95% gasoline, and a corrosion inhibitor.

Both opponents and proponents of methanol blends agree there should be uniform labeling requirements for pumps that dispense methanol blends. Not all states require that the pump be labeled with the specific type of alcohol; only the word *alcohol* is required. As a result, consumers in many states are not aware that the gasoline they are buying contains methanol. Most assume the label *alcohol* refers to ethanol or gasohol, since ethanol was the first alcohol used in gasoline blends.

Methanol is more soluble in water (recall hydrogen bonding) than in hydrocarbons. Hence, water will extract the methanol into a two-layered system.

NATURAL GAS

Natural gas is a mixture of gases trapped with petroleum in the Earth's crust and is recoverable from oil wells or gas wells where the gases have migrated through the rock. The natural gas found in North America is a mixture of C_1 to C_4 alkanes, methane (60–90%), ethane (5–9%), propane (3–18%), and butane (1–2%), with a number of other gases, such as CO_2, N_2, H_2S, and noble gases present in varying amounts. In Europe and Japan the natural gas is essentially all methane. Most natural gas is used as a fuel, but it is also an important source of raw materials for the organic chemical industry (see Fig. 13–1).

Energy is obtained by burning the constituents of natural gas in air:

$$CH_4 + 2\ O_2 \longrightarrow CO_2 + 2\ H_2O + 213\ \text{kcal/mol}\ CH_4$$
$$2\ C_2H_6 + 7\ O_2 \longrightarrow 4\ CO_2 + 6\ H_2O + 372.8\ \text{kcal/mol}\ C_2H_6$$

COAL

The use of coal as an energy source was discussed in Chapter 8. Heating coal at high temperatures in the absence of air produces a mixture of coke, coal tar, and coal gas. The process, called **pyrolysis,** is represented by

$$\text{Coal} \longrightarrow \text{Coke} + \text{Coal tar} + \text{Coal gas}$$

One ton of bituminous (soft) coal yields about 1500 lb of coke, 8 gal of coal tar, and 10,000 ft³ of coal gas. Coal gas is a mixture of H_2, CH_4, CO, C_2H_6, NH_3, CO_2, H_2S, and other gases. At one time coal gas was used as a fuel. Coal tar can be distilled to yield the fractions listed in Table 13–5. Note the predominance of aromatic compounds. Some uses of these compounds as raw materials are shown in Figure 13–1.

Synthesis gas, a mixture of CO and H_2 obtained by treating pulverized coal with superheated steam, is also used as a starting material for the production of organic chemicals.

$$C + H_2O \longrightarrow CO + H_2$$

TABLE 13–5 Fractions from Distillation of Coal Tar

Boiling Range (°C)	Name	Tar, Mass %	Primary Constituents
Below 200	Light oil	5	Benzene, toluene, xylenes
200–250	Middle oil (carbolic oil)	17	Naphthalene, phenol, pyridine
250–300	Heavy oil (creosote oil)	7	Naphthalenes and methylnaphthalenes, cresols
300–350	Green oil	9	Anthracene, phenanthrene
Residue		62	Pitch or tar

SELF-TEST 13-A

1. The fractions of petroleum are separated by _____ .
2. The principal component in natural gas is _____ .
3. A widely used octane enhancer is _____ .
4. The _____ process is used to produce branched-chain and aromatic hydrocarbons from straight-chain hydrocarbons.
5. The major source of organic chemicals used as starting materials in the synthesis of a wide range of commercial products is _____ .
6. The _____ process is used in refining petroleum to convert molecules in the higher boiling fractions to molecules in the gasoline fraction.
7. Heating coal in the absence of air is called _____ .
8. Coal tar is the principal source of what major class of hydrocarbons? _____ .
9. Which of the following hydrocarbons would be expected to have the highest octane number?

a. $CH_3CH_2CH_2CH_2CH_2CH_2CH_3$ b. $CH_3CH_2\overset{\overset{\displaystyle CH_3}{|}}{C}HCH_2CH_2CH_3$

c. $CH_3-\overset{\overset{\displaystyle CH_3}{|}}{\underset{\underset{\displaystyle CH_3}{|}}{C}}-\overset{\overset{\displaystyle CH_3}{|}}{\underset{\underset{\displaystyle H}{|}}{C}}-CH_3$

ORGANIC SYNTHESIS

The preparation of new and different organic compounds through chemical reactions is called **organic synthesis.** Millions of organic compounds have been synthesized in the laboratories of the world during the past 150 years. Organic compounds were obtained originally from plants, animals, and fossil fuels, and these are still direct sources for many important chemicals such as sucrose from sugar cane or ethanol from fermented grain mash. However, the development of organic chemistry led to cheaper methods for the synthesis of both naturally occurring substances and new substances. Prior to 1828, it was widely believed that chemical compounds synthesized by living matter could not be made without living matter—a "vital force" was necessary for the synthesis. In 1828, a young German chemist, Friedrich Wöhler (Fig. 13–5), destroyed the vital force myth and opened the door to modern organic syntheses. Wöhler heated a solution of silver cyanate and ammonium chloride, neither of which had been derived from any living substance. From these he prepared urea, a major animal waste product found in urine.

Figure 13–5   Friederich Wöhler (1800–1882) was professor of chemistry at the University of Berlin and later at Göttingen. His preparation of the organic compound urea from the inorganic compound ammonium cyanate did much to overturn the theory that organic compounds must be prepared in living organisms. He was also one of the first to study the properties of aluminum, and the first to isolate the element beryllium, among many other outstanding contributions to chemistry.

$$AgOCN + NH_4Cl \longrightarrow AgCl + NH_4OCN$$

Silver Cyanate   Ammonium   Silver Chloride   Ammonium
                   Chloride   (Precipitate)   Cyanate

$$NH_4OCN \xrightarrow{\text{Heat}} H_2N\overset{\displaystyle O}{\overset{\displaystyle \|}{C}}NH_2$$

Ammonuim Cyanate   Urea

The notion of a mysterious vital force declined as other chemists began to synthesize more and more organic chemicals without the aid of a living system. Soon it was shown that chemists could do more than imitate the products of living tissue; they could form unique materials of their own design.

Organic synthesis is based on an understanding of the classification of organic compounds by functional groups, a concept introduced in Chapter 12, and the study of the chemical reactions of those groups. Advances in understanding the structure of organic compounds also gave organic synthesis a tremendous boost. The organic chemist who knows the structure of compounds can predict, by analogy with simpler molecules, what reactions might take place when organic reagents are used. Very elegant and reliable schemes of synthesis can then be constructed that lead to the synthesis of new and useful compounds or to the more economical synthesis of known compounds; both are central functions of modern organic chemistry.

See Chapter 12 Essay.

Acquaintance with a few representative compounds of each functional group class will give you a feel for the subject of organic chemistry. Keep in mind that the functional group classes represented in Table 12–5 and discussed here are examples of the most important functional groups, but there are many other functional groups.

ALKYL HALIDES

The halogens are group VIIA elements (F, Cl, Br, I).

Alkyl halides are molecules in which one or more hydrogen atoms are replaced by halogen atoms. In the IUPAC system for naming these compounds, the halogen is specified as fluoro, chloro, bromo, or iodo. Structural formulas and names, including familiar common names, of some alkyl halides are:

$$Cl-\underset{\underset{\displaystyle Cl}{|}}{\overset{\overset{\displaystyle H}{|}}{C}}-Cl \qquad Cl-\underset{\underset{\displaystyle Cl}{|}}{\overset{\overset{\displaystyle Cl}{|}}{C}}-Cl \qquad Cl-\underset{\underset{\displaystyle Cl}{|}}{\overset{\overset{\displaystyle Cl}{|}}{C}}-\underset{\underset{\displaystyle H}{|}}{\overset{\overset{\displaystyle H}{|}}{C}}-H$$

Trichloromethane   Tetrachloromethane   1,1,1-trichloroethane
(Chloroform)   (Carbon Tetrachloride)   (Methylchloroform)

$$Cl-\underset{\underset{\displaystyle Cl}{|}}{\overset{\overset{\displaystyle H}{|}}{C}}-H \qquad H-\underset{\underset{\displaystyle Br}{|}}{\overset{\overset{\displaystyle H}{|}}{C}}-\underset{\underset{\displaystyle Br}{|}}{\overset{\overset{\displaystyle H}{|}}{C}}-H$$

Dichloromethane   1,2-dibromoethane
(Methylene Chloride)   (Ethylene Dibromide)

At one time chloroform and carbon tetrachloride were widely used as solvents in the laboratory and in industrial cleaning, but their toxicity and carcinogenicity have led them to be removed from use. Many chlorinated hydrocarbons are on either the carcinogen or the suspected carcinogen list of the Environmental Protection Agency. Only 1,1,1-trichloroethane appears to be safe, and it has replaced the others in many solvent applications.

Substituted alkanes that contain both fluorine and chlorine are called chlorofluorocarbons (CFCs). Chlorofluorocarbons are relatively nontoxic, nonflammable, noncorrosive, odorless gases or liquids. These nonreactive properties led to the use of CFC-11 and CFC-12 as propellants in aerosol spray cans and as refrigerants in refrigerators

See Chapter 16 for a discussion of the toxicity and carcinogenicity of these compounds.

CFCs are perhaps better known as Freons®, (the Du Pont trade name for a variety of CFCs used as refrigerants in refrigerators and air conditioning units and propellants in aerosol cans).

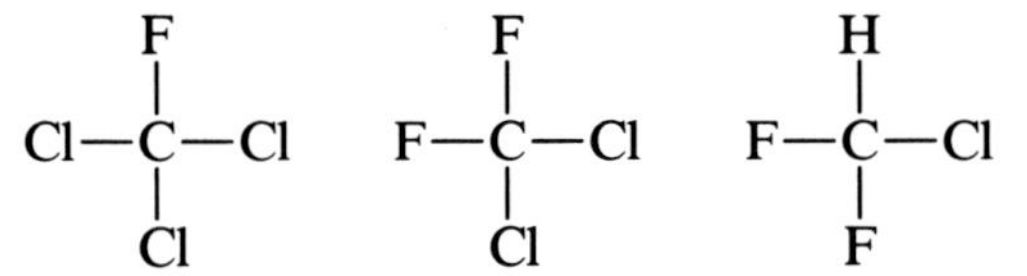

CFC-11   CFC-12   CFC-22

The numbers in CFC-11, CFC-12, and CFC-22 are industrial code numbers.

and air conditioning units for buildings and vehicles. However, recent studies have shown that CFCs do react in the upper atmosphere, decreasing statospheric ozone that is essential for filtering out harmful ultraviolet radiation. The reactions of ozone with CFCs and proposed reductions in CFC production are discussed in Chapter 18.

Several other halogenated hydrocarbons have also caused environmental problems because of their toxicity or carcinogenicity (Table 13–6). Environmental problems associated with these compounds are discussed in chapters 17 and 18.

TABLE 13–6 Some Organic Chlorine Compounds

Name	Formula	Uses
Dichlorodiphenyltrichloroethane DDT	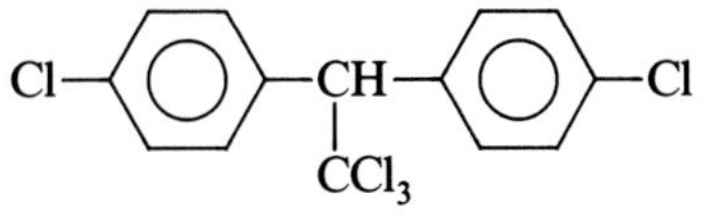	Insecticide
2,4-dichlorophenoxyacetic acid (2,4-D)	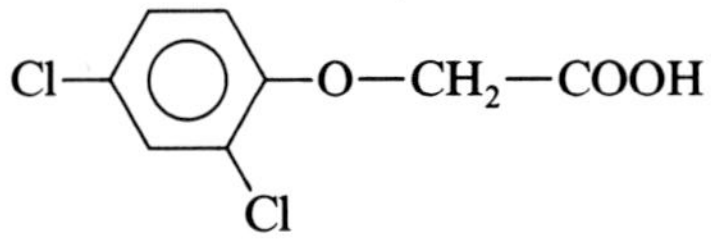	Herbicide
3,4,3′,4′,5′-pentachlorobiphenyl (a typical polychlorinated biphenyl —PCB)	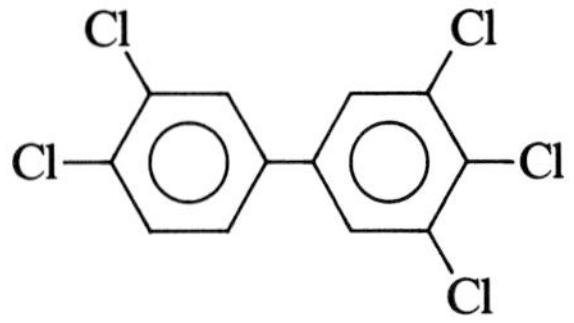	Plasticizer, solvent, coolant
Vinyl chloride	$CH_2{=}CHCl$	Plastics
2,3,7,8-tetrachlorodibenzo-p-dioxin	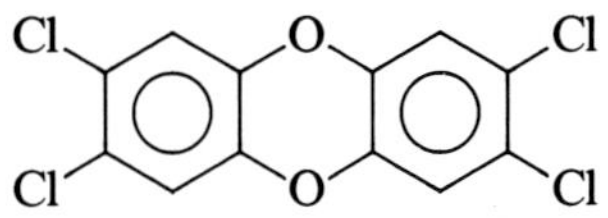	Impurity in herbicides

TABLE 13–7 Some Important Alcohols

Formula	IUPAC Name	Common Name	Typical Use
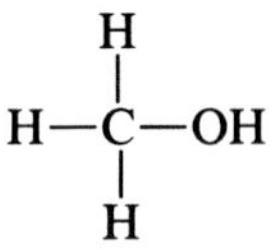 H │ H—C—OH │ H	Methanol	Methyl alcohol (wood alcohol)	Industrial solvent
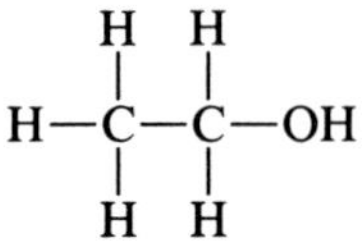 H H │ │ H—C—C—OH │ │ H H	Ethanol	Ethyl alcohol (grain alcohol)	Beverage
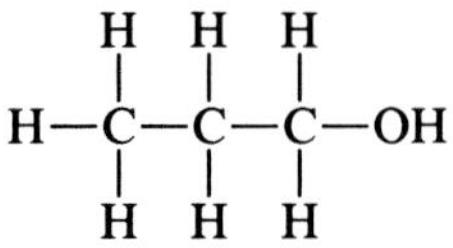 H H H │ │ │ H—C—C—C—OH │ │ │ H H H	1-propanol	*n*-propyl alcohol	Chemical intermediate

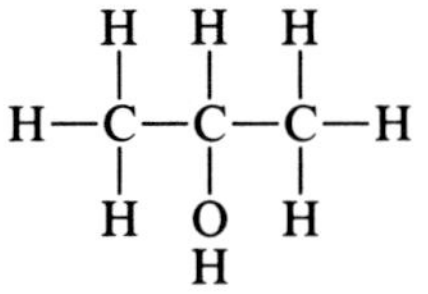 H H H
│   │   │
H—C—C—C—H
│   │   │
H O H
   │
   H
	2-propanol	Isopropyl alcohol	Rubbing alcohol

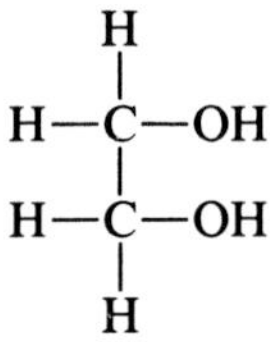 H │ H—C—OH │ H—C—OH │ H	1,2-ethanediol	Ethylene glycol	Permanent antifreeze
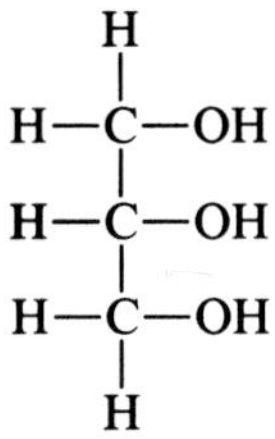 H │ H—C—OH │ H—C—OH │ H—C—OH │ H	1,2,3-propanetriol	Glycerol (glycerin)	Manufacture of drugs and cosmetics

ALCOHOLS

Alcohols among the top 50 chemicals produced in the United States (Table 13–1) include methanol (22), ethylene glycol (29), and isopropyl alcohol (48). These along with some of the other important alcohols are listed in Table 13–7.

Methanol

The production of synthesis gas, a mixture of carbon monoxide and hydrogen, from coal was discussed earlier in this chapter and in Chapter 8.

Methanol, CH_3OH, is the simplest of all alcohols. Over 7 billion pounds of methanol are produced each year from synthesis gas. High pressure, high temperature, and a mixture of catalysts are used to increase the yield.

$$C + H_2O \longrightarrow CO + H_2$$
Coal   Steam   Synthesis Gas

$$CO + 2\,H_2 \xrightarrow[300°C]{ZnO,\,Cr_2O_3} CH_3OH$$

An old method of producing methanol involved heating of a hardwood such as beech, hickory, maple, or birch in a retort in the absence of air. For this reason methanol is sometimes called *wood alcohol.*

About 50% of the methanol produced in United States is used in the production of formaldehyde (used in plastics, embalming fluid, germicides, and fungicides); 30% is used in the production of other chemicals; and the remaining 20% is used for jet fuels, antifreeze mixtures, solvents, as a gasoline additive, and as a denaturant (a poison added to make ethanol unfit for beverages). Methanol is a deadly poison that causes blindness in less than lethal doses. Many deaths and injuries have resulted from the accidental substitution of methanol for ethanol in beverages.

Methanol will likely move upward in the ranking of top 50 chemicals when petroleum and natural gas become too expensive as sources of both energy and chemicals. Although most of the world's methanol currently comes from synthesis gas made from natural gas, coal gasification will become a more important source of methanol as the natural gas reserves are used up. Since methanol is relatively cheap, its potential as a fuel and as a starting material for the synthesis of other chemicals is receiving more attention.

Methanol as a Fuel

Methanol is being considered as a replacement for gasoline, especially in urban areas with extremely high levels of air pollution caused by motor vehicles. For example, methanol-powered cars and buses have been tested in southern California since 1980. The positive results with these vehicles and the high levels of air pollutants in southern California have led to a proposal for the gradual elimination of gasoline-powered vehicles in southern California over the next 20 years. The first five-year step of this proposal is to require owners of fleet vehicles to replace existing vehicles with electric- or methanol-powered vehicles. This would involve an estimated 1 million vehicles.

What are the advantages and disadvantages of switching to methanol-powered vehicles? Since methanol burns more completely, levels of troublesome pollutants such as carbon monoxide, unreacted hydrocarbons, nitrogen oxides, and ozone would be reduced. Estimates are that these reductions would be to levels below those recommended by the EPA (see Chapter 18). Levels of methanol and formaldehyde would be higher. Present projections indicate acceptable levels of methanol and formaldehyde, but careful monitoring of these would be necessary.

The technology for methanol-powered vehicles has existed for many years, particularly for racing cars, which burn methanol because of its high-octane rating. Since methanol has about one-half the energy content of gasoline, it takes almost 2 gal of methanol to go as far as 1 gal of gasoline. Methanol costs about half as much as gasoline so the price per mile would be competitive. However, the size of fuel tanks will need to be doubled. In addition, methanol corrodes regular steel so the fuel system will need to be

Methanol is the main ingredient in many windshield washer fluids.

Methanol model. (Charles D. Winters.)

Cars at the Indianapolis 500 are powered by methanol.

made out of stainless steel or have a methanol-resistant coating. Until sufficient numbers of methanol-powered vehicles are on the road, cars equipped to run on either methanol or gasoline will be necessary because of the lack of service stations selling methanol. Ford Motor Company is producing such a vehicle, and California plans to purchase 5000 of them. ARCO has also announced plans to build 20 service stations in southern California for marketing methanol. As the problems of distribution and storage are solved, better engineered methanol-fueled engines will be designed and produced, which will lead to higher efficiency utilization of methanol as a fuel.

Methanol to Gasoline

Another option is to use methanol to make gasoline. Mobil Oil Company has developed a methanol-to-gasoline process, which is currently not competitive with refined gasoline prices in the United States, but is competitive in those regions of the world, such as New Zealand, where the price of gasoline is much higher. In fact, the production of 92-octane gasoline from methanol is now taking place in New Zealand.

New Zealand Synthetic Fuels Company is operating a plant based on a methanol-to-gasoline process developed by Mobil Oil Company. The process starts with the production of synthesis gas from natural gas and then uses the synthesis gas to make methanol. The key reaction for the production of gasoline from methanol is the dehydration of methanol with a clay catalyst developed by Mobil; this catalyst is known as the ZSM-5 zeolite catalyst. The catalyst aids the dehydration to yield short-chain alkenes, which then cyclize and polymerize to give a mixture of C_5 to C_{12} hydrocarbons made up of branched chains, straight chains, and aromatics, similar to the 92-octane gasoline currently obtained by the refinement of straight-run gasoline from petroleum refining. The dehydration and subsequent polymerization can be represented by the following reactions:

Polymerization is discussed in Chapter 14.

$$2\ CH_3OH \xrightarrow[\text{Catalyst}]{\text{ZSM-5}} (CH_3)_2O + H_2O$$
$$\text{Dimethyl Ether}$$

$$2\ (CH_3)_2O \xrightarrow[\text{Catalyst}]{\text{ZSM-5}} 2\ C_2H_4 + 2\ H_2O$$
$$\text{Ethylene}$$

$$C_2H_4 \xrightarrow[\text{Catalyst}]{\text{ZSM-5}} \text{Hydrocarbon mixture in the } C_5\text{–}C_{12} \text{ range}$$
$$\text{Gasoline}$$

The New Zealand plant is currently producing 14,000 barrels per day of gasoline with an octane rating of 92 to 94. This is about one-third the amount of gasoline used in New Zealand.

Ethanol

Ethanol can be prepared by the fermentation of grains.

Ethanol, also called ethyl alcohol or grain alcohol, can be obtained by the fermentation of carbohydrates (starch, sugars). For example, glucose is converted into ethanol and carbon dioxide by the action of yeast in the absence of oxygen.

$$C_6H_{12}O_6 \xrightarrow{\text{Yeast}} 2\ C_2H_5OH + 2\ CO_2$$
$$\text{Glucose} \qquad\qquad \text{Ethanol}$$

Wine (far right) is produced from the glucose in grape juice (left) by fermentation. The fermentation jug (center) has a bubble chamber to allow CO_2 to escape but to prevent oxygen from entering and oxidizing ethanol to acetic acid (vinegar). (Charles D. Winters.)

A mixture of 95% ethanol and 5% water can be recovered from the fermentation products by distillation. Ethanol is the active ingredient of alcoholic beverages. Some of the most commonly encountered alcoholic beverages and their characteristics are presented in Table 13–8. The "proof" of an alcoholic beverage is twice the volume percent of ethanol; 80 proof vodka, for example, contains 40% ethanol.

95% ethanol (190 proof) is a strong dehydrating agent. Never drink it straight.

Although ethanol is not as toxic as methanol, 1 pint of pure ethanol, rapidly ingested, would kill most people. Ethanol is a depressant for nonalcoholics. The effects of different blood levels of alcohol are shown in Table 13–9. Rapid consumption of two 1-oz "shots" of 90-proof whiskey or of two 12-oz beers can cause one's blood alcohol level to reach 0.05%.

The breathalyzer test used to detect drunken drivers is based on the color change that occurs when ethanol is oxidized to acetic acid by dichromate anion ($Cr_2O_7^{2-}$) in acidic solution.

$$16 \text{ H}^+ + 2 \text{ Cr}_2\text{O}_7^{2-} + 3 \text{ CH}_3\text{CH}_2\text{OH} \longrightarrow 3 \text{ CH}_3\text{COOH} + 4 \text{ Cr}^{3+} + 11 \text{ H}_2\text{O}$$

Yellow-orange   Green

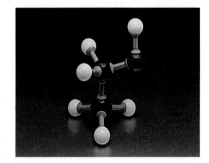

Ethanol model. (Charles D. Winters.)

Ethanol is quickly absorbed by the blood and metabolized by enzymes produced in the liver. The rate of detoxification is about 1 oz of pure alcohol per hour. The ethanol is oxidized to acetaldehyde, which is further oxidized

TABLE 13–8 Common Alcoholic Beverages

Name	Source of Fermented Carbohydrate	Amount of Ethyl Alcohol	Proof
Beer	Barley, wheat	5%	10
Wine	Grapes or other fruit	12% maximum, unless fortified*	20–24
Brandy	Distilled wine	40–45%	80–90
Whiskey	Barley, rye, corn, etc.	45–55%	90–110
Rum	Molasses	~45%	90
Vodka	Potatoes	40–50%	80–100

* The growth of yeast is inhibited at alcohol concentrations over 12%, and fermentation comes to a stop. Beverages with a higher concentration are prepared either by distillation or by fortification with alcohol that has been obtained by the distillation of another fermentation product.

TABLE 13–9 Alcohol Blood Level and Effect

Blood Alcohol Level (Percentage by Volume)	Effect
0.05–0.15	Lack of coordination
0.15–0.20	Intoxication
0.30–0.40	Unconsciousness
0.50	Possible death

to acetic acid; eventually CO_2 and H_2O are produced and eliminated through the lungs and kidneys.

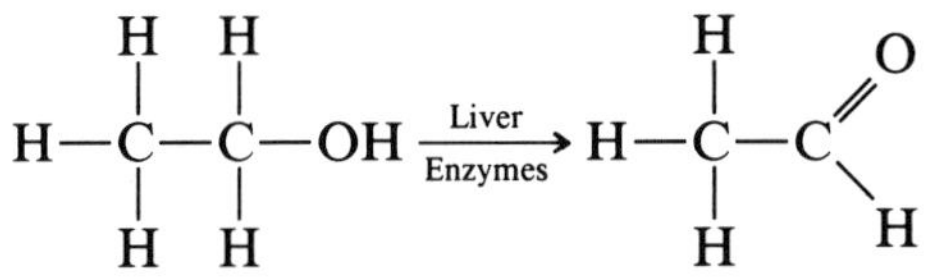

Ethanol   Acetaldehyde

Industrial Use of Ethanol

The federal tax on alcoholic beverages is about $20 per gallon. Since the cost of producing ethanol is only about $1 per gallon, ethanol intended for industrial use must be *denatured* to avoid the beverage tax. **Denatured alcohol** contains small amounts of a toxic substance, such as methanol or gasoline, that cannot be removed easily by chemical or physical means.

Apart from being used in the alcoholic beverage industry, ethanol is used widely in solvents and in the preparation of many other organic compounds. Over 1 billion lb of ethanol are used for these purposes each year, and this is produced synthetically rather than by a fermentation process. The reaction involves the addition of water vapor to ethylene under high pressure in the presence of a catalyst.

Where does the ethylene come from?

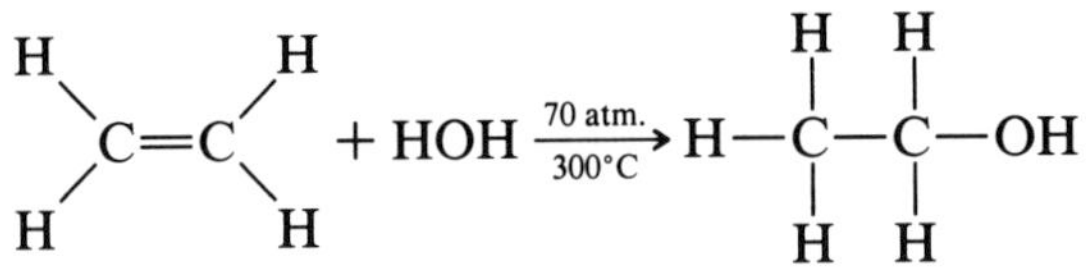

Ethylene   Ethanol

Propanols

When one considers the possible structures for propyl alcohol (propanol), it is apparent that two isomers are possible. The most common one is 2-propanol (rubbing alcohol).

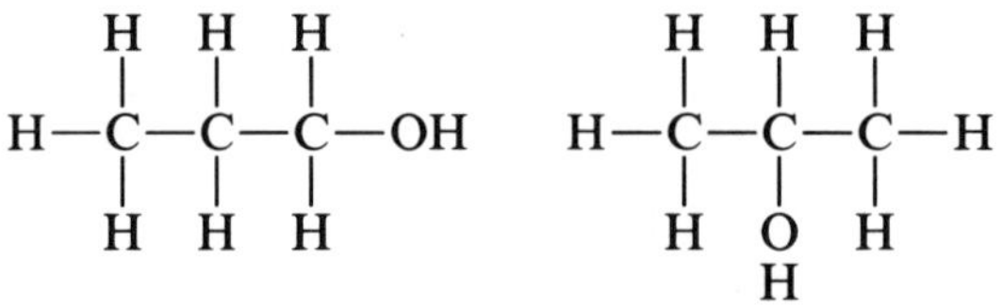

1-propanol   2-propanol
n-propyl Alcohol   Isopropyl Alcohol
                                (Rubbing Alcohol)

Ethylene Glycol and Glycerol

More than one alcohol group (—OH) can be present in a single molecule. Glycerol and ethylene glycol, the base of permanent antifreeze, are examples of such compounds. Ethylene glycol is made in a two-step synthesis, starting with ethene.

Permanent antifreeze is ethylene glycol.

$$CH_2{=}CH_2 + \tfrac{1}{2}O_2 \xrightarrow[300°C]{Ag} H_2C\underset{O}{\diagdown\underset{}{\diagup}}CH_2 + H_2O \xrightarrow{Acid} \begin{array}{c} H \\ | \\ H{-}C{-}OH \\ | \\ H{-}C{-}OH \\ | \\ H \end{array}$$

| Ethylene | Ethylene Oxide | 1,2-ethanediol Ethylene Glycol |

In organic reactions, catalysts, reagents, and reaction conditions are often listed above and below the arrow.

Glycerol is a byproduct in the manufacture of soaps. Because of its moisture-holding properties, glycerol has many uses in foods and tobacco as a digestible and nontoxic humectant (gathers and holds moisture), and in the manufacture of drugs and cosmetics. It is also used in the production of nitroglycerin and numerous other chemicals. Perhaps the most important compounds of glycerol are its natural esters (fats and oils), which we shall discuss later in this chapter.

$$\begin{array}{c} H \\ | \\ H{-}C{-}OH \\ | \\ H{-}C{-}OH \\ | \\ H{-}C{-}OH \\ | \\ H \end{array}$$

1,2,3-propanetriol
Glycerol
(Glycerin)

Hydrogen Bonding in Alcohols

The physical properties of water, methanol, ethanol, the propanols, ethylene glycol, and glycerol offer another interesting example of the effects of **hydrogen bonding** between molecules in liquids. In Table 13–10 the boiling points for these compounds are listed.

See Chapter 5 for a discussion of hydrogen bonding in water.

Since boiling involves overcoming the attractions between liquid molecules as they pass into the gas phase, a higher boiling point indicates stronger intermolecular forces holding the molecules together. Another factor is also present: as the molecules become larger, higher boiling points result because more energy is required to change the longer chain molecules from the liquid to the gaseous phase, owing in part to the larger London forces. A graph showing the boiling points of the normal alcohols (straight carbon chains

TABLE 13–10 Boiling Points for Some —OH Compounds

Name	Formula	Boiling Point (°C)
Water	HOH	100
Methanol	CH_3OH	65.0
Ethanol	CH_3CH_2OH	78.5
1-propanol	$CH_3CH_2CH_2OH$	97.4
2-propanol	$CH_3CHOHCH_3$	82.4
Ethylene glycol	CH_2OHCH_2OH	198
Glycerol	$CH_2OHCHOHCH_2OH$	290

Note: The parent hydrocarbon of methanol is methane (bp −164°C); of ethanol and ethylene glycol, ethane (bp −88.6°C); and of the propanols and glycerol, propane (bp −42.1°C).

Products which contain ethylene glycol or glycerin. (Charles Steele.)

Figure 13–6   Boiling points of straight-chain alcohols.

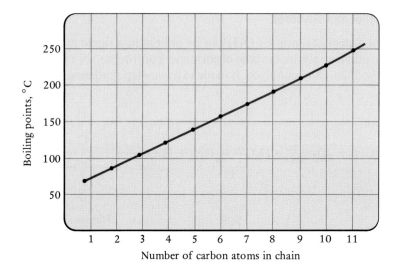

Number of carbon atoms in chain

with the —OH group on an end carbon) as a function of chain length is given in Figure 13–6.

Methanol, like water, has an —OH group, and some hydrogen bonding is to be expected, as shown in Figure 13–7. Hydrogen bonding explains why methanol (molecular weight 32) is a liquid, whereas propane (C_3H_8, molecular weight 44), an even heavier molecule, is a gas at room temperature. Methanol has only one hydrogen through which it can hydrogen bond, but water can hydrogen bond from either of its two hydrogen atoms. Thus, water, with more extensive intermolecular bonding, has the higher boiling point even though water has lighter molecules.

The higher boiling point of ethylene glycol is readily explained in terms of the two —OH groups per molecule and the enhanced possibility for hydrogen bonding. Glycerol, with three —OH groups per molecule, has an even higher boiling point, as well as a very high viscosity (resistance to flow).

Since the —OH group in an organic molecule causes at least that area of the molecule to be polar, such molecules will be attracted to other polar molecules such as water molecules. As a result of these attractions, the lower molecular weight alcohols are quite soluble in water. In the higher molecular

Figure 13–7   Hydrogen bonding in methanol.

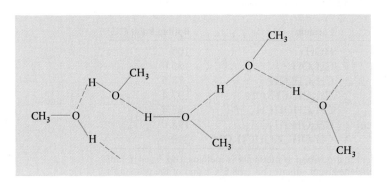

weight alcohols, the nonpolar hydrocarbon chain significantly decreases their solubility in water.

Synthesis with Alcohols

Alcohols can serve as the starting substances for the synthesis of many other types of organic compounds. Oxidation of alcohols may yield aldehydes, ketones, or acids, depending on the starting compound and the amount of oxygen added.

   If the starting compound is a primary alcohol, the first oxidation product will be an aldehyde. Secondary alcohols are oxidized to give ketones. For example, the oxidation of ethanol, a primary alcohol, can be used to make acetaldehyde and acetic acid:

$$\underset{\text{Ethanol}}{\overset{\displaystyle H}{\underset{\displaystyle H}{CH_3\overset{|}{\underset{|}{C}}-OH}}} \xrightarrow{\text{Oxidation}} \underset{\text{Acetaldehyde}}{\overset{\displaystyle H}{CH_3\overset{|}{C}=O}} \xrightarrow{\text{Oxidation}} \underset{\text{Acetic Acid}}{\overset{\displaystyle O}{CH_3\overset{\|}{C}-OH}}$$

In organic compounds, oxidation is often accompanied by the loss of hydrogen or the addition of oxygen to a molecule.

The oxidation of 2-propanol, a secondary alcohol, provides the ketone, acetone:

$$\underset{\text{2-propanol}}{\overset{\displaystyle H}{\underset{\displaystyle OH}{CH_3\overset{|}{\underset{|}{C}}CH_3}}} \xrightarrow{\text{Oxidation}} \underset{\text{Acetone}}{\overset{\displaystyle }{\underset{\displaystyle O}{CH_3\overset{}{\underset{\|}{C}}CH_3}}}$$

Workable oxidants are hot copper oxide (CuO), potassium dichromate ($K_2Cr_2O_7$) with sulfuric acid, or potassium permanganate ($KMnO_4$).

   Alcohols dehydrate to form alkenes or ethers. Two important dehydration reactions of ethanol illustrate how temperature can be used to determine the product. Sulfuric acid is the dehydrating agent.

Dehydration reactions involve the formation and removal of water molecules.

$$2\ CH_3CH_2OH \xrightarrow[H_2SO_4]{140°C} CH_3CH_2OCH_2CH_3 + H_2O$$
$$\underset{\text{Ethanol}}{} \qquad\qquad \underset{\text{Diethyl Ether}}{}$$

$$\underset{\text{Ethanol}}{CH_3CH_2OH} \xrightarrow[H_2SO_4]{180°C} \underset{\text{Ethene}}{\overset{\displaystyle H\ \ H}{H-\overset{|}{C}=\overset{|}{C}-H}} + H_2O$$

ETHERS

Ethers, which contain the $R—O—R'$ linkage, are formed by the dehydration of alcohols. The most common ether is diethyl ether, which for many years was used as an anesthetic. It produces unconsciousness by depressing the activity of the central nervous system. It is no longer used because it irritated the respiratory system and caused postanesthetic nausea and vomiting. Methyl propyl ether, $CH_3OCH_2CH_2CH_3$, known as neothyl, is currently used as an anesthetic because it is relatively free of side effects.

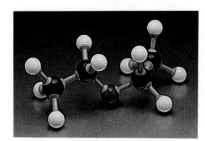

Diethyl ether model. (Charles D. Winters.)

$$\underset{\text{Diethyl Ether}}{CH_3CH_2—O—CH_2CH_3}$$

Formaldehyde model.

O
‖
H—C—H
Formaldehyde

Formaldehyde is a suspected carcinogen.

Cinnamaldehyde is a *trans* isomer.

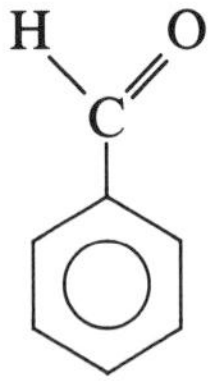

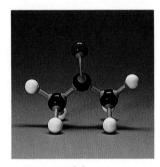

Acetone model.

Oxidation of ethanol to acetic acid by oxygen in air is responsible for the souring of wine.

Organic acids are compounds of the type R—C
O
OH
. They are generally weak acids.

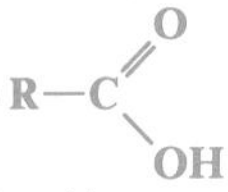

ALDEHYDES AND KETONES

Aldehydes and ketones contain a **carbonyl group,** $>$C=O, and are generally obtained by the oxidation of alcohols. In aldehydes the carbonyl group is on an end carbon, whereas in ketones the carbonyl group is bonded to two carbon atoms.

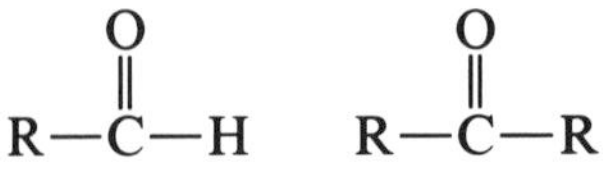

O   O
‖   ‖
R—C—H   R—C—R
Aldehyde   Ketone

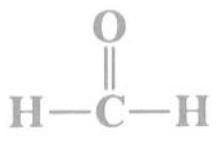

Formaldehyde, the simplest aldehyde, has a foul odor. It is the starting material in the production of several plastics and is used in the laboratory as a preservative for dead animals. Aldehydes with an aromatic ring have pleasant odors, and some are used in food flavors and perfumes.

Benzaldehyde
(Bitter Almonds)

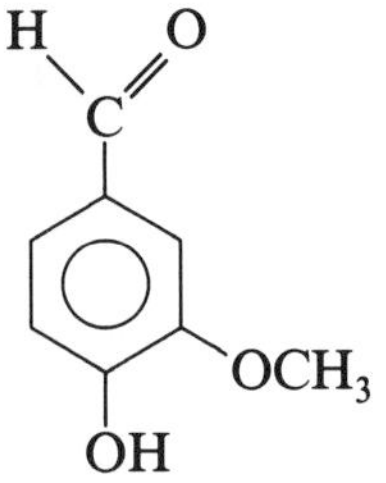

Vanillin
(Vanilla Bean)

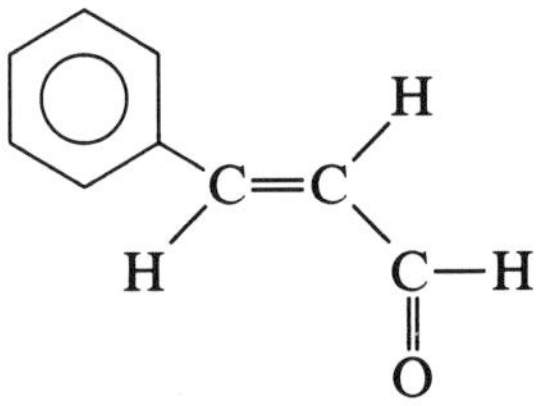

Cinnamaldehyde
(Cinnamon)

The simplest ketone is acetone, an important commercial solvent. Methyl ethyl ketone is a solvent in model-airplane glue.

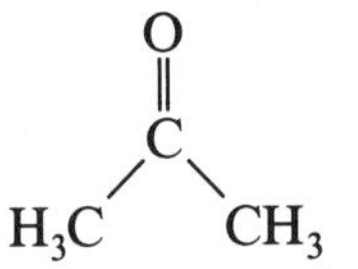

O   O
‖   ‖
C   C
H₃C   CH₃   H₃C   CH₂CH₃
Acetone 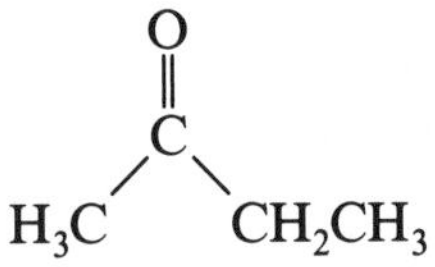   Methyl Ethyl Ketone

CARBOXYLIC ACIDS

Organic, or carboxylic, acids contain the **carboxyl group,** —COOH, and can be prepared by the oxidation of alcohols or aldehydes. These reactions occur quite readily, as evidenced by the souring of wine, which is the oxidation of ethanol to acetic acid in the presence of oxygen from the air.

Carboxylic acids are found in both the plant and animal kingdoms. The first six carboxylic acids, with their sources, common names, and odors, are given in Table 13–11. Longer chain carboxylic acids do not smell as bad, in part because they are less volatile. Some of the other common carboxylic acids found in nature are given in Table 13–12. As can be seen in the table, some organic acids have more than one carboxyl group as well as other groups, usually hydroxyl groups.

TABLE 13–11 First Six Carboxylic Acids

Formula	Source	Common Name	IUPAC Name	Odor
HCOOH	Ants (Latin, *formica*)	Formic acid	Methanoic acid	Sharp
CH_3COOH	Vinegar (Latin, *acetum*)	Acetic acid	Ethanoic acid	Sharp
CH_3CH_2COOH	Milk (Greek, *protos pion,* "first fat")	Propionic acid	Propanoic acid	Swiss cheese
$CH_3(CH_2)_2COOH$	Butter (Latin, *butyrum*)	Butyric acid	Butanoic acid	Rancid butter
$CH_3(CH_2)_3COOH$	Valerian root (Latin, *valere,* "to be strong")	Valeric acid	Pentanoic acid	Manure
$CH_3(CH_2)_4COOH$	Goats (Latin, *caper*)	Caproic acid	Hexanoic acid	Goat

In Chapter 9 an acid was defined as a species that has a tendency to donate hydrogen ions. The electronegative character of the $C=O$ group in

$$-C\overset{\displaystyle O}{\underset{\displaystyle OH}{\big\|}}$$

tends to drain electron density away from the region between the oxygen and hydrogen atoms. The partial positive charge assumed by the hydrogen then makes it possible for polar water molecules to remove hydrogen ions from some of the carboxyl groups. The strength of an organic acid depends on the group that is attached to the carboxyl group. If the attached

Condensed formulas are often used to save space. For example, $CH_3(CH_2)_3COOH$ is the same as $CH_3CH_2CH_2CH_2COOH$.

TABLE 13–12 Some Other Naturally Occurring Carboxylic Acids

Name	Structure	Natural Source
Citric acid	$HOOC-CH_2-\overset{\displaystyle OH}{\underset{\displaystyle COOH}{C}}-CH_2-COOH$	Citrus fruits
Lactic acid	$CH_3-\underset{\displaystyle OH}{CH}-COOH$	Sour milk
Malic acid	$HOOC-CH_2-\underset{\displaystyle OH}{CH}-COOH$	Apples
Oleic acid	$CH_3(CH_2)_7-CH=CH-(CH_2)_7-COOH$	Vegetable oils
Oxalic acid	$HOOC-COOH$	Rhubarb, spinach, cabbage, tomatoes
Stearic acid	$CH_3(CH_2)_{16}-COOH$	Animal fats
Tartaric acid	$HOOC-\underset{\displaystyle OH}{CH}-\underset{\displaystyle OH}{CH}-COOH$	Grape juice, wine

Sources of some naturally occurring carboxylic acids. (Charles D. Winters.)

group has a tendency to pull electrons away from the carboxyl group, the acid is a stronger acid. For example, trichloroacetic acid is much stronger than acetic acid:

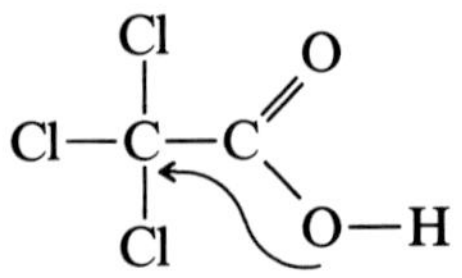
Trichloroacetic Acid

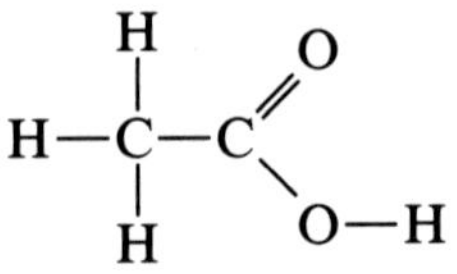
Acetic Acid

As shown by the arrow, the highly electronegative chlorine atoms draw away electron density from the region of the carboxyl group and make the loss of the hydrogen ion even easier than in acetic acid. However, trichloroacetic acid is still weaker than a strong mineral acid such as sulfuric acid.

The ionization of carboxylic acids in water is simply:

The unequal double arrows ($\leftharpoondown\!\!\rightleftharpoons$) indicate that the equilibrium favors the un-ionized acid molecule; that is, the acid is weak.

$$R{-}\overset{\overset{\displaystyle O}{\|}}{C}{-}OH + H_2O \rightleftharpoons R{-}\overset{\overset{\displaystyle O}{\|}}{C}{-}O^- + H_3O^+$$

Carboxylic acids are *neutralized* by bases to form salts.

$$R{-}\overset{\overset{\displaystyle O}{\|}}{C}{-}OH + Na^+ + OH^- \longrightarrow \underset{\text{A salt}}{R{-}\overset{\overset{\displaystyle O}{\|}}{C}{-}O^-Na^+} + H_2O$$

An example of an organic salt is sodium benzoate, a food preservative:

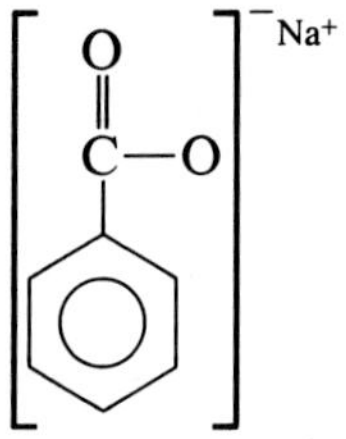
Sodium Benzoate

Methanoic Acid

The simplest organic acid is methanoic acid, also called formic acid, in which the carboxyl group is attached directly to a hydrogen atom.

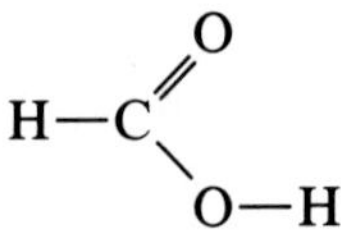
Methanoic Acid
(Formic Acid)

Weak bases such as ammonia (NH_3), neutralize formic acid and are used in the treatment of insect bites.

This acid is found in ants and other insects and is part of the irritant that produces itching and swelling after a bite.

Formic acid may be prepared from its sodium salt, which is readily prepared by heating carbon monoxide (CO) with sodium hydroxide (NaOH):

$$CO + NaOH \xrightarrow[6-10 \text{ min}]{200°C} HCOO^-Na^+$$
Sodium Formate

If the resulting salt is mixed with a mineral acid, formic acid can be distilled from the mixture:

$$HCOO^-Na^+ + H_3O^+ + Cl^- \longrightarrow HCOOH + Na^+Cl^- + H_2O$$

| Sodium Formate | Hydrochloric Acid | Formic Acid | Sodium Chloride |

Ethanoic Acid

Ethanoic (acetic) acid is the most widely used of the organic acids. It is found in vinegar, an aqueous solution containing 4% to 5% acetic acid. Flavor and colors are imparted to vinegars by the constituents of the alcoholic solutions from which they are made. Ethanol in the presence of certain bacteria and air is oxidized to acetic acid:

$$CH_3CH_2OH + O_2 \xrightarrow{\text{Bacteria}} CH_3COOH + H_2O$$
Ethanol   Oxygen   Ethanoic Acid   Water
(Acetic Acid)

Acetic acid model.

The bacteria, called mother of vinegar, form a slimy growth in a vinegar solution. The growth of bacteria can sometimes be observed in a bottle of commercially prepared vinegar after it has been opened to the air.

Acetic acid is an important starting substance for making textile fibers, vinyl plastics, and other chemicals and is a convenient choice when a cheap organic acid is needed.

Fatty Acids

A fatty acid is a naturally occurring organic acid with a long hydrocarbon chain. The chain often contains only C—C single bonds, but several fatty acids also contain some C=C double bonds. Stearic acid and palmitic acid are examples of saturated fatty acids, and oleic acid is an example of an unsaturated fatty acid.

Fatty acids may be obtained from animal and vegetable fats or oils. The carbon chain in fatty acids is generally 8 to 18 carbon atoms in length.

Recall from Chapter 12 that *saturated* compounds contain the maximum number of hydrogen atoms per carbon atom.

$$CH_3CH_2CH_2CH_2CH_2CH_2CH_2CH_2CH_2CH_2CH_2CH_2CH_2CH_2CH_2CH_2CH_2C\overset{O}{\underset{OH}{<}}$$
Stearic Acid, $CH_3-(CH_2)_{16}-COOH$

$$CH_3CH_2CH_2CH_2CH_2CH_2CH_2CH_2CH_2CH_2CH_2CH_2CH_2CH_2CH_2C\overset{O}{\underset{OH}{<}}$$
Palmitic Acid, $CH_3-(CH_2)_{14}-COOH$

$$CH_3CH_2CH_2CH_2CH_2CH_2CH_2CH_2CH=CHCH_2CH_2CH_2CH_2CH_2CH_2C\overset{O}{\underset{OH}{<}}$$
Oleic Acid, $CH_3(CH_2)_7CH=CH(CH_2)_7COOH$

Stearic acid is obtained by the hydrolysis of animal fat, palmitic acid results from the hydrolysis of palm oil, and oleic acid is obtained from olive oil.

Stearic and palmitic acids are especially important in the manufacture of soaps.

It has been known for about 60 years that the human body has a small requirement for certain types of fatty acids (called **essential fatty acids**). Until recently, essential fatty acids were thought to be **linoleic, linolenic,** and **arachidonic** acids, but it has been determined that the human body can produce the latter two from linoleic acid.

$$CH_3CH_2CH_2CH_2CH_2CH = CHCH_2CH = CHCH_2CH_2CH_2CH_2CH_2CH_2C \overset{O}{\underset{OH}{\big<}}$$

Linoleic Acid, $(C_{18}\Delta_{9,12})$

$$CH_3CH_2CH = CHCH_2CH = CHCH_2CH = CHCH_2CH_2CH_2CH_2CH_2CH_2C \overset{O}{\underset{OH}{\big<}}$$

Linolenic Acid, $(C_{18}\Delta_{9,12,15})$

$$CH_3CH_2CH_2CH_2CH_2CH = CHCH_2CH = CHCH_2CH = CHCH_2CH = CHCH_2CH_2CH_2C \overset{O}{\underset{OH}{\big<}}$$

Arachidonic Acid, $(C_{20}\Delta_{5,8,11,14})$

Δ indicates the positions of the double bonds.

Medical evidence links the decrease in peptic ulcer disease in the United States and Britain to an increase in the consumption of polyunsaturated fats, which provide linoleic acid for the synthesis of prostaglandins. The prostaglandins protect the stomach and the intestinal tract from ulcers.

The presence of fatty acids in the diet permits the body to synthesize a very important group of compounds, the prostaglandins. The key compound here again is linoleic acid.

SELF-TEST 13–B

1. Gin that is 84 proof contains what percentage of ethanol?
2. Ethanol is quickly absorbed by the blood and oxidized to _____ in the liver.
3. Ethanol intended for industrial use is _____ by the addition of small amounts of a toxic substance.
4. 2-Propanol is commonly known as _____ alcohol.
5. A long-chain organic acid found in nature is called a _____ acid.
6. The organic acid found in vinegar is _____.
7. The simplest aldehyde is _____.
8. The alcohol being considered as a replacement for gasoline in heavily polluted urban areas is _____.
9. One-third of the gasoline used in _____ is produced by a methanol-to-gasoline process developed by Mobil Oil Company.

ESTERS

In the presence of strong mineral acids, organic acids react with alcohols to form compounds called **esters**. For example, when ethyl alcohol is mixed with acetic acid in the presence of sulfuric acid, ethyl acetate is formed. This

TABLE 13–13 Some Alcohols, Acids, and Their Esters

Alcohol	Acid	Ester	Odor of the Ester

$CH_3CHCH_2CH_2OH$
    $\mid$
   CH_3
Isopentyl Alcohol
	CH_3COOH
	Acetic Acid
		$CH_3CHCH_2CH_2-O-\overset{\displaystyle \\ }{\underset{\displaystyle O}{C}}-CH_3$
		    $\mid$
		   CH_3
		Isopentyl Acetate
			Banana

$CH_3CHCH_2CH_2OH$
    $\mid$
   CH_3
Isopentyl Alcohol
	$CH_3CH_2CH_2CH_2COOH$
	Pentanoic Acid
		$CH_3CHCH_2CH_2-O-\overset{\displaystyle \\ }{\underset{\displaystyle O}{C}}-CH_2CH_2CH_3$
		    $\mid$
		   CH_3
		Isopentyl Pentanoate
			Apple

$CH_3CH_2CH_2CH_2OH$ *n*-Butyl Alcohol	$CH_3CH_2CH_2COOH$ Butanoic Acid	$CH_3CH_2CH_2CH_2-O-\overset{\displaystyle \\ }{\underset{\displaystyle O}{C}}-CH_2CH_2CH_3$ Butyl Butanoate	Pineapple
$\langle\bigcirc\rangle-CH_2-OH$ Benzyl Alcohol	$CH_3CH_2CH_2COOH$ Butanoic Acid	$\langle\bigcirc\rangle-CH_2-O-\overset{\displaystyle \\ }{\underset{\displaystyle O}{C}}-CH_2CH_2CH_3$ Benzyl Butanoate	Rose

reaction is a dehydration in which sulfuric acid acts as a catalyst and dehydrator.

$$CH_3CH_2O\underbrace{+H + HO+}CCH_3 \underset{H_2SO_4}{\rightleftarrows} CH_3CH_2O\underset{O}{C}CH_3 + H_2O$$
                                 Ethyl Acetate

Organic esters are compounds of the type

$$R-O-\underset{O}{C}-R'$$

formed by the reaction of organic acids and alcohols.

Ethyl acetate is a common solvent for lacquers and plastics and is often used as fingernail polish remover.

Some odors of common fruits are due to the presence of mixtures of volatile esters (Table 13–13). In contrast, esters of higher molecular weight often have a distinctly unpleasant odor.

Fats and Oils

Fats and oils are esters of glycerol (glycerin) and a fatty acid. R, R′, and R″ stand for the hydrocarbon chains of the acids in the following equation:

Fats and oils are esters of fatty acids and glycerol. Fats are solids, and oils are liquids.

$$\begin{array}{c}CH_2-OH \\ \mid \\ CH-OH \\ \mid \\ CH_2-OH\end{array} \quad + \quad \begin{array}{c}HO-\overset{O}{C}-R \\ \\ HO-\overset{O}{C}-R' \\ \\ HO-\overset{O}{C}-R''\end{array} \quad \rightleftharpoons \quad \begin{array}{c}CH_2-O-\overset{O}{C}-R \\ \mid \\ CH-O-\overset{O}{C}-R' \\ \mid \\ CH_2-O-\overset{O}{C}-R''\end{array} \quad + \quad 3\,H_2O$$

Glycerol (One Molecule)   Fatty Acid (Three Molecules That May or May Not Be the Same)   Fat or Oil (One Molecule)   Water (Three Molecules)

Lipids are soluble in fats and oils.

Chapter 20 describes the importance of reducing saturated fats in your diet.

The term *fat* is usually reserved for solid glycerol esters (butter, lard, tallow) and *oil* for liquid esters (castor, olive, linseed, tung, and so forth). The term *lipid* includes fats, oils, and fat-soluble compounds.

Saturated fatty acids (which contain all single bonds with maximum hydrogen content) are usually found in solid or semisolid fats, whereas *unsaturated* fatty acids (containing one or more double bonds) are usually found in oils. Hydrogen can be catalytically added to the double bonds of an oil to convert it into a semisolid fat. For example, liquid soybean and other vegetable oils are **hydrogenated** to produce cooking fats and margarine.

Consumers in Europe and North America have historically valued butter as a source of fat. As the population of these parts of the world increased, the advantages of a substitute for butter became apparent, and efforts to prepare such a product began about 100 years ago. One initial problem was that common fats are almost all *animal* products with very pronounced tastes of their own. Analogous compounds from vegetable oils, which are bland or have mixed flavors, were generally *unsaturated* and consequently *oils*. A solid fat could be made from the much cheaper vegetable oils if an inexpensive way could be discovered to add hydrogen across the double bonds. After extensive experiments, many catalysts were found, of which finely divided nickel is among the most effective. The nature of the process can be illustrated by the following reaction:

Catalytic hydrogenation can convert a liquid oil into a solid fat.

$$
\begin{array}{l}
H_2C\!-\!O\!-\!\underset{\underset{O}{\|}}{C}\!-\!(CH_2)_7CH\!=\!CH(CH_2)_7CH_3 \\[6pt]
HC\!-\!O\!-\!\underset{\underset{O}{\|}}{C}\!-\!(CH_2)_7CH\!=\!CH(CH_2)_7CH_3 \xrightarrow[200^\circ C]{H_2,\ Ni} \\[6pt]
H_2C\!-\!O\!-\!\underset{\underset{O}{\|}}{C}\!-\!(CH_2)_7CH\!=\!CH(CH_2)_7CH_3
\end{array}
\qquad
\begin{array}{l}
H_2C\!-\!O\!-\!\underset{\underset{O}{\|}}{C}\!-\!(CH_2)_7CH_2CH_2(CH_2)_7CH_3 \\[6pt]
HC\!-\!O\!-\!\underset{\underset{O}{\|}}{C}\!-\!(CH_2)_7CH_2CH_2(CH_2)_7CH_3 \\[6pt]
H_2C\!-\!O\!-\!\underset{\underset{O}{\|}}{C}\!-\!(CH_2)_7CH_2CH_2(CH_2)_7CH_3
\end{array}
$$

Triolein (A Liquid Oil)   Tristearin (A Solid Fat)

Oils commonly subjected to this process include those from cottonseed, peanuts, corn germ, soybeans, coconuts, and safflower seeds. In recent years, as it has become apparent that saturated fats may encourage diseases of the heart and arteries, soft margarines and cooking oils (which still contain some of the unhydrogenated fatty acid) have been placed on the market.

Hydrogenation of the double bonds in vegetable oils converts the liquid oil into a solid fat. (J. Morgenthaler.)

AMINES AND AMIDES

Organic amines can be considered derivatives of ammonia (NH_3). The nitrogen in the amine may be attached to R groups or may be part of a ring.

$$CH_3\!-\!NH_2$$

Methyl Amine
(Fish Odor)

Pyridine
(Foul Odor)

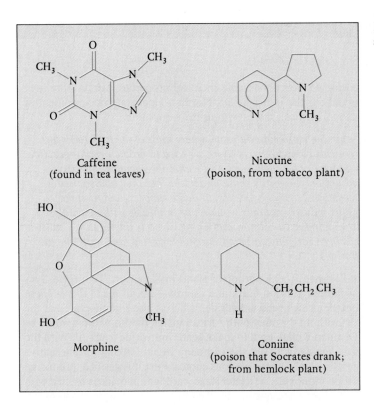

Figure 13–8   Some common alkaloids.

Most amines have unpleasant odors. The stench of decaying protein, for example, is due to some of the compounds listed below.

$$H_2N—CH_2—CH_2—CH_2—CH_2—NH_2$$ PUTRESCINE

$$H_2N—CH_2—CH_2—CH_2—CH_2—CH_2—NH_2$$ CADAVERINE

SKATOLE (IN FECES)

Alkaloids (alkalilike) are amines derived from plants. The amine nitrogen is usually part of a ring. Caffeine, nicotine, morphine, and coniine are examples of alkaloids (Fig. 13–8). ·

The functional group in amides is $—\overset{\displaystyle O}{\overset{\|}{C}}—NH_2$. The functional groups of amines and amides are found in many important biological compounds (discussed in Chapter 15).

Urea

Nicotinamide

A CATALYST IN ACTION

Because of their remarkable chemical properties, catalysts are used throughout the industry. Dr. Norman Hochgraf, vice-president of Exxon Chemical, shares with us his viewpoint:

Within the petroleum industry, where most of the feedstocks for chemicals come from, catalysts are used to produce motor gasoline, heating oil, and feedstocks for chemicals. And within the chemical industry, catalysts are used not only to purify those feedstocks, but they're also used to polymerize the feedstocks, to make plastics, to make rubbers, and to make synthetic fibers. This industry not only wouldn't be profitable without catalysts, but in a sense, it wouldn't exist. Almost everything we do is the result of catalytic activity, which has been carefully designed, carefully selected, and built into our commercial operations.

Let's look at Eastman Kodak's group of plants in Kingsport, Tennessee. Synthetic products—fabrics, plastics, films, and aspirin—are all made there. These products are all made using acetic anhydride, which is also manufactured at the same location. Yet without the rare South African metal, rhodium, there would be no plant at all. Eastman Kodak used to make acetic anhydride from oil. With the rise in oil prices in the early 1970s, the company began looking for alternative inexpensive materials. Coal was the obvious choice. First it is gasified, producing hydrogen and carbon monoxide. At a later stage, carbon monoxide is reacted with methyl acetate to produce acetic anhydride. This crucial step requires rhodium as a catalyst.

If rhodium is so expensive, how can it be profitable to use the catalyst in such a large-scale reaction? The answer lies in the fact that catalysts are not used up in reactions. Each catalyst molecule may react with thousands and thousands of molecules of reactant. So this catalyst need only be present in tiny amounts. In addition, as the mixture is drawn off, the catalyst is carefully separated from the acetic anhydride and recycled back into the reactor. Very little rhodium is lost.

Without the development of a viable catalyst system for producing acetic anhydride from methyl acetate and carbon monoxide, the chemicals-from-coal complex at Kingsport, Tennessee would not have been built.

The World of Chemistry (Program 14) "Molecules in Action."

Solution of rhodium chloride catalyst. (*The World of Chemistry,* Program 14, "Molecules in Action.")

WILL COAL BECOME THE MAJOR SOURCE OF CHEMICALS?

Although petroleum is now the source of over 90% of the organic chemicals used to synthesize consumer products, the projected depletion of petroleum reserves mentioned in Chapter 8 is focusing more attention on coal.

Synthesis gas from coal is already receiving increased attention as a starting material for the production of organic chemicals that are among the top 50 produced in the United States (Table 13–1). For example, the first complete chemicals-from-coal plant, built by Eastman Kodak in Kingsport, Tennessee, started production in 1983. Figure 13–9a is a schematic

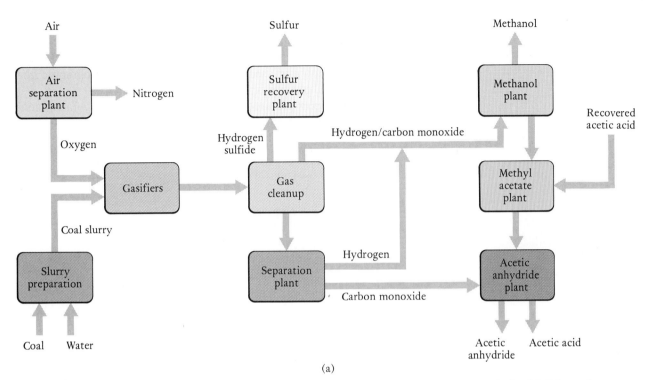

(a)

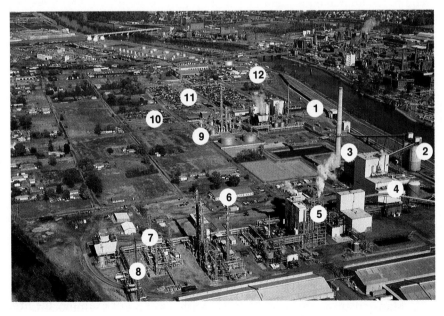

Figure 13-9 **(a) Schematic drawing showing the production of acetic anhydride from coal. (b) Eastman Kodak's chemicals-from-coal facility in Kingsport, Tennessee. Numbers on the photograph represent different parts of the plant: 1, coal unloading; 2, coal silos; 3, steam plant; 4, slurry preparation; 5, coal gasification plant; 6, gas cleanup and separation; 7, sulfur recovery plant; 8, gas flare stack; 9, chemical storage; 10, methanol plant; 11, methyl acetate plant; 12, acetic anhydride plant. (Courtesy of Tennessee Eastman.)**

drawing of the various components of the plant, which is pictured in Figure 13–9b. The basic reactions are to produce synthesis gas from coal, to use the synthesis gas to make methanol, and to use the methanol in the synthesis of acetic anhydride. Acetic anhydride is used by Eastman Kodak to make cellulose acetate, a polymer used in the manufacture of photographic film base, synthetic fibers, plastics, and other products.

Within the complex pictured in Figure 13–9b are nine separate plants, four related to the gasification of coal, two for synthesis-gas preparation, and three for the synthesis of methanol, methyl acetate, and acetic anhydride. The main chemical reactions used in the process are shown below:

1. $\underset{\text{Coal}}{C} + \underset{\text{Steam}}{H_2O} \longrightarrow \underset{\text{Synthesis Gas}}{CO + H_2}$

2. $CO + 2\,H_2 \longrightarrow \underset{\text{Methanol}}{CH_3OH}$

3. $CH_3OH + \underset{\text{Acetic Acid}}{CH_3\overset{\displaystyle O}{\overset{\|}{C}}OH} \longrightarrow \underset{\text{Methyl Acetate}}{CH_3\overset{\displaystyle O}{\overset{\|}{C}}OCH_3} + H_2O$

4. $CH_3\overset{\displaystyle O}{\overset{\|}{C}}OCH_3 + CO \longrightarrow \underset{\text{Acetic Anhydride}}{CH_3-\overset{\displaystyle O}{\overset{\|}{C}}-O-\overset{\displaystyle O}{\overset{\|}{C}}-CH_3}$

About 900 tons per day of high-sulfur coal from nearby Appalachian coal mines are ground in water to form a slurry of 55% to 65% by weight of coal in water. The slurry is fed into two gasifiers to make synthesis gas. To produce the same amounts of these chemicals by conventional means would require the annual equivalent of 1 million barrels of oil.

Plant design uses the latest environmental control technologies to protect the environment. For example, the sulfur recovery unit converts the hydrogen sulfide gas that was removed during the gasification of coal into free sulfur. This process removes over 99% of the sulfur from the coal, and this sulfur is sold to chemical companies.

Synthesis gas will become increasingly important as a raw material, since it can be used to make a variety of hydrocarbons that can then be converted into other commercially important organic chemicals.

DESIGN OF ORGANIC SYNTHESIS REACTIONS

In this chapter we have given examples of useful compounds in some major functional group classes. The flow diagram in Figure 13–10 reviews the general reactions for obtaining compounds with these functional groups. A knowledge of these reactions along with the ability to recognize the functional groups in illustrations of organic molecules will help you understand both organic chemistry and biochemistry.

Many chemists are engaged in the synthesis of organic compounds. In educational, government, and industrial laboratories throughout the world, they prepare new and different compounds on a small scale. If the new

Acetic anhydride reacts with water to give acetic acid (number 36 in the top 50 chemical list).

Over half of the estimated 140,000 chemists in United States are working in the organic chemical industry.

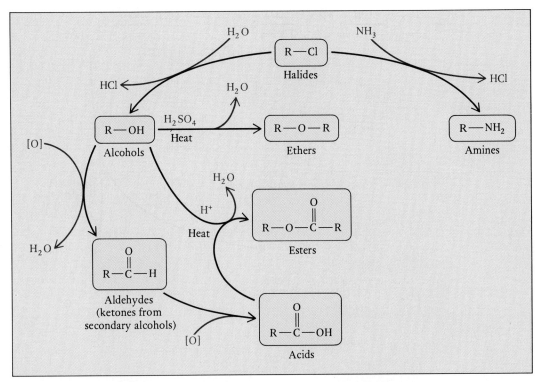

Figure 13–10   Flow diagram of the synthesis of compounds with different functional groups.

compound has commercial value, the preparation is subsequently adapted for full-scale plant operations. Since multistep syntheses are often used to make important commercial products, an example will be given here for the synthesis of aspirin.

Aspirin (acetylsalicylic acid) was first synthesized for medical use in 1893 by Felix Hofmann, a German chemist working for the F. Bayer Company. Aspirin is still the leading pain killer and the standard treatment to reduce fever and swelling. Over 30 million lb of aspirin, or 150 tablets per person, are consumed in the United States each year, and worldwide use exceeds 100,000 tons per year.

The starting point for the synthesis of aspirin is benzene, which is obtained from coal tar. The steps in the synthesis are shown in Figure 13–11. Only the principal organic substance is shown for each step. Other products such as sodium chloride (a coproduct with phenol) and water (a coproduct with sodium phenoxide) are sometimes important in the synthesis because they have to be removed to avoid interference with subsequent steps. However, in a broad outline of the synthetic process, the coproducts are generally omitted; only those products made in a previous step and required for subsequent steps are included. In Figure 13–11 the step-by-step structural changes can be followed by noting groups in color. Conditions and additional reactants for each step are written with the arrow. These conventions are generally used to summarize organic syntheses.

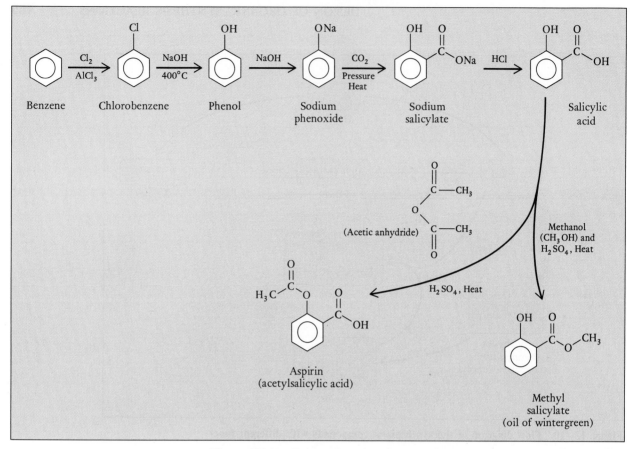

Figure 13–11 **Preparation of aspirin and oil of wintergreen. A discussion of the syntheses is given in the text.**

Any given organic compound can usually serve as the starting substance for the synthesis of many other organic compounds.

Some intermediates are useful compounds in their own right. Phenol, commonly called carbolic acid, is used to prepare plastics such as Bakelite®, drugs, dyes, and other compounds. Phenol also has medical application as a topical anesthetic for some types of lesions and in the treatment of mange and colic in animals. Methyl salicylate, or oil of wintergreen, is used as a flavoring agent and as a component of rubbing alcohol mixtures for sore muscles.

The conversion of phenol into sodium phenoxide is an acid–base reaction. Phenol, with an acidic hydrogen in the hydroxyl group, reacts with a base (sodium hydroxide) to give a salt (sodium phenoxide) and water. The reaction of salicylic acid to form oil of wintergreen is an **esterification.** The organic acid reacts with an alcohol in the presence of strong mineral acid to produce the ester, methyl salicylate, and water.

Obviously, it is beyond the scope of this text to give an extensive overview of organic synthesis. You can be assured, however, that if the "miracle cancer drug" is found or whenever any other new and useful compounds are discovered, their production will involve step-by-step molecular modifications to produce the desired product. An exciting aspect of organic synthesis is the prediction of desired properties of new molecular arrangements and then the testing of the theoretical properties. With so many possibilities yet to be discovered, one can only guess at the potential power of organic synthesis.

SELF-TEST 13-C

1. Fats and oils are esters obtained from the reaction of fatty acids with _____.

2. The class of organic compounds which are likely to be used as fragrances is _____.

3. Alcohols are oxidized to _____, which are oxidized to _____, which react with _____ to give esters.

4. Amines derived from plants are known as _____. An example of a compound in this class is _____.

5. Aspirin and oil of wintergreen
   a. are structurally similar
   b. are both acids
   c. can be made from chlorobenzene

6. Complete the following equation:

$$CH_3CH_2CH_2OH + CH_3CH_2COOH \xrightarrow{H_2SO_4}$$

7. a. When referring to edible lipids, what is the difference between a fat and an oil?
   b. How can the melting point of an edible oil be increased?

MATCHING SET

_____ 1. Synthesized from NH_4OCN by Wöhler
_____ 2. Odor of banana
_____ 3. Alkaloid
_____ 4. Found in sour milk
_____ 5. Fuel for racing cars
_____ 6. Linoleic acid
_____ 7. Saturated fatty acid
_____ 8. Octane rating
_____ 9. Present primary source of organic chemicals used as raw materials
_____ 10. Likely future primary source of organic chemicals used as raw materials
_____ 11. Catalytic cracking
_____ 12. Catalytic re-forming

a. Stearic acid
b. Essential fatty acid
c. Ethanol
d. Coal
e. Petroleum
f. Represents knocking properties of gasolines
g. Isopentyl acetate
h. Methanol
i. Urea
j. Lactic acid
k. Caffeine
l. Converts straight-chain hydrocarbons to their branched-chain isomers
m. Converts long-chain hydrocarbons into short-chain hydrocarbons
n. Acetic acid
o. Amide

QUESTIONS

1. What is the difference between natural gas found in North America and that found in Europe?
2. Make a drawing that illustrates the petroleum refinement process and label the fractions that are separated.
3. List four gasoline additives that will increase the octane rating of straight-run gasoline.
4. Explain the octane rating scale.
5. What is gasohol?
6. What is catalytic re-forming?
7. Why is coal receiving increased attention as a source of organic compounds?
8. Why and how are methanol blends used in gasoline?
9. What is meant by the following terms?
   a. Proof rating of an alcohol
   b. Denatured alcohol
10. Explain how gasoline can be made from methanol.
11. Why is a fatty acid so named?
12. After reading this chapter and the previous one, what new thoughts do you have when you view a lump of coal or a drop of petroleum?
13. What is synthesis gas? How can it be used to produce chemicals?
14. Explain the common names of *wood alcohol* for methanol and *grain alcohol* for ethanol.
15. Which propanol is used as rubbing alcohol?
16. Pure ethanol is what proof?
17. What structural features make ethylene glycol a desirable antifreeze agent?
18. Indicate what products would be formed in the reaction of
    a. Methanol and acetic acid
    b. 1-propanol and stearic acid
    c. Ethylene glycol and acetic acid
19. What reaction is used for the breathalyzer test?
20. Which would you expect to boil at a higher temperature, $CH_3CH_2CH_2CH_3$ or CH_3CH_2OH? Why?

21. Would you say that hydrogen bonding is stronger in ethanol or water?
22. Draw structural formulas for:
    a. Aspirin
    b. Oil of wintergreen
    c. Phenol
23. What chemical reactions can be used to distinguish between:
    a. C_2H_5OH and CH_3COOH
    b. $CH_3COOC_2H_5$ and CH_3COOH
24. The product of oxidation of a primary alcohol is a(n) _____, which can be oxidized to a(n) _____.
25. How many hydrogen bonds are possible per molecule of methanol? Of ethylene glycol? Of glycerol?
26. Would you expect glycerin to be water soluble? Explain.
27. Outline the steps used by Eastman Kodak to make acetic anhydride from coal.
28. What are two advantages and two disadvantages of using methanol as a fuel for vehicles?
29. Methanol is now 22nd in the list of top 50 chemicals produced in the United States. What factors are likely to lead to an increased demand for methanol in the next decade?
30. Biphenyl, ⬡—⬡, is a compound that can be chlorinated to make chlorinated biphenyl. The family of chlorinated biphenyls is called polychlorinated biphenyls, or PCBs. Draw the structures of several polychlorinated biphenyls. There are 209 possible structures.

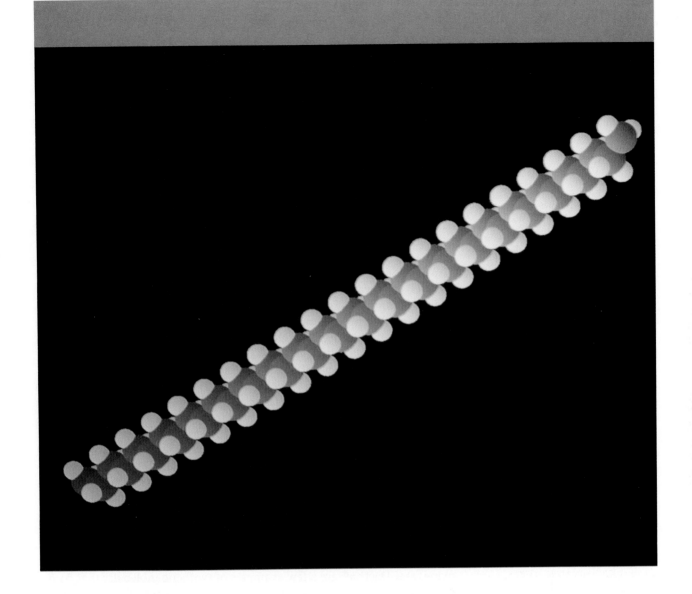

14

Giant Molecules —The Synthetic Polymers

Model of polyethylene polymer. (Courtesy of Phillips Petroleum.)

t is impossible for us to get through a day without using a dozen or more synthetic **polymers**. The word *polymer* means "many parts" (Greek, *poly* meaning "many," *meros* meaning "parts"). Polymers are *giant molecules* with molecular weights ranging from thousands to over a million. Our clothes are polymers; our food is packaged in polymers; our appliances and cars contain a number of polymer components. However, polymers were not invented by humans. Both *inorganic* and *organic* polymers exist in nature. The mineral silicates and silica sand (see Chapter 11) are examples of inorganic polymers. The structural materials of all forms of life are organic polymers—proteins, nucleic acids (DNA and RNA), cellulose, lignin, and starch are examples.

Many synthetic polymers are **plastics** of one sort or another. Examples include plastic dishes and cups; plastic containers; telephones; plastic bags for packaging foods and trash; plastic pipes and fittings; plastic water-dispersed paints; automobile steering wheels and seat covers; and cabinets for appliances, radios, and television sets. In fact, these plastics along with textile fibers and synthetic rubbers are so widely used, they are usually taken for granted. The prominence of synthetic polymers in consumer products is indicated by the fact that 24 of the top 50 chemicals (Table 13–1) are used in the production of plastics, fibers, and rubbers.

This "flood of plastic objects" did not arise accidentally; it slowly became a reality during the last 50 years because (1) natural resources like wood were dwindling, (2) with rising labor costs, many items could be made less expensively and more uniformly by molding than by sawing, shaping, sanding, and gluing, and (3) for many applications, the properties of the new synthetic materials were superior to those of metals, wood, or other natural materials. We have moved from the Iron Age to the Plastics Age, as evidenced by the fact that since 1976, plastic has exceeded steel as the nation's most widely used material.

Some of our most useful polymer chemistry has resulted from copying giant molecules in nature. Rayon is remanufactured cellulose; synthetic rubber is copied from natural latex rubber. As useful as they may be, however, polymer chemistry is not restricted to nature's models. Polystyrene, nylon, and Dacron® are a few examples of synthetic molecules that do not have exact duplicates in nature. We have gone to school on nature and extended our knowledge to produce polymers that are more useful than natural ones.

The purpose of this chapter is to investigate the structural chemistry of polymers to see just why they have such useful properties. Are these properties the result of stronger bonds or of groups of molecules acting together, or is there some other explanation? In the next chapter we shall study some of nature's polymers.

Natural polymers are essential to life.

A plastic is a substance that will flow under heat and pressure, and hence is capable of being molded into various shapes. All plastics are polymers, but not all polymers are plastic.

The average production of synthetic polymers in the United States exceeds 200 lb per person annually.

Over half of the industrial chemists in United States are polymer chemists (100,000 people).

WHAT ARE GIANT MOLECULES?

In the early 20th century many chemists were reluctant to accept the concept of giant molecules, but in the 1920s a persistent German chemist, Hermann Staudinger (1881–1965; he won the Nobel Prize in chemistry in 1953),

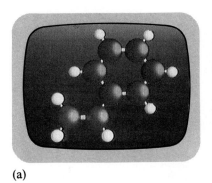

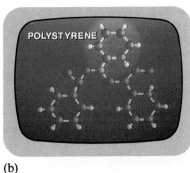

Figure 14–1 (a) Model of styrene molecule. (b) Model of the polystyrene molecule. (*The World of Chemistry,* Program 22, "The Age of Polymers.")

(a)　　　　　　　　(b)

championed the idea and introduced a new term, **macromolecule,** for these giant molecules. Staudinger devised experiments that yielded accurate molecular weights, and he synthesized "model compounds" to test his theory. One of his first model compounds was prepared from styrene, a chemical made from ethylene and benzene.

A *macromolecule* is a molecule with a very high molecular weight.

Styrene is 21st on the list of the top 50 commercial chemicals.

$$H_2C{=}CH$$

Styrene

Under the proper conditions, styrene molecules use the "extra" electrons of the double bond to undergo a **polymerization** reaction to yield polystyrene, a material composed of giant molecules (Fig. 14–1). The molecules of styrene are the **monomers** (Greek, *mono* meaning "one"); they provide the repeating units in the giant molecule analogous to identical railroad cars coupled together to make a long train.

The macromolecule polystyrene is represented as a long chain of monomer units bonded to each other. Each unit is bonded to the next by a covalent bond. The polymer chain is not an endless one; some polystyrenes made by Staudinger were found to have molecular weights of about 600,000, corresponding to a chain of about 5700 styrene units. The polymer chain can be indicated as

$$R{-}CH_2{-}CH{-}\left(CH_2{-}CH{-}\right)_n CH_2{-}CH{-}R$$

where R represents some terminal group, often an impurity, and *n* is a large number, in this case $n = 5700$.

Polystyrene is a clear, hard, colorless solid at room temperature that can be molded easily at 250°C. Commercial production of polystyrene began in Germany in 1929, and today its U.S. production exceeds 2 million tons per year. Polystyrene is used to make food containers, toys, electrical parts, insulating panels, appliance and furniture components, and many other

A polystyrene cup. (Beverly March.)

Figure 14-2   Large piece of styrofoam. (*The World of Chemistry,* Program 22, "The Age of Polymers.")

items. The variation in properties shown by polystyrene products is typical of synthetic polymers. For example, a clear polystyrene drinking glass that is brittle and breaks into sharp pieces somewhat like glass is much different from the styrofoam coffee cup that is soft and pliable.

Styrofoam is produced by "expansion molding." In this process, polystyrene beads are placed in a mold and heated with steam or hot air. The beads, 0.25 to 1.5 mm in diameter, contain 4-7% by weight of a low-boiling liquid such as pentane. The steam causes the low-boiling liquid to vaporize and this expands the beads and the foamed particles are then molded in the shape of the mold cavity. Styrofoam (Fig. 14-2) is used for egg cartons, meat trays, coffee cups, and packing material.

There are two broad categories of plastics. One, when heated repeatedly, will soften and flow; when it is cooled, it hardens again. Materials that undergo such reversible changes when heated and cooled are called **thermoplastics;** polystyrene is a thermoplastic. The other type is plastic when first heated, but when heated further it forms a set of interlocking bonds. When reheated, it cannot be softened and reformed without extensive degradation. These materials are called **thermosetting plastics** and include such familiar names as Bakelite® and rigid-foamed polyurethane, a polymer that is finding many new uses as a construction material.

To gain a better understanding of polymers, we must look at representative examples of the different types of polymerization processes. We shall see in the sections that follow that synthetic polymers can be **addition polymers,** in which monomer units are joined directly, or **condensation polymers,** in which monomer units combine by splitting out a small molecule, usually water.

Thermoplastic polymers can be repeatedly softened by merely heating.

Thermosetting polymers form cross-linking bonds when heated, resulting in a rigid structure.

ADDITION POLYMERS

In the previous section it was noted that some polymers, such as polystyrene, are made by adding monomer to monomer to form a polymer chain of great length. Perhaps the addition reactions that are easiest to understand chemically are those involving monomers containing double bonds. The simplest monomer of this group is ethene (C_2H_4). When ethene (ethylene) is heated under pressure in the presence of oxygen, polymers with molecular weights of up to 50,000 are formed (Fig. 14–3). To enter into reaction, the carbon-carbon double bond in the ethene molecule must be partially broken. This forms **reactive sites** composed of unpaired electrons at either end of the molecule.

$$
\begin{array}{c}
\text{H} \quad \text{H} \\
| \quad\quad | \\
\text{C}{=}\text{C} \xrightarrow{\text{Energy}} \cdot\text{C}{-}\text{C}\cdot \\
| \quad\quad | \\
\text{H} \quad \text{H}
\end{array}
\qquad \text{Reactive Site}
$$

Ethylene (ethene) is the number four commercial chemical.

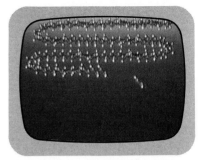

Figure 14–3 Model of polyethylene chain. (*The World of Chemistry*, Program 22, "The Age of Polymers.")

The partial breaking of the double bond can be accomplished by physical means such as heat, ultraviolet light, X rays, and high-energy electrons. This *initiation* of the polymerization reaction can also be accomplished with chemicals such as organic peroxides. These initiators, which are very unstable, break apart into pieces with unpaired electrons. These fragments (called **free radicals**) are ravenous in trying to find a "buddy" for their unpaired electrons. They react readily with molecules containing carbon-carbon double bonds.

An organic peroxide, RO—OR′, produces free radicals, RO·, each with an unpaired electron.

$$
\text{CH}_2{:}\text{CH}_2 \xrightarrow[\substack{\text{A}\\ \text{Peroxide}\\ \text{Free}\\ \text{Radical}}]{\cdot\,\text{OR}} \dot{\text{C}}\text{H}_2{-}\text{CH}_2\text{OR} \xrightarrow{n\text{CH}_2{=}\text{CH}_2} (\text{CH}_2{-}\text{CH}_2)_{n+1}\text{OR}
$$

(*n* RANGES FROM 500 TO 5,000)

The extension of the polyethylene chain shown above comes from the unpaired electron bonding to an electron in an unreacted ethylene molecule. This leaves another unpaired electron to bond with yet another ethylene molecule. For example,

$$
\text{ROCH}_2{-}\text{CH}_2\cdot + \text{CH}_2{:}\text{CH}_2 \longrightarrow \text{ROCH}_2{-}\text{CH}_2{-}\text{CH}_2{-}\text{CH}_2\cdot
$$

Polyethylenes formed under various pressures and catalytic conditions have different molecular structures and hence different physical properties. For example, chromium oxide as a catalyst yields almost exclusively the linear polyethylene shown below. A kinked structure represents more closely the tetrahedral carbon in the saturated polyethylene chain (Fig. 14–4).

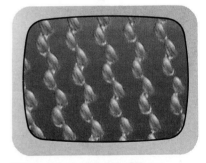

Figure 14–4 Model of linear chains of polyethylene. (*The World of Chemistry*, Program 22, "The Age of Polymers.")

$$
\begin{array}{c}
\cdots\text{C}{-}\text{C}{-}\text{C}{-}\text{C}{-}\text{C}{-}\text{C}{-}\text{C}{-}\text{C}\cdots
\end{array}
$$

A Portion of a Polyethylene Molecule

or

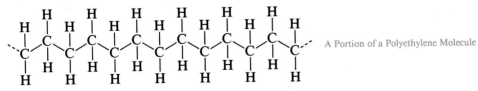

Where Each Point is a CH_2 Group.

Figure 14–5 **(a) Model of branched-chain polyethylene. (b) Model of cross-linked polyethylene.** (*The World of Chemistry,* **Program 22, "The Age of Polymers."**)

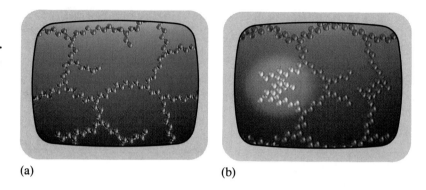

(a)   (b)

If ethylene is heated to 230°C at a pressure of 200 atm, irregular branches result (Fig. 14–5). Under these conditions, free radicals attack the chain at random positions, causing the irregular branching.

$$-CH_2-CH_2-CH_2-CH_2- \longrightarrow -CH_2-CH-CH_2-CH_2- + H\cdot$$

$$RCH_2CH_2\cdot$$

with branches:
$$\begin{array}{c} | \\ CH_2 \\ | \\ CH_2 \\ | \\ R \end{array}$$

Branched Polymer Chains

Uses of Polyethylene

Polyethylene is the world's most widely used polymer. Over 8 million tons were produced in 1987 in the United States alone. What are some of the reasons for this popularity? The wide range of properties of polyethylene leads to its variety of uses (Fig. 14–6). The key to the range of properties is the variety of structures and molecular weights that are possible. The formation of linear, branched, and cross-linked polyethylene gives polymer materials with widely different properties. Long, linear chains of polyethylene ($n = 10,000$) can pack closely together (Fig. 14–4) and give a material with high density and high molecular weight, referred to as high-density polyethylene (HDPE). This material is hard, tough, and rigid. The plastic milk carton is a good example of an application of HDPE. Short, branched chains of polyethylene ($n = 500$) cannot be packed closely together (Fig. 14–5); so the resulting material is low-density polyethylene (LDPE). This material is soft and flexible. Sandwich bags are made from LDPE. If the linear chains of polyethylene are treated in a way that causes cross-links between chains to form cross-linked polyethylene (CLPE), a very tough form of polyethylene is produced. The plastic caps on soft-drink bottles are made from CLPE. These examples of the influence of polyethylene structure on the properties of polyethylene illustrate how polymer chemists and engineers can obtain polymers with differing properties by varying the temperature, pressure, catalyst, time, and the order of mixing the various reactants.

The bottle on the left is made of flexible, low-density branched polyethylene. The one on the right is made of rigid, high-density, linear polyethylene. (Marna G. Clarke.)

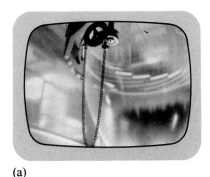

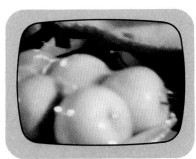

(a)

(b)

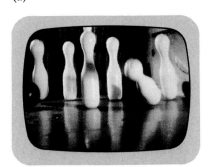

(c)

(d)

Figure 14–6 (a) Production of polyethylene. (b), (c), (d) The wide range of properties of different structural types of polyethylene lead to a variety of applications. (*The World of Chemistry,* Program 22, "The Age of Polymers.")

A Variety of Addition Polymers

A large number of addition polymers can be made by using monomers that are derivatives of ethylene, in which one or more of the hydrogen atoms have been replaced with either atoms of halogens or organic groups. If the formation of polyethylene is represented as

$$n \quad \overset{H}{\underset{H}{>}}C=C\overset{H}{\underset{H}{<}} \longrightarrow \left(\begin{array}{cc} H & H \\ | & | \\ C{-}C \\ | & | \\ H & H \end{array} \right)_n$$

then the general reaction is

$$n \quad \overset{H}{\underset{H}{>}}C=C\overset{H}{\underset{X}{<}} \longrightarrow \left(\begin{array}{cc} H & H \\ | & | \\ C{-}C \\ | & | \\ H & X \end{array} \right)_n$$

where X is Cl, F, or an organic group that can be used to represent a number of other important addition polymers.

Polypropylene, used in making indoor–outdoor carpeting, bottles, and battery cases, is made from propylene.

$$n \quad \underset{H}{\overset{H}{>}} C = C \underset{CH_3}{\overset{H}{<}} \quad \longrightarrow \quad \left(\underset{\underset{H}{|}}{\overset{\overset{H}{|}}{C}} - \underset{\underset{CH_3}{|}}{\overset{\overset{H}{|}}{C}} \right)_n$$

Propylene   Polypropylene

Polyvinyl chloride (PVC), used for making floor tile, garden hoses, plumbing pipes, and trash bags, has a chlorine atom substituted for one of the hydrogen atoms in ethene.

$$n \quad \underset{H}{\overset{H}{>}} C = C \underset{Cl}{\overset{H}{<}} \quad \longrightarrow \quad \left(\underset{\underset{H}{|}}{\overset{\overset{H}{|}}{C}} - \underset{\underset{Cl}{|}}{\overset{\overset{H}{|}}{C}} \right)_n$$

Vinyl Chloride   Polyvinylchloride

Polystyrene, first synthesized by Staudinger, is made from the monomer styrene, which has a benzene ring substituted for one of the hydrogen atoms in ethene.

Polyacrylonitrile, the acrylic plastic used to make fibers for rugs and fabrics, is an addition polymer of acrylonitrile. (See box on p. 418.)

$$n \quad \underset{H}{\overset{H}{>}} C = C \underset{CN}{\overset{H}{<}} \quad \longrightarrow \quad \left(\underset{\underset{H}{|}}{\overset{\overset{H}{|}}{C}} - \underset{\underset{CN}{|}}{\overset{\overset{H}{|}}{C}} \right)_n$$

Acrylonitrile   Polyacrylonitrile

Although the representation

$$\left(\underset{\underset{H}{|}}{\overset{\overset{H}{|}}{C}} - \underset{\underset{H}{|}}{\overset{\overset{H}{|}}{C}} \right)_n$$

saves space, keep in mind how large the polymer molecules are. Generally n is 500 to 10,000, and this gives molecules with molecular weights ranging from 10,000 to over a million. The molecules that make up a given polymer sample are of different lengths and thus are not all of the same molecular weight. As a result, only the average molecular weight can be determined.

In summary, the variation in substituents, length, branching, and cross-linking will give a variety of properties for each addition polymer. These examples illustrate why the uses of polymers continue to increase. The chemists and chemical engineers can fine-tune the properties of the polymer to match desired properties by appropriate selection of monomer and reaction conditions for making the polymer. Table 14–1 summarizes information about common addition polymers.

Teflon-coated pans. (Beverly March.)

TABLE 14–1 Ethylene Derivatives That Undergo Addition Polymerization

Formula	Monomer Common Name (Top 50 Rank)	Polymer Name (Trade Names)	Uses	Polymer U.S. Production (Tons/Yr)
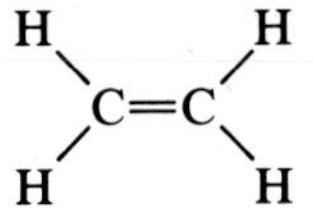 $\begin{array}{c} H \\ \!\!\!\! \end{array}$ C=C with H, H, H, H	Ethylene (4)	Polyethylene (Polythene)	Squeeze bottles, bags, films, toys and molded objects, electrical insulation	8 million
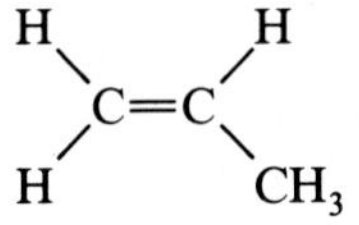 C=C with H, H, H, CH_3	Propylene (10)	Polypropylene (Vectra, Herculon)	Bottles, films, indoor–outdoor carpets	2.7 million
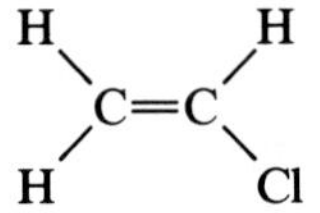 C=C with H, H, H, Cl	Vinyl chloride (20)	Polyvinyl chloride (PVC)	Floor tile, raincoats, pipe, phonograph records	3.5 million
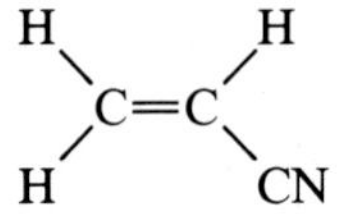 C=C with H, H, H, CN	Acrylonitrile (40)	Polyacrylonitrile (Orlon, Acrilan)	Rugs, fabrics	920,000
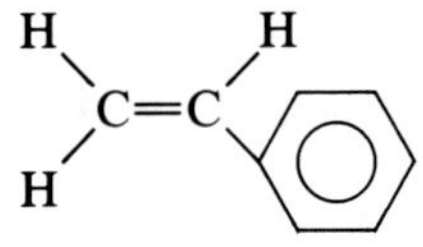 C=C with H, H, H, (phenyl)	Styrene (21)	Polystyrene (Styrene, Styrofoam, Styron)	Food and drink coolers, building material insulation	2 million
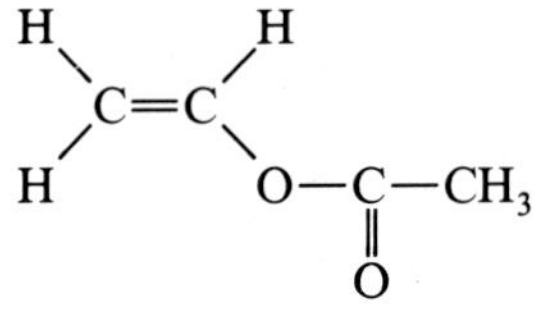 C=C with H, H, H, O—C(=O)—CH_3	Vinyl acetate (41)	Polyvinylacetate (PVA)	Latex paint, adhesives, textile coatings	500,000
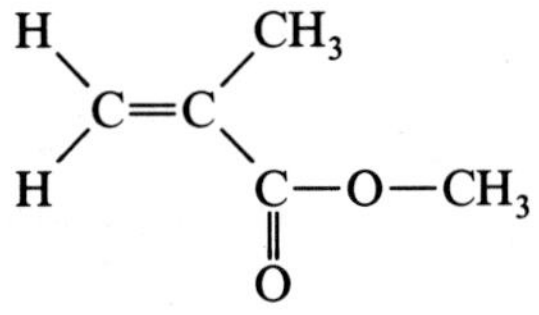 C=C with H, CH_3, H, C(=O)—O—CH_3	Methyl methacrylate	(Plexiglas, Lucite)	High-quality transparent objects, latex paints, contact lenses	450,000
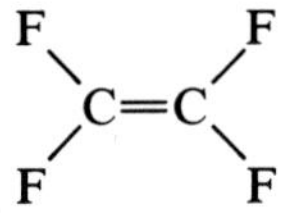 C=C with F, F, F, F	Tetrafluoroethylene	(Teflon)	Gaskets, insulation, bearings, pan coatings	7,000

DISCOVERY OF A CATALYST FOR POLYACRYLONITRILE PRODUCTION

Oil companies invest considerable sums in equipment and human effort to develop new catalysts to make fuels and chemicals from petroleum. Research and development in this area is a multimillion-dollar gamble. There is no guarantee the money spent will produce anything useful. But if it does, the payoff can be enormous. Just one catalyst breakthrough made more than half a billion dollars for Standard Oil of Ohio, now part of BP America. In the late 1950s, SOHIO researchers came up with a new catalytic process that soon dominated all others in the production of poly-acrylonitrile, the polymer used to make textiles, tires, and car bumpers. Oddly enough, SOHIO researchers weren't even trying to produce acrylonitrile at first. They simply wanted to make a metal oxide catalyst to convert waste propane gas from petroleum refining into something more valuable. As Dr. Jeanette Grasselli said:

The theory, the hypothesis at the time, was that we could take the oxygen from the catalyst and insert it into the propane, a relatively unreactive molecule. So this was a tough technical objective. And, in turn, we wanted to generate or take the catalyst back to its original oxidized form by using oxygen from the air.

But the theory didn't hold up. Propane was too stable to react, and the catalyst particles broke down in service. Management gave them three more months to show results, so they made some changes. They replaced propane with a more reactive refinery gas, propylene. They made their catalyst out of different metal oxides, and they added ammonia to promote, or speed up, the reaction.

To their surprise, ammonia reacted. Rather than just encouraging the reaction to go faster, as a promoter, oxygen reacted and became part of the reaction sequence, and acrylonitrile was made in one step. The researchers had struck pay dirt. Their new catalyst, combined with ammonia, had made a valuable product out of a cheap gas. Management quickly saw the value of the new process and wasted no time building a plant to use it. As Dr. Grasselli recalls,

Jeanette Grasselli. (*The World of Chemistry,* Program 20, "On the Surface.")

In 1960, when our plant came on stream, acrylonitrile was selling for 28¢/lb. We were making it for 14¢/lb. And we shut down every other commercial process. Today 90% of the world's acrylonitrile is manufactured by the SOHIO process.
The World of Chemistry (Program 20) "On the Surface."

Natural and Synthetic Rubbers Are Addition Polymers

Natural rubber, a product of the *Hevea brasiliensis* tree, is a hydrocarbon with the composition C_5H_8; when decomposed in the absence of oxygen, the monomer isoprene is obtained:

$$CH_2{=}\underset{\underset{CH_3}{|}}{C}{-}CH{=}CH_2 \quad \text{or}$$

Isoprene

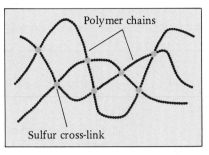
(a) Before stretching

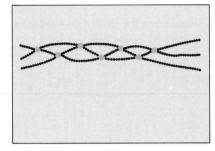
(b) Stretched

Figure 14–7 Stretched vulcanized rubber will spring back to its original structure, an elastomeric property.

Natural rubber occurs as latex (an emulsion of rubber particles in water) that oozes from rubber trees when they are cut. Precipitation of the rubber particles yields a gummy mass that is not only elastic and water-repellent but also very sticky, especially when warm. In 1839, after ten years' work on this material, Charles Goodyear (1800–1860) discovered that the heating of gum rubber with sulfur produced a material that was no longer sticky but was still elastic, water-repellent, and resilient.

Vulcanized rubber, as Goodyear called his product, contains short chains of sulfur atoms that bond together the polymer chains of the natural rubber and reduce its unsaturation. The sulfur chains help align the polymer chains, so the material does not undergo a permanent change when stretched but springs back to its original shape and size when the stress is removed (Fig. 14–7). Substances that behave this way are called **elastomers.**

In later years chemists searched for ways to make a synthetic rubber so we would not be completely dependent on imported natural rubber during emergencies, such as during the first years of World War II. In the mid-1920s, German chemists polymerized butadiene (obtained from petroleum and structurally similar to isoprene, but without the methyl group side chain). The product was buna rubber, so named because it was made from butadiene (Bu—) and catalyzed by sodium (—Na).

The behavior of natural rubber (polyisoprene), it was learned later, is due to the specific arrangement within the polymer chain. We can write the formula for polyisoprene with the CH_2 groups on opposite sides of the double bond (the *trans* arrangement):

Poly-*trans*-isoprene (The —CH_2—CH_2— Groups Are *trans*)

The formula can also be written with the CH_2 groups on the same side of the double bond (the *cis* arrangement, from Latin meaning "on this side").

Poly-*cis*-isoprene (The —CH_2—CH_2— Groups Are *cis*)

Vulcanization

TABLE 14–2 A Rubber Formulation

Ingredient	Name	Percentage	Formula	Function
Rubber	Poly-*cis*-isoprene	62.0	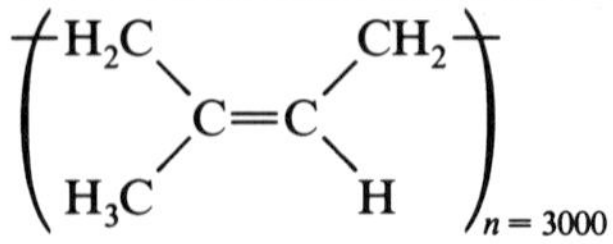 $\left(\begin{array}{c} H_2C \quad\quad\quad CH_2 \\ \backslash\quad\quad\quad / \\ C=C \\ /\quad\quad\quad \backslash \\ H_3C \quad\quad\quad H \end{array}\right)_{n=3000}$	Elastomer
Activators	Zinc oxide	2.7	ZnO	Activates vulcanizing
	stearic acid	0.6	$C_{17}H_{35}COOH$	agents; stearic acid acts as a lubricant in processing
Vulcanizing agent	Sulfur	1.5	S_8	Cross-links polymer chains
Filler	Carbon black	30.5	C	Provides strength and abrasion resistance
Accelerator	Dibenzthiozole disulfide	1.1	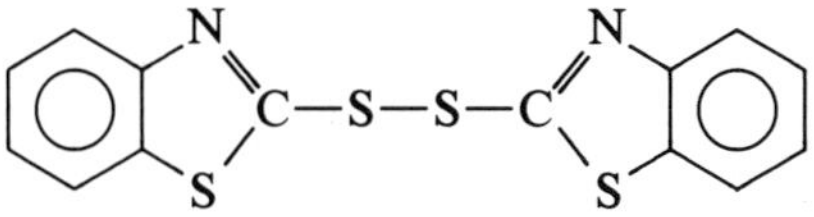 (benzothiazole)C—S—S—C(benzothiazole)	Catalyzes vulcanization
Antioxidant	Alkylated diphenylamine	1.1	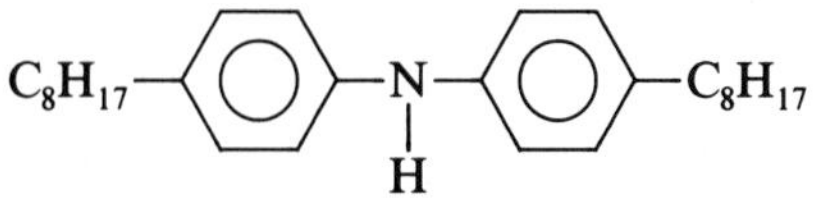 C_8H_{17}—⬡—N(H)—⬡—C_8H_{17}	Inhibits attack by oxygen or ozone in the air
Processing oil	Hydrocarbon oil	0.5	C_nH_{2n+2}	Plasticizer

Natural rubber is poly-*cis*-isoprene. However, the *trans* material also occurs in nature in the leaves and bark of the sapotacea tree and is known as *gutta-percha.* It is used as a thermoplastic for golf ball covers, electrical insulation, and other such applications. Without an appropriate catalyst, polymerization of isoprene yields a solid that is like neither rubber nor gutta-percha. Neither the *trans* polymer nor the randomly arranged material is as good as natural rubber *(cis)* for making automobile tires.

In 1955, chemists at the Goodyear and Firestone companies almost simultaneously discovered how to use stereoregulation catalysts to prepare synthetic poly-*cis*-isoprene. This material is structurally identical to natural rubber. Today, synthetic poly-*cis*-isoprene can be manufactured cheaply and is used almost equally well (there is still an increased cost) when natural rubber is in short supply. More than 2.4 million tons of synthetic rubber are produced in the United States yearly. Table 14–2 gives a typical rubber formulation as it might be used in a tire.

Neoprene

One of the first synthetic rubbers produced in the United States was neoprene, an addition polymer of the monomer 2-chlorobutadiene:

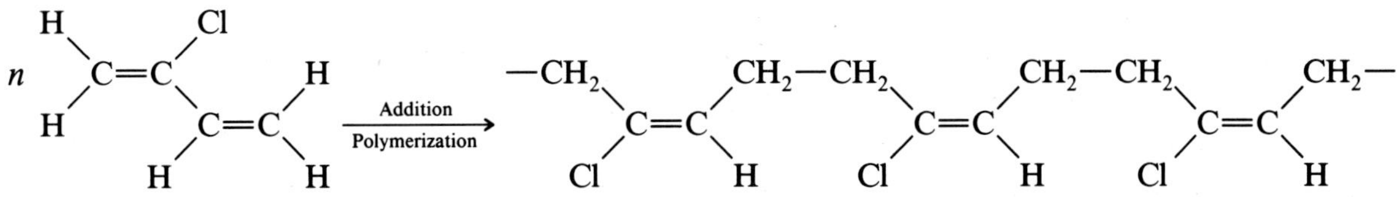

2-chlorobutadiene

which has a chlorine atom substituted for the methyl group in isoprene. Neoprene is used in the production of gaskets, garden hoses, and adhesives.

Polybutadiene

Polybutadiene, a synthetic rubber used in the production of tires, hoses, and belts, is an addition polymer of the monomer 1,3-butadiene:

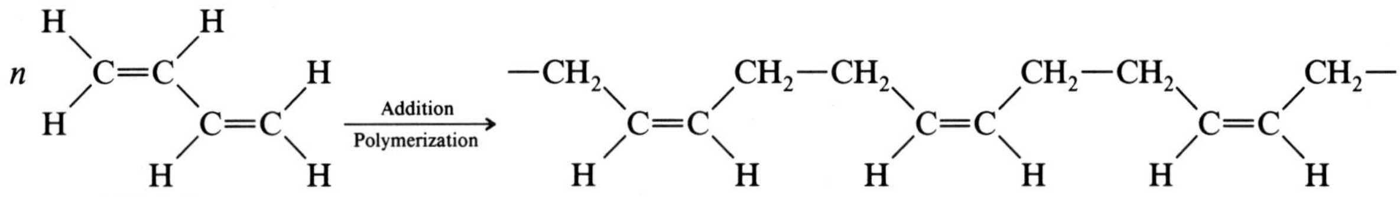

1,3-butadiene

Many other synthetic rubbers are copolymers.

COPOLYMERS

After examining Table 14–1, you might well wonder what would happen if a mixture of two monomers is polymerized. If we polymerize pure monomers A or B, we get a **homopolymer,** poly A or poly B:

—AAAAAAAAAAAAAAAAAA—

or

—BBBBBBBBBBBBBBBBBBBB—

In contrast, if the monomers A and B are mixed and then polymerized, we get **copolymers** such as the following:

—ABABABABAB—   ALTERNATING COPOLYMER

or

—AABABAAABB—
—AABABABABB—   RANDOM COPOLYMERS
—BABABBAABA—

In such polymers the order of the units is either alternating or completely random, in which case the properties of the copolymer will be determined by the ratio of the amount of A to the amount of B and the reaction conditions during polymerization.

It is possible to produce copolymers that have long chains of similar monomers in their structures. These are called **block copolymers** and can be represented as:

$$-AAAAAABBBAAAAAA-$$ BLOCK COPOLYMER

A **graft copolymer** is produced by grafting blocks of monomers to the backbone of a linear polymer:

$$-AAAAAAAAAAAAAAAAAAAAAAAAAAAAAAAA-$$ GRAFT COPOLYMER
     | |
     BBBBBBB   BBBBBBBBBBBB

Both random and block copolymers of ethylene and polypropylene have important uses. The block copolymers are more resistant to impact. They are molded into articles such as toys by a process called **injection molding,** in which the molten plastic flows under pressure into the mold. Random copolymers are more transparent. Some uses of these copolymers are given in Table 14–3 and illustrated in Figure 14–8.

A copolymer of styrene with butadiene (SBR) is the most important synthetic rubber produced in the United States. It is used primarily for making tires. This copolymer was developed during World War II in a search for a synthetic product that would be as good as natural rubber. American chemists found that a block copolymer with a 1-to-3 ratio of styrene to butadiene possessed properties that were better than those of butyl rubber.

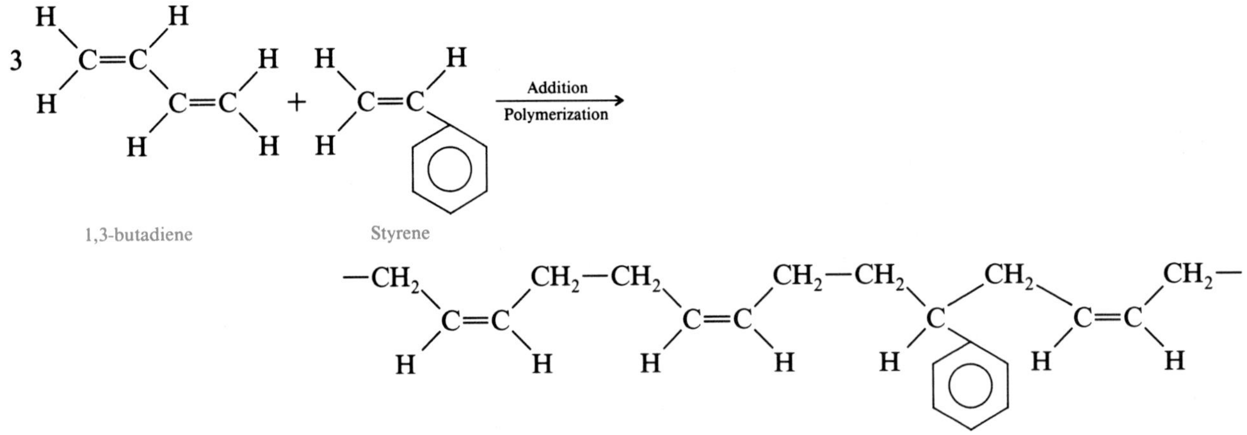

1,3-butadiene   Styrene

Styrene-butadiene Rubber (SBR)

The double bonds in SBR molecules can be cross-linked by vulcanization similar to the process used for poly-*cis*-isoprene.

Another widely used copolymer (ABS) is made from the three monomers acrylonitrile, butadiene, and styrene. This polymer is not an elastomer, but has desirable molding properties, and about 600,000 tons per year are used in the United States to make automotive parts (grills, instrument panels, and exterior decorative trim), power-tool housings, business machines, television cabinets, and appliance housings. Adding acrylonitrile changes the polymer from an elastomer to a plastic.

TABLE 14–3 Examples of Elastomer Copolymers

Name	Uses	U.S. Annual Production (tons)
ethylene-propylene polymer	radiator and heater hoses, appliance parts, car bumpers, body and chassis parts, wire and cable insulation, appliance parts, coated fabrics	150,000
polybutadiene	tires, metal can coatings, hoses, belts	400,000
polychloroprene	bridge mounts, automotive belts, hoses, conveyor belts, wire and cable jacketing	120,000
polyisoprene	sporting goods, sealants, caulking compounds, car and truck tires, footwear	75,000
SBR	tires	1.4 million

(a)

(b)

Figure 14–8 Tires and car bumpers are examples of products made from synthetic rubber polymers. (*The World of Chemistry,* Program 22, "The Age of Polymers.")

Table 14–3 lists a variety of copolymers along with some of their uses and annual production estimates.

ADDITION POLYMERS AND PAINTS

All paints involve polymers in one form or another (Fig. 14–9). Popular latex and acrylic paints contain addition polymers that serve as **binders**. A paint binder forms a molecular network to hold the **pigment** (coloring agent) in place and to hold the paint to the painted surface. In oil-based paints, a drying oil (such as linseed oil) or a resin is the binder. All paints also have a volatile solvent or thinner; this is water in water-based paints and turpentine (or mineral spirits, or both) in oil-based paints.

Early latex paints were emulsions of partly polymerized styrene and butadiene in water. Some type of emulsifying agent (such as soap) was present to keep the small drops of nonpolar styrene and butadiene dispersed in the polar water.

The first commercial water-based latex paint was Glidden's Spred Satin, introduced in 1948.

Figure 14–9 Paints are a $10 billion per year business in the United States. (Courtesy of Porter Paint Company.)

(Courtesy of Porter Paint Company.)

Acrylic polymers have a sheen that allows latex paint to compete in the exterior gloss market traditionally monopolized by oil-based coatings. Acrylics adhere well and control corrosion. Acrylics are polymers of acrylonitrile,

$$\underset{H}{\overset{H}{\diagdown}}C=C\underset{H}{\overset{CN}{\diagup}}$$

Mineral spirits are petroleum fractions of moderate volatility.

Immediately after the application of a latex paint, the water begins to evaporate. When some of the water is gone, the emulsion breaks down, and the remaining water evaporates quickly, leaving the paint film. Further polymerization of the styrene and butadiene follows slowly, but the paint appears to be dry in a few minutes. The pigment is trapped in the network of the polymer. If the paint is white, the pigment is probably titanium dioxide (TiO_2), which has replaced the poisonous compound "white lead" [$Pb(OH)_2 \cdot 2PbCO_3$] that was widely used in paints until it was banned in 1977.

The styrene-butadiene resin is the least expensive binder material used, but it has a relatively long curing period, relatively poor adhesion, and a tendency to yellow with age. Polyvinylacetate is only a little more expensive and is an improvement over the styrene-butadiene resin. It quickly captured 50% of the latex market for interior paints. Another type with rapidly growing popularity, though about one-third more expensive, includes the acrylic resins and the "acrylic latex" paints. These are more washable and much more resistant to light damage. They are especially useful as exterior paints.

The fluoropolymers, similar to Teflon®, are especially promising as surface coatings because of their great stability. Fluorine atoms are substituted for hydrogen atoms in the organic structure. Metals covered with polyvinylidene fluoride carry up to a 20-year guarantee against failure from exposure.

The drying of modern oil-based paints involves much more than the evaporation of the mineral spirits or turpentine solvent. The chemical reaction between a drying oil and oxygen from the air completes the drying process. Common drying oils are soybean, castor, coconut, and linseed oils; the most widely used is linseed, which comes from the seed of the flax plant. All of these oils are glyceryl esters of fatty acids, as discussed in Chapter 13.

The chemical action of oxygen on a drying oil is to replace a hydrogen atom on a carbon atom next to a C=C double bond in an unsaturated fatty-acid chain. When oxygen reacts with two fatty acids on two oil molecules, the result is cross-linking between the two molecules.

Part of One Molecule

$$-CH_2-CH_2-CH=CH-CH_2- \qquad\qquad -CH_2-CH-CH=CH-CH_2-$$

$$+ O_2 \rightarrow \qquad\qquad \underset{|}{O}\ \text{Ether Linkage} \qquad\qquad +H_2O$$

$$-CH_2-CH_2-CH=CH-CH_2- \qquad\qquad -CH_2-CH-CH=CH-CH_2-$$

Part of Another Molecule

The polymeric network produced by the cross-linking hardens the paint, traps the pigment, and secures the paint in the crevices of the painted surface.

SELF-TEST 14-A

1. The individual molecules from which polymers are made are called
   _____.

2. Plastics that undergo reversible changes when heated and cooled are
   ( ) thermosetting ( ) thermoplastic.

3. Draw the formulas of the monomers used to prepare the following polymers.
   a. Polypropylene
   b. Polystyrene
   c. Teflon
   d. Polyvinyl chloride
4. Draw the repeating unit for the following polymers.
   a. Polyethylene
   b. Polyacrylonitrile
5. Natural rubber is a polymer of _____ .
6. When styrene and butadiene are polymerized together, the product is a type of _____ .

CONDENSATION POLYMERS

Polyesters

A chemical reaction in which two molecules react by splitting out or eliminating a small molecule is called a **condensation reaction.** For example, acetic acid and ethyl alcohol will react, splitting out a water molecule, to form ethyl acetate, an **ester.**

$$\underset{\text{Acetic Acid}}{CH_3\overset{\overset{\displaystyle O}{\|}}{C}-OH} + \underset{\text{Ethanol}}{HOCH_2CH_3} \xrightarrow[\text{Catalyst}]{H^+} \underset{\substack{\text{Ethyl Acetate}\\\text{(An Ester)}}}{CH_3\overset{\overset{\displaystyle O}{\|}}{C}-OCH_2CH_3} + H_2O$$

This important type of chemical reaction does not depend on the presence of a double bond in the reacting molecules. Rather, it requires the presence of two kinds of functional groups on two different molecules. If each reacting molecule has *two* functional groups, both of which can react, it is then possible for condensation reactions to lead to a long-chain polymer. If we take a molecule with two carboxyl groups, such as terephthalic acid, and another molecule with two alcohol groups, such as ethylene glycol, each molecule can react at each end. The reaction of one acid group of terephthalic acid with one alcohol group of ethylene glycol initially produces an ester molecule with an acid group left over on one end and an alcohol group left over on the other:

$$\underset{\text{Terephthalic Acid}}{HO-\overset{\overset{\displaystyle O}{\|}}{C}-\underset{}{\bigcirc}-\overset{\overset{\displaystyle O}{\|}}{C}-OH} + \underset{\text{Ethylene Glycol}}{HO-CH_2-CH_2-OH} \longrightarrow$$

$$\underset{\text{(An Ester)}}{HO-\overset{\overset{\displaystyle O}{\|}}{C}-\bigcirc-\overset{\overset{\displaystyle O}{\|}}{C}-OCH_2-CH_2-OH} + H_2O$$

Subsequently, the remaining acid group can react with another alcohol group, and the alcohol group can react with another acid molecule. The

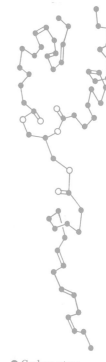

Typical molecule in linseed oil.

● Carbon atom
○ Oxygen atom

The esterification of a dialcohol and a diacid involves two positions on each molecule.

Figure 14-10 A garment made of Dacron (Courtesy of Du Pont de Nemours and Company.)

Terephthalic acid is the number 18 commercial chemical.

process continues until an extremely large polymer molecule, known as a **polyester,** is produced with a molecular weight in the range of 10,000 to 20,000.

$$HO-\overset{\overset{O}{\|}}{C}-\bigcirc-\overset{\overset{O}{\|}}{C}-O-CH_2-CH_2-O\left(\overset{\overset{O}{\|}}{C}-\bigcirc-\overset{\overset{O}{\|}}{C}-O-CH_2-CH_2-O\right)_n\overset{\overset{O}{\|}}{C}-\bigcirc-\overset{\overset{O}{\|}}{C}-O-CH_2-CH_2-OH$$

A typical polyester is produced from a dialcohol and a diacid.

Fifty percent of all synthetic fiber is Dacron®.

Over 2 million tons of poly(ethylene terephthalate), commonly referred to as PET, are produced in the United States each year for use in making apparel, tire cord, film for photography and magnetic recording, food packaging, beverage bottles, coatings for microwave and conventional ovens, and home furnishings. A variety of trade names are associated with the various applications. Polyester textile fibers are marketed under such names as Dacron® or Terylene® (Fig. 14-10). Films of the same polyester, when magnetically coated, are used to make audio and TV tapes. This film, Mylar®, has unusual strength and can be rolled into sheets one-thirtieth the thickness of a human hair. The inert, nontoxic, nonallergenic, noninflammatory, and non-blood-clotting natures of Dacron polymers make Dacron tubing an excellent substitute for human blood vessels in heart bypass operations (Fig. 14-11) and as a skin substitute for burn victims.

INVENTOR OF THE POLY(ETHYLENE TEREPHTHALATE) BOTTLE

The inventor of the poly(ethylene terephthalate) soft-drink bottle is Nathaniel Wyeth, who comes from the internationally famous family of artists. His brother, Andrew Wyeth, expresses his creativity on canvas, but Nat Wyeth expresses his through chemical engineering. He has an intriguing story:

I got to thinking about the work that Wallace Carothers did for Du Pont way back in the days when nylon was born, where he found that, if you took a thread of nylon when it was cold, that is, below the melt point, and stretched it, it would orient itself. That is, the molecules of the polymer would align themselves. This is what you're doing to the molecules when you orient them, you're lining them up so they can give you the most strength. They're all pulling in the direction you want them to pull in.

But the bottles kept splitting. Wyeth estimates that he made 10,000 tries and 10,000 failures before he made a simple observation.

Well, then I realized what we've got to do now is to align these molecules in the sidewall of the bottle; not only in one direction, but in two directions. So I thought I'd play a trick on this mold, on this problem. I took two pieces of poly-ethylene and turned one of them ninety degrees with the other. So then I had one that would split in this direction, and one that would split in that direction. Well, one piece reinforced the other. As soon as I did that, I could blow bottles. That seems almost dirt simple. But as I've often said, quoting Einstein, the biggest part of a problem and the easiest way to solving a problem is to understand it, have the problem in a form you can understand what's going on. And what I was doing here was learning about what was going on. Once I knew, it was simple to solve.

Nathaniel Wyeth. (*The World of Chemistry,* Program 22, "The Age of Polymers.")

The World of Chemistry (Program 22) "The Age of Polymers."

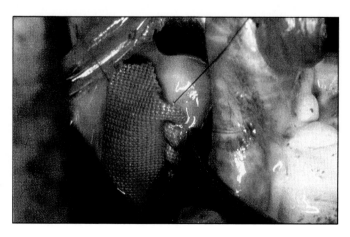

Figure 14–11 A Dacron patch is used to close an atrial septal defect in a heart patient. (Courtesy of Drs. James L. Monro and Gerald Shore and the Wolfe Medical Publications, London, England.)

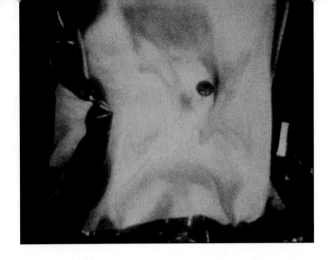

This vest, made of Kevlar fiber, has deflected a bullet. Kevlar is a polyamide made from *p*-phenylenediamine and terephthalic acid. It is also used in aircraft construction and tires. (DuPont/Conoco.)

$$\underset{\text{Acid}}{-\overset{\overset{\displaystyle O}{\|}}{C}+OH} + \underset{\text{Amine}}{H+\overset{\overset{\displaystyle H}{|}}{N}-}$$

$$\downarrow$$

$$\underset{\text{Amide}}{-\overset{\overset{\displaystyle O}{\|}}{C}-\underset{\underset{\displaystyle H}{|}}{N}-} + HOH$$

Common nylon can be made by the reaction of adipic acid and hexamethylenediamine.

Polyamides (Nylons)

Another useful condensation reaction is that occurring between an acid and an amine to split out a water molecule to form an **amide**. Reactions of this type yield a group of polymers that perhaps have had a greater impact on society than any other type. These are the **polyamides,** or nylons.

In 1928, the Du Pont Company embarked on a program of basic research headed by Dr. Wallace Carothers (1896–1937), who came to Du Pont from the Harvard University faculty. His research interests were high-molecular-weight compounds, such as rubber, proteins, and resins, and the reaction mechanisms that produced these compounds. In February 1935, his research yielded a product known as nylon 66 (Fig. 14–12), prepared from adipic acid (a diacid) and hexamethylenediamine (a diamine):

$$\underset{\text{Adipic Acid}}{HO-\overset{\overset{\displaystyle O}{\|}}{C}-(CH_2)_4-\overset{\overset{\displaystyle O}{\|}}{C}-OH} + \underset{\text{Hexamethylenediamine}}{H_2N-(CH_2)_6-NH_2} \longrightarrow$$

$$-\overset{\overset{\displaystyle O}{\|}}{C}-(CH_2)_4\boxed{\overset{\overset{\displaystyle O}{\|}}{C}-\underset{\underset{\displaystyle H}{|}}{N}}-(CH_2)_6\boxed{\underset{\underset{\displaystyle H}{|}}{N}-\overset{\overset{\displaystyle O}{\|}}{C}}-(CH_2)_4\boxed{\overset{\overset{\displaystyle O}{\|}}{C}-\underset{\underset{\displaystyle H}{|}}{N}}-(CH_2)_6-\overset{\overset{\displaystyle H}{|}}{N}- + xH_2O.$$

Nylon 66
(*The amide groups are outlined for emphasis.*)

This material could easily be extruded into fibers that were stronger than natural fibers and chemically more inert. The discovery of nylon jolted the American textile industry at almost precisely the right time. Natural fibers were not meeting the needs of 20th-century Americans. Silk was not durable and was very expensive, wool was scratchy, linen crushed easily, and cotton did not lend itself to high fashion. All four had to be pressed after cleaning. As women's hemlines rose in the mid-1930s silk stockings were in great demand, but they were very expensive and short-lived. Nylon changed all that almost overnight. It could be knitted into the sheer hosiery women wanted, and it was much more durable than silk. The first public sale of nylon hose took place in Wilmington, Delaware (the hometown of Du Pont's main office), on October 24, 1939. The stockings were so popular they

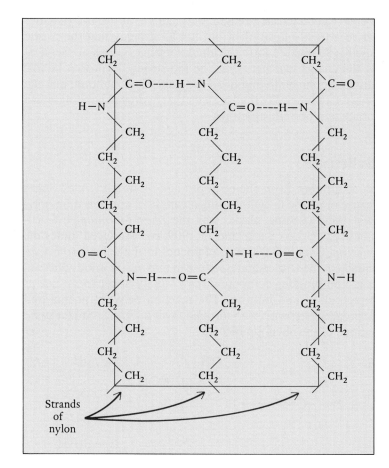

Strands
of
nylon

Figure 14–12   Nylon 66. Hexamethylenediamine is dissolved in water (bottom layer), and a derivative of adipic acid (adipoyl chloride) is dissolved in hexane (top layer). The two compounds mix at the interface between the two layers to form nylon, which is being wound onto a stirring rod. (Charles D. Winters.)

Figure 14–13   Structure and hydrogen bonding in nylon 6.

had to be rationed. World War II caused all commercial use of nylon to be abandoned until 1945, as the industry turned to making parachutes and other war materials. Not until 1952 was the nylon industry able to meet the demands of the hosiery industry and to release nylon for other uses as a fiber and as a thermoplastic.

Many kinds of nylon have been prepared and tried on the consumer market, but two, nylon 66 and nylon 6, have been most successful. Nylon 6 is prepared from caprolactam, which comes from aminocaproic acid. Notice how aminocaproic acid contains an amine group on one end of the molecule and an acid group on the other end.

$$H_2C \underset{H_2C-NH}{\overset{CH_2}{\underset{\diagdown}{\bigcirc}}} \overset{CH_2}{\underset{C=O}{}}$$

Caprolactam

$$H_2N-(CH_2)_5-\overset{\overset{\displaystyle O}{\|}}{C}-OH \xrightarrow[\text{Polymerization}]{-H_2O} -\overset{\overset{\displaystyle H}{|}}{N}-(CH_2)_5-\overset{\overset{\displaystyle O}{\|}}{C}-\overset{\overset{\displaystyle H}{|}}{N}-(CH_2)_5-\overset{\overset{\displaystyle O}{\|}}{C}-$$

Aminocaproic
Acid   Portion of Nylon 6

Figure 14–13 illustrates another facet of the structure of nylon—hydrogen bonding. This type of bonding explains why the nylons make such

The amide linkage in nylon is the same linkage found in proteins where it is called the peptide linkage.

429

Hair, wool, and silk are examples of nature's version of nylon. However, these natural polymers have only one carbon

$$\begin{array}{c} O \\ \parallel \\ \text{between each pair of}\ -C-N-\ \text{units} \\ | \\ H \end{array}$$

instead of the half dozen or so found in synthetic nylons.

good fibers. To have good tensile strength, the chains of atoms in a polymer should be able to attract one another, but not so strongly that the plastic cannot be initially extended to form the fibers. Ordinary covalent chemical bonds linking the chains together would be too strong. Hydrogen bonds, with a strength about one tenth that of an ordinary covalent bond, link the chains in the desired manner. We shall see later that hydrogen bonding is also of great importance in protein structures.

Formaldehyde Resins

Formaldehyde is number 23 in chemical production in the United States primarily because of its use in synthesizing a variety of condensation polymers. The first thermosetting plastic was the *phenol-formaldehyde* copolymer synthesized by Leo Baekeland in 1909 and produced under the tradename Bakelite. Over 700,000 tons of phenol-formaldehyde resins are produced annually in the United States for use in making plywood adhesive, glass fiber resin, and molding compound for a variety of products such as distributor caps, radios, and buttons. The reaction between phenol and formaldehyde proceeds stepwise, with formaldehyde adding first to the *ortho* and *para* positions of the phenol molecule.

Figure 14–14   Preparation of phenol-formaldehyde polymer. (Charles D. Winters.)

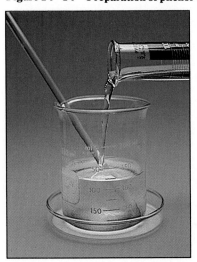

During the hardening process the substituted phenol molecules react to split out water and form a three-dimensional network of cross-linked units.

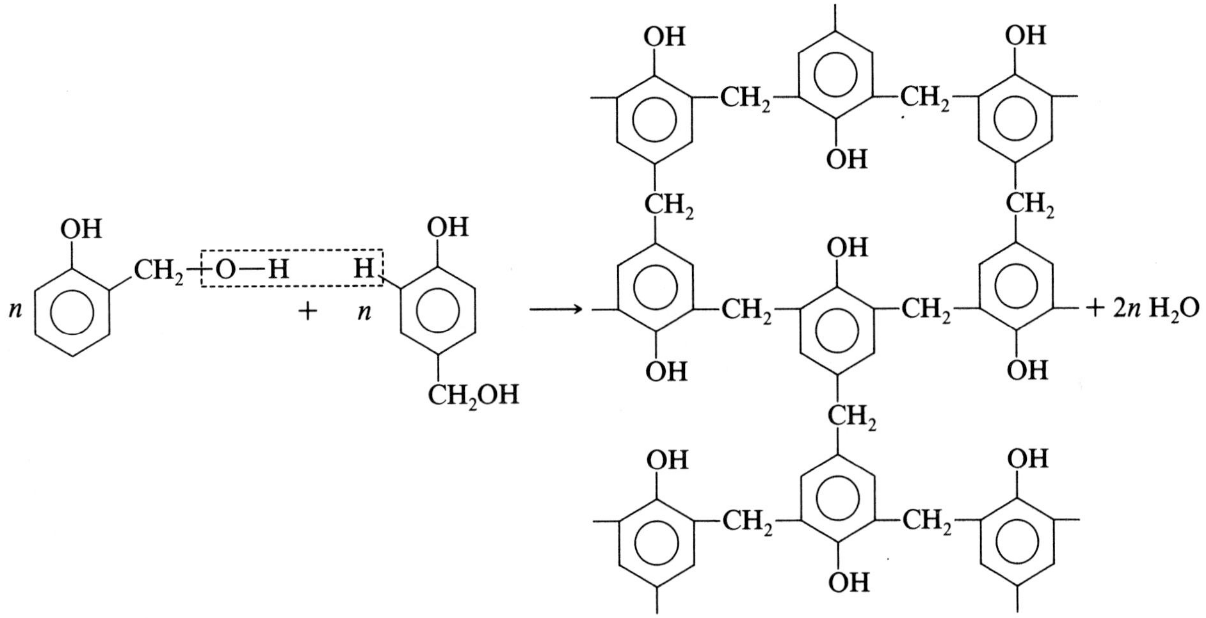

$$n \; \text{(phenol-CH}_2\text{-O-H)} + n \; \text{(phenol-CH}_2\text{OH)} \longrightarrow \text{cross-linked network} + 2n \; H_2O$$

The *melamine-formaldehyde* resin is used for dinnerware (Melmac®), coatings, table tops (Formica®). Melamine is a cyclic molecule formed by the condensation of three molecules of urea.

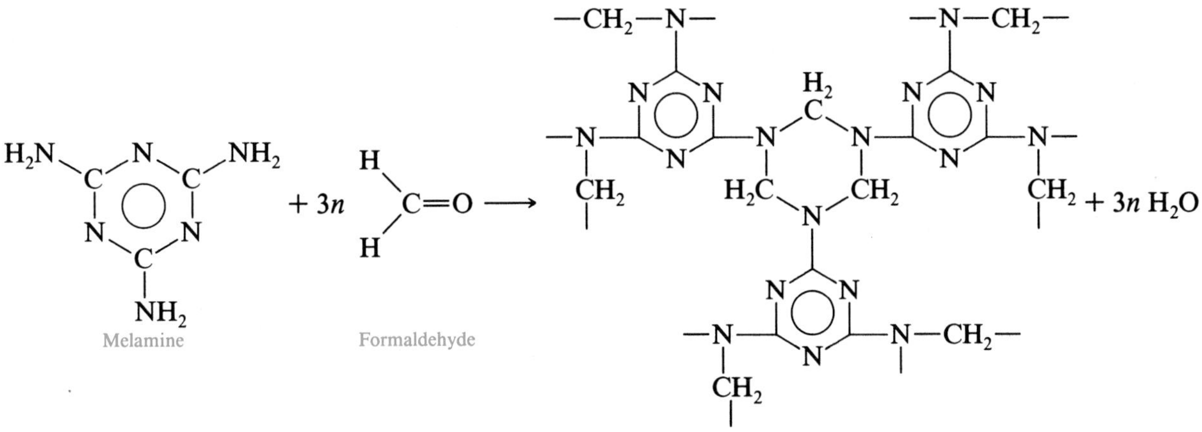

$$\text{Melamine} + 3n \; \text{Formaldehyde} \longrightarrow \text{cross-linked network} + 3n \; H_2O$$

Moon walk. The visor in the astronaut's helmet contains Lexan, a polycarbonate. (Courtesy of NASA.)

Urea and formaldehyde also form an important condensation polymer. Over 600,000 tons of the *urea-formaldehyde* resin are produced annually in the United States for use as particle-board binder resin, molding compound, coatings, and paper and textile treatment.

$$n \; H_2N-\underset{\underset{O}{\parallel}}{C}-NH_2 + 2n \; \underset{\underset{H}{|}}{\overset{\overset{H}{|}}{C}}=O \longrightarrow$$

$$-H_2CNH\overset{\overset{O}{\parallel}}{C}-NH-CH_2-\underset{\underset{CH_2}{|}}{N}-\overset{\overset{O}{\parallel}}{C}-NH-$$

$$-NH\overset{\overset{O}{\parallel}}{C}-\underset{\underset{CH_2}{|}}{N}$$

$$NH-\underset{\underset{O}{\parallel}}{C}-\underset{\underset{CH_2}{|}}{N}-CH_2- \quad + \; 2n \; H_2O$$

Urea   Formaldehyde

Although formaldehyde is an important starting material for these condensation polymers, it presents a number of health hazards because of its toxicity and carcinogenicity (See Chapter 16).

Polycarbonates

The tough, clear polycarbonates constitute another important group of condensation plastics. One type of polycarbonate, commonly called Lexan® or Merlon®, was first made in Germany in 1953. It is as "clear as glass" and nearly as tough as steel. A 1-in. sheet can stop a .38-caliber bullet fired from 12 ft away. Such unusual properties have resulted in Lexan's use in "bulletproof" windows and as visors in astronauts' space helmets. More than 115 million lb of polycarbonates were produced in the United States in 1980.

The polycarbonates are formed by condensing phosgene with a substance containing two phenol structures. A molecule of HCl is condensed out in the formation of a C—O bond to complete the ester ($-\overset{\overset{\text{O}}{\|}}{\text{C}}-\text{O}-$) linkage. Other chlorines then react with other alcohol groups to give a polymer chain containing $-\text{O}-\overset{\overset{\text{O}}{\|}}{\text{C}}-\text{O}-$ functional groups in the backbone of the chain. A representative portion of Lexan is made as follows:

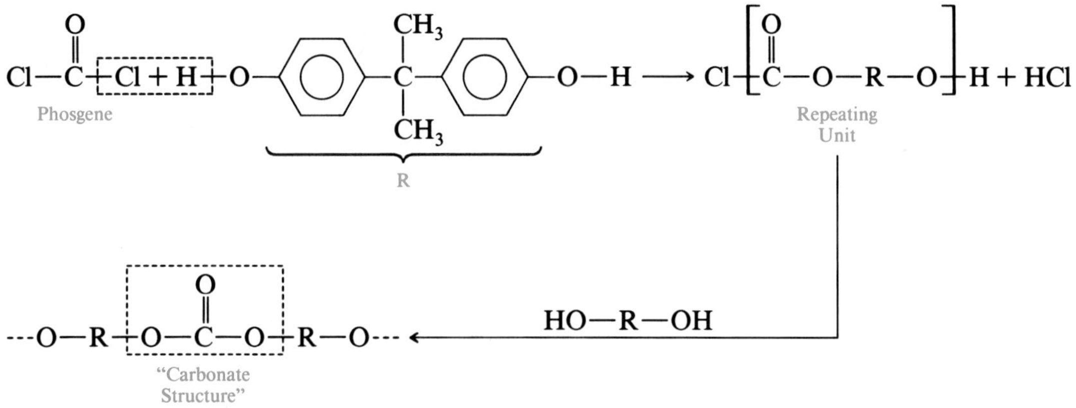

The name *polycarbonate* comes from the linkage's similarity to an inorganic carbonate ion, CO_3^{2-}.

Rearrangement Polymers

Some molecules polymerize by **rearrangement reactions** to yield very useful products. Molecules containing the isocyanate group (—NCO), for example, will react with almost any other molecule containing an active hydrogen atom (such as in an —OH or —NH_2 group) in a rearrangement process. An example is the reaction of hexamethylene diisocyanate and butanediol. The urethane linkage $\left(\begin{array}{c} -\text{N}-\text{C}-\text{O}- \\ | \| \\ \text{H} \text{O} \end{array} \right)$ is produced by a shift (or rearrangement) of a hydrogen atom, moving it from the alcohol (butanediol) to a nitrogen atom on the isocyanate group; the linkage (—N—C—O—) is similar to, but not the same as, the amide bond (—N—C—C—) in nylons.

The continued reaction of the other groups gives rise to a polymer chain—a polyurethane.

$$\underset{\text{A Portion of Polyurethane}}{-\overset{\overset{\text{O}}{\|}}{\text{C}}-\overset{\overset{\text{H}}{|}}{\text{N}}-(\text{CH}_2)_6-\overset{\overset{\text{H}}{|}}{\text{N}}-\overset{\overset{\text{O}}{\|}}{\text{C}}-\text{O}-(\text{CH}_2)_4-\text{O}-\overset{\overset{\text{O}}{\|}}{\text{C}}-\overset{\overset{\text{H}}{|}}{\text{N}}-(\text{CH}_2)_6-\overset{\overset{\text{H}}{|}}{\text{N}}-\overset{\overset{\text{O}}{\|}}{\text{C}}-\text{O}-(\text{CH}_2)_4-\text{O}-}$$

Polyurethanes are structurally similar to many polyamides.

A polyurethane is structurally similar to a polyamide (nylon). In Europe polyurethanes have applications similar to those of nylon in this country. Polyurethanes have viscosities and melting points that make them useful for foam applications. Foamed polyurethanes are known as "foam rubber" and "foamed hard plastics," depending on the degree of cross-linking.

Condensation Polymers and Baked-On Paints

When General Motors lacquered the 1923 Oakland with a nitrocellulose lacquer, the protective coatings industry first began its expansion into the use of a wide variety of materials instead of a few naturally occurring oils and minerals.

If you have ever had a car repainted, perhaps you have seen the baking oven in which the paint is dried (Fig. 14–15). Automobile finishes and those on major appliances (such as refrigerators, washing machines, and stoves) require very tough, adherent paints in order to withstand abuse. The tough coating is produced by extensively cross-linked condensation polymers.

A popular type of baked-on paint is the **alkyd** variety. The term comes from a combination of the words *alcohol* and *acid*. Alkyds, then, are polyesters with extensive cross-linking. One of the simpler alkyds is formed from the diacid, phthalic acid, and the trialcohol, glycerol.

The —OH and —COOH groups continue to react with more and more reactant molecules until extensive cross-linking occurs. Heating to about

Figure 14–15 Spray-painting a car. (*The World of Chemistry,* Program 15, "The Busy Electron.")

130°C for about 1 h causes maximum cross-linking. A portion of the resin's structure is shown as follows:

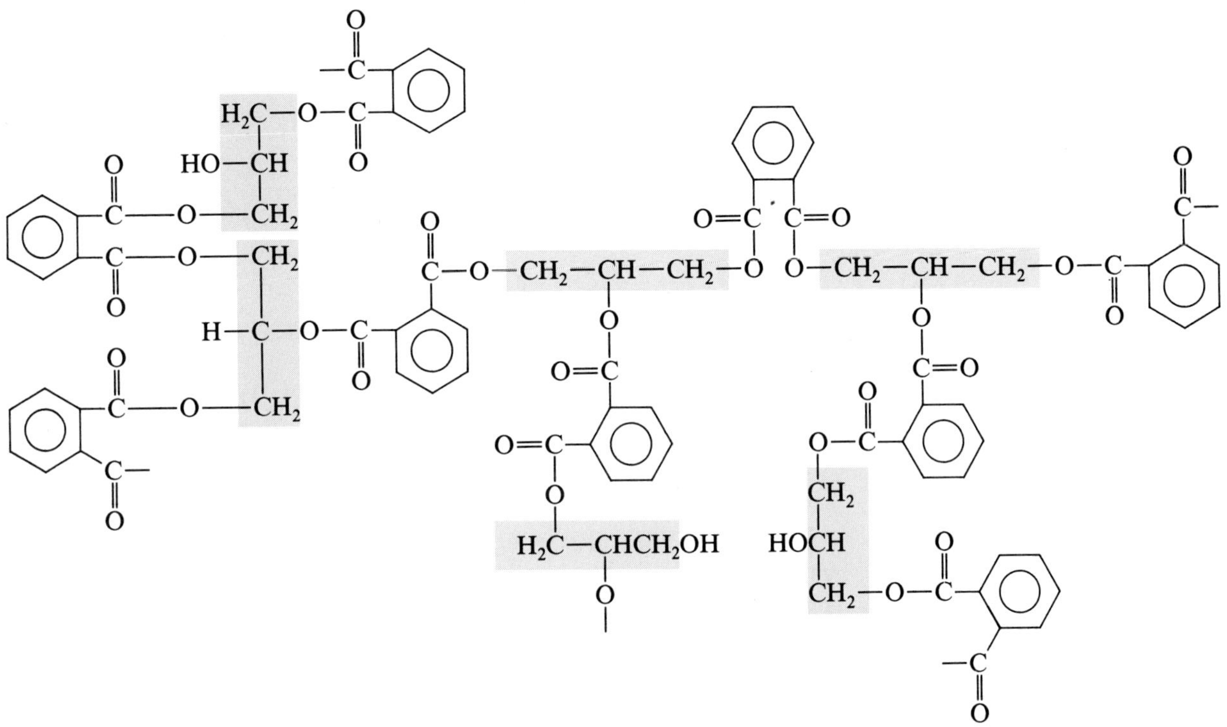

SILICONES

The element silicon, in the same chemical family as carbon, also forms many compounds with numerous Si—Si and Si—H bonds, analogous to C—C and C—H bonds. However, the Si—Si bonds and the Si—H bonds react with both oxygen and water; hence, there are no useful silicon counterparts to most hydrocarbons. However, silicon does form stable bonds with carbon, and especially oxygen, and this fact gives rise to an interesting group of condensation polymers containing silicon, oxygen, carbon, and hydrogen (bonded to carbon).

In 1945, at the General Electric Research Laboratory, E. G. Rochow discovered that a silicon-copper alloy will react with organic chlorides to produce a whole class of reactive compounds, the **organosilanes.**

Silane (SiH_4) is structurally like methane (CH_4), in that both are tetrahedral.

$$2 \ CH_3Cl + \ Si(Cu) \ \rightarrow \ (CH_3)_2SiCl_2 \ + Cu$$

Methyl Chloride   Silicon-Copper Alloy   Dimethyldichlorosilane

The chlorosilanes readily react with water and replace the chlorine atoms with hydroxyl (—OH) groups. The resulting molecule is similar to a dialcohol.

$$(CH_3)_2SiCl_2 + 2 \ H_2O \rightarrow (CH_3)_2Si(OH)_2 + 2 \ HCl$$

Two dihydroxysilane molecules undergo a condensation reaction in which a water molecule is split out. The resulting Si—O—Si linkage is very strong; the same linkage holds together all the natural silicate rocks and minerals. Continuation of this condensation process results in polymer molecules with molecular weights in the millions:

Silicones are polymers held together by a series of covalent Si—O bonds.

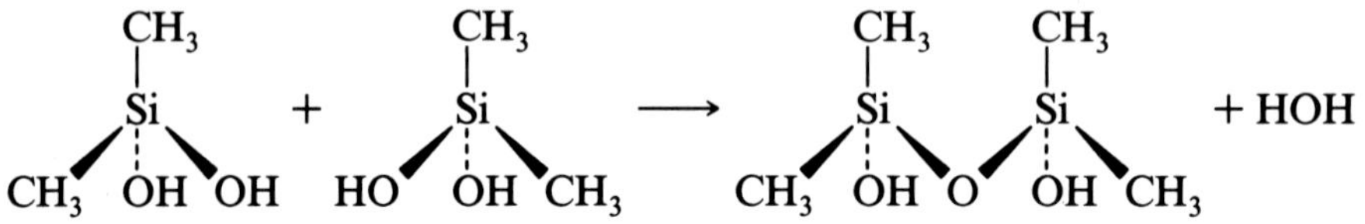

Further reaction yields:

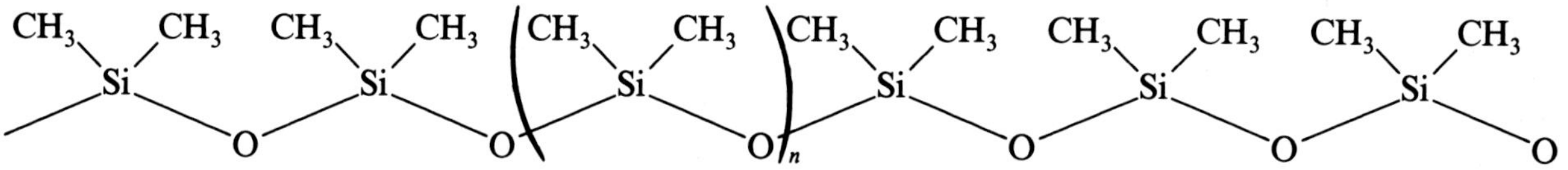

By using different starting silanes, polymers with different properties result. For example, two methyl groups on each silicon atom result in **silicone oils,** which are more stable at high temperatures than hydrocarbon oils and also have less tendency to thicken at low temperatures.

Silicone rubbers are very high molecular weight chains cross-linked by Si—O—Si bonds. Silicone rubbers that vulcanize at room temperature are commercially available; they contain groups that readily cross-link in the presence of atmospheric moisture. The —OH groups are first produced, and then they condense in a cross-linking "cure" similar to the vulcanization of organic rubbers.

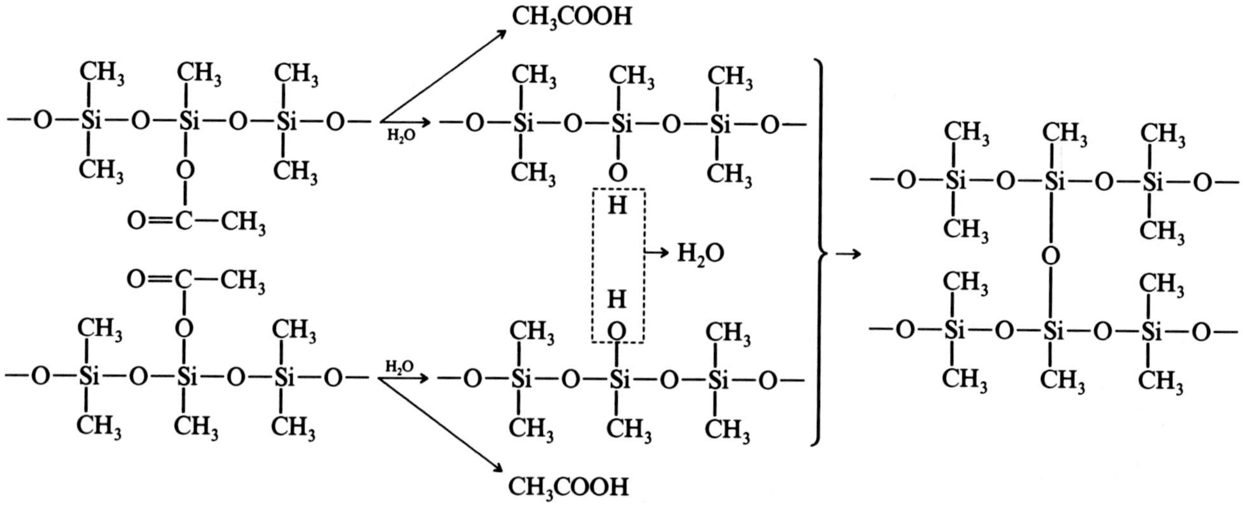

Silicone oils and rubbers find many medicinal uses.

Over 3 million lb of silicone rubber are produced each year in the United States. The uses include window gaskets; O-rings; insulation; sealants for buildings, space ships, and jet planes; and even some wearing apparel

(a)

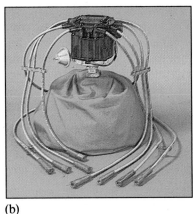

(b)

Figure 14–16 (a) Examples of some consumer products that contain silicone polymers. (b) Silicone rubber is used in automotive ignition systems because of its superior electrical properties and heat resistance. (Courtesy of Stauffer-Wacker Silicones Corporation.)

(Fig. 14–16). The first footprints on the moon were made with silicone rubber boots, which readily withstood the extreme surface temperatures.

Silly Putty®, a silicone widely distributed as a toy, is intermediate between silicone oils and silicone rubber. It is an interesting material with elastic properties on sudden deformation, but its elasticity is quickly overcome by its ability to flow like a liquid when allowed to stand.

POLYMER ADDITIVES—TOWARD AN END USE

Few plastics produced today find end uses without some kind of modification. For example, body panels for the GM Corvette and Pontiac Fiero are made of **reinforced plastics,** which contain fibers embedded in the matrix of a polymer. These are often referred to as **composites.** The strongest geometry for a solid is a wire or a fiber, and the use of a polymer matrix prevents the fiber from bending or buckling. As a result, reinforced plastics are stronger than steel. In addition, the composites have a low specific gravity—1.5 to 2.25 compared with 2.7 for aluminum, 7.9 for steel, and 2.5 for concrete. The only structural material with a lower specific gravity is wood, which has an average value of 0.5. In addition, polymers do not corrode. The low specific gravity, high strength, and high chemical resistance of composites are the basis for their increased use in automobile and airplane construction and sporting goods industries.

Glass fibers currently account for over 90% of the fibrous material used in reinforced plastics because glass is inexpensive and glass fibers possess high strength, low density, good chemical resistance, and good insulating properties. In principle, any polymer can be used for the matrix material. Polyesters are the number one polymer matrix at the present time. Glass-reinforced polyester composites have been used in structural applications such as boat hulls, airplanes, missile casings, and automobile body panels.

Other fibers and polymers have been used, and the trend is toward increased utilization of composites in automobiles and aircraft. For example, a composite of graphite fibers in an epoxy matrix is used in the construction of the Lear jet. Graphite–epoxy composites are used in a number of sporting goods such as golf-club shafts, tennis racquets, fishing rods, and skis. The F-16 military aircraft was the first to contain graphite–epoxy composite material, and the technology has advanced to the point where many aircraft, such as the F-18, use graphite composites for up to 26% of the aircraft's structural weight. This percentage is projected to increase to 40 or 50% in future aircraft.

Although few automobiles contain exterior body panels made of plastics, a number of components are plastic. Examples include car bumpers, trim, light lenses, grilles, dashboards, seat covers, steering wheels—enough plastics to account for an average of 250 lb per car. The increased emphasis on improving fuel efficiency will lead to higher amounts of plastics in the construction of automobiles, both in interior components and exterior body

TABLE 14–4 Polymer Additives

Additive	Structure	Use/Comments
Foaming agent Pentane	$CH_3CH_2CH_2CH_2CH_3$	Used to foam polyurethane and polystyrene.
Plasticizers Dioctyl phthalate (DOP)	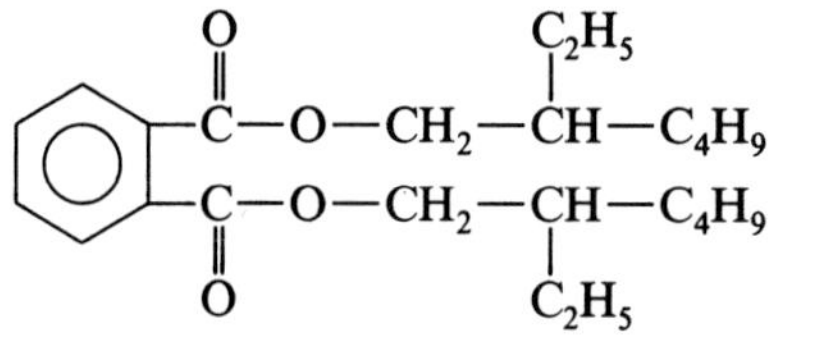	Plasticizer in polyvinylchloride to lend flexibility. Gets into the enviroment.
Dioctyl adipate (DOA)	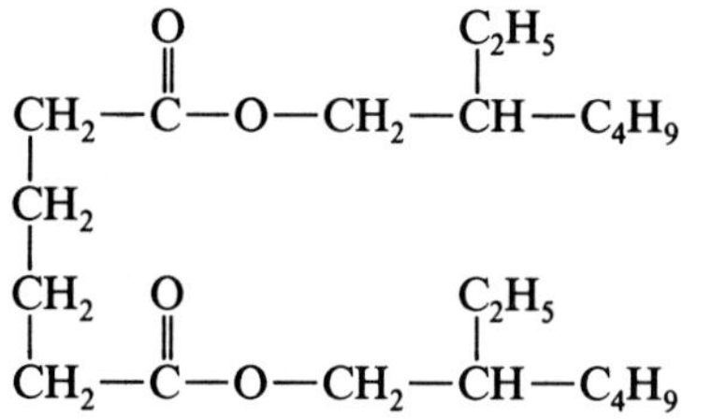	Used in plastic films to make them flexible. Has Food and Drug Adminstration approval for food contact.
UV stabilizers Phenylsalicylate	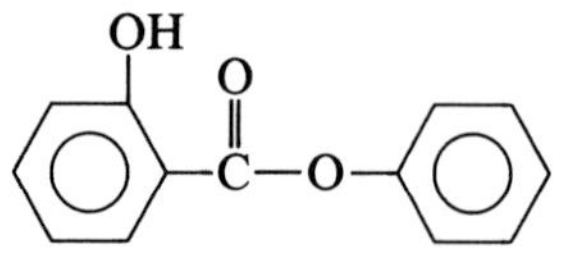	Absorbs UV light very efficiently.
Carbon black	Similar to graphite below, but smaller particles, less-organized structure	Absorbs UV light and radiates energy as heat. Fine for all-black articles.
Reinforcing agents Glass fibers Boron fibers Graphite fibers	SiO_2 units (see Fig. 11-18) Clusters of B_{12} units Hexagonal rings of carbon atoms, joined on all sides in a layered arrangement	Used in polyesters and other plastics to improve strength. Found in car bodies, boats, fishing poles, tennis racquets, bicycle frames, radio antennas, and so on.

panels. General Motors predicts that it will be manufacturing one million plastic-body automobiles per year in the 1990s, up from the current 150,000 per year.

The use of a **foaming agent** in the production of foamed polystyrene was described earlier. Foaming agents are also used with polyurethanes. Foamed polyurethane is used for insulation in refrigerators, refrigerated trucks, and railroad cars, and as construction insulation. It is also used in bedding and car upholstery.

Fillers are small particles of low-cost materials that provide bulk for composites, and they enhance hardness, abrasion resistance, color, flame resistance, and chemical resistance. Up to half the volume of a composite may be filler. Fillers include clays, silica, calcium carbonate, diatomaceous earth, alumina, carbon black, and titanium dioxide. Note that a typical rubber formulation (Table 14–2) has 30.5% carbon black as filler.

UV stabilizers (Table 14–4) prevent light from damaging plastics. Compounds such as phenylsalicylate and carbon black absorb photons in the 290- to 400-nm range of the spectrum (photons with such energy as to break the chemical bonds found in polymers). If enough bonds are broken, the polymer becomes brittle and will break under stress. Of course, with time, even the stabilizers decay, and polymer articles may begin to show visible signs of sunlight degradation.

Plasticizers, such as dioctyl phthalate or dioctyl adipate (Table 14–4), improve the pliability of plastics such as polystyrene, polyethylene, or polypropylene.

DISPOSAL OF PLASTICS

Discarded plastic is contributing to the mounting garbage problem (Fig. 14–17) since most of the present plastic materials are not biodegradable. One method of reducing the decomposition time from decades to around five years is to add cornstarch to the plastics in the production process. Bacteria eat the starch and cause the plastic to break apart into small pieces.

Figure 14–17 **The stability of plastics and their increased use contribute to the mounting problem of garbage disposal. (*The World of Chemistry,* Program 22, "The Age of Polymers.")**

Some plastics can be incinerated without polluting the atmosphere, but others, such as polyvinyl chloride or polyurethane, give off poisonous gases such as HCl and HCN when burned.

Plastics make up 7% of solid wastes by weight but are 30% of the volume of municipal wastes. At present, only about 1% of plastics are being recycled; compared with 30% of aluminum and 20% of paper. Federal legislation is being considered that would authorize a comprehensive recycling program on a national level. The legislation would require the EPA and the Commerce Department to identify and list all consumer items deemed recyclable. Consumer items not identified as recyclable would have to be biodegradable. Many states have already developed recycling programs that offer incentives to bring the recycling of plastics up to a level comparable with that for aluminum cans, glass bottles, and paper. Thermoplastics, which account for 87% of the plastics sold, are the most recyclable form of plastics because they can be remelted and reprocessed. The plastics that make up consumer packaging applications include high-density polyethylene (HDPE), low-density polyethylene (LDPE), polypropylene, poly(ethylene terephthalate) [PET], polystyrene, and polyvinyl chloride (PVC). Because several of these substances are incompatible, recycling them together gives a low-quality product often called "plastic lumber," which can substitute for wood in landscaping timbers, boat docks, and benches. However, to obtain a higher quality material for reuse, plastics must be separated. For example, a 2-L soft-drink bottle has a clear or green body of PET, an opaque base of HDPE, a paper label with adhesive, and an aluminum cap. The bottles are shredded and ground into chips for processing. The processing includes removing the labels and adhesive from the chips and then separation of the lighter HDPE from PET. The remaining PET is then separated from the aluminum. The price of the resulting recycled plastic chips is about half the cost of the original plastic.

THE FUTURE OF POLYMERS

As we have seen in this chapter, the development and use of synthetic polymers is quite recent. Polyethylene, for example, was not discovered until 1933, yet by 1981, its production in the United States amounted to almost 12 billion lb. Chemists are constantly synthesizing new polymers and finding applications for them. The space age has brought with it the need for new polymers, especially in electronics and as special coatings that can withstand high temperatures without breaking down. Among the newcomers are the polyimides, prepared from the polycondensation of a diacid anhydride and a diamine. Some of these polymers have very high service temperatures.

Pyromellitic Dianhydride   1,2-diaminoethane   A Polyimide

Plastic materials are being improved constantly. Some have been made with the strength and rigidity of steel while having only 15% to 20% of the density of steel. The structural strength of such plastics offers the possibility of self-supporting domes for buildings and automobiles that contain more plastic than metal. New low-temperature polymers that don't require the use of a solvent are being developed. An application is to make "spray-on" clothes. Simply spray the monomers onto a mannequin — a little more here, a little less there. Then cut along desired lines, add buttons and zippers, and wear. Some plastics are being developed to replace wood fiber in paper for the printed page. These papers offer smooth surfaces without the graininess of paper, an improvement especially in the quality of microfilming.

Developing plastics that conduct electricity is one of the most exciting recent developments. A conducting polymer is made by doping polymers with small amounts of certain chemicals, similar to the concept used to make semiconductors by doping silicon. A polymer of acetylene shows the most promise as a conducting polymer.

See Chapter 8 for a discussion of doping.

Polyacetylene, a Plastic That Conducts Electricity

Acetylene (C_2H_2), can be polymerized in the presence of a catalyst to produce a polymer with a conjugated double–single bond system, a chain of carbon atoms in which every other bond is a double bond and every other bond is a single bond. This addition polymerization reaction can be represented as follows:

$$2n \ H-C \equiv C-H \longrightarrow \left[\begin{array}{cccc} H & H & H & H \\ | & | & | & | \\ C & = C - C & = C \end{array} \right]_n$$

This polymer appears as a black powder in the usual laboratory preparation and received little attention prior to 1970. In that year a Korean university student, having trouble understanding his Japanese instructor, Hideki Shirakawa, prepared the polymer using an excessive amount of the catalyst. The result was a silver film that looked more like a metal than anything else (Fig. 14–18). Furthermore, the film conducted electricity, which was a first for plastic materials.

The metal-like polyacetylene can be explained in terms of very long conjugated polymer molecules that fit nicely together in a crystalline structure. Also, it has long been known that conjugated carbon systems are adept at passing electric charge from atom to atom along the system.

In 1975, at the University of Pennsylvania, Alan MacDiarmid began a systematic study of this new form of polyacetylene. It was soon learned that the electric conductivity of the plastic could be increased a trillionfold (10^{12}) by the introduction of iodine into the polymer during its formation. The resulting conductivity rivaled that of metals! Recall that iodine atoms have an attraction for one additional electron per atom. If an iodine atom removed an electron from a double bond at one end of the polyacetylene molecule, the entire molecule would simply pass negative charge along the conjugated system and into the "positive hole" if an electron were available at the other end of the molecule. This flow of electric charge is electric

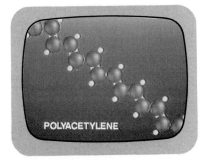

Model of polyacetylene. (*The World of Chemistry*, Program 22, "The Age of Polymers.")

Figure 14–18 A piece of electrically conducting polyacetylene film. (*The World of Chemistry*, Program 22, "The Age of Polymers.")

TABLE 14–5 Energy Requirements to Produce Some Plastics and Metals (Including the Fuel Equivalent of the Monomers)

	Million Btu/ton*
Aluminum	244
Copper	112
Low-density polyethylene	106
High-density polyethylene	96
Polystyrene	64
Polyvinyl chloride	49
Steel	24

* Burning 1 ton of coal produces about 25 million British thermal units of heat energy.

conduction. An applied electric potential could subsequently remove the newly received electron from the iodine atom, and the process could be repeated over and over again.

Following polyacetylene, many similar plastics with useful electrical properties can be conceived theoretically, and many are now being made. Obviously, the applications are very exciting to the many workers in this field. For example, lightweight batteries with plastic components have been made and appear to be commercially exploitable. Think what this might mean since the one great fault of the electric car has been the weight of the lead electrodes in batteries. Another area of intense research is the conversion of sunlight into electric energy; the plastics may offer a much cheaper conversion surface where the energy transfer takes place. Motors, generators, wires, magnets, and electronic parts are all possible applications of the new materials. Will plastic conductors replace metal conductors as synthetic fibers have partially replaced silk and wool? Time will tell.

The long-range future of plastics looks dismal unless we curtail our ravenous appetite for petroleum. Most of the raw materials for plastics come from petroleum and less from coal. Petroleum and coal are the principal sources of energy in this country. Not only are plastics and energy linked through the raw materials–fossil fuels relationship, but also considerable energy is required to purify starting materials and to change them into the desired plastic in the preferred shape. It is the age-old principle that we cannot continue to eat our cake (burn petroleum and coal) and have it, too (use petroleum and coal to make plastics, fibers, and medicines). Table 14–5 illustrates the point that when we consider polymers in the broadest sense in terms of their energy costs of production, we must include their energy value as if they were used as fuel instead of for some object. Considered this way, plastics, fibers, and other items made of polymers derived from petroleum may have only a short history in human existence. Wood, paper, and mineral products such as metals and cement appear either renewable or present in the Earth's crust in far greater abundance than petroleum.

We hope that these and similar problems will be solved as we begin to understand more fully how to use what we have on this planet and how to live in greater harmony with nature.

In the United States, 90% of our petroleum is used as fuel; only 10% is used to make products such as medicines, textiles, and plastics.

SELF-TEST 14–B

1. Nylon is an example of a _____ polymer.
2. Polyamides are formed when _____ is split out from the reaction of many organic acid groups and many amine groups.
3. When an acid such as terephthalic acid

reacts with ethylene glycol ($HOCH_2CH_2OH$), what is the structure of the resulting polymer?
4. Polyesters are formed by ( ) addition ( ) condensation reactions.
5. Many molecules of a carboxylic diacid reacting with many molecules of a dialcohol produce a _____ .
6. Formaldehyde resins are examples of ( ) thermosetting ( ) thermoplastic polymers.
7. The electric conductivity of polyacetylene can be increased a trillionfold by adding small amounts of _____ to the polymer.
8. When $(CH_3)_2SiCl_2$ reacts with water, what is a representative portion of the structure of the polymer?
9. Stabilizers protect plastics against the action of
_____ .
10. Baked-on paints are usually alkyds. This means that they are usually made from polyfunctional _____ and _____ .
11. A plastic that is too stiff can be rendered more flexible by the addition of a _____ .
12. A silicone polymer contains Si— _____ bonds.
13. The burning of plastics containing chlorine, such as polyvinyl chloride, produces what toxic gas?
14. ( ) Ultraviolet ( ) visible light is more destructive to plastics.
15. Which requires more energy per ton to produce? ( ) polyethylene ( ) steel

MATCHING SET

_____ 1. Nylon	a. Plastic that forms interlocking bonds when heated
_____ 2. Block copolymer	b. Rubber
_____ 3. Monomer	c. Cross-linking via reaction with sulfur
_____ 4. Thermoplastic	d. —AAAAABBBBBBAAAAA—
_____ 5. Thermosetting plastic	e. Causes room-temperature-vulcanizing silicone to cross-link
_____ 6. Homopolymer	f. Forms polymers of desired structure
_____ 7. Polymer with a memory	g. —AAAAAAAAAAAAAAAA—
	h. Conducting polymer
_____ 8. Vulcanize	i. A synthetic rubber
_____ 9. Stereochemical control	j. Building unit for a polymer
	k. Plastic softened by heat

_____ **10.** Styrene-butadiene copoly-
mer

_____ **11.** Poly-*cis*-iso-
prene

_____ **12.** Polyester

_____ **13.** Moisture

_____ **14.** Dioctyl phthalate

_____ **15.** Polyacetylene

l. Graphite
m. Formed from a dialcohol and a diacid
n. Possibly harmful plasticizer
o. Natural rubber
p. A polyamide

QUESTIONS

1. In what ways is a railroad train like polystyrene?

2. Where do you suppose the first chemist who prepared a polymer got the idea for giant molecules?

3. What property does a polymer have when it is extensively cross-linked?

4. Describe on the molecular level the end result of the vulcanization process.

5. What is the origin of the word *polymer?*

6. Is polystyrene a thermoplastic or a thermosetting plastic?

7. What property of the molecular structure of rubber allows it to be stretched?

8. Explain how polymers could be prepared from each of the following compounds. (Other substances may be used.)

a. $CH_3-\underset{\underset{H}{|}}{C}=\underset{\underset{H}{|}}{C}-CH_3$

b. $HO-\overset{\overset{O}{\|}}{C}-CH_2-CH_2-\overset{\overset{O}{\|}}{C}-OH$

c. $\underset{\underset{OH}{|}}{CH_2}-\underset{\underset{OH}{|}}{CH}-\underset{\underset{OH}{|}}{CH_2}$

d. $H_2N-CH_2-\bigcirc-CH_2-NH_2$

9. What are the monomers used to prepare the following polymers?

a. $-CH_2CH_2CH_2CH_2CH_2CH_2CH_2CH_2CH_2-$

b. $-\underset{\underset{CH_3}{|}}{CH}CH_2\underset{\underset{CH_3}{|}}{CH}CH_2\underset{\underset{CH_3}{|}}{CH}CH_2-$

c. $-CH_2-\underset{\underset{\bigcirc}{\overset{\overset{H}{|}}{C}}}{}CH_2-\underset{\underset{\bigcirc}{\overset{\overset{H}{|}}{C}}}{}CH_2-\underset{\underset{\bigcirc}{\overset{\overset{H}{|}}{C}}}{}CH_2-\underset{\underset{\bigcirc}{\overset{\overset{H}{|}}{C}}}{}-$

10. Write equations showing the formation of polymers by the reaction of the following pairs of molecules:

a. $\underset{\underset{COOH}{|}}{\overset{\overset{COOH}{|}}{\bigcirc}}$ and $HOCH_2CH_2OH$

b. $HOOCCH_2CH_2COOH$ and $H_2NCH_2CH_2NH_2$

c. $\underset{\underset{CH_2OH}{|}}{\overset{\overset{CH_2OH}{|}}{HCOH}}$ and $\underset{\underset{C-OH}{\overset{\overset{O}{\|}}{}}}{\overset{\overset{C-OH}{\overset{\overset{O}{\|}}{}}}{\bigcirc}}$

11. Is a small molecule eliminated when each monomer unit is added to the chain in addition polymers?

12. Give an example of a copolymer.

13. What structural features must a molecule have in order to undergo addition polymerization?

14. What is meant by the term *macromolecule?*

15. Orlon has a polymeric chain structure of

$-CH_2-\underset{\underset{CN}{|}}{CH}-CH_2-\underset{\underset{CN}{|}}{CH}-CH_2-\underset{\underset{CN}{|}}{CH}-$

What is the monomer from which this structure can be made?

16. Which white pigment is banned in interior paints? Explain.

17. What feature do all condensation polymerization reactions have in common?

18. What type of chemical change takes place during the drying of oil paints?

19. What are the starting materials for nylon 66?

20. Suggest a major difference in the bonding of thermosetting and thermoplastic polymers. Which is more likely to have an interlacing (cross-linking) of covalent bonds throughout the structure? Which is more likely to have weak bonds between large molecules?

21. What is a major difference between silicone oils and silicone rubbers?

22. Explain how a plasticizer can make a polymer more flexible.

23. Name one commercial plasticizer found in food wraps.

24. Describe the properties and structure of Silly Putty.

25. Which is more likely to produce a thermosetting polymer, the monomers of Question 10a or 10c?

26. Draw representative portions of Acrilan® and polyvinyl-chloride.

27. In what way is the structure of ice like that of a cross-linked polymer? How is it different?

28. What single property must a molecule possess in order to be a monomer?

29. Which do you think is the source of most polymers used today, green plants or petroleum? Do you think this will ever change? Explain.

30. Would isoprene make a good motor fuel? Explain.

31. What properties of plastics make them superior to metals? What properties of plastics make them inferior to metals?

32. To what would you attribute the superior thermal stability of silicone polymers?

33. A tiny sample of rubber, held in the flame of a match, burns with a small bright flame and gives a *white* flame in contrast to the black smoke of burning tires. Explain.

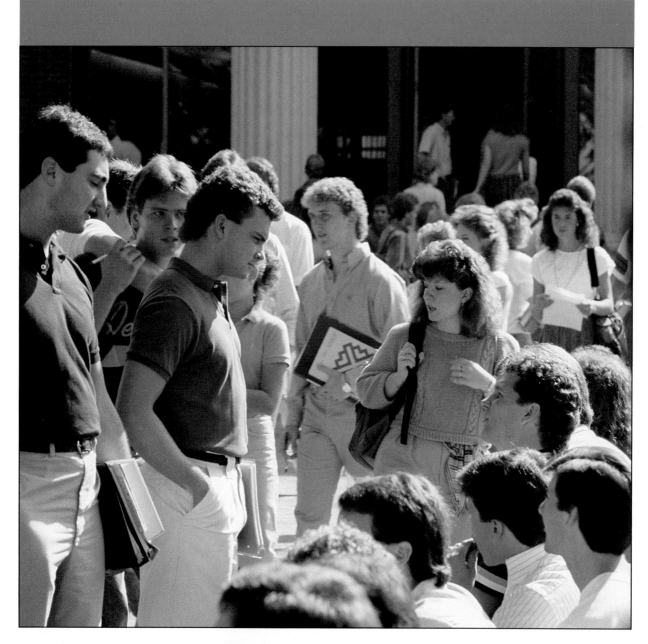

15

Chemistry of Living Systems

The continuance of life requires many chemicals and chemical reactions at all stages of life. (David England, David Lipscomb University.)

Have you ever marveled at the vastness of the universe or the intricacies of the atom? Have you wondered how a huge airplane flies or a computer works? The most marvelous and intricate wonderment of them all is a human being and other living organisms. One cannot help but be amazed at the complicated chemical sequences that occur when, for example, we eat, digest, use, and eliminate a burger, french fries, and a milkshake. **Biochemistry,** the science of life, has helped us make great strides in medicine and health care, and thus in our general well-being.

The purposes of this chapter are to acquaint you with some of the biochemicals in our bodies, to discuss how biochemicals give us energy, and to help you understand genetics from a chemical point of view. Biochemicals are not life inherent; they are simply part of living systems and follow the same laws as other kinds of matter.

Biochemicals common to all living systems are fats and oils, carbohydrates (sugars and starches), proteins, enzymes, vitamins, hormones, nucleic acids, and compounds for the storage and exchange of energy, such as adenosine triphosphate (ATP). In addition to these biochemicals, certain minerals are required for proper functioning of living organisms. Vitamins and minerals are discussed in Chapter 20 on nutrition. Medicines, the chemicals frequently necessary to sustain life and to make life more bearable are discussed in Chapter 21.

Some biochemicals are polymers. Starches are condensation polymers of simple sugars (the monomers); sucrose (table sugar) is composed of only two simple sugars. Proteins are condensation polymers of amino acids (the monomers). Nucleic acids are condensation polymers of simple sugars, nitrogenous bases, and phosphoric acid species.

Other biochemicals are composed of two or more smaller molecular structures. Recall from Chapter 13 that a fat molecule is composed of one glycerol and three fatty acid molecules bonded by ester linkages. Enzymes are constructed of a protein alone or a protein bonded to a metal ion or a vitamin.

Biochemistry embodies the relationships between chemicals and life forms.

Starch, glycogen, cellulose, and proteins are condensation polymers (Chapter 14).

A nitrogenous base is basic because hydrogen ions are attracted to the nonbonding pairs of electrons on nitrogen atoms:

$$-\overset{..}{\underset{|}{N}}-$$

CARBOHYDRATES

Carbohydrates have the three elements carbon, hydrogen, and oxygen arranged primarily into three structural groups: alcohol ($-OH$), aldehyde

$(-\overset{\displaystyle O}{\overset{\displaystyle \|}{C}}H)$, and ketone ($-\overset{\displaystyle O}{\overset{\displaystyle \|}{C}}-$). The hydrogen and oxygen atoms are generally in the ratio of 2 to 1. Carbohydrates are divided into three groups on the basis of the degree of condensation polymerization: monosaccharides (Latin *saccharum,* "sugar"), oligosaccharides, and polysaccharides. Monosaccharides are simple sugars that cannot be dissociated into smaller units by acid hydrolysis. Hydrolysis of a molecule of an oligosaccharide yields two to six molecules of a simple sugar; complete hydrolysis of a polysaccharide molecule produces many (sometimes thousands) monosaccharide monomers.

mono—one
oligo—few
poly—many

Hydrolysis is a water-splitting reaction in which H· bonds with one fragment of the attacked molecule and ·OH bonds with the other fragment.

Oligosaccharides and polysaccharides are composed of a few or many of the ring structures of monosaccharides, whereas monosaccharides alone exist in an equilibrium between the linear structure and a ring structure (Figs. 15–1 through 15–5). The equilibrium lies far toward the ring structure.

Glucose ($C_6H_{12}O_6$) and some of the other simple sugars are quick sources of energy for cells. Large amounts of energy are stored in polysaccharides, such as starch. The stored energy is usable by living cells only if polysaccharides are hydrolyzed into monosaccharides. Some complex polysaccharides are used by some organisms for structural purposes. Cellulose, for example, is partially responsible for plant support.

Monosaccharides

D-glucose has a relative sweetness of 74.3, compared with sucrose, which has an assigned value of 100.0. The sweetness value of fructose is 173.3.

Sweetness is judged by taste testers.

Approximately 70 monosaccharides are known; 20 occur naturally. The most common simple sugar is D-glucose (Fig. 15–1), which is found in fruit, blood, and living cells. As can be seen from the structure of D-glucose, the great solubility of monosaccharides in water is caused by the numerous —OH groups, which hydrogen-bond with water. An aqueous solution of D-glucose contains all three structures shown in the figure in dynamic equilibria involving mostly the two ring forms in the presence of a relatively small amount of the straight-chain form. The aldehyde group in the straight-chain structure of D-glucose qualifies this sugar as an **aldose** monosaccharide.

Also known as dextrose, grape sugar, and blood sugar, D-glucose is used in the manufacture of candy and in commercial baking. A solution of D-glucose is often given intravenously when a quick source of energy is needed to sustain life.

Figure 15–1   The structures of D-glucose; d and e are two-dimensional representations of b and c, respectively. Note the difference in the positions of the —OH groups *(color)* in the α and β forms of glucose: the —OH groups on the 1 and 4 carbons are *trans* when the structure is β, and the —OH groups are *cis* when the structure is α. In both α and β glucose, the —OH group on the number-4 carbon atom must be in the same position.

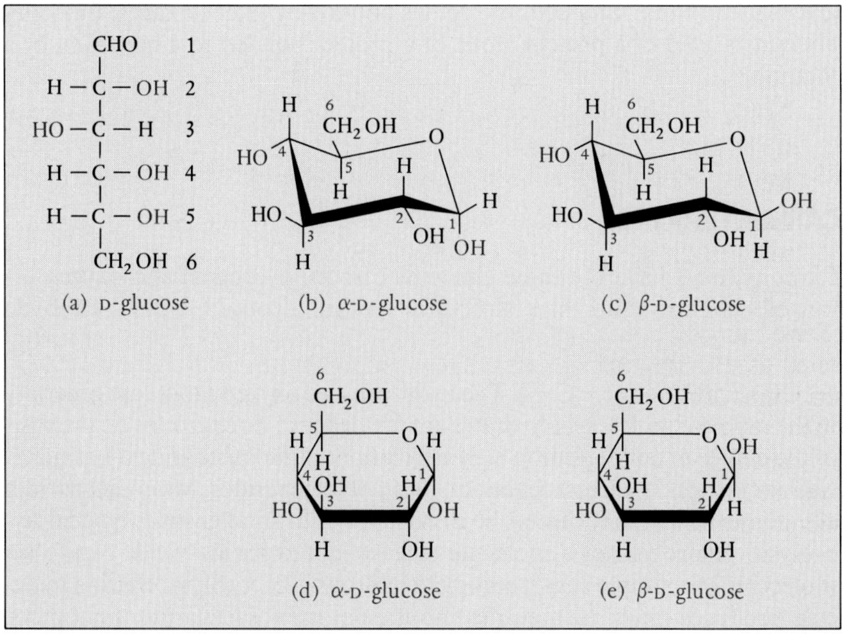

(a) D-glucose   (b) α-D-glucose   (c) β-D-glucose

(d) α-D-glucose   (e) β-D-glucose

Recall from Chapter 12 that a carbon atom is at each vertex not occupied by another atom.

Refer to Chapter 12 for a discussion of *cis* and *trans* (geometric) isomers.

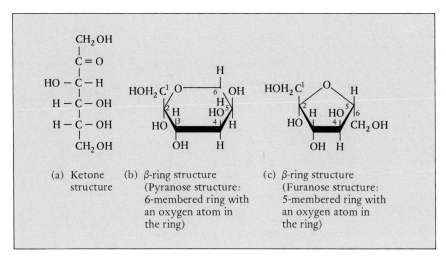

(a) Ketone structure

(b) β-ring structure (Pyranose structure: 6-membered ring with an oxygen atom in the ring)

(c) β-ring structure (Furanose structure: 5-membered ring with an oxygen atom in the ring)

Because D-fructose, a monosaccharide found in many fruits and table sugar, has a ketone group in its straight-chain form (Fig. 15-2), it is classified as a **ketose** monosaccharide.

Glucose and fructose are hexose monosaccharides because they have six carbons in a row in the chain form. Two essential pentoses (five carbon atoms in a row), are part of the heredity material, DNA, and are discussed later in this chapter.

Oligosaccharides

The most commonly encountered oligosaccharides are the disaccharides, which have two simple sugar monomers per molecule. Examples include three widely used disaccharide sugars:

Sucrose (from sugar cane or sugar beets), which consists of a glucose monomer and a fructose monomer

Maltose (from starch), which consists of two glucose monomers

Lactose (from milk), which consists of a glucose monomer and a galactose (an optical isomer of glucose) monomer

The formula for these disaccharides ($C_{12}H_{22}O_{11}$), is not simply the sum of two monosaccharides, $C_6H_{12}O_6 + C_6H_{12}O_6$. A water molecule is eliminated as two monosaccharides are united to form the disaccharide. The structures of sucrose, maltose, and lactose, along with their hydrolysis reactions, are shown in Figure 15-3.

Sucrose is produced in a high state of purity on an enormous scale— over 80 million tons per year. Originally produced in India and Persia, sucrose is now used universally as a sweetener. About 40% of the world sucrose production comes from sugar beets and 60% from sugar cane. Sucrose provides a high caloric value (1794 kcal/lb); it is also used as a preservative in jams, jellies, and candied fruit.

Bags of dextrose. (*The World of Chemistry,* Program 11, "The Mole.")

The structure of galactose is shown on page 450.

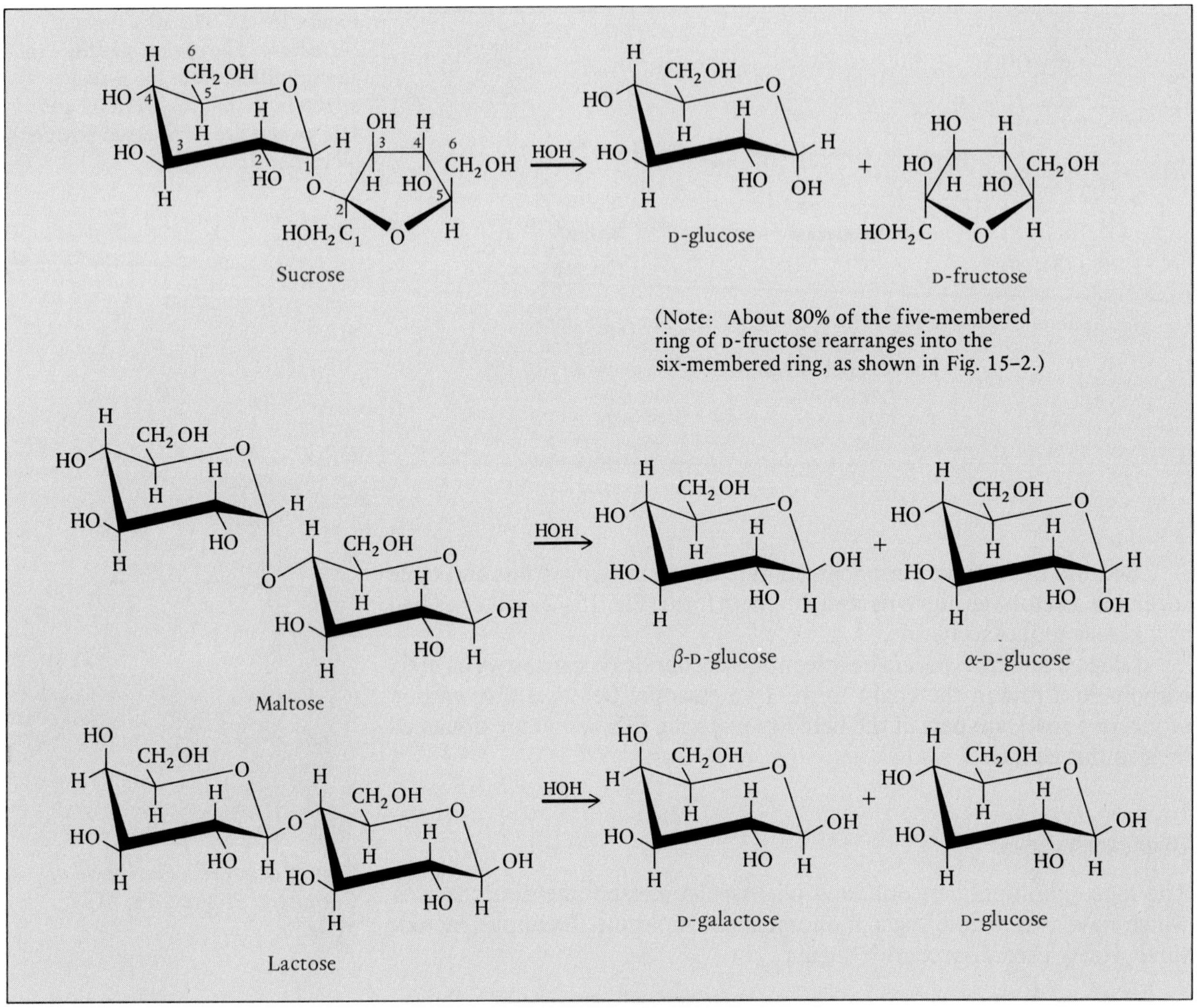

Figure 15–3   Hydrolysis of disaccharides (sucrose, maltose, and lactose).

The polar alcohol (—OH) groups on the disaccharides and monosaccharides cause hydrogen bonding with water. This is why table sugar dissolves readily in coffee and tea, and why glucose is transported easily by the blood.

Polysaccharides

Polysaccharides are condensation polymers.

Nature's most abundant polysaccharides are the starches, glycogen, and cellulose. Some molecular structures are known to combine more than 5000 monosaccharide monomers into molecules with molecular weights of over 1 million. The monosaccharide most commonly used to build polysaccharides is D-glucose.

Starches and Glycogen

Plant starch is found in protein-covered granules. If these granules are ruptured by heat, they yield a starch that is soluble in hot water, **amylose,** and an insoluble starch, **amylopectin.** Amylose constitutes about 25% of most natural starches. When tested with iodine solution, amylose turns blue-black, whereas amylopectin turns red.

Structurally, amylose is a straight-chain condensation polymer with an average of about 200 α-D-glucose monomers per molecule. Each monomer is bonded to the next with the loss of a water molecule, just as the two units are bonded in maltose (Fig. 15–3). A representative portion of the structure of amylose is shown in Figure 15–4.

A typical amylopectin molecule has about 1000 α-D-glucose monomers arranged into branched chains (Fig. 15–5). Complete hydrolysis yields D-glucose; partial hydrolysis produces mixtures called **dextrins.** Dextrins are used as food additives and in mucilage, paste, and finishes for paper and fabrics.

Glycogen is an energy reservoir in animals, just as starch is in plants. The α-glucose chains in glycogen are more highly branched than the chains in amylopectin.

Starch molecules consist of many glucose monomers bonded together.

Cellulose

Cellulose is the most abundant polysaccharide in nature. Like amylose, it is composed of D-glucose units. The difference between the structures of cellulose and amylose lies in the bonding between the D-glucose units; in cellulose all of the glucose units are in the β-ring form, whereas in amylose they are in the α-ring form. (Review the ring forms in Fig. 15–1 and compare the structures in Figs. 15–4 and 15–6). This subtle structural difference between starch and cellulose causes their differences in digestibility. Human beings and carnivorous animals do not have the necessary enzymes (biochemical catalysts) to hydrolyze cellulose, as do numerous microorganisms (such as bacteria in the digestive tracts of termites).

D-glucose can be obtained from cellulose by heating a suspension of the polysaccharide in the presence of a strong acid. At present wood cannot be hydrolyzed into food (D-glucose) economically enough to satisfy the world's growing need for an adequate food supply.

Human beings do not have an enzyme to hydrolyze cellulose into its glucose monomers.

Figure 15–4   Amylose structure. From 60 to 300 α-D-glucose units are bonded together by α linkages to form amylose molecules. In α linkages, only the α structure of glucose is used. The bonding is between monomers at the 1 and 4 carbon atoms (see Fig. 15–1). The —OH groups on the 1 and 4 carbon atoms are *cis* in α-glucose, leading all bonds between α-glucose monomers ($\diagdown$O$\diagup$) to point in the same direction.

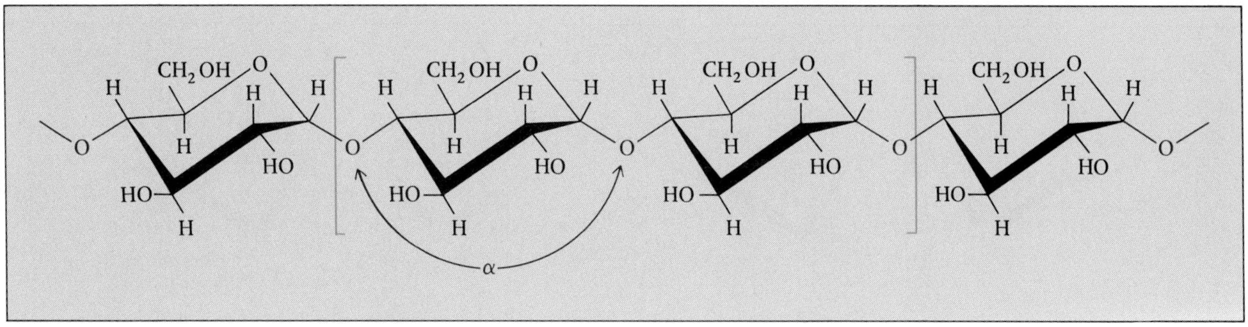

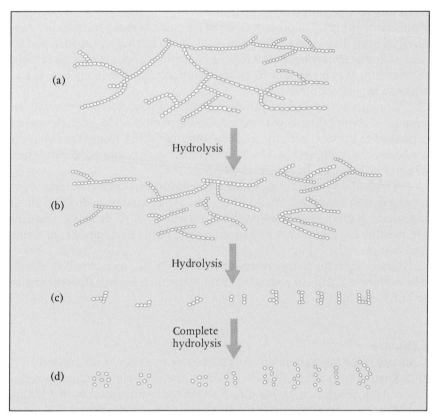

Figure 15–5 (a) Partial schematic amylopectin structure. (b) Dextrins from incomplete hydrolysis of a. (c) Oligosaccharides from hydrolysis of dextrins. (d) Final hydrolysis product: D-glucose. Each circle represents a glucose unit.

Figure 15–6 Cellulose structure. About 2800 β-D-glucose units are bonded together by β linkages to form an unbranched cellulose structure. Cellulose contains only the β form of glucose. The —OH groups on the 1 and 4 carbon atoms (see Fig. 15–1) are *trans* in β glucose, leading the bonds between β-glucose monomers (⌒O⌒ and ⌒O⌒) to alternate in direction. Compare cellulose with amylose (Fig. 15–4). Note in cellulose that every other β-glucose monomer is turned over. In amylose all α-glucose monomers are in the same position.

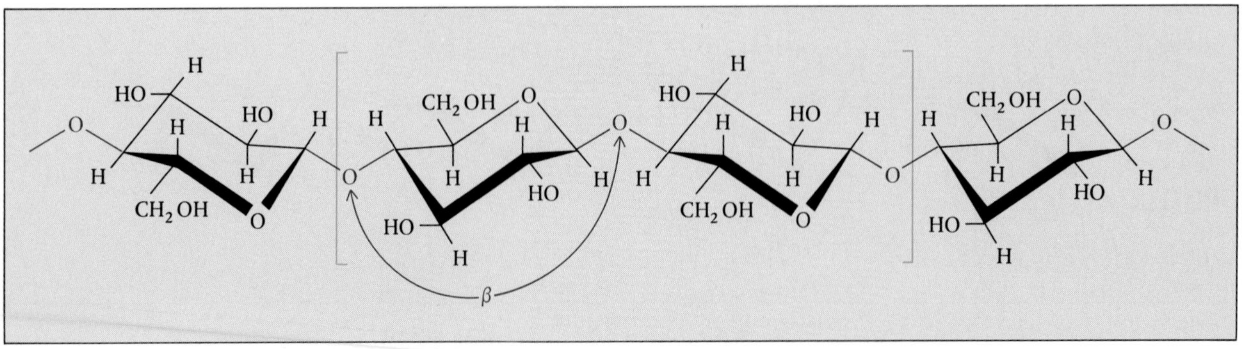

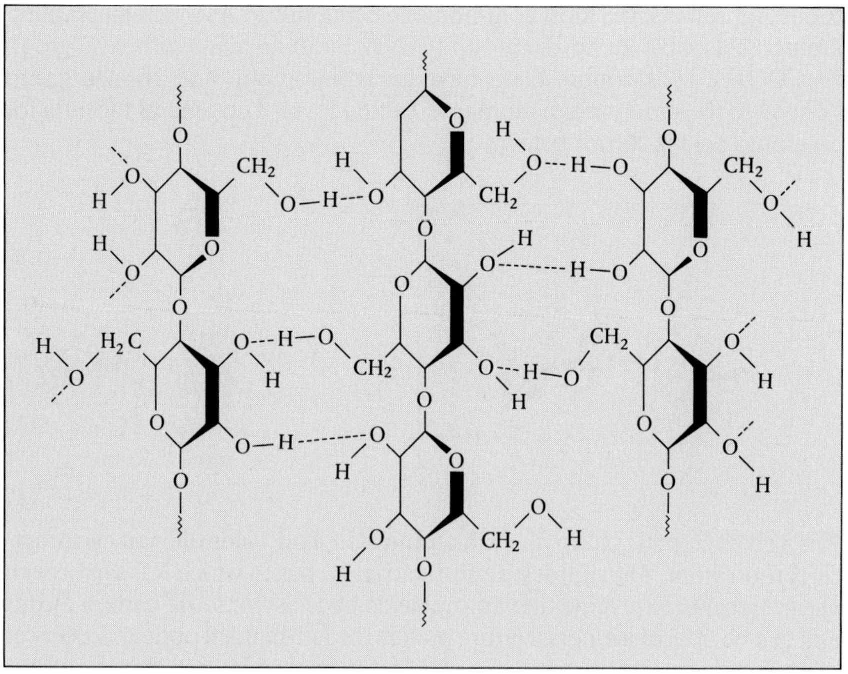

Figure 15–7 The properties of cotton, which is about 98% cellulose, can be explained in terms of this submicroscopic structure. A small group of cellulose molecules, each with 2000 to 9000 units of D-glucose, are held together in an approximately parallel fashion by hydrogen bonding (-----). When several of these *chain bundles* cling together in a relatively vast network of hydrogen bonds, a *microfibril* results; the microfibril is the smallest microscopic unit that can be seen. The macroscopic *fibril* is a collection of numerous microfibrils. The absorbent nature of cotton results from the numerous capillaries wherein the smaller water molecules are held by hydrogen bonds.

Paper, rayon, cellophane, and cotton are principally cellulose. A representative portion of the structure of cotton is shown in Figure 15–7. Note the hydrogen bonding between cellulose chains.

SELF-TEST 15–A

1. Carbohydrates contain the elements _____, _____, and _____.
2. The complete hydrolysis of a polysaccharide yields _____.
3. When a molecule of sucrose is hydrolyzed, the products are one molecule each of the monosaccharides _____ and _____.
4. The sugar referred to as blood sugar, grape sugar, or dextrose is actually the compound _____.
5. Starch is a condensation polymer built of _____ monomers.
6. What type of bonding holds polysaccharide chains together, side by side, in cellulose?
7. What kind of bonding enables sugar to dissolve in coffee?

PROTEINS

Proteins are condensation polymers of **amino acids.** The twenty different amino acids that can be found in proteins are made primarily from carbon, oxygen, hydrogen, and nitrogen. Small amounts of other elements are also

Proteins are high-molecular-weight compounds made up of amino acid monomers.

found in proteins, the most common one being sulfur. As the name implies, amino acids have an amino group ($-NH_2$) and an acid (carboxyl) group ($-COOH$). Most amino acids have an amine group and an acid group bonded to the same carbon atom (see Table 15–1). The general formula for an amino acid is shown below:

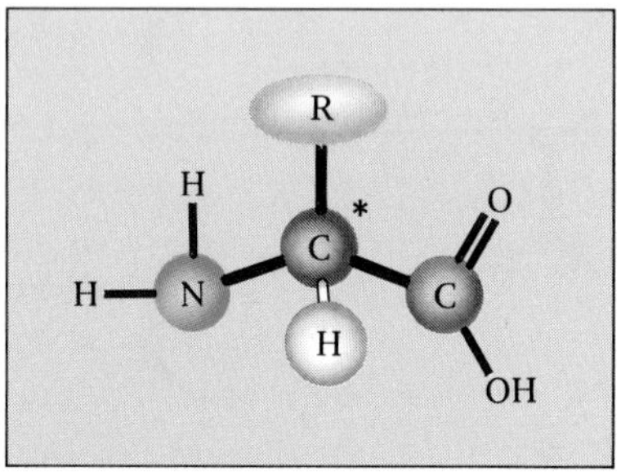

R is a characteristic group for each amino acid, and * identifies an asymmetric carbon atom. The simplest amino acid is **glycine,** in which R is a hydrogen atom. Except for glycine, the amino acids have asymmetric carbon atoms and can be optical isomers. Nature prefers the left-handed optical isomers of amino acids in protein synthesis, which is discussed later in this chapter.

The close relationship between proteins and living organisms was first noted in 1835 by the German chemist G. T. Mulder. He named proteins from the Greek **proteios** ("first"), thinking that proteins are the starting point for a chemical understanding of life. Proteins play a role in a wide variety of functions, including motion of organisms, defense mechanisms against foreign substances, makeup of enzymes, and makeup of the all-important cell wall. Each unique kind of protein is composed of several specific amino acids arranged in a definite molecular structure. In a few proteins the major fraction is only one kind of amino acid; the protein in silk, for example, is 44% glycine.

The **essential amino acids** must be ingested from food; they are indicated by asterisks in Table 15–1. The other amino acids can be synthesized by the human body.

Amino acid monomers are bonded together by **peptide bonds.** The chemical reaction is an acid–base reaction in which two monomers bond and water is split out. For example, when two glycine molecules react, a peptide bond is formed and a water molecule is produced:

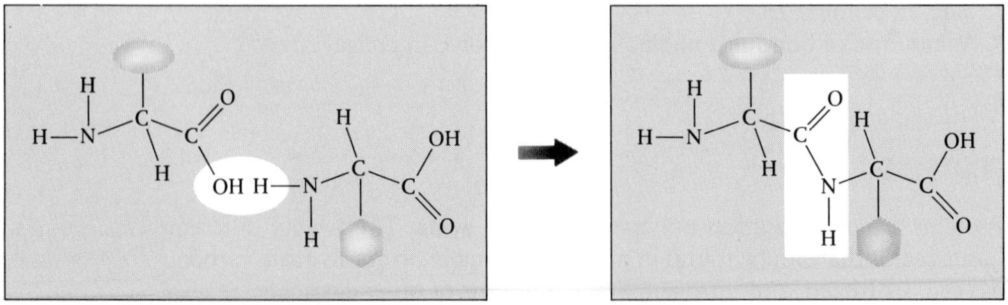

Review the discussion of optically active amino acids in Chapter 12.

For good nutrition we require *all* of the essential amino acids in our daily diet, but the amount required does not exceed 1.5 g per day for any of them.

Peptide bonds form polyamides like nylon 66 (Chapter 14).

TABLE 15–1 Common L-amino acids found in proteins. In some listings, two others are included: the amide of aspartic acid, asparagine (R group is —CH_2CONH_2) and the amide of glutamic acid, glutamine (R group is —$CH_2CH_2CONH_2$).

All of the amino acids except proline and hydroxyproline have the general formula

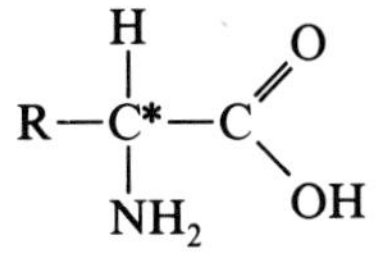

$$R-\overset{\overset{\displaystyle H}{|}}{\underset{\underset{\displaystyle NH_2}{|}}{C^*}}-C\underset{\displaystyle OH}{\overset{\displaystyle O}{\diagup}}$$

in which R is the characteristic group for each acid. The R groups are as follows.

1. Glycine—H
2. Alanine—CH_3
3. Serine—CH_2OH
4. Cysteine—CH_2SH
5. Cystine—CH_2—S—S—CH_2—
*6. Threonine—$\underset{\displaystyle OH}{\overset{\displaystyle |}{CH}}$—$CH_3$

*7. Valine CH_3—$\overset{\displaystyle |}{CH}$—$CH_3$

*8. Leucine—CH_2—$\underset{\underset{\displaystyle CH_3}{|}}{CH}$—$CH_3$

*9. Isoleucine—$\underset{\underset{\displaystyle CH_2-CH_3}{|}}{\overset{\overset{\displaystyle CH_3}{|}}{CH}}$

*10. Methionine—CH_2—CH_2—S—CH_3
11. Aspartic acid—CH_2CO_2H
12. Glutamic acid—CH_2—CH_2—CO_2H
*13. Lysine—CH_2—CH_2—CH_2—CH_2—NH_2

*14. Arginine—CH_2—CH_2—CH_2—$NH\overset{\overset{\displaystyle NH}{||}}{C}NH_2$

*15. Phenylalanine—CH_2—⬡

16. Tyrosine—CH_2—⬡—OH

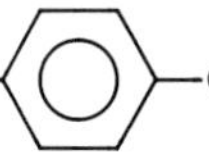

*17. Tryptophan—CH_2—[indole ring]

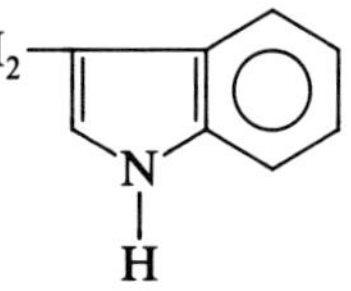

*18. Histidine—CH_2—[imidazole ring]

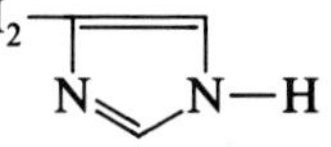

The structures for the other two are:

19. Proline H_2C——CH_2
   H_2C ╲   ╱$CHCO_2H$
        N
        |
        H

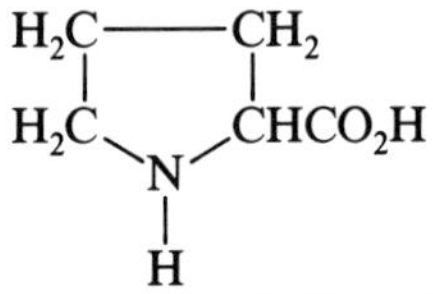

20. Hydroxyproline   HOHC——CH_2
   H_2C ╲   ╱$CHCO_2H$
        N
        |
        H

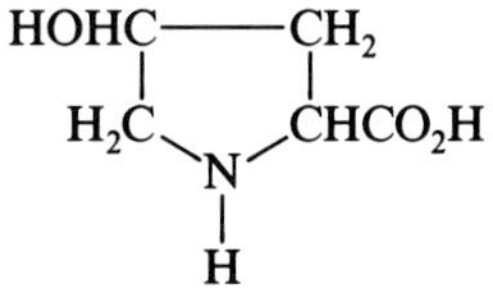

* Essential amino acids; arginine and histidine are essential for children but may not be essential for adults.

Part of the uniqueness of each human being is caused by the uniqueness of some of that person's protein structures. For billions of people, this implies a tremendously large number of protein structures.

When two different amino acids are bonded, two different combinations are possible, depending on which amine reacts with which acid group. For example, when glycine and alanine react, both glycylalanine and alanylglycine can be formed.

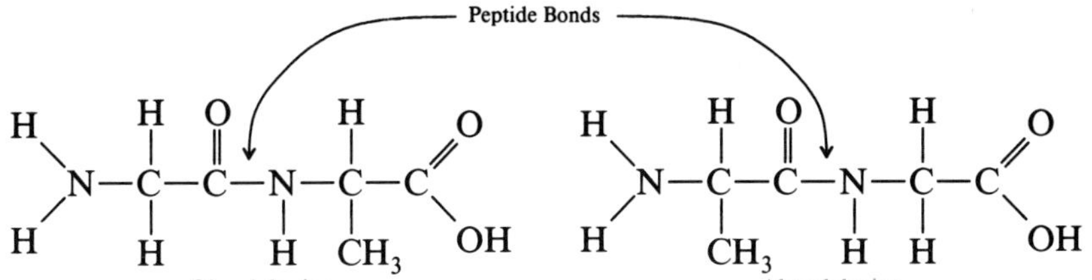

Glycylalanine   Alanylglycine

Figure 15–8 **(a) Helical structure for a polypeptide in which each oxygen atom can be hydrogen-bonded to a nitrogen atom in the third amino acid unit down the chain. (b) α-helix structure of proteins. The sketch represents the actual position of the atoms and shows where intrachain hydrogen bonds occur (dotted line).**

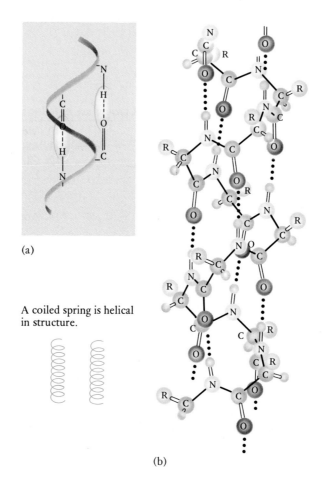

(a)

A coiled spring is helical in structure.

(b)

A very large number of different proteins can be prepared from a small number of different amino acids.

Twenty-four **tetra**peptides are possible if four amino acids (for example, glycine, Gly; alanine, Ala; serine, Ser; and cystine, Cys) are linked in all possible combinations.* They are:

Gly-Ala-Ser-Cys	Ala-Gly-Ser-Cys	Ser-Ala-Gly-Cys	Cys-Ala-Gly-Ser
Gly-Ala-Cys-Ser	Ala-Gly-Cys-Ser	Ser-Ala-Cys-Gly	Cys-Ala-Ser-Gly
Gly-Ser-Ala-Cys	Ala-Ser-Gly-Cys	Ser-Gly-Ala-Cys	Cys-Gly-Ala-Ser
Gly-Ser-Cys-Ala	Ala-Ser-Cys-Gly	Ser-Gly-Cys-Ala	Cys-Gly-Ser-Ala
Gly-Cys-Ser-Ala	Ala-Cys-Gly-Ser	Ser-Cys-Ala-Gly	Cys-Ser-Ala-Gly
Gly-Cys-Ala-Ser	Ala-Cys-Ser-Gly	Ser-Cys-Gly-Ala	Cys-Ser-Gly-Ala

If 17 different amino acids are bonded, the sequences alone make 3.56×10^{14} (356 trillion) uniquely different 17-monomer molecules. Many more combinations can be made when more than one molecule of each amino acid is taken. However, of the many different proteins that could be

---

* If the amino acids are all different, the number of arrangements in $n!$ (n factorial). For five different amino acids, the number of different arrangements is 5! (or $5 \times 4 \times 3 \times 2 \times 1 = 120$).

made from a set of amino acids, a living cell will make only the relatively small, select number it needs.

There are several distinguishing characteristics in the structures of proteins. The sequence of amino acids bonded to one another in a chain is the **primary structure.** The twisting of the amino acid chain into a helical shape is a **secondary structure** (Fig. 15–8). Hydrogen bonds hold the helices in place as a nitrogen atom hydrogen-bonds with the oxygen atom in the third amino acid down the chain.

Another secondary structure of proteins is like a sheet in which several chains of amino acids are joined side to side by hydrogen bonds (Fig. 15–9). Most of the properties of silk can be explained by its sheetlike structure.

In addition to hydrogen bonding between two different primary chains of amino acids and hydrogen bonding between an amine group and a carboxyl group at separated sites on a single twisted, curled primary chain, there are three other major types of interactions between R groups on the amino acids in a primary protein chain. London forces can hold two nonpolar hydrocarbon groups together. A disulfide bridge (—S—S—) bonds two sites together if the amino acid cystine is included in the sequence of the

chains. An ionic bond such as $-NH_3^+ \ ^-O-\overset{\overset{\displaystyle O}{\displaystyle \|}}{C}-O^-$ is effective in holding two chains together if the pH is such to form the ionic groups. The disulfide bridge and the ionic bond are structural features of hair and are important in permanent-waving. (See Chapter 22.)

All of these interactions between R groups on the amino acids in a primary chain contribute to the third level of protein structure, the **tertiary structure.** One type of tertiary structure is found in **collagen,** a fibrous protein tissue. Three amino acid chains twisted into left-handed helices are then

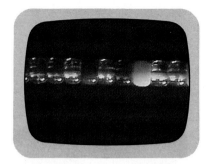

The primary structure of a protein is the sequence of amino acids in a chain like the sequence of beads on a string. (*The World of Chemistry,* Program 23, "Proteins: Structure and Function.")

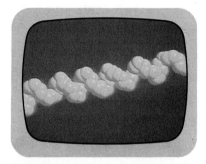

An α-helix structure of protein. (*The World of Chemistry,* Program 23, "Proteins: Structure and Function.")

Figure 15–9   An example of the secondary sheet structure of proteins as found in silk. The primary strands of amino acids are held together by hydrogen bonds.

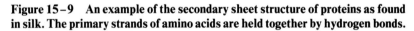
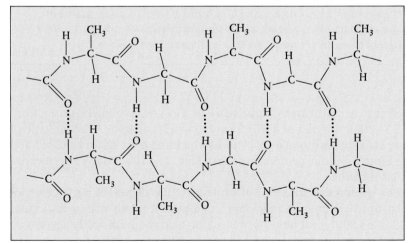

UNRAVELING THE PROTEIN STRUCTURE

One of the key steps in unraveling the mystery of hydrogen bonds in protein structure involved Linus Pauling, a cold, and a Nobel Prize. Pauling presently lives in the Big Sur region of Northern California. His living room is his office. There he spends a large part of each day at a simple desk, working on a new research interest, metals. Earlier in his career, Pauling had another interest, the structure of protein molecules. At that time there were several conflicting theories. Pauling and his colleagues thought that the first level of protein structure was a polypeptide chain. Then they asked themselves a fundamental question.

We asked: How is the polypeptide chain folded? We couldn't answer the question, but we said it's probably held together by hydrogen bonds. The conclusion we reached was that there are . . . polypeptide chains in the protein, which, far longer if they were stretched out than the diameter of the molecule, are coiled back and forth; and that they are coiled into a very well defined structure, configuration, with the different part of the chain held together by hydrogen bonds. In 1937, I spent a good bit of the summer with models for—I assumed that I knew what a polypeptide chain looks like except for the way in which it's folded. And I wanted to fold it to form the hydrogen bonds. I didn't succeed. The fact is, I thought that there was something about proteins that perhaps I didn't know.

Pauling continued to work on this problem, but the solution eluded him. Then one day he had a crucial insight in a completely different and unexpected setting.

I had a cold. I was lying in bed for two or three days, and I read detective stories, light reading, for awhile, and then I got sort of bored with that. So I said to my wife, "Bring me a sheet of paper, and I'm going to—I think I'll work on that problem of how polypeptide chains are folded in proteins. So she brought me a sheet of paper and the slide rule and pencil, and I started working.

Using the knowledge gained from his years of model building, he drew the backbone of a polypeptide chain on a piece of paper. Then it occurred to him to try to fold the paper to see how hydrogen bonds could form along the polypeptide chain. The result was a structure that twisted around like a spring.

Well, I succeeded. It only took a couple of hours of work that day, March of 1948, for me to find the structure, called the alpha helix.

The World of Chemistry (Program 23) "Proteins: Structure and Function."

Linus Pauling (b. 1901), along with R. B. Corey, proposed the helical and sheetlike secondary structures for proteins. For his bonding theories and for his work with proteins, Pauling was awarded the Nobel Prize in 1954. For his fight against nuclear danger, he received the 1963 Nobel Peace Prize. (*The World of Chemistry*, Program 23, "Proteins: Structure and Function.")

twisted into a right-handed superhelix to form an extremely strong fibril (Fig. 15–10a). Bundles of fibrils form the tough collagen. A second type of tertiary structure is globular protein in which the helix chain is folded and twisted into a definite geometric pattern (Fig. 15–10b). Many enzymes are globular proteins.

The **quaternary structure** of proteins is the degree of aggregation of protein units. Human hemoglobin, a globular protein with a molecular weight of 68,000, must have its four amino acid chains properly aggregated

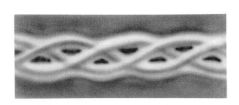

(a)

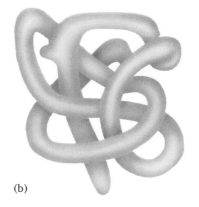

(b)

Figure 15–10   Tertiary molecular structures of proteins. (a) The imaginary twisted structure of collagen. (b) The imaginary folded structure of the helix in a globular protein.

in order to form active hemoglobin. Insulin is also composed of subunits of protein properly arranged into its quaternary structure.

How important is structure to protein? If hemoglobin, for example, has an abnormal primary, secondary, tertiary, or quaternary structure because of a wrong amino acid in a given position, it may be unable to transfer oxygen through the bloodstream. The cause of **sickle cell anemia** is the alteration of only one specific amino acid of the 146 amino acid units in a single hemoglobin chain.

Hemoglobin carries oxygen and carbon dioxide in the bloodstream and helps control pH.

(a)

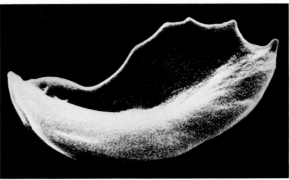

(b)

Figure 15–11   (a) Normal red blood cells. (b) Sickle red blood cells. (c) Sickle cells are caused by the substitution of the nonpolar amino acid valine for the negatively charged amino acid glutamate in the protein structure of hemoglobin. This substitution produces a crucial alteration in the tertiary structure, which causes the sickling. (a and b from J. R. Holum: *Fundamentals of General, Organic, and Biological Chemistry,* 2nd ed., p. 487. New York, John Wiley & Sons, Inc., 1982. c is from *The World of Chemistry,* Program 24, "Genetic Code.")

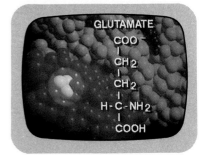

(c)

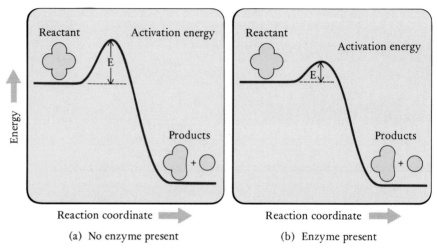

Figure 15–12   Effect of enzyme activity on activation energy. The vertical axis represents energy, and the horizontal axis represents time. For energy-producing reactions, the reactant molecules are at a higher energy than the product molecules. It is necessary for the reactant molecules to "get over" the energy barrier (acquire the activation energy, E) in going from being reactants to being products. The function of an enzyme is to lower the activation energy as illustrated in (b), and thereby to speed up the reaction.

ENZYMES

In 1926 at Cornell, James B. Sumner (1887–1955) separated, crystallized, identified, and characterized the first enzyme, urease. Sumner had been advised not to enter the field of chemistry because he had only one arm. In 1946 he won the Nobel Prize.

The names of most enzymes end in -ase.

Enzymes function as catalysts for chemical reactions in living systems. Each enzyme performs a specific catalytic task. As we shall discuss later, a major part of the structure of an enzyme is globular protein. Like all catalysts, enzymes increase the rate of a reaction by weakening bonds and causing a lowering of the **energy of activation** (Fig. 15–12). The action of an enzyme on a chemical reaction is similar to the effect of a key opening a lock (Fig. 15–13). The lock can be opened without the key by using more energy (i.e., the lock can be broken). Similarly, the reaction will occur without the enzyme, but at a much slower rate. The enzyme makes the procedure easier and faster. For example, enzyme-catalyzed action allows a single molecule of β-amylase to catalyze the breaking of bonds between the α-glucose monomers in amylose at the rate of 4000 per second.

Most enzymes are very specific. The enzyme maltase hydrolyzes maltose into two molecules of D-glucose. This is the only function of maltase, and no other enzyme can substitute for it. Sucrase, another enzyme, hydrolyzes only sucrose. Some enzymes are less specific. The digestive enzyme trypsin, for example, primarily hydrolyzes peptide bonds in proteins. However, the structure and polarity of trypsin are such that it can also catalyze the hydrolysis of some esters.

Some enzymes are globular proteins only. Other enzymes are globular proteins plus either a metal ion (e.g., Co^{3+}, Fe^{3+}, Mg^{2+}, or another essential

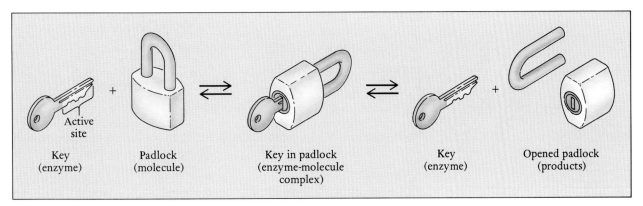

Figure 15–13   Lock-and-key theory for enzymatic catalysis. Although the analogy is an oversimplification, one very important point is made; the enzyme makes a difficult job easy by reducing the energy required to get the job started. The analogy also suggests that the enzyme has a particular structure at an active site that will allow the enzyme to work only for certain molecules, similar to a key that fits the shape of a particular keyhole and a particular sequence of tumblers inside the lock.

mineral) or a vitamin. The vitamin or the mineral is the **coenzyme,** and the protein is the **apoenzyme.** Both parts are needed for enzymatic activity, just as two keys are required to open a bank lock-box. Neither your key nor the bank's key alone will open the box; both are needed. The B vitamins are coenzymes in various oxidative processes in the human body. For example, niacin (vitamin B_3) is part of a larger molecule, **nicotinamide adenine dinucleotide (NAD^+)**; NAD^+ serves as a coenzyme in concert with the apoenzyme, a globular protein. Riboflavin (vitamin B_2) is part of the coenzyme **flavin adenine dinucleotide (FAD).**

The action of an enzyme is shown in Figure 15–14. The reactant molecule is the **substrate.** Enzymes and substrates have electrically polar regions, partially charged groupings, or ionic sections that attract and guide the enzyme and substrate together; these regions of chemical activity are the **active sites.** Substrates sit down on active sites on enzymes in assembly-line fashion at a remarkably fast rate. For example, enzymes renew 3 million red blood cells in the human body *every second.*

Genetic effects are often observed in the pattern of enzymes produced by individuals or races. An example of this is found in "lactose intolerance," common in certain peoples of Asia (e.g., Chinese and Japanese) and Africa (many black tribes), whose diets have traditionally contained little milk after the age of weaning. While infants, such people manufacture the enzyme **lactase** that is necessary to digest lactose, a sugar occurring in all mammals' milk. As they grow older, their bodies stop producing this enzyme because their diets normally contain no milk, and the ingestion of milk products containing lactose can lead to considerable discomfort in the form of stomach aches and diarrhea. People whose ancestral adult diets contained substantial amounts of milk or milk products (African tribes such as the Masai), continue to produce lactase as adults and can eat such foods and digest the

Why must we have minerals and vitamins? Answer: In part, because vitamins and minerals serve as coenzymes.

Niacin prevents pellagra.

Riboflavin promotes growth, healthy eyes and skin, and the oxidation of foods.

Enzyme structure is the key to specific catalytic activity.

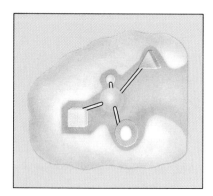

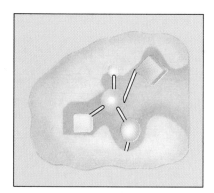

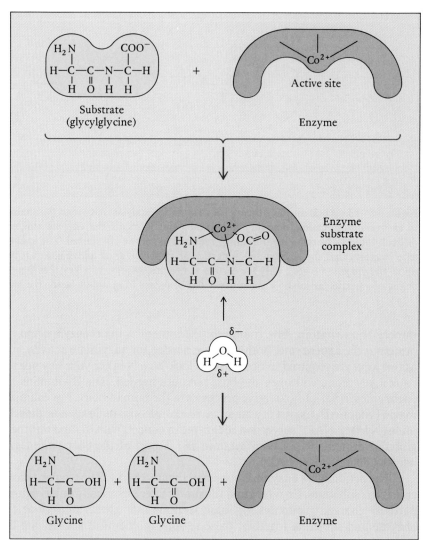

Figure 15–14 (a) The enzyme is active when it fits the substrate at the active sites. (b) The enzyme is inactive when there is no fit with the substrate at the active sites. (c) Action of an enzyme. The substrate molecule is chemically bonded to the enzyme (glycylglycine dipeptidase). The negative oxygen and the nitrogen atoms of the substrate bond to the positive cobalt ion (Co^{2+}) in the enzyme. The bonding of the substrate makes it more susceptible to attack by water. *Hydrolysis* occurs and the glycine molecules are released by the enzyme, which is then ready to play its catalytic role again.

lactose they contain. It is quite possible that this is only one of several similar cases in which a traditional tribal diet has altered the pattern of production of digestive enzymes.

SELF-TEST 15–B

1. The fundamental building units in proteins are the _____.
2. Amino acids that the body cannot synthesize from other molecules are called _____.

3. The peptide linkage that bonds amino acids together in protein chains has the structure _____.

4. The basic structure present in almost all of the amino acids can be represented as _____.

5. **a.** The primary structure of a protein refers to its _____;
   b. the secondary structure refers to its _____;
   c. its tertiary structure refers to _____;
   d. and its quaternary structure refers to _____.

6. **a.** If we have three different amino acids and can use each three times in any given tripeptide, we can make a total of _____ different tripeptides.
   b. If we can use each amino acid only once, there are still _____ possible different tripeptides.

7. Describe how hydrogen bonding is involved in the secondary structures of proteins.

8. The best term to describe the general function of enzymes is ( ) *catalyst,* ( ) *intermediate,* ( ) *oxidant.*

9. In the lock-and-key analogy of enzyme activity, the enzyme functions as the _____, and the substrate molecule serves as the _____.

10. The activation energy of many biological reactions is decreased if a(n) _____ is present.

11. Apoenzyme + coenzyme $\rightarrow$ _____

12. That portion of the enzyme at which the reaction is catalyzed is called the _____.

13. The four major types of interactions between chains of R groups on amino acids in a primary protein chain are _____, _____, _____, and _____.

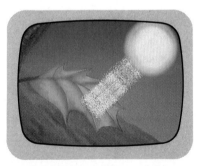

Photosynthesis requires sunlight, chlorophyll, carbon dioxide, and water. (*The World of Chemistry,* Program 17, "The Precious Envelope.")

ENERGY AND BIOCHEMICAL SYSTEMS

Energy for life's processes comes from the Sun. During photosynthesis, green plants absorb energy from the Sun to make glucose and oxygen from carbon dioxide and water. The energy stored in glucose is transferred eventually to the bonds in molecules such as ATP. When needed, the ATP molecules release energy to drive other chemical reactions.

Photosynthesis

In the complex process of photosynthesis, carbon dioxide is reduced to make sugar,

Photosynthesis involves a number of different steps and is a very complex process.

$$6\ CO_2 + 24\ H^+ + 24\ e^- \longrightarrow C_6H_{12}O_6 + 6\ H_2O$$

Reduction is the gain of electrons or hydrogen. *Oxidation* is the loss of electrons or hydrogen.

and water is oxidized to oxygen:

$$12\ H_2O \longrightarrow 6\ O_2 + 24\ H^+ + 24\ e^-$$

The oxidation and reduction reactions added together give the overall reaction:

$$6\ CO_2 + 6\ H_2O + 688\ kcal \longrightarrow C_6H_{12}O_6 + 6\ O_2$$

Carbon Dioxide Water Energy (Sunlight) Glucose Oxygen

It was discovered in 1985 that a blue-green protozoa, *Stentor coeruleus,* contains a light-absorbing substance, stentorin, that allows it to undergo a unique type of photosynthesis.

The oxygen produced in photosynthesis is the source (and only present source) of all of the oxygen in our atmosphere. Only this life-giving gas, given off by trees, grass, greenery, and even by algae in the sea, makes possible human life and most animal life on Earth. We are dependent on the plant life of our planet, and we must live in balance with the oxygen output of that plant life, as well as with the food output of the same plant life. Photosynthesis is thus absolutely vital to life on Earth.

Photosynthesis is generally considered a series of **light reactions,** which occur only in the presence of light energy, and a series of **dark reactions,** which can occur in the dark. The dark reactions feed on high-energy compounds (such as ATP) produced by the light reactions. During the light reactions, green pigments such as the chlorophylls (either A or B, Fig. 15–15) absorb photons of light and raise electrons within these structures to higher energy levels. As electrons move back to the ground state, chloroplasts absorb this energy. Through a series of reactions, water is oxidized to oxygen, and energy is stored in the bonds of energy-bank compounds such as ATP (see Fig. 15–16). ATP stores energy in two high-energy phosphate bonds, shown as wiggle lines in Figure 15–16.

The energy of a photon is captured by chlorophyll when an electron is raised to a higher energy state.

Refer to atomic theory in Chapter 3.

Chlorophyll is green because violet light and red light are absorbed and green light is reflected.

By the combination of several diverse fields of scientific investigation, more details have been discovered of how the excited electron transfers its energy into molecular bonds (Fig. 15–17). Within about 4×10^{-12} s after an

Figure 15–15 The two structures of chlorophyll.

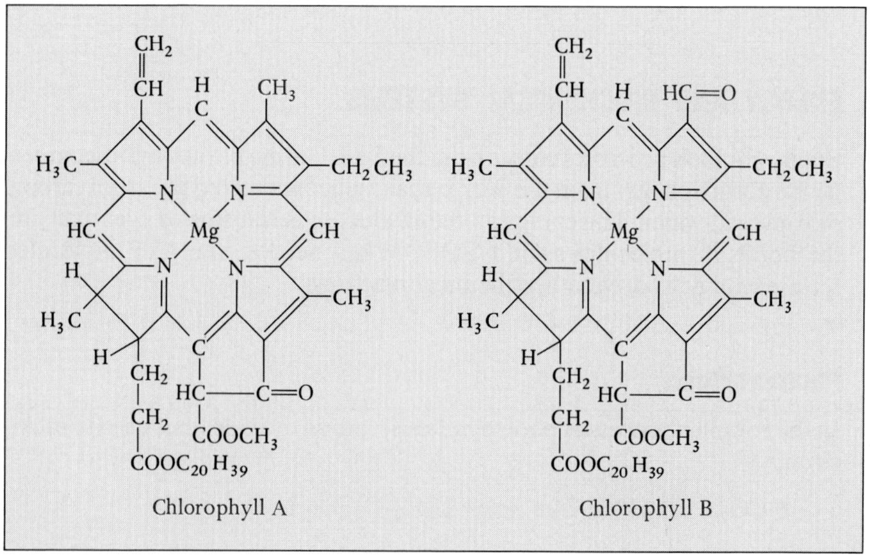

Chlorophyll A   Chlorophyll B

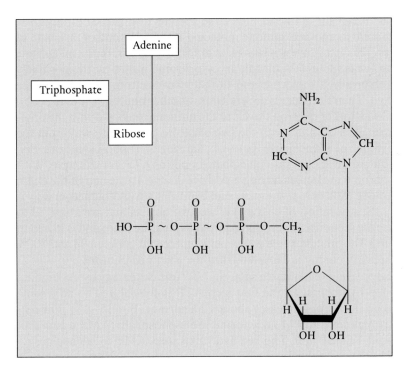

Figure 15–16   Molecular structure of adenosine triphosphate (ATP).

electron in chlorophyll absorbs a photon, the electron moves to a neighboring molecule, pheophytin, which is attached to chlorophyll and is very similar in composition and structure to chlorophyll. The chlorophyll molecule is left with an excess positive charge. The electron then moves from the pheophytin to a molecule of quinone, which is at the end of a spiraling chain of

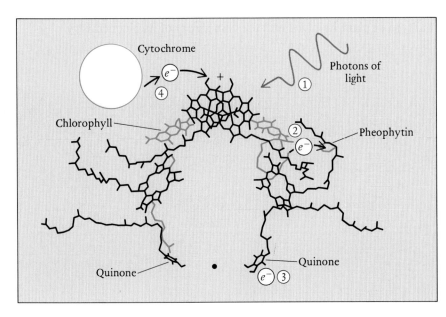

Figure 15–17   Part of a mechanism of how chlorophyll and associate molecules store photons of light. (1) A photon excites an electron in chlorophyll. (2) The electron transfers to pheophytin. (3) The electron moves to the quinone part of the pheophytin. (4) A cytochrome molecule relinquishes an electron to the positively charged chlorophyll, and chlorophyll is ready to repeat the process.

molecules. From the quinone molecule, the electron passes through the central protein to another quinone molecule at the other end of the chain of molecules. This last process is slower, at about 1×10^{-4} s. In the meantime, a globular, water-soluble molecule of cytochrome passes by the positively charged chlorophyll molecule and donates an electron, neutralizing the chlorophyll. Then the process occurs again: another photon strikes the chlorophyll, and another electron travels to the quinone molecule with the original electron. The quinone with the two electrons then pulls away from the chain of molecules and participates in the stages of photosynthesis that transfer the stored energy into phosphate bonds of ATP, for example. With two cytochrome molecules having a positive charge at one end of the chain and a quinone molecule at the other end being negatively charged, energy is stored by the separation of charge. (When two objects that are attracted to each other are separated, energy is required to separate them and is stored in the system.) The process is remarkably efficient in that between 98 and 100% of the necessary energy is captured from the absorbed photons.

In the presence of a suitable catalyst, ATP releases energy by undergoing a three-step hydrolysis. In the first step ATP is hydrolyzed to adenosine diphosphate (ADP) and releases about 12 kcal/mol (Fig. 15–18). The second hydrolysis step, ADP to adenosine monophosphate (AMP), also produces about 12 kcal/mol. The last hydrolysis step, AMP to adenosine, releases only about 2.5 kcal/mol. The hydrolysis of ATP releases energy (is *exothermic*); the synthesis of ATP from AMP or ADP requires energy (is *endothermic*). It is the synthesis of ATP that occurs during the light reactions of photosynthesis, and it is this process that stores the Sun's energy in chemical compounds.

During the dark reactions, hydrolysis of the P—O bonds of ATP provide the energy to convert CO_2 and hydrogen (from water) into glucose through a series of chemical reactions.

After photosynthesis the living plant may convert glucose to oligosaccharides, starches, cellulose, proteins, or oils. The end-product depends on the type of plant involved and the complexity of its biochemistry.

Exothermic means energy is released.

Endothermic means energy is required.

Figure 15–18 Hydrolysis of ATP to ADP.

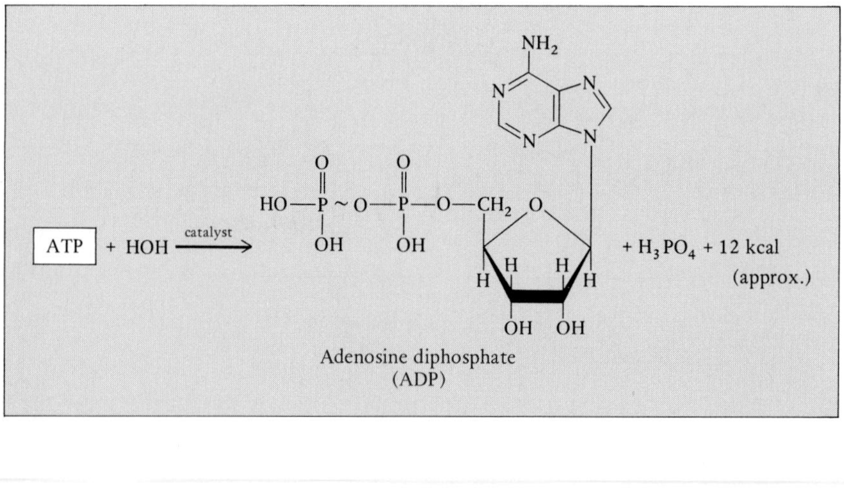

Adenosine diphosphate
(ADP)

The next steps involved in use of the energy stored in high-energy compounds are for the compounds to be eaten, digested, transported to the cells of the body, and metabolized.

Digestion

From a chemical point of view, digestion is the breakdown of ingested foods by hydrolysis. The products of digestion are relatively small molecules that can be absorbed through the intestinal walls. The hydrolytic reactions of digestion are catalyzed by enzymes, there being a specific enzyme for the hydrolysis of each type of substance. The hydrolysis of carbohydrates ultimately yields simple sugars, proteins yield amino acids, and fats and oils yield fatty acids and glycerol.

> Digestion is the hydrolysis of carbohydrates, fats, and proteins to provide small molecules that can be absorbed.

In our food, carbohydrates requiring digestion are polysaccharides such as starch and disaccharides such as sucrose and lactose. The digestion process begins in the mouth with salivary amylase, or ptyalin. Starch is partially hydrolyzed into the disaccharide maltose by ptyalin, which is later rendered inactive by the high acidity of the stomach. No more digestion of carbohydrates occurs in the stomach. When the food passes from the stomach into the small intestine, the acidity is neutralized by a secretion from the pancreas. Enzymes from the pancreas complete the hydrolysis of carbohydrates into simple sugars such as glucose, fructose, and galactose. These simple sugars are then absorbed into the bloodstream. The hormone insulin (a protein) escorts simple sugars through the cell membranes and into the cells. There, in the mitochondria, these simple sugars are oxidized for their energy content.

> Acidic solutions: pH below 7
>
> Basic solutions: pH above 7
>
> Human blood normally contains between 0.08% and 0.1% glucose.
>
> Insulin is a protein.

If the sugar level in the bloodstream becomes too high, the simple sugars are converted into glycogen in the liver. If the sugar level is too low, stored glycogen is hydrolyzed to raise it. Malfunctions in these processes can lead to too much blood sugar, **hyperglycemia,** or too little blood sugar, **hypoglycemia.** Either condition, if sustained, produces a type of **diabetes.**

> Types of diabetes are discussed in Chapter 20.

The normal fasting level of glucose in blood occurs after 8 to 12 h without food, which is just before most people eat breakfast. The blood sugar level for normal adults during the normal fasting level is between 70 and 100 mg of glucose for each 100 mL of blood.

The digestion of fats and oils, such as the triesters of fatty acids and glycerol, occurs primarily in the small intestine. The enzyme that catalyzes the hydrolysis of fatty acid esters is water-soluble, but the fats and oils themselves are not. Bile salts, secreted by the liver, emulsify the oil by forming an interface between the nonpolar oil and the polar water, thereby making it possible for the oil to "dissolve" in water. For a molecule to be an emulsifier between polar and nonpolar molecules, it must have both polar and nonpolar structures. The sodium salt of glycocholic acid, a bile salt, contains the bulky nonpolar hydrocarbon groups, which are compatible with fat or oil, and the —OH and ionic groups, which attract water molecules (Fig. 15–19).

> Bile salts act chemically much like detergent and soap molecules (Chapter 22).

The digestion of proteins begins in the stomach and is completed in the small intestine. Many enzymes are known to be involved. In the stomach

Figure 15–19 The sodium salt of glycocholic acid, a bile salt.

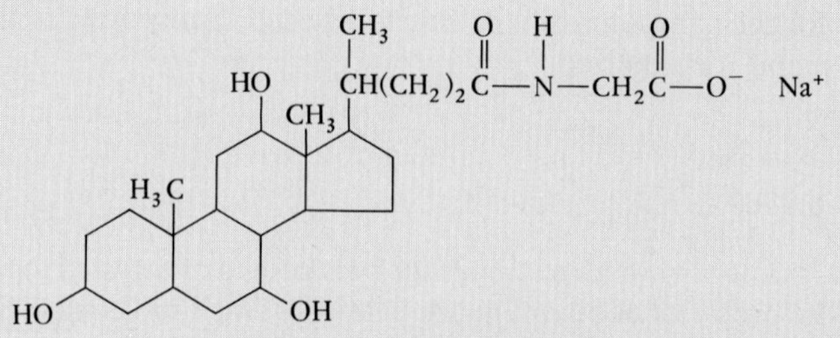

Sodium salt of glycocholic acid

Some protein enzymes are sold commercially as meat tenderizers and stain removers. Some are used to free the lens of the eye before cataract surgery.

pepsin catalyzes the hydrolysis of only about 10% of the bonds in a typical protein, leaving protein fragments with molecular weights of 600 to 3000. In the small intestine hydrolysis is completed to amino acids, which are absorbed through the intestinal wall.

The stomach is protected from protein-splitting enzymes by a mucous lining. The mucus is mostly protein. Although the lining is being digested slowly, it is also constantly being renewed.

The Liver: The Nutrient Bank of the Body

After digestion most food nutrients pass directly to the liver for distribution to the body. Glucose is used for energy in the liver and to prepare glucose phosphate as the first step in the preparation of glycogen (the storage carbohydrate); in addition, about one third goes on in the bloodstream to nourish the cells. From the liver a fraction of the amino acids is sent to the cells to build proteins. In the liver amino acids are used to form enzymes, and some are oxidized to obtain energy. The liver is thus the central nutrient bank, or warehouse, of the body in that it stores, converts, and classifies nutrients.

Glucose Metabolism

Aerobic reactions use elemental oxygen. Anaerobic reactions use no elemental oxygen.

The sequence of reactions by which energy is obtained from glucose and similar compounds begins with an **anaerobic** (without elemental oxygen) series of reactions followed by a series of **aerobic** (with elemental oxygen) reactions.

The anaerobic process was discovered by the German chemist Otto Fritz Meyerhof (1884–1951). More details were discovered by Gustav Embden (1884–1933). In 1918, Meyerhof showed that animal cells break down sugar in much the same way as does yeast, a plant. This work made clear for the first time that, with only minor differences, glucose metabolism follows the same sequences in all creatures. Details of the individual steps were discovered between 1932 and 1933 by Embden and between 1937 and 1941 by Carl Ferdinand Cori (1896–1984) and his wife Gerty Theresa Cori (1896–1957). The Coris shared a Nobel Prize in 1947. Meyerhof shared the Nobel Prize in physiology and medicine in 1922 with Archibald Vivian Hill (1886–1977), who had investigated muscle from the point of view of heat

Gerty Cori was the third woman to receive a Nobel Prize. The first two were Marie Curie and her daughter, Irène Joliot.

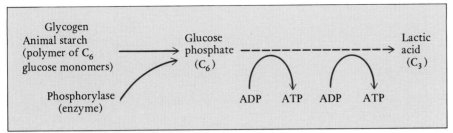

Figure 15–20   The Embden-Meyerhof anaerobic pathway begins with glycogen and after conversion into 11 successive compounds (4 C$_6$ compounds and 7 C$_3$ compounds) becomes lactic acid, another C$_3$ compound. In the process, two molecules of ADP are converted into two molecules of ATP for each glucose monomer carried through the pathway.

production. The anaerobic sequence of reactions is known as the **Embden-Meyerhof pathway** (Fig. 15–20).

The aerobic sequence of reactions was discovered by the German biochemist Sir Hans Adolf Krebs (1900–1981). In 1933, Krebs fled from Hitler's Germany to England, where he studied at Cambridge, later joined the faculty at Oxford, and was knighted in 1958. In 1953, he shared the Nobel Prize in physiology and medicine with Fritz Albert Lipmann (1899–1986), who discovered the roles of ATP and other such compounds in the storing of energy in phosphorus-oxygen bonds. The aerobic sequence of reactions is known as the **Krebs cycle** (Fig. 15–21).

When a muscle is used, glucose is converted anaerobically to lactic acid by a series of 11 steps in the Embden-Meyerhof pathway. The overall reaction can be represented by the following equation:

$$C_6H_{12}O_6 + 2\ ADP + 2\ H_3PO_4 \longrightarrow 2\ CH_3-\overset{\overset{\displaystyle H}{|}}{\underset{\underset{\displaystyle OH}{|}}{C}}-\overset{\overset{\displaystyle O}{\|}}{C}-OH + 2\ ATP + 2\ H_2O$$

Glucose   Lactic Acid

Note that the energy of one molecule of glucose is transferred to two molecules of ATP (two P—O bonds), and the remainder resides in two molecules of lactic acid. Since the process is anaerobic, elemental oxygen (O_2) is not a reactant.

If muscle is used strenuously for a sufficiently long period of time, the lactic acid buildup produces tiredness and a painful sensation. The bloodstream carries the lactic acid away eventually, but time and oxygen are needed to convert it to carbon dioxide and water, which are excreted.

The conversion of lactic acid to CO_2 and H_2O is accomplished by the Krebs cycle. The equation for the aerobic conversion of lactic acid into carbon dioxide and water contains the number of ATP molecules formed in the process. Note that it includes elemental oxygen (O_2), which is required for an aerobic process:

$$C_3H_6O_3 + 18\ ADP + 18\ H_3PO_4 + 3\ O_2 \longrightarrow 3\ CO_2 + 21\ H_2O + 18\ ATP$$

Lactic Acid

Muscular activity converts glucose to lactic acid, which produces fatigue in muscles as the lactic acid accumulates.

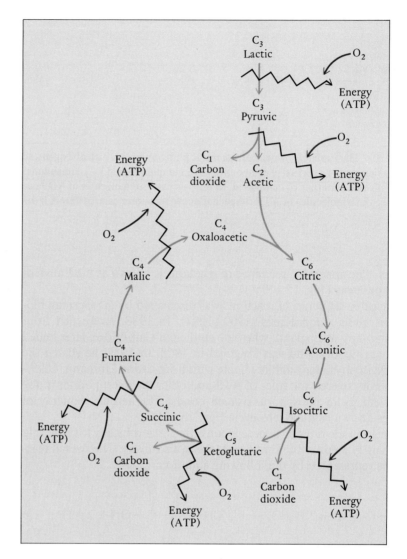

The energy available to the body from one glucose molecule is now stored in 38 ATP molecules, 2 from the Embden-Meyerhof pathway and 36 from two lactic acid molecules. The complete oxidation of glucose into CO_2 and H_2O with energy stored in ATP bonds is 41% efficient. This is remarkable when you consider that the efficiency of the automobile engine is only about 20% and that the efficiency of a heat engine of any size seldom goes above 35%.

When ATP hydrolyzes back to ADP by losing a phosphate group, the energy is used to drive chemical reactions that cause muscles to move, such as the heart to pump, the diaphragm to breathe, the eyelid to twitch, the jaw to chew, the throat to swallow, and so on.

You now have seen some of the detailed chemistry involved in simply raising your arm, and you are now aware of what happens to some of the sugars and starches you ingest. As far as our bodies are concerned, this is the fate of the quantity of the Sun's energy absorbed by the green plant. Of course, there is much more known than is presented here, and there appears to be no end to what is left to be discovered.

CHEMISTRY OF VISION

In addition to chlorophyll's absorption of light, which leads to the production of food and fuel, many species have rhodopsin, which absorbs light and leads to sight. The very complex eye–brain coordination required for us to see is now beginning to be understood.

> Visible light photons correspond to colors or wavelengths of light from violet, with wavelength = 400 nm, to deep red, with wavelength = 700 nm.

The outer parts of the eye, particularly the lens, focus the photons on the retina, a light-sensitive material with the thickness of tissue paper. The retina has two types of light-sensitive substances or **photoreceptor cells, rods** and **cones.** The human eye contains about 1 billion rods and 3 million cones. Cones work in bright light and are sensitive to color, whereas rods function in dim light and are unable to distinguish colors. This is why only shades of gray and not colors are distinguished in moonlight. The sensitivity and adaptability of the eye is phenomenal. A person can sense as few as five photons of light in a darkened room and then adjust within a minute to the myriad of photons in bright light without being blinded.

Rhodopsin, the photosensitive material in the photoreceptor cells includes the protein **opsin** and the compound **11-*cis*-retinal.** A schematic of the complete visual cycle is shown in Figure 15–22. Photons focused on the retina isomerize the 11-*cis*-retinal to **all-*trans*-retinal** (Fig. 15–23). This isomerization causes a series of other molecular changes, which result in the dissociation of rhodopsin into opsin and all-*trans*-retinal. These structural changes trigger an electrical signal, which is carried by the optic nerve to the brain. Later in the visual cycle all-*trans*-retinal is converted back to 11-*cis*-retinal, which then combines with opsin to yield rhodopsin.

Retinal is derived from vitamin A, which is why a shortage of vitamin A in the diet can lead to night blindness. Eating carrots is good for your eyes because β-carotene is converted to vitamin A in your body (Fig. 15–24).

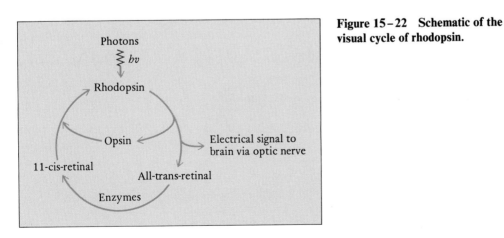

Figure 15–22   Schematic of the visual cycle of rhodopsin.

Figure 15–23 (a) Isomerization of 11-*cis*-retinal to all-*trans*-retinal by a photon of light. (b) Structural change in rhodopsin results in the separation of all-*trans*-retinal from opsin.

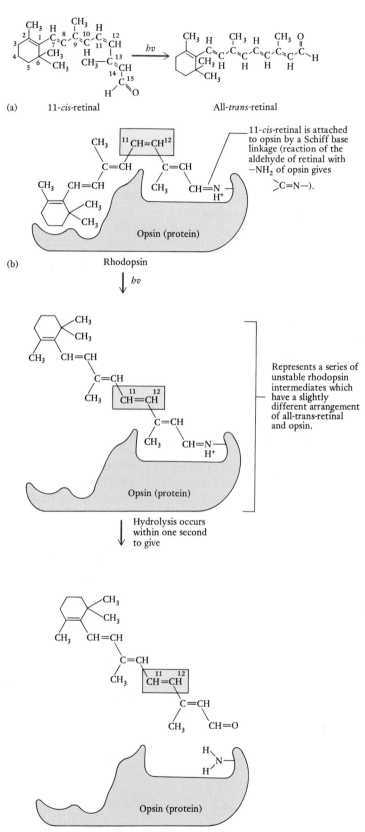

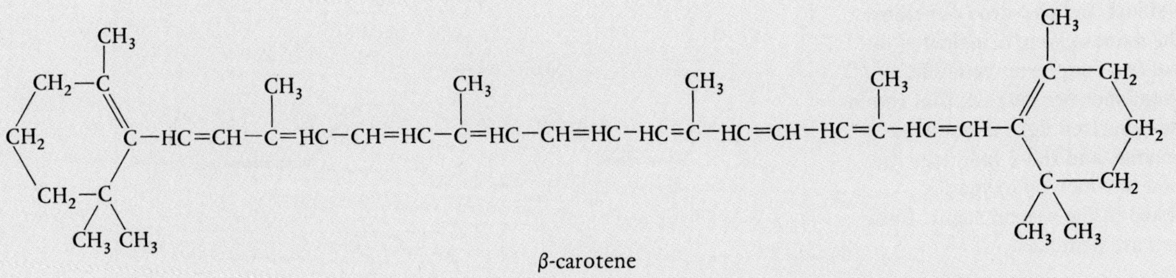

β-carotene

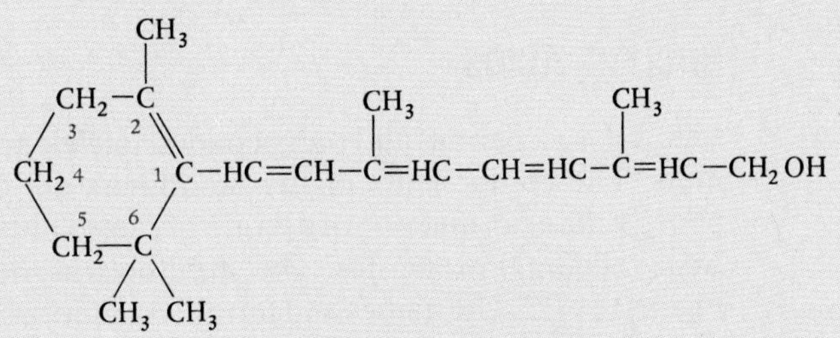

Vitamin A or retinol

Figure 15–24   Structures of β-carotene and vitamin A.

SELF-TEST 15–C

1. The source of energy for photosynthesis is the _____.
2. Most of the energy obtained by the oxidation of food is used immediately to synthesize the molecule _____.
3. The hydrolysis of ATP results in the molecules _____ and _____; The other "product" is _____.
4. The reactants in the photosynthesis process are _____ and _____; _____ must also be supplied.
5. Energy from the Sun is absorbed by _____ in the green cells of a plant.
6. Digestion is the breakdown of foodstuffs by _____.
7. Bile salts act as ( ) catalysts, ( ) emulsifying agents, ( ) enzymes.
8. The two products of the Embden-Meyerhof pathway are _____ and _____.
9. The end-products of the Krebs energy cycle are _____, _____, and _____.
10. The photosensitive material in the photoreceptor cells of the eye retina is _____.
11. Photoreceptor cells in the retina are referred to as _____ and _____.
12. _____ are sensitive to different colors; _____ are not sensitive to colors.
13. _____ in carrots is converted to vitamin A in your body.

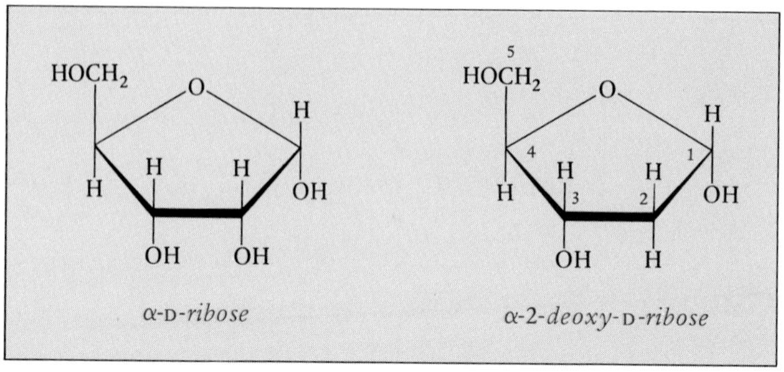

Figure 15–25 The structure of α-D-ribose and α-2-deoxy-D-ribose. In the names given, α indicates the one of two ring forms possible, D distinguishes the isomers that rotate plane-polarized light in opposite directions, and the 2 indicates the carbon to which no oxygen is attached in the second sugar. Both sugars are pentoses.

α-D-*ribose*   α-2-*deoxy*-D-*ribose*

NUCLEIC ACIDS

Like polysaccharides and polypeptides, **nucleic acids** are condensation polymers. The components of the monomers are one of two simple sugars, phosphoric acid, and one of a group of ringed nitrogen compounds that have basic (alkaline) properties. The structures of the two sugars are shown in Figure 15–25. The names and formulas of the basic nitrogen compounds are given in Figure 15–26.

Nucleic acids are **deoxyribonucleic acids (DNA)** if they contain the sugar **α-2-deoxy-D-ribose,** or **ribonucleic acids (RNA)** if they contain the sugar **α-D-ribose.** DNA is found primarily in the nucleus of the cell (Fig. 15–27), whereas RNA is found mainly in the cytoplasm, outside of the nucleus. Nucleic acids are found in all living cells, with the exception of the red blood cells of mammals.

Three major types of RNA have been identified. They are messenger RNA (mRNA), transfer RNA (tRNA), and ribosomal RNA (rRNA). Each has a characteristic molecular weight and base composition. Messenger RNAs are generally the largest, with molecular weights between 25,000 and 1 million. They contain from 75 to 3000 mononucleotide units. Transfer RNAs have molecular weights in the range of 23,000 to 30,000 and contain 75 to 90 mononucleotide units. Ribosomal RNAs, which have molecular weights between those of mRNAs and tRNAs, make up as much as 80% of the total cell RNA. Besides having different molecular weights, the three types of RNA differ in function. One difference in function is described in the discussion of natural protein synthesis.

Figure 15–26 Some nitrogenous bases obtained from the hydrolysis of nucleic acids.

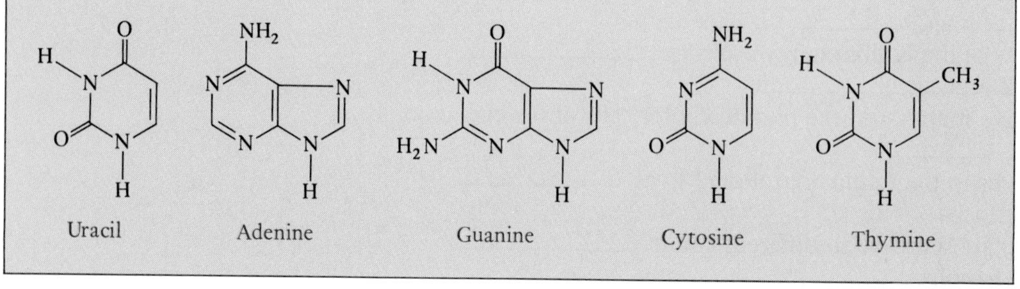

Uracil   Adenine   Guanine   Cytosine   Thymine

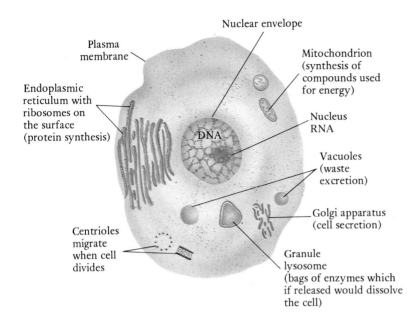

Figure 15–27 **Diagrammatic generalized cell showing the relationships among the various components of the cell. Cytoplasm is the material of the cell, exclusive of the nucleus. Many of the components shown are not visible through an ordinary optical microscope.**

The monomers of both DNA and RNA contain a simple sugar, one of the nitrogenous bases, and one or two phosphoric acid units. The structure of a monomer, a **nucleotide,** is shown in Figure 15–28a. The nucleotides of DNA and RNA have two structural differences: (1) the sugar (Fig. 15–25) and (2) the use of uracil base only in RNA, whereas thymine base is found only in DNA. The other bases, adenine, guanine, and cytosine, are found in both DNA and RNA. Three monomers condensed into an oligonucleotide can be seen in Figure 15–28b.

When the phosphate group is absent, one of the pentoses bonded to a nitrogeneous base is a **nucleoside.**

Polynucleotides with molecular weights up to several million are known. The sequence of nucleotides in the polynucleotide chain is its **primary structure.**

In 1953, James D. Watson and Francis H. C. Crick (Fig. 15–29) proposed a **secondary structure** for DNA that has since gained wide acceptance. Figure 15–30 illustrates a small portion of the structure, in which two polynucleotides are arranged in a double helix stabilized by hydrogen bonding between the base groups opposite each other in the two chains. RNA is generally a single strand of helical polynucleotide.

The function of polynucleotides is to transcribe cellular and organism information so that like begets like. The almost infinite variety of primary structures of polynucleotides allows an almost infinite variety of information to be recorded in the molecular structures of the strands of nucleic acids. The different arrangements of just a few different bases give the large variety of structures. In a somewhat similar fashion, the multiple arrangements of just a few language symbols convey the many ideas in this book. The coded information in the polynucleotide is believed to control the inherited characteristics of the next generation as well as most of the continuous life processes of the organism.

One nucleotide is joined to another by an ester-forming reaction:

$$-P-OH + HO-C-$$

$$\underset{\underset{O}{\parallel}}{\overset{}{}}$$

$$\longrightarrow P-O-C-+ H_2O$$

The inherited traits of an organism are controlled by DNA molecules.

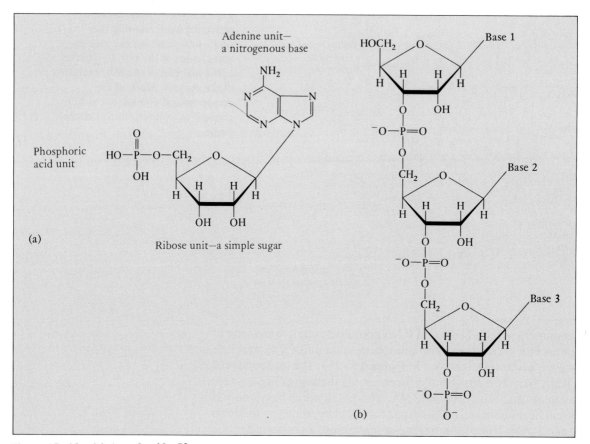

Figure 15–28   (a) A nucleotide. If other bases are substituted for adenine, several nucleotides are possible for each of the two sugars shown in Figure 15–25. (b) Bonding structure of a trinucleotide. Bases 1, 2, and 3 represent any of the nitrogenous bases obtained in the hydrolysis of DNA and RNA (Fig. 15–26). The primary structure of both DNA and RNA is an extension of this structure and produces molecular weights as high as a few million.

Double stranded DNA forms the 46 human chromosomes, which have specialty heredity areas called genes. The approximately 100,000 genes in a human genome consist of 3 billion pairs of bases, with each gene having as few as 1000 or as many as 100,000 base pairs. Each gene is sandwiched between "junk" or noncoding DNA sequences, which do not code anything. There are also short segments that act as switches to signal where the coding sequence begins.

The total sequence of the base pairs of a cell has been coined a **genome.** Until 1986, the experimental determination of the sequence of base pairs in a DNA strand was laboriously slow, taking place at a speed of about 200,000 base pairs per year. With the invention of an automatic DNA sequencer by a team headed by Professor Leroy Hood at California Institute of Technology in Pasadena, the complete sequencing of a genome is now possible. The process is rather expensive at about $3 billion for the complete mapping and sequencing of a human genome. This is a cost of about $1 per base. At the time of this writing, eight chromosomes had been mapped and reported. The complete sequencing of a human genome is expected to be complete by the year 2000.

A complete mapping of an individual's genome would improve knowledge of the about 4000 known hereditary diseases, provide genetic guidance to better health, and give a key to understanding diversity among individ-

Figure 15–29 Francis H. C. Crick (b. 1916) (right) and James D. Watson (b. 1928) (left), working in the Cavendish Laboratory at Cambridge, built scale models of the double helical stucture of DNA based on the X-ray data of Maurice H. F. Wilkins (b. 1916). Knowing distances and angles between atoms, they compared the task to the working of a three-dimensional jig-saw puzzle. Watson, Crick, and Wilkins received the Nobel Prize in 1962 for their work relating to the structure of DNA.

uals. The specific base-pair sequences and genes that cause cystic fibrosis, manic depression, Down's syndrome, and muscular dystrophy, for example, could theoretically be spliced out and not transmitted to future generations. Medical treatment could be tailored to an individual's genetic makeup rather than to some "average" patient. The applications of genome mapping are diverse, exciting, and technically possible. Before the general population of the 5 billion here on Earth can benefit from the specific information in

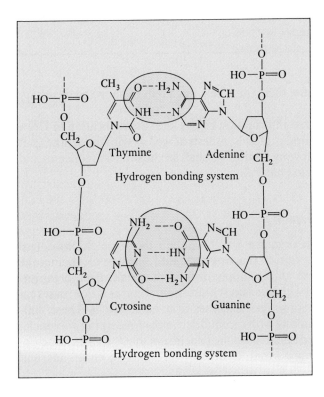

Figure 15–30 Double helix structure proposed by Watson and Crick for DNA. Hydrogen bonds in the thymine-adenine and cytosine-guanine pairs stabilize the double helix. Adenine also pairs with uracil in mRNA, which contains no thymine.

HEREDITARY DISEASES AND THE GENETIC CODE

A change in the genetic code is passed on from generation to generation. That is why sickle cell anemia is called a hereditary disease. Sickle cell anemia is just one of almost 4,000 genetic diseases caused by a change in a single gene. Hemophilia, cystic fibrosis, and Tay-Sachs disease are other familiar examples, though none is as common as sickle cell anemia. As yet, there is no cure for any genetic disease, for we have no way of correcting the defective gene, though some, like sickle cell anemia, can be successfully treated. But the future holds great promise, thanks to a new technique called **recombinant DNA technology.** Scientists have discovered a special group of enzymes, restriction enzymes, which recognize specific base sequences in DNA and cut the strands at that point. This has made it possible to remove a gene from one DNA molecule and insert it into another. The genetic code can now be edited. This technology has already made important contributions to medical science. Genes coding for human insulin have been inserted into the DNA of bacteria, which are then grown in huge fermenting tanks. Here billions of bacteria act like human insulin factories, producing a virtually limitless supply of this important hormone. More and more diabetics are now injecting themselves with this human insulin rather than the animal insulin alternative, which can produce undesirable side effects. There are many who view recombinant DNA technology as the beginning of a new age in science. Laboratories throughout the world are already genetically modifying crop species, introducing genes that code for proteins that improve growth and protect the plant against insect pests. And scientists are hopeful that children with life-threatening genetic diseases may someday be cured by introducing normal copies of the defective gene into their cells. But recombinant DNA technology has even more to offer. It is allowing scientists to unravel chemistry's most remarkable secret, the molecular basis of life.

The World of Chemistry (Program 24) "Genetic Code."

The billions of bacteria in the tank have DNA coded to produce human insulin. The bacteria act like human insulin factories and produce a virtually limitless supply of this hormone. (*The World of Chemistry,* Program 24, "Genetic Code.")

each individual's genome, there must be a huge reduction in the monetary and time costs.

The transfer of coded information begins with the replication of DNA and continues with natural protein synthesis as well as with the synthesis of body tissues. In the following section we shall see how DNA replicates and how protein is synthesized naturally.

Replication of DNA: Heredity

Almost all nuclei in an organism's cells contain the same chromosomal composition. This composition remains constant regardless of whether the cell is starving or has an ample supply of food materials. Each organism begins life as a single cell with this same chromosomal composition; in sexual reproduction half of a chromosome comes from each parent. These well-known biological facts, along with recent discoveries concerning polynucleotide structures, have led scientists to the conclusion that the DNA structure is faithfully copied during normal cell division (**mitosis** — both strands) and

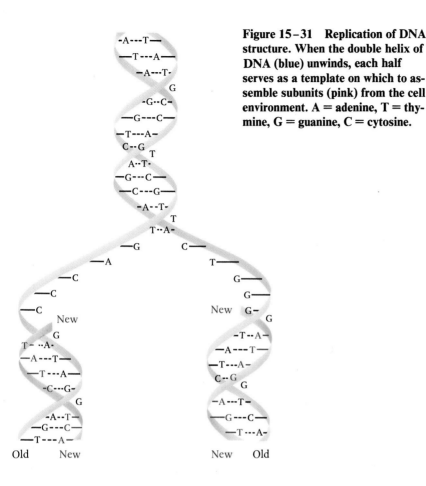

Figure 15-31 Replication of DNA structure. When the double helix of DNA (blue) unwinds, each half serves as a template on which to assemble subunits (pink) from the cell environment. A = adenine, T = thymine, G = guanine, C = cytosine.

that only half is copied in cell divison producing reproductive cells (**meiosis** — one strand).

In replication the double helix of the DNA structure unwinds and each half of the structure serves as a template, or pattern, from which the other complementary half can be reproduced from the molecules in the cell environment (Fig. 15-31). Replication of DNA occurs in the nucleus of the cell.

Natural Protein Synthesis

The proteins of the body are continually being replaced and resynthesized from the amino acids available to the body.

The use of isotopically labeled amino acids has made possible studies of the average lifetimes of amino acids as constituents in proteins — that is, the time it takes the body to replace a protein in a tissue. For a process that must be extremely complex, replacement is very rapid. Only minutes after radioactive amino acids are injected into animals, radioactive protein can be found. Although all the proteins in the body are continually being replaced, the rates of replacement vary. Half of the proteins in the liver and plasma are replaced in 6 days; the time needed for replacement of muscle proteins is

Figure 15–32 A schematic illustration of the role of DNA and RNA in protein synthesis. A, C, G, T, and U are nitrogen bases characteristic of the individual nucleotides. See Figure 15–26 for structures of the bases and Table 15–2 for abbreviations of the amino acids used.

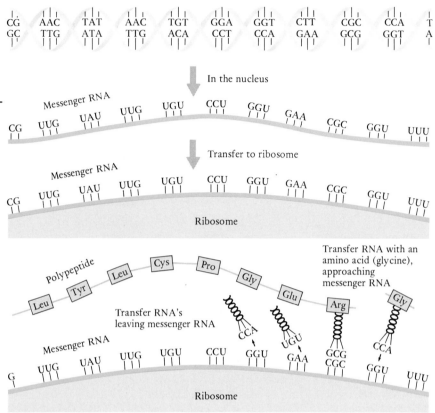

about 180 days, and replacement of protein in other tissues, such as bone collagen, takes even longer.

Recall that each organism has its own kinds of proteins. The number of possible arrangements of 20 amino acid units is 2.43×10^{18}, yet proteins characteristic of a given organism can be synthesized by the organism in a matter of a few minutes.

The DNA in the cell nucleus holds the code for protein synthesis. Messenger RNA, like all forms of RNA, is synthesized in the cell nucleus. The sequence of bases in one strand of the chromosomal DNA serves as the template from which a single strand of a messenger ribonucleotide (mRNA) is made (Fig. 15–32). The bases of the mRNA strand complement those of the DNA strand. A pair of complementary bases is structured such that each one fits the other and forms one or more hydrogen bonds. Messenger RNA contains only the four bases adenine (A), guanine (G), cytosine (C), and uracil (U). DNA contains principally the four bases adenine (A), guanine (G), cytosine (C), and thymine (T). The base pairs are as follows:

DNA	mRNA
A	U
G	C
C	G
T	A

This means that, provided the necessary enzymes and energy are present, wherever a DNA has an adenine base (A), the mRNA will transcribe a uracil base (U).

Before the mRNA, referred to as pre-mRNA, leaves the nucleus, quality control is applied by **SNURPS,** small nuclear ribonucleoproteins (Fig. 15–33). SNURPS help to remove meaningless sections of the transcribed mRNA. The meaningless sections are **introns;** the meaningful sections are **exons.** About 14 SNURPS have been identified, but only 4 are very abundant. Each SNURP attacks genetic noncoding at a different point along the sequence of the mRNA. When the SNURPS attach to the unwanted mRNA, a spliceosome is formed, and the relatively large knot on the mRNA makes it difficult for the mRNA to pass through the nuclear membrane into the cytoplasm. Hence, the mRNA is not released from the nucleus until the cleanup procedure is complete. The unwanted introns are removed, and the wanted exons are spliced back together, an amazing sequence of cellular proofreading.

After transcription and cleanup, mRNA passes from the nucleus of the cell to a ribosome, where mRNA serves as the template for the sequential ordering of amino acids during protein synthesis. As its name implies, messenger RNA contains the sequence message, in the form of a three-base code, for ordering amino acids into proteins. Each of the thousands of different proteins synthesized by cells is coded by a specific mRNA or segment of an mRNA molecule.

Transfer RNAs carry the specific amino acids to the mRNA. Each of the 20 amino acids found in proteins has at least one corresponding tRNA, and some have multiple tRNAs (Table 15–2). For example, there are five distinctly different tRNA molecules specifically for the transfer of the amino

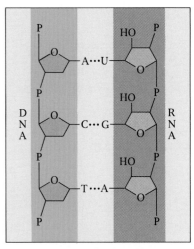

RNA is transcribed from DNA by proper base pairing. (*The World of Chemistry,* Program 24, "Genetic Code.")

The code on tRNA is the second genetic code; the first code is on DNA.

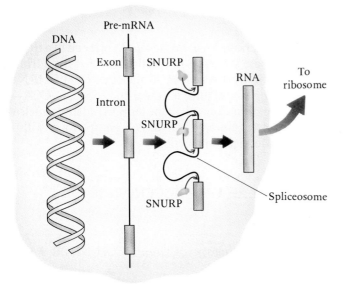

Nucleus of a cell

Figure 15–33 SNURPS, small nuclear ribonucleoproteins, exercise quality control on meaningless messenger RNA by orchestrating the removal of the meaningless sections, the introns.

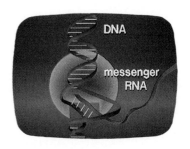

After transcription, RNA leaves DNA and the nucleus of the cell. (*The World of Chemistry,* **Program 24, "Genetic Code.")**

TABLE 15-2 Messsenger RNA Codes for Amino Acids*

Amino Acid	Shortened Notation Used for Amino Acids in Fig. 15-32	Base Code on mRNA
Alanine	Ala	GCA, GCC, GCG, GCU
Arginine	Arg	AGA, AGG, CGA, CGG, CGC, CGU
Asparagine	Asp-NH$_2$	AAC, AAU
Aspartic acid	Asp	GAC, GAU
Cysteine	Cys	UGC, UGU
Glutamic acid	Glu	GAA, GAG
Glutamine	Glu-NH$_2$	CAG, CAA
Glycine	Gly	GGA, GGC, GGG, GGU
Histidine	His	CAC, CAU
Isoleucine	Ileu	AUA, AUC, AUU
Leucine	Leu	CUA, CUC, CUG, CUU, UUA, UUG
Lysine	Lys	AAA, AAG
Methionine	Met	AUG
Phenylalanine	Phe	UUU, UUC
Proline	Pro	CCA, CCC, CCG, CCU
Serine	Ser	AGC, AGU, UCA, UCG, UCC, UCU
Threonine	Thr	ACA, ACG, ACC, ACU
Tryptophan	Try	UGG
Tyrosine	Tyr	UAC, UAU
Valine	Val	GUA, GUG, GUC, GUU

*In groups of three (called codons), bases of mRNA code the order of amino acids in a polypeptide chain. A, C, G, and U represent adenine, cytosine, guanine, and uracil, respectively. Some amino acids have more than one codon, and hence more than one tRNA can bring the amino acid to mRNA. The research on this coding was initiated by Marshall Warren Nirenberg. (Adapted from J. I. Routh, D. P. Eyman, and D. J. Burton: *Essentials of General, Organic, and Biochemistry,* 3rd ed. Philadelphia, Saunders College Publishing, 1977.)

The bases in groups of three are called codons.

acid leucine in cells of the bacterium *Escherichia coli.* At one end of a tRNA molecule is a trinucleotide base sequence (the **anticodon**) that fits a trinucleotide base sequence on mRNA (the **codon**). At the other end of a tRNA molecule is a specific base sequence of three terminal nucleotides — CCA — with a hydroxyl group on the sugar exposed on the terminal adenine nucleotide group. With the aid of enzymes, this hydroxyl group reacts with a specific amino acid by an esterification reaction.

$$\text{(Mononucleotides)}_{75-90}\text{CCA}-\text{OH} + \text{HOCCH(NH}_2)\text{R} \longrightarrow$$
tRNA   Amino Acid

$$\text{(Mononucleotides)}_{75-90}\text{CCA}-\text{OCCH(NH}_2)\text{R} + \text{H}_2\text{O}$$
tRNA-Amino Acid

The P—O bonds of ATP provide energy for the reaction between an amino acid and its tRNA. A molecule of ATP first activates an amino acid.

ATP + Amino acid $\longrightarrow$ ATP-Amino acid activated species

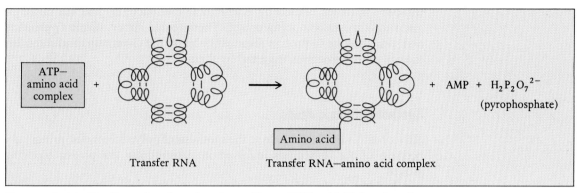

Figure 15–34 Bonding of activated amino acid to tRNA. AMP is adenosine monophosphate.

The activated complex then reacts with a specific tRNA and forms the products shown in Figure 15–34.

The tRNA and its amino acid migrate to the ribosome, where the amino acid is used in the synthesis of a protein. The tRNA is then free to migrate back to the cell cytoplasm and repeat the process.

The ribosome is the part of the cell in which protein synthesis takes place.

Messenger RNA is used at most only a few times before being depolymerized. Although this may seem to be a terrible waste, it allows the cell to produce different proteins on very short notice. As conditions change, different types of mRNA come from the nucleus, different proteins are made, and the cell responds adequately to a changing environment.

Cell Communicators — the G-Proteins

Cells have a common need to sense the environment in order to carry on the functions of the cell. Bacteria move toward food and away from danger. When environmental conditions are right, cells grow and divide. In a different environment, cells will not multiply. Thousands of signals can activate cells. The signals are usually chemical, and are often cell-produced proteins. One type of cell communicator or activator are the **G-proteins,** guanine-nucleotide-binding proteins. Seven different G-proteins have been identified. Their molecular weights are in the 100,000 range, and they are composed of three distinct protein subunits.

G-proteins function primarily by causing the production of second messengers such as enzymes. The G-protein attaches to a cell membrane and stimulates the membrane to produce the enzyme needed for a particular metabolic function. In essence, G-proteins are the amplifiers of the communication process by triggering the production of second transmitters.

The G-proteins operate in such diverse systems as vision, smell, control of cell growth and division, cellular regulation of a host of hormones and neurotransmitters, as well as the production of enzymes.

Several diseases either interfere or compete with the function of G-proteins. Some examples are the cholera toxin and its effects on intestinal cells, traveler's diarrhea caused by *Escherichia coli* bacteria, and the bacterial toxin that causes whooping cough. These toxins either mimic G-proteins and stimulate the G-protein response or tie up the G-proteins and diminish the activation processes triggered by them.

Synthetic Nucleic Acids

Slow progress has been made in the synthesis of polynucleotides, principally because of the difficulties involved in determining the proper blocking groups.

In 1959, Arthur Kornberg synthesized a DNA type of polynucleotide, for which he received a Nobel Prize. He used natural enzymes as templates to arrange the nucleotides in the order of the desired polynucleotide. His product was not biologically active. In 1965, Sol Spiegelman synthesized the polynucleotide portion of an RNA virus. This polynucleotide was biologically active and reproduced itself readily when introduced into living cells. In 1967, Mehran Goulian and Arthur Kornberg synthesized a fully infectious virus of a more complicated DNA type.

In 1970, Gobind Khorana synthesized a complete, double-stranded, 77-nucleotide gene. He, too, used natural enzymes to join previously synthesized, short, single-stranded polynucleotides into the double-stranded gene.

The ability to alter genes has led to a new field of science known as **biogenetic engineering.** This is a very active field of research that has developed (among other accomplishments) bacteria that can clean up oil spills. In this case a patent was granted to the General Electric Company for the production of life — a unique patent. Other bacteria have been produced that can synthesize protein, human growth hormone, and insulin. The method of producing bacteria for a particular function involves removing a gene from the bacterium, splicing in part of a gene from a human or other organism (the part that produces human insulin, for example), placing the spliced gene back into the bacterium, and letting the bacterium make millions of other insulin-producing bacteria. This process of splicing and recombining genes is referred to as recombinant DNA technology. The implications of gene splicing are tremendous — for both good and bad — and will demand responsible human decision making for guidance toward the common good.

A **mutation** occurs whenever an individual characteristic appears that has not been inherited but is duly passed along as an inherited factor to the next generation. A mutation can readily be accounted for in terms of an alteration in the DNA genetic code; that is, some force alters the nucleotide structure in a reproductive cell. Some sources of energy, such as gamma radiation, are known to produce mutations. This is entirely reasonable because certain kinds of energy can disrupt some bonds, which can re-form in another sequence.

If scientists can control the genetic code, can they control hereditary diseases such as sickle cell anemia, gout, some forms of diabetes, and mental retardation? If our understanding of detailed DNA structure and the enzymatic activity required to build these structures continues to grow, it is

Biogenetic engineering is the alteration of genes for a desired purpose.

The process of forming and cloning recombinant DNA is discussed in Chapter 1. See Figure 1–9 for an outline of this process.

A mutation results when there has been an alteration of the genetic code contained within the DNA molecule.

Ethics and risks were discussed in Chapter 1.

reasonable to believe that some detailed relationships between structure and gross properties will emerge. If this happens, it may be possible to build compounds that, when introduced into living cells, can combat or block inherited characteristics.

SELF-TEST 15–D

1. **a.** The basic code for the synthesis of protein is contained in the _____ molecule.
   b. The synthesis of a protein is carried out when _____ molecules bring up the required amino acids to mRNA.
2. When DNA replicates itself, each nitrogenous base in the chain is matched to another one via _____ bonds.
3. The energy for DNA replication and natural protein synthesis is supplied by substances such as _____.
4. What nitrogenous base complements (matches through hydrogen bonding)
   a. adenine (A)?
   b. cytosine (C)?
   c. guanine (G)?
   d. thymine (T)?
   e. uracil (U)?
5. A gene has been synthesized from individual nucleotides in the laboratory without the aid of natural enzymes. True ( ) False ( )
6. Replication means the same as *duplication.* True ( ) False ( )
7. The sugar in RNA is _____, whereas the one in DNA is _____.
8. A nucleotide contains _____, _____, and _____.
9. The secondary structure of DNA is in the shape of a(n) _____.

MATCHING SET

_____	**1.** Energy "cash" in the living cell	**a.** Ptyalin
		b. ADP + energy
_____	**2.** Mutation	**c.** Krebs cycle
_____	**3.** Enzyme that splits polysaccharides in the mouth	**d.** Chlorophylls
		e. Structure determined by DNA and RNA
		f. Altered DNA
_____	**4.** Natural protein	**g.** ATP
_____	**5.** Occurs under aerobic (with air) conditions	**h.** Embden-Meyerhof pathway
		i. Polymer consisting of α-D-glucose monomers
_____	**6.** Product of ATP hydrolysis	**j.** Proteins
		k. Sugar present in the blood

_____ 7. Molecules that absorb light energy

_____ 8. D-glucose

_____ 9. Methionine

_____ 10. Enzymes

_____ 11. Carbohydrate stored in animals

_____ 12. Starch

_____ 13. Polypeptides

_____ 14. DNA

_____ 15. Fibrous protein

_____ 16. Cellulose

_____ 17. Vitamins

_____ 18. Enzyme that splits sucrose into fructose and glucose

l. Amino acid
m. Sucrase
n. Polynucleotide
o. Biochemical catalysts
p. Coenzymes
q. Glycogen
r. Collagen
s. Polymer consisting of β-D-glucose monomers

QUESTIONS

1. Show the structure of the product that would be obtained if two alanine molecules (Table 15–1) were to react to form a dipeptide.

2. What is an essential amino acid?

3. Name a polysaccharide that yields only α-D-glucose upon complete hydrolysis. Name a disaccharide that yields the same hydrolysis product.

4. What is the chemical difference between the starch amylopectin and the "animal starch" glycogen?

5. What is the chief function of glycogen in animal tissue?

6. Explain the basic differences between amylose and cellulose.

7. Why does cotton, a cellulose material, absorb moisture so much better than nylon 66? (The structure of nylon 66 is given in Chapter 14.)

8. What functional groups are always present in each molecule of an amino acid?

9. a. What element is necessarily present in proteins that is not present in either carbohydrates or fats?
   b. Name another element that is probably present in proteins but is not present in either carbohydrates or fats.

10. What are the meanings of the terms *primary, secondary,* and *tertiary structures of proteins?*

11. In a protein, what type of bond holds the helical structure in place?

12. a. Which of the following biochemicals are polymers: starch, cellulose, glucose, fats, proteins, DNA, and RNA?
    b. What are the monomer units for those that are polymers?

13. What is the chemical function of many vitamins? Give some examples.

14. Why are carbohydrates considered "energy-rich"?

15. Why can humans not digest cellulose?

16. The molecular structures of enzymes (particularly apoenzymes) are most closely related to which of the following structures: proteins, fats, carbohydrates, or polynucleic acids?

17. What is the basic nature of the digestion processes for large molecules?

18. What compound produces soreness in the muscles after a period of vigorous exercise?

19. If protein digestion is facilitated by enzymes and these enzymes are produced in body organs made of proteins, why do the enzymes not cause rapid digestion of the organs themselves?

20. What are the end-products in the digestion of carbohydrates? Of fats? Of proteins?

21. How do living beings store and transfer energy?

22. What is the role of enzymes in digestion?

23. What is the purpose of ATP?

24. What important types of chemicals can function as coenzymes?

25. What three molecular units are found in nucleotides?

26. What are the basic differences between DNA and RNA structures?

27. What stabilizing forces hold the double helix together in the secondary structure of DNA proposed by Watson and Crick?

28. What is recombinant DNA?

29. Does a strand of DNA actually duplicate itself base for base in the formation of a strand of mRNA? Explain.

30. A mutation can be explained in terms of a change in which chemical in the cell?

31. The replication of DNA occurs in what part of the cell?

32. What is meant by a base *pair* in protein synthesis? What type of bond holds base pairs together?

33. Why is the liver called the central nutrient bank of the body?

34. What is the function of G-proteins? Give three types of systems in which G-proteins function.

35. What are SNURPS and what is their function?

36. What is a genome? Why would knowing one's genome help to improve health? How?

37. Visible light causes a reversible chemical change in the eye. However, ultraviolet light can cause permanent blindness. Explain on the basis of the energy of the photons of visible light (smaller) and the photons of ultraviolet light (larger).

38. Use Figure 15–1 to help decide which structure below represents α-D-glucose and which represents β-D-glucose. (Courtesy Phillips Petroleum.)

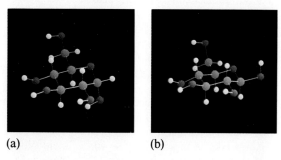

(a)   (b)

39. Which of the structures below is ribose and which is deoxyribose? Figure 15–25 may help. (Courtesy Phillips Petroleum.)

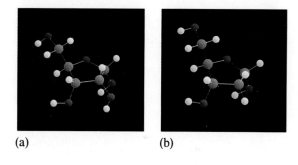

(a)   (b)

ESSAY by ROALD HOFFMANN

There seems to be nothing beautiful about molecule **1**. It looks like a clump of tangled spaghetti or a nest of worms. The molecule's function cannot be guessed (unless you already know what it is!). It is a far more complicated structure than those we examined in prior essays.

Complexity poses problems in any field, in chemistry as in the visual arts or music. In science, simplicity and complexity always coexist. The world of real phenomena is infinitely detailed, while the underlying principles are simpler (although not as simple as our naive minds imagine them to be). In chemistry more than in most other sciences, complexity is at the core. I call it a richness of possibilities.

Chemistry is not as much the science of a hundred or so elements, but the science of the infinite variety of compounds that can be built from them. If you want a simple molecule — one shaped like a tetrahedron or a cube — we have it for you. If you want a complicated one — intricate enough to work efficiently in a living body with its 10,000 simultaneous reactions — we have that, too. Do you want to change a male hormone into a female one, or convert the blue of

cornflowers into the red of poppies? No problem; merely change a CH_3 group or a proton, respectively.

Beautiful molecule **1** is **hemoglobin,** the protein that carries oxygen throughout the mammalian body. Like many proteins, it is built from several chunks or subunits. In this case, there are two pairs of subunits, called α (alpha) and β (beta). The way the subunits fit together is critical for hemoglobin's task, which is to take oxygen from the lungs to all the cells in the body.

One of the hemoglobin subunits is shown in picture **2**. It is a single chain, many hundreds of atoms long, that is twisted into a springlike helix. (The "tube" is the outline of the helix; the actual molecule is just the thin coil within the outline.) A "heme" molecule is nestled into a pocket within the curves of the chain.

All proteins contain such chains, called **polypeptides.** A schematic formula of part of a polypeptide is shown in **3**. The building blocks of proteins are **amino acids,** molecules whose general structure is shown in **4**.

In a reaction known as a condensation, a hydrogen atom from the left end of one amino acid molecule combines with the —OH group from the right end of another amino acid to form water. Simultaneously, the nitrogen on the first molecule attaches to the end carbon of the other to form a bond called a "peptide link." This reaction occurs again and again, each time lengthening the chain by one amino acid.

The "side chain" represented by R in **4** is different for each kind of amino acid. Fragment **3** shows three different amino acids, with side chains R, R′, and R″, linked together.

In all the thousands of proteins in living systems, only about 20 different amino acids occur. However, that supply of building blocks is quite rich enough for nature's purposes. A typical protein contains a hundred or more amino acid groups — for example, the hemoglobin subunit is made of 146 of them. Each position along the chain could (in principle) have any one of the 20 distinct side chains. Thus, we can imagine 20^{146} possible

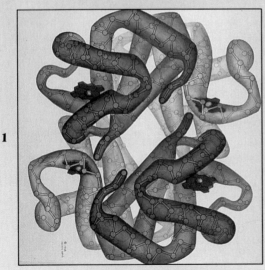

Illustration copyright by Irving Geis.

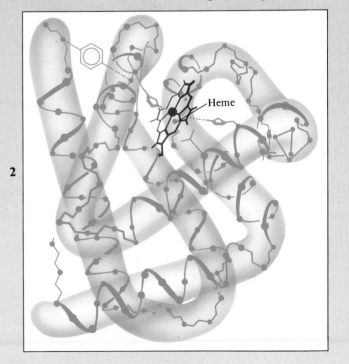

2

Heme

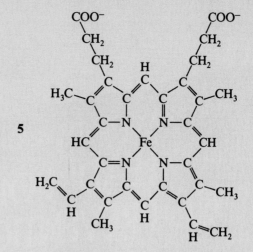

3
$$\text{—N—C—C—N—C—C—N—C—C—}$$
(with R, O, R', O, R'', O substituents; H atoms on N and C)

5 (heme structure)

proteins with the same length as the hemoglobin subunit—an astronomical number! Only a very small fraction of all the possible proteins exist in living organisms.

The various side chains have specific effects on the shape of the protein, and hence on its function. As you can see in **2**, large sections of the hemoglobin subunit curl into a helix. At certain places the chain kinks sharply; at these points we find the amino acid called proline. Other amino acid side chains tend to form bonds with partners in other parts of the protein, stabilizing the positions of the helical sections. The seeming jumble of spaghetti is not at all random, but a direct consequence of the order in which the amino acids are assembled into the polypeptide.

The hemoglobin subunit has two functions. The first is to hold the flat, disklike heme molecule that binds oxygen. The second is to change shape, in a certain way, once the oxygen is bound. The structure of heme is shown in **5**.

The oxygen molecule binds to the iron atom at the center of the heme. As it does, the iron changes its position a little, the heme flexes, and the surrounding protein moves. In a cascade of molecular motions, the fact that one hemoglobin subunit has captured an oxygen molecule is communicated to another subunit. As a result, the second subunit changes shape in such a way that it is easier for it, too, to bind an oxygen. Drawing **6** shows an interpretation of the changes in the geometry of all four subunits that accompany oxygenation of hemoglobin.

That bizarre sculpted folding has a purpose related to the function of a

4
(amino acid structure with H, N, R, C, C, O, OH)

molecule that is critical to life. Suddenly we see its dazzling beauty that cries out, "For this task, I am the best that can be." If you are so inclined, it testifies to a Designer.

Beautiful, certainly. The best, fashioned according to a plan? Hardly. It takes only a few bubbles of carbon monoxide, the lethal product of fires and car exhausts, to bring us back to reality. CO fits into the same protein pocket, and it binds to heme several hundred times better than does oxygen. This would hardly be the case if hemoglobin was the perfect solution to the oxygen transport problem.

As F. Jacob wrote in *Science*, Nature is a tinkerer. It has a wonderful evolutionary mechanism for exploring the effects of chance variations in molecular structure. Also, until we humans came along, changes in the chemical makeup of the environment

were gradual. While organisms were evolving hemoglobin, there wasn't much CO around, so there was no need for protection against it.

Actually, the story of molecular evolution is still more complicated and wonderful. We might wonder why the little heme molecule needs the enormous hemoglobin protein to carry it around. It seems that a bit of CO is always present in the body as a result of cellular processes. Heme that is removed from its protein binds CO much better than heme within hemoglobin. Apparently, one of the functions of the protein is to discourage CO bonding a little. This is not enough to protect against massive doses of external carbon monoxide, but enough to allow the heme to bind O_2 in the presence of naturally produced CO. Chemical complexity makes it possible.

Illustration copyright by Irving Geis.

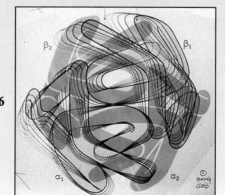

6

16   Toxic Substances

Toxic substances upset the incredibly complex system of chemical reactions occurring in the human body. Sometimes toxic substances cause mere discomfort; sometimes they cause illness, disability, or even death. Toxic symptoms can be caused by very small amounts of extremely toxic materials (an example is sodium cyanide) or larger amounts of a less toxic substance. The term *toxic substances* usually is limited to materials that are dangerous in small amounts. However, as most of us know, ill effects can be caused by excessive intake of substances normally considered harmless (eating too much candy, for example). Fortunately, in most cases the human body is capable of recognizing "foreign" chemicals and ridding itself of them. In this chapter, we shall focus on the chemical mechanisms by which toxic substances act.

A large enough dose of any compound can result in poisoning.

DOSE

Lethal doses of toxic substances are customarily expressed in milligrams (mg) of substance per kilogram (kg) of body weight of the subject. For example, the cyanide ion (CN^-) is generally fatal to human beings in a dose of 1 mg of CN^- per kg of body weight. For a 200-lb (90.7 kg) person, about 0.1 g of cyanide is a lethal dose. Examples of somewhat less toxic substances and the range of lethal doses for human beings follow:

"Dosis sola facit venenum"—the dose makes the poison.

Morphine	1–50 mg per kg
Aspirin	50–500 mg per kg
Methyl alcohol	500–5000 mg per kg
Ethyl alcohol	5000–15,000 mg per kg

A quantitative measure of toxicity is obtained by introducing into laboratory animals (such as rats) various dosages of substances to be tested. The dosage found to be lethal in 50% of a large number of the animals under controlled conditions is called the LD_{50} (lethal dosage—50%) and is reported in milligrams of poison per kilogram of body weight. Thus, if a

The death of Socrates by hemlock (coniine), as painted by Jacques David (1748–1825). (Courtesy of the Metropolitan Museum of Art, New York.)

TABLE 16-1 LD$_{50}$ Values for Dioxin (2,3,7,8-TCDD)

Species*	LD$_{50}$ (mg/kg)
Guinea pig	0.0006
Rat	0.04
Monkey	0.07
Rabbit	0.115
Dog	0.150
Mouse	0.200
Hamster	3.5
Bullfrog	>1.0

* No known human deaths have been reported for this compound.

TABLE 16-2 Approximate Comparison of LD$_{50}$ Values with Lethal Doses for Human Adults

Oral LD$_{50}$ for Any Animal (mg/kg)	Probable Lethal Oral Dose for Human Adult
<5	A few drops
5-50	"A pinch" to 1 tsp
50-500	1 tsp-2 T
500-5000	1 oz-1 pt (1 lb)
5000-15,000	1 pt-1 qt (2 lb)

See Table 13-6 for the structure of a dioxin isomer.

statistical analysis of data on a large population of rats showed that a dosage of 1 mg/kg was lethal to 50% of the population tested, the LD$_{50}$ for this poison would be 1 mg/kg. Obviously, metabolic variations and other differences between species will produce different LD$_{50}$ values for a given poison in different kinds of animals. For some toxic chemicals, the effect between animal species can be very great. For example the toxicity of dioxin, a multi-ring chlorinated compound produced when chlorinated compounds are incinerated, and also found as an impurity in some defoliants, such as Agent Orange, varies over an extremely large range (Table 16-1). The extremely low LD$_{50}$ for the guinea pig is unusual. Some have classed dioxin as one of the most toxic compounds known. Dioxin certainly is toxic to animals, but it does not rank as a potent human toxin.

Due to species differences in LD$_{50}$ values, toxicity cannot be extrapolated to human beings with any assurance, but it is generally safe to assume that a chemical with a low LD$_{50}$ value for several species will also be quite toxic to human beings (Table 16-2).

Toxic substances can be classified according to the way in which they disrupt the chemistry of the body. Some modes of action of toxic substances can be described as **corrosive, metabolic, neurotoxic, mutagenic, teratogenic,** and **carcinogenic,** and these will serve as the bases of our discussion.

Metabolism (from the Greek, *metaballein,* meaning "to change or alter") is the sum of all the physical and chemical changes by which living organisms are produced and maintained.

CORROSIVE POISONS

Toxic substances that actually destroy tissues are corrosive poisons. Examples include strong acids and alkalies and many oxidants, such as those found in laundry products, which can destroy tissues. Sulfuric acid (found in auto batteries) and hydrochloric acid (also called muriatic acid, used for cleaning purposes) are very dangerous corrosive poisons. So is sodium hydroxide, used in clearing clogged drains. Death has resulted from the swallowing of 1 oz of concentrated (98%) sulfuric acid, and much smaller amounts can cause extensive damage and severe pain.

Strong acids and bases destroy cell protoplasm.

Concentrated mineral acids such as sulfuric acid act by first dehydrating cellular structures. The cell dies because its protein structures are destroyed by the acid-catalyzed hydrolysis of the peptide bonds.

$$R \underset{\text{Peptide Link}}{\overset{\overset{\displaystyle O \quad H}{\|\quad|}}{-C-N}}-R' + H_2O \xrightarrow[\text{From Acid}]{H^+} R-\overset{\overset{\displaystyle O}{\|}}{C}-OH + H-\overset{\overset{\displaystyle H}{|}}{N}-R'$$

Peptide Link (In Protein)   Carboxyl End of Smaller Peptide or Amino Acid   Amine End of Smaller Peptide or Amino Acid

In the early stages of this process there will be a large proportion of larger fragments present. Subsequently, as more bonds are broken, smaller and smaller fragments result, leading to the ultimate disintegration of the tissue.

Some poisons act by undergoing chemical reaction in the body to produce corrosive poisons. Phosgene, the deadly gas used during World War I, is an example. When inhaled, it is hydrolyzed in the lungs to hydrochloric acid, which causes pulmonary edema (a collection of fluid in the lungs) due to the dehydrating effect of the strong acid on tissues. The victim "drowns" because oxygen cannot be absorbed effectively by the flooded and damaged tissues.

Chemical "warfare gases," such as phosgene, were outlawed by an international conference in 1925.

$$\underset{\text{Phosgene}}{\overset{\overset{\displaystyle O}{\|}}{\underset{Cl \quad\quad Cl}{C}}} + H_2O \longrightarrow \underset{\substack{\text{Hydrochloric}\\\text{Acid}}}{2\ HCl} + \underset{\substack{\text{Carbon}\\\text{Dioxide}}}{CO_2}$$

Sodium hydroxide, NaOH (caustic soda—a component of drain cleaners), is a very strongly alkaline, or basic, substance that can be just as corrosive to tissue as strong acids. The hydroxide ion also catalyzes the splitting of peptide linkages:

$$R-\overset{\overset{\displaystyle O}{\|}}{C}-\overset{\overset{\displaystyle H}{|}}{N}-R' + H_2O \xrightarrow[\text{Base}]{OH^-} R-\overset{\overset{\displaystyle O}{\|}}{C}-OH + H-\overset{\overset{\displaystyle H}{|}}{N}-R'$$

Both acids and bases, as well as other types of corrosive poisons, continue their action until they are consumed in chemical reactions.

Some corrosive poisons destroy tissue by oxidizing it. This is characteristic of substances such as ozone, nitrogen dioxide, and possibly iodine, which destroy enzymes by oxidizing their functional groups. Specific groups, such as the —SH and —S—S— groups in the enzyme, are believed to be converted by oxidation to nonfunctioning groups; alternatively, the oxidizing agents may break chemical bonds in the enzyme, leading to its inactivation.

A summary of some common corrosive poisons is presented in Table 16–3.

METABOLIC POISONS

Metabolic poisons are more subtle than the tissue-destroying corrosive poisons. In fact, many of them do their work without actually indicating their presence until it is too late. Metabolic poisons can cause illness or death by

TABLE 16–3 Some Corrosive Poisons

Substance	Formula	Toxic Action	Possible Contact
Hydrochloric acid	HCl	Acid hydrolysis	Tile and concrete floor cleaner; concentrated acid used to adjust acidity of swimming pools
Sulfuric acid	H_2SO_4	Acid hydrolysis, dehydrates tissue—oxidizes tissue	Auto batteries
Phosgene	ClCOCl	Acid hydrolysis	Combustion of chlorine-containing plastics (PVC or Saran)
Sodium hydroxide	NaOH	Base hydrolysis	Caustic soda, drain cleaners
Trisodium phosphate	Na_3PO_4	Base hydrolysis	Detergents, household cleaners
Sodium perborate	$NaBO_3 \cdot 4H_2O$	Base hydrolysis—oxidizing agent	Laundry detergents, denture cleaners
Ozone	O_3	Oxidizing agent	Air, electric motors
Nitrogen dioxide	NO_2	Oxidizing agent	Polluted air, automobile exhaust
Iodine	I_2	Oxidizing agent	Antiseptic
Hypochlorite ion	OCl^-	Oxidizing agent	Bleach
Peroxide ion	O_2^{2-}	Oxidizing agent	Bleach, antiseptic
Oxalic acid	$H_2C_2O_4$	Reducing agent, precipitates Ca^{2+}	Bleach, ink eradicator, leather tanning, rhubarb, spinach, tea
Sulfite ion	SO_3^{2-}	Reducing agent	Bleach
Chloramine	NH_2Cl	Oxidizing agent	Produced when household ammonia and chlorinated bleach are mixed
Nitrosyl chloride	NOCl	Oxidizing agent	Mixing household ammonia and bleach

interfering with a vital biochemical mechanism to such an extent that it ceases to function or is prevented from functioning efficiently.

Carbon Monoxide

The interference of carbon monoxide with extracellular oxygen transport is one of the best understood processes of metabolic poisoning. As early as 1895, it was noted that carbon monoxide deprives body cells of oxygen (asphyxiation), but it was much later before it was known that carbon monoxide, like oxygen, combines with hemoglobin:

$$O_2 + \text{hemoglobin} \longrightarrow \text{oxyhemoglobin}$$

$$CO + \text{hemoglobin} \longrightarrow \text{carboxyhemoglobin}$$

Laboratory tests show that carbon monoxide reacts with hemoglobin to give a compound (carboxyhemoglobin) that is 140 times more stable than the compound of hemoglobin and oxygen (oxyhemoglobin) [Fig. 16–1]. Since hemoglobin is so effectively tied up by carbon monoxide, it cannot perform its vital function of transporting oxygen.

An organic material that undergoes incomplete combustion will always liberate carbon monoxide. Sources include auto exhausts, smoldering

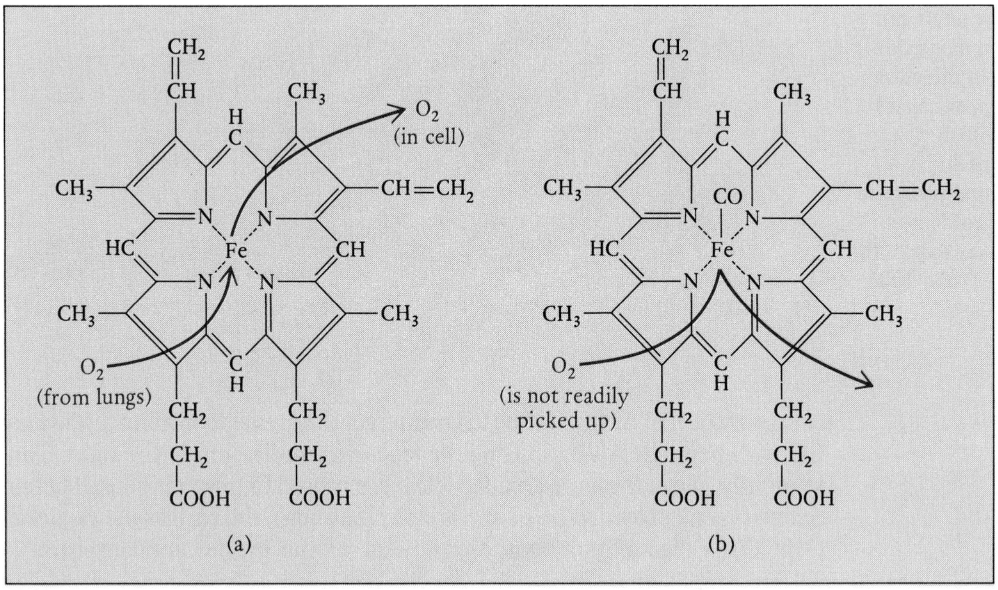

Figure 16–1 **Structure of the heme portion of hemoglobin. (a) Normal acceptance and release of oxygen. (b) Oxygen blocked by carbon monoxide.**

leaves, lighted cigars or cigarettes, and charcoal burners. In the United States alone, combustion sources of all types dump about 200 million tons of carbon monoxide per year into the atmosphere.

While the best estimates of the maximum global background level of carbon monoxide are of the order of 0.1 ppm, the background concentration in cities is higher. In heavy traffic, sustained levels of 100 or more ppm are common; for offstreet sites an average of about 7 ppm is typical for large cities. A concentration of 30 ppm for 8 h is sufficient to cause headache and nausea. Breathing an atmosphere that is 0.1% (1000 ppm) carbon monoxide for 4 h converts approximately 60% of the hemoglobin of an average adult to carboxyhemoglobin (Table 16–4), and death is likely to result (Fig. 16–2).

Since both the carbon monoxide and oxygen reactions with hemoglobin involve easily reversed reactions, the concentrations, as well as relative strengths of bonds, affect the direction of the reaction. In air that contains 0.1% CO, oxygen molecules outnumber CO molecules 200 to 1. The larger

ppm — parts per million — a measure expressing concentration. 50 ppm CO means 50 ml CO for every million ml of air.

To convert ppm to percent, divide by 10,000.

TABLE 16–4 Concentration of CO in Atmosphere versus Percentage of Hemoglobin (Hb) Saturated*

CO concentration in air	0.01% (100 ppm)	0.02% (200 ppm)	0.10% (1000 ppm)	1.0% (10,000 ppm)
Percentage of hemoglobin molecules saturated with CO†	17	20	60	90

* A few hours of breathing time is assumed.
† Normal human blood contains up to 5% of the hemoglobin as carboxyhemoglobin (HbCO).

Figure 16–2 A healthy adult can tolerate 100 ppm carbon monoxide in air without suffering ill effect. A 1-hr exposure to 1000 ppm causes a mild headache and a reddish coloration of the skin develops. A 1-hr exposure to 1300 ppm turns the skin cherry red and a throbbing headache develops. A 1-hr exposure to concentrations greater than 2000 ppm will likely cause death.

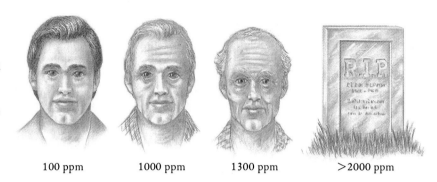

| 100 ppm | 1000 ppm | 1300 ppm | >2000 ppm |

Air is 21% O_2 by volume; in 1 million "air molecules" there would be 210,000 O_2 molecules.

concentration of oxygen helps to counteract the greater combining power of CO with hemoglobin by shifting the reaction equilibrium to the right. Consequently, if a carbon monoxide victim is exposed to fresh air or, still better, pure oxygen (provided he or she is still breathing), the carboxyhemoglobin (HbCO) is gradually decomposed, owing to the greater concentration of oxygen:

$$HbCO + O_2 \rightleftharpoons HbO_2 + CO$$

Equilibrium Shifted to Right Because
of Greater Concentration of Oxygen

Although carbon monoxide is not a cumulative poison, permanent damage can occur if certain vital cells (e.g., brain cells) are deprived of oxygen for more than a few minutes.

Individuals differ in their tolerance of carbon monoxide, but generally those with anemia or an otherwise low reserve of hemoglobin (e.g., children) are more susceptible. No one is helped by carbon monoxide, and smokers suffer chronically from its effects. There is also a strong relationship between low infant birth weight and mothers who smoke.

Finally it should be noted that carbon monoxide is a subtle poison, since it is colorless, odorless, and tasteless.

Cyanide

The cyanide ion (CN^-) is the toxic agent in cyanide salts such as sodium cyanide used in electroplating. Since the cyanide is a relatively strong base, it reacts easily with many acids (weak and strong) to form volatile hydrogen cyanide (HCN):

$$CH_3COOH + Na^+CN^- \rightleftharpoons HCN + Na^+CH_3COO^-$$

| Acetic Acid | Sodium Cyanide | Hydrogen Cyanide | Sodium Acetate |

Since HCN boils at a relatively low temperature ($26°C$), it is a gas at temperatures slightly above room temperature. It is often used as a fumigant in storage bins and holds of ships because it is toxic to most forms of life and, in gaseous form, can penetrate into tiny openings, even into insect eggs.

Natural sources of cyanide ions include the seeds of the cherry, plum, peach, apple, and apricot fruits. Hydrogen cyanide is produced by hydrolysis of certain compounds, such as amygdalin, contained in the seeds:

The cyanide is not toxic as long as it is tied up in the amygdalin, but presumably if enough apple or peach seeds were hydrolyzed in warm acid, sufficient HCN would result to cause considerable danger. There are a few recorded instances of humans poisoned by eating large numbers of apple seeds. Amygdalin is not confined to the seeds; amounts as high as 66 mg/100 g have been reported in peach leaves.

The cyanide ion is one of the most rapidly working poisons. Lethal doses taken orally act in minutes. Cyanide poisons by asphyxiation, as does carbon monoxide, but the mechanism of cyanide poisoning is different (Fig. 16–3). Instead of preventing the cells from getting oxygen, cyanide interferes with oxidative enzymes, such as cytochrome oxidase. Oxidases are enzymes containing a metal, usually iron or copper. They catalyze the oxidation of substances such as glucose:

$$\text{Metabolite (H)}_2 + \tfrac{1}{2}\,O_2 \xrightarrow{\text{Oxidase}} \text{Oxidized metabolite} + H_2O + \text{energy}$$

The iron atom in cytochrome oxidase is oxidized from Fe^{2+} to Fe^{3+} to provide electrons for the reduction of O_2. The iron regains electrons from other steps in the process. The cyanide ion forms stable cyanide complexes with the metal ion of the oxidase and renders the enzyme incapable of reducing oxygen or oxidizing the metabolite.

$$\text{Cytochrome oxidase (Fe)} + CN^- \longrightarrow \text{Cytochrome oxidase } \underbrace{\text{(Fe)} \cdots CN^-}_{\text{Complex}}$$

In essence, the electrons of the iron ion are "frozen"—they cannot participate in the oxidation–reduction processes. Plenty of oxygen gets to the cells, but the mechanism by which the oxygen is used in the support of life is stopped. Hence, the cell dies, and if this occurs fast enough in the vital centers, the victim dies.

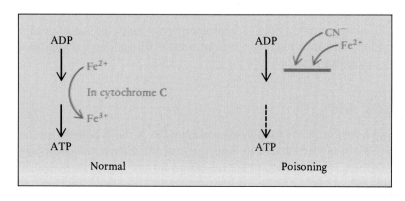

Figure 16–3 The mechanism of cyanide (CN^-) poisoning. Cyanide binds tightly to the enzyme cytochrome C, an iron compound, thus blocking the vital ADP–ATP reaction in cells.

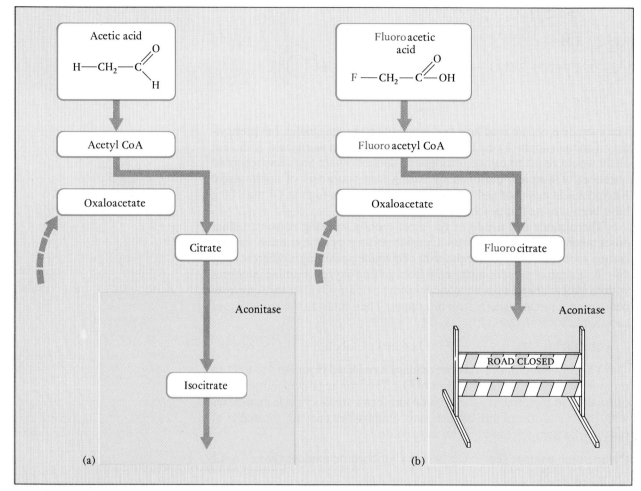

Figure 16–4   Fluoroacetic acid is converted into fluorocitrate, which then forms a stable bond with the enzyme aconitase. This blocks the normal Krebs cycle, a portion of which is shown.

The body has a mechanism for ridding itself slowly of cyanide ions. The cyanide-oxidative enzyme reaction is reversible, and other enzymes such as rhodanase, found in almost all cells, can convert cyanide ions to relatively harmless thiocyanate ions. For example,

$$CN^- + S_2O_3^{2-} \xrightarrow{\text{Rhodanase}} SCN^- + SO_3^{2-}$$

Thiosulfate   Thiocyanate

This mechanism is not as effective in protecting a cyanide-poisoning victim as it might appear, since only a limited amount of thiosulfate is available in the body at a given time.

The body can rid itself of many toxic substances if the dose is small enough and sufficient time is allowed.

Fluoroacetic Acid

Nature has used the synthesis of fluoroacetic acid as part of a defense mechanism for certain plants. Native to South Africa, the *gilbaar* plant contains lethal quantities of fluoroacetic acid. Cattle that eat these leaves usually sicken and die.

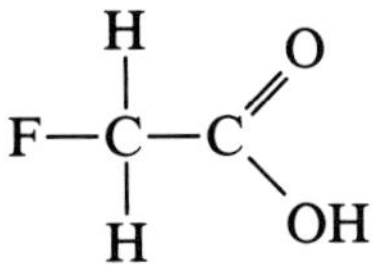

$$F-\overset{\overset{\displaystyle H}{|}}{\underset{\underset{\displaystyle H}{|}}{C}}-\overset{\displaystyle O}{C}\diagdown_{OH}$$

Fluoroacetic Acid

Sodium fluoroacetate, the sodium salt of this acid (Compound 1080), is a potent rodenticide (rat poison). Because it is odorless and tasteless it is especially dangerous, and its sale in this country is strictly regulated by law.

Fluoroacetate is toxic because it enters the Krebs cycle, producing fluorocitric acid, which in turn blocks the Krebs cycle (Fig. 16–4). The C—F linkage apparently ties up the enzyme aconitase, thus preventing it from converting citrate to isocitrate.

In this instance, the poison is similar enough to the normal substrate to compete effectively for the active sites on the enzyme. If a poison has sufficient affinity for the active site on the enzyme, it blocks the normal function of the enzyme. The blocking of the Krebs cycle by fluorocitrate is a typical example of this affinity. If fluoroacetate is not present in excessive amounts, its action can be reversed simply by increasing the concentration of available citrate.

Some poisons can act by mimicking other compounds.

HEAVY METALS

Heavy metals are perhaps the most common of all the metabolic poisons. These include such common elements as lead and mercury, as well as many less common ones such as cadmium, chromium, and thallium. In this group we should also include the infamous poison, arsenic, which is really not a metal but is metal-like in many of its properties, including its toxic action.

Most heavy metals are cumulative poisons.

Arsenic

Arsenic, a classic homicidal poison, occurs naturally in small amounts in many foods. Shrimp, for example, contain about 19 ppm arsenic; corn may contain 0.4 ppm arsenic. Some agricultural insecticides contain arsenic (Table 16–5), and so some arsenic is observed in very small amounts on

TABLE 16–5 Some Arsenic-Containing Insecticides

Name	Formula
Lead arsenate	$Pb_3(AsO_4)_2$
Monosodium methanearsenate	$CH_3-\overset{\overset{\displaystyle O}{|}}{\underset{\underset{\displaystyle OH}{|}}{As}}-O^-Na^+$
Paris green (copper acetoarsenite)	$3\,CuO \cdot 3\,As_2O_3 \cdot Cu(C_2H_3O_2)_2$

Figure 16–5 Glutathione reaction with a metal (M).

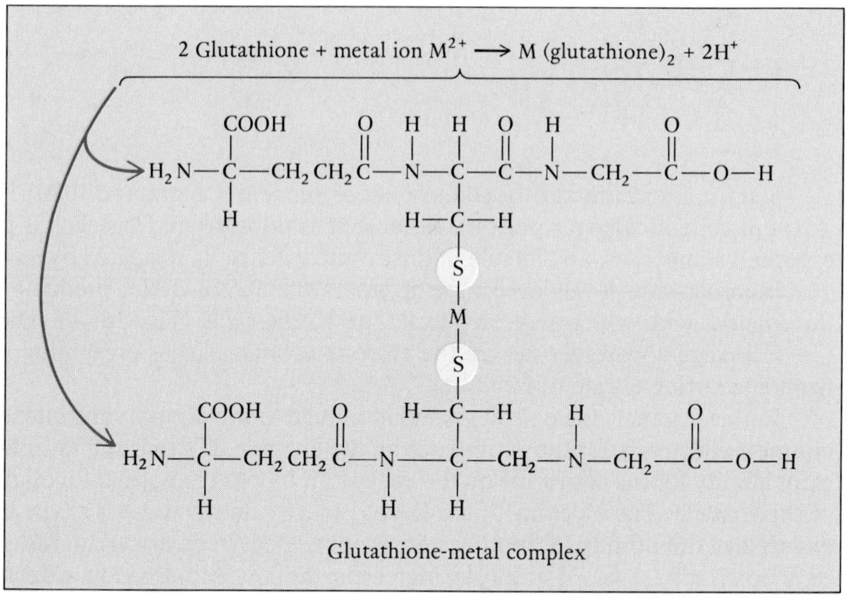

2 Glutathione + metal ion M^{2+} $\longrightarrow$ M (glutathione)$_2$ + 2H$^+$

Glutathione-metal complex

some fruits and vegetables. The Federal Food and Drug Administration (FDA) has set a limit of 0.15 mg of arsenic per pound of food, and this amount apparently causes no harm. Several drugs, such as arsphenamine, which has found some use in treating syphilis, contain covalently bonded arsenic. In its ionic forms, arsenic is much more toxic.

Arsenic and heavy metals owe their toxicity primarily to their ability to react with and inhibit sulfhydryl (—SH) enzyme systems, such as those involved in the production of cellular energy. For example, glutathione (a tripeptide of glutamic acid, cysteine, and glycine) occurs in most tissues; its behavior with metals illustrates the interaction of a metal with sulfhydryl groups. The metal replaces the hydrogen on two sulfhydryl groups on adjacent molecules (Fig. 16–5), and the strong bond that results effectively eliminates the two glutathione molecules from further reaction. Glutathione is involved in maintaining healthy red blood cells.

The typical forms of toxic arsenic compounds are inorganic ions such as arsenate (AsO_4^{3-}) and arsenite (AsO_3^{3-}). The reaction of an arsenite ion with sulfhydryl groups results in a complex in which the arsenic unites with two sulfhydryl groups, which may be on two different molecules of protein or on the same molecule:

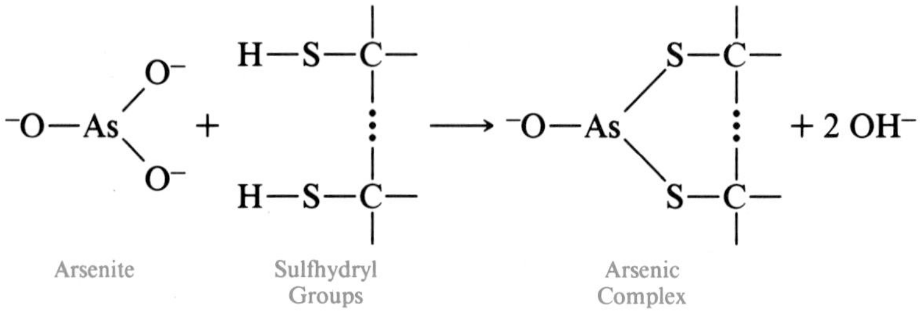

Arsenite   Sulfhydryl   Arsenic
                  Groups   Complex

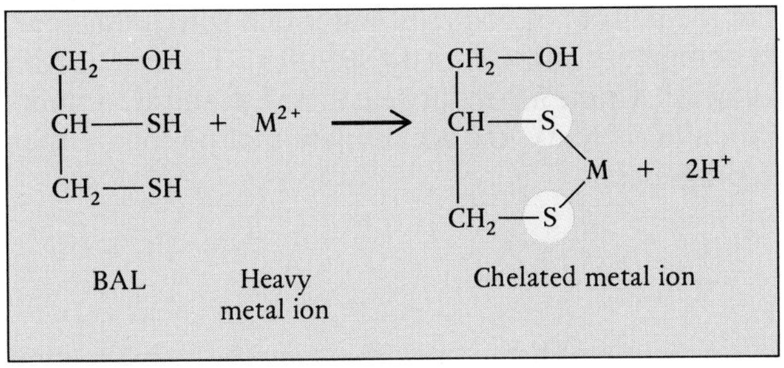

Figure 16-6 BAL chelation of arsenic or a heavy metal ion such as lead.

The problem of developing a compound to counteract *Lewisite,* an arsenic-containing poison gas used in World War I, led to an understanding of how arsenic acts as a poison and subsequently to the development of an antidote. Once it was understood that Lewisite poisoned people by the reaction of arsenic with protein sulfhydryl groups, British scientists set out to find a suitable compound that contained highly reactive sulfhydryl groups that could compete with sulfhydryl groups in the natural substrate for the arsenic, and thus render the poison ineffective. Out of this research came a compound now known as British anti-Lewisite (BAL).

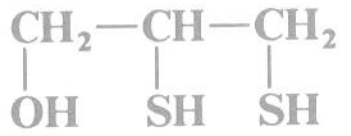

The BAL, which bonds to the metal at several sites, is called a **chelating agent** (Greek, *chela,* meaning "claw"), a term applied to a reacting agent that envelops a species such as a metal ion. BAL is one of many compounds that can act as chelating agents for metals (Fig. 16-6).

With the arsenic or heavy metal ion tied up, the sulfhydryl groups in vital enzymes are freed and can resume their normal functions. BAL is a standard therapeutic item in a hospital's poison emergency center and is used routinely to treat heavy-metal poisoning.

$$CH_2 - CH - CH_2$$

OH   SH   SH

BAL
(British anti-Lewisite)

A chelating agent encases an atom or ion like a crab or an octopus surrounds a bit of food.

Mercury

Mercury deserves some special attention because it has a rather peculiar fascination for some people, especially children, who love to touch it. It is poisonous and, to make matters worse, mercury and its salts accumulate in the body. This means the body has no quick means of ridding itself of this element and there tends to be a buildup of the toxic effects leading to *chronic* poisoning.

Although mercury is rather unreactive compared with other metals, it is quite volatile and easily absorbed through the skin. In the body, the metal atoms are oxidized to Hg_2^{2+} [mercury (I) ion] and Hg^{2+} [mercury (II) ion]. Compounds of both Hg_2^{2+} and Hg^{2+} are known to be toxic.

Today mercury poisoning is a potential hazard to those working with or near this metal or its salts, such as dentists (who use it in making amalgams for fillings), various medical and scientific laboratory personnel (who routinely use mercury compounds or mercury pressure gauges), and some agricultural workers (who employ mercury salts as fungicides).

Mercury can also be a hazard when it is present in food (see Chapter 1). It is generally believed that mercury enters the food chain through small

A vivid description of the psychic changes produced by mercury poisoning can be found in the Mad Hatter, a character in Lewis Carroll's *Alice in Wonderland.* The fur felt industry once used mercury (II) nitrate, $Hg(NO_3)_2$, to stiffen the felt. Chronic mercury poisoning accounted for the Mad Hatter's odd behavior; it also gave the workers in hat factories symptoms known as "hatter's shakes."

Amalgam: Any mixture or alloy of metals of which mercury is a constitutent.

organisms that feed at the bottom of bodies of water that contain mercury from industrial waste or mercury minerals in the sediment. These in turn are food for bottom-feeding fish. Game fish in turn eat these fish and accumulate the largest concentration of mercury, the accumulation of poison building up as the food chain progresses.

Lead

Lead is another widely encountered heavy-metal poison. The body's method of handling lead provides an interesting example of a "metal equilibrium" (Fig. 16–7). Lead often occurs in foods (100–300 μg/kg), beverages (20–30 μg/L), public water supplies (100 μg/L, from lead-sealed pipes), and even air (up to 2.5 μg/m^3 from lead compounds in auto exhausts). With this many sources and contacts per day, it is obvious that the body must be able to rid itself of this poison; otherwise everyone would have died long ago of lead poisoning! The average person can excrete about 2 mg of lead a day through the kidneys and intestinal tract; the daily intake is normally less than this. However, if intake exceeds this amount, accumulation and storage result. In the body lead not only resides in soft tissues but also is deposited in bone. In the bones lead acts on the bone marrow, in tissues it behaves like other heavy-metal poisons, such as mercury and arsenic. Lead, like mercury and arsenic, can also affect the central nervous system.

1μg (microgram) $= 10^{-6}$ g

1 mg (milligram) $= 10^{-3}$ g $= 1000$ μg

At the height of the Roman empire, lead production worldwide was about 80,000 tons per year. Today it is about 3 million tons per year.

Figure 16–7 Lead equilibrium in humans. Figures chosen for intake are probable upper limits.

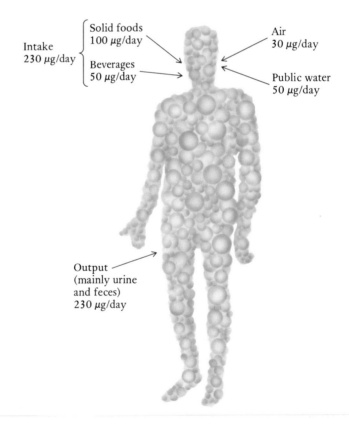

Intake 230 μg/day

Solid foods 100 μg/day

Beverages 50 μg/day

Air 30 μg/day

Public water 50 μg/day

Output (mainly urine and feces) 230 μg/day

Unless they are very insoluble, lead salts are always toxic, and their toxicity is directly related to the salt's solubility. But even metallic lead can be absorbed through the skin; cases of lead poisoning have resulted from repeated handling of lead foil, bullets, and other lead objects.

One of the truly tragic aspects of lead poisoning is that even though lead-pigmented paints have not been used for interior painting in this country during the past 30 years, children are still poisoned by lead from old paint. Health experts estimate that up to 225,000 children become ill from lead poisoning each year, with many experiencing mental retardation or other neurological problems. The reason for this is twofold. Lead-based paints still cover the walls of many older dwellings. Coupled with this is the fact that many children in poverty-stricken areas are ill fed and anemic. These children develop a peculiar appetite trait called **pica,** and among the items that satisfy their cravings are pieces of flaking paint, which may contain lead. Lead salts also have a sweet taste, which may contribute to this consumption of lead-based paint.

If a child consumes as little as 200 mg (200,000 μg) of older, lead-based paint, he or she may ingest as much as 2600 μg of lead, of which 550 μg will be absorbed. Since 1977, lead in paints has been limited to 600 ppm (dry weight). For every gallon of paint, about 8 lb, there should be no more than 0.0048 lb, or 2,100,000 μg, of lead. If this paint is spread uniformly over 400 ft^2 or about 371,600 cm^2, a paint chip 1 cm square will contain less than 6 μg of lead. Some older paints contained up to 50,000 ppm lead (or about 175 μg of lead/cm^2 of paint). Lead compounds have been used in paints as pigments, such as yellow lead chromate, and as drying agents, such as lead naphthenate.

Children retain a larger fraction of absorbed lead than do adults, and children do not immediately tie up absorbed lead in their bones as adults do. This inability to quickly absorb lead into their bones means the lead stays in a child's blood longer, where it can exert its toxic effects on various organs. Table 16–6 shows the effects of lead in children's blood.

Gamma ray fluorescent device for determining lead in paint. (*The World of Chemistry,* Program 6, "The Atom.")

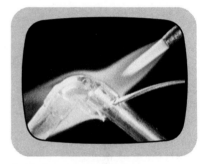

Soldering copper water pipes together with lead-based solder has been the cause of lead in drinking water. (*The World of Chemistry,* Program 19, "Metals.")

TABLE 16–6 Effects of Lead in Children's Blood

Pb Levels (μg/dL)*	Effect
~5	Elevated blood pressure
~10	Lowered intelligence
15–25	Decreased heme synthesis, decreased vitamin-D and calcium metabolism
25–40	Impaired central nervous system functions, delayed cognitive development, reduced IQ scores, impaired hearing, reduced hemoglobin formation
40–80	Peripheral neuropathy, anemia
80+	Coma, convulsions, irreversible mental retardation, possible death

* μg/dL = micrograms per deciliter of blood.
Source: Agency for Toxic Substance and Disease Registry, U.S. Department of Health and Human Services, Atlanta, 1988.

Figure 16–8   **The office of the Lead Poisoning Prevention Program in Chatham County (Savannah), Georgia.**

Studies in the mid-1980s by the U.S. Department of Health showed that about 1.5 million children have blood levels above 15 μg/dL. About 900,000 children have blood levels above 20 μg/dL (Fig. 16–8).

Toxicologists have discovered an effective chelating agent for removing lead from the human body—ethylenediaminetetraacetic acid, also called EDTA (Fig. 16–9).

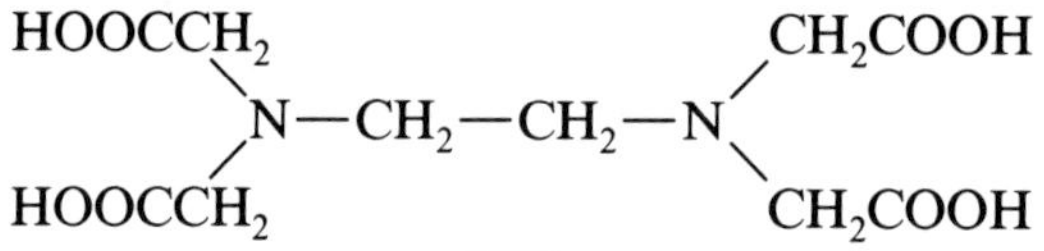

$$\text{HOOCCH}_2 \qquad\qquad\qquad \text{CH}_2\text{COOH}$$
$$\text{N}-\text{CH}_2-\text{CH}_2-\text{N}$$
$$\text{HOOCCH}_2 \qquad\qquad\qquad \text{CH}_2\text{COOH}$$

EDTA
(Ethylenediaminetetraacetic Acid)

The calcium disodium salt of EDTA is used in the treatment of lead poisoning because EDTA by itself would remove too much of the blood serum's calcium. In solution, EDTA has a greater tendency to complex with lead (Pb^{2+}) than with calcium (Ca^{2+}). As a result, the calcium is released and the lead is tied up in the complex:

$$[\text{CaEDTA}]^{2-} + Pb^{2+} \longrightarrow [\text{PbEDTA}]^{2-} + Ca^{2+}$$

The lead chelate is then excreted in the urine.

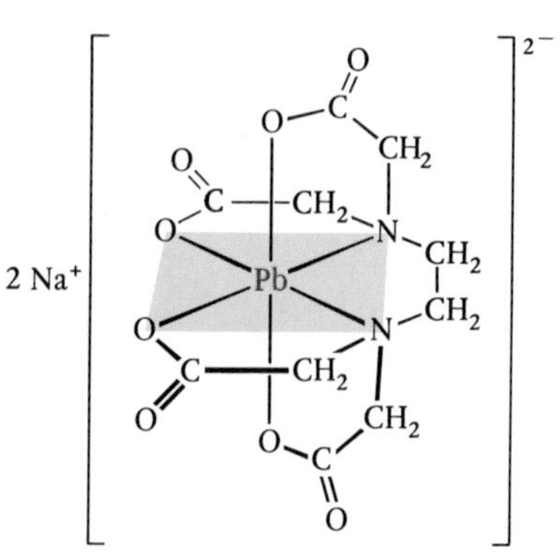

Figure 16–9   **The structure of the chelate formed when the anion of EDTA envelops a lead (II) ion.**

SELF-TEST 16–A

1. Corrosive poisons such as sulfuric acid destroy tissue by _____ followed by _____ of proteins.
2. Corrosive poisons, such as ozone, nitrogen dioxide, and iodine, destroy tissue by _____ it.

3. Carbon monoxide poisons by forming a strong bond with iron in _____ and thus preventing the transport of _____ from the lungs to the cells throughout the body.
4. CO is a cumulative poison. ( ) True ( ) False
5. The cyanide ion has the formula _____. It poisons by complexing with iron in the enzyme _____, thus preventing the use of _____ in the oxidative processes in the cells.
6. Give an example of a metabolic poison that is toxic because its structure is so similar to a useful substance that it can mimic the useful substance. _____
7. BAL is an antidote for _____. BAL is effective because its sulfhydryl (—SH) groups _____ arsenic and heavy metals and render them ineffective toward enzymes.
8. Mercury is a cumulative poison. ( ) True ( ) False
9. Name three ways children can be exposed to lead. _____, _____, and _____.

NEUROTOXINS

Anticholinesterase Poisons

Some metabolic poisons are known to limit their action to the nervous system. These include poisons such as strychnine and curare (a South American Indian dart poison), as well as the dreaded nerve gases developed for chemical warfare. The exact modes of action of most neurotoxins are not known for certain, but investigations have discovered the action of a few.

A nerve impulse or stimulus is transmitted along a nerve fiber by electric impulses. The nerve fiber connects either with another nerve fiber or with some other cell (such as a gland or cardiac, smooth, or skeletal muscle) capable of being stimulated by the nerve impulse (Fig. 16–10). Neurotoxins often act at the point where two nerve fibers come together, called a **synapse.** When the impulse reaches the end of certain nerves, a small quantity of **acetylcholine** is liberated. This activates a receptor on an adjacent nerve or organ. The acetylcholine is thought to activate a nerve ending by changing the permeability of the nerve cell membrane. The method of increasing membrane permeability is not clear, but it may be related to an ability to dissociate fat–protein complexes or to penetrate the surface films of fats. Such effects can be brought about by as little as 10^{-6} mol of acetylcholine, which could alter the permeability of a cell so ions can cross the cell membrane more freely.

To enable the receptor to receive further electrical impulses, the enzyme **cholinesterase** breaks down acetylcholine into acetic acid and choline (Fig. 16–11):

Investigations of the actions of neurotoxins have provided insight into how the nervous system works.

$$CH_3COCH_2CH_2N^+CH_3, OH^-$$

Acetylcholine

Permeability: The ability of a membrane to let chemicals pass through it.

10^{-6} of a mole of acetylcholine is 6×10^{17} molecules.

$$CH_3COCH_2CH_2N^+—CH_3, OH^- + H_2O \xrightarrow{\text{Cholinesterase}} CH_3COH + HOCH_2CH_2N^+—CH_3, OH^-$$

Acetylcholine   Water   Acetic Acid   Choline

Figure 16–10 "Cholinergic" nerves, which transmit impulses by means of acetylcholine, include nerves controlling both voluntary and involuntary activities. Exceptions are parts of the "sympathetic" nervous system that utilize norepinephrine instead of acetylcholine. Sites of acetylcholine secretion are circled in color; poisons that disrupt the acetylcholine cycle can interrupt the body's communications at any of these points. The role of acetylcholine in the brain is uncertain, as is indicated by the broken circles.

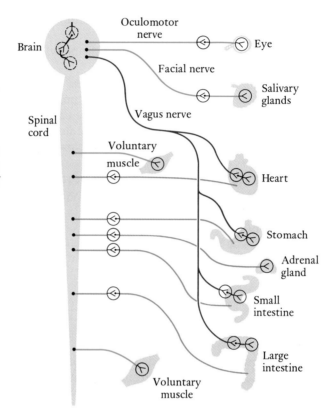

In the presence of potassium and magnesium ions, other enzymes such as acetylase resynthesize new acetylcholine from the acetic acid and the choline within the incoming nerve ending:

$$\text{Acetic acid} + \text{Choline} \xrightarrow{\text{Acetylase}} \text{Acetylcholine} + H_2O$$

The new acetylcholine is available for transmitting another impulse across the gap.

Neurotoxins can affect the transmission of nerve impulses at nerve endings in a variety of ways. The **anticholinesterase poisons** prevent the breakdown of acetylcholine by deactivating cholinesterase. These poisons are usually structurally analogous to acetylcholine, so they bond to the enzyme cholinesterase and deactivate it (Fig. 16–12). The cholinesterase molecules bound by the poison are held so effectively that the restoration of proper nerve function must await the manufacture of new cholinesterase. In the meantime, the excess acetylcholine overstimulates nerves, glands, and muscles, producing irregular heart rhythms, convulsions, and death. Many of the organic phosphates widely used as insecticides are metabolized in the body to produce anticholinesterase poisons. For this reason, they should be treated with extreme care. Some poisonous mushrooms also contain an anticholinesterase poison. Figure 16–13 contains the structures of some anticholinesterase poisons.

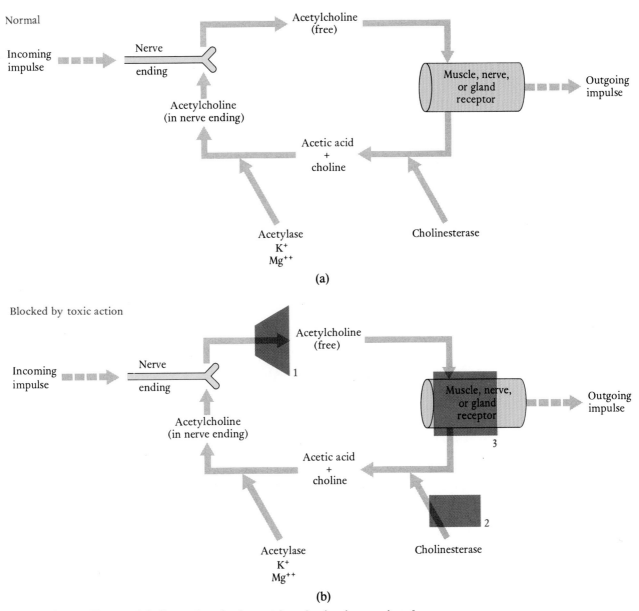

Figure 16–11   The acetylcholine cycle, a fundamental mechanism in nerve impulse transmission, is affected by many poisons. An impulse reaching a nerve ending in the normal cycle (a) liberates acetylcholine, which then stimulates a receptor. To enable the receptor to receive further impulses, the enzyme cholinesterase breaks down acetylcholine into acetic acid and choline; other enzymes resynthesize these into more acetylcholine. (b) Botulinus and dinoflagellate toxins inhibit the synthesis, or the release, of acetylcholine (1). The "anticholinesterase" poisons inactivate cholinesterase and therefore prevent the breakdown of acetylcholine (2). Curare and atropine desensitize the receptor to the chemical stimulus (3).

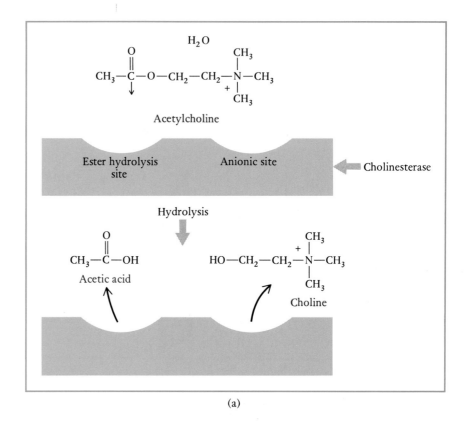

(a)

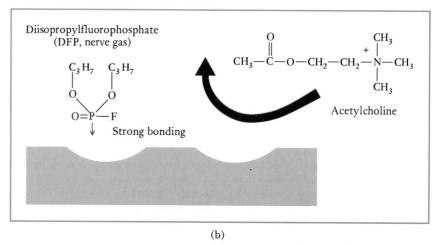

(b)

**Figure 16–12   (a) The mechanism of cholinesterase breakdown of acetylcholine. (b)
The tie-up of cholinesterase by an anticholinesterase poison like the nerve gas DFP
blocks the normal hydrolysis of acetylcholine since the acetylcholine cannot bind to
the enzyme.**

Name	Structure	LD_{50} (rat;oral), mg/kg	Use
Sarin	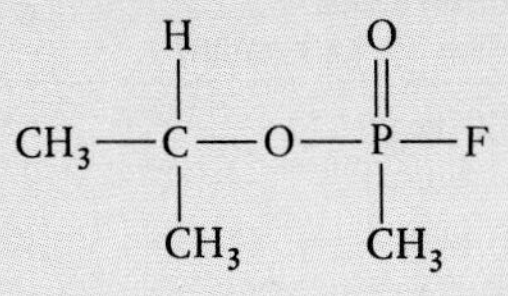	0.55	Chemical warfare agent first produced in World War II
Tabun	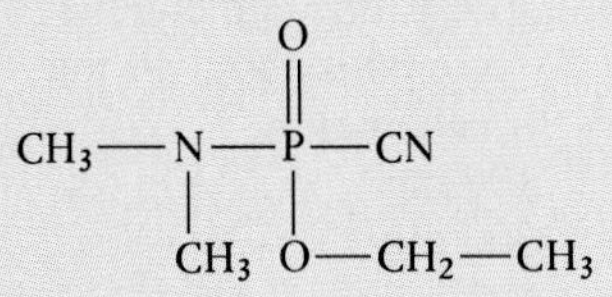	3.7	Chemical warfare agent first produced in World War II
Parathion	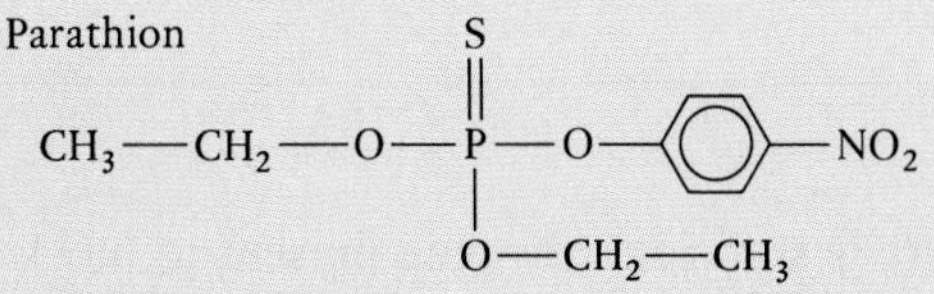	20	Insecticide
Paraoxon	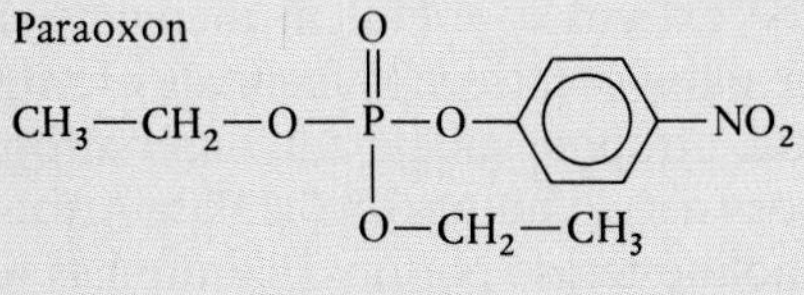	1.8	Insecticide
Malathion	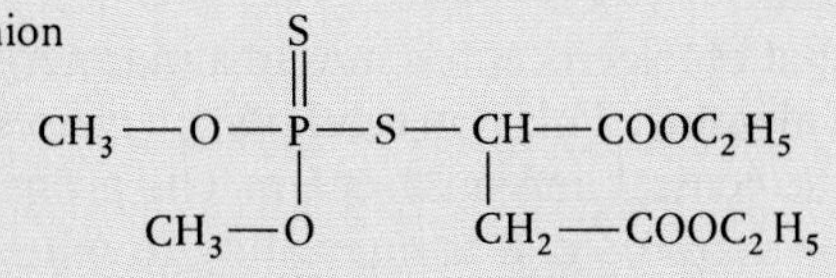	885	Insecticide
Carbaryl (sevin)	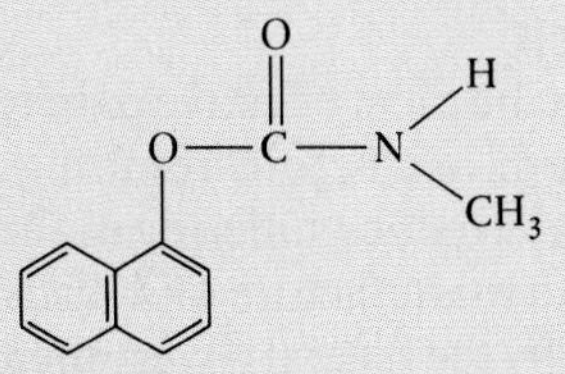	400	Insecticide

Figure 16–13 Some anticholinesterase poisons. In animals, parathion is converted into paraoxon in the liver. Carbaryl and malathion do not bind to cholinesterase as strongly. Malathion was the insecticide used in California in July 1981 to eradicate Medflies.

Neurotoxins such as **atropine** and **curare** (Table 16–7) are able to occupy the receptor sites on nerve endings of organs that are normally occupied by the impulse-carrying acetylcholine. When atropine or curare occupies the receptor site, no stimulus is transmitted to the organ. Acetylcholine in excess causes a slowing of the heartbeat, a decrease in blood pressure, and excessive saliva, whereas atropine and curare produce excessive thirst and dryness of the mouth and throat, a rapid heartbeat, and an increase in blood pressure. The normal responses to acetylcholine activation are absent, and the oppo-

TABLE 16-7 Alkaloid Neurotoxins That Compete with Acetylcholine for the Receptor Site

Name	Normal Contact	Lethal Dose (for a 70 kg human)	Discussed in
Atropine	Dilation of the pupil of the eye	100 mg	Chapter 13
Curare	Muscle relaxant	20 mg	Chapter 13
Nicotine	Tobacco, insecticides	75 mg	Chapter 13
Caffeine	Coffee, tea, cola drinks	13.4 g (one cup of coffee contains about 40 mg caffeine)	Chapter 13
Morphine	Opium—pain killer	100 mg	Chapter 21
Codeine	Opium—pain killer	300 mg	Chapter 21
Cocaine	Leaves of *Erythroxylon coca* plant in South America	1 g	Chapter 21

site responses occur when there is sufficient atropine present to block the receptor sites.

Neurotoxins of this kind can be extremely useful in medicine. For example, atropine is used to dilate the pupil of the eye to facilitate examination of its interior. Applied to the skin, atropine sulfate and other atropine salts relieve pain by deactivating sensory nerve endings on the skin. Atropine is also used as an antidote for anticholinesterase poisons. Curare has long been used as a muscle relaxant.

Alkaloids are discussed in Chapter 13.

A well-known alkaloid that blocks receptor sites in a manner similar to that of curare and atropine is **nicotine.** This powerful poison causes stimulation and then depression of the central nervous system. The probable lethal dose for a 70-kg person is less than 0.3 g. It is interesting to note that pure nicotine was first extracted from tobacco and its toxic action observed *after* tobacco use was established as a habit.

Morphine is the most effective pain killer known.

Natural or synthetic **morphine** is the most effective pain reliever known. It is widely used to relieve short-term acute pain resulting from surgery, fractures, burns, and so on, as well as to reduce suffering in the later stages of terminal illnesses such as cancer. The manufacture and distribution of narcotic drugs are stringently controlled by the federal government through laws designed to keep these products available for legitimate medical use. Under Federal law, some preparations containing small amounts of narcotic drugs may be sold without a prescription (for example, cough mixtures containing codeine), but not many.

In spite of stringent controls, drugs such as **morphine, heroin, meperidine,** and **methadone** are abused and illicitly used. Heroin is prepared from morphine, which is derived from sap in the opium poppy. It takes about 10 lb of opium to prepare 1 lb of morphine. Morphine reacts with acetic anhydride in a one-to-one reaction to form heroin. Street-grade heroin is only 9% to 10% pure.

Meperidine and **methadone** are products of chemical laboratories rather than of poppy fields. Meperidine was claimed to be nonaddictive when first

Meperidine

produced. Experience, however, proved otherwise (as it did with morphine and heroin). A major difference between methadone and morphine and heroin is that when methadone is taken orally, under medical supervision, it prevents withdrawal symptoms for approximately 24 h.

Some other natural products that affect the central nervous system and can be neurotoxic in comparatively small amounts are listed in Table 16–7.

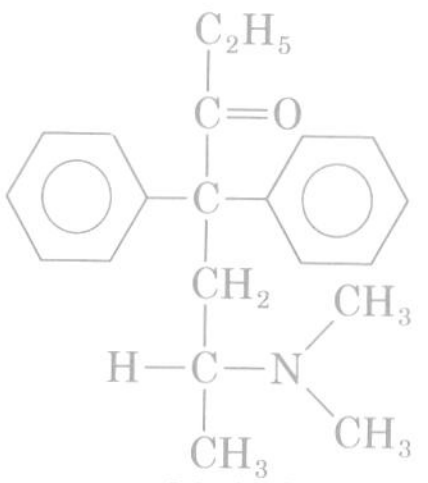

Methadone

CHEMICAL WARFARE AGENTS

Chemical warfare is the use of toxic chemicals to kill and incapacitate enemy troops. The Greeks used choking clouds of sulfur dioxide (SO_2) gas caused by burning sulfur and pitch during the Peloponnesian War between Sparta and Athens (413–404 B.C.) Modern chemical warfare began in 1915 when the Germans released chlorine gas on Allied troops at Ypres, Belgium during World War I. After the initial use of chlorine, various other gases were developed and used (Table 16–8).

In general, the World War I war gases caused death if the victim was exposed to high enough doses, but their most significant contribution to warfare was their effect on dispersing unprotected troops as they ran from the areas of highest concentration. Because most of the early war gases were strongly irritating, their use always caused confusion and disorder among troop concentrations. The actual number of deaths due to chemical warfare agents during World War I was fairly small. This was probably due to technical problems of delivering the toxic chemical so as to produce consistently a lethal concentration exactly where the enemy troops were located. In addition, gas masks were quickly issued to troops of all belligerent nations. These

TABLE 16–8 Chemical Warfare Agents

Type	Example	Action	History
Mustard gas	Bis (2-chloroethyl) sulfide ($Cl—CH_2—CH_2)_2—S$	Strong blisters, strong irritant	WW I
Choking gas	Phosgene Cl_2CO Chlorine Cl_2	Lung damage	WW I
Vomit gas	Diphenylamine chloroarsine (Adamsite)	Nausea and vomiting	WW I
Blood gas	Hydrogen cyanide HCN	Cell Death	WW I
	Arsine AsH_3	Vital Enzymes	WW I
Nerve gas	Tabun (see text)	Anticholinesterase poisons	WW II
	Saran (see text)		WW II

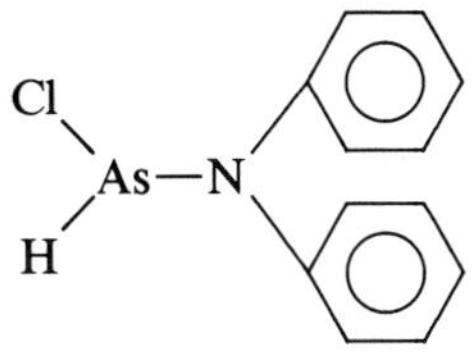

gas masks offered sufficient protection to prevent death from exposures except in cases where wounded troops could not put on their masks as the toxic cloud approached.

After World War I, most nations agreed to never use toxic chemicals in warfare — yet development of these agents continued. During World War II, the Germans developed Tabun and Saran (Table 16 – 7), two nerve gases that are anticholinesterase poisons. Their discovery led to our present-day phosphate ester insecticides such as Parathion® and Malathion®. Throughout World War II, war gases were available, but were never used.

Recently, in the 1980s, chemical agents were used in the Iran – Iraq war against both troops and civilians. Against civilians, chemical warfare agents are especially devastating since civilians are not only untrained and uninformed about the effects of these chemicals, but are unprepared to protect themselves. Modern concern regarding chemical warfare agents centers on protecting civilians, especially against terrorist attacks using weapons of this sort.

TERATOGENS

The effects of chemicals on human reproduction are a frightening aspect of toxicity. The study of birth defects produced by chemical agents is the discipline of **teratology.** The root word *terat* comes from the Greek word meaning "monster." There are three known classes of **teratogens:** radiation, viral agents, and chemical substances.

Birth defects occur in 2% to 3% of all births. About 25% of these occur from genetic causes, some possibly due to contact with mutagens, and 5% to 10% are the result of teratogens. The remaining 60% or so result from unknown causes.

In the development of the newborn, there are three basic periods during which the fetus is at risk. For a period of about 17 days between conception and implanatation in the uterine wall, a chemical "insult" will result in cell death. The rapidly multiplying cells often recover, but if a lethal dose is administered, death of the organism occurs followed by spontaneous abortion or reabsorption. The chemical in the so-called abortion pill, RU-486 (discussed in Chapter 21), developed in France in 1988 and being tested in the United States, works in this way.

During the critical embryonic stage (18 to 55 days) organogenesis, or development of the organs, occurs. At this time the fetus is extremely sensitive to teratogens. During the fetal period (56 days to term), the fetus is less sensitive. Contact with teratogens results in reduction of cell size and number. This is manifested in growth retardation and failure of vital organs to reach maturity.

The horrible thalidomide disaster in 1961 focused worldwide attention on chemically induced birth defects. Thalidomide®, a tranquilizer and sleeping pill, caused gross deformities (flipperlike arms, shortened arms, no arms or legs, and other defects) in children whose mothers used this drug during the first two months of pregnancy. The use of this drug resulted in more than 4000 surviving malformed babies in West Germany, more than

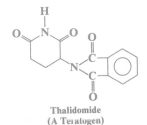

Thalidomide
(A Teratogen)

TABLE 16–9 Teratogenic Substances

Substances	Species	Effects on Fetus
Metals		
Arsenic	Mice Hamsters	Increase in males born with eye defects, renal damage
Cadmium	Mice Rats	Abortions Abortions
Cobalt	Chickens	Eye, lower limb defects
Gallium	Hamsters	Spinal defects
Lead	Humans Rats Chickens	Low birth weights, brain damage, stillbirth, early and late deaths
Lithium	Primates	Heart defects
Mercury	Humans Mice Rats	Minamata disease (Japan) Fetal death, cleft palate Brain damage
Thallium	Chickens	Growth retardation, abortions
Zinc	Hamsters	Abortions
Organic Compounds		
DES (diethylstilbestrol)	Humans	Uterine anomalies
Caffeine (15 cups per day equivalent)	Rats	Skeletal defects, growth retardation
PCBs (polychlorinated biphenyls)	Chickens Humans	Central nervous system and eye defects Growth retardations, stillbirths

1000 in Great Britain, and about 20 in the United States. With shattering impact, this incident demonstrated that a compound can appear to be remarkably safe on the basis of animal studies (so safe, in fact, that thalidomide was sold in West Germany without prescription) and yet cause catastrophic effects in humans. Although the tragedy focused attention on chemical mutagens, thalidomide presumably does not cause genetic damage in the germinal cells and is really not mutagenic. Rather, thalidomide, when taken by a woman during early pregnancy, causes direct injury to the developing embryo.

Any chemical substance that can cross the placenta is a potential teratogen, and any activity resulting in the uptake of these chemicals into the mother's blood might prove a dangerous act for the health and well-being of the fetus. Smoking a cigarette results in higher than normal blood levels of such substances as carbon monoxide, hydrogen cyanide, cadmium, nicotine, and benzo(α)pyrene. Of course, many of these substances are present in polluted air as well. Table 16–9 lists a number of chemical substances known to be teratogenic in humans and laboratory animals.

A pregnant woman should always be advised to limit her exposure throughout the term of her pregnancy to chemicals of unknown toxicity, any of which could be harmful to the developing child. This is especially true during the 18th through 55th days. She should take no drugs nor medicines except on the advice of her physician, and she should avoid the use of alcohol and tobacco.

MUTAGENS

A mutagen is a chemical that can change the hereditary pattern of a cell.

Mutagens are chemicals capable of altering the genes and chromosomes sufficiently to cause abnormalities in offspring. Chemically, mutagens alter the structures of DNA and RNA, which compose the genes (and, in turn, the chromosomes) that transmit the traits of parent to offspring. Mature sex, or germinal, cells of humans normally have 23 chromosomes; body, or somatic, cells have 23 *pairs* of chromosomes.

Although many chemicals are under suspicion because of their mutagenic effects on laboratory animals, it should be emphasized that no one has yet shown conclusively that any chemical induces mutations in human germinal cells. Part of the difficulty of determining the effects of mutagenic chemicals in humans is the extreme rarity of mutation. A specific genetic

TABLE 16–10 Mutagenic Substances as Indicated by Experimental Studies on Plants and Animals

Substance	Experimental Results
Aflatoxin (from mold, *Aspergillus flavus*)	Mutations in bacteria, viruses, fungi, parasitic wasps, human cell cultures, mice
Benzo(α)pyrene (from cigarette and coal smoke)	Mutations in mice
Caffeine	Chromosome changes in bacteria, fungi, onion root tips, fruit flies, human tissue cultures
Captan (a fungicide)	Mutagenic in bacteria and molds; chromosome breaks in rats and human tissue cultures
Chloroprene	Mutagenic in male sex cells; results in spontaneous abortions
Dimethyl sulfate (used extensively in chemical industry to methylate amines, phenols, and other compounds)	Methylates DNA base guanine; potent mutagen in bacteria, viruses, fungi, higher plants, fruit flies
LSD (lysergic acid diethylamide)	Chromosome breaks in somatic cells of rats, mice, hamsters, white blood cells of humans and monkeys
Maleic hydrazide (plant growth inhibitor; trade names Slo-Gro®, MH-30®)	Chromosome breaks in many plants and in cultured mouse cells
Mustard gas (dichlorodiethyl sulfide)	Mutations in fruit flies
Nitrous acid (HNO_2)	Mutations in bacteria, viruses, fungi
Ozone (O_3)	Chromosome breaks in root cells of broadleaf plants
Solvents in glue (glue sniffing) (toluene, acetone, hexane, cyclohexane, ethyl acetate)	4% more human white blood cells showed breaks and abnormalities (6% versus 2% normal)
TEM (triethylenemelamine) (anticancer drug, insect chemosterilants)	Mutagenic in fruit flies, mice

disorder may occur as infrequently as only once in 10,000 to 100,000 births. Therefore, to obtain meaningful statistical data, a carefully controlled study of the entire population of the United States would be required. In addition, the very long time between generations presents great difficulties, and there is also the problem of tracing a medical disorder to a single specific chemical out of the tens of thousands of chemicals with which we come in contact.

If there is no direct evidence for specific mutagenic effects in human beings, why, then, the interest in the subject? The possibility of a deranged, deformed human race is frightening; the chance for an improved human body is hopeful; and the evidence for chemical mutation in plants and lower animals is established. A wide variety of chemicals is known to alter chromosomes and to produce mutations in rats, worms, bacteria, fruit flies, and other plants and animals. Some of these are listed in Table 16–10.

Experimental work on the chemical basis of the mutagenic effects of nitrous acid (HNO_2) has been very revealing. Repeated studies have shown that nitrous acid is a potent mutagen in bacteria, viruses, molds, and other organisms. In 1953, at Columbia University, Dr. Stephen Zamenhof demonstrated experimentally that nitrous acid attacks DNA. Specifically, nitrous acid reacts with the adenine, guanine, and cytosine bases of DNA by removing the amino group of each of these compounds. The eliminated group is replaced by an oxygen atom (Fig. 16–14). The changed bases may

Figure 16–14   Reaction of nitrous acid (HONO) with nitrogenous bases of DNA. Nitrogen and water are also products of each reaction.

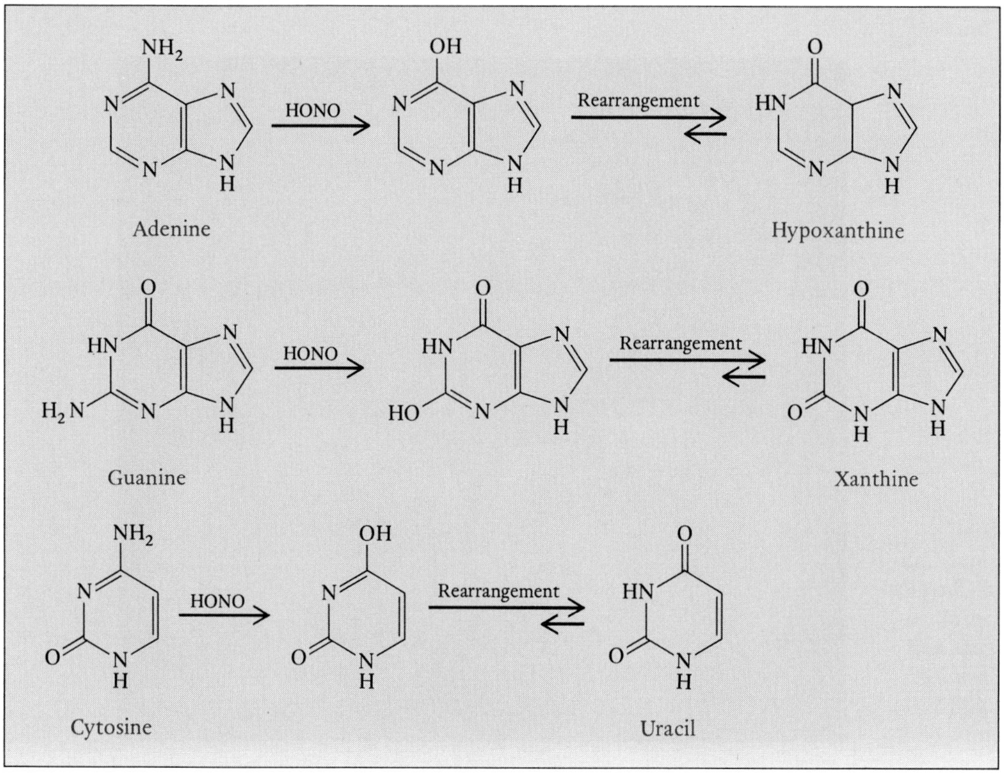

garble a part of DNA's genetic message, and in the next replication of DNA, the new base may not form a base pair with the proper nucleotide base.

For example, adenine (A) typically forms a base pair with thymine (T) (Fig. 16–14). However, when adenine is changed to hypoxanthine, the new compound forms a base pair with cytosine (C). In the second replication, the cytosine forms its usual base pair with guanine (G). Thus, where an adenine–thymine (A–T) base pair existed originally, a guanine–cytosine (G–C) pair now exists. The result is an alteration in the DNA's genetic coding, so that a different protein is formed later.

Do all of these findings mean that nitrous acid is mutagenic in humans? Not necessarily. We do know that **sodium nitrite** has been widely used as a preservative, color enhancer, or color fixative in meat and fish products for at least the past 30 years. It is currently used in such foods as frankfurters, bacon, smoked ham, deviled ham, bologna, Vienna sausage, smoked salmon, and smoked shad. The sodium nitrite is converted to nitrous acid by hydrochloric acid in the human stomach:

Sodium nitrite produces nitrous acid in the stomach.

$$NaNO_2 + HCl \longrightarrow HNO_2 + NaCl$$

Sodium   Nitrous
Nitrite   Acid

The FDA now considers the mutagenic effects of nitrous acid in lower organisms sufficiently ominous to suggest strongly that the use of sodium nitrite in foods be severely curtailed, and a complete ban of this use of sodium nitrite is being considered. Several European countries already restrict the use of sodium nitrite in foods. The concern is that this compound, after being converted in the body to nitrous acid, may cause mutation in

The process of frying bacon produces carcinogenic nitrosamines.

$$HNO_2 + R_2NH \longrightarrow$$

Nitrous   Amines
Acid   Found in
          Bacon

$$R_2N-N=O + H_2O$$

Nitrosamine

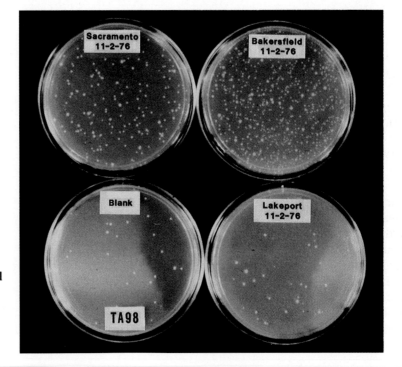

Figure 16–15 The effect of mutagens in the air in various California cities. (© American Chemical Society, From P. Flessel, et al.: "Ames Testing for Mutagens and Carcinogens in Air," *J. Chemical Education*, **64**, 391–395, 1987. Reproduced by permission.)

somatic cells (and possibly in germinal cells) and thus could possibly produce cancer in the human stomach. Other scientists doubt that nitrous acid is present in germinal cells and, therefore, seriously question whether this compound could be a cause of genetically produced birth defects in humans. The uncertainty of extrapolating results obtained in animal studies to human beings hovers over the mutagenic substances.

In the early 1980s, Professor Bruce Ames and his colleagues at the University of California, Berkeley, developed a simple test that can identify chemicals capable of causing mutations in sensitive strains of bacteria. In this test, about 100 million bacteria unable to synthesize histidine are mixed in an agar suspension along with a suspected mutagenic chemical. This mixture is then added to a hard agar gel containing salts and glucose, and incubated in a petri dish for several days. If the suspected chemical is a mutagen, some of the histidine-requiring cells mutate and the biosynthesis of histidine resumes. The growth of these bacterial colonies can be seen in the petri dish (Fig. 16–15).

The Ames test has utility in identifying not only mutagenic chemicals, but potential carcinogenic chemicals as well, since mutagenic chemicals are often carcinogenic. In studies involving hundreds of chemicals, nearly four out of every five animal carcinogens have been found to be mutagenic in the Ames test.

CARCINOGENS

Carcinogens are chemicals that cause cancer. **Cancer** is an abnormal growth condition in an organism that manifests itself in at least three ways. The rate of cell growth (that is, the rate of cellular multiplication) in cancerous tissue differs from the rate in normal tissue. Cancerous cells may divide more rapidly or more slowly than normal cells. Cancerous cells spread to other tissues; they know no bounds. Normal liver cells divide and remain a part of the liver. Cancerous liver cells may leave the liver and be found, for example, in the lung. Most cancer cells show partial or complete loss of specialized functions. Although located in the liver, cancer cells no longer perform the functions of the liver.

Attempts to determine the cause of cancer have evolved from early studies in which the disease was linked to a person's occupation. Dr. Percivall Pott, an English physician, first noticed in 1775 that persons employed as chimney sweeps had a higher rate of skin cancer than the general population. It was not until 1933 that **benzo(α)pyrene**, $C_{20}H_{12}$ (a five-ringed aromatic hydrocarbon), was isolated from coal dust and shown to be metabolized in the body to produce one or more carcinogens. In 1895, the German physician Rehn noted three cases of bladder cancer, not in a random population, but in employees of a factory that manufactured dye intermediates in the Rhine Valley. Rehn attributed these cancers to his patients' occupation. These and other cases confirmed that at times as many as 30 years passed between the time of the initial employment and the occurrence of bladder cancer. The principal product of these factories was aniline. Although ani-

Benzo(α)pyrene

TABLE 16–11 Some Industrial Chemicals That are Carcinogenic for Humans

Compound	Formula	Use or Source	Site Affected	Confirming Animal Tests*
Inorganic Compounds				
Arsenic (and compounds)	As†	Insecticides, alloys	Skin, lungs, liver	−
Asbestos	$Mg_6(Si_4O_{11})(OH)_6$	Brake linings, insulation	Respiratory tract	+
Beryllium	Be	Alloy with copper	Bone, lungs	+
Cadmium	Cd	Metal plating	Kidney, lungs	+
Chromium	Cr†	Metal plating	Lungs	+
Nickel	Ni†	Metal plating	Lungs, nasal sinus	+
Organic Compounds				
Benzene		Solvent, chemical intermediate in syntheses	Blood (leukemia)	+
Acrylonitrile	$CH_2{=}CH(CN)$	Monomer	Colon, lungs	+
Carbon tetrachloride	CCl_4	Solvent	Liver	+
Diethylstilbestrol		Hormone	Female genital tract	+
Benzo(α)pyrene		Cigarette and other smoke	Skin, lungs	+
Benzidine		Dye manufacture, rubber compounding	Bladder	+
Ethylene oxide		Chemical intermediate used to make ethylene glycol, surfactants	Gastrointestinal tract	±
Soots, tar, and mineral oils		Roofing tar, chimney soot, oils of hydrocarbon nature	Skin, lungs, bladder	+
Vinyl chloride	$CH_2{=}CHCl$	Monomer for making PVC	Liver, brain, lungs, lymphatic system	+

* For animal tests, (+) means positive supporting data, (−) means a lack of supporting data, (±) means conflicting data.
† Certain compounds or oxidation states only.

line was first thought to be the carcinogenic agent, it was later shown to be noncarcinogenic. It was not until 1937 that continuous long-term treatment with **2-naphthylamine,** one of the suspected dye intermediates, in dosages of up to 0.5 g per day produced bladder cancer in dogs. Since then other dye intermediates have been shown to be carcinogenic.

A vast amount of research has verified the carcinogenic behavior of a large number of diverse chemicals. Some of these are listed in Table 16–11. This research has led to the formulation of a few generalizations concerning the relationship between chemicals and cancer. For example, carcinogenic effects on lower animals are commonly extrapolated to humans. The mouse has come to be the classic animal for studies of carcinogenicity. Strains of inbred mice and rats have been developed that are genetically uniform and show a standard response.

Some carcinogens are relatively nontoxic in a single, large dose, but may be quite toxic, often increasingly so, when administered continuously. Thus, much patience, time, and money must be expended in carcinogen studies. The development of a sarcoma in humans, from the activation of the first cell to the clinical manifestation of the cancer, takes from 20 to 30 years. With life expectancy of an average person in the United States now set at about 70 years, it is not surprising that the number of deaths due to cancer is increasing.

Cancer does not occur with the same frequency in all parts of the world. Breast cancer occurs less frequently in Japan than in the United States or Europe. Cancer of the stomach, especially in males, is more common in Japan than in the United States. Cancer of the liver is not widespread in the western hemisphere but accounts for a high proportion of the cancers among the Bantu in Africa and in certain populations in the Far East. The widely publicized incidence of lung cancer is higher in the industrialized world and is increasing at an appreciable rate.

Some compounds cause cancer at the point of contact. Other compounds cause cancer in an area remote from the point of contact. The liver, the site at which most toxic chemicals are removed from the blood, is particularly susceptible to such compounds. Since the original compound does not cause cancer on contact, some other compound made from it must be the cause of cancer. For example, it appears that the substitution of an $>$NOH group for an $>$NH group in an aromatic amine derivative produces at least one of the active intermediates for carcinogenic amines. If R denotes a two- or three-ring aromatic system, then the process can be represented as follows:

Aniline   2-naphthylamine

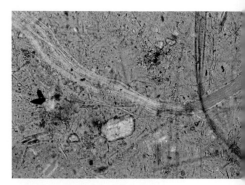

Asbestos sample as seen under a polarizing light microscope. This sample contains 60% chrysotile asbestos. (Courtesy of Particle Data Laboratories.)

An abnormal growth is classified as cancerous or malignant when examination shows it is invading neighboring tissue. A growth is benign if it is localized at its original site.

$$\underset{\substack{\text{Inactive} \\ \text{on Contact}}}{\overset{\text{H}}{\underset{|}{\text{RNCOCH}_3}}} \longrightarrow \underset{\substack{\text{Active on} \\ \text{Contact}}}{\overset{\text{OH}}{\underset{|}{\text{RNCOCH}_3}}} \longrightarrow \underset{\substack{\text{Other Unknown} \\ \text{Intermediates}}}{\text{RX?}} \longrightarrow \text{RY?} \xrightarrow{\text{Tissue}} \text{Tumor cell}$$

As indicated by the variety of chemicals in Table 16–11, many molecular structures produce cancer, whereas ones closely related to them do not. The 2-naphthylamine mentioned earlier is carcinogenic, but repeated testing gives negative results for 1-naphthylamine.

1-naphthylamine
(Noncarcinogenic)

2-naphthylamine
(Carcinogenic)

Almost all human cancers caused by chemicals have a long induction period, which makes it extremely difficult for researchers to obtain meaningful interpretations of exposure data. For example, asbestos particles are known to cause cancer, but only those particles in a narrow size range — that is, $5-100$ μm long and less than 2 μm wide seem to be capable of actually causing cancer under controlled conditions involving lab animals. Neither lung cancer nor mesothelioma (a very malignant tumor of the linings of the chest and abdominal cavities) commonly occurs less than 20 years after the first exposure. Since very few precise measurements of asbestos in the air at workplaces were made prior to 1950, and it wasn't until 1965 when the smaller-sized asbestos particles could be measured, it is not surprising that we know very little about the concentrations of asbestos particles that caused cancer in so many factory and construction workers.

Asbestos is thought to be responsible for over 10,000 cancer deaths per year — second only to tobacco-caused cancer deaths.

DIETARY CARCINOGENS

Apart from industrial chemicals with which we may come in contact in the workplace and which may contaminate our atmosphere (see Chapter 18) and our drinking water (see Chapter 17), our everyday diets contain a great variety of natural carcinogens. Some of these chemicals are also mutagens and teratogens. Plants produce these chemicals as defense mechanisms, or natural pesticides, and often produce more of them when diseased or stressed. Celery, for example, contains isoimpinellin, a member of a chemical family called psoralens, at a level of 100 μg/100 g. This level increases by 100 fold if the celery is diseased. This compound is a potent light-activated carcinogen. Psoralens, when activated by sunlight, damage DNA. Oil of bergamont, found in citrus fruits, contains a psoralen that was once used by a French manufacturer of suntan oil. Sunlight caused the psoralens to enhance tanning.

Black pepper contains small amounts of saffrole, a known carcinogen, and large amounts (up to 10% by weight) of piperine, a compound related to saffrole. Black pepper extracts cause tumors in mice at a variety of sites at dose rates equivalent to 4 mg of dried pepper per day for 3 months. Many people consume over 140 mg of black pepper per day for life.

Most hydrazines that have been tested are carcinogenic. This includes industrial products such as rocket fuels, and hydrazines produced by plants. The widely eaten false morel *(Gyromitra esculenta)* contains eleven hydrazines, three of which are known carcinogens. One of these is *N*-methyl-*N*-formylhydrazine, which is present at concentrations of 50 mg/100 g and causes lung tumors in mice at very low dietary doses of 20 μg per mouse per day. The common mushroom, *Agaricus bisporus,* contains about 300 mg of

Isoimpinellin

Hydrazine

Sym-dimethylhydrazine

TABLE 16–12 Some Natural Carcinogens Found in Foods and Beverages

Compound	Formula	Food
Ethyl alcohol	CH_3CH_2OH	Alcoholic beverages
Threobromine	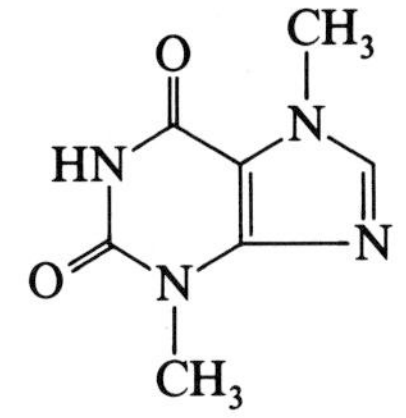	Cocoa-based drinks, tea
Methylgyloxal		Coffee
Prunasin	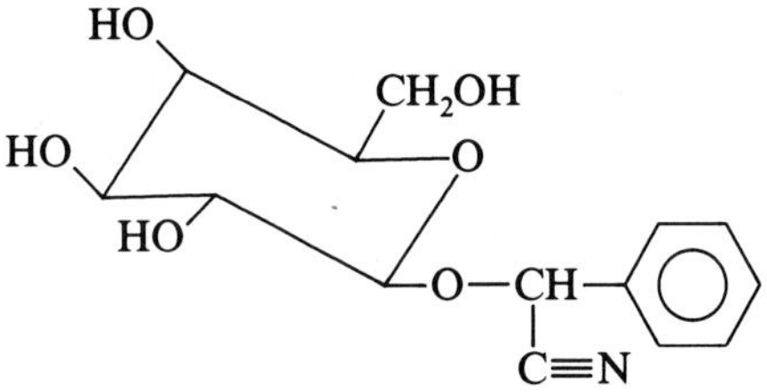	Lima beans
D-Limonene	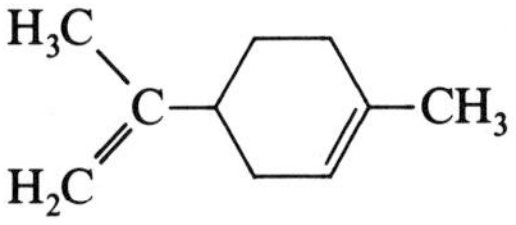	Citrus fruits
Serotonin	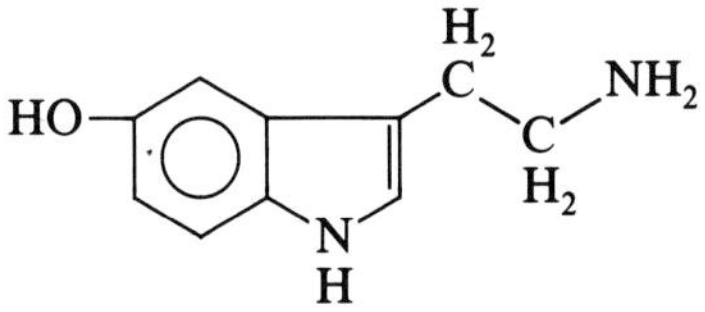	Tomatoes, pineapples, bananas
Aflatoxin B₁	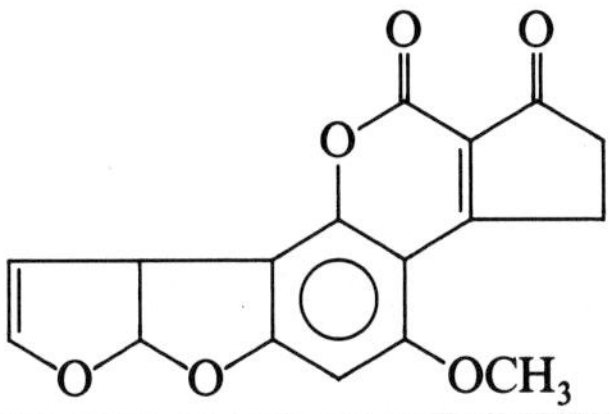	Moldy foods

agaritine per 100 g. Agaritine is a mutagen and is metabolized to a compound that is extremely carcinogenic.

Allyl isothiocyanate, the main flavor ingredient in oil of mustard and horseradish, has been shown to cause cancers in rats. It also causes chromosome damage in hamster cells at low concentration.

Table 16–12 contains information on several carcinogens found in foods.

It has been estimated that we consume about 10,000 times more natural carcinogens than man-made pesticides. This means that many of the cancers

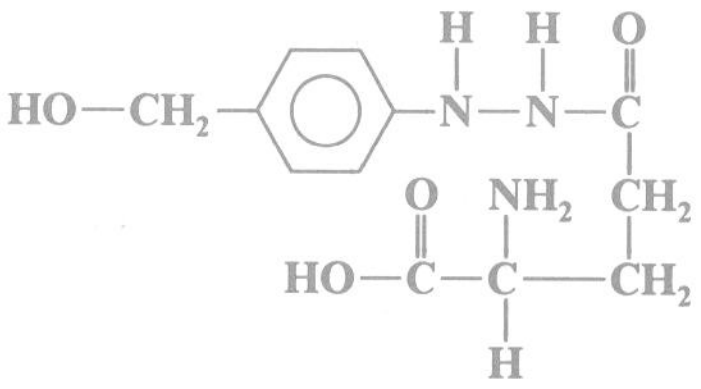

Agaritine

$CH_2{=}CH{-}CH_2N{=}C{=}S$

Allyl Isothiocyanate

TOXIC SUBSTANCES

How do scientists perceive the risk from chemicals and the environment? Interestingly most chemicals, even vitamins, can be toxic at large doses. Most chemicals, in fact, have a threshold dose of toxicity. A dose above the threshold is dangerous, a dose below it is not. But there's a great deal of debate over whether potential carcinogens have such thresholds, that is, concentrations below which they won't cause cancer. Thus, given conflicting and imprecise evidence, some scientists talk about carcinogens in terms of risk or probabilities. Risk is relative. For example, some think the cancer risk from hazardous waste is negligible compared with the naturally occurring carcinogens we successfully resist every day.

The originator of a test for determining whether a chemical has the potential to cause cancer is Dr. Bruce Ames, biochemist at the University of California, Berkeley. He claims:

People have been very worried about toxic waste dumps but, in fact, the evidence that they're really causing any harm is really minimal, there's not very much evidence. And the levels of chemicals are very tiny, so we don't really know whether there's no hazard or a little bit of hazard.

The world is full of carcinogens, because half the natural chemicals they've tested have come out as carcinogens. Some plants have toxic chemicals to keep off insects, and we are eating those every time we eat a tomato or potato. Mushrooms have carcinogens, celery has carcinogens, and an apple has formaldehyde in it. So there is an incredible number of carcinogens in nature; we're getting much more of those than man-made chemicals.

Dr. Halina Brown, Professor of Toxicology, Clark University has stated:

If we accept those risks, why can't we accept small risks from chemical carcinogens in the environment? It's a valid argument. But then there is, of course, the counter argument. The counter argument is we cannot do much about trace amount of carcinogens that are present in food, why should we add to this burden that we already have by increasing the amount of exposure to carcinogens. But then it boils down to money. Unfortunately, it takes tremendous resources to reduce the levels of exposure in the environment to carcinogens, especially when you get to very low levels. Reducing it by another order of magnitude may take millions of dollars at one hazardous waste site. And the pie is not unlimited. Even those who don't consider hazardous waste dumps a health threat think the money should be spent to clean them up.

Dr. Bruce Ames also states:

I mean if Congress has put $10 billion for cleaning up toxic waste dumps, you might as well find the worst ones and clean them up. Now, whether you're getting anything—whether you're gaining much in public health for cleaning them up is something one could argue about. I think probably very little. But, in any case, you can—you might as well spend the money cleaning up the worst dumps.
The World of Chemistry (Program 25) "Chemistry and the Environment."

Dr. Bruce Ames. (*The World of Chemistry*, Program 25, "Chemistry and the Environment.")

we may get as we grow older may have been caused not by the chemicals getting all of the publicity, but by the very foods that we have been eating all along. We are not likely to eliminate all carcinogens either from our diets or from our general environment.

HOW CARCINOGENS WORK

Every cancer comes from a single cell—one that is a modification of a normal cell. A normal cell functions according to directions stored in its genetic data bank, the complex DNA molecule. Normally, when a cell divides, each new cell gets its own exact copy of the parent DNA. When anything disrupts this DNA replication process, the genetic code found in one of the descendent cells may cause that cell to act differently from a normal cell. For example, the cell may become distorted, or its growth sensors may not respond to controls from neighboring cells telling it to stop dividing. If this happens, the cell will not confine its growth within tissue boundaries and remain in one place.

DNA is discussed in Chapter 15.

Carcinogenesis is often a two-stage process. The first stage is that of **initiation.** A chemical, physical, or viral agent alters the cell's DNA. Sometimes a single exposure to some carcinogen causes a rapid onset of a tumor that is composed of rapidly growing, uncontrolled cells, but usually the abnormal cells continue to reproduce in about the same way as normal cells around them. Then a **promotion** occurs. This is the second stage and may occur days, months, or years after the initiation causing DNA alteration. This promotion may be a physical irritation or exposure to some toxic chemical that is itself not a carcinogen. In either case, the promotion results in the killing of a large number of cells. The destruction of cells is almost always compensated for by a sudden growth of new cells, and the abnormal cells begin to grow in ways the original DNA coding never intended. The cancer has started.

When a cancer spreads from one site to another, the process is called metastasis.

In 1947, Dr. Phillipe Shubik and Dr. Isaac Berenblum at Oxford University in England, applied very small doses of dimethylbenzanthracene (DMBA), a known carcinogenic component of coal tar, to the skin of a group of mice (Fig. 16–16). They then separated these mice into two groups. In one group they immediately daubed the exposed skin with croton oil, a strongly irritating natural oil. The other group of mice had their skin daubed four months later with croton oil. Almost every mouse in both groups developed a tumor where the DBMA had been applied. In other groups of mice tested, neither croton oil nor DMBA alone produced any tumors, and if croton oil was applied first to the skin, followed by the DBMA, tumors failed to appear.

Smoking is thought to play both an initiation and promotion role in cancer causation.

The cell's disruption by cancer is concentrated largely in the epithelial cells. These cells cover the skin and other tissues, make up the glandular organs such as the breasts, and line the lungs and gastrointestinal tract. Cancers in the epithelial cells are known as **carcinomas** and account for about 85% of all cancers. The epithelial cells are those most likely to be exposed to external chemical, physical, and viral agents. **Sarcomas are**

Cancers of the epithelial tissue— carcinoma. Cancers of the connective tissue—sarcoma.

Figure 16–16 A second chemical can promote tumor growth in mice after an initiation period. Treatment of mice with croton oil produces no tumor nor does treatment with small quantities of benzo(α)pyrene alone. Croton oil is an irritant oil similar to castor oil. Both are derived from plants.

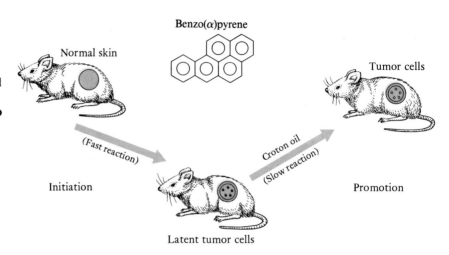

Benzo(α)pyrene

Normal skin

Tumor cells

(Fast reaction)

Croton oil (Slow reaction)

Initiation

Latent tumor cells

Promotion

Cancers of the lymph system— lymphomas. Cancers of the blood— leukemias.

cancers of the connective and supportive tissue, **lymphomas** are cancers of the lymph system, and **leukemias** are cancers of the blood.

SELF-TEST 16–B

1. Substrates that poison the nervous system are called _____.
2. Most neurotoxins affect chemical reactions that occur in the opening between two nerve cells. These openings are called _____.
3. The electric impulse is carried across a synapse by the chemical
   _____.
4. Mutagens alter the structures of _____ or _____.
5. If a substance is mutagenic in test animals, particularly dogs, it must necessarily be mutagenic in human beings. ( ) True ( ) False
6. The first occupation definitely linked to cancer was _____.
7. Two dangers associated with smoking are _____ and
   _____.
8. A chemical that can cross the placenta and harm the fetus is called a
   _____.

MATCHING SET

____ 1. Metabolic poison	a. Diethylstilbestrol
____ 2. Metabolism	b. Cyanide ion
____ 3. Corrosive poison	c. Cancer in connective tissue
	d. Benzo(α)pyrene metabolite
____ 4. Neurotoxin	e. Cancerous growths of lung tissue located in liver
____ 5. Mutagen	f. Use of chemicals in the body
____ 6. Carcinogen	g. Cancer of skin

_____ **7.** Carcinoma
_____ **8.** Metastases
_____ **9.** Sarcoma
_____ **10.** Chelating agent
_____ **11.** Teratogen

h. Sodium hydroxide (caustic soda)
i. Atropine
j. Alters DNA
k. EDTA
l. Acetylcholine
m. Pica

QUESTIONS

1. Give an example of a toxic substance that is toxic as a result of:
  a. Binding to an oxygen-carrying molecule
  b. Disguising itself as another compound
  c. Attack on an enzyme
  d. Hydrolysis

2. True or False. Explain each answer concisely.
  a. Lead is a corrosive poison.
  b. Carbon monoxide and the cyanide ion poison in the same way.
  c. There are no known chemical compounds that cannot be toxic under some circumstances.

3. The application of a single, minute dose of a fused-ring hydrocarbon such as benzo(α)pyrene fails to produce a tumor in mice. Does this mean this compound is definitely noncarcinogenic? Give a reason for your answer.

4. Describe the chemical mechanism by which the following substances show their toxic effects.
  a. Fluoroacetic acid
  b. Phosgene
  c. Curare
  d. Carbon monoxide

5. Read a recent magazine article on the chemical Alar. What has it been used for? Why is it dangerous?

6. Discuss some of the pros and cons of testing toxic substances on animals.

7. Give chemical reactions in words for
  a. Action of NaOH on tissue
  b. Action of carbon monoxide in blood
  c. Reaction of EDTA and lead ion

8. What questions do you think need to be answered before the action of ethyl alcohol is understood?

9. Assume a normal diet has the quantity of lead in a given quantity of food stated in the text. What would a person's total food intake of lead be per day?

10. What is the meaning of the symbol LD_{50}?

11. Write chemical equations for:
  a. The hydrolysis of acetylcholine
  b. Acid hydrolysis of a protein having a glycyl–glycine primary structure

12. Describe how the corrosive poisons lye (NaOH), NO_2, and the hypochlorite ion (OCl^-) destroy tissue.

13. What are some common sources of carbon monoxide?

14. If a relatively small amount of carbon monoxide is inhaled, are the chemical reactions reversible, or is carbon monoxide a cumulative poison?

15. What poisons can be rendered ineffective by wrapping a large molecule around them?

16. What type of poisoning is associated with the term _pica_?

17. What is a common structural feature of alkaloids?

18. Give two examples of each: (a) corrosive poison, (b) metabolic poison, (c) neurotoxin, (d) mutagen, (e) carcinogen.

19. Phosgene hydrolyzes in the lungs to produce what acid?

20. What concentration of carbon monoxide in the air is likely to cause death in 1 h?

21. Is it possible that one molecule of a mutagen could cause a problem to human life?

22. Two new chemicals are prepared in the lab. One is a relatively simple acid with corrosive properties. The other chemical has carcinogenic properties. Which chemical's toxic property is more likely to be discovered?

23. An old laboratory chemical is discovered to have a new property. It reacts with amine groups to produce —OH groups. Could this chemical be a possible mutagen?

24. If a poisoning victim has pinpoint pupils and is salivating excessively, what type of poison was the probable cause of these symptoms?

25. If your drinking water contains numerous toxic substances, why are you not normally harmed by drinking it?

17 Water — Plenty of It, But of What Quality?

Testing stream water for pollutants. (Courtesy of Carolina Biological Supply Company.)

Water is an unusually unique substance (see Chapters 5 and 6). Without it life on this planet would not be possible. There is plenty of water, but it is unevenly distributed. Parts of our world have ample supplies, and others have virtually none. Water's unique properties result from its molecular polarity and hydrogen bonding, which enable it to carry dissolved chemicals, some of them quite toxic, as well as harmful bacteria and viruses. In developed countries throughout the world, pure water is often taken for granted, but producing water that is both clean and pure enough for use by man, animals, and plants is not easy. The job of purifying water is becoming more difficult as water becomes contaminated by the chemical byproducts of mining, farming, industry, and household activities. Serious water supply problems exist in some states and communities, and this has resulted in recent years in the rationing of all types of water use. For nearly two decades, the U.S. Environmental Protection Agency (EPA) has issued standards for toxic contaminants commonly found in our drinking water. Therefore, we often do not have enough water, and what we have is at risk of being contaminated by chemicals that can cause harm. In this chapter we shall examine these problems by looking at some of their causes and the possible solutions to them.

WATER—THE MOST ABUNDANT COMPOUND

Water is the most abundant substance on the Earth's surface. Oceans (with an average depth of 2.5 miles) cover about 72% of the Earth. They are the reservoir of 97.2% of the Earth's water. The rest consists of 2.16% in glaciers, 0.0197% fresh water in lakes and rivers, 0.61% groundwater (water underground), 0.01% in brine wells and brackish waters, and 0.001% in atmospheric water.

Water is the major component of all living things. For example, the water content of human adults is 70%—the same proportion as the Earth's surface (Table 17–1).

It is estimated that an average of 4350 billion gal of rain and snow fall on the contiguous United States each day. Of this amount, 3100 billion gal return to the atmosphere by evaporation and transpiration. The discharge to the sea and to underground reserves amounts to 800 billion gal daily, leaving 450 billion gal of surface water each day for domestic and commercial use. The 48 contiguous states withdrew 40 billion gal per day from natural sources in 1900 and 394 billion gal in 1985, and it is estimated that the demand will be at least 900 billion gal per day by the year 2000. The demand for water by our growing population is already greater than the resupply by natural resources in many parts of the country.

ARE WE FACING A WATER CRISIS?

Yes, we are entering a water crisis era that may be as serious as the highly publicized energy crisis. We are not running out of water because the 394 billion gal we use every day is only one tenth of the daily supply. However, we

TABLE 17–1 Water Content

Marine invertebrates	97%
Human fetus (1 month)	93%
Adult human	70%
Body fluids	95%
Nerve tissue	84%
Muscle	77%
Skin	71%
Connective tissue	60%
Vegetables	89%
Milk	88%
Fish	82%
Fruit	80%
Lean meat	76%
Potatoes	75%
Cheese	35%

Brackish water contains dissolved salts but at a lower level than sea water.

Transpiration is the release of water by leaves of plants. An acre of corn is estimated to release 3000 gal per day, while a large oak tree releases 110 gal per day.

(*The World of Chemistry,* **Program 12, "Water."**)

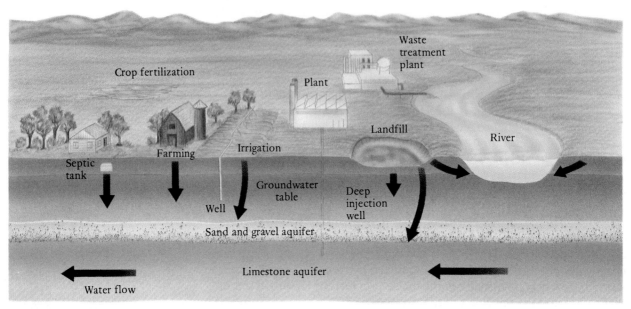

Figure 17–1   How groundwater gets polluted.

An aquifer is a water-bearing stratum of permeable rock, sand, or gravel, as illustrated in Figure 17–1.

need to examine the sources of usable water before we can understand the seriousness of the water crisis.

The two sources of usable water are **surface water** (lakes, rivers) and **groundwater.** Groundwater is that part of underground water that is below the water table. Figure 17–1 shows the various parts of the water cycle and the flow of groundwater. About 90 billion gal of the 394 billion gal per day of water usage come from groundwater supplies. These groundwater supplies are from wells drilled into the **aquifers.** The supply and demand for surface water and groundwater are uneven across the country, and in many areas the quantity and quality of the withdrawn water are not being resupplied to the lakes, rivers, and aquifers at the rate needed.

In the arid West, wells used to pump water for irrigation either are going dry or are requiring drilling so deep that irrigation is no longer economically feasible. The huge Ogallala aquifer that stretches from South Dakota to Texas has 150,000 wells tapping it for irrigation of 10 million acres. As a result, the Ogallala aquifer is being drawn down at a rate that has reduced the average thickness of the aquifer from 58 ft in 1930 to 8 ft today. At the current rate, the Ogallala aquifer will be used up as a source of groundwater in 20 to 30 years!

The depletion of the Ocala aquifer along the eastern seaboard has caused large sinkholes in Georgia and Florida when the limestone rock strata of the aquifer collapse as the water is withdrawn. Many coastal cities are also experiencing problems with brackish drinking water that comes from aquifers where the fresh water has been drawn off, causing sea water to flow into the depleted aquifer.

Depletion of underground sources has also caused sinkholes in Texas. Houston has sunk several feet as the result of extensive use of the underground water sources in that area! Figure 17–2 is a graphic illustration of the change in surface level in California's San Joaquin Valley as a result of groundwater depletion. So much groundwater has been pumped out for irrigation that the land sunk 29 ft between 1925 and 1977.

The disputes between Arizona and California about Colorado River water and between the Great Lakes states and states along the Mississippi River are other examples of how water shortages can affect neighbors. The U.S. Supreme Court ruled in Arizona's favor, setting California's allotment of the Colorado River at 4.4 million acre-feet per year, which is less than California's previous intake of 5 million acre-feet per year. The potential dispute about diverting Great Lakes water down the Mississippi River instead of allowing it to flow out the St. Lawrence River to the Atlantic Ocean came up during the summer of 1988 when a midwest drought severely limited the Mississippi flow. The initial reaction from states bordering the Great Lakes was against diversion of this water from its normal path.

These water shortages together with the problems of polluted water require careful study of possible solutions and widespread cooperation if we are to solve the water crisis. There are no quick fixes. A reliable long-term solution will require conservation, reuse, and less reliance on fresh water sources.

An acre-foot is the volume that would cover 1 acre to the depth of 1 foot.

WATER USE AND REUSE

Water reuse will be a major consideration in meeting the growing demand for water.

Who's using the water? Table 17–2 shows the breakdown for 1950, 1965, 1980, and 1985. Industry is the major user and also accounts for most of the increase since 1950. Table 17–3 gives an idea of how much water is needed for different products.

TABLE 17–2 Water Use in the United States

Use	Billions of Gallons per Day			
	1950	*1965*	*1980*	*1985*
Public supplies	14	26	36	40
Agricultural irrigation	110	115	140	137
Industry	77	165	255	217
Total	201	306	431	394

From W. W. Hales: Use and Reuse of Water. *Chemtech,* Vol. 12, pp. 532–537, 1982. Updated from Statistical Abstracts of the U.S., 1988.

TABLE 17–3 Sample of Water Usage by Industry in 1980

Industry	Water Used	
	Per Unit Production	*Per Finished Product*
Paper	20,000 gal/ton	1 gal/8 sheets typing paper
Oil refinery	20,000 gal/barrel crude oil	80 gal/gal gasoline
Steel	50,000 gal/ton	25 gal/1-lb box nails
Power	360 gal/min/Mw	51 gal/100-W bulb burning for 24 h

From W. W. Hales: Use and Reuse of Water. *Chemtech,* Vol. 12, pp. 532–537, 1982.

Figure 17–2 Markers on a utility pole in California's San Joaquin Valley indicate the large drop in surface level caused by withdrawal of groundwater for irrigation. (Courtesy of U.S. Geological Survey.)

Figure 17–3   **Industrial cooling towers. (Courtesy of Betz Laboratories, Inc.)**

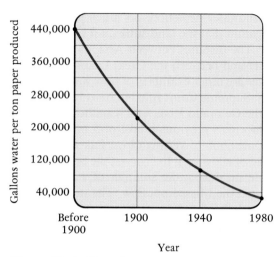

Figure 17–4   **Drop in water usage by paper industry.**

However, it is important to note that much of the industrial usage now involves recycled water. For example, the largest single industrial use of water is in plant cooling systems. Recirculating cooling water systems, which often use cooling towers (Fig. 17–3), are an important means of water reuse. Such reuse also helps to reduce thermal pollution of the river or lake where the used water is discharged. In addition, the high heat capacity of water enables the industrial user to recycle this important source of heat energy during the reuse water cycle.

An indication of how the paper industry has reduced the amount of water by reuse is shown in Figure 17–4. The steel industry, another large water user, uses 50,000 gal per ton of steel produced. However, 31,200 gal of this are recycled water, 1200 gal are vaporized, and the rest is discharged as wastewater.

These examples illustrate the difference between water use and water consumption. Much of the industrial use is temporary. The water is used to cool equipment or to provide steam, and then some fraction of it is discharged (treated if necessary) back to the river or lake source.

PUBLIC USE OF WATER

Where do we get the water we drink? Groundwater and surface water sources each provide half of the more than 40 billion gal of potable (drinkable) water that is used each day in the United States. In the United States, the average use of potable water per day is 90 gal per person. Table 17–4 gives the average amounts for various personal uses. One of the contributing factors to the water crisis is the fact that only *one-half gallon* of water per person per day must be of drinkable quality. Although urban water delivery systems are not set up to deliver two types of water—drinkable water and water for other purposes—dual water supply systems are feasible. Only a small fraction of municipal water would have to be of drinking water quality. The largest portion of nonpotable water supply would be disinfected and bacteriologically safe, but would not meet drinking water regulations. This second source would be suitable for irrigation of parks and golf courses, air conditioning, industrial cooling, and toilet flushing.

Table 17–4 illustrates the inefficiency of residential water systems. We use 33% of residential water for flushing toilets and 25% for bathing. Conventional showers use up to 10 gal/min, which can be reduced by as much as 70% by the installation of inexpensive water-saving shower heads.

Residential water conservation is a way to cut demand for fresh water supplies. It has been effective during drought conditions. This was illustrated during the California drought of 1976–1977, when individual usage was reduced by 63% in Marin County. Although Table 17–2 shows that residential use is a small part of the total, home water conservation is an important step in cutting demand for fresh water supplies in large urban areas. Home conservation will also help focus attention on the need for a total and continuing analysis of the problems relating to water supply.

Direct reuse of water is even possible for drinkable water. The idea of obtaining potable water from sewage is psychologically difficult for many persons to accept, but the technology has been developed. The rate of depletion of aquifers in the southwestern United States has led to use of water recycling plants in several cities. For example, El Paso, Texas, uses a water recycling plant to obtain 10 million gal of pure water per day from sewage effluent. The recycled water is pumped into the underground aquifer that is the main source of water for El Paso.

TABLE 17–4 Average Water Usage per Person per Day (Gallons)

Flushing toilets	30
Bathing	23
Laundering	11
Dishwashing	6
Drinking and cooking	10
Miscellaneous	10
Total	90

WHAT IS CLEAN WATER?—WHAT IS POLLUTED WATER?

Water that is judged unsuitable for drinking, washing, irrigation, or industrial uses is **polluted water**. The pollution may be heat, radioisotopes, toxic metal ions, organic solvent molecules, acids, or alkalies. Water suitable for some uses might be considered polluted and therefore unsuitable for other uses. Water that is unsuitable for use has generally been polluted by human activity, although natural leaching of some metal ions from rocks and soil, organic substances like tannins from decaying leaves and animal wastes, and silt can pollute clean water. As human activities have continued to pollute water, various governments have passed laws designed to cause us to keep our waters clean and unpolluted.

The Clean Water Act of 1977 shifted the burden of producing water suitable for reuse from the user to the wastewater discharger. This action was a crucial step in improving the quality of our rivers and lakes, since it is easier

Even drinking water has enough impurities to build up in water pipes over time. (*The World of Chemistry*, Program 12, "Water.")

TABLE 17–5 Classes of Water Pollutants, with Some Examples

Oxygen-demanding wastes	Plant and animal material
Infectious agents	Bacteria, viruses
Plant nutrients	Fertilizers, such as nitrates and phosphates
Organic chemicals	Pesticides such as DDT,* detergent molecules
Other minerals and chemicals	Acids from coal mine drainage, inorganic chemicals such as iron from steel plants
Sediment from land erosion	Clay silt on stream bed, which may reduce or even destroy life forms living at the solid–liquid interface
Radioactive substances	Waste products from mining and processing of radioactive material, used radioactive isotopes
Heat from industry	Cooling water used in steam generation of electricity

* Banned in the U.S., but still produced and exported.

to clean the wastewater prior to dumping than to clean the river water after the untreated waste has been discharged. In addition, the quality of the wastewater effluent is often high enough to be used as a resource of water for other purposes, such as irrigation or cooling towers.

As our industrial and commercial use of water has increased, water pollution has become more diversified. The U.S. Public Health Service now

TABLE 17–6 Maximum Contaminant Levels for Drinking Water Supplies

Contaminant	Maximum Concentration (mg/L)
Inorganics	
Arsenic	0.05
Barium	1
Cadmium	0.01
Chromium	0.05
Lead	0.05
Mercury	0.002
Nitrate	10
Selenium	0.01
Silver	0.05
Organics	
Endrin	0.0002
Lindane	0.004
Methoxychlor	0.1
Toxaphene	0.005
2,4-D	0.1
2,4,5-TP (Silvex)	0.01
Total trihalomethanes (includes bromotrichloromethane, dibromochloromethane, tribromomethane, trichloromethane (chloroform)	0.1

Source: EPA, 1988.

Figure 17–5 A water quality laboratory. This atomic absorption instrument can analyze both drinking water and treated waste water samples for trace amounts of various metals. (Courtesy of Metro Water Services, Nashville, Tennessee)

classifies water pollutants into eight broad categories (Table 17–5). The very fact that the EPA has published limits for chemical contaminants in drinking water (Table 17–6), is evidence enough that our water supplies are polluted. The EPA requires that municipal water supplies be monitored constantly (Fig. 17–5).

SELF-TEST 17–A

1. Approximately what percentage of the human body is water?
2. What is the major reservoir of water on the planet Earth?
3. The average person in the United States uses _____ gallons of water per day. Based on the U.S. population being _____ persons, this becomes _____ gallons for personal use.
4. The actual amount of potable (drinkable) water a person needs is _____ per day.
5. Three common water pollutants are _____, _____, and _____.
6. Water held in a stratum of porous rock is called _____.
7. A water bearing stratum of porous rock, sand, or gravel is called a(n) _____.
8. What happens to most of the water that falls on the United States each day?

BIOCHEMICAL OXYGEN DEMAND

The way in which organic materials are oxidized in the natural purification of water deserves special attention. The process opposes **eutrophication.** Even in the natural state, living organisms found in natural waters are constantly discharging organic debris into the water. To change this organic material into simple inorganic substances (such as CO_2 and H_2O) requires oxygen. The amount of oxygen required to oxidize a given amount of organic material is called the **biochemical oxygen demand (BOD).** The oxygen is required by microorganisms, such as many forms of bacteria, to metabolize the organic matter that constitutes their food. Ultimately, given near normal conditions and enough time, the microorganisms will convert huge quantities of organic matter into the following end products:

Organic carbon $\longrightarrow CO_2$

Organic hydrogen $\longrightarrow H_2O$

Organic oxygen $\longrightarrow H_2O$

Organic nitrogen $\longrightarrow NO_3^-$ or N_2

One way to determine the amount of organic pollution is to determine how much oxygen a given sample of polluted water will require for complete oxidation. For example, a known volume of the polluted water is diluted with a known volume of standardized sodium chloride solution of known oxygen content. This mixture is then held at 20°C for 15 days in a closed bottle. At the end of this time the amount of oxygen that has been consumed is taken to be the biochemical oxygen demand.

Highly polluted water often has a high concentration of organic material, with resultant large biochemical oxygen demand (Fig. 17–6). In extreme cases, more oxygen is required than is available from the environment, and putrefaction results. Fish and other freshwater aquatic life can no longer survive. The aerobic bacteria (those that require oxygen for the decomposition process) die. As a result of the death of these organisms, even more lifeless organic matter results and the BOD soars. Nature, however, has a backup system for such conditions. A whole new set of microorganisms (anaerobic bacteria) takes over; these organisms take oxygen from oxygen-containing compounds to convert organic matter to CO_2 and water. Organic

A quantitative relationship exists between oxygen needs and organic pollutants to be destroyed. This is BOD.

A standardized solution is one of known concentration.

Fish cannot live in water that has less than 0.004 g_{O_2}/L (4 ppm).

Low available oxygen in a stream results in dead fish. (*The World of Chemistry*, Program 12, "Water.")

Figure 17–6 Graph showing oxygen content and oxidizable nutrients (BOD) as a result of sewage introduced by a city. The results are approximated on the basis of a river flow of 750 gal/s. Note that it takes 70 miles for the stream to recover from a BOD of 0.023 g oxygen per liter.

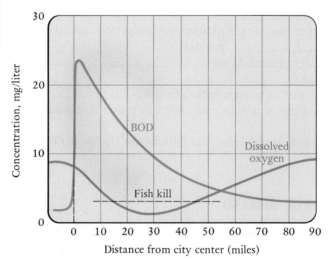

nitrogen is converted to elemental nitrogen by these bacteria. Given enough time, enough oxygen may become available, and aerobic oxidation will then return.

A stream containing 10 ppm by weight (just 0.001%) of an organic material, the formula of which can be represented by $C_6H_{10}O_5$, will contain 0.01 g of this material per liter. The calculation used to obtain this is:

$$?g = 1\,L \times \frac{1000\ ml}{1\ L} \times \frac{1\ g}{ml} = 1000\ g$$

0.001% of this is the pollutant:

0.001% of 1000 g = (0.00001)(1000 g) = 0.010 g

To transform this pollutant to CO_2 and H_2O, the bacteria present use oxygen as described by the equation:

$$C_6H_{10}O_5\ +\ 6\ O_2\ \longrightarrow\ 6\ CO_2 + 5\ H_2O$$

Relative Weight 162   Relative Weight 192

The 0.010 g of pollutant requires 0.012 g of dissolved oxygen.

$$\frac{?g\ oxygen}{L} = \frac{0.010\ g\ pollutant}{L} \times \frac{192\ g\ oxygen}{162\ g\ pollutant} = \frac{0.012\ g\ oxygen}{L}$$

At 68°F (20°C), the solubility of oxygen in water under normal atmospheric conditions is 0.0092 g of oxygen per liter (Table 17–7.)

Because the BOD (0.012 g/L) is greater than the equilibrium concentration of dissolved oxygen (0.0092 g/L), as the bacteria utilize the dissolved oxygen in this stream, the oxygen concentration of the water will soon drop too low to sustain any form of fish life. Life forms can survive in water where the BOD exceeds the dissolved oxygen if the water is flowing vigorously in a shallow stream (this facilitates the absorption of more oxygen from the air via aeration).

BOD values can be greatly reduced by treating industrial wastes and sewage with oxygen and/or ozone. Numerous commercial cleanup operations now being developed and used employ this type of "burning" of the organic wastes. Another benefit of treating waste water with oxygen is that some of the nonbiodegradable material becomes biodegradable as a result of partial oxidation.

Characteristic BOD Levels	g_{o_2}/L
Untreated municipal sewage	0.1–0.4
Runoff from barnyards and feed lots	0.1–10
Food-processing wastes	0.1–10

TABLE 17–7 Solubility of Oxygen in Water at Various Temperatures

Temperature °C	Solubility of O_2 (g_{O_2}/L H_2O)
0	0.0141
10	0.0109
20	0.0092
25	0.0083
30	0.0077
35	0.0070
40	0.0065

These data are for water in contact with air at 760 mm mercury pressure.

High concentration of organic pollutants
↓
Low oxygen concentration
↓
Dead organisms
↓
Higher concentration of organic pollutants
↓
Lower oxygen concentration
↓
Anaerobic conditions

IMPACT OF HAZARDOUS WASTES ON WATER QUALITY

Industrial wastes can be an especially vexing sort of pollution problem because often they either are not removed or are removed very slowly by naturally occurring purification processes, and are generally not removed at all by a typical municipal water treatment plant. Table 17–8 lists some of the industrial pollutants that result from products important to us.

Disposal of hazardous wastes in landfills has been the principal method of disposal for industries, agriculture, and municipalities for decades. Inci-

TABLE 17–8 Important Industrial Products and Consequent Hazardous Wastes

The Products We Use	The Potentially Hazardous Waste They Generate
Plastics	Organic chlorine compounds
Pesticides	Organic chlorine compounds, organic phosphate compounds
Medicines	Organic solvents and residues, heavy metals like mercury and zinc, for example
Paints	Heavy metals, pigments, solvents, organic residues
Oil, gasoline, and other petroleum products	Oils, phenols, and other organic compounds, lead, salts, acids, alkalies
Metals	Heavy metals, fluorides, cyanides, acids, and alkaline cleaners, solvents, pigments, abrasives, plating salts, oils, phenols
Leather	Chromium, zinc
Textiles	Heavy metals, dyes, organic chlorine compounds, organic solvents

dents such as the Love Canal disaster drew attention to the serious contamination of groundwater by hazardous wastes. Action on local, state, and federal levels began in the 1970s to solve problems caused by past disposal and to develop workable methods for future disposal of hazardous wastes. In 1980, Congress established the "Superfund," a $1.6 billion program designed to clean up hazardous waste sites that were threatening to contaminate the nation's underground water supplies. In 1985, the Office of Technology Assessment estimated that the number of hazardous waste sites requiring cleanup will increase, perhaps to as high as 10,000, and the cost of cleanup may reach $100 billion. By June 1988, the Environmental Protection Agency (EPA) had placed 1177 hazardous waste sites on its National Priorities List for cleanup under the Superfund Law.

Although only 1 or 2% of the aquifers are known to be polluted by hazardous wastes, many of these aquifers are near large population centers, so the problem is a serious one. The basic problem with land disposal of hazardous wastes is the contamination of groundwater as it moves through the disposal area (Fig. 17–1). Water pollution from these sites generally occurs as seepage into an underlying aquifer.

In 1976, the federal government passed the Resource Conservation and Recovery Act (RCRA). This law is designed to give "cradle-to-grave" (origin to disposal) responsibility to generators of hazardous wastes. The RCRA regulations cover generation, transportation, storage treatment, and disposal of hazardous wastes.

Considerable attention has been given to safe disposal of hazardous wastes, monitoring groundwater near hazardous waste sites and reducing the quantity of hazardous wastes by recycling chemicals. The technology for safe disposal exists, but the costs are high. Data reported by the EPA in 1988 (Table 17–9) indicate that 7% of hazardous wastes are being disposed of by environmentally unsound methods. The effect of the present government regulations and the greater public awareness are making current disposal

Field sampling of industrial waste water. (Courtesy of Isco, Inc.)

TABLE 17–9 Hazardous Waste Disposal Methods in 1988

Method	Percentage of Total
Unacceptable	
Unlined surface impoundment	5
Land disposal	1
Uncontrolled incineration	1
Acceptable	
Controlled incineration	55
Secure landfills	20
Recovered	25

Source: EPA, 1989.

Figure 17–7   Waste drums containing hazardous wastes. Their cleanup is both time-consuming and costly. (John Cunningham/Visuals Unlimited.)

methods safer, but the cleanup of Superfund sites and other landfills that are contaminating groundwater will take time.

The following are examples of groundwater contamination by seepage from hazardous waste sites.

In 1978, the area around an old chemical dumpsite in Love Canal, a community in southeastern Niagara Falls, New York, was declared a disaster area by President Carter. Record rainfall caused leaching of chemicals from corroding waste-disposal barrels buried in the old chemical dumpsite. Over 230 families were relocated, and the area was fenced off. In 1980, new boundaries that affected an additional 800 families were established. The emergency declarations by President Carter in 1978 and 1980 provided federal funds to assist the state in relocating families. This was the first use of federal emergency funds for something other than a "natural" disaster.

Groundwater supplies in Toone and Teague, Tennessee, were contaminated by organic wastes from a nearby landfill in 1978. The landfill, closed 16 years earlier, held 350,000 drums, and pesticide wastes were leaking from many of them. The towns must pump water from other locations.

Groundwater in a 30-mile^2 area near Denver was contaminated from disposal of pesticide waste between 1943 and 1957 in unlined disposal ponds.

At least 1500 drums containing wastes from a metal-finishing operation were buried near Bryon, Illinois, until 1972. Surface water, soil, and groundwater were contaminated with cyanide, heavy metals, and organic toxic compounds.

About 17,000 waste drums littered a 7-acre site in Kentucky that became known as the "Valley of the Drums" (Fig. 17–7). Many drums have been leaking their contents onto the ground. In 1979, an EPA survey identified about 200 toxic organic chemicals and 30 heavy metals in the soil and in water samples near the dump.

Workers in protective suits at a waste site. (*The World of Chemistry,* Program 25, "Chemistry and the Environment.")

Number of hazardous
waste sites in each state:

none	11–15
1–5	16–40
6–10	over 60

Hawaii: 6
Alaska: none
Puerto Rico: 8
Guam: 1

Figure 17–8 Hazardous waste sites in the United States as designated by EPA Superfund.

The EPA has listed over 2000 sites in the United States where toxic wastes have been stored and should be cleaned up (Fig. 17–8). This cleanup will cost billions of dollars.

It was once considered good engineering practice to put all wastes into landfills. There, many of these wastes leached into the groundwater causing serious water pollution problems. Today, industrial wastes categorized as **hazardous wastes** (see Table 17–10) must be placed into secure landfills, incinerated, or treated in some way to render them nonhazardous. Secure

The cleanup of hazardous waste sites is costly and time-consuming. (*The World of Chemistry*, Program 25, "Chemistry and the Environment.")

TABLE 17–10 Hazardous Wastes as Defined by EPA

Wastes containing the following metals and pesticides:
    Arsenic, Barium, Cadmium, Chromium, Lead, Mercury, Selenium, Silver, Endrin, Lindane, Methoxychlor, Toxaphene, 2,4-D, 2,4,5,-TP (Silvex)
Wastes that have the following characteristic properties*:
    Ignitible, Corrosive, Reactive, Acutely toxic
Twenty-one wastes from nonspecific sources: (such as)
    Wastes containing the cyanide ion, distillation residues, used halogenated solvents like carbon tetrachloride
Eighty-nine wastes from specific sources: (such as)
    Wastewater sludges from chloride production, wastewater from pesticide manufacture, sludges from the production of petroleum products
A large number of various discarded and off-specification chemicals, many of which are used in the chemical industry to manufacture pharmaceuticals, polymers, paints, dyes, automotive products, cosmetics, etc.

Note: Shipments of hazardous wastes are carefully monitored by EPA and state governments. In addition, all facilities receiving these wastes are permitted and licensed.
Source: EPA.
* Detailed definitions apply to these waste characteristics.

Figure 17–9 Pits for holding containers of hazardous wastes must be lined with thick polymer liners that help prevent the escape of wastes into the groundwater. (Courtesy of Chemical Waste Management, Inc.)

landfills generally have plastic linings (see Fig. 17–9) and carefully spaced monitoring wells so any leaching from the landfill's contents may be monitored for any leakage. Other wastes from industry, not considered hazardous but still containing potentially harmful chemicals (Table 17–11), may go into public landfills, or may go to sewers or receiving bodies of water after some form of pretreatment. The amounts of wastes disposed of this way are very large and the potential for polluting ground and surface waters is considerable.

TABLE 17–11 Some Additional Hazardous Chemicals Not Considered Hazardous Wastes

Chemical	Use
Ammonium nitrate (solution)	Fertilizer, explosives
Beryllium nitrate	Chemical manufacture
Biphenyl	Fungistat for citrus fruit
Cobalt oxide	Pigment
Copper nitrate	Electroplating, light-sensitive papers
Ethylene glycol	Antifreeze
Manganese dioxide	Battery manufacture
Nickel nitrate	Metal plating
Quinoline	Pharmaceuticals, flavorings

Note: Chemicals on this list must be reported when released into the environment by any means according to the Superfund Reauthorization and Amendments Act of 1986.
Source: EPA.

**An industrial waste discharge.
(Karen Roeder.)**

TABLE 17–12 Releases of Hazardous Chemicals by Industry to Water. The Top 10 States for 1987.

State	Quantity Released (lb)
1. California	3.8×10^9
2. Louisiana	7.0×10^8
3. Texas	6.0×10^8
4. Alabama	6.0×10^8
5. Mississippi	4.0×10^8
6. Georgia	4.0×10^8
7. South Carolina	3.0×10^8
8. Washington	3.0×10^8
9. Virginia	2.0×10^8
10. North Carolina	2.0×10^8

Source: EPA, 1989.
Note: These release data are now available to the public from the National Library of Medicine as an on-line database. This release reporting is a part of the "Community Right-to-Know" provisions of the Superfund Reauthorization and Amendments Act of 1986.

In 1988, new reporting regulations regarding releases of certain listed hazardous chemicals went into effect for manufacturing industries. These releases must be reported each year. Data for 1987 (Table 17–12) indicate that the state of California had the largest total amount of hazardous chemicals released.

Because of known releases of harmful chemicals into water, states like California have taken drastic steps to protect their ground and surface waters. In California, Proposition 65, the Safe Drinking Water and Toxic Enforcement Act of 1986 lists approximately 200 chemicals or classes of chemicals known to cause cancer or reproductive toxicity and prohibits their discharge into drinking water supplies.

Proposition 65:
"No person in the course of doing business shall knowingly discharge or release a chemical known to cause cancer or reproductive toxicity into water or onto or into land where such chemical passes or probably will pass into any source of drinking water."

HOUSEHOLD WASTES AS HAZARDOUS WASTES

Often we do not think about the things we discard in our garbage, but what we throw away and how we do it can affect our groundwater purity. Household wastes that are incinerated can contribute to air pollution (See Chapter 18), but since the bulk of our household waste goes to landfills, we too, can be responsible for causing pollution of groundwater. Table 17–13 lists some common household products and the kinds of chemicals they contain. Because we are the consumers of industrial products, we can put the very same chemicals into our groundwater as industry can. Although the individual amounts of harmful chemicals used in a household are less than those used in a large industry, the total amounts of harmful chemicals disposed of daily by all households can be very large, even for a medium-sized city.

TABLE 17–13 Some Household Hazardous Wastes and Recommended Disposal

Type of Product	Harmful Ingredients	Disposal*
Bug sprays	Pesticides, organic solvents	Special
Oven cleaner	Caustics	Drain
Bathroom cleaners	Acids or caustics	Drain
Furniture polish	Organic solvents	Special
Aerosol cans (empty)	Solvents, propellants	Trash
Nail-polish remover	Organic solvents	Special
Nail polish	Solvents	Trash
Antifreeze	Organic solvents, metals	Special
Insecticides	Pesticides, solvents	Special
Auto battery	Sulfuric acid, lead	Special
Medicine (expired)	Organic compounds	Drain
Paint (latex)	Organic polymers	Drain
Gasoline	Organic solvents	Special
Motor oil	Organic compounds, metals	Special
Drain cleaners	Caustics	Drain
Shoe polish	Waxes, solvents	Trash
Paints (oil-based)	Organic solvents	Special
Mercury batteries	Mercury	Special
Moth balls	Chlorinated organic compound	Special

* Special: Professional disposal as a hazardous waste. Drain: disposal down the kitchen or bathroom drain. Trash: Treat as normal trash—no harm to the groundwater. In most households, the items marked special are disposed of as normal trash, which results in groundwater pollution.
Source: "Household Hazardous Waste: What You Should and Shouldn't Do," Water Pollution Control Federation, 1986.

Households have a greater problem disposing of chemicals that are potentially harmful to the groundwater than industry does. Even if a city has an active recycling project for glass, paper, metals, and plastics, most municipalities provide no means of pickup of those chemicals separated from the ordinary trash destined for the landfill. If these chemicals are mixed with ordinary garbage, all of it goes to the city landfill or incinerator. Professional hazardous waste disposers seldom offer their services to homeowners at a cut rate, so their high prices deter most households from disposing of these chemicals in the proper way.

How can we dispose of hazardous household wastes without danger to the groundwater supply? We can ask our city's municipal waste authorities to provide disposal sites for these wastes. As an example of how this can be done, in some European countries (such as the Netherlands) special trucks pick up paints, oils, batteries, and other products for disposal. Another approach is to purchase products with their ingredients in mind. Ordinary alkaline batteries often work just as well as mercury batteries, for example. We should also purchase the quantity of a product we believe we can use instead of buying larger quantities simply because it is priced more economically in larger containers. When we purchase more of a product than we can reasonably expect to use, we often allow the unused portion to sit on a shelf until it is eventually discarded.

Recycling of materials like glass, paper, metals, and plastics helps conserve resources like raw materials and energy. Recycling also conserves valuable landfill space and keeps some otherwise harmful chemicals from the groundwater.

Ordinary garbage costs about $27/ton for disposal, whereas hazardous waste costs about $1000/ton for proper disposal.

Recycling of household wastes can be important in lessening the burden on our municipal landfills and incinerators. Some communities in the United States are more actively involved in recycling than others. Household items like waste paper (newspaper, cereal boxes, and junk mail), plastics (milk jugs, plastic wrap, and other containers), and metals (aluminum cans, old lawn furniture, bicycles, old tools) unnecessarily contribute to overburdened landfills, causing them to fill up prematurely. When metals and plastics are mixed with incinerated wastes, they add pollutants to the atmosphere. Used motor oil and discarded automobile batteries can add metals, acids, and unwanted hydrocarbons to groundwater and therefore should be recycled whenever possible. Most communities have companies that specialize in recycling one or more of these household waste products, but it is up to each homeowner to seek out alternatives to indiscriminate mixing of garbage and other household wastes.

SELF-TEST 17–B

1. The amount of oxygen required to oxidize a given amount of organic material is called the _____, which is abbreviated _____.
2. Which can hold more dissolved oxygen, 1 L of water at 5°C, or 1 L of water at 40°C?
3. Name two industrial products whose manufacture introduces heavy metals into groundwater.
4. Name two industrial products whose manufacture introduces chlorinated organic compounds into groundwater.
5. Name three household wastes that can contaminate groundwater with the same harmful chemicals as industrial wastes. Beside each list the harmful chemical.

Household waste	Harmful chemical
_____	_____
_____	_____
_____	_____

6. List four household waste types that lend themselves to recycling.

WATER PURIFICATION IN NATURE

Water is a natural resource that, within limitations, is continuously renewed. The familiar water cycle (Fig. 17–1) offers a number of opportunities for nature to purify its water. The worldwide **distillation** process results in rain water containing only traces of nonvolatile impurities, along with gases dissolved from the air. **Crystallization** of ice from ocean saltwater results in relatively pure water in the form of icebergs. **Aeration** of groundwater as it trickles over rock surfaces, as in a rapidly running brook, allows volatile

Rain water in clean air is very pure.

Volatile means going easily into the gaseous state.

impurities to be released to the air. **Sedimentation** of solid particles occurs in slow-moving streams and lakes. **Filtration** of water through sand rids the water of suspended matter such as silt and algae. Next, and of very great importance, are the **oxidation processes.** Practically all naturally occurring organic materials—plant and animal tissue, as well as their waste materials—are changed through a complicated series of oxidation steps in surface waters to simple substances common to the environment. Finally, another process used by nature is **dilution.** Most, if not all, pollutants found in nature are rendered harmless if reduced below certain levels of concentration by dilution with pure water.

Before the explosion of the human population and the advent of the Industrial Revolution, natural purification processes were quite adequate to provide ample water of very high purity in all but desert regions. Nature's purification processes can be thought of as massive but somewhat delicate. In many instances the activities of humans push the natural purification processes beyond their limit, and polluted water accumulates.

A simple example of nature's inability to handle increased pollution comes from dragging gravel from stream beds. This excavation leaves large amounts of suspended matter in the water. For miles downstream from a source of this pollutant, aquatic life is destroyed. Eventually, the solid matter settles, and normal life can be found again in the stream.

A more complex example, and one for which there is not nearly so much reason to hope for the eventual solution by natural purification, is the degradability of organic materials. A **biodegradable** substance is composed of molecules that are broken down to simpler ones in the natural environment by microorganisms. For example, cellulose suspended in water will eventually be converted to carbon dioxide and water. Some organic compounds, notably some of those synthetically produced, are not easily biodegradable; these substances simply stay in the natural waters or are absorbed by life forms and remain intact for a long time. An example is DDT.

Even nature's pure rain water is in jeopardy. If the acidic air pollutants, such as the oxides of sulfur, are concentrated enough, the absorbing rain water will become acidic enough to harm life forms and mar metal and stone structures. The government of Canada has complained to the United States because of acid rains arising from the industrial Northeast. Acid rain is discussed in more detail in Chapter 18. In areas in which heavy concentrations of automobile fumes collect, poisonous lead compounds have been found in rain water in concentrations many times higher than the 0.01 ppm generally allowed in drinking water. The concentration of the lead can be correlated with the concentration of exhaust fumes in the air. Fortunately, lead does not long remain in water, since it generally forms insoluble compounds.

Pure water:
Chemist: "Pure H_2O—no other substance."
Parent: "Nothing harmful to human beings."
Game and Fish Commission: "Nothing harmful to animals."
Sunday boater: "Pleasing to the eye and nose, no debris."
Ecologist: "Natural mixture containing necessary nutrients."

Only about 1% of groundwater supplies in the United States are now considered unsafe.

Some synthetic compounds are not biodegradable and therefore are very persistent in natural waters.

Acid rain in Pasadena, California (1976–1977), contained:
1. Acid (pH = 4.06)
2. NH_4^+
3. K^+
4. Ca^{2+}
5. Mg^{2+}
6. Cl^-
7. NO_3^-
8. SO_4^{2-}

Concentrations in 10^{-5} to 10^{-6} M range.

WATER PURIFICATION: CLASSICAL AND MODERN PROCESSES

The outhouses of some rural dwellers had their counterparts in city cesspools. The terrible job of cleaning led to the development of cesspools that could be flush-cleaned with water, followed by a connecting series of such

Cesspools were an early and crude form of the modern activated sludge process.

PLANT FOR PRIMARY SEWAGE TREATMENT

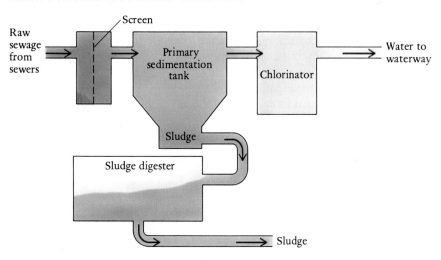

PLANT FOR SECONDARY SEWAGE TREATMENT

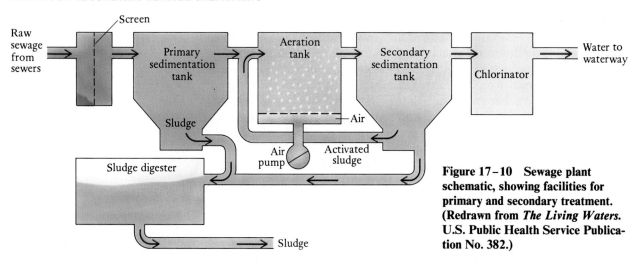

Figure 17–10 Sewage plant schematic, showing facilities for primary and secondary treatment. (Redrawn from *The Living Waters*. U.S. Public Health Service Publication No. 382.)

Sewage is still 99.9% water!

pools that could be flushed from time to time. City sewer systems with no holding of the wastes were the next step.

Since there were not enough pure wells and springs to serve the growing population, water purification techniques were developed. The classical method, which is now termed **primary water treatment,** involved settling and filtration (Fig. 17–10).

In the settling stage, calcium hydroxide and aluminum sulfate are added to produce aluminum hydroxide. Aluminum hydroxide is a sticky, gelatinous precipitate that settles out slowly, carrying suspended dirt particles and bacteria with it.

$$3\ Ca(OH)_2 + Al_2(SO_4)_3 \longrightarrow 2\ Al(OH)_3 + 3\ CaSO_4$$

(a)

(b)

Figure 17–11   Water treatment. (a) Aerobic. (b) Anaerobic. (*The World of Chemistry,* Programs 12 and 16, "Water," and "The Proton in Chemistry.")

If the intake water is polluted enough with biological wastes, the primary treatment, even with chlorination, cannot render the water safe. To be sure, enough chlorine or other oxidizing agents could be added to kill all life forms, but the result would be water loaded with a wide variety of noxious chemicals, especially chlorinated organics, many of which are suspected carcinogens. Some way had to be found to coagulate and separate out the organic material that passed through the primary filters.

Secondary water treatment revives the old cesspool idea under a more controlled set of conditions and acts only on the material that will not settle or cannot be filtered. Modern secondary treatment operates in an oxygen-rich environment (aerobic), whereas the cesspool operates in an oxygen-poor environment (anaerobic) (Fig. 17–11). The results are the same: the organic molecules that will not settle are consumed by organisms; the resulting sludge will settle. Bacteria and even protozoa are introduced into the oxygen-rich environment for this purpose. Two techniques, the trickle filter and the activated sludge method have been widely used in secondary water treatment.

Primary and secondary water treatment systems will not remove dissolved inorganic materials such as poisonous metal ions or even residual amounts of organic materials. These materials are removed by a variety of **tertiary water treatments.**

Two technologies are now being used for the removal of toxic materials from wastewater; these are carbon adsorption and activated sludge. Carbon black has been used for many years for adsorbing vapors and solute materials from liquid streams. Many toxic organic materials can be removed with activated or baked carbon granules that have been activated by high-temperature baking. This activated carbon has a high surface area that readily adsorbs chemicals from the wastewater. Activated sludge is a hurry-up ver-

Americans spend about $350 million a year on bottled water. Buyer beware: a very wide variety of standards exist for bottled water.

Primary, secondary, and tertiary water treatment methods can be used in both the purification of water to be consumed and the preparation of sewage to be sent back into a stream.

TABLE 17-14 Ions Present in Sea Water at Concentrations Greater Than 0.001 g/kg

Ion	g/kg Sea Water
Cl^-	19.35
Na^+	10.76
SO_4^{2-}	2.71
Mg^{2+}	1.29
Ca^{2+}	0.41
K^+	0.40
HCO_3^-, CO_3^{2-}	0.106
Br^-	0.067
$H_2BO_3^-$	0.027
Sr^{2+}	0.008
F^-	0.001
Total	35.129

sion of natural stream purification. Bacteria and other microorganisms degrade the water pollutants in the sludge medium.

FRESH WATER FROM THE SEA

Since sea water covers 72% of the Earth, it is not surprising that this source would be a major consideration for areas where fresh water supplies aren't sufficient to meet the demand. The oceans contain an average 3.5% dissolved salts by weight, a concentration too high for most uses. The solvent properties of water are illustrated by the average composition of sea water in Table 17-14. If you add these up in terms of the number 0.001 g/kg, you have over 35,000 ppm of dissolved ions. The total must be reduced to below 500 ppm before the water is suitable for human consumption.

The technology has been developed for the conversion of sea water to fresh water. The extent to which this technology is actually put to use depends on the availability of fresh water and the cost of the energy for the conversion. Over 2200 desalination plants were in operation throughout the world in the 1980s. Two methods used to purify sea water are reverse osmosis and solar distillation.

Reverse Osmosis

When a membrane is permeable to water molecules but not to ions or molecules larger than water, it is called a **semipermeable membrane.** If a semipermeable membrane is placed between sea water and pure water, the pure water will pass through the membrane to dilute the sea water. This is **osmosis.** The liquid level on the sea water side rises as more water molecules enter than leave, and pressure is exerted on the membrane until the rates of

Figure 17-12 Normal osmosis is represented by a and b. Water molecules create osmotic pressure by passing through the semipermeable membrane to dilute the brine solution. Reverse osmosis, represented in c, is the application of an external pressure in excess of osmotic pressure to force water molecules to the pure water side.

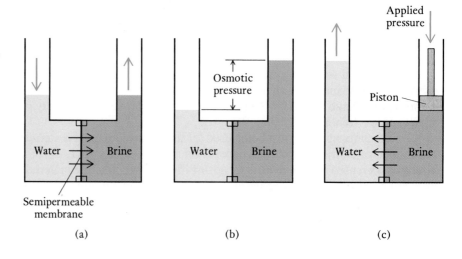

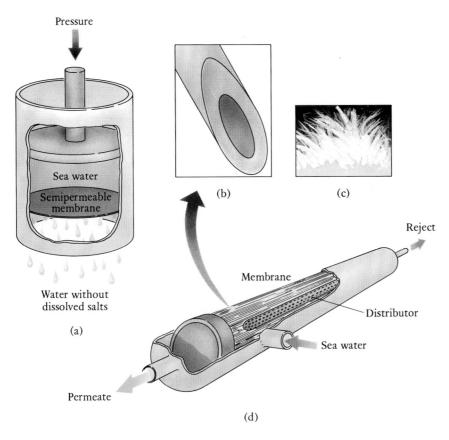

Pressure

Sea water

Semipermeable membrane

Water without dissolved salts

(a)

(b)

(c)

Reject

Membrane

Distributor

Sea water

Permeate

(d)

Figure 17–13   Commercial reverse osmosis units. (a) Mechanical pressure forces water against osmotic pressure to region of pure water. (b) Enlargement of individual membrane. (c) Mass of many membranes. (d) Industrial unit; feed water (salt) that passes through membranes collects at the left end (permeate). The more concentrated salt solution flows out to the right as the reject.

diffusion of water molecules in both directions are equal. **Osmotic pressure** is defined as the external pressure required to prevent osmosis. Figure 17–12 illustrates the concept of osmosis and osmotic pressure.

Reverse osmosis is the application of pressure to cause water to pass through the membrane to the pure-water side (Fig. 17–13). The osmotic pressure of normal sea water is 24.8 atm (atmospheres). As a result, pressures greater than 24.8 atm must be applied to cause reverse osmosis. Pressures up to 100 atm are used to provide a reasonable rate of filtration and to account for the increase in salt concentration that occurs as the process proceeds.

The most common semipermeable membrane used in reverse osmosis is a modified cellulose acetate polymer, although several polyamide polymers also have been used. The largest reverse osmosis plant in operation today is the Yuma Desalting Plant in Arizona. This plant, which began operation in 1982, can produce 100 million gal/day. The plant was built to reduce the salt concentration of irrigation wastewater in the Colorado River from 3200 ppm to 283 ppm. This project is part of a U.S. commitment to supply Mexico with a sufficient quantity of water suitable for irrigation.

Sea water can also be purified economically using reverse osmosis. One reverse osmosis plant, built in 1983 on the Mediterranean island of Malta, can purify 5.3 million gal/day, lowering the total dissolved solids from about 38,000 ppm to between 400 and 500 ppm — well within the World Health

Irrigation water of desert fields dissolves about 2 tons of salt per acre per year. Irrigation wastewater carries the salt back to the Colorado River.

TABLE 17–15 Large Solar Stills

Location	Country	Area (m²)	Year Built	Feed Type
Gwadar	Pakistan	9072	1972	Sea water
Patmos	Greece	8667	1967	Sea water
Las Salinas	Chile	4757	1872	Sea water
Coober Pedy	Australia	3160	1966	Brackish
Megisti	Greece	2528	1973	Sea water
Kimolos	Greece	2508	1968	Sea water
Klonlon	Greece	2400	1971	Sea water
Fiskardo	Greece	2200	1971	Sea water
Nisiros	Greece	2005	1969	Sea water
Awania	India	1867	1978	Brackish
St. Vincent	West Indies	1710	1967	Brackish
Mahdia	Tunisia	1300	1968	Brackish

Organization's limits for drinking water. Malta now uses four reverse osmosis plants that produce a total of 12 million gal of fresh water per day from the sea. On Florida's Sanibel Island, increasing salinity in its wellwater led to the installation of a reverse osmosis system. This facility has a design capacity of 3.6 million gal/day and has one of the lowest energy consumptions per 1000 gal of potable water of any comparably sized system in commercial use.

Solar Distillation

Solar evaporation units can be used to purify sea water in areas that receive a lot of sunlight. The main disadvantage of solar units is the amount of land required to produce appreciable amounts of fresh water. Table 17–15 lists the location and size of large solar stills. The output of these units is about 3 L/m²/day or 7000 gal/day for the larger units. Figure 17–14 shows a basic

Figure 17–14 Principle of solar still. Radiation heats salt water in black trough. Vapor condenses on sloping glass surfaces and runs off into distilled water troughs.

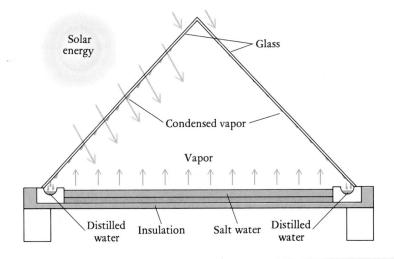

design used for solar stills. Smaller units can be constructed that will provide enough fresh water for homes.

SOFTENING OF HARD WATER

The presence of Ca^{2+}, Mg^{2+}, Fe^{3+}, or Mn^{2+} will impart "hardness" to waters. Hardness in water is objectionable because (1) it causes precipitates (scale) to form in boilers and hot water systems, (2) it causes soaps to form insoluble curds (this reaction does not occur with some synthetic detergents), and (3) it can impart a disagreeable taste to the water.

Hardness due to calcium or magnesium, present as bicarbonates, is produced when water containing carbon dioxide trickles through limestone or dolomite:

$$CaCO_3 + CO_2 + H_2O \longrightarrow Ca^{2+} + 2\ HCO_3^-$$
Limestone

$$CaCO_3 \cdot MgCO_3 + 2\ CO_2 + 2\ H_2O \longrightarrow Ca^{2+} + Mg^{2+} + 4\ HCO_3^-$$
Dolomite

Such "hard water" can be softened by removing these compounds. The principal methods for softening water are (1) the lime–soda process and (2) ion exchange processes.

The lime–soda process is based on the fact that calcium carbonate ($CaCO_3$) is much less soluble than calcium bicarbonate [$Ca(HCO_3)_2$] and that magnesium hydroxide is much less soluble than magnesium bicarbonate. The raw materials added to the water in this process are hydrated lime [$Ca(OH)_2$] and soda (Na_2CO_3). In the system, several reactions take place, which can be summarized:

$$HCO_3^- + OH^- \longrightarrow CO_3^{2-} + H_2O$$

$$Ca^{2+} + CO_3^{2-} \longrightarrow CaCO_3\downarrow$$

$$Mg^{2+} + 2\ OH^- \longrightarrow Mg(OH)_2\downarrow$$

The overall result of the lime–soda process is to precipitate almost all the calcium and magnesium ions and to leave sodium ions as replacements.

Iron present as Fe^{2+} and manganese present as Mn^{2+} can be removed from water by oxidation with air (aeration) to higher oxidation states. If the pH of the water is 7 or above (either naturally or by adding lime), the insoluble compounds $Fe(OH)_3$ and $MnO_2(H_2O)_x$ are produced and precipitate from solution.

The desire for and achievement of soft water for domestic use has sparked a rather heated health debate during the past two decades. Soft water is usually acidic and contains Na^+ ions in the place of di- and trivalent metal ions. An increased intake of Na^+ is known to be related to heart disease. Also, the acidic soft water is more likely to attack metallic pipes, resulting in the solution of dangerous ions such as Pb^{2+}. One way to avoid sodium ions in drinking water and to use less soap when washing would be to drink naturally hard water and wash in soft water.

Hard water contains metal ions that react with soaps and give precipitates.

Water softeners that act like ion-exchange resins are used to make soft water. They remove the hard water ions, Ca^{2+}, Mg^{2+}, and Fe^{3+}, and put Na^+ ions in the water in exchange.

Soft water: <65 mg of metal ion/gal
Slightly hard: 65–228 mg
Moderately hard: 228–455 mg
Hard: 455–682 mg
Very hard: >682 mg

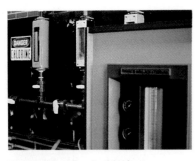

Figure 17–15   A chlorinator apparatus for a 60 million gal/day water plant. (Courtesy of the Robert L. Lawrence, Jr. filtration plant.)

CHLORINATION

With the advent of chlorination of water supplies in the early 1900s, the number of deaths in the United States caused by typhoid and other water-borne diseases dropped from 35/100,000 population in 1900 to 3/100,000 population in 1930.

Chlorine is introduced into water as the gaseous free element (Cl_2) (Fig. 17–15) and it acts as a powerful oxidizing agent for the purpose of killing bacteria that remain in water after preliminary purification. The principal water-borne diseases spread by bacteria include cholera, typhoid, paratyphoid, and dysentery.

Most city water supplies are not bacteria-free. Surviving bacteria will usually produce counts numbering in the tens of thousands but only rarely do these surviving bacteria cause disease. Today the most common water-borne bacterial disease is giardiasis, a gastrointestinal disorder. Most often this disease comes from surface water but, on occasion, it can be traced to city water systems.

Chlorination of industrial wastes and city water supplies presents a potential threat because of the reaction of chlorine with residual concentrations of organic compounds. Traditional purification methods do not remove chlorinated hydrocarbons, or for that matter, most organics. The chlorinated hydrocarbons, which may be present at levels of a few parts per million or less, include dichloromethane, chloroform, trichloroethylene, and chlorobenzene, all suspected carcinogens.

A number of these chemicals in the same concentration range have been shown to be mutagenic to salmonella bacteria. Studies show an increased risk of 50% to 100% in rectal, colon, and bladder cancers in individuals who drink chlorinated water. According to the EPA, mutagenic or carcinogenic chemicals have been found in 14 major river basins in the United States. It is estimated that more than 500 water systems in the United States exceed EPA's maximum of 0.1 ppm for chlorinated hydrocarbons. The presence of these chlorinated hydrocarbons can be prevented either by using another disinfectant or by removing the low-level organic compounds before chlorination.

An efficient process for reducing the level of organic compounds in water is to pass the water through biologically activated carbon. (See p. 545 for a description of the use of activated carbon.)

Ton cylinders of chlorine. Each cylinder holds 2000 lb of gaseous chlorine for final treatment of drinking water. (Courtesy of the Robert L. Lawrence, Jr. filtration plant.)

OZONE TREATMENT

Ozone, a disinfectant that does not produce harmful compounds when it reacts with low-level organic compounds in water, is used in over 1000 water treatment plants, mostly in Europe. Ozone is produced on-site by passing oxygen or air through an electric discharge. This process normally gives about a 20% ozone–oxygen mixture. Although the use of ozone in the United States has been minimal, 20 of the 25 ozone plants in the United States have been built in the last decade. Table 17–16 shows some data on several of these plants in the United States.

TABLE 17-16 Some Ozonation Plants in the United States

Location	Flow (million gal/day)	Purpose of Ozonation
Bay City, MI	14	Taste and odor control
Monroe, MI	6.1	Taste and odor control
Hackensack, NJ	33.8	Color removal, iron and manganese removal
Los Angeles, CA	196.6	Coagulation aid, chlorinated chemical control
Myrtle Beach, SC	10	Color removal, chlorinated chemical control

Source: "Drinking Water Treatment with Ozone," W. H. Glaze, *Environmental Science and Technology,* March, 1987, p. 224.

Ozone gas bubbling through water. (*The World of Chemistry,* Program 12, "Water.")

WHAT ABOUT THE FUTURE?

Our water quality in the United States has improved over the past two decades. Municipal water is more carefully monitored for heavy metals, pesticides, and chlorinated organic molecules. Yet we still have hazardous waste sites that pollute groundwater, and the cleanup of those sites will require decades and perhaps hundreds of billions of dollars. Industry is complying more with hazardous waste and pollution-discharge regulations, and much research is going on to try and find ways to process hazardous wastes, but nagging problems still exist. Household wastes continue to be a major problem. Recycling of wastes is growing, but the recycling of some items like automobile batteries, mercury batteries, unused pesticides, solvents, and used lubricating oil is still being severely limited by economic factors that overshadow a strong desire by many for purer groundwater.

The politics of groundwater protection must improve in the future. Many communities are taking rather short-sighted approaches to hazardous waste disposal. In effect, these communities are saying "not in my backyard!" States that allow communities veto power over the location of hazardous waste sites have been singled out for retaliation by other states having active hazardous waste disposal sites located within their borders. These states are passing laws effectively banning another state's hazardous wastes if that state doesn't allow hazardous waste disposal in its own borders, or if it allows communities to have veto power over the location of disposal sites within their city or community limits. Political problems like these can be solved when everyone recognizes the importance of the proper disposal of hazardous industrial wastes.

The loss of massive supplies of cheap irrigation water will require changes to more efficient irrigation techniques, such as drip and trickle irrigation. Industrial reuse is already a major factor in water and energy conservation and will continue to be a part of the long-term solution to the water crisis. Residential conservation such as was practiced in the California drought of 1977 should be evaluated in every community. Communities should consider dual water supply systems for delivery of drinkable water and water for other purposes. Expansion of the capacity for desalination of sea water, particularly for industrial and agricultural uses, will help ease the

A recycling plant for industrial solvents. Reuse of solvents prevents their entry into water supplies. (Courtesy of Chemical Waste Management, Inc.)

demand for groundwater. Large cities such as Boston and New York are considering replacement of leaky plumbing, which accounts for up to one third of their water use. Pumping sludge from sewage plants directly onto fields offers a way to recycle both water and nutrients to the soil.

The discussion in this chapter has focused on water quality in the United States. A combined program of water conservation, protection of water quality, and water recycling will help to alleviate the water crisis in the United States and other industrialized nations. However, contaminated water is still a serious problem for 75% of the world's population. It has been estimated that 80% of the sickness in the world is caused by contaminated water. For years, many countries and international organizations have provided financial and technical aid to help improve the water quality in developing countries. However, much work remains to be done to reduce sickness caused by contaminated water.

SELF-TEST 17 – C

1. The four metal ions present in sea water at concentrations of 400 ppm or higher are: _____, _____, _____, and _____.
2. Heavy-metal ions in more than trace concentrations are usually _____ to life forms.
3. Select the ions that may cause water to be hard: ( ) sodium, ( ) calcium, ( ) magnesium, ( ) potassium.
4. The element _____ is added to water to kill microbes.
5. Primary water treatment involves _____ and _____ of particles.
6. Tertiary water treatment removes _____ ions and trace amounts of _____.

MATCHING SET

_____	1. Sedimentation	a. A measure of organic material in water
_____	2. Biodegradable	
_____	3. Detergent	b. Widely used as a detergent builder
_____	4. BOD	c. California
_____	5. Proposition 65	d. Caused by metal ions such as Ca^{2+} and Mg^{2+} in solution
_____	6. Phosphate	
_____	7. Activated sludge process	e. Puts less burden on landfills
		f. Hazardous waste cleanup
_____	8. Reverse osmosis	g. Alternative to landfills
_____	9. Water hardness	h. Chemicals banned from discharge into drinking water sources
_____	10. Ozone	
_____	11. Recycling	i. Removes toxic chemicals from water
_____	12. Superfund sites	j. Kills bacteria

_____ **13.** Incineration
_____ **14.** Beryllium
nitrate
_____ **15.** Carbon adsorption
_____ **16.** Chlorination
_____ **17.** Auto battery

k. Common household hazardous waste
l. Primary purification process
m. Disinfectant used in water treatment plants in Europe
n. Secondary purification process
o. Naturally reducible to simpler compounds
p. Soap substitute
q. Requires high pressure
r. Hazardous chemical not considered a hazardous waste
s. Clean Water Act

QUESTIONS

1. If four fifths of the Earth is covered with water, why is there a problem with water supply for humans?
2. Which can dissolve more oxygen to support marine or aquatic life, cold or warm water?
3. Find out what industrial wastes are produced in your community. Are you satisfied with the way these wastes are handled? Explain.
4. Name two ways to obtain fresh water from sea water.
5. What are some processes that _decrease_ the amount of dissolved oxygen in a stream? What are some processes that _increase_ the amount of dissolved oxygen in a stream? Which ones are most readily subject to human control?
6. Explain why each of the following introduces a pollution problem when its wastes are emptied into a stream:
   a. A chlorine-producing plant
   b. A steel mill
   c. A neighborhood laundry.
   d. An agricultural area that is intensively cultivated
7. An old rule of thumb is, "Water purifies itself by running 2 miles from the source of incoming waste." What processes are active in purifying the water? Is this adage foolproof? Explain.
8. What is meant by a biodegradable substance?
9. Obtain some distilled water and evaluate its taste. What can you conclude about drinking pure water?
10. How is water purified by ozone?
11. Debate the topic: Since water pollution is a national problem, the federal government should license water districts to supply the water for U.S. citizens.
12. Describe how these processes purify water.
    a. Solar distillation
    b. Reverse osmosis
13. What is the chemical cause of hard water? Describe how hard water can be made soft.
14. What is the intent of California's Proposition 65 Law?
15. What pertinent facts would you try to gather if it were your responsibility to vote on a bill to regulate water pollution?
16. At what point should pollutants be removed from used water? Who should be responsible for this removal? Would you distinguish between industrial wastes and household wastes?
17. The most abundant elements in organic compounds are carbon, hydrogen, oxygen, and nitrogen. What are the oxidation products for these elements in the decomposition that occurs in nature?
18. Classify water pollutants into as few major groups as you can. Describe some effects of each group and a removal process.
19. In your judgment, what are the most serious pollution problems? Be ready to defend your points in class discussion.
20. What is natural osmosis? Explain the significance of the word _reverse_ in reverse osmosis.

18

Clean Air—Should It Be Taken for Granted?

Smog in New York City. (Photograph by John Cunningham/Visuals Unlimited.)

P lanet Earth is enveloped by a few vertical miles of chemicals that compose the gaseous medium in which we exist—the atmosphere. Close to the Earth's surface and near sea level, the atmosphere is mostly nitrogen (80%) and life-sustaining oxygen (20%). It is the few little fractions of a percentage point of other chemicals that make a difference in the quality of life in various spots on Earth. Extra water in the atmosphere can mean a rain forest; a little less water produces a balanced rainfall; and practically no water results in a desert.

Urbanization created an unhealthful, unpleasant medium for the existence of human life. With its consequent vast number of vehicles and increase in industrialization, urbanization produced an unwanted (and for a while ignored) increase in some of the pesky, naturally occurring "minor" chemicals in the atmosphere (nitrogen oxides, sulfur dioxide, carbon monoxide, carbon dioxide, and ozone). An atmosphere containing such unwanted and harmful ingredients is called **polluted.**

Air pollution is nothing new. Shakespeare wrote about it in the 17th century.

this most excellent canopy, the air,
look you, this excellent o'erhanging
firmament, this magestical roof fretted
with golden fire, why, it appears
no other thing to me but a foul
and pestilent congregation of vapors.

Hamlet *(act II, scene 2)*

Nature pollutes the air on a massive scale with volcanic ash, mercury vapor, and hydrogen sulfide from volcanoes, and reactive and odorous organic compounds from coniferous plants such as pine trees. But automobiles, power plants, smelting plants, other metallurgical plants, and petroleum refineries add significant quantities of toxic chemicals to the atmosphere, especially in heavily populated areas. Atmospheric pollutants cause burning eyes, coughing, acid rain, smog, the destruction of ancient monuments, and even the destruction of the atmosphere itself.

Prior to 1960, there was little concern about air pollution. Most smoke, carbon monoxide, sulfur dioxide, nitrogen oxides, and organic vapors were emitted into the air with little apparent thought of their harmful nature as long as they were scattered into the atmosphere and away from human smell and sight (Fig. 18–1).

A few decades ago, we operated on the principle that "Dilution is the solution to pollution."

Early in the 1960s, air pollution became generally recognized as a problem and caused widespread concern, although devastating air pollution was prevalent earlier in certain geographical areas such as London, England, and where volcanic eruptions and burning of large areas occurred.

In many parts of the world, smog- and air-pollution-control devices are now a part of our daily life, and this new awareness of how easy it is to pollute our atmosphere has resulted in pollution-control measures that have improved the atmosphere in many locations. In spite of greater industrialization and increasing population pressures, in most areas of the United States,

Figure 18-1 Coppertown Basin (Ducktown), Tennessee, as photographed in 1943. Copper ore (principally copper sulfide, Cu_2S) had been mined and smelted in this area since 1847. In the early years, large quantities of sulfur dioxide, a by-product, were discharged directly into the atmosphere and killed all vegetation for miles around the smelter. Today the sulfur is reclaimed in the exhaust stacks to make sulfuric acid, but the denuded soil remains a monument to the misuse of the atmosphere.

The stratosphere is that part of the atmosphere between 40 and 80 km above the Earth.

the air is cleaner now than it was 10 years ago (see Table 18-1), and the trend seems to be toward even cleaner air. But, as we shall see in this chapter, several disturbing trends lead scientists to conclude that our very existence on this planet may be in danger from such problems as increased ultraviolet radiation caused by a depletion of the ozone layer in the stratosphere and increasing carbon dioxide that may be causing a generalized global warming.

Air pollution knows no political boundaries (since molecules and weather do not know where the borders are located). Progress in cleaner air is dependent on efforts at the local, regional, national, and international levels.

TABLE 18-1 Trends in Some Air Pollutants, 1977 to 1986

Pollutant	Change	Comments
Lead (particles)	Decreased 87%	35% decrease from 1985 to 1986 as leaded gasoline phased out
Sulfur dioxide	Decreased 37%	Related to the "acid rain" problem
Ozone (at ground level)	Decreased 21%	Ozone is harmful in the air we breathe, but beneficial in the stratosphere
Carbon monoxide	Decreased 32%	Primarily a product of the automobile
Nitrogen dioxide	Decreased 14%	Primarily a product of the automobile
Particulates	Decreased 23%	Primarily a product of industry. Carry many other harmful chemicals

Source: U.S. EPA, "National Air Quality and Emissions Trends Report," 1988.

Air pollution has caused international controversy and concern over acid rain that is largely produced in the United States and falls in Canada.

Now, let's put air pollution in context, understand it a bit better, and focus on the sources, reactions, and removal of the polluting chemicals.

AIR POLLUTANTS—PARTICLE SIZE MAKES A DIFFERENCE

Pollutants may exist in particle sizes from fly ash particles big enough to see down to single, isolated molecules, ions, or atoms. Often, because of their polar nature, many pollutants are attracted into water droplets and form **aerosols,** or onto larger particles called **particulates.** The solids in an aerosol or particulate may be various dusts, consisting of metal oxides and soil particles, sea salt, fly ash from electric generating plants and incinerators, elemental carbon, or even small metal particles. Aerosols range upward from a diameter of 1 nm (nanometer) to about 10,000 nm and may contain as many as a trillion atoms, ions, or small molecules. Aerosol particles are small enough to remain suspended in the atmosphere for long periods. Such particles are easily breathable and can cause lung diseases. They may also contain mutagenic or carcinogenic compounds. Smoke, dust, clouds, fog, mist, and sprays are typical aerosols. Because of their vast combined surface area, aerosol particles have great capacities to *adsorb* and concentrate chemicals on the surfaces of the particles. Liquid aerosols or particles covered with a thin coating of water may *absorb* air pollutants, thereby concentrating them and providing a medium in which reactions may occur. A typical urban aerosol is shown schematically in Figure 18–2.

Adsorption is the attachment of particles to a surface.

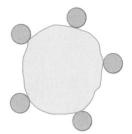

1 nanometer $= 10^{-9}$ m.

Absorption is pulling particles inside.

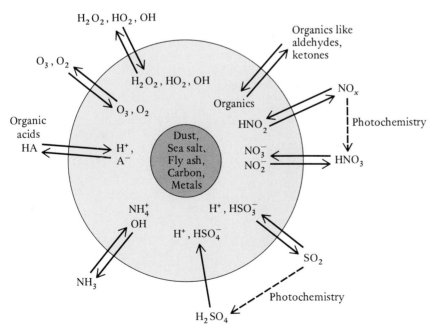

Figure 18–2 Schematic of an aerosol particle and some of its chemical reactions involving urban air pollutants.

1 μm = 10^{-6} m, or 1000 nm.

Major contributors to the amount of atmospheric particulates are volcanic eruptions by: Krakatoa, Indonesia, 1883; Mt. Katmai, Alaska, 1912; Hekla, Iceland, 1947; Mt. Spurr, Alaska, 1953; Bezymyannaya, U.S.S.R., 1956; Mt. St. Helens, Washington, 1980.

Ammonia may concentrate in this aerosol as ammonium hydroxide, sulfur dioxide may react to form sulfurous acid, nitric oxide may form nitrous acid, and many other reactions may occur.

Many aerosols are large enough to be seen; these are the particulates. Particles of sizes below 2 μm (microns) in diameter are largely responsible for the deterioration of visibility often observed in highly populated cities like Los Angeles and New York.

Millions of tons of soot, dust, and smoke particles are emitted into the atmosphere of the United States each year. The average suspended particulate concentrations in the United States vary from about 0.00001 g/m³ of air in rural areas to about six times as much in urban locations. In heavily polluted areas, concentrations of particulates may increase to 0.002 g/m³.

Particulates in the atmosphere can cool the Earth by partially shielding it from the Sun. Large volcanic eruptions such as that from Mt. St. Helens in 1980 have a cooling effect on the Earth.

Particulates and aerosols are removed naturally from the atmosphere by gravitational settling and by rain and snow. They can be prevented from entering the atmosphere by treating industrial emissions by one or more of a variety of physical methods such as filtration, centrifugal separation, spraying, and electrostatic precipitation. A method often used is electrostatic

Figure 18–3   An electrostatic precipitator. The central electrode is negatively charged and imparts a negative charge to particles in the gases. These are attracted to the positively charged walls.

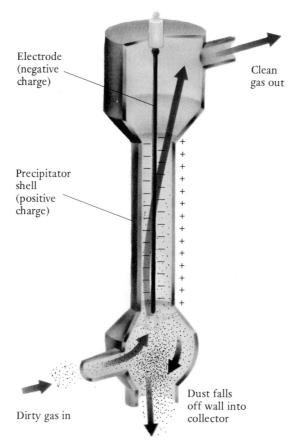

Electrode (negative charge)

Clean gas out

Precipitator shell (positive charge)

Dirty gas in

Dust falls off wall into collector

precipitation, which is better than 98% effective in removing aerosols and dust particulates even smaller than 1 μm from exhaust gases of industrial plants. A diagram of a Cottrell electrostatic precipitator is shown in Figure 18–3. The central wire is connected to a source of direct current at high voltage (about 50,000 V). As dust or aerosols pass through the strong electric field, the particles attract ions that have been formed in the field, become strongly charged, and are attracted to the electrodes. The collected solid grows larger and heavier and falls to the bottom, where it is collected.

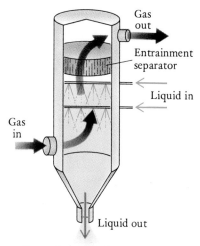

Removing particulates and aerosols by scrubbing. Schematic drawing of a spray collector, or scrubber.

SMOG—INFAMOUS AIR POLLUTION

The poisonous mixture of smoke, fog, air, and other chemicals was first called **smog** in 1911 by Dr. Harold de Voeux in his report on a London air pollution disaster that caused the deaths of 1150 people. Through the years, smog has been a technological plague in many communities and industrial regions.

Two general kinds of smog have been identified. One is the chemically reducing type that is derived largely from the combustion of coal and oil and contains sulfur dioxide mixed with soot, fly ash, smoke, and partially oxidized organic compounds. This is the **London type,** which is diminishing in intensity and frequency as less coal is burned and more controls are installed.

The type of smog formed in London and around some industrial and power plants is thought to be caused by sulfur dioxide. Laboratory experiments have shown that sulfur dioxide increases aerosol formation, particularly in the presence of mixtures of hydrocarbons, nitrogen oxides, and air energized by sunlight (Fig. 18–4). For example, mixtures of 3 ppm olefin,

Industrial, or London-type, smog: fog + SO_2
Photochemical smog: fog + NO_x + hydrocarbons.

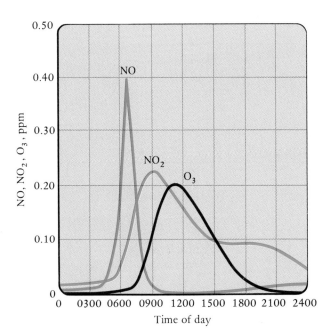

Figure 18–4 Average concentrations of the pollutants, NO, NO_2, and O_3 on a smoggy day in Los Angeles, California. The NO concentration builds up first during the morning rush hour. Later in the day the concentrations of NO_2 and O_3 build up.

Relative humidity is a measure of the amount of water vapor air contains compared with the maximum amount it can contain.

"Olefin" is another name for an unsaturated hydrocarbon.

Organic peroxides contain the R—O—O—R′ structure and are produced by ozone reacting with organic molecules. Hydrogen peroxide is H—O—O—H.

Thermal inversion: mass of warmer air over a mass of cooler air.

1 ppm NO_2, and 0.5 ppm SO_2 at 50% relative humidity form aerosols that have sulfuric acid as a major product. Even with 10 to 20% relative humidity, sulfuric acid is a major product. Sulfuric acid, which is formed in this kind of smog, is very harmful to people suffering from respiratory diseases such as asthma or emphysema. At a concentration of 5 ppm for 1 h, SO_2 can cause constriction of bronchial tubes. A level of 10 ppm for 1 h can cause severe distress. In the 1962 London smog, readings as high as 1.98 ppm of SO_2 were recorded. The sulfur dioxide and sulfuric acid are thought to be the primary causes of deaths in the London smogs.

A second type of smog is the chemically oxidizing type, typical of Los Angeles and other cities where exhausts from internal combustion engines are highly concentrated in the atmosphere. This type is called **photochemical smog** because light—in this instance sunlight—is important in initiating the photochemical process. This smog is practically free of sulfur dioxide but contains substantial amounts of nitrogen oxides, ozone, ozonated olefins, and organic peroxide compounds, together with hydrocarbons of varying complexity.

What general conditions are necessary to produce smog? Although the chemical ingredients of smogs often vary, depending on the unique sources of the pollutants, certain geographical and meteorological conditions exist in nearly every instance of smog.

There must be a period of windlessness so that pollutants can collect without being dispersed vertically or horizontally. This lack of movement in the ground air can occur when a layer of warm air rests on top of a layer of cooler air. This sets the conditions for a **thermal inversion,** which is an abnormal temperature arrangement for air masses (Fig. 18–5). Normally the warmer air is on the bottom nearer the warm Earth, and this warmer, less dense air rises and transports most of the pollutants to the upper troposphere where they are dispersed. In a thermal inversion the warmer air is on top, and

Figure 18–5   A diagram of a temperature inversion over a city. Warm air over a polluted air mass effectively acts as a lid, holding the polluted air over the city until the atmospheric conditions change. The line on the left of the diagram indicates the relative air temperature.

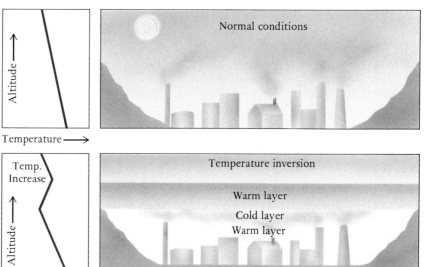

The brown haze shown above this city is caused by NO_2. (National Center for Atmospheric Research.)

the cooler, more dense air retains its position nearer the Earth. The air becomes stagnated. If the land is bowl shaped (surrounded by mountains, cliffs, or the like), this stagnant air mass can remain in place for quite some time.

When these natural conditions exist, humans supply the pollutants by combustion and evaporation in automobiles, electric power plants, space heating, and industrial plants. The chief pollutants are sulfur dioxide (from burning coal and some oils), nitrogen oxides, carbon monoxide, and hydrocarbons (chiefly from the automobile). Add to these ingredients the radiation from the Sun, and a massive smog is in the offing.

Photochemical Smog

A city's atmosphere is an enormous mixing bowl of frenzied chemical reactions. Ferreting out the exact chemical reactions that produce smog has been a tedious job, but in 1951, insight into the formation process was gained when smog was first duplicated in the laboratory. Detailed studies have subsequently revealed that the chemical reactions involved in the smog-making process are photochemical and that aerosols serve to keep the reactants together long enough to form **secondary pollutants.** Ultraviolet radiation from the sun is the energy source for the formation of this photochemical smog.

The exact reaction scheme by which primary pollutants are converted into the secondary pollutants found in smog is still not completely understood (Fig. 18–6). The process is thought to begin with the absorption of a quantum of light by nitrogen dioxide, which causes its breakdown into nitrogen oxide and atomic oxygen, a chemical radical. The very reactive atomic oxygen reacts with molecular oxygen to form ozone (O_3), which is then consumed by reacting with nitrogen oxide to form the original reactants—nitrogen dioxide and molecular oxygen. Atomic oxygen, however, also reacts with reactive hydrocarbons—olefins and aromatics—to form other chemical radicals. These radicals, in turn, react to form other radicals and secondary pollutants such as aldehydes (e.g., formaldehyde). About 0.2 ppm of nitrogen oxides and 1 ppm of reactive hydrocarbons are sufficient to initiate these reactions. The hydrocarbons involved come mostly from unburned petroleum products like gasoline.

Primary pollutants: pollutants emitted into the air.

Secondary pollutants: pollutants formed in the air by chemical reaction.

A chemical radical is a species with an unpaired valence electron. They are usually very reactive and short lived.

561

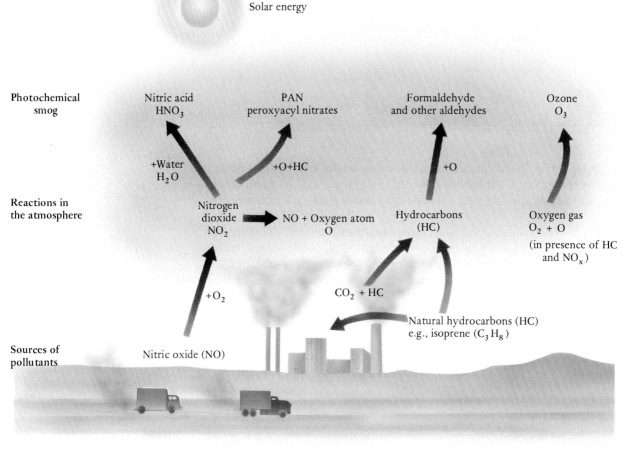

Solar energy

Photochemical smog

| Nitric acid HNO_3 | PAN peroxyacyl nitrates | Formaldehyde and other aldehydes | Ozone O_3 |

+Water H_2O

+O+HC

+O

Reactions in the atmosphere

Nitrogen dioxide NO_2 → NO + Oxygen atom O

Hydrocarbons (HC)

Oxygen gas $O_2 + O$

(in presence of HC and NO_x)

$+O_2$

CO_2 + HC

Natural hydrocarbons (HC) e.g., isoprene (C_3H_8)

Sources of pollutants

Nitric oxide (NO)

Figure 18–6   The formation of photochemical smog. The details of these processes are discussed in this chapter.

In the following sections, we shall look at the major ingredients of photochemical smog, the primary pollutants, the oxides of nitrogen and hydrocarbons, and the secondary pollutant ozone, and see how they interact with oxygen in our atmosphere to produce urban pollution.

NITROGEN OXIDES

About 97% of the nitrogen oxides in the atmosphere are naturally produced and only 3% result from human activity.

Radicals are shown with a "·" as part of the formula. Both NO and NO_2 molecules have an unpaired valence electron so they are shown as NO· and NO_2·.

There are eight known oxides of nitrogen, two of which are recognized as important components of the atmosphere: dinitrogen oxide (N_2O) and nitrogen dioxide (NO_2).

Most of the nitrogen oxides emitted are in the form of NO, a colorless reactive gas. In a combustion process involving air, some of the atmospheric nitrogen reacts with oxygen to produce NO:

$$N_2 + O_2 + \text{heat} \longrightarrow 2\,NO\cdot$$

TABLE 18–2 Emissions of NO$_x$

Source	Emissions (millions of tons)	
	United States	*Global*
Fossil fuel combustion	66	231
Biomass burning	1.1	132
Lightning	3.3	88
Microbial activity in soil	3.3	88
Input from the stratosphere	0.3	5.5
Total (uncertainty in estimates)	74 ($\pm$1)	544.5 ($\pm$275)

Note: The large uncertainty for global emissions is due to incomplete data for much of the world.
Source: Stanford Research Institute, 1983.

Nitrogen oxide is formed in this manner during electrical storms. Since the formation of nitrogen oxide requires heat, it follows that a higher combustion temperature would produce relatively more NO.

In the atmosphere NO reacts rapidly with atmospheric oxygen to produce NO$_2$:

$$2\,NO\cdot + O_2 \longrightarrow 2\,NO_2\cdot$$
Nitrogen
Dioxide

Normally the atmospheric concentration of NO$_2$ is a few parts per billion (ppb) or less. In the United States, most oxides of nitrogen (NO$_x$) are produced from fossil fuel combustion such as automobile engines burning gasoline, with significantly less coming from natural sources like lightning (Table 18–2).

In laboratory studies, nitrogen dioxide in concentrations of 25 to 250 ppm inhibits plant growth and causes defoliation. The growth of tomato and bean seedlings is inhibited by 0.3 to 0.5 ppm of NO$_2$ applied continuously for 10 to 20 days.

In a concentration of 3 ppm for 1 h, nitrogen dioxide causes bronchioconstriction in humans, and short exposures at high levels (150–220 ppm) cause changes in the lungs that produce fatal results. A seemingly harmless exposure one day can cause death a few days later.

Nitrogen dioxide reacts with photons of light in a **photodissociation** reaction that produces nitric oxide and free oxygen atoms (O$\cdot$, oxygen radicals) that can react with molecular oxygen to produce ozone.

Ozone is discussed in the next section.

$$NO_2\cdot + h\nu \longrightarrow NO\cdot + O\cdot$$

$$O_2 + O\cdot \longrightarrow O_3$$

The photon, hν, has a wavelength between 280 and 430 nm. The nitric oxide then reacts with an ozone molecule to regenerate NO$_2$:

$$NO\cdot + O_3 \longrightarrow NO_2\cdot + O_2$$

At night, another oxide of nitrogen, the nitrogen trioxide radical, NO$_3\cdot$, is produced by the reaction

$$NO_2\cdot + O_3 \longrightarrow NO_3\cdot + O_2$$

Nitric acid is a source of acid rain, which is discussed later in this chapter.

In daylight the nitrogen trioxide radical would quickly be photodissociated, but in the absence of photons it accumulates and reacts with nitric oxide to form N_2O_5, which in turn reacts with water to form nitric acid.

$$NO_3\cdot + NO_2\cdot \longrightarrow N_2O_5$$

$$N_2O_5 + H_2O \longrightarrow 2\ HNO_3$$

Nitrogen dioxide can also react with water, such as in aerosol particles, to form nitric acid and nitrous acid.

$$2\ NO_2 + H_2O \longrightarrow \underset{\substack{\text{Nitric}\\\text{Acid}}}{HNO_3} + \underset{\substack{\text{Nitrous}\\\text{Acid}}}{HNO_2}$$

In addition, nitrogen dioxide and oxygen yield nitric acid:

$$4\ NO_2 + 2\ H_2O + O_2 \longrightarrow 4\ HNO_3$$

These acids in turn can react with ammonia or metallic particles in the atmosphere to produce nitrate or nitrite salts. For example,

Nitrates are important components of fertilizers.

$$\underset{\text{Ammonia}}{NH_3} + HNO_3 \longrightarrow \underset{\substack{\text{Ammonium Nitrate}\\\text{(A Salt)}}}{NH_4NO_3}$$

The acids or the salts, or both, ultimately form aerosols, which eventually settle from the air or dissolve in raindrops. Nitrogen dioxide, then, is a primary cause of haze in urban or industrial atmospheres because of its participation in the process of aerosol formation. Normally nitrogen dioxide has a lifetime of about three days in the atmosphere.

OZONE AND ITS ROLE IN AIR POLLUTION

Allotropes are different forms of the same element that differ significantly in physical and chemical properties. Diatomic oxygen (O_2) and triatomic ozone (O_3) are allotropes. Carbon has three common allotropes; graphite, diamond, and carbon black.

Ozone is an allotropic form of oxygen consisting of three oxygen atoms bound together in a molecule with the formula O_3. Ozone has a pungent odor that can be detected at concentrations as low as 0.02 ppm. We often smell the ozone produced by sparking electric appliances, or after a thunderstorm when lightning-caused ozone washes out with the rainfall.

As we shall see in this chapter, there is "good" and "bad" ozone. The bad ozone is that found in the air we breathe, whereas the good ozone is found in the stratosphere, where it forms a protective blanket, absorbing harmful ultraviolet radiation.

The only significant chemical reaction producing ozone in the atmosphere is one involving molecular oxygen and atomic oxygen.

$$O_2 + O\cdot + M \longrightarrow O_3 + M$$

In the reaction above, M stands for a third molecule, like a nitrogen or an oxygen molecule. This third molecule takes away some of the energy of the reaction and keeps the ozone molecule from dissociating immediately. At high altitudes oxygen atoms are produced by ultraviolet photons striking oxygen molecules, and most of these high-energy photons are absorbed before they get to the lower atmosphere (see the discussion of the protective "good" ozone layer later in this chapter), where only photons with wavelengths greater than 280 nm are present. The low-energy photons are sufficiently energetic to react with nitric oxide,

$$NO_2 \cdot + h\nu \longrightarrow NO \cdot + O \cdot$$

so there can be plenty of oxygen atoms, especially if there is a ready source of nitric oxide, like automobile exhaust or other high-temperature combustion sources.

Ozone also photodissociates when struck by photons with wavelengths less than 320 nm to produce oxygen atoms that can react with water to produce hydroxyl radicals ($OH \cdot$), an important species for other reactions taking place in the atmosphere.

$$O_3 + h\nu \longrightarrow O_2 + O \cdot$$

$$O \cdot + H_2O \longrightarrow 2\, OH \cdot$$

In the daytime, when they are produced in large numbers, hydroxyl radicals can react with nitric oxide to produce nitric acid.

$$NO_2 \cdot + OH \cdot \longrightarrow HNO_3$$

As we shall see later, hydroxyl radicals play an important role in the oxidation of hydrocarbons found in the atmosphere.

Ozone is a secondary air pollutant, and is the most difficult pollutant to control. According to EPA, during 1983 to 1985, the standard for ozone of 0.12 ppm was exceeded in 76 urban areas of the United States. This high ozone was primarily related to excess nitrogen oxides emissions from automobiles, buses, and trucks. Most major urban areas have vehicle inspection centers for passenger automobiles in an effort to control nitrogen oxides emissions as well those of carbon monoxide and unburned hydrocarbons (Fig. 18–7).

Automobile exhaust contains nitrogen oxides, carbon monoxide, and unburned hydrocarbons, in addition to water vapor and carbon dioxide. (*The World of Chemistry*, Program 6, "The Atom.")

Figure 18–7   At this test station, automobiles are tested for carbon monoxide, nitrogen oxides, and unburned hydrocarbons. Those failing the standards established by EPA must be repaired.

NET REACTIONS

We can write the three reactions just discussed as a net, or overall reaction. To do this, we first need to double the third reaction by writing twos in front of each of the reactants and products. When we do that, we add the reactants together in all three reactions, set them opposite the sum of the products of the three reactions, and then cancel out any species appearing in equal numbers as both products and reactants. The result is a **net reaction.**

$$O_3 + h\nu \longrightarrow O_2 + O\cdot$$

$$O\cdot + H_2O \longrightarrow 2\ OH\cdot$$

$$2\ NO_2 + 2\ OH\cdot \longrightarrow 2\ HNO_3$$

$$O_3 + h\nu + \cancel{O\cdot} + H_2O + 2\ NO_2 + \cancel{2\ OH\cdot} \longrightarrow O_2 + \cancel{O\cdot} + \cancel{2\ OH\cdot} + 2\ HNO_3$$

or,

$$O_3 + h\nu + 2\ NO_2 + H_2O \longrightarrow 2\ HNO_3 + O_2$$

The net reaction is what scientists can often directly measure, but the individual reactions, or steps, are usually of equal or sometimes greater importance in understanding what is going on. In the reaction above, the hydroxyl radical cancels out, so their role cannot be appreciated unless the individual steps are known. Most chemical reactions that involve more than two reactants are actually net reactions involving two or more steps.

As difficult as it is to attain, the ozone standard may not be low enough for good health. Exposure to concentrations of ozone at or near 0.12 ppm lowers the volume of air a person breaths out in 1 s (the forced expiratory volume, or FEV_1). Children studied who were exposed to ozone concentrations close to the EPA standard, but not exceeding it, showed a 16% decrease in the FEV_1. Some scientists have been urging EPA to lower the standard to

Lower FEV_1 accelerates the aging of the lungs.

Ozone can affect the breathing of children at play in urban environments. (*The World of Chemistry,* Program 25, "Chemistry and the Environment.")

0.08 ppm. If that is done even some midsized cities would probably fail to meet the EPA standards.

No matter what the standard becomes, present ozone concentrations in many urban areas represent health hazards to children at play, joggers, others doing outdoor exercise, and older persons who may have diminished respiratory capabilities.

Trees in urban environments may emit as many reactive hydrocarbons as automobiles do.

HYDROCARBONS AND AIR POLLUTION

Hydrocarbons occur in the atmosphere from both natural sources and human activities. Isoprene and α-pinene are produced in large quantities by both coniferous and deciduous trees. Methane gas is produced by such diverse sources as ruminant animals, termites, ants, and decay-causing bacteria acting on dead plants and animals. Human activities such as the use of industrial solvents, petroleum refining and its distribution, and unburned gasoline and diesel fuel components account for a large amount of hydrocarbons in the atmosphere. The types of hydrocarbons found in urban air (Table 18–3) reads like something you saw in the organic chemistry chapter (Chapter 12).

The hydroxyl radical (OH·), produced indirectly from ozone, helps in the oxidation of hydrocarbons in the atmosphere. Like the reactions we have seen earlier, several steps are involved. Using RH as the general formula for a hydrocarbon, we can write the first step as

$$RH + O_2 + OH\cdot \longrightarrow RO_2\cdot + H_2O$$

where atmospheric oxygen is the oxidant. The second step involves nitric oxide and produces aldehydes and ketones and the hydroperoxide radical (HO$_2$·).

$$RO_2\cdot + NO\cdot \longrightarrow (\text{aldehydes and ketones}) + NO_2\cdot + HO_2\cdot$$

In September, 1988, William Chameides, of Georgia Tech in Atlanta, published a report in *Science* magazine, in which he stated that in some cities trees may account for more hydrocarbons in the atmosphere than that produced from human activities. The EPA has since found this to be true. This fact is causing a rethinking about how to control urban pollution.

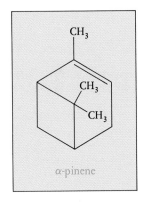

TABLE 18–3 The ten most abundant ambient air hydrocarbons found in cities. Results from 800 air samples taken in 39 different cities.

Compound	Concentration Median	Maximum ppb of Carbon
Isopentane	45	3393
n-Butane	40	5448
Toluene	34	1299
Propane	23	399
Ethane	23	475
n-Pentane	22	1450
Ethylene	21	1001
m-Xylene, *p*-Xylene	18	338
2-Methylpentane	15	647
Isobutane	15	1433

Source: Air Pollution Control Association, 1988.

The third step is the photodissociation of the NO_2 radical. This is followed quickly by the production of ozone.

$$NO_2 \cdot + h\nu \longrightarrow NO \cdot + O \cdot$$

$$O_2 + O \cdot \longrightarrow O_3$$

Adding all of these steps together to get the net reaction, we can see how the presence of hydrocarbons in the atmosphere can produce ozone as they are oxidized by oxygen in the air.

$$RH + OH \cdot + (3\ O_2) \longrightarrow O_3 + HO_2 \cdot + \text{(aldehydes and ketones)} + H_2O$$

The hydroperoxide radical can react with NO to produce NO_2 and more hydroxyl radicals.

$$HO_2 \cdot + NO \cdot \longrightarrow NO_2 \cdot + OH \cdot$$

Hydroxyl radicals also are involved in the oxidation of aldehydes that can react with NO_2 and cause it to be transported over great distances. The simplest aldehyde is acetaldehyde (CH_3CHO). It reacts with the hydroxyl radical and oxygen in the following way.

$$CH_3CHO + OH \cdot + O_2 \longrightarrow CH_3C(O)\!-\!O\!-\!O \cdot + H_2O$$
$$\text{Peroxoaldehyde}$$

The peroxoaldehyde can then react with NO_2 to produce peroxyacetylnitrate (PAN):

$$CH_3C(O)\!-\!O\!-\!O \cdot + NO_2 \cdot \longrightarrow CH_3C(O)\!-\!O\!-\!O\!-\!NO_2$$
$$\text{PAN}$$

The average lifetime of NO_2 before it reacts with something is about 1 h.

PAN stabilizes NO_2 and allows it to be carried over great distances with prevailing winds. Eventually PAN decomposes and releases NO_2. In this way urban pollution in the form of NO_2 may be carried to outlying areas where it may do additional damage to vegetation, human tissue, and fabrics —a direct result of an unfortunate combination of events involving hydrocarbons emitted into the atmosphere.

In addition to simpler hydrocarbons like alkanes, alkenes (olefins), and alkynes, a large number of polynuclear aromatic hydrocarbons (PAH) are released into the atmosphere, primarily from motor vehicle exhaust. These chemicals can react with hydroxyl radicals and oxygen much like simpler hydrocarbons, but their greatest danger is their toxic properties. One PAH, benzo(α)pyrene (BAP), is a known carcinogen. Concentrations as high as 60 $\mu g/m^3$ of air have been found in urban air. Coal smoke contains about 300 ppm benzo(α)pyrene. Measurements have shown that for every million tons of coal burned, about 750 tons of benzo(α)pyrene can be produced. British researchers reported that a typical London resident in the 1950s inhaled about 200 mg of BAP a year. Heavy smokers (those who smoke about two packs a day without filters) receive an additional 150 mg a year. This is about 40,000 times the amount necessary to produce cancer in mice. Extracts of urban air taken at various times during the past decade have in fact produced cancer in mice, but not all of these cancers were caused by PAHs like benzo(α)pyrene; other organic chemicals were present as well (see later in this chapter.)

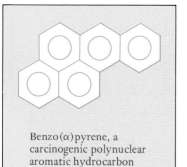

Benzo(α)pyrene, a carcinogenic polynuclear aromatic hydrocarbon found in smoke.

SELF-TEST 18–A

1. Parts per million is abbreviated ——————. To change from ppm to percent, divide by ——————. Thus, 10 ppm would be —————— percent.
2. What has been the general trend in air pollutants for approximately the past decade? ( ) increase ( ) decrease.
3. Name a chemical that is considered an air pollutant and a beneficial chemical.
4. Because of their large surface areas, aerosol particles can (absorb/adsorb) chemicals onto their surfaces.
5. A liquid aerosol particle will likely ( ) adsorb ( ) absorb a chemical.
6. Electrostatic fields can be used to remove which pollutant?
   ( ) oxides of nitrogen ( ) ozone ( ) particulates.
7. A thermal inversion occurs when ( ) warm ( ) cool air is above ( ) warm ( ) cool air below.
8. London type smog is often associated with ( ) coal burning ( ) sunlight.
9. In all combustion processes in air, some nitrogen —————— are formed.
10. What are the products of the photodissociation of nitrogen dioxide?
11. What species reacts with molecular oxygen to form ozone?
12. What two classes of compounds are formed when hydrocarbon molecules react with hydroxyl radicals in air?
13. Name the chemical that can transport nitrogen dioxide.

SULFUR DIOXIDE—A MAJOR PRIMARY POLLUTANT

Sulfur dioxide is produced when sulfur or sulfur-containing compounds are burned in air.

$$S + O_2 \longrightarrow SO_2$$

Most of the coal burned in the United States contains sulfur in the form of the mineral pyrite (FeS_2). The weight percent of sulfur in this coal ranges from 1 to 4%. The pyrite is oxidized as the coal is burned.

$$4 FeS_2 + 11 O_2 \longrightarrow 2 Fe_2O_3 + 8 SO_2$$

Large amounts of coal are burned in this country to generate electricity. A 1000 MW (megawatt) coal-fired generating plant can burn about 700 tons of coal an hour. If the coal contains 4% sulfur, that equals 56 tons of SO_2 an hour, or 490,560 tons of SO_2 every year. Oil-burning electric generating plants can also produce comparable amounts of SO_2 since fuel oils can contain up to 4% sulfur. The sulfur in the oil is in the form of mercapto compounds in which sulfur atoms are bound to carbon and hydrogen atoms. Eight states have the highest SO_2 emissions in the United States. Table 18–4 shows the number of coal-fired power plants and the amounts of SO_2 emitted in those eight states. Operators of these facilities are making efforts to eliminate most of the SO_2 before it reaches the stack.

Electric power plants are discussed in Chapter 8.

—SH
Mercapto group.

TABLE 18–4 Characteristics of Coal-Fired Power Plants in Eight States

State	Plants	SO_2 Emissions (thousand tons/yr)	Capacity (gigawatts)
Ohio	99	2,221	22.31
Indiana	66	1,588	14.58
Pennsylvania	70	1,427	17.93
Missouri	41	1,214	9.97
Illinois	59	1,136	15.75
West Virginia	33	966	14.46
Tennessee	37	950	9.41
Kentucky	54	947	11.82

Note: Some of these plants have emissions controls installed; others do not.
Source: EPA, 1988.

Most low-sulfur coals are mined far from the major metropolitan areas where they are most needed for power generation. The cleansing of sulfur from closer, high-sulfur coal is costly and incomplete. One method is to pulverize the coal to the consistency of talcum powder and remove the pyrite (FeS_2) by magnetic separation. Technology is available to decrease the sulfur content of fuel oil to 0.5%, but this process, too, is costly. It involves the formation of hydrogen sulfide (H_2S) by bubbling hydrogen through the oil in the presence of metallic catalysts, such as a platinum-palladium catalyst.

Several efficient methods are available to trap SO_2. In one method, limestone is heated to produce lime. The lime reacts with SO_2 to form calcium sulfite, a solid particulate, which can be removed from an exhaust stack by an electrostatic precipitator.

$$CaCO_3 \xrightarrow{\text{Heat}} CaO + CO_2$$
$$\text{Limestone} \qquad \text{Lime}$$

$$CaO + SO_2 \longrightarrow CaSO_3\ (s)$$
$$\text{Calcium Sulfite}$$

Another trapping method involves the passage of SO_2 through molten sodium carbonate. Solid sodium sulfite is formed.

$$SO_2 + Na_2CO_3 \xrightarrow{800°C} Na_2SO_3 + CO_2$$
$$\text{Sodium Carbonate} \qquad \text{Sodium Sulfite}$$

The less desirable method of dissipating SO_2 is by tall stacks. Although tall stacks emit SO_2 into the upper atmosphere away from the immediate vicinity and give SO_2 a chance to dilute itself on the way down, the fact remains that SO_2 will come down, and the longer it stays up the greater chance it has to become sulfuric acid. A 10-year study in Great Britain showed that although SO_2 emissions from power plants increased by 35%, the construction of tall stacks decreased the ground level concentrations of SO_2 by as much as 30%. The question is, who got the SO_2? In this case, Britain's solution was others' pollution. In the United States, the EPA may have added to a pollution problem unwittingly with rules in 1970 that caused plants to increase the height of smokestacks and caused pollutants to be carried longer distances by winds. There are about 179 stacks in the United States that are 500 ft or higher, and 20 stacks are 1000 ft or more tall.

TABLE 18–5 Physiological and Corrosive Effects of SO_2

SO_2 Exposure (ppm)	Duration	Effect	Comment
0.03–0.12	Annual average	Corrosion	Moist temperate climate with particulate pollution
0.3	8 h	Vegetation damage (bleached spots, suppression of growth, leaf drop, and low yield)	Laboratory experiment; other environmental factors optimal. Field studies are consistent but dose is difficult to estimate.
0.47	<1 h	Odor threshold (50% of subjects detect)	May be higher for many persons or when other methods are used
0.2	Daily average	Respiratory symptoms	Community exposure exceeding 0.2 ppm more than 3% of the time
>0.05	Long-term average	Respiratory symptoms	With particulates > 100 $\mu g/m^3$
0.2	Daily average	Respiratory symptoms	With particulates
0.9	Hourly average	Respiratory symptoms	With particulates
>0.05	Monthly average	Respiratory symptoms, including impairment of lung function in children	With particulates

Most of the SO_2 in the atmosphere reacts with oxygen to form sulfur trioxide (SO_3). Several reactions are possible. SO_2 may react with atomic oxygen:

$$SO_2 + O\cdot \longrightarrow SO_3$$

It may react with molecular oxygen:

$$2\,SO_2 + O_2 \longrightarrow 2\,SO_3$$

Or, it may react with hydroxyl radicals:

$$SO_2 + 2\,OH\cdot \longrightarrow SO_3 + H_2O$$

The SO_3 formed has a strong affinity for water and will dissolve in aqueous aerosol particles forming sulfuric acid, a strong acid.

$$SO_3 + H_2O \longrightarrow H_2SO_4$$

Sulfur dioxide can corrode metals and decay building stones, in particular marble and limestone. Both marble and limestone are forms of calcium carbonate ($CaCO_3$), which reacts readily with acid (H^+) and with SO_2 and water.

$$CaCO_3 + 2\,H^+ \longrightarrow Ca^{2+} + H_2O + CO_2$$

$$CaCO_3 + SO_2 + 2\,H_2O \longrightarrow \underset{\substack{\text{Calcium Sulfite}\\ \text{(Soluble)}}}{CaSO_3\cdot 2H_2O} + CO_2$$

Sulfur dioxide in the air is harmful to people, animals, plants, and buildings.

An alarming example is the disintegration of marble statues and buildings on the Acropolis in Athens, Greece. As all coatings have failed to protect the marble adequately, the only known solution is to bring the prized objects into air-conditioned museums protected from SO_2 and other corroding chemicals.

Sulfur dioxide is physiologically harmful to both plants and animals. Table 18–5 shows a summary of the effects. Most healthy adults can tolerate fairly high levels of SO_2 without apparent lasting ill effects. Individuals with chronic respiratory difficulties such as bronchitis or asthma tend to be much more sensitive to SO_2.

ACID RAIN

The term **acid rain** was first used in 1872 by Robert Angus Smith, an English chemist and climatologist. He used the term to describe the acidic precipitation that fell on Manchester just at the start of the Industrial Revolution. Although neutral water has a pH of 7, rainwater becomes naturally acidified from dissolved carbon dioxide, a normal component of the atmosphere. The

Figure 18–8 The pH of acid rain as compared with the pH of other mixtures.

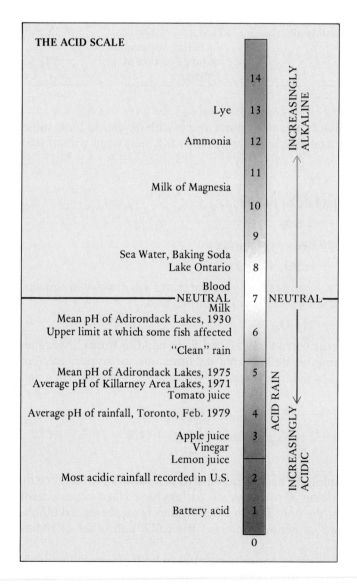

carbon dioxide reacts reversibly with water to form a solution of the weak acid, carbonic acid.

$$H_2O + CO_2 \rightleftharpoons H^+ + HCO_3^-$$

At equilibrium, the pH of this solution is 5.6. Thus, natural rainfall will be slightly acidic from carbon dioxide (Fig. 18–8). Any precipitation with a pH below 5.6 is considered excessively acidic.

As we have seen earlier in this chapter, NO_2 and SO_2 can react with chemicals in the atmosphere to produce acids. These gases can dissolve in water droplets or aerosol particles where they greatly lower the pH, and when conditions are favorable, this moisture precipitates as rain or snow. NO_2 yields nitric acid (HNO_3) and nitrous acid (HNO_2); SO_2 yields sulfuric acid (H_2SO_4) and sulfurous acid (H_2SO_3). Ice core samples taken in Greenland and dating back to 1900 contain sulfate (SO_4^{2-}) and nitrate (NO_3^-) ions. This indicates that at least from 1900 onward, acid rain has been commonplace, and more importantly, has been deposited far from where the oxides of nitrogen and sulfur were formed.

Acid rain is a problem today due to the large annual amounts of these acidic oxides being produced by human activities and put into the atmosphere (Fig. 18–9). When this acidic precipitation falls on natural areas that cannot easily tolerate all of this abnormal acidity, serious environmental problems occur. The average annual pH of precipitation falling on much of the northeastern United States and northeastern Europe is between 4 and 4.5. Specific storms in some areas like West Virginia have been recorded with rain having pH values as low as 1.5. Further complicating matters is the fact that acid rain is an international problem—rain and snow don't observe borders. Many Canadian residents are offended by the government of the United States because some of the acid rain produced in the United States falls on Canadian cities and forests (Fig. 18–10).

A more adequate term for acid rain might be acid precipitation. Some scientists use acid deposition.

Figure 18–9   Major source and components of acid rain.

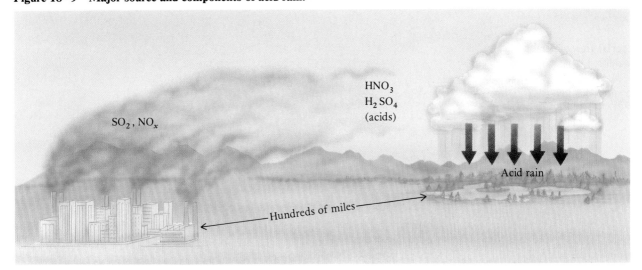

Figure 18–10   Most of the oxides of sulfur responsible for acid rain come from Midwestern states. Prevailing winds carry the acid droplets formed over the Northeast and into Canada. Oxides of nitrogen also contribute to acid rain formation.

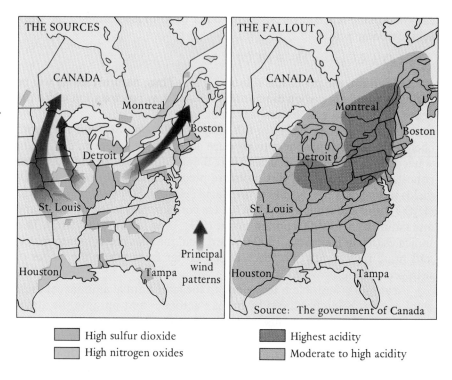

Source: The government of Canada

High sulfur dioxide
High nitrogen oxides

Highest acidity
Moderate to high acidity

Acid rain destroys lakes.

The extent of the problems with acid rain can be seen in dead (fishless) ponds and lakes, dying or dead forests, and crumbling buildings. Because of wind patterns, Norway and Sweden have received the brunt of western Europe's emission of sulfur oxides and nitrogen oxides as acid rain. As a result, of the 100,000 lakes in Sweden, 4000 have become fishless, and 14,000 other lakes have been acidified to some degree. In the United States, 6% of all ponds and lakes in the Adirondack Mountains of New York are now fishless, and 200 lakes in Michigan are dead. For the most part, these "dead" lakes are still picturesque, but no fish can live in the acidified water.

Aerial view of dead trees at a lake acidified by acid rain. (*The World of Chemistry,* Program 16, "The Proton in Chemistry.")

Lake trout and yellow perch die at pH below 5.0, and smallmouth bass die at pH below 6.0. Mussels die at pH below 6.5.

Acid rain damages trees in several other ways. It disturbs the stomata (openings) in tree leaves and causes increased transpiration and a water deficit in the tree. Acid rain can acidify the soil, damaging fine root hairs, and thus diminish nutrient and water uptake. The acid can leach out needed minerals in the soil, and the minerals are carried off in the groundwater. The surface structures of the bark and the leaves can be destroyed by the acid in the rain. The effects of acid rain have been so severe on some forests already that experts predict that they will be lost, possibly forever. This is the case with Germany's famous Black Forest.

In addition to destroying vegetation, acid rain also damages forests by leaching minerals from the soil. Many of these minerals contain metal ions that are toxic to plant life. These minerals can exist in natural soil indefinitely because they are insoluble in groundwater and surface waters of normal pH. However, acid solutions can increase the solubility of many minerals. For example, protons in acid rain can react with insoluble aluminum hydroxide in the soil, causing aluminum ions (Al^{3+}) to be taken up by the roots of plants where they have toxic effects.

$$Al(OH)_3 + 3 H^+ \longrightarrow Al^{3+} + 3 H_2O$$

The effects of acid rain and other pollution on stone and metal structures are more subtle. These effects are especially devastating because of their irreversibility. By damaging stone buildings in Europe, acid rain is slowly but surely dissolving the continent's historical heritage. The bas-reliefs on the Cologne (West Germany) cathedral are barely recognizable. The Tower of London, St. Paul's Cathedral, and the Lincoln Cathedral in London (Fig. 18–11) have suffered the same fate. Other beautifully carved statues and bas-reliefs on buildings throughout Europe and the eastern part of the United States and Canada are slowly passing into oblivion by the action of pollutants, in particular, acid rain.

Dead trees caused by acid rain. (*The World of Chemistry,* Program 16, "The Proton in Chemistry.")

The leaching of toxic metal ions into groundwater by acid rain may also increase groundwater pollution.

Acid rain crumbles buildings.

Figure 18–11 The photo on the left was taken at the Lincoln Cathedral in London in 1910. The photo on the right was taken in 1984. (Dean and Chapter of London.)

(a)

(b)

Loading a statue onto a boat in Venice, Italy to take it to a cleaner environment. (*The World of Chemistry,* Program 18, "The Chemistry of Earth.")

What can be done about acid rain? Some stopgap measures are being taken, such as spraying lime, $Ca(OH)_2$, into acidified lakes to neutralize at least some of the acid and raise the pH toward 7 (Fig. 18–12).

$$Ca(OH)_2 + 2\ H^+ \longrightarrow Ca^{2+} + 2\ H_2O$$

Sweden is spending $40 million a year to neutralize the acid in some of its lakes. Some lakes in the problem areas have their own safeguard against acid rain by having limestone-lined bottoms, which supply calcium carbonate ($CaCO_3$) for neutralizing the acid from acid rain (just as an antacid tablet relieves indigestion). Statues and bas-reliefs have been coated with a variety of plastics and other materials. None of these materials appear to be long-range protectors.

The ultimate answers to acid rain problems lie with governments. Twenty-one European countries agreed in 1985 to lower their SO_2 emissions by 30% or more over a 10-year period. By 1989, more than half of those countries had already reached that goal. In 1988, the Canadian government set a goal of lowering its SO_2 emissions by half by 1994. In the United States, progress has not been as rapid.

Figure 18–12   Calcium carbonate (limestone) is being dispersed over forests affected by acid rain. (Courtesy of Ohio Edison.)

CARBON MONOXIDE

Carbon monoxide (CO) is the most abundant and widely distributed air pollutant found in the atmosphere. Like ozone, carbon monoxide is one of the most difficult pollutants to control. Cities like Los Angeles with high densities of automobiles tend to be repeatedly cited by EPA for not attaining the required ambient air quality for carbon monoxide.

Carbon monoxide is always produced when carbon or carbon-containing compounds are oxidized by insufficient oxygen.

$$2\,C + O_2 \longrightarrow 2\,CO$$

For every 1000 gal of gasoline burned, 2300 lb of carbon monoxide are formed in the combustion gases. Modern catalytic converters on car mufflers convert much of this carbon monoxide to carbon dioxide, but the amounts that are not converted make being near a heavily traveled street dangerous because of the carbon monoxide concentrations. At peak traffic times, concentrations as high as 50 ppm are common (refer to Chapter 16 on toxic substances). In the countryside, carbon monoxide levels are closer to the global average of 0.1 ppm.

At least ten times more carbon monoxide enters the atmosphere from natural sources than from all industrial and automotive sources combined. Of the 3.8 billion tons of carbon monoxide emitted every year, about 3 billion tons are emitted by the oxidation of decaying organic matter in the topsoil.

A bit of a mystery concerning carbon monoxide is that its global level does not seem to be changing as is the case with some pollutants. Although polar carbon monoxide molecules dissolve readily in water, they react very slowly with oxygen to form carbon dioxide. Where carbon monoxide goes is the subject of ongoing research in atmospheric chemistry.

Heavy traffic contributes to high CO levels. (*The World of Chemistry,* Program 2, "Color.")

About 10^{14} g of CO are released each year in the U.S., that's about 1000 lb for every person.

CHLOROFLUOROCARBONS AND THE OZONE LAYER

Most pollutants are adsorbed onto surfaces, absorbed into water droplets and react, or they react in the gas phase with other pollutants in the lower atmosphere (troposphere) and eventually wash out in precipitation. There is one class of industrial pollutants, the halogenated hydrocarbons collectively called **chlorofluorocarbons,** or **CFCs,** which are relatively unreactive and are not eliminated in the troposphere. Instead, many of them eventually mix with air in the stratosphere where they reside for many years. As we shall see, it is the series of reactions between CFCs and ozone that cause so much concern about the presence of CFCs in our atmosphere.

The common halogenated hydrocarbons were discussed in Chapter 13. These chemicals have found uses as refrigeration fluids and degreasing solvents in the manufacture of such diverse products as electronic parts and machined metallic objects, and even as fire extinguishers because of their highly unreactive nature. The lower molecular weight CFCs were at one time used as propellants for aerosol products like hair sprays, deodorants, medicines, and foods, but those uses were banned in the United States in 1978. But CFCs are still used worldwide in both developed and developing countries.

The dangers of CFCs were announced in 1974, when M. J. Molina and F. S. Rowland of the University of California, Irvine, published a scientific paper in which they predicted that continued use of CFCs would lead to a serious depletion of the Earth's protective stratospheric ozone layer. The reason the depletion of the ozone layer is important is that for every 1% decrease in stratospheric ozone, an additional 2% of the sun's most damaging ultraviolet radiation reaches the Earth's surface resulting in increased skin cancer, damage to plants, and possibly other effects we know little about now. Let's examine how these CFCs destroy the ozone layer.

In the stratosphere there is an abundance of ozone being produced because high-energy ultraviolet photons with wavelengths below 280 nm readily photodissociate oxygen molecules.

$$O_2 + h\nu \longrightarrow 2\ O\cdot$$

$$O\cdot + O_2 \longrightarrow O_3$$

Recall that these same reactions take place to a lesser extent in the troposphere with the ozone becoming involved in producing much of the urban secondary pollutants.

This stratospheric ozone is so abundant (about 10 ppm) that it absorbs between 95 and 99% of the sunlight in the 200 to 300 nm wavelength range (the ultraviolet range). Light in this wavelength range is especially damaging to living organisms, so it is important that this protective ozone layer remain intact if we are to avoid drastic harmful effects on our planet.

The carbon-chlorine bond in a CFC molecule is broken easily by an ultraviolet photon, especially those available in the stratosphere. When the CFC molecule photodissociates, an active chlorine atom ($Cl\cdot$) is produced. The reaction below shows one of the most common CFCs, CFC-11, undergoing photodissociation.

$$\begin{array}{ccc} & Cl & & Cl \\ & | & & | \\ F-&C&-Cl + h\nu \longrightarrow F-&C&\cdot + Cl\cdot \\ & | & & | \\ & Cl & & Cl \end{array}$$

The chlorine atom then combines with an ozone molecule, producing a chlorine oxide ($ClO\cdot$) radical and an oxygen molecule.

$$Cl\cdot + O_3 \longrightarrow ClO\cdot + O_2$$

Thus, an ozone molecule has been destroyed. If this were the only reaction the single CFC molecule caused, there would be little danger to the ozone layer. However, the $ClO\cdot$ radical can react with oxygen atoms and produce the chlorine atom again, which is ready to react with yet another ozone molecule.

$$ClO\cdot + O\cdot \longrightarrow O_2 + Cl\cdot$$

$$Cl\cdot + O_3 \longrightarrow O_2 + ClO\cdot$$

The net reaction from these steps is the destruction of an ozone molecule for every reaction cycle in which the chlorine atom participates.

$$O_3 + O\cdot \longrightarrow 2 O_2$$

A single chlorine atom may undergo reaction thousands of times. Eventually the chlorine atom reacts with a water molecule to form HCl, which mixes into the troposphere and washes out in acidic rainfall. This chlorine atom chain is thought to account for about 80% of the loss of ozone observed.

The bromine oxide radical can cause ozone depletion in a manner similar to chlorine oxide. Bromine oxide $(BrO\cdot)$, is produced from halons, a class of compounds structurally similar to the CFCs (Table 18–6), but containing a carbon-bromine bond. When a halon like Halon-1301 reacts with a photon, a bromine atom is produced.

$$\underset{\underset{F}{|}}{\overset{\overset{F}{|}}{F-C-Br}} + h\nu \longrightarrow \underset{\underset{F}{|}}{\overset{\overset{F}{|}}{F-C\cdot}} + Br\cdot$$

The bromine atom then reacts with ozone to form $BrO\cdot$ and O_2.

$$Br\cdot + O_3 \longrightarrow BrO\cdot + O_2$$

Nitric oxide can also deplete ozone. Nitric oxide, also a radical species and written as $NO\cdot$, is formed from microbially produced nitrous oxide (N_2O) reacting with oxygen atoms.

$$O\cdot + N_2O \longrightarrow NO\cdot + NO\cdot$$

For $NO\cdot$ the reactions are

$$NO\cdot + O_3 \longrightarrow NO_2\cdot + O_2$$

$$NO_2\cdot + O\cdot \longrightarrow NO\cdot + O_2$$

At one time it was feared that high-flying supersonic aircraft might destroy ozone by the nitric oxides they produced, but this has not proved to be the case.

Molina and Rowland's warning in 1974 regarding CFCs has proven correct, however. Satellite and ground-based measurements since 1978 indicate that global concentrations of ozone in the stratosphere have been decreasing.

Recently NASA scientists published measurements that show an average of about 2.5% decrease in the ozone layer worldwide in the decade 1978–1988. Local decreases in ozone may be even larger. In a latitude band

TABLE 18–6 Halons — Halogenated Hydrocarbons Containing Carbon-Bromine Bonds

Name	Formula	Uses
Halon-1211	CF_2BrCl	Portable fire extinguishers
Halon-1301	CF_3Br	Fire extinguisher systems

Figure 18–13   The hole in the ozone layer over the Antarctic continent. The purple region has the lowest ozone concentration. (Courtesy of NASA.)

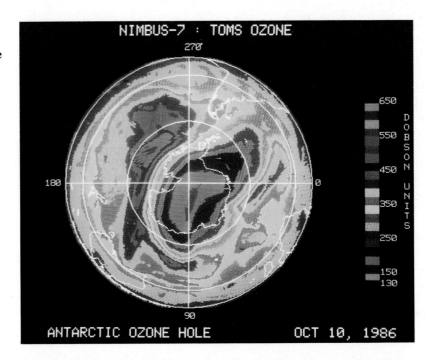

that includes Dublin, Ireland, Moscow, USSR, and Anchorage, Alaska, ozone decreased 8% from January 1969 to January 1986.

Near the North and South Poles ozone losses have been between 1 and 2.5%/year (Fig. 18–13). Recently, massive losses, termed "holes," have been observed. These ozone holes are of special concern since many scientists believe that they may happen at the mid-latitudes in the future.

Intense studies during the winters of 1987 and 1988 have shown that other factors lower the ozone concentration in addition to the gas-phase reactions mentioned earlier. In an Antarctic winter, a vortex of intensely cold air containing ice crystals builds up. On the surfaces of these crystals additional reactions take place that produce two chemical species not ordinarily involved in chlorine atom production. These species are hydrogen chloride (HCl) and chlorine nitrate ($ClONO_2$).

On the surface of these ice crystals, HCl and $ClONO_2$ react to form chlorine molecules (Cl_2). These are readily photodissociated into chlorine atoms, which become involved in the ozone destruction reactions.

$$HCl + ClONO_2 \longrightarrow Cl_2 + HNO_3$$

$$Cl_2 + h\nu \longrightarrow 2\ Cl\cdot$$

In an effort to reduce the use of CFCs and other compounds that harm the stratospheric ozone layer, some state legislatures have begun to pass laws restricting CFC use. Some states have passed laws requiring automobile service centers to reclaim automobile air-conditioning refrigerant gases. Vermont will ban CFC refrigerants in 1990, and the state of Connecticut will ban CFCs from automobile air conditioners by 1993 (Fig. 18–14). Nations are also acting. In January, 1989, the Montreal Protocol on Substances That Deplete the Ozone Layer went into effect. Signed by 24 nations, the protocol

Figure 18–14 CFCs like those used in automotive air conditioners can be recycled during repairs. At many service centers the CFCs are simply vented into the atmosphere prior to repair work. Many states and localities are beginning to require recycling as a means of lowering the amounts of CFCs escaping into the atmosphere. (Courtesy of Robinair.)

calls for reductions in production and consumption of several of the long-lived CFCs. There is a likelihood that there will be a total ban on CFCs by the year 2000. In addition, the use of two halons is to remain at 1986 levels. CFC and halon manufacturers are seeking alternatives to these compounds. One such possible substitute is 1,1,1-trifluoro-2-fluoroethane, which contains no chlorine.

$$\begin{array}{ccc} & F & H \\ & | & | \\ F- & C-C & -F \\ & | & | \\ & F & H \end{array}$$

SELF-TEST 18–B

1. When coal and fuel oil are burned, what two primary pollutants are formed?
2. When sulfur dioxide reacts with oxygen, what oxide of sulfur is formed? When this oxide reacts with water, what acid is formed?
3. Which chemical would most likely react with sulfur dioxide and remove it from combustion gases? (   ) sodium chloride (NaCl), (   ) lime (CaO), (   ) nitric oxide (NO).
4. Name two acids found in acid rain. One must contain sulfur, the other nitrogen.
5. What is the pH of normal rainfall? What dissolved chemical causes this pH to be below pH 7?
6. Approximately when was acid rainfall first observed?
7. Name the chemical bond broken by a photon of light in a typical CFC molecule.
8. Write the reaction producing ozone from oxygen.
9. What is the chemical species containing chlorine that destroys ozone molecules?
10. What type of chemical bond does a typical halon molecule contain?
11. Over what continent have scientists found an ozone "hole"?
12. If ozone in the stratosphere is destroyed, what form of radiation will pass through to the Earth below?

The ocean dissolves carbon dioxide to form bicarbonates and carbonates. (Charles D. Winters.)

CARBON DIOXIDE AND THE GREENHOUSE EFFECT

How can carbon dioxide (CO_2) be considered a pollutant when it is a natural product of respiration and fossil fuel burning and is a required reactant for photosynthesis? Actually, CO_2 is not a pollutant, but the fact that it is increasing in the Earth's atmosphere is cause for deep concern. Consequently, it is treated as a pollutant. Without human influences, the flow of carbon between the air, plants, animals, and the oceans would be roughly balanced. However, between 1900 and 1970, the global concentration of CO_2 increased from 296 ppm to 318 ppm, an increase of 7.4%. By 1985, the concentration was 350 ppm and expectations are that the CO_2 concentration will continue to increase (Fig. 18–15). For example, since the end of World War II, a world energy growth rate of about 5.3% per year took place, until the OPEC oil embargo in the mid-1970s. Rates of energy use have actually decreased since then.

To see how easily our everyday activities affect the amount of CO_2 being put into the atmosphere, consider a round-trip flight from New York to Los Angeles. Each passenger pays for about 200 gal of jet fuel, which weighs 1400 lb. When burned, each pound of jet fuel produces about 3.14 lb of carbon dioxide. So 4400 lb, or 2.2 tons of carbon dioxide are produced per passenger during that trip.

Population pressures are also contributing heavily to increased CO_2 concentrations. In the Amazon region of Brazil, for example, extensive cut-and-burn practices are being used to create cropland. This is causing a tremendous burden on the natural CO_2 cycle, since CO_2 is being added to the atmosphere during burning while there are fewer trees present to photosynthesize this additional CO_2 into plant nutrients. Counting all forms of fossil fuel combustion worldwide, the amount of CO_2 added to the atmosphere is about 50 billion tons a year. About half of this remains in the atmosphere to increase the global concentration of CO_2. The other half is taken up by plants during photosynthesis and by the oceans where CO_2

Figure 18–15 Atmospheric carbon dioxide is up more than 10% since 1960. (Adapted from *Chemical and Engineering News*, March 13, 1989.)

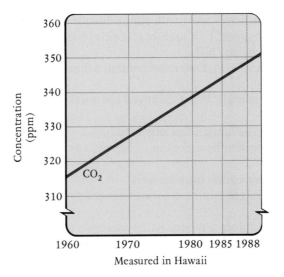

Measured in Hawaii

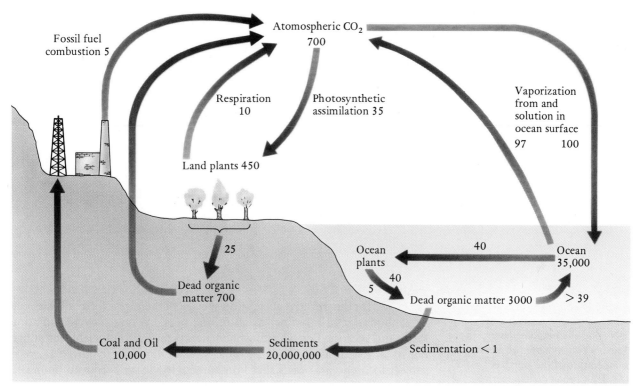

Figure 18-16 **The carbon cycle in the biosphere. The numbers are in units of 10^9 tons.**

dissolves to form carbonic acid, which then can form bicarbonates and carbonates (Fig. 18-16).

$$CO_2 + H_2O \;\rightleftharpoons\; H^+ + \underset{\text{Bicarbonate}}{HCO_3^-} \;\rightleftharpoons\; H^+ + \underset{\text{Carbonate}}{CO_3^{2-}}$$

It seems reasonable that if we are rapidly burning fossil fuels that took millions of years to form, we are then going to be adding back into the atmosphere CO_2 at a more rapid pace than it can be used up in natural processes.

How does increasing global CO_2 concentrations produce a greenhouse effect? Consider solar radiation as it arrives at the Earth's atmosphere. About half of the visible light (400-700 nm) striking the Earth is reflected back into space. The remainder reaches the Earth's surface and causes warming (Fig. 18-17). These warmed surfaces then re-radiate this energy as heat. Water vapor and CO_2 readily absorb some of this radiated energy coming from the Earth's surface, and in turn warm the atmosphere. A botanical greenhouse works on the same principle. Glass transmits visible light, but blocks infrared radiation trying to leave. The effect is a warming of the air inside the greenhouse. In warm weather, the windows of a greenhouse must be opened or the plants inside will overheat and die.

The Earth, with an average surface temperature of about 300 K (27°C), radiates energy in the spectral region of 3,000 to 30,000 nm. CO_2, water vapor, and ozone all absorb in various portions of this infrared region (see

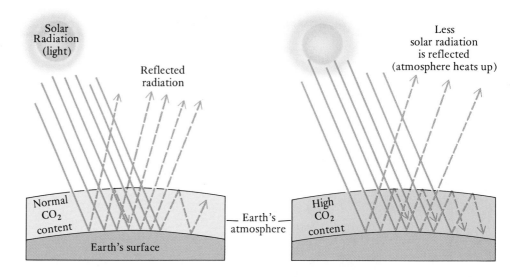

Figure 18–17   The greenhouse effect. Owing to a balance of incoming and outgoing energy in the Earth's atmosphere, the mean temperature of the Earth is 14.4°C (58°F). Carbon dioxide permits the passage of visible radiation from the Sun to the Earth but traps some of the heat radiation attempting to leave the Earth.

Fig. 5–22 for the infrared absorption spectra of these gases). So all three are "greenhouse gases" and act as absorbing blankets that prevent radiation losses and keep the Earth's atmospheric temperature comfortable (although not in all locations at the same time!). Water vapor in the atmosphere is subject to such vast cycles that human activity doesn't seem to bother it. Since ozone is present in relatively low concentrations, our attention is focused on CO_2.

Recently, Russian scientists took ice core samples dating back 160,000 years. In these ice samples were tiny pockets of air that could be analyzed for CO_2 content. They found a direct correlation between CO_2 and geologic temperatures known by other means. As the CO_2 increased, the global temperature increased and as the CO_2 decreased, the global temperature decreased. It is generally agreed that rising CO_2 concentrations will probably lead to increasing global temperatures and corresponding changes in climates. If predictions by the National Academy of Sciences prove correct, when and if the global concentration of CO_2 reaches 600 ppm, the average global temperature could rise by 1.5 to 4.5°C (2.7–8.1°F). Even a 1.5°C warming would produce the warmest climate seen on Earth in the past 6000 years, and a 4.5°C warming would produce world temperatures higher than any since the Mesozoic era—the time of dinosaurs.

Clearly, global warming is a major potential problem the world faces. Controlling CO_2 emissions worldwide will undoubtedly prove to be more difficult than controlling CFCs or the precursors to acid rain, but if we do not do something in this regard, the future may be bleak indeed.

INDUSTRIAL AIR POLLUTION

Besides CO_2, CFCs, NO_x, and SO_2, industry pollutes the atmosphere by emitting a wide variety of solvents, metal particulates, acid vapors, and unreacted monomers. The extent to which this takes place became evident

TABLE 18–7 The Top Ten States by Toxic Chemicals Released to Air in 1988

State	Emissions (millions of lbs)
1. Texas	229
2. Louisiana	134
3. Tennessee	132
4. Virginia	131
5. Ohio	122
6. Michigan	106
7. Indiana	103
8. Illinois	103
9. Georgia	94
10. North Carolina	92

Note: The total emissions for all 50 states in 1988 was 2,396,915,248 lb.
Source: U.S. EPA, 1989.

A chemical plant. Possible source for releases of toxic chemicals into the atmosphere. (*The World of Chemistry,* Program 17, "The Precious Envelope.")

in 1989 when the first summary of annual releases was published from data received by the EPA. This report was a part of the Superfund Reauthorization and Amendments Act of 1986, regulations which resulted in part from the tragedy in Bhopal, India (see Chapter 1). These release-reporting regulations were placed on manufacturers who use any of a group of about 320 chemicals and classes of compounds representing special health hazards. The reporting was divided into releases to air, water, and land. (Water and land releases directly affect surface and groundwater purity and are discussed in Chapter 17).

These regulations have been called "Community Right-to-Know" regulations because they inform communities about releases of harmful chemicals in their areas.

In the report, an industrial facility was asked to count all releases to the atmosphere, regardless of type. This meant that leaky valves, fittings, accidental spills, vapor losses while filling tank trucks and rail tank cars, emissions at stacks, and so forth, were all added together. As expected, heavily industrialized states and states with a lot of chemical industry had high releases (Table 18–7), but the amounts of some chemicals released were also

Testing for leaks at a valve in a chemical plant. (*The World of Chemistry,* Program 25, "Chemistry and the Environment.")

TABLE 18–8 Top Ten Chemicals Released into Air in 1988.

Chemical	Emissions (millions of lbs)	Uses
Toluene	235	Gasoline, solvent
Ammonia	233	Refrigerant, reactant*
Acetone	186	Solvent in paints
Methanol	182	Solvent, reactant
Carbon disulfide	137	Solvent, reactant
1,1,1-Trichloroethane	130	Degreasing operations
Methyl ethyl ketone	124	Solvent in paints
Xylene (mixed isomers)	120	Gasoline, solvents
Dichloromethane	112	Solvent, reactant†
Chlorine	103	Reactant, bleach‡

* Ammonia's use as a fertilizer not reported.
† A known carcinogen.
‡ Chlorine's use to disinfect water and wastewater not reported.
Source: U.S. EPA, 1989.

This form of release reporting could be termed a pollution "glasnost."

surprisingly large (Table 18–8). In the future, the EPA expects to make these data available by means of a publicly accessible computerized database.

INDOOR AIR POLLUTION

We shouldn't be surprised that air in our homes is contaminated by industrial chemicals—after all, we bring industry's products into our homes.

As if the data about pollutants in the outside air were not enough to concern us, the air inside our homes and workplaces is also contaminated and usually by the same chemicals emitted by industry.

TABLE 18–9 Some Common Household Products and the Chemicals They Contribute to Indoor Air Pollution.

Product	Major Organic Chemicals
Silicone caulk	Methyl ethyl ketone, butyl propionate, 2-butoxyethanol, butanol, benzene, toluene
Floor adhesive	Nonane, decane, undecane, xylene
Particleboard	Formaldehyde, acetone, hexanal, propanal, butanone, benzaldehyde, benzene
Moth crystals	p-dichlorobenzene
Floor wax	Nonane, decane, dimethyloctane, ethylmethylbenzene
Wood stain	Nonane, decane, methyloctane, trimethylbenzene
Latex paint	2-propanol, butanone, ethylbenzene, toluene
Furniture polish	Trimethylpentane, dimethylhexane, ethylbenzene, limonene
Room freshener	Nonane, limonene, ethylheptane, various substituted aromatics (as fragrances)

Source: EPA, 1988.

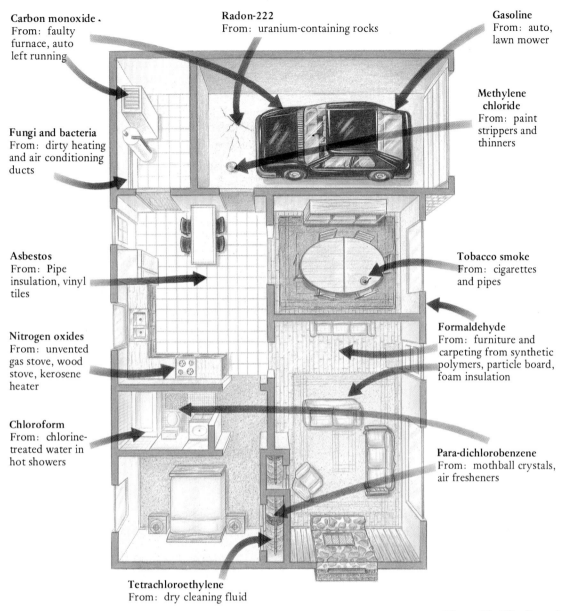

Carbon monoxide
From: faulty furnace, auto left running

Radon-222
From: uranium-containing rocks

Gasoline
From: auto, lawn mower

Methylene chloride
From: paint strippers and thinners

Fungi and bacteria
From: dirty heating and air conditioning ducts

Asbestos
From: Pipe insulation, vinyl tiles

Tobacco smoke
From: cigarettes and pipes

Nitrogen oxides
From: unvented gas stove, wood stove, kerosene heater

Formaldehyde
From: furniture and carpeting from synthetic polymers, particle board, foam insulation

Chloroform
From: chlorine-treated water in hot showers

Para-dichlorobenzene
From: mothball crystals, air fresheners

Tetrachloroethylene
From: dry cleaning fluid

Figure 18–18   Some indoor air pollutants and their sources.

At Home

Some scientists have concluded that air in our homes may be more harmful than the air outdoors, even in heavily industrialized areas. A study by EPA indicated that indoor pollution levels in rural homes were about the same as for homes in industrialized areas. One cause for this is the emphasis on tighter, more energy-efficient homes, which tend to trap air inside for long periods.

What are the sources of home air pollution (see Fig. 18–18)? Tobacco smoke, if present, is an obvious source. Benzene, a known carcinogen, occurs at 30 to 50% higher levels in homes of smokers than homes of non-smokers. Building materials and other consumer products are also sources of pollutants (Table 18–9). Entire buildings can acquire a "sick building syndrome" when a particular chemical or group of chemicals is found in sufficiently high concentration to cause headaches, nausea, stinging eyes, itching nose, or some combination of these symptoms. Usually, the best cure for all forms of indoor air pollution is to limit the introduction of the offending chemicals and to have better exchange with the outside air.

Radon, which was discussed in Chapter 7, is also a major potential air pollutant inside the home.

SELF-TEST 18-C

1. Name three different human activities that produce large amounts (millions of tons/year) of CO_2.
2. What are two principal processes whereby CO_2 is consumed?
3. Name three major greenhouse gases found in the atmosphere. Which is most closely associated with human activities?
4. Global temperature seems to follow carbon dioxide concentrations. ( ) True ( ) False.
5. Approximately what is the current global CO_2 concentration?
6. What state had the highest air releases of toxic chemicals in 1988, the first year such data were reported?
7. What chemical was released in greatest amount nationwide in 1988, according to the EPA release report?
8. What single activity inside the home can account for increased concentrations of benzene, a known human carcinogen?

MATCHING SET

_____ 1. Smog	a.	Heat absorbing
_____ 2. Primary pollutant	b.	Mixture of fog, SO_2, and (or) hydrocarbons and NO_2
_____ 3. Secondary pollutant	c.	Parts per million
	d.	Disease of lungs
_____ 4. Aerosol	e.	Hydrocarbon with double bond
_____ 5. Micron	f.	Eradication
_____ 6. ppm	g.	CO
_____ 7. Unsaturated hydrocarbon	h.	Destroys stratospheric ozone
	i.	Process of decreasing
_____ 8. Photochemical	j.	Intermediate in size between individual molecules and particulates
_____ 9. Abatement		
_____ 10. Emphysema	k.	Benzo(α)pyrene
_____ 11. Brown gas	l.	Peroxyacetyl nitrate
_____ 12. Endothermic	m.	Chemical reaction energized by light
_____ 13. Polynuclear hydrocarbon	n.	10^{-6} m
	o.	NO_2
_____ 14. Chlorofluorocarbon		

QUESTIONS

1. The formation of photochemical smog involves very reactive chemical species. What structural feature makes these species reactive?
2. Write a balanced chemical equation for the burning of iron pyrite (FeS_2) in coal to sulfur dioxide and Fe_2O_3, iron (III) oxide.
3. What conditions are necessary for thermal inversion?
4. What are the basic chemical differences between industrial smog and photochemical smog?
5. What are the major sources of the following pollutants?
   a. Carbon monoxide
   b. Sulfur dioxide
   c. Nitrogen oxides
   d. Ozone
6. What is a photochemical reaction? Give an example.
7. If air pollutants rise from the Earth into the atmosphere, why do they not continue on into space?
8. If nature emits more than 90% of the particulates (volcanic eruptions), nitrogen oxides (lightning), and carbon monoxide (decaying organic matter), why the concern about air pollution caused by people?
9. What effects does weather have on local air pollution problems? On regional air pollution problems?
10. Knowing the chemistry of photochemical smog formation, list some ways to prevent its occurrence.
11. What part do aerosols play in the formation of smogs?
12. Which substance, naturally found in the atmosphere, seems to be increasing in concentration over the years?
13. Describe an air pollution problem in your community. How can this problem be solved?
14. Discuss the merits of abatement versus eradication of air pollution.
15. Of the following air pollutants—particulates, sulfur dioxide, carbon monoxide, ozone, and nitrogen dioxide—
   a. Which is normally a secondary pollutant?
   b. Which is emitted almost exclusively by human-controlled sources?
   c. Which can be removed from emissions by centrifugal separators?
   d. Which two react with water to form acids?
16. Define:
   a. Micron
   b. ppm

17. Should human beings make a major effort to alter their activities in order to stop the increase of carbon dioxide in the atmosphere? What would be some of the costs to achieve this goal?
18. Why are small particulates in the atmosphere dangerous to human beings?
19. Which are more effective in producing smog, hydrocarbons with all single bonds or hydrocarbons with some double bonds?
20. Why is natural gas a good substitute for oil and coal as far as air pollution is concerned? What problem is related to its substitution?
21. What is the approximate concentration of air pollutants in the atmosphere during smoggy conditions?
22. What brown oxide of nitrogen is a necessary component of photochemical smog?
23. What is the chief source of air pollution in the United States?
24. Why should CO_2, SO_2, and CO be more soluble in water than N_2 and O_2?
25. What is a PAH? How is this class of compound produced?
26. Sketch these generalized (no actual data required) graphs:
   a. y axis: SO_2 emissions; x axis: concentration of S in coal
   b. y axis: CO emissions; x axis: amount of oxygen present
   c. y axis: CO emissions; x axis: gasoline consumed
   d. y axis: CO emissions; x axis: controls applied
27. Why are our desires for cheap energy and a clean environment in conflict?
28. The Cottrell precipitator makes use of what property of particulate particles?
29. Why does a temperature inversion tend to act as a lid over polluted air?
30. Does sulfur dioxide have a long or short atmospheric life?
31. What are some sources of indoor pollutants not necessarily found outdoors?
32. How does acid affect trees, buildings, and lakes?

19

Chemistry of Food Production

The green plant leaf is the best solar collector. Under ideal conditions the energy of the sun is captured in the biochemicals of life. Human interest in understanding the optimum growth conditions for plants and the modification of the plants for greater yields is intense. (Mark Boulton/Photo Researchers, Inc.)

T hrough the 8,000 to 10,000 years of recorded human history, food production techniques have developed enormously. It is generally estimated that 90% of the U.S. population worked to provide food and fiber during most of the 19th century. Sophistication in agriculture has reduced this figure to 4% and it is likely to go lower! However, we are still playing catch-up in supplying the human appetite for food. Today's efficient farming methods are required and will need to be improved if we are to feed adequately the approximately 5 billion people who are now alive in an expanding world population (Fig. 19 – 1).

To help grow the enormous amount of food we need, the chemical industry supplies modern scientific agriculture with a large assortment of chemicals — the **agrichemicals** — including fertilizers, medicine for livestock, chemicals to destroy unwanted pests and plant diseases, food supplements, and many others. However, current events tell us that even with the use of new agrichemicals and improved plant species, an improper climate or political-economic conditions that deter the flow of food still does result in areas of massive starvation.

A number of the ancient civilizations apparently developed good farming practices that are not directly recorded in history. However, in Roman

The British Isles recorded 201 famines between 10 A.D. and 1846, with none since. In China there were 1846 famines between 108 B.C. and 1828 A.D.

Annual U.S. agricultural exports (in billions of dollars): $43.3 in 1981; $36.6, 1982; $36.1, 1983; $37.8, 1984; $29.0, 1985; $26.2, 1986; $28.6, 1987. As much as 40% of our corn, 45% of our soybeans, 30% of our wheat, and 65% of our rice have been exported in a given year.

Figure 19 – 1   World population from 500 B.C. to 2000 A.D. The curve is shaped like a J because slow growth over a long period of time was followed by rapid growth over a relatively short period. It is projected that 6 billion humans will inhabit the earth by the year 2000.

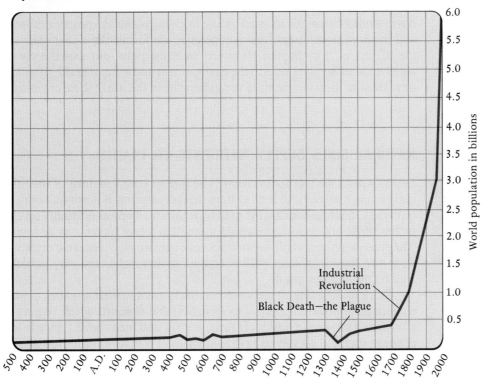

George Washington had marl, an alkaline mixture of limestone and clay, dug from the Potomac River bed for application to his fields.

"No man qualifies as a statesman who is entirely ignorant of the problems of wheat." — Socrates

times, Cato the Elder described seed selection, green manuring with legumes, testing of the soil for acidity, the use of marl, the value of alfalfa and clover, composting, the preservation and use of animal manures for soil fertilization, pasture management, early-cut hay, and the importance of livestock in any general farming operation. Such progress based on practical experience was apparently gained and lost a number of times in the course of history.

In 18th-century England, Arthur Young set the stage for modern scientific agriculture. As the first noted agricultural extension worker, he set in motion a system of dissemination of agricultural information that has now been developed on a worldwide basis. In 1840, Justus von Liebig published his *Organic Chemistry in Its Applications to Agriculture and Physiology.* Liebig began the description of the chemicals required by plants, and his work set the stage for the chemical fertilizer industry. As a result of his work, Liebig has been called the "father of modern soil science." The Industrial Revolution provided better tools and the machines that made the physical tasks of agriculture relatively easy. Even now, however, the best estimate is that two thirds of the world's agriculture is "backward."

The use of too much of a chemical or the use of the wrong chemical produces unwanted side effects that negate effective food production. Two notable examples are the phosphate runoff from fertilized fields, which can pollute streams, and the inclusion of DDT in bird eggs, which interferes with calcium metabolism and causes eggshells to be too thin and therefore to crack prematurely.

Benefits and problems related to DDT are discussed later in this chapter.

Our fundamental food is plants. Either we eat plants or animals that eat plants. In order to grow, plants require the proper temperature, nutrients, air, water, and freedom from disease, weeds, and harmful pests. Chemistry has provided chemicals to assist nature in giving plants the proper nutrients and freedom from disease and competetive life forms; these chemicals are the subject of this chapter. However, we must point out that the use of agrichemicals involves risks to the environment and human health; it will be important to measure the risks versus benefits in the use of these chemicals.

Figure 19–2   Through photosynthesis powered by energy from the sun, the earth's green plant population provides the food to sustain life on earth. (*The World of Chemistry*, Program 23, "Proteins: Structure and Function.")

NATURAL SOILS SUPPLY NUTRIENTS TO PLANTS

Layers within the soil are **horizons.** The **topsoil** contains most of the presently living material and humus from dead organisms. It is not uncommon to find as much as 5% organic matter in topsoil. Topsoil is usually several inches thick, and in some locations more than 3 ft of topsoil can be found. The **subsoil,** up to several feet in thickness, contains the inorganic materials from the parent rocks as well as organic matter, salts, and clay particles washed out of the topsoil.

Since healthy topsoil has abundant life forms and their remains, it must contain an abundant supply of oxygen. Soil that supports a rich vegetative growth and serves as a host for insects, worms, and microbes is typically full of pores; such soil is likely to have as much as 25% of its volume occupied by air. The ability of soil to hold air depends on soil particle size and how well the particles pack and cling together to form a solid mass. The particle size

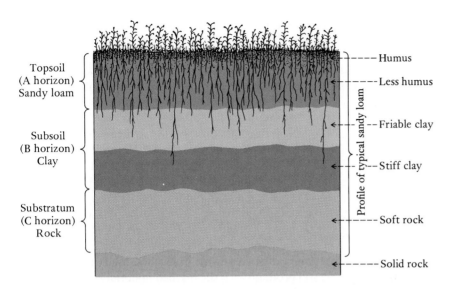

Figure 19–3   Structure of a typical sandy soil. Note that the soil is composed of layers or horizons that can be built up through the weathering of the parent rock below and the interaction of the weathered material with air and plant life from above. (U.S. Department of Agriculture.)

groups in soils, called **separates,** vary from clays (the finest) through silt and sand to gravel (the coarsest). The particle size of a clay is 0.005 mm or less. The small particles in a clay deposit pack closely together to eliminate essentially all air and thus support little or no life. A typical soil horizon is composed of several separates. A **loam,** for example, is a soil consisting of a friable mixture of varying proportions of clay, sand, and organic matter; a loam is rich in air content.

Friable material crumbles easily under slight pressure.

Air in soil has a different composition from the air we breathe. Normal dry air at sea level contains about 21% oxygen (O_2) and 0.03% carbon dioxide (CO_2). In soil the percentage of oxygen may drop to as low as 15%, and the percentage of carbon dioxide may rise above 5%. This results from the partial oxidation of organic matter in the closed space. The carbon in the organic material uses oxygen to form carbon dioxide. This increased concentration of carbon dioxide tends to cause groundwater to become acidic; acidic soils are described as *sour* soils because of the sour taste of aqueous acids.

$$CO_2 + H_2O \rightleftharpoons H^+ + HCO_3^-$$

Crushed limestone ($CaCO_3$) applied to soil combines with hydrogen ions to form bicarbonate ions, thus raising the pH.

$$H^+ + CO_3^{2-} \rightleftharpoons HCO_3^-$$

If enough limestone is added to neutralize the acid in the soil and leave an excess of limestone, the pH of the soil is raised to slightly basic.

A slightly basic soil is a sweet soil.

WATER IN SOIL: TOO MUCH, TOO LITTLE, OR JUST RIGHT

Water can he held in soil in three ways: it can be *absorbed* into the structure of the particulate material, it can be *adsorbed* onto the surface of the soil particles, and it can occupy the pores ordinarily filled with air.

Labels on figure:
Topsoil (A horizon) Sandy loam
Subsoil (B horizon) Clay
Substratum (C horizon) Rock
Profile of typical sandy loam
Humus
Less humus
Friable clay
Stiff clay
Soft rock
Solid rock

It takes several hundred pounds of water for the typical food crop to make 1 lb of food.

Water is removed from soil in four ways: plants transpire it while carrying on life processes, soil surfaces evaporate water, water is carried away in plant products, and gravity pulls it to the subsoil and rock formations below in a process called **percolation.** Percolation is the ability of a solid material to drain a liquid from the spaces between the solid particles. Soils with good percolation drain water from all but the small pores in the natural flow of the water.

The percolation of a soil depends on the soil particle size and on the chemical composition of the soil material. Because of the small particle sizes involved, clays, and to a lesser degree silts, tend to pack together in an impervious mass with little or no percolation. Of course, sand, gravel, and rock pass water readily. Waterlogged soils that will not percolate support few crops because of their lack of air and oxygen. Rice is an important exception. A negative aspect of the massive flow of water through soil is the **leaching effect.** Water, known as the universal solvent because of its ability to dissolve so many different materials, dissolves away, or leaches, many of the chemicals needed to make a soil productive. If the leached material is not replaced, the soil becomes increasingly unproductive.

In some arid regions, calcium collects as calcium carbonate just under the solum, or true soil.

Soils become acidic, or sour, not only because of the oxidation of organic matter but also because of *selective leaching* by the passing groundwater. Salts of the alkali and alkaline-earth metals are more soluble than salts of the group III and transition metals. For example, a soil containing calcium, magnesium, iron, and aluminum is likely to be slightly alkaline, or sweet, before leaching with water. If calcium and magnesium are removed and iron and aluminum salts remain, the soil becomes acidic. The iron and

Figure 19–4   In addition to the food manufacturing process, a leaf must transport large masses of materials from one place to another and exchange gases and water vapor with the surrounding air. All of these processes depend on an abundant supply of water. Desert plants have protective mechanisms to protect a high water content for chemically active cells. (*The World of Chemistry,* Program 12, "Water.")

Figure 19–5   It is estimated that 2 million gal of water per person per year are used in the United States and that 80% of this water is for agriculture. We want to grow crops where rainwater is not sufficient and as we become more adept at using fresh groundwater, it is becoming increasingly evident that the fresh water supply will continue to be the limiting factor in food production. Note the prospects for using salt water for agriculture at the end of this chapter. (*The World of Chemistry,* Program 12, "Water.")

aluminum ions each tie up hydroxide ions from water and release an excess of hydrogen ions:

$$Fe^{3+} + H_2O \longrightarrow FeOH^{2+} + H^+$$

$$Al^{3+} + H_2O \longrightarrow AlOH^{2+} + H^+$$

HUMUS

Organic matter varies in soil from the relatively fresh remains of leaves, twigs, and other plant and animal parts to peat, which results from old, decayed animal and vegetable matter. Peat is the precursor of coal and oil. Humus is not far removed in time from the living debris. However, it is well decomposed, dark-colored, and rather resistant to further decomposition. As a source of nutrients for plants, humus is almost like a time-release capsule, taking considerable time to release its contents while holding them in an insoluble form. Eventually, humus is decomposed into minerals and inorganic oxides.

Humus releases its nutrients to plants slowly.

In addition to being a source of plant nutrients, humus is important in maintaining good soil structure, often keeping the soil friable in a soil rich in clay. Soil rich in humus may contain as much as 5% organic matter. Soils in the grasslands of North America are rich in humus to a considerable depth, in contrast to rain-forest regions, where there is only a thin film of humus on the ground surface.

Maintaining humus in the soil is of major concern to the agriculturist. Humus such as peat moss or organic fertilizer can be added. However, there is no real substitute for natural plant growth that is returned to the ground for humus formation. Clover is often grown for this purpose and plowed under at the point of its maximum growth. The compost pile of the gardener is another effort to maintain humus for a productive soil.

If large amounts of organic matter such as leaves or sawdust are added to soil to promote humus formation, it should be remembered that the oxidation of this material produces acids. Acid soils require treatment with lime if the soil is to remain sweet.

CHEMICAL COMPOSITION OF SOIL

The chemical compositions of soils reflect the Earth's crust composition, the composition of the parent rocks, and chemical and physical activity during and after soil formation. Even though it is second to oxygen in percent composition, silicon is the central element in explaining soil chemicals. Sand is silicon dioxide (SiO_2), clays are mixtures of silicates, and the different kinds of silicate rock fragments are numerous. The bulk of most soil horizons is composed of silicate materials. A wide variety of other elements is present in any given soil sample.

Ask your county extension office for advice on how to take soil samples and where to send the samples for analysis. The cost of analysis is generally under $5 per sample. The analysis will tell you the amount of lime in the soil and the amount and type of fertilizer to apply.

Soils black in color are usually rich in organic matter and consequently contain the elements required for plant life. Red soils are likely to be rich in iron, and soils that are nearly white have been heavily leached and are likely

Iron oxide is red.

to be of poor quality. Poorly drained soils are likely to be of uneven chemical texture, with several colors, such as gray, brown, and yellow, appearing in a spadeful of the material.

NUTRIENTS

At least 18 known elemental nutrients are required for normal green plant growth (Table 19–1). Three of these, the **nonmineral nutrients,** carbon, hydrogen, and oxygen, are obtained from air and water. The **mineral nutrients** must be absorbed through the plant root system as solutes in water. The 15 known mineral nutrients fall into three groups: **primary nutrients, secondary nutrients,** and **micronutrients,** depending on the amounts necessary for healthy plant growth.

Nonmineral Nutrients

Sir Humphrey Davies argued the humus theory. "Carbon for plants came from humus." A Swiss, de Saussure, showed the carbon came from carbon dioxide.

Carbon, hydrogen, and oxygen are available from the air and water. Carbon comes to plants as carbon dioxide, and hydrogen and oxygen come as water; in addition, plant roots absorb some free oxygen dissolved in water. During photosynthesis, green plants produce an excess of oxygen, which is released through the leaves and other green tissue.

Primary Nutrients

The air above each acre of earth's surface contains 36,000 tons of nitrogen.

Nitrogen species ranked by degree of oxidation: NO_3^-, NO_2, NO, N_2O, N_2, NH_4^+.

The primary nutrients are nitrogen, phosphorus, and potassium. Although bathed in an atmosphere of nitrogen, most plants are unable to use the air as a supply of this vital element. **Nitrogen fixation** is the process of changing atmospheric nitrogen into the compounds of this element that can be dissolved in water, absorbed through the plant's roots, and assimilated by the plant (Fig. 19–6). Most plants thrive on soils rich in nitrates, but many plants that grow in swamps, where there is a lack of oxidized materials, can use reduced forms of nitrogen such as the ammonium ion. The nitrate ion is the most highly oxidized form of combined nitrogen, and the ammonium ion is the most reduced form of nitrogen.

TABLE 19–1 Essential Plant Nutrients

Nonmineral	Primary	Secondary	Micronutrients
Carbon	Nitrogen	Calcium	Boron
Hydrogen	Phosphorus	Magnesium	Chlorine
Oxygen	Potassium	Sulfur	Copper
			Iron
			Manganese
			Molybdenum
			Sodium
			Vanadium
			Zinc

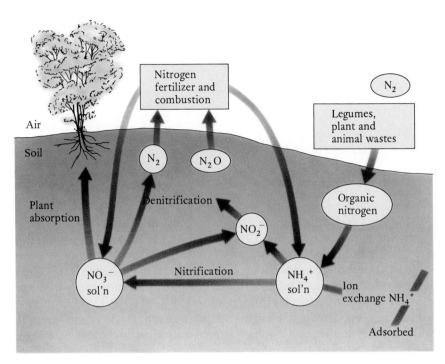

Figure 19-6   Nitrogen pathways through the soil.

Nature fixes nitrogen on a massive scale in two ways. Nitrogen is oxidized under highly energetic conditions, such as in the discharge of lightning or to a lesser extent, such as in a fire. The initial reaction is the reaction of nitrogen and oxygen to form nitric oxide (NO):

$$N_2 + O_2 \longrightarrow 2\ NO$$

Nitric oxide is easily oxidized in air to nitrogen dioxide (NO_2), which dissolves in water to form nitric acid (HNO_3) and nitrous acid (HNO_2):

$$2\ NO + O_2 \longrightarrow 2\ NO_2$$

$$H_2O + 2\ NO_2 \longrightarrow HNO_3 + HNO_2$$

Nitric acid is readily soluble in rain, clouds, or ground moisture and thus increases nitrate concentration in soil.

Nitrogen is an important element in amino acids and proteins.

A German, Hellriegel, showed in 1886 that leguminous plants "fix" nitrogen. Under ideal conditions, legume fixation can add more than 100 lb of nitrogen per acre of soil in one growing season.

A vast amount of atmospheric nitrogen is fixed as a result of natural electric discharges in the atmosphere. The energy in a bolt of lightning is sufficient to disrupt the very stable triple bond in a nitrogen molecule. Oxidation of nitrogen results. (Gary Ladd.)

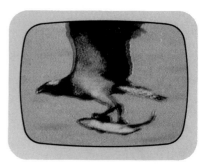

Figure 19–7 The control of all life forms is dependent on fixed nitrogen in the form of protein. Animals can obtain the building blocks for proteins either by eating plants, plant products, or other plant-eating animals. Upon death and decay, the fixed nitrogen may be recycled in fixed form through the soil or waters to other plants or it may be returned to the air in the elemental form. (*The World of Chemistry*, Programs 23 and 24, "Proteins: Structure and Function" and "Genetic Code.")

Potassium is absorbed as the free ion, $K^+(aq)$.

Another major source of nitrogen replenishment in soil is dead organisms and animal wastes (Fig. 19–7). Even in the absence of legumes, this can be an adequate source of nitrogen.

Like nitrogen, phosphorus must be in mineral or inorganic form before it can be used by plants. Unlike nitrogen, phosphorus comes totally from the mineral content of the soil. Orthophosphoric acid (H_3PO_4) loses hydrogen ions to form the dihydrogen and monohydrogen ions ($H_2PO_4^-$ and HPO_4^{2-}), which are the dominant phosphate ions in soils of normal pH (Fig. 19–8). Because of the great concentration of electric charge associated with the trivalent phosphate ion, phosphates tend to be held to positive centers in the soil structure and are not as easily leached by groundwater as are nitrate ions. The nitrate ion has only one negative charge; nitrate salts are generally soluble in water.

Potassium is a key element in the enzymatic control of the interchange of sugars, starches, and cellulose. Although potassium is the seventh most abundant element in the Earth's crust, soil used heavily in crop production can be depleted in this important metabolic element, especially if it is regularly fertilized with nitrate with no regard to potassium content. Some fungus plants in the soil produce chemicals that cause bound potassium to be released into a soluble form that can be taken in through the plant root system in excessive amounts or simply leached out by the flow of soil water.

Figure 19–8 The availability of phosphate in the soil is a function of pH. The dominant species present for phosphoric acid at pH 5 to pH 8 are $H_2PO_4^-$ and HPO_4^{2-}. At very low pH values, the phosphorus is in the form of the acid H_3PO_4 (all three protons on the acid structure). At a very high pH, all three protons are removed, and the phosphorus is in the form of the PO_4^{3-} ion. Low soil temperatures in temperate regions significantly reduce phosphorus uptake by plants.

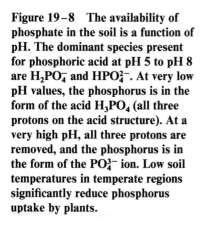
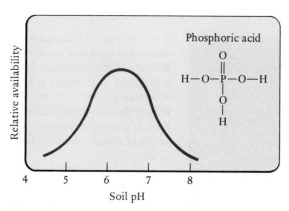

Secondary Nutrients

Calcium and magnesium are available as Ca^{2+} and Mg^{2+} ions in small amounts as well as in complex ions and crystalline formations. These abundant elements are bound tightly enough so they are not readily leached yet loosely enough to be available to plants. When in the soil as sulfate (SO_4^{2-}), sulfur is readily available to plants.

Chlorophyll requires nitrogen and magnesium from the soil. Magnesium deficiencies, like nitrogen deficiencies, cause chlorosis, *a condition of low chlorophyll content.*

Micronutrients

Only very small amounts of micronutrients are required by plants; therefore, unless extensive cropping or other factors deplete the soil of these nutrients, sufficient quantities are usually available.

Iron is also an essential component of the catalyst involved in the formation of chlorophyll, the green plant pigment. When the soil is iron deficient or when too much lime, $Ca(OH)_2$, is present in the soil, iron availability decreases. This condition is usually indicated by plant leaves that are light or yellowish in color. Often a gardener or lawn worker will apply phosphate and lime to adjust soil acidity, only to see green plants turn yellow. What happens in such cases is that both phosphate and the hydroxide from the lime tie up the iron and make iron unavailable to the plants:

Lime, as CaO or Ca(OH)$_2$, is the number-six commercial chemical. See listing inside front cover.

$$Fe^{3+} + 2\ PO_4^{3-} \longrightarrow Fe(PO_4)_2^{3-}$$
Phosphate   Tightly Bound Complex

$$Fe^{3+} + 3\ OH^- \longrightarrow Fe(OH)_3$$
Insoluble Hydroxide

Boron is absolutely necessary in trace amounts, but there is a relatively narrow concentration range above which boron is toxic to most plants.

SELF-TEST 19–A

1. Rank the following types of soils from those with the smallest soil particles to those with the largest: silts, sandy soils, loams, clays.
2. Acidic soils are described as _____ because of the common taste of aqueous acids.
3. Carbon dioxide causes soils to be (   ) acidic (   ) basic. Limestone ($CaCO_3$) causes soils to be (   ) acidic (   ) basic.
4. The two factors that determine the percolation of a soil are _____ and _____.
5. Which is more acidic, a monovalent ion like Na^+ or a trivalent ion like Fe^{3+}?
6. A well-decomposed, dark-colored plant residue that is relatively resistant to further decomposition is known as _____.
7. The bulk of most soil horizons is composed of _____ materials.
8. The secondary elemental plant nutrients are _____, _____, and _____.

Atmospheric nitrogen is fixed in root nodules on leguminous plants such as the soybean pictured here. Rather than requiring fertilization for nitrogen, such plants can be grown to enrich the soil with nitrogen compounds and produce a valuable crop at the same time. (Metcalfe, Williams, and Castka and Walter O. Scott.)

9. Nitrogen fixation involves breaking a nitrogen-nitrogen triple bond and combining nitrogen with another _____.
10. Some micronutrients can poison plants. True ( ) False ( )

FERTILIZERS SUPPLEMENT NATURAL SOILS

Primitive people raised crops on a cultivated plot until the land lost its fertility; then they moved to a virgin piece of ground. In many cases, the slash-burn-cultivate cycle was no more than a year in length, and few found a piece of ground anywhere that could support successful cropping for more than 5 years without fertilization. Farming villages, developed in ancient times and prevalent throughout the Middle Ages, demanded innovation in fertilization, since they had to use the same land for many years. With the use of legumes in crop rotations, manures, dead fish, or almost any organic matter available, the land was kept in production.

An estimated 4 billion acres are used worldwide in the cultivation of crops for food, less than 0.8 acre per person. This acreage would likely be sufficient if modern chemical fertilization were employed on all of it. If about

In 1988, vast amounts of acres were torched in the Amazon basin.

Chinese farmers added calcined bones to their soil 2000 years ago.

Crop yield explosions: (1) U.S. corn—25 bushels per acre in 1800, 110 bushels per acre in the 1980s; (2) English wheat—below 10 bushels per acre from 800 A.D. to 1600, above 75 bushels per acre in the 1980s. (3) Rice in Japan, Korea, and Taiwan—fourfold increase in the last 40 years.

As chemical fertilizers became widely available during the early part of the 20th century, farmers were eager to use them because of the explosive increase in crop yields obtained. Tractor-powered spreading equipment can now easily fertilize a 40-ft path across the field in a side-dressing to the growing plant, a general application to the soil surface or subsurface. (*The World of Chemistry*, Program 2, "Color.")

TABLE 19–2 Some Chemical Sources for Plant Nutrients

Element	Source Compound(s)
Nonmineral Nutrients	
C	CO_2 (carbon dioxide)
H	H_2O (water)
O	H_2O (water)
Primary Nutrients	
N	NH_3, (ammonia), NH_4NO_3 (ammonium nitrate), H_2NCONH_2 (urea)
P	$Ca(H_2PO_4)_2$ (calcium dihydrogen phosphate)
K	KCl (potassium chloride)
Secondary Nutrients	
Ca	$Ca(OH)_2$ (calcium hydroxide [slaked lime]), $CaCO_3$, (calcium carbonate [limestone]), $CaSO_4$ (calcium sulfate [gypsum])
Mg	$MgCO_3$ (magnesium carbonate), $MgSO_4$ (magnesium sulfate [epsom salts])
S	Elemental sulfur, metallic sulfates
Micronutrients	
B	$Na_2B_4O_7 \cdot 10H_2O$ (borax)
Cl	KCl (potassium chloride)
Cu	$CuSO_4 \cdot 5H_2O$ (copper sulfate pentahydrate)
Fe	$FeSO_4$ (iron(II) sulfate, iron chelates)
Mn	$MnSO_4$ (manganese(II) sulfate, manganese chelates)
Mo	$(NH_4)_2 MoO_4$ (ammonium molybdate)
Na	NaCl (sodium chloride)
V	V_2O_5, VO_2 (vanadium oxides)
Zn	$ZnSO_4$ (zinc sulfate, zinc chelates)

$40 were spent on fertilizer for each cultivated acre, world crop production would increase by 50%, the equivalent of having 2 billion more acres under cultivation. However, the cost to produce this additional food would approach a prohibitive $160 trillion.

Fertilizers that contain only one nutrient are called **straight fertilizers;** those containing a mixture of the three primary nutrients are called **complete,** or **mixed, fertilizers.** Urea for nitrogen, and potassium chloride for potassium are examples of straight products. The macronutrients are absorbed by plant roots as simple inorganic ions: nitrogen in the form of nitrates (NO_3^-), phosphorus as phosphates ($H_2PO_4^-$ or HPO_4^{2-}), and potassium as the K^+ ion. Organic fertilizers can supply these ions, but only when used in large quantities over a long time. For example, a manure might be a 0.5-0.24-0.5 fertilizer, in contrast to a typical chemical fertilizer, which might carry the numbers 6-12-6. These numbers indicate the **grade,** or **analysis,** in order, of the percentage of nitrogen, phosphorus as P_2O_5, and potassium as K_2O in the fertilizer (Fig. 19–9). In addition to having the desired ion, the chemical fertilizer places the ion in the soil in a form that can be absorbed directly by plants. The problem is that these inorganic ions are relatively easily leached from the soil and may pose pollution problems if not contained. The much slower organic fertilizer tends to say put. **Quick-release fertilizers** are water-soluble, as opposed to **slow-release products,** which require days or weeks for the material to dissolve completely. Table 19–2 lists the necessary plant nutrients and suitable chemical sources of each.

Table 19–3 lists the quantities of nutrients known to be necessary to produce 150 bushels of corn.

Mixed fertilizers were first produced in Baltimore in 1850.

Peruvian guano, a natural source of nitrates, was first imported into the United States in 1824.

Manure releases about half of its total nitrogen in the first growing season. Manure as a fertilizer is graded less than 1-2-1.

TABLE 19–3 Approximate Amounts of Nutrients Required to Produce 150 Bushels of Corn

Nutrient	Approximate Amount (Pounds per Acre)*	Source
Oxygen	10,200	Air
Carbon	7800	Air
Water	3225–4175 tons	29–36 in. of rain
Nitrogen	310	1200 lb of high-grade fertilizer
Phosphorus	120 (as phosphate)	1200 lb of high-grade fertilizer
Potassium	245 (as K_2O)	1200 lb of high-grade fertilizer
Calcium	58	150 lb of agricultural limestone
Magnesium	50	275 lb of magnesium sulfate
Sulfur	33	33 lb of powdered sulfur
Iron	3	15 lb of iron sulfate
Manganese	0.45	13 lb of manganese sulfate
Boron	0.05	1 lb of borax
Zinc	Trace amounts	Small amount of zinc sulfate
Copper	Trace amounts	Small amount of copper sulfate
Molybdenum	Trace amounts	Trace of ammonium molybdate

* Except for water.

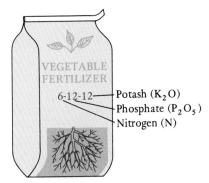

Figure 19–9 Fertilizer analysis numbers refer to the percentage by weight of N (nitrogen), P_2O_5 (phosphate), and K_2O (potash). Following the lead of Liebig, Samuel William Johnson, an American student of Liebig and the author of *How Crops Grow*, burned plants and analyzed their ashes. He expressed the nutrient concentrations in the oxide form present in the ashes as P_2O_5, K_2O, and so on, a practice that has continued to this day.

AGRICULTURE

When any covalent bond forms, energy is released. In turn, the same amount of energy is needed to break a bond.

Consider nitrogen. There is nitrogen in all living things. Muscles, hair, and DNA all contain nitrogen bonded to other elements. But 80% of the atmosphere is nitrogen molecules held together by strong triple bonds. How do living things get the form of nitrogen they need? Lightning helps. The electrical flash in the sky has enough energy to break apart nitrogen molecules, which then react with oxygen in the air, eventually forming nitric acid. The natural acid dissolves in rain and falls to earth as a dilute solution. There it is absorbed and metabolized by plants.

Some plants, though, convert molecular nitrogen in a different way. Soybeans and other legumes, like peas and peanuts, host a unique bacterium in their roots. It is this bacterium that converts the nitrogen molecule into a nitrogen compound, ammonia, which the plant can then use to make amino acids.

Exactly how the bacterium works is the subject of vigorous research. Don Keister, of the U.S. Department of Agriculture, claims, "This is one of the very unique enzymes in all of nature, because it is the only solution that nature has evolved for biologically reducing nitrogen."

The soybean and the bacterium have a symbiotic relationship. The plant houses and feeds the bacterium and, in turn, it receives the nitrogen it needs. But not all plants can host these nitrogen fixers. They have to rely on rain, natural fertilizers, as well as expensive manufactured fertilizers, like ammonium nitrate.

Keister goes on to point out, "We are currently using something like 300 million barrels of oil per year in the United States alone to produce nitrogen fertilizers. We forget sometimes, that we're going to need to double the food supply over the next 20 years." He further asks the unanswered questions, "Where is that energy going to come from? Where is the fertilizer going to come from?"

For feeding the world, there are two basic options: We can either produce more fertilizer at greater cost and some risk to the environment, or we can create new varieties of nitrogen-fixing plants. Both options are being pursued worldwide.

The World of Chemistry (Program 8), "Chemical Bonds."

Nitrogen

Fixed nitrogen refers to nitrogen present in chemical compounds, that is, in combinations other than N_2.

The first commercial U.S. ammonia plant began production in 1921.

The production of nitrogen through the distillation of air is described in Chapter 11. The reaction of steam with hydrocarbons is discussed in Chapter 13.

The fixed nitrogen in commercial fertilizers is obtained by the direct reaction of nitrogen with hydrogen to produce ammonia by the Haber process (see Chapter 11).

$$N_2 + 3\,H_2 \rightleftharpoons \underset{\text{Ammonia}}{2\,NH_3}$$

Pure nitrogen is obtained for the process by distillation of oxygen and other gases from liquid air. Hydrogen is more difficult to obtain. At present, petroleum products such as propane (C_3H_8) are reacted with steam in the presence of catalysts to produce hydrogen:

$$C_3H_8 + 6\,\underset{\text{Steam}}{H_2O} \xrightarrow{\text{Catalyst}} 3\,CO_2 + 10\,H_2$$

Figure 19-10 At normal temperatures ammonia can be maintained as a liquid under tank pressure. On release from pressure the ammonia turns to a gas and is readily dissolved in the moisture of the soil. If the soil pH is slightly acid, the ammonia is immediately converted to the ammonium ion and enters into the natural nitrogen pathways of the soil (Fig. 19-6). (Grant Heilman.)

This is one of the principal reasons why ammonia fertilizer costs are so closely tied to petroleum prices. More expensive hydrogen can also be prepared by the electrolysis of water:

$$2\ H_2O \xrightarrow[\text{KOH}]{\text{Electricity}} 2\ H_2 + O_2$$

Several other methods are also available, but all of them require considerable energy per pound of hydrogen produced. So as energy costs rise, the costs of fertilizers and therefore also the costs of food will necessarily rise.

At normal temperatures and pressures, ammonia is a gas. At higher pressures, ammonia is a liquid, known as **anhydrous ammonia.** Anhydrous ammonia can be injected directly into the soil; it is retained there because the polar ammonia molecules have a very high affinity for moisture in the soil. Some danger is involved if liquid ammonia is released from a pressure tank into the atmosphere; there it will immediately evaporate and can create a toxic atmosphere prior to its dispersal through the denser air.

Water solutions of ammonia, containing as much as 30% nitrogen by weight, can also be applied successfully to the soil. Again, special equipment and training are required to prevent the loss of ammonia through volatilization.

Ammonia to Nitrates

Nitrogen in the form of soil ammonia is readily available for plant growth. Under usual soil conditions, the ammonia molecules pick up a hydrogen ion to form the ammonium ion:

$$NH_3 + H^+ \longrightarrow NH_4^+$$

Soil, rich in oxygen, is an oxidizing medium, so that the ammonium ion, through a process called **nitrification,** can be oxidized to the nitrate ion (NO_3^-). The nitrate ion is then taken up and used by the plants.

Major energy problems in the fertilizer industry began with the energy crisis in the 1970s.

Anhydrous ammonia was first applied commerically to a crop in 1944.

Research shows solid and liquid fertilizers to be equally effective.

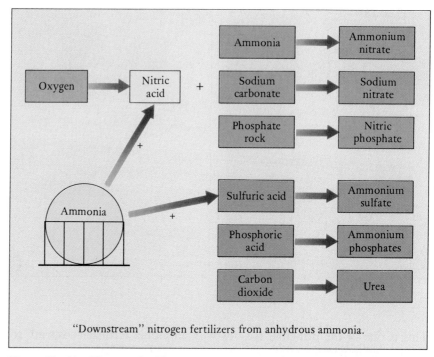

"Downstream" nitrogen fertilizers from anhydrous ammonia.

Figure 19–11   Nitrogen fertilizers, produced from anhydrous ammonia.

Solid nitrate fertilizers are prepared from ammonia (Fig. 19–11). Ammonia from the Haber process is burned in oxygen over a platinum catalyst to obtain nitric oxide (NO):

$$4\ NH_3 + 5\ O_2 \longrightarrow 4\ NO + 6\ H_2O$$

The NO reacts readily with O_2 from the air to form NO_2:

$$2\ NO + O_2 \longrightarrow 2\ NO_2$$

NO_2 in turn reacts with water to yield nitric acid and nitrous acid. The nitrous acid, being unstable, is decomposed with heat, and the resulting NO can be volatilized and recycled:

$$H_2O + 2\ NO_2 \longrightarrow HNO_3 + HNO_2$$

$$3\ HNO_2 \longrightarrow H_2O + 2\ NO + HNO_3$$

Additional ammonia then reacts with the nitric acid to produce ammonium nitrate (NH_4NO_3). Inorganic nitrate fertilizers are water soluble and contribute significantly to nitrate pollution in the groundwater of farming regions. Since excess nitrates in drinking water and food supplies cause blood disorders and contribute to the formation of carcinogenic nitrosamines (Chapter 16), the present concern is to keep the nitrate out of the groundwater or to remove it in the purification of drinking water. Solid ammonium nitrate should be handled with caution. While pure ammonium nitrate is quite stable in industrial and farming operations, it is an oxidizing agent and can explode on heating when mixed in bulk storage with reducing materials. Pure ammonium nitrate applied to the soil poses no explosive threat at all.

A flooded soil quickly becomes a reducing medium as the air supply of oxygen is cut off. **Denitrification** occurs when conditions are such that nitrate is reduced to elemental nitrogen, which escapes into the atmosphere. It is estimated that soil fertilized with a soluble nitrate and then flooded for 3 to 5 days will lose 15% to 30% of its nitrogen as a result of denitrification.

Urea

Urea (NH_2CONH_2) is one of the world's most important chemicals because of its wide use as a fertilizer and as a feed supplement for cattle. Ammonia and carbon dioxide react under high pressure near 200°C to produce ammonium carbamate, which then decomposes into urea and water.

Solid urea was placed on the market in 1935.

$$2\,NH_3 + CO_2 \longrightarrow H_2N-C\overset{O}{\underset{O^-NH_4^+}{\diagdown}} \longrightarrow H_2N-\overset{O}{\overset{\|}{C}}-NH_2 + H_2O$$

A slurry of water, urea, and ammonium nitrate is often applied to crops under the name of "liquid nitrogen." Such a solution can contain up to 30% nitrogen and is easy to store and apply.

When applied to the surface of the ground around plants, urea is subject to considerable nitrogen loss unless it is washed into the soil by rain or irrigation. When urea hydrolyzes (is decomposed by water), ammonia is formed, and some ammonia is lost to the air and some is absorbed by moist soil particles. As much as half of the nitrogen applied can be lost in this way.

Sulfur-coated urea (SCU) slows the release of nitrogen from urea.

Phosphorus and Potassium (Phosphate Rock and Potash)

Phosphate rock and **potash** are two minerals that can be mined, pulverized, and dusted directly onto deficient soil. Often they are specially treated to produce desirable mixing properties. Phosphorus, for example, is found scattered throughout the world in deposits of phosphate rock. The phosphate rock itself is not very useful to a growing plant because of its very low solubility. When treated with sulfuric acid, however, mined phosphate becomes more soluble and is called "superphosphate."

The world has limited deposits of phosphate rock, which are essential to the manufacture of fertilizers.

$$\underset{\text{Phosphate Rock}}{Ca_3(PO_4)_2} + 2\,H_2SO_4 \longrightarrow \underset{\text{Superphosphate}}{Ca(H_2PO_4)_2} + 2\,CaSO_4$$

Within a few years, large new deposits of phosphate rock will be needed to meet the demand. A major problem is an increasing concentration of impurities in the available phosphate rock. Of particular concern is the natural cadmium content in phosphate rock, which may range from 3 mg of cadmium per kg of phosphate (P_2O_5) for deposits in Finland, the Soviet Union, and South Africa to a high of 280 mg in Senegal. U.S. phosphate contains from 20 to 120 mg cadmium per kg of phosphate. Cadmium, highly toxic to many life forms, follows the phosphoric acid into the solid phosphate fertilizers, and is subsequently taken up by the plants. It is esti-

Superphosphate fertilizer (0-45-0). Recall that the middle number measures the amount of phosphorus in the form of P_2O_5 that is available from the fertilizer. (Charles Steele.)

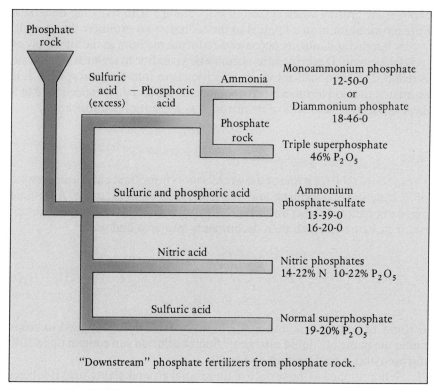

Figure 19–12   Fertilizers produced from phosphate rock.

mated that removal of the cadmium by ion exchange or calcination (trapping the cadmium in the flue gases) may add approximately 10% to the fertilizer cost. See Figure 19–12 for products produced from phosphate rock.

Potassium in the form of potash (K_2CO_3) exists in enormous quantities throughout the world. A major problem involved in obtaining potash is the depth of mining required. Another soluble form of potassium is its chloride (KCl), called muriate of potash. Because KCl is often found in combination with sodium chloride, which is toxic to plants, some process (such as recrystallization) must be used to separate the two compounds.

Calcium, Magnesium, and Sulfur

Calcium is rarely in short supply as a plant nutrient. In addition to its widespread occurrence in clays and minerals, it is a major component of lime, which is used to make soil alkaline. Calcium hydroxide is insoluble enough in neutral and alkaline media to prevent dangerously excessive uptake by plants.

Magnesium deficiencies occur in acidic sandy soils and in other soils in which the magnesium is bound so strongly that an insufficient amount of it is released in solution form. Magnesium sulfate or a double salt of magnesium and potassium sulfate is effective in supplying the element. Liming with dolomitic limestone has the advantage of adding magnesium from the dolomite while adjusting the pH upward and, at the same time, adding calcium.

Sulfur deficiencies usually occur in sandy and well-drained soils and have been most often noted in the southeastern and northwestern parts of the United States. Positive crop responses to added sulfate have been documented in 29 of the states. Several sulfur chemicals are effective, including ammonium sulfate, ammonium thiosulfate, potassium sulfate, potassium magnesium sulfate, and ordinary superphosphate.

Micronutrients

The biggest difficulty in fertilizing with micronutrients is knowing what element is needed and in what amount. Precise analytical methods will tell the analyst the concentration of each element in the soil, but these tests will not indicate how much of the elements are available for plant absorption. Such analytical tests can be calibrated in terms of crop response in greenhouse and field studies, but this is a costly and laborious process, and the results are likely to vary from one type of soil to another and from crop to crop.

Micronutrients can be applied to the soil, to the plant, or to the seed before it is planted. Iron is often applied to the leaves of plants, and molybdenum compounds are dusted onto the seeds. Micronutrients are mixed with macronutrients in many applications, but in these cases one must allow for possible chemical reactions among the ingredients in the mixture. For example, manganese sulfate is effective as a source of manganese, but not in the presence of more soluble polyphosphate fertilizers. Another complication of micronutrient fertilization is the need for uniform application. In the case of boron, which is toxic in higher concentrations, a nonuniform application of the correct amount per acre could result in areas that are deficient in boron as well as other areas in the same plot that contain poisonous concentrations of this necessary nutrient.

Considerable chemical innovation has been required to address some of the micronutrient deficiencies that have been observed in crop production. Iron is a good example. In the 1950s severe iron deficiencies appeared on thousands of acres of both acid and alkaline soils in the Florida citrus groves. Extensive applications of iron sulfate (100 lb per acre) were ineffective in correcting the problem, since the iron was bound by soil chemicals and was therefore unavailable to the plants. It occurred to researchers to introduce into the soil another complexing agent that would hold the iron with just the right tenacity; these new complexes were meant to keep the iron from the "gripping" complexing agents already in the soil and at the same time to release enough iron to the roots. Ethylenediaminetetraacetic acid (EDTA) was tried, and it worked! The iron is encircled by EDTA, which chelates the iron ion from six different directions (Fig. 19–13). The word *chelate* comes from the Greek *chela,* meaning "claw." (The iron is bonded at six different bonding sites.) A number of micronutrients are now introduced into plants by this technique.

A dolomite is a soil mixture containing both calcium and magnesium carbonate.

It is a problem to keep micronutrients uniform when they are mixed with granulated fertilizers.

Waxes and oils have been used to bind micronutrients to the surface of fertilizer particles.

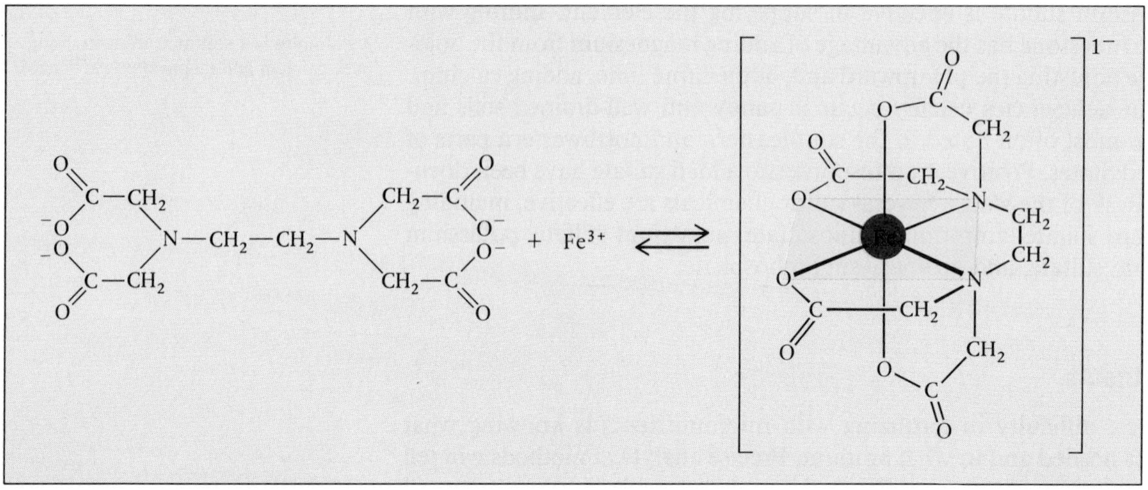

Figure 19–13 The EDTA anion
formula and the structure of the
iron-EDTA complex ion.

PROTECTING PLANTS IN ORDER TO PRODUCE MORE FOOD

The natural enemies of plants include over 80,000 diseases brought on by
viruses, bacteria, fungi, algae, and similar organisms: 30,000 species of
weeds; 3,000 species of nematodes; and about 10,000 species of plant-eating
insects. One third of the food crops in the world are lost to pests each year,
with the loss going above 40% in some developing countries. Crop losses to
pests amount to $20 billion per year.

Pesticides are the chemical answer to pest control. The 18 classes of
pesticides (Table 19–4) are fortified with more than 2600 active ingredients
to fight the battle with pests. The agricultural market consumes about three
fourths of the nearly $6.0 billion spent annually on pesticides in the United
States. Although the dollar cost is up and expected to reach $7.5 billion in
1995, the actual poundage use began a decline in 1987. There are three
reasons for a slowing in the demand for pesticides: (1) Cropland planted is
below 330 million U.S. acres, down from a high of 383 million acres in 1982.
(2) Farming is becoming more cost effective as farmers learn to use the least
amount of pesticide possible for the desired effect. (3) Farmers are becoming
more concerned with environmental and health issues. Twenty-two farmers
tested negative for insecticides in 1986 by the University of Iowa prior to the
corn season, but a number of them showed low levels of insecticides in their
urine afterwards. On a national level complacency in the use of agrichemicals is turning into concern for the environment as well as the individual. In a
1987 study by *U.S. News and World Report,* 51 food samples in each of three
cities were tested with 87 tests for pesticides. Forty-six of the samples showed
no traces of pesticides while the other five did test positive but below legal
limits set by the Environmental Protection Agency. Three of these five failed
a new negligible-risk standard that has been proposed by the National Academy of Sciences.

Without pesticides, crop production, on average, would be 20% lower
than at present. **Organic farming,** a new name for farming without chemical
fertilizers and pesticides, is gaining a significant number of followers. Or-

Figure 19–14 Unchecked insects
will consume on average up to one
third of the grower's efforts, and in
extreme cases all of the food
produced. (*The World of Chemistry,*
Program 24, "Genetic Code.")

TABLE 19–4 Classes of Pesticides

Class	Kills	Word Origin (Latin [L] or Greek [G] or English [E])
Acaricide	Mites	(G) *akari,* mite or tick
Algicide	Algae	(L) *alga,* seaweed
Avicide	Birds	(L) *avis,* bird
Bactericide	Bacteria	(L) *bacterium*
Fungicide	Fungi	(L) *fungus*
Herbicide	Plants	(L) *herbum,* grass or plant
Insecticide	Insects	(L) *insectum*
Larvicide	Larvae	(L) *lar,* mask
Molluscicide	Snails, slugs, etc.	(L) *molluscus,* soft or thin shell
Nematicide	Round worms	(G) *nema,* thread
Ovicide	Eggs	(L) *ovum,* egg
Pediculicide	Lice	(L) *pedis,* louse
Piscicide	Fish	(L) *piscis,* fish
Predicide	Predators (coyotes, wolves, etc.)	(L) *praeda,* prey
Rodenticides	Rodents	(L) *rodere,* to gnaw
Silvicide	Trees and brush	(L) *silva,* forest
Slimicide	Slimes	(E)
Termiticide	Termites	(L) *termes,* wood-boring worm

ganic farming uses only about 40% of the energy required for modern farming with synthetic chemicals and produces about 90% of the yield. The costs of energy saved in organic farming are offset by the costs of human labor, which is required by the use of natural fertilizers. Many claims are made that organic farming produces a better product for human consumption. However, there is no real evidence that these claims are generally true. Organic farming does have one clear advantage, however; it is definitely less of a threat to the environment compared with regular farming if agrichemicals are not very carefully controlled.

Insecticides

Before World War II, the list of insecticides included only a few compounds of arsenic, petroleum oils, nicotine, pyrethrum, rotenone, sulfur, hydrogen cyanide gas, and cryolite. DDT, the first of the chlorinated organic insecticides, was originally prepared in 1873, but it was not until the beginning of World War II that it was recognized as an insecticide.

The use of synthetic insecticides increased enormously on a worldwide basis after World War II. As a result, insecticides such as DDT have found their way into our environment. There is a great variety of pesticides, and their use frequently leads to severe damage to other forms of animal life, such as fish and birds. The toxic reactions and peculiar biological side effects of many of the pesticides were often not thoroughly studied or understood prior to their widespread use.

A case in point is DDT (Fig. 19–15). This insecticide, which has not been shown to be toxic to humans in doses as high as those received by

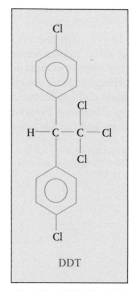

Figure 19–15 DDT [1,1-bis(*p*-chlorophenyl)-2,2,2-trichloroethane]

Figure 19–16 Aerial application of an insecticide to a field crop. This once widely used method has been largely discarded for local farming operations because of the lack of control over the spread of the expensive insecticide. (*The World of Chemistry,* Program 25, "Chemistry and the Environment.")

Half-life is the time required for half of the substance to disappear.

Figure 19–17 The massive use of DDT for insect control nearly wiped out several species of birds due to the interference of this insecticide in the formation of egg shells. (*The World of Chemistry,* Program 25, "Chemistry and the Environment.")

Dieldrin and aldrin were banned by the courts from their major uses (such as on corn) in 1974.

factory workers involved in its manufacture (400 times the average exposure), does have peculiar biological consequences. The structure of DDT is such that it is not metabolized (broken down) very rapidly by animals; instead it is deposited and stored in the fatty tissues. The biological half-life of DDT is about 8 years; that is, it takes about 8 years for an animal to metabolize half of the amount it assimilates. If ingestion continues at a steady rate, DDT will build up within the animal over time.

For many animals this buildup of DDT is not a problem, but for some predators, such as the eagle and osprey, that feed on other animals and fish, the consequences are disastrous. The DDT in the fish eaten by a bird is concentrated in the bird's fatty tissue; the bird then attempts to metabolize the insecticide by altering its normal metabolic pattern. This alteration involves the use of compounds that normally regulate calcium metabolism in the bird and are vital to the female's ability to lay eggs with thick shells. When these compounds are diverted to their new use, they are chemically modified and are no longer available for the egg-making process. As a consequence, the eggs the bird does lay are easily damaged, and the survival rate of the species decreases drastically. This process has led to the nearly complete extinction of eagles and ospreys in some parts of the United States.

The buildup of DDT in natural waters is not an irreversible process; the Environmental Protection Agency reported a 90% reduction of DDT in Lake Michigan fish by 1978 as a result of a ban on the use of the insecticide.

DDT and other insecticides such as **dieldrin** and **heptachlor** (Fig. 19–18), are referred to as **persistent pesticides.** Other substances with biodegradable structures are now substituted as much as possible; the compound **chlordan** is an example of just such a substitution. Heptachlor, because of its persistent nature, and chlordan, because of its short-term toxic effects on test animals, were banned for most garden and home use in December 1975. It is interesting to note that the structural differences between heptachlor

---

* The use of DDT was banned in the United States in 1973, although it is still in use in some other parts of the world. Over 1.8 billion kg of DDT have been produced and used.

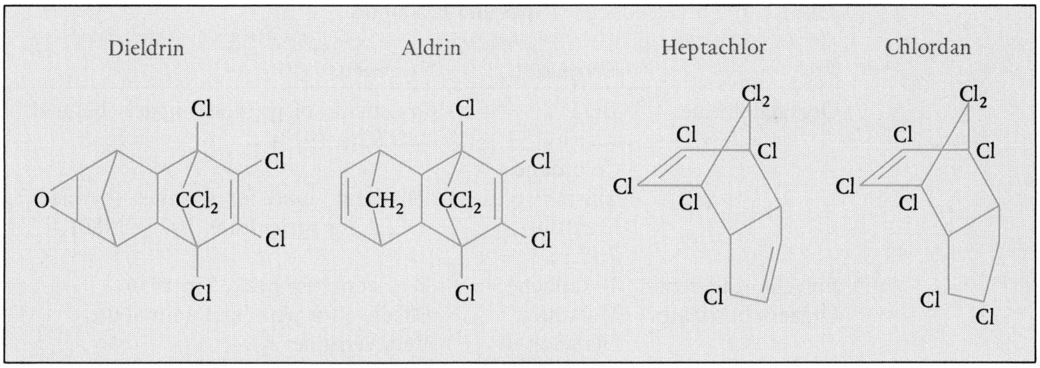

Figure 19-18   Dieldrin, Aldrin, Heptachlor, and Chlordan.

(persistent) and chlordan (short-lived) are relatively slight (look at the chlorine atom on the lower five-membered ring in Fig. 19-18).

Many insecticides are much more toxic to humans than is DDT. These include inorganic materials based on arsenic compounds, as well as a wide variety of phosphorus derivatives based on the structure indicated in part (a) of Figure 19-19. In these structures the central phosphorus atom has a double bond to Z, which is either oxygen or sulfur, and to X, which is a group that is easily split from the phosphorus. This relatively unstable P—X bond accounts for the short life of the insecticide in the environment. R and R′ represent alkyl, alkoxy, alkylthio, or amide groups. Insecticides of this type include **parathion** and **malathion,** which are effective against a large number of insects but are also very toxic to human begins. Called anticholinesterase poisons (Chapter 16), these compounds are readily hydrolyzed to less toxic substances that are not residual poisons.

The choice of solutions to the problems of pesticides is not an easy one. By using insecticides, we introduce them into our environment and our water supplies. If we fail to use them, we must tolerate malaria, plague, sleeping sickness, and the consumption of a large part of our food supply by insects. Continuing research in the development of more effective and safer insecticides is intense and new products are introduced each year. For

The goal of the insecticide quest: a selectively toxic chemical that is quickly biodegradable.

Figure 19-19   General formula for anticholinesterase poisons, and structures for commercial products parathion and malathion.

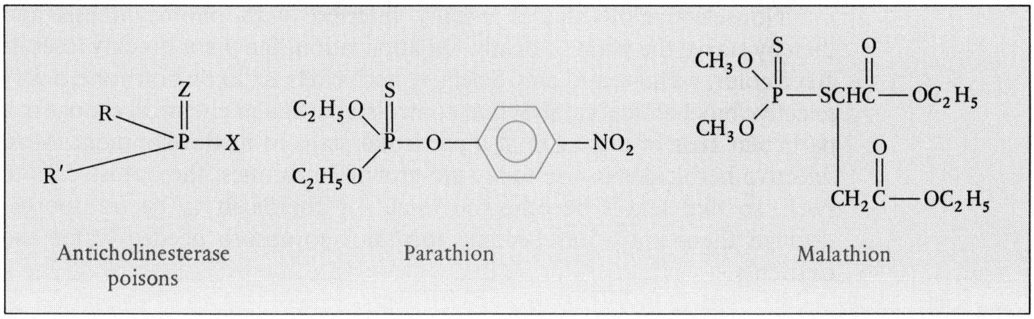

TABLE 19–5 Insecticide Types and Examples

Type	Example(s)	Comment(s)
Organochlorines	DDT	Insecticide of greatest impact (banned in U.S. in 1973)
	Chlordane	
	Aldrin	Persistent, used for termites (banned by EPA for agriculture, 1975–1980)
	Dieldrin	
	Endrin	
Polychloroterpenes	Toxaphene	Persistent, use peaked in 1976
Organophosphates	Malathion	Effective for plant and animal use
	Diazinon	Very versatile
Organosulfurs	Aramite	Choice for mite control
Carbamates	Carbaryl (Sevin)	Lawn and garden choice, low toxicity
Formamidines	Chlordimeform	Promising new group for resistant insects
Thiocyanates	Lethane	Limited use in aerosols
Dinitrophenols	Dinoseb	Choice for mildew fungi
Organotins	Cyhexatin	Most selective acaricide
Botanicals	Pyrethrum	Extracted from chrysanthemum
Synergists	Sesamex	Increases insecticide activity
Inorganics	Sulfur	Oldest insecticide
	Arsenicals	Stomach poisons
Fumigants	Methyl bromide	Kills insects in stored grain
Microbials	Heliothis virus	Specific for corn earworm and cotton ball-worm
Insect growth regulators	Ecdysone	Environmentally sound
Insect repellents	Delphene	Superior to other repellents

example, Du Pont is touting Asana® as the pyrethroid insecticide of the 1990s. The active ingredient in this new product is a chemical derivative of pyrethrum, one of the oldest natural insecticides, which was originally extracted from chrysanthemum plants (Table 19–5). The search goes on!

Herbicides

Because there are so many varieties and such large quantities of herbicides produced, (about 300,000 tons yearly in the U.S.), they are considered a separate category from pesticides.

Herbicides kill plants. They may be **selective** and kill only a particular group of plants, such as the broad-leaved plants or the grasses, or they may be **nonselective,** making the ground barren of all plant life.

Nonselective herbicides usually interfere with photosynthesis and thereby starve the plant to death. On application, the plant quickly loses its green color, withers, and dies. Selective herbicides act like a hormone, a very selective biochemical catalyst that controls a particular chemical change in a particular type of organism at a particular stage in its development. Most selective herbicides in use today are growth hormones; they cause cells to swell, so that leaves become too thick for chemicals to be transported through them and roots become too thick to absorb needed water and nutrients.

PHEROMONES

One important area in insect research is the development of pheromones, the sex attractants used by female insects to entice males. They offer a safe, nontoxic way to lure insects into traps and avoid the use of pesticides. Dr. Meyer Schwartz makes such synthetic insect pheromones in his laboratory.

An important part of Dr. Schwartz's work is confirming that he has made exact copies of the natural pheromones. He uses infrared spectroscopy to verify the structure of the molecules he's tried to duplicate in the lab.

Once an accurate copy is made, it's tried out to see if it works. A miniature wind tunnel (Fig. 19–20) lets Agriculture Department scientists see just how tantalizing their creation is. The pheromone is placed at one end of the tunnel, and a love-struck male gypsy moth flies against the wind to get it. It's bad enough that he's going to all this trouble for a synthetic chemical, but the Agriculture Department scientists have another trick up their sleeve. They can control the speed of a striped conveyor belt on the wind tunnel floor. The distracted moth sees the belt moving by and thinks he's flying full tilt toward a female. By carefully adjusting the belt speed, the researchers can stop the moth's forward progress. The speed required to stop the moth gives them a good measure of how strong a sex lure their pheromone is. When he finally is allowed to reach the end, all he finds are synthetic pheromones in a cold steel cage.

The World of Chemistry (Program 10), "Signals From Within."

Figure 19–20   A male moth is attracted to a synthetic pheromone. By placing the bait in a wind tunnel with a moving floor, the moth can be quantitatively measured in its effort to reach the "female." Thus, the efficacy of the chemical is measured. (*The World of Chemistry*, Program 10, "Signals from Within.")

The traditional method for the control of weeds in agriculture was **tillage.** Only in the early 1900s was it recognized that some fertilizers were also weed killers. For example, when calcium cyanamide (CaNCN) was used as a source of nitrogen, it was found to retard the growth of weeds. Arsenites, arsenates, sulfates, sulfuric acid, chlorates, and borates have also found use as weed killers. A typical product still in commercial use contains 40% sodium chlorate ($NaClO_3$), 50% sodium metaborate ($NaBO_2$), and 10% inert filler. These herbicides are nonselective and must be used with considerable care to protect the desired plants.

Nitrophenol was used in 1935 as the first selective, organic herbicide, Figure 19–21. It was also in the 1930s that work began on the auxins, or hormone-type weed killers. The most widely used herbicide today, 2,4-D (Fig. 19–22) came out of this work. The corresponding trichloro-compound (common name: 2,4,5-T) has also been shown to be highly effective. The only difference between it and 2,4-D is the additional chlorine atom on the benzene ring in the fifth position. Agent Orange, widely used as a defoliant during the Vietnam War, is a mixture of these two compounds. The second compound, 2,4,5-T, has been banned by the EPA because of a number of health problems associated with its use. It is probable that most of the problems were caused by the presence of an impurity, dioxin, described in Chapter 16. Dioxin is a severe poison with a toxicity equaled by few compounds. It will be interesting to see whether 2,4,5-T, which is now commer-

Tillage is land prepared for agricultural use by plowing, harrowing, etc.

Figure 19–21 Organic herbicides. (a) The first organic herbicides used were the three nitrophenols: *ortho, meta,* and *para.* (b) A widely used organic herbicide today is 2,4-dichlorophenoxyacetic acid (2,4-D).

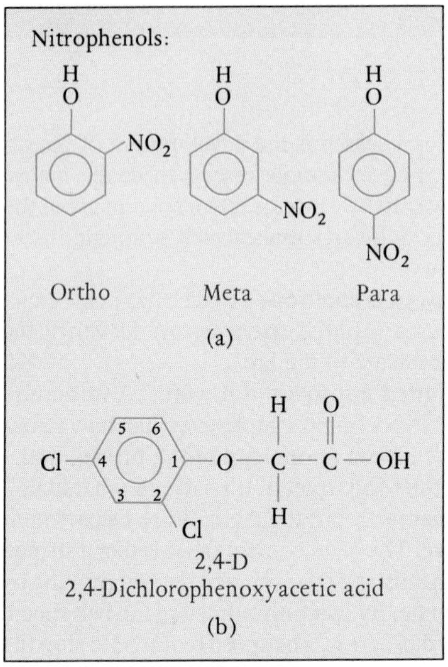

Nitrophenols:

Ortho

Meta

Para

(a)

2,4-D
2,4-Dichlorophenoxyacetic acid

(b)

No-till farming is the control of weeds by herbicides without cultivation.

cially produced free of dioxin, will be reestablished as a herbicide. Both of these compounds result in an abnormally high level of RNA in the cells of the affected plants, causing the plants to grow themselves to death.

Several different triazines have been effective as herbicides, the most famous one being atrazine (Fig. 19–22). Atrazine is widely used in no-till corn production or for weed control in minimum tillage. Atrazine is a poison to any green plant if it is not quickly changed into another compound. Corn and certain other crops have the ability to render atrazine harmless, which weeds cannot do. Hence, the weeds die and the corn shows no ill effect.

Paraquat is also used as a contact herbicide. When applied directly to susceptible plants, it quickly causes a frostbitten appearance and death. Paraquat has received considerable attention because it was used to spray illegal poppy and marijuana fields in Mexico and elsewhere. Like atrazine,

Figure 19–22 (a) 1,3,5-Triazine; (b) 2-Chloro-4-ethylamino-6-isopropyltriazine.

1,3,5-triazine

(a)

Atrazine
2-chloro-4-ethylamino-6-isopropyltriazine

(b)

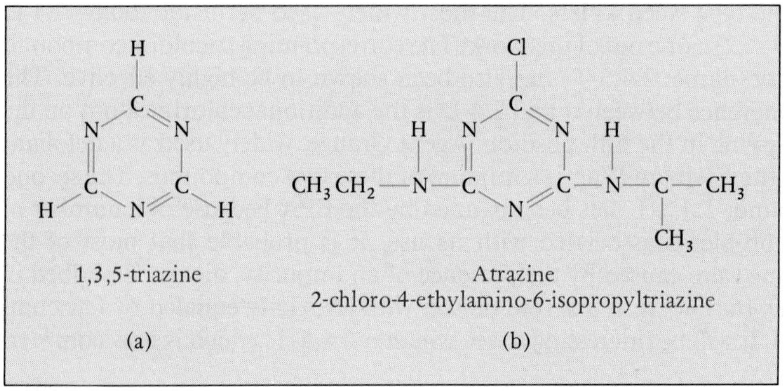

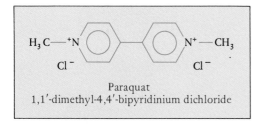

Figure 19–23 Paraquat [1,1'-dimethyl-4-4'-bipyridinium dichloride].

Paraquat
1,1'-dimethyl-4,4'-bipyridinium dichloride

paraquat has a nitrogen atom in each aromatic ring of the two-ring system (Fig. 19–23).

A typical fact sheet by a state agricultural extension service lists six formulations using atrazine for no-till corn and nine mixtures for preplant and pre-emergence applications on tilled ground. For example, one formulation for no-till corn calls for a mixture of 2 to 3 lb of atrazine, 0.25 to 0.50 lb of paraquat, and a detergent. The surfactant lowers the surface tension of the liquid spray and makes it easier for the liquid to wet and penetrate the plant surface.

The amount of energy saved by herbicides used in no-till farming is enormous. The saving of topsoil is also considerable, since the cover from the previous crop holds the soil against wind and water runoff. However, agriculturists who use herbicides are highly dependent on agricultural research institutions for the selection of herbicides that will do the desired job without harmful side effects. Such selections depend on considerable research, much of which is carried out on a trial-and-error basis on test plots. A procedure that is recommended today may be outdated by the next growing season.

GENETIC ENGINEERING FOR PEST AND DISEASE CONTROL

Armed with the ability to insert genes into organisms (see Chapters 1 and 15), it follows that we might be able to introduce genes into food plants in order to fight pests and control disease without interfering with the production or

Figure 19–24 After the laboratory work of gene modification, the test plants must be evaluated in greenhouse and field tests to see if a sought-for characteristic is obtained. (*The World of Chemistry,* Programs 26, 2, and 24, "Futures," "Color," and "Genetic Code").

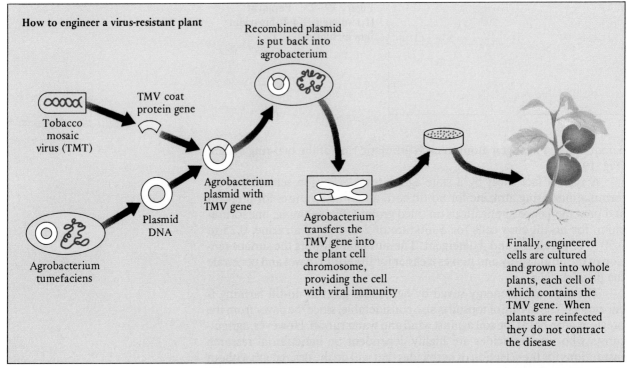

How to engineer a virus-resistant plant

Recombined plasmid
is put back into
agrobacterium

Tobacco
mosaic
virus (TMT)

TMV coat
protein gene

Agrobacterium
plasmid with
TMV gene

Plasmid
DNA

Agrobacterium
tumefaciens

Agrobacterium
transfers the
TMV gene into
the plant cell
chromosome,
providing the cell
with viral immunity

Finally, engineered
cells are cultured
and grown into whole
plants, each cell of
which contains the
TMV gene. When
plants are reinfected
they do not contract
the disease

**Figure 19–25   Steps necessary to insert a gene from tobacco mosaic virus into a
tomato plant in order to make the tomato plant resistant to a particular viral disease.**

quality of the food from that plant. Progress has been made in three areas:
growing plants that produce their own insecticide, making plants resistant to
viral infections, and matching the chemistry of a plant with a protecting
herbicide.

Monsanto Company scientists have modified tomato plants to produce
a protein characteristic of the bacterium *Bacillus thuringiensis*. This protein
is a naturally occurring insecticide. The protein has been widely used to
control the gypsy moth through external application to the affected plants.
Using recombinant DNA techniques, the gene for the production of the
protein was copied from the bacterium and spliced into the genetic structure
of the tomato plant. The modified tomato plants then produced the protein
and effectively poisoned the insects that ate the plant tissue. The obvious
follow-up problems to be studied are the extent of soil penetration by the
protein and any possible effect on the quality of the food produced.

Natural breeding methods require up to 10 years to produce plants
suitable for field testing that show resistance to a particular virus. The genetic
engineer has accomplished this in just 1 year! DNA segments from the
tobacco mosaic virus have been introduced into the genetic code of tomato
plants (Fig. 19–25). The implants caused the tomato plants to be strongly
resistant to attack by this virus. Will it be possible for the geneticists to
modify food plants for effective viral protection as nature constantly produces new viral strains? Perhaps so.

SELF-TEST 19-B

1. Most virgin soils can support crop production for a decade or more before fertilization is needed. True ( ) False ( )
2. What do the numbers 6-12-8 on a fertilizer mean? The 6 is the percentage of _____; the 12 is the percentage of _____; and the 8 is the percentage of _____ in the fertilizer.
3. Which would be properly termed a quick-release fertilizer, potassium nitrate for potassium or manure for nitrogen?
4. Would the nitrogen in ammonia be considered "fixed" nitrogen?
5. Pure ammonia under ordinary conditions is a ( ) solid, ( ) liquid, ( ) gas.
6. If ammonia is oxidized to NO_2 and dissolved in water, what two acids are produced?
7. Flooded soils cause ( ) nitrification ( ) denitrification.
8. The chemical formula for potash is _____.
9. Which is more likely to be in short supply as a plant nutrient in soils, calcium or magnesium?
10. Approximately what percentage of the food crops of the world is lost to pests each year?
11. The first chlorinated organic insecticide was _____.
12. Which of the following is not a persistent insecticide: DDT, dieldrin, heptachlor, or chlordan?
13. Which is more likely to be a hormone, a selective or a nonselective herbicide?
14. The most widely used herbicide today is _____.

MATCHING SET

_____ 1.	Humus	**a.** Solid nitrogen compound
_____ 2.	Loam	**b.** Friable material
_____ 3.	Sweet soil	**c.** Result of dissolution
_____ 4.	Percolation	**d.** Herbicide
_____ 5.	Leaching	**e.** Source of humus
_____ 6.	Compost pile	**f.** Insecticide
_____ 7.	Silicon	**g.** Provides nitrogen to soil
_____ 8.	Zinc	**h.** Central element in soil
_____ 9.	Lime	**i.** Likely in sandy soils
_____ 10.	Ammonia	**j.** Applied organic chemistry to agriculture
_____ 11.	Haber process	**k.** Calcium compound that sweetens soil
_____ 12.	Urea	**l.** Alkaline
_____ 13.	Sulfur deficiency	**m.** Micronutrient
_____ 14.	EDTA	**n.** Fortunately leached
_____ 15.	DDT	**o.** Gas; fixed nitrogen
_____ 16.	2,4-D	**p.** Passage of water
		q. Ammonia production
		r. Complexing agent for micronutrients

QUESTIONS

1. Describe the horizons that would be found in a typical soil.
2. Why is the air in soil of a different chemical composition than the air around us?
3. Which will contain more air per soil volume, a clay or a loam? Give a reason for your answer.
4. What causes a soil to be sour? Sweet?
5. If crushed limestone is spread on soil, will it raise or lower the pH of the soil? Explain.
6. How many pounds of water are typically required to produce 1 lb of food?
7. Give the approximate dates for the beginning of the world population explosion and the beginning of the application of scientific information to agriculture. In your opinion, is there a direct cause-and-effect relationship between these phenomena?
8. Guano was used as a fertilizer in colonial America. If you do not know the meaning of this word, look it up in the dictionary. What nutrients does guano add to the soil?
9. What are three principal elements in the soils of the Earth? Contrast the elements in this list to the elemental composition of the crust of the Earth.
10. Which groups of elements are first leached from soils, the alkali and alkaline-earth metals or the group III and transition metals? What is the effect of this selective leaching on soil pH?
11. What are two important roles of humus in the soil? Do leaves turned into the soil to produce humus raise or lower the soil pH?
12. If a soil is very light in color, what plant nutrient is likely to be absent?
13. The oxide of what element predominates in the soils of the Earth?
14. What three elements are obtained from the air and water as nonmineral plant nutrients?
15. What are the three primary mineral plant nutrients that are considered first in fertilizer formulations?
16. Explain the necessity of nitrogen fixation for plant growth. Give a physical, a biological, and a chemical method of nitrogen fixation.
17. Are nitrates or phosphates more easily leached from the soil? Explain.
18. Soil phosphates are in different ionic forms, depending on the pH. What are the predominant forms of ionic phosphate in sweet soils?
19. Which is more likely to be a problem in farming, a soil shortage of N, P, and K, or a shortage of Ca, Mg, and S? Give a reason for your answer.
20. Explain the numbers 6-12-6 as found on a fertilizer bag.
21. What is the danger in the use of anhydrous ammonia as a chemical fertilizer?
22. What is superphosphate? How is it made?
23. What two herbicides were formulated to produce Agent Orange? Which of these herbicides is presently banned in the United States for agricultural use?
24. How is it possible that your great-grandfather might have been happy with 25 bushels of corn per acre, whereas farmers today who cannot average over 100 bushels per acre cannot adequately compete in the corn market?
25. In the period after World War II, most farmers fertilized "enough to be sure." Farmers today are likely to have the soil analyzed and have a fertilizer formulated on prescription. What is the cause of this change? Can you see how this change in farming practice might have an effect on water pollution?
26. Investigate a no-till farming operation. What herbicides are used? How is energy saved and, at the same time, how is additional energy required? What is the effect of no-till farming on the conservation of topsoil?
27. Trace the rise and fall of the use of DDT in agriculture. Debate whether it has been more good than bad for the human race.
28. Debate the proposition that herbicides such as paraquat should be sprayed from airplanes to destroy crops grown to produce illegal drugs.
29. Contact the U.S. Department of Agriculture through the Soil Conservation Service in your area. Find out if there is documentation of a micronutrient problem in the agriculture in your state. Define the problem if one is found, and outline a chemical solution.

20

Nutrition: The Basis of Healthy Living

Both athletes and spectators require good nutrition for a chance at a healthful life. (David England, David Lipscomb University.)

Chapter 15 and part of Chapter 13 (fats) are basic to material in this chapter.

Nutrition is the science that deals with diet and health. The old saying "we are what we eat" is true in the sense that we are continually replacing parts of our bodies and that the material to make these replacements comes from our food. The skin that covers us now is not the same skin that covered us 7 years ago. The fat beneath our skin is not the same fat that was there just a year ago. Our oldest red blood cells are 120 days old. The entire linings of our digestive tracts are renewed every 3 days. Many chemical reactions are required to replace these tissues, and all of these reactions are supplied ultimately by what we eat.

Nutrition, then, is concerned with the chemical requirements of the body—the nutrients. The six classes of nutrients are carbohydrates, fats, proteins, vitamins, minerals, and water. The preparation, molecular structures, and fundamental properties of these nutrients were discussed in Chapters 13 and 15. In this chapter we focus on the health effects of too much or too little of these nutrients, why we need these nutrients (their physiological functions), general ways to assess nutritional status, and the recommended intake of nutrients.

In our society, a discussion of needed food would be incomplete without mention of food additives. Although the vitamins, minerals, and sugars added to natural foods often have nutritional value, chemicals added to preserve food, to make it taste and look better, or to give it a certain consistency, often have no nutritional value and sometimes can even be dangerous to our health. It is difficult to find food that has not been altered with palate-pleasing food additives. We shall discuss how these additives act chemically and why they are added to food.

THE SETTING OF MODERN NUTRITION

Concern about nutrition differs in different parts of the world. In the United States we no longer have diseases such as beriberi, scurvy, and rickets, which are due to the lack of a particular nutrient, but we do have cancer, heart and circulatory disorders, and so on, some forms of which are perhaps caused by an excess of certain nutrients.

By the beginning of the 20th century, the existence of carbohydrates, fats, and proteins was well recognized, and the heat-producing values of the various foodstuffs had been determined. The role of iron and several other minerals in human nutrition had also been recognized. At the turn of the century, an increasing amount of food was obtained from stores rather than from fresh farm products. In the store-bought food there were often appreciable amounts of dirt and other filth; furthermore, false or misleading labels frequently described the food. As a result of these problems, a federal food and drug law was passed in 1906. This law was replaced in 1938 by a more comprehensive act, known as the Pure Food and Drug Law. Although these laws cleaned up food and made labeling more reliable, they also caused food to be too clean chemically. In the process of food purification, nutrients such as vitamins and minerals were removed or destroyed. The replacement of needed nutrients began the food additive process.

As early as 1915, E. V. McCollum had discovered the fat-soluble group of closely related organic compounds known as vitamin A and a water-soluble collection of organic compounds known as the B vitamins. Thereafter other investigators discovered other vitamins, essential minerals, essential fats, and essential amino acids.

The usual method of testing the effect of a particular nutrient on the human body is to give an excess of the nutrient and observe its effect or to withhold the nutrient and observe the effect of its absence. However, there are large individual variations with respect to the need for many of the essential nutrients. There are also anatomical and physiological differences among animals—such as rats, dogs, cats, guinea pigs, and monkeys—used for nutritional studies. For example, human beings require vitamin C in their diets, whereas rats and dogs, which can produce vitamin C in their bodies, do not. Although valuable information can be gained from animal studies, final conclusions can be reached only about the effects of the ingested food on the species involved and, to some extent, only about the effects on the individual involved.

GENERAL NUTRITIONAL NEEDS—SOME CHEMICAL ELEMENTS

Why do we need carbohydrates, fats, proteins, vitamins, minerals, and water to sustain life? Basically, we need the elements in the nutrients to compose and supply the tissues, organs, and systems of the body. The major elements in the human body, on a weight percentage basis, are oxygen (65%), carbon (18%), hydrogen (10%), and nitrogen (3%). Practically all of these elements are in the form of water or organic compounds. Other elements present (and required) are calcium (2%), phosphorus (1%), potassium (0.35%), sulfur (0.25%), sodium (0.15%), chlorine (0.15%), and magnesium (0.05%). The total thus far equals over 99.9% of the total body weight. The rest of the body is composed of trace amounts of other elements. Table 20–1 lists all of the elements found in the human body, the total amount of each element in a 70-kg man, and average daily intake. Either a lack of proper nutrients or an excess of improper nutrients can produce **malnutrition.**

General nutritional needs are determined by one's nutritional status. There are many ways to assess the nutritional status of a human being: (1) size and weight—a person should be neither too thin nor too fat; (2) effect of stress—if a person is well nourished, stress is more bearable; (3) intelligence— undernourished and malnourished individuals are dull and unresponsive; (4) ability to reproduce—undernourished individuals are sometimes unable to reproduce; and (5) biochemical and clinical analysis—analysis of urine, blood, appearance, weight change, posture, and other chemical and medical tests and observations. Items 1 and 5 are used most often to determine nutritional status because they are most reliable (do not involve many variables) and are determined quickly.

How is nutritional status assessed?

Elements in the body are combined into many chemical compounds in a precise fashion to build the various tissues and organs of the body. Knowledge of the composition and chemistry of these organs and tissues is needed

TABLE 20-1 Chemical Elements in the Adult Human Body*

Element	Total Amount in Body (mg)	Daily Intake (mg)	Element	Total Amount in Body (mg)	Daily Intake (mg)
Aluminum	61	34	Manganese	12	3.7
Antimony	*ca.* 8	*ca.* 50	Mercury	15	0.015
Arsenic	18	1	Molybdenum	9	0.3
Barium	22	*ca.* 0.8	Nickel	10	0.4
Beryllium	0.036	0.012	Niobium	*ca.* 120	0.6
Bismuth	*ca.* 0.2	0.02	Nitrogen	1,800,000	16,000
Boron	20	1.3	Oxygen	43,000,000	3,500,000
Bromine	200	7.5	Phosphorus	780,000	1400
Cadmium	50	0.15	Potassium	140,000	3300
Calcium	1,000,000	1000	Radium	3×10^{-8}	$(1-7) \times 10^{-15}$
Carbon	16,000,000	300,000	Rubidium	680	2.2
Cesium	1.5	0.01	Selenium	15	0.15
Chlorine	95,000	5000	Silver	0.8	0.07
Chromium	*ca.* 6	0.15	Sodium	100,000	*ca.* 5000
Cobalt	*ca.* 1.5	0.30	Strontium	32	1.9
Copper	72	3.5	Sulfur	140,000	850
Fluorine	2600	1.8	Tellurium	9	0.6
Gold	*ca.* 9	—	Tin	*ca.* 16	4
Hydrogen	7,000,000	3,500,000	Titanium	9	0.9
Iodine	13	0.20	Uranium	0.09	0.002
Iron	4200	15	Vanadium	*ca.* 10	2
Lead	120	0.44	Zinc	2300	13
Lithium	80	2	Zirconium	*ca.* 450	4.2
Magnesium	19,000	340			

* The approximate values given here refer to a 70-kg man. (Adapted from W. S. Snyder, M. J. Cook, E. S. Nasset, L. R. Karhausen, G. P. Howells, and I. H. Tipton: *Report of the Task Group on Reference Man,* ICRP Pub. 23. Oxford, England, Pergamon Press, 1975.)

to understand the nutritional requirements of humans. For example, a person's nutritional status can be determined by examination of that person's liver, fat tissue, and muscle tissue. If one consumes an excess of alcohol and has a poor diet, the liver grows and the cells become fatty. Fat tissue in humans can vary from a low of 13% to a high of 70% of body weight; the amount of fat tissue depends on the amount and type of food consumed, the age of the person, and certain inherited traits. Muscle tissue ranges from 25% to 45% by body weight. A weightlifter may possess a large percentage of muscle tissue. As one grows older, the amount of muscle tissue generally decreases. Despite common belief, the amount of muscle tissue is not determined by the amount of protein eaten but rather by the amount of exercise.

A set of dietary goals proposed by the U.S. Senate Select Committee on Nutrition and Human Needs in 1976 is popularized and emphasized by many nutritional groups of today. The goals proposed are 30% fat, 12% protein, and 58% carbohydrate (38% starch, 10% naturally occurring sugars, and 10% sucrose). These goals are significantly lower in fat and sucrose and higher in starches than the average diet of U.S. citizens of today which is 42% fat, 12% protein, and 46% carbohydrate (20% starch, 6% naturally occurring

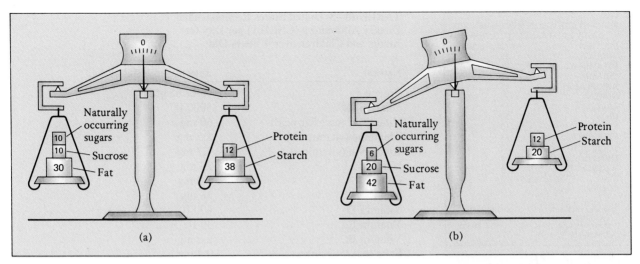

(a)   (b)

Figure 20-1   Successful nutrition means eating a balanced diet. Scientific studies recommend the percentages shown in a. However, the average citizen of the United States consumes the percentages shown in b, which are low in naturally occurring sugars and complex starches, high in fats and refined sugar.

sugars, and 20% sucrose). Why do we need less fat and sucrose (refined sugar) and more starch or complex carbohydrates? Do we need less of all types of fats? These kinds of questions will be answered as we discuss the individual nutrients.

Nutrient Requirements — RDA and USRDA

Recommended Dietary Allowance (RDA) per day

A specific list of recommendations for nutrient intake is the Recommended Dietary Allowance (RDA), published by the Food and Nutrition Board of the National Academy of Sciences and the National Research Council. Sample data are shown in Table 20-2. RDAs are given according to age, sex, energy requirements, pregnancy, and lactation (milk production). The RDA is the intake level of a nutrient that should ensure adequate nutrition among as large a percentage of the population as possible, with account being taken of the effects of some stress and biochemical differences. The listing was first published in 1968 and was revised in 1974 and 1980. For most nutrients, each revision produced slightly lower values.

TABLE 20-2 A Selection of Recommended Dietary Allowances (RDA), 1980

Group	Age (Years)	Weight (Pounds)	Height (Inches)	Protein (g)	Vitamins				Calcium (mg)
					D (μg)	C (mg)	E (mg)	B_6 (mg)	
Infants	0.5–1	20	28	18	10	4	35	0.6	540
Children	4–6	44	44	30	10	6	45	1.3	800
Males	19–22	154	70	56	7.5	10	60	2.2	800
Females	19–22	120	64	44	7.5	8	60	2.0	800
Pregnant women				+30	+5	+2	+20	+0.6	+400
Lactating women				+20	+5	+3	+40	+0.5	+400

PERCENTAGE OF U.S. RECOMMENDED
DAILY ALLOWANCES (U.S. RDA)

PROTEIN	2	10
VITAMIN A	15	20
VITAMIN C	25	25
THIAMIN	35	40
RIBOFLAVIN	35	45
NIACIN	35	35
CALCIUM	**	15
IRON	10	10
VITAMIN D	10	25
VITAMIN B_6	35	35
FOLIC ACID	35	35
PHOSPHORUS	4	15
MAGNESIUM	2	6
ZINC	2	6
COPPER	2	4

*WHOLE MILK SUPPLIES AN ADDITIONAL 30
CALORIES, 4 g FAT, AND 15 mg CHOLESTEROL
**CONTAINS LESS THAN 2% OF THE U.S. RDA OF
THIS NUTRIENT.

The numbers on one cereal box for the percentages of the USRDA requirements per serving of the cereal. The left column is for 1 cup of cereal; the right column is for cereal plus ½ cup of skim milk.

TABLE 20–3 United States Recommended Dietary Allowances (USRDA) per Day for Adults and Children over 4 Years Old

Nutrient	Amount
Protein	45 or 65 g*
Vitamin A	5000 IU
Vitamin C (ascorbic acid)	60 mg
Thiamine (vitamin B_1)	1.5 mg
Riboflavin (vitamin B_2)	1.7 mg
Niacin	20 mg
Calcium	1.0 mg
Iron	18 mg
Vitamin D	400 IU
Vitamin E	30 IU
Vitamin B_6	2.0 mg
Folic acid (folacin)	0.4 mg
Vitamin B_{12}	6 μg
Phosphorus	1.0 g
Iodine	150 μg
Magnesium	400 mg
Zinc	15 mg
Copper	2 mg
Biotin	0.3 mg
Pantothenic acid	10 mg

* 45 g if protein quality is equal to or greater than milk protein, 65 g if protein quality is less than milk protein.

1 μg (microgram) = 10^{-6} g
1 mg (milligram) = 10^{-3} g

The set of recommendations used for labeling products is the United States Recommended Dietary Allowances (USRDA), published by the Food and Drug Administration (FDA). The USRDA is based on the 1968 version of the RDA but lists slightly higher values than the RDA and does not give ranges of recommendations (Table 20–3). On packages of products, such as cereal boxes, the datum for each nutrient is usually the percentage of the USRDA recommendation contained in one serving of the product. International Units (IUs) are commonly listed for vitamins A, D, and E on packaged, enriched foods. A remnant from the past, IUs were employed before chemical analyses on the specific vitamins were possible. In modern units, 1 IU of vitamin A is 0.344 μg of crystalline vitamin A acetate; therefore, 5000 IU of vitamin A is the same as 1.72 mg of vitamin A acetate. Another unit sometimes used for vitamin A is the **retinal equivalent** (RE); 1 RE equals 5 IU. For vitamin D, 1 IU is 0.025 μg of cholecalciferol; the USRDA of 400 IU is the same as 0.01 mg. For vitamin E, 1 IU is 1 mg; the USRDA of 30 IU is the same as 30 mg of vitamin E. The more important information on a cereal box is the percentage of the USRDA provided by one serving of the cereal.

Dietary guidelines for Americans

A generalized summary of dietary recommendations was proposed in 1980 by the Department of Health and Human Services and the United States Department of Agriculture:

1. Eat a variety of foods.
2. Maintain ideal body weight.

3. Avoid too much saturated fat and cholesterol.
4. Eat foods with adequate starch and fiber.
5. Avoid excess sugar.
6. Avoid excess sodium.
7. Drink alcoholic beverages in moderation.

On July 27, 1988, Dr. C. Everett Koop, then surgeon general of the United States, released the *Surgeon General's Report on Nutrition and Health*. The 712-page report presents scientific evidence and conclusions relating diet and specific chronic diseases. The recommendations of the 1980 report were reiterated and emphasized in the 1988 report; there was no disagreement between the two reports. Highest priority in the 1988 report was placed on lessening fat in our diets. In addition to the points listed above, the 1988 report recommends:

1. Fluoride should be available in water supplies or in other sources for prevention of tooth decay.
2. Calcium intake should be increased for adolescent girls and adult women.
3. Iron intake should be improved, especially for children, adolescents, and women of child-bearing age, by eating lean red meats, fish, certain beans, iron-rich cereals, and whole-grain products. Improved iron intake is of special concern for low-income families.

Other recommendations of the 1988 report include improvements in food labels, food service programs, and food products; and expanded nutrition research and monitoring. An interesting chapter deals with diet fads and fraud.

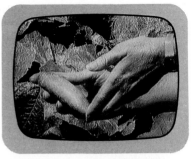

Vegetables are a good source of complex starch. (*The World of Chemistry*, Program 10, "Signals from Within.")

Figure 20–2 A pole vaulter needs high strength, speed, and energy to soar. (*The World of Chemistry*, Program 14, "Molecules in Action.")

Caloric Needs

Heat is needed by humans to maintain body temperature at about 37°C (98.6°F under the tongue) and to energize endothermic chemical reactions. The principal source of this heat is the oxidation of some fats and carbohydrates. The oxidation of proteins and various other exothermic reactions provides the rest of the heat for the body.

Since fat contains less oxygen per gram than do carbohydrates, the exothermic reaction of a fat with oxygen to form carbon dioxide and water produces more heat per gram of fat. Some specific oxidations representative of the three major sources of energy from food are the oxidation of glucose (a sugar),

Some fats, such as those made from linoleic acid, are required for chemical reactions in the body other than oxidations.

$$C_6H_{12}O_6 + 6\ O_2 \longrightarrow 6\ CO_2 + 6\ H_2O + 670\ \text{kcal (3.7 kcal/g glucose)}$$
Glucose Oxygen Carbon Water
       Dioxide (Liquid)

A food, or dietary, calorie is a kilocalorie (kcal).

the oxidation of a fatty acid (representing a fat),

$$C_{16}H_{32}O_2 + 23\ O_2 \longrightarrow 16\ CO_2 + 16\ H_2O + 2385\ \text{kcal (9.3 kcal/g palmitic acid)}$$
Palmitic Acid

and the oxidation of an amino acid (representing a protein),

$$2\ C_3H_7O_2N + 6\ O_2 \longrightarrow CO(NH_2)_2 + 5\ CO_2 + 5\ H_2O + 416\ \text{kcal (2.3 kcal/g alanine)}$$
Alanine   Urea

TABLE 20-4 Calorie Data for Fats, Carbohydrates, and Proteins

Foodstuff	kcal/g	RDA	Actually Consumed in U.S. Daily Diet	kcal Produced by Daily Intake	Percentage of Daily Calorie Output
Fat	9	—	100–150 g	900–1350	30–50
Carbohydrate	4	—	300–400 g	1200–1600	35–45
Protein	4	46–56 g (10 oz meat)	80–120 g	320–480	10–15

Calorie values for specific foods are listed in Table 20–6.

For the purposes of comparison, the oxidation of ethanol is shown below:

$$C_2H_5OH + 3\,O_2 \longrightarrow 2\,CO_2 + 3\,H_2O + 327\ \text{kcal} \ (7.1\ \text{kcal/g ethanol})$$

Ethanol

The values from Table 20–4 can be used to calculate the calorie value of a food, if the composition of the food is known and if the complete quantity of each food is oxidized. For example, if a steak is 49% water, 15% protein, 0% carbohydrate, 36% fat, and 0.7% minerals, then 3.5 oz (about 100 g) would produce about 384 kcal, or 384 food Cal.

Nutrient	Weight kcal/g	Total
Water	49 g × 0 kcal/g =	0 kcal
Protein	15 g × 4 kcal/g =	60
Carbohydrate	0 g × 4 kcal/g =	0
Fat	36 g × 9 kcal/g =	324
Minerals	0.7 g × 0 kcal/g =	0
	Total	384 kcal

Calorie values of most foods are calculated by this method, and these are the values that are listed in diet books.

A 150-lb person who skis at 10 mph requires 600 kcal/h. (Gordon Wiltsie.)

TABLE 20-5 Approximate Energy Expenditure by a 150-lb Person in Various Activities

Activity	Energy (kcal/h)	Activity	Energy (kcal/h)
Bicycling, 5.5 mph	210	Roller skating	350
13 mph	660	Running, 10 mph	900
Bowling	270	Skiing, 10 mph	600
Domestic work	180	Square dancing	350
Driving an automobile	120	Squash and handball	600
Eating	150	Standing	140
Football, touch	530	Swimming, 0.25 mph	300
tackle	720	Tennis	420
Gardening	220	Volleyball	350
Golf, walking	250	Walking, 2.5 mph	210
Lawn mowing (power mower)	250	3.75 mph	300
Lying down or sleeping	80	Wood chopping or sawing	400

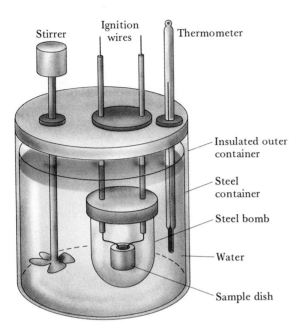

Figure 20–3 A diagram of a bomb calorimeter, which can be used to determine the caloric content of food. A sample of food is weighed and placed in a heavy steel, closed vessel, the bomb. The bomb is filled with oxygen to about 30 times atmospheric pressure. The food is ignited and the heat produced is measured by the rise in temperature of the water surrounding the bomb. From the temperature rise and the known heat absorption of the equipment, the heat emitted per gram of food ignited can be calculated.

Stirrer

Ignition wires

Thermometer

Insulated outer container

Steel container

Steel bomb

Water

Sample dish

Physical activity is one way to consume the foods that would be stored as fat (use them to produce heat and energy). Some average calorie values for various activities are listed in Table 20–5.

Energy spent for normal maintenance activities of the body is the **basal metabolic rate (BMR)**. These maintenance activities include the beating of the heart, breathing, maintenance of life in each cell, maintenance of body temperature, and the sending of nerve impulses from the brain to direct these automatic activities. Energy for these activities must be supplied before energy can be taken for digesting food, running, walking, talking, or other activities. BMR, usually expressed as kilocalories per hour, is defined as the energy spent by a body at rest after a 12-h fast. To get a rough estimate of your BMR (kcal/day), multiply your weight (in pounds) by 10.

The BMR can be affected by many factors. An increased BMR can come from anxiety, stress, lack of sleep, low food intake, congestive heart failure, fever, increased heart activity, and the ingestion of drugs, including caffeine, amphetamine, and epinephrine. A decreased BMR can result from malnutrition, menopause, inactive tissue due to obesity, and low-functioning adrenal glands.

Basal metabolic rate is energy required to do nothing willfully.

SELF-TEST 20–A

1. The six classes of nutrients are _____, _____, _____, _____, _____, and _____.

2. The most abundant element in the human body is _____, and the most abundant mineral is _____.

3. Which set of recommendations for nutrient intake is printed on packaged products?

4. The following statement is ( ) completely true, ( ) partially true, ( ) completely false: All humans are sufficiently alike for nutrition studies to be extrapolated from one to the other.

5. The nutrient that produces the most heat per gram is _____.

6. A food calorie is the same value as _____ kilocalorie(s), and it will raise the temperature of 1000 g of water _____ degree(s) C.

7. The amount of heat required to operate the body at rest is _____, which in kcal is about ten times your _____.

8. Which nutrient does the Surgeon General's report strongly recommend be decreased in the American diet?

INDIVIDUAL NUTRIENTS—WHY WE NEED THEM IN A BALANCED AMOUNT

Proteins

Properties and structures of proteins are given in Chapter 15.

Histidine is required for wound healing.

See Table 15–1, and relate the essential amino acids to the letters TV Till PM HA.

Of the some 22 amino acids identified in human protein, 10 are considered essential, meaning that the human body cannot synthesize them and therefore must obtain them from ingested food. Infants require arginine because they cannot make it fast enough to have a supply for both protein synthesis and urea synthesis. The lack of an essential amino acid in one meal is not supplied by an excess of the amino acid in another, since excess amino acids are not stored very long except in functioning proteins. If proteins are eaten at only one meal per day, the liver must store a full day's supply from that one meal.

Functions in the Human Body

Hormone (Greek hormaein, "to set in motion, spur on"): a chemical substance, produced by the body, that has a specific effect on the activity of a certain organ.

Humans must have proteins to provide the structural compounds for repairing and maintaining muscles and most organs. Proteins are part (the apoenzyme) of the some 80,000 known enzymes. Some hormones, transport molecules (such as hemoglobin and transferrin), antibodies, and fibrinogin (for blood clotting) contain proteins.

Daily Needs

The USRDA requirement for protein is 46 g for young female adults and 56 g for adult males.

Proteins are nearly the only source of nitrogen in the diet. An adult male has about 10 kg of protein, about 300 g of which is replaced daily. Part of the 300 g is recycled, and part comes from intake. Various studies indicate that, on the average, 25 to 38 g of high-quality protein (as in meat, chicken eggs, and cow's milk) or 32 to 42 g of lower quality proteins (as in corn and wheat) are required in the daily diets of healthy adult humans in order to maintain nitrogen equilibrium in the body. The average daily intake has remained near 100 g of protein per person since 1910, although there was a small drop in protein intake during the Depression of the 1930s. Methionine is the essential amino acid required in the greatest amount (2 g of the total of 7.1 g of all of the essential amino acids). Protein is lost in urine (as urea, a by-product of protein metabolism), fecal material, sweat, hair and nail cuttings, and sloughed skin.

Food Sources

Table 20–6 lists some foods that are relatively high in protein content. Generally speaking, persons who are reasonably well fed and eat meat, fish, eggs, or dairy products every day have no worry about their protein intake.

Protein-Related Problems

If the diet does not contain the proper balance of the essential amino acids, protein synthesis is curtailed. **Kwashiorkor** (pronounced: kwash-ee-OR-core) is a protein-deficiency disease. To Ghanaians, who named the disease, *kwashiorkor* originally meant "the evil spirit that infects the first child when the second child is born." Traditionally, a first Ghanaian child would nurse until a second child was born, at which time the first child would be weaned from the mother's protein-rich milk to a starchy, protein-poor sustenance of gruel. The first child would then begin to sicken and would often die within a few years. If kwashiorkor set in around the age of 2, by the time the child was 4, growth would be stunted, hair would have lost its color, skin would be patchy and scaly with sores, the belly, limbs, and face would be swollen by the collection of fluid in intercellular spaces (edema), and the child would sicken easily (because of a lowered supply of antibodies) and would be weak, fretful, and apathetic. If a child with kwashiorkor is given nutritional therapy before the disease has progressed to its last stages, the chances of recovery are good.

If proteins occupy too large a proportion of dietary intake (and carbohydrates too low a proportion), some of the excess amino acids are consumed for energy or are converted into glucose and then glycogen in the liver. If the condition continues, **uremia,** a form of blood poisoning resulting from liver failure, can occur. Uremia is marked by nausea, vomiting, headache, vertigo, dimness of vision, coma or convulsions, and a urinous odor of the breath and perspiration. Protein metabolized completely forms ammonia. The liver converts the ammonia to urea, some of which is used to make "nonessential" amino acids. Excess urea is excreted in the urine. If insufficient carbohydrates are available to be oxidized for energy, too much urea from the oxidation of proteins is sent to the kidneys, which may become overworked, and uremia sets in.

Athletes often believe that their diets should be extremely high in protein. Athletes do, in fact, use a little more protein than nonathletes, but only a little more—perhaps 10%. Most young people's diets already contain about twice as much protein as they can possibly use to build muscle; the excess is used for energy—a purpose any other energy nutrient could serve just as well, and less expensively. Cells do not respond to what is given to them, but rather they select from what is offered when they need nutrients in order to perform. So the way to make muscle cells grow is to put a demand on them, that is, to make them work.

Meats such as beef and poultry are the primary sources of protein in the diets of most human beings. (*The World of Chemistry*, Programs 5 and 13, "A Matter of State" and "The Driving Forces.")

TABLE 20–6 The Approximate Percentages of Carbohydrates, Fats, Proteins, and Water in Some Whole Foods as Normally Eaten

Food	Water	Protein	Fat	Carbohy-drates	kcal/100 g
Vegetables					
Spinach, raw	90.7	3.2	0.3	4.3	26
Collard greens, cooked	89.6	3.6	0.7	5.1	33
Lettuce, Boston, raw	91.1	2.4	0.3	4.6	25
Cabbage, cooked	93.9	1.1	0.2	4.3	20
Potatoes, cooked	75.1	2.6	0.1	21.1	93
Turnips, cooked	93.6	0.8	0.2	4.9	23
Carrots, raw	88.2	1.1	0.2	19.7	42
Squash, raw summer	94.0	1.1	0.1	4.2	19
Tomatoes, raw	93.5	1.1	0.2	4.7	22
Corn kernels, cooked on cob	74.1	3.3	1.0	21.0	91
Snap beans, cooked	92.4	1.6	0.2	5.4	25
Green peas, cooked	81.5	5.4	0.4	12.1	71
Lima beans, cooked	70.1	7.6	0.5	21.1	111
Red kidney beans, cooked	69.0	7.8	0.5	21.4	118
Soybeans, cooked	73.8	9.8	5.1	10.1	118
Meats and Fish					
Lean beef, broiled	61.6	31.7	5.3	0	183
Beef fat, raw	14.4	5.5	79.9	0	744
Lean lamb chops, broiled	61.3	28.0	8.6	0	197
Lean pork chops, broiled	69.3	17.8	10.5	0	171
Lard, rendered	0	0	100.0	0	902
Calf's liver, cooked	51.4	29.5	13.2	4.0	261
Beef heart, cooked	61.3	31.3	5.7	0.7	188
Brains	78.9	10.4	8.6	0.8	125
Chicken, whole, broiled	71.0	23.8	3.8	0	136

Food	Water	Protein	Fat	Carbohy-drates	kcal/100 g
Meats and Fish (cont'd)					
Cod, raw	81.2	17.6	0.3	0	78
Salmon, broiled	63.4	27.0	7.4	0	182
Freshwater perch, raw	79.2	19.5	0.9	0	91
Oysters, raw	84.6	8.4	1.8	3.4	66
Grains and Grain Products					
Wheat grain, hard	13.0	14.0	2.2	69.1	330
Brown rice, dry	12.0	7.5	1.9	77.4	360
Brown rice, cooked	70.3	2.5	0.6	25.5	119
Whole-wheat bread	36.4	10.5	3.0	47.7	243
White bread	35.8	8.7	3.2	50.4	269
Whole-wheat flour	12.0	14.1	2.5	78.0	361
White cake flour	12.0	7.5	0.8	79.4	364
Dairy Products and Eggs					
Milk, whole	87.4	3.5	3.5	4.9	65
Yogurt, whole-milk	89.0	3.4	1.7	5.2	50
Ice cream	62.1	4.0	12.5	20.6	207
Cottage cheese	79.0	17.0	0.3	2.7	86
Cheddar cheese	37.0	25.0	32.2	2.1	398
Eggs	73.7	12.9	11.5	0.9	163
Fruits, Berries, and Nuts					
Apples, raw	84.4	0.2	0.6	14.5	58
Pears, raw	83.2	0.7	0.4	15.3	61
Oranges, raw	86.0	1.0	0.2	12.2	49
Cherries, sweet	80.4	1.3	0.3	17.4	70
Bananas, raw	75.7	1.1	0.2	22.2	85
Blueberries, raw	83.2	0.7	0.5	15.3	62
Red raspberries, raw	84.2	1.2	0.5	13.6	57
Strawberries, raw	89.9	0.7	0.5	8.4	37
Almonds	4.7	18.6	54.2	19.5	598
Pecans	3.4	9.2	71.2	14.6	689
Walnuts	3.5	14.8	64.0	15.8	651

$$CH_3CH_2CH_2CH_2CH_2CH = CHCH_2CH = CHCH_2CH_2CH_2CH_2CH_2CH_2CH_2C \diagdown_{OH}^{O}$$

Linoleic acid ($C_{18}\Delta_{9,12}$)

$$CH_3CH_2CH = CHCH_2CH = CHCH_2CH = CHCH_2CH_2CH_2CH_2CH_2CH_2C \diagdown_{OH}^{O}$$

Linolenic acid ($C_{18}\Delta_{9,12,15}$)

$$CH_3CH_2CH_2CH_2CH_2CH = CHCH_2CH = CHCH_2CH = CHCH_2CH = CHCH_2CH_2CH_2C \diagdown_{OH}^{O}$$

Arachidonic acid ($C_{20}\Delta_{5,8,11,14}$)

Figure 20-4 **The essential fatty acid, linoleic acid. The two other fatty acids are required by the human body but can be synthesized from linoleic acid. The symbol Δ indicates the positions of the double bonds in the unsaturated acids.**

Fats

A **lipid** is an organic substance that has a greasy feel and is insoluble in water but soluble in organic solvents. Lipids include neutral fats and oils, waxes, steroids, phospholipids, and similar compounds. When we refer to fats, we are usually referring to triglycerides, composed of one glycerol molecule esterified by three fatty acid molecules. Ninety-five percent of the lipids in the diet are triglycerides. The other 5% are phospholipids (lecithin is an example) and steroids (cholesterol is the major one in food).

The only truly essential fatty acid is **linoleic acid;** it cannot be synthesized in the body and therefore must be eaten in the diet. Arachidonic and linolenic fatty acids were thought to be essential until it was discovered that they can be synthesized in the body from linoleic acid (Fig. 20-4).

Properties and structures of fats and fatty acids are given in Chapter 13.

Triacylglycerol is used by some sources for the older term triglyceride.

Structures of glycerol and triglycerides are in Chapter 13, and the structure of cholesterol is in Chapter 21.

Functions in the Human Body

Fats are essential structural parts of cell membranes. They are the most concentrated source of food energy in our diets; they furnish 9000 cal/g when oxidized for energy, compared with glucose, which furnishes about 3800 cal/g. Stored fat is a potential energy source for the body. Fats insulate thermally, pad the body, and are packing material for various organs. Fatty, or adipose, tissue is composed mainly of specialized cells, each featuring a relatively large globule of triglycerides.

Fatty acids are precursors of prostaglandins; the oxidation of arachidonic acid yields several possible prostaglandins. The prostaglandins function as bioregulators to influence the action of certain hormones and nerve transmitters. They inhibit high blood pressure, ulcer formation, and inflammation.

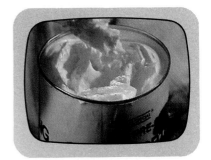

Solid animal fat. (*The World of Chemistry*, Program 9, "Molecular Architecture.")

Daily Needs

The daily consumption of fat in the United States has risen continuously from about 125 g per person in 1910 to about 155 g per person today. The fat in today's diet is about 40% saturated, 40% monounsaturated, and 20% polyunsaturated. There are no RDA or USRDA recommendations for fats. Dietary Guidelines for Americans recommend that we avoid too much fat, saturated fat, and cholesterol:

Choose low-fat protein sources such as lean meats, fish, poultry, dry peas and beans; use eggs and organ meats in moderation; limit intake of fats on and in foods; trim fats from meats; broil, bake, or boil—don't fry; read labels for fat contents.

Other recommendations suggest a decrease in saturated and monounsaturated fats and an increase in polyunsaturated fats so that each type of fat will represent one-third of the total consumed.

Food Sources

Table 20–7 lists the fat content of some foods. The fat in the edible portions of the food, as normally eaten, is difficult to ascertain, since people trim varying amounts of fat off their food before eating it. Animal fats (oils) are high in saturated and monounsaturated fatty acids. Vegetable fats (oils) have a high percentage of polyunsaturated fatty acids. The ratios of saturated and unsaturated fatty acids in common fats and oils are listed in Table 20–7. Note that coconut oil is very high in saturated fatty acids although coconut oil is a vegetable oil. Since coconut oil is cheap, it is used in some foods. Those people who tend toward artery blockage from atherosclerotic plaque (see below and Chapter 21) should avoid foods that have coconut oil.

TABLE 20–7 Ratios of Saturated and Unsaturated Fatty Acids from Common Fats and Oils*

Oil or Fat	Percentage of Total Fatty Acids by Weight		
	Saturated	*Monounsaturated*	*Polyunsaturated*
Coconut oil†	93	6	1
Corn oil	14	29	57
Cottonseed oil	26	22	52
Lard	44	46	10
Olive oil	15	73	12
Palm oil	57	36	7
Peanut oil†	21	49	30
Safflower oil	10	14	76
Soybean oil	14	24	62
Sunflower oil	11	19	70

* Recall that *saturated* means a full complement of hydrogen (only C—C single bonds); *monounsaturated* means one C=C double bond per fatty acid molecule; *polyunsaturated* means two or more C=C double bonds per molecule of fatty acid. The chief unsaturated fatty acid is linoleic acid.
† Although derived from vegetable rather than animal fats, both coconut oil and peanut oil have been associated recently with hardening of the arteries when combined with a high cholesterol intake.

Nearly all diets supply enough linoleic acid to meet the needs of the human body. Pork (lard) and chicken fat contain mostly linoleic acid. Even in a totally fat-free diet, 1 tsp (5 g) of corn oil supplies the daily need of linoleic acid. Two of the highest sources of cholesterol are brains (2.5 g/100 g) and egg yolk (1.15 g/100 g).

Problems Associated with Eating Fats

Too much fat in the diet can lead to obesity. After digestion of a fat, if the components glycerol and fatty acids are not used otherwise, they are re-synthesized in the liver into fats and stored as such.

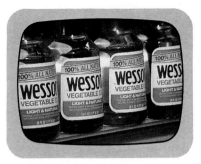

Liquid vegetable oil. (*The World of Chemistry*, Program 9, "Molecular Architecture.")

According to standards set by life insurance companies, anyone over the "ideal" weight for his or her age group is considered obese. According to these standards, 10% to 25% of teenagers and 25% to 50% of adults in the United States are overweight. One test for obesity is the skinfold test, in which one measures the thickness of a big pinch of skin (skinfold) on the back of the upper arm, the back, or the waist; in this test fatness is defined as a skinfold thicker than 1 in. The problem with both of these criteria and other such tests is that each individual has his or her own "set point," or ideal weight for optimum health, which may appear fat (or skinny) to others. The set-point weight depends on heredity, bone density, muscle conformations, occupation, number of fat cells, and other such factors and may well differ from other general recommendations.

The number of adipose (fat) cells is fixed by adulthood. The more adipose cells one has, the harder it is to lose weight.

When dieting, there is one immutable law: for each pound of body weight lost, there must be an expenditure of 3500 kcal. There is no magic escape from this principle. The kilocalories can be used by the body for the BMR or for additional activities, but activity is required to use the energy. A person or a machine moving your muscles does not expend nearly as much of your energy as you do moving yourself; you must move your muscles to expend the energy. Before participating in a diet plan, it may be wise to consult an authority who can examine the plan to verify that it is based on sound, scientific studies. One such authority is the Committee on Nutritional Misinformation of the Food and Nutrition Board, National Academy of Sciences, National Research Council.

If there is too much fat in the diet and too little carbohydrate, **ketosis** can develop. Ketosis is the combination of high blood-ketone levels **(ketonemia)** and ketones in the urine **(ketonuria)**. Ketones are formed when fats are broken down to form glucose when no glucose is readily available to the body. Glycerol derived from fat destruction forms pyruvate, then glucose. The fatty acids form ketones (the simplest being acetone [CH_3COCH_3]) and keto-acids (such as acetoacetic acid [CH_3COCH_2COOH] and the substance in largest amount, 3-hydroxybutanoic acid, [$CH_3CH(OH)CH_2COOH$]). One noticeable characteristic of ketosis is "acetone" breath. Although some cells in the body can use ketones for fuel, other cells must have glucose. There is a small amount of ketones in the blood normally. Excess ketones and keto-acids lead to **ketoacidosis,** a potentially fatal condition.

If too little fats are eaten, especially the essential fatty acid, one can develop coarsened, sparse hair and eczema (a skin disease characterized by lesions, watery discharge, and crusts and scales).

Ingestion of hydrogenated polyunsaturated fats (oils) causes problems. Hydrogenation is used to convert oils into solid fats to make margarine, cooking fats, and similar products. However, hydrogenation of vegetable oils (liquid fats) decreases some of the double bonds, forms unnatural *trans*-fatty acids from natural *cis*-fatty acids, and moves double bonds around to form conjugated structures. The *trans*-fatty acids are not metabolized in the human system, but they can be stored for the life of the individual because the *trans*-fatty acids are "straight" molecular structures and pack together into solids like the saturated acids. Both stack similar to sticks of wood. By contrast, the *cis*-fatty acids are bent like broken but attached sticks of wood and do not pack into solid masses easily. Another problem with *trans*-fatty acids is their effect on cholesterol. *Cis*-fatty acids participate but *trans*-fatty acids do not participate in the formation of cholesterol esters, the principal storage form of cholesterol. By the inactivity of *trans*-fatty acids, cholesterol is free to roam through the bloodstream and become entangled in a blockage in an artery, **atherosclerosis.** Both cholesterol and straight-structured fatty acids (saturated and *trans*) are implicated as the ingredients for forming the solid blockages in an artery, particularly damaging to the small arteries serving the heart. The site of blockage is often a damaged or rough lining of the artery caused by heredity, chemicals (smoking is implicated), or physical injury.

Problems with saturated fatty acids can be decreased by eating less animal fat, butter, and lard. Problems with *trans*-fatty acids are reduced by avoiding processed vegetable fats. Although the cause is not understood, the effect of the use of fish oils as a source for oils and fats is well documented. Eskimos and persons living in fishing villages in Japan have much less arterial blockage and consequently fewer heart attacks than are common in other parts of the world. On the down side, those people who consume large amounts of fish oil bruise and bleed easily.

Another solution to problems with fats is to use "fake fats," substances that give the fat or oil taste and consistency, but are not fats or oils. Some of the fake fats either now on the market or about ready to be sold commercially are Simplesse, Olestra, emulsified starch, and emulsified protein. Simplesse was developed by the makers of NutraSweet, G. D. Searles Co., and is a butter substitute with only 15% of the real fat calories in butter. An ounce of cheese made from Simplesse instead of butter fat drops from 82 Cal to 36. Simplesse is made from egg white or milk proteins. It feels creamy on the tongue. However, Simplesse is not suitable for cooking because heat makes it tough. Olestra, developed by Proctor and Gamble Co., is a sucrose polyester

Figure 20–5   Saturated fatty acids (a) and *trans*-unsaturated fatty acids (top structure in b) are linear and tend to pack into solid masses. *Cis*-unsaturated fatty acids (c and bottom structure in b) are bent, do not pack as well as straight structures, and tend to be liquid at normal temperatures. (*The World of Chemistry*, Program 9, "Molecular Architecture.")

(a)

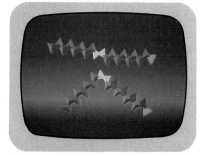

(b)

(c)

made from sugar and fatty acids. Olestra contributes no calories because it is indigestible, and it can be used in cooking. Emulsified starch is now used in Hellman's light mayonnaise and salad dressing. It is not used for cooking, but can be used in ice cream and yogurt. Emulsified protein (Unilever) is an emulsion of gelatin and water that cuts margarine calories in half. It can be used for baking and light frying.

When saturated or unsaturated fats are used in cooking, they should not be heated to temperatures at which they smoke. Under these conditions fats produce toxic peroxides, and unsaturated fats can polymerize.

All fats and oily foods should be smelled for rancidity when the package or bottle is opened and should be returned to the store for credit if there is any evidence of rancidity.

Carbohydrates

Carbohydrates in foods include digestible simple sugars (glucose, fructose, galactose), disaccharides (sucrose, maltose, lactose), and polysaccharides (amylose, amylopectin, glycogen). Indigestible carbohydrates consumed include cellulose, hemicellulose, lignin, plant gums, sulfated polysaccharides, carrageenan, and cutin.

Properties and structures of carbohydrates are given in Chapter 15.

Functions in the Human Body

The only beneficial function of digestible carbohydrates is to provide energy at the rate of approximately 4 kcal/g of glucose oxidized. Excess digestible carbohydrates are stored first as glycogen, principally in the liver; further excesses are converted into fats and stored as such. The indigestible carbohydrates serve as roughage in the diet, along with bran and fruit pulp.

Table 20–6 lists the carbohydrate contents of some foods.

Daily Needs

A daily caloric intake of 2000 kcal would require the ingestion of about 500 g of glucose (or its equivalent). Fats and proteins are also oxidized for energy, so less digestible carbohydrate is required. Daily consumption of digestible carbohydrate has declined from 500 g per person in 1910 to 380 g per person in the 1980s. The decline in total amount of carbohydrates includes a rise in the amounts of refined sugars: 150 g in 1910 to 200 g in the 1980s. There is no RDA or USRDA for carbohydrates. The 1980 Dietary Guidelines for Americans and the 1988 Surgeon General's Guidelines recommend that we decrease our ingestion of concentrated sweets (candy, soft drinks, cookies, etc.) and substitute starches (complex carbohydrates), fresh fruits, and fiber.

Problems Associated with Carbohydrates

Problems can be encountered when the glucose level is too low in the blood and when there is too little roughage in the diet; the general medical term for too low a concentration of glucose in the blood is **hypoglycemia**. Persons with hypoglycemia need a regular intake of sugar to avoid lows, characterized by an inability to think clearly, emotional disturbances, and a feeling of general indisposition.

Diseases associated with lack of dietary fiber (roughage) are appendicitis, diverticular disease (herniation of the mucous membrane lining of a tubular organ), and benign or malignant tumors of the colon and rectum. In

One source of roughage in the diet is fruit pulp. Eating whole oranges provides delectable juice, vitamin C, and needed roughage. (James W. Morgenthaler.)

Dyspepsia is the name applied to chronic indigestion.

Insulin escorts glucose to the fatty cell membrane.

Diabetes:
Type I: insulin deficient
Type II: plenty of insulin but the insulin receptors on the cells are inactive

Insulin, a protein, is hydrolyzed (digested) in the gastrointestinal tract if taken orally.

Some substances such as insulin have to be injected to save life (shown above). Other substances are injected daily by millions for a short-term "high" and sooner or later result in the destruction of the human system. (*The World of Chemistry*, Program 24, "Genetic Code.")

the gastrointestinal tract, fiber absorbs water, swells, facilitates regular bowel movements, and prevents the stagnation of foods, particularly refined foods, in the intestines. Bran is a good source of dietary fiber. Fruit pulp is another good source if the whole fruit is eaten. Some bakers incorporate wood or cotton cellulose into their high-fiber foods.

Clinical problems associated with a diet rich in carbohydrates, particularly refined sugar, are obesity, dyspepsia, atherosclerosis, and diabetes mellitus.

Diabetes mellitus is characterized by elevated blood glucose levels, multiple hormonal and metabolic disturbances in the secretion of insulin and growth hormone, thirst, hunger, weakness, low resistance to infection, slowness to heal, and, in later stages, blindness and coma. About 15% of older people have diabetes, but the disease is generally rare in parts of the world where the people eat no refined or processed food. Although several types of diabetes are due to a decrease in glucose metabolism, diabetes mellitus is caused by too little insulin due to defective *islets of Langerhans* in the pancreas. In this condition the pancreas does not produce sufficient insulin because of hereditary incapacity, a disease or injury that destroyed part or all of the pancreas, interruption of the mechanism of insulin production by lack of proper reactants or the presence of toxins, or too much demand for insulin invoked by too high glucose intake. If the pancreas produces even a small amount of insulin, diabetes can be held at bay by a diet low or absent in sugar. If the pancreas produces no insulin, daily injections of insulin are required.

The yearly sugar consumption in the United States in 1750 was 2 lb per person; today it is 110 to 135 lb per person. Sixty percent of the sugar comes from sugar cane; the other 40% comes from sugar beets. The sugar content of some commercially processed foods is given in Table 20–8.

Some problems with refined sugar are due to the removal of required nutrients during the refining process and the dumping of too much refined sugar into the bloodstream too quickly. The production of white sugar (almost pure sucrose) removes all other nutrients, such as B vitamins, manganese, and chromium, which generally coexist in natural foods with sucrose in the appropriate amounts for proper metabolism in the human body.

TABLE 20–8 Refined Sugar Added to Some Commercially Processed Foods*

Food	Sugar (%)
Cherry Jello	82.6
Coffeemate	65.4
Shake'N Bake, Barbecue Style	50.9
Wishbone Russian Dressing	30.2
Heinz Ketchup	28.9
Sealtest Chocolate Ice Cream	21.4
Libby's Peaches (in Heavy Syrup)	17.9
Skippy Peanut Butter	9.2
Coca Cola	8.8

* According to *Consumer Reports,* "Too Much Sugar" (March 1978) pp. 136–142; percents by weight.

Therefore, a large intake of refined sugar means that the B vitamins and certain minerals must be obtained from another food source. Brown sugar supplies more minerals than white sugar because brown sugar is darkened with molasses, the residue from sugar cane that is rich in essential minerals.

Unrefined and unprocessed sugar is often contained in cellular structures, which are not easily digestible. Sugar such as from sugar cane or from apples goes to the bloodstream much more slowly than refined sugar. The slower transfer allows the body to metabolize the sugar for energy more efficiently and avoids both the buildup of glucose in the bloodstream and the consequent storage of excess glucose as fat. If we obtain our sugar from an unrefined, unprocessed source, we are likely to eat less sugar; for example, it takes four apples to supply the same amount of sugar present in one 12-oz cola drink.

Milled white flour constitutes about two thirds of all of the grains consumed by humans in the United States. The milling of flour removes some lysine and fat and reduces the fiber content to 10% of that found in wheat grains. Vitamins are reduced to between 10% and 50% of the original content, mostly by removal of wheat germ. Enriched flour has some of the removed vitamins and minerals added back. Some millers add white paper pulp to replace the roughage provided by the removed bran. Some people supplement white flour with wheat germ, though this practice is inadvisable, since wheat germ separated from wheat degrades nutritionally and becomes rancid easily.

Refined flour, like refined sugar, is digested more quickly and more completely than unrefined (whole-wheat) flour. Therefore, digested, refined flour and refined sugar have less bulk. This causes the stomach to empty more slowly, and the digested food stays in the intestines longer. As a result, wastes in the intestines have more time to effect any toxicity they might have. Studies have shown that it takes 40 to 140 h for refined carbohydrates to pass through the human body and only 15 to 45 h for a traditional diet (not supplemented by Western foods) to pass through the body.

The process of bleaching to make white flour also destroys vitamins. Commonly used bleaching agents are chlorine dioxide and benzoyl peroxide. The destruction of vitamins can be averted by the use of unbleached flour.

Unrefined carbohydrates contain factors that destroy the bacteria that consume carbohydrates and produce acid and storage carbohydrates (tooth plaque).

SELF-TEST 20-B

1. Proteins compose the _____ part of enzymes.
2. Kwashiorkor is a disease caused by _____ deficiency.
3. Muscles are built primarily by eating excess proteins. True (   ) False (   )
4. Uremia is caused by excess protein intake and the excretion of _____ .
5. A common phospholipid in food is _____ , and a common steroid in food is _____ .
6. Most of the lipids in the diet are _____ .
7. The one essential fatty acid is _____ .

8. Natural sources rich in linoleic acid are _____ and _____ .

9. Ketosis is caused by too much _____ and too little _____ in the diet.

10. Eczema is caused by too little _____ in the diet.

11. Heating fats until they smoke can produce _____ .

12. All ingested carbohydrates are digestible. True ( ) False ( )

13. Refined _____ and refined _____ are absorbed more quickly than unrefined carbohydrates.

14. ( ) *Trans-* ( ) *Cis-* fatty acids have straight structures and stack together to make a solid mass.

Minerals

Carbon, hydrogen, oxygen, and nitrogen are supplied by organic fats, carbohydrates, and proteins.

As nutrients, minerals are substances that are needed for good health and that contain elements other than C, H, O, and N. On vitamin and mineral supplement labels and elsewhere, nutrient elements are called minerals, the two terms being used interchangeably in nutrition. Most of the elements needed for nutrition are obtained from soluble inorganic salts either in foods or in food supplements. Magnesium is an exception in that it is obtained primarily from organic chlorophyll.

The required inorganic nutrients can be grouped into two classes. Calcium, phosphorus, and magnesium are required in amounts of 1 g or more per day. Trace elements such as chromium, chlorine, cobalt, copper, fluorine, iodine, iron, manganese, molybdenum, nickel, selenium, sulfur, vanadium, and zinc are needed in milligram or microgram quantities each day.

That the human body needs minerals is borne out both by the functions of the minerals and by the effects of mineral deficiencies. Table 20–9 lists the functions and deficiency effects of some minerals, as well as a few food sources for each.

The nutrient minerals have varied functions, including as components of enzymes, as structural components (calcium and phosphorus in bones and teeth), in electrolyte balance in body fluids, and as transport vehicles (iron in hemoglobin transports oxygen; iron and cobalt transport electrons in electron transport cycles). Not only does the human body need minerals for its functions, but the minerals must be maintained in balanced amounts, with no deficiencies and no excesses. Many of the body's minerals are excreted daily in the feces, urine, and sweat and must therefore be replenished. For most of the elements, the amount excreted each day is very nearly the amount ingested.

One way to ensure ingestion of an ample supply of each mineral nutrient, particularly the trace nutrients, is to eat a variety of whole foodstuffs grown in different places. Mineral supplements are also available.

The role of calcium in regulating the heartbeat is discussed in Chapter 21.

Calcium slows down the heartbeat by increasing electrical resistance across nerve membranes. The movement of potassium and sodium ions across the membrane is constrained, and the nerve impulse rate is thus decreased. Calcium is metabolized in the body by a hormone synthesized from calciferol (vitamin D). The calciferol also brings about synthesis of a

TABLE 20-9 Need for and Sources of Some Nutrient Minerals

Element (Amount Ingested per Day)	USRDA*	Function	Deficiency Effects	Food Sources
Calcium (Ca) (1.0 g)	1 g	In bone apatite, collagen	Bone dissolution	Milk products
Chromium (Cr) (0.15 mg)	—	In collagen, glucose tolerance factor	Increase in cholesterol	Honey
Cobalt (Co) (0.30 mg)	—	In vitamin B_{12}	Wasting disease, pernicious anemia	Meat, eggs
Copper (Cu) (3.5 mg)	2 mg	Coenzymes (4)	Anemia, infertility	Seafood, spinach, molasses (copper water pipes)
Fluorine (F) (1.8 mg)	—	Fluorapatite	—	Toothpaste, drinking water
Iodine (I) (0.20 mg)	0.15 mg	In thyroxine hormone	Goiter, cretin children	Kelp, seafood, iodized salt
Iron (Fe) (15 mg)	18 mg	In hemoglobin, myoglobin	Anemia	Meat, eggs, raisins
Magnesium (Mg) (340 mg)	400 mg	In bone, dentine, coenzyme	Circulatory and mental problems, red nose of alcoholics	Green vegetables
Manganese (Mn) (3.7 mg)	—	In melanin (skin pigment) coenzymes	Growth, skeletal, reproduction abnormalities	Spinach, beans, grains
Molybdenum (Mo) (0.3 mg)	—	Coenzymes (2)	Cancer of esophagus, sexual impotency	Legumes, liver
Phosphorus (P) (1.4 g)	1 g	In bone, teeth, ATP	Weak bones, lack of energy	Beans, grain, meat, milk
Selenium (Se) (0.15 mg)	—	Coenzymes, growth stimulator	Degeneration of skeletal muscles	Wheat
Sulfur (S) (0.85 g)	—	In amino acids	Unhealthy hair, muscles	Eggs, meat, mustard
Zinc (Zn) (13 mg)	15 mg	In insulin, coenzymes; heals wounds	Dwarfism, stretch marks, painful joints, finickiness in appetite	Oysters, meat, nuts

* USRDA values for adults and children over 4 years old.

substance called calcium-binding protein (CBP), which carries calcium through the small intestine wall. Fat slows down the transfer, and lactose speeds up calcium absorption.

We could not support ourselves nor eat without calcium. The hard part of bone is a calcium compound, and the structure of teeth is mostly hydroxy-apatite, a calcium compound.

Excess calcium may lead to the formation of kidney stones, but the body has a protein, **calmodulin,** that collects excess calcium and then binds to a number of enzymes to mediate their activity. By the use of calmodulin, the body monitors the amount of calcium in the bloodstream. A possible benefit of excess calcium is that it will make a person taller.

A deficiency in calcium can occur in postmenopausal women, who produce less estrogen than premenopausal women. The estrogen suppresses bone dissolution. Further bone dissolution can be suppressed by long-term medication with estrogen, but this has produced some toxic side effects in

The structure of apatite is given and the role of fluoride in tooth decay is discussed in Chapter 22.

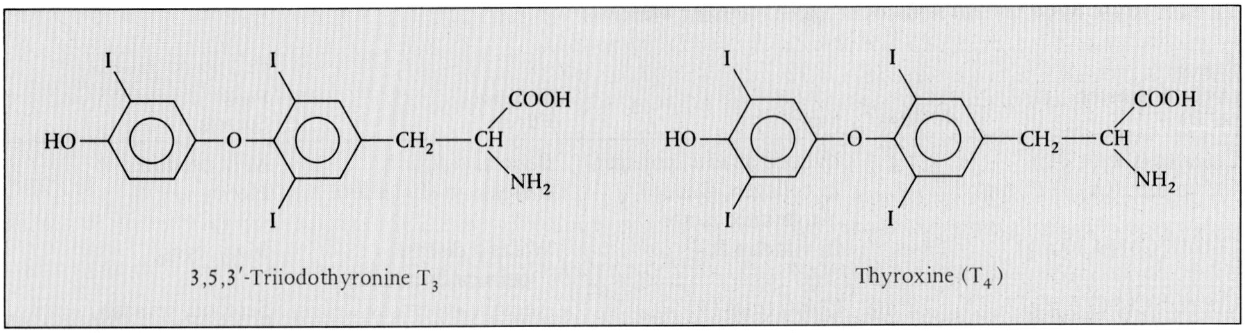

Figure 20–6   Two hormones of the thyroid gland.

some women and therefore seems unwise. Taking calcium supplements may help.

The principal function of *iodine* in the human body is the proper operation of the thyroid glands located at the base of the neck. Two of the thyroid hormones are T_3 and T_4 (Fig. 20–6). These hormones and other similar ones, collectively known as **thyroxine,** go into every cell and regulate the rate at which the cell uses oxygen. Thyroxine thus regulates the BMR and the Krebs cycle. Iodine is absolutely necessary to the production of thyroxine. If there is a deficiency of iodine, the thyroid glands sometimes swell to as large as a person's head; this swelling is called a goiter. In 1960 it was estimated that 7% of the world's population (200 million) had goiters. Treatment with iodized salt (0.1% KI) with the hormone thyroxine decreases the size of or even eliminates small goiters. Larger goiters may require surgery. Since an excess amount of iodine also causes goiters, balance is the key to health.

The lack of either vitamins B_9 or B_{12} can cause pernicious anemia.

Anemia can be caused by a deficiency of *iron,* but it can also be due to heredity, an improper level of vitamin B_6, lack of folic acid (vitamin B_9), and lack of vitamin B_{12} (pernicious anemia). Iron-deficient anemia is not necessarily fatal; a person with only 20% of the normal amount of hemoglobin still has the energy and strength to walk.

In the case of some nutrient elements, good health depends on the element's being present in the proper amount and in the *proper ratio* to one or more other elements. An example of an important ratio is the potassium/sodium ratio (K/Na ratio).

Typical values of the **K/Na ratio** are greater than 1. Some K/Na ratios for specific tissues follow: muscle, 4; liver, 2.5; heart, 1.8; brain, 1.7; and kidney, 1.0. For individual cells, potassium ions (K^+) concentrate inside the cell, whereas sodium ions (Na^+) concentrate outside in the fluid that bathes the cell. Natural, unprocessed food has high K/Na weight ratios. Fresh, leafy vegetables average a K/Na ratio of 35. Fresh, nonleafy vegetables and fruits average a ratio of 360, with extreme values of 3 for beets and 840 for bananas. K/Na ratios in meats range from 2 to 12. Thus, when such foods are eaten, the body has K/Na ratios greater than 1. However, problems occur with processed and cooked foods. Potassium and sodium compounds are quite soluble in water. During processing (and cooking, if foods are boiled), both potassium and sodium compounds are dissolved by water and discarded. The sodium is replenished by "salting" of the food (addition of sodium

Foods high in potassium (mg K/3.5 oz of food):

Raw prunes	940
Raw raisins	763
Banana (1)	740
Turkey breast	411
Boiled potato	404
Orange (1)	300
Apple (1)	165
Cottage cheese	81

chloride). Potassium is usually not added to the food. One solution to the problem is to eat unprocessed, natural food, which "naturally" has the proper K/Na ratios. Another solution is to "salt" food with a commercial product that contains both potassium and sodium, such as Morton's Lite Salt.® In summary, do not add much NaCl, if any, to food, and eat fresh vegetables and fruits high in potassum.

Normal daily urinary excretion of *sodium* is in the range of 1.4 to 7.8 g for adults. If excess sodium is not eliminated, water is retained, which may lead to edema (swollen legs and ankles). Various clinical studies have shown that increased levels of sodium raise the blood pressure of some individuals but have no effect on the blood pressure of others. The high-salt diets of 70 g NaCl per day in certain areas of Japan have traditionally produced an unusually high frequency of heart attacks. Sodium levels in the bloodstream are regulated by **aldosterone,** which is secreted from the adrenal gland. Aldosterone works in the kidney to reabsorb sodium from the urine. The secretion of aldosterone is controlled by receptors that measure salt concentration in the blood. If the blood sodium concentration is too high, less aldosterone is excreted and less sodium is reabsorbed from the urine.

Sodium is also excreted in sweat as sodium chloride. Salt concentration in sweat depends on dietary sodium intake, environmental temperature, the amount of sweating, and the degree of acclimation to the environment. Abrupt overheating is a problem for the body, involving an increase in skin and rectal temperatures, rapid beating of the heart, and a greatly increased sweat rate. After about a week of working at high temperatures, the body adapts by lowering the pulse rate and body temperature to normal levels. A high rate of sweating continues, but the concentration of NaCl in the sweat is decreased. As a person becomes acclimated to the heat, he or she needs more water but no more salt than at normal temperatures. Salt tablets may be helpful during the acclimation period but are not advisable after the body becomes acclimated. It should be noted that some persons never adapt to heat and should avoid overheating.

The word *salary* derives from the Latin *sal,* for "salt." Roman soldiers were given an allowance for salt.

Vitamins

A vitamin is an organic constituent of food that is consumed in relatively small amounts (less than 0.1 g/kg of body weight per day) and is essential to the maintenance of life. Vitamins are not synthesized in the cells of human beings; they are synthesized by plants, our principal natural source of them. Of the some million organic compounds eaten in a normal diet, only about 100 are of proper size and stability to be absorbed from the digestive tract into the bloodstream without digestion or breakdown. Vitamins are included in this group of compounds.

USRDA amounts of vitamins are given in Tables 20–3 and 20–10.

The structures of vitamins divide them into two classes: oil-soluble and water-soluble. The oil-soluble vitamins—A, D, E, F, and K—tend to be stored in the fatty tissues of the body (especially the liver). The structures of oil-soluble vitamins have nonpolar hydrocarbon chains and rings that are compatible with nonpolar oil and fat. For good health and nutrition, it is important to store enough oil-soluble vitamins, but not too much.

TABLE 20–10 Vitamin Summary Chart

Name	USRDA*	Deficiency Effect	Sources
Water-Soluble			
Thiamine (B$_1$)	1.5 mg	Beriberi	Seeds, pork, whole-wheat bread
Riboflavin (B$_2$)	1.7 mg	Cheilosis (shark skin)	Organ meats, yeast, wheat germ
Niacin (B$_3$) (nicotinic acid)	2 mg	Pellagra	Meat, yeast, legumes
Pantothenic acid (B$_5$)	10 mg	Neuromotor disturbance	Yeast, liver, eggs
Vitamin B$_6$ (pyridoxine)	2 mg	Skin lesions, anemia	Liver, nuts, wheat germ
Biotin (B$_7$)	0.3 mg	Dermatitis	Liver, yeast, grains
Folic acid (B$_9$) (folacin)	0.4 mg	Anemia, gastrointestinal changes	Green leafy vegetables, liver
Vitamin B$_{12}$ (cobalamin)	6 μg	Pernicious anemia	Intestinal bacteria, organ meats
Ascorbic acid (vitamin C)	60 mg	Scurvy	Fruits, vegetables
Oil-Soluble			
Vitamin A (retinol)	5000 IU	"Night blindness," (xerophthalmia)	Liver, fruits, vegetables
Vitamin D (calciferol)	400 IU	Rickets	Fish liver oil
Vitamin E (tocopherol)	30 IU	Lack of hemoglobin in blood cells (hemolysis)	Plant oils
Vitamin F (linoleic acid)	—	Lesions, scales (eczema)	Pork lard, fatty foods
Vitamin K (phylloquinone)	†	Blood loss	Green leafy vegetables

* USRDA values are for adults and children over 4 years old.
† There is no USRDA value for vitamin K, but an estimated need is 0.1 mg per day.

The water-soluble vitamins tend to pass through the body and are not stored readily. Water-soluble vitamins are the B group (called vitamin B complex) and C. The structures of these vitamins have polar hydroxy (—OH) and carboxyl (—COOH) groups, which are attracted to polar water. Fewer problems are caused by excessive intake of water-soluble than of oil-soluble vitamins.

Vitamin D is the most toxic of all of the vitamins; avoid excesses.

Retinol
(Oil-Soluble Vitamin A)

Ascorbic Acid
(Water-Soluble Vitamin C)

Table 20–10 lists the vitamins, their USRDAs, some food sources, and their deficiency effects. Since vitamins are synthesized by plants, a good natural source is plants that are not overcooked.

Contrary to popular belief, carrots provide the **provitamin** β-carotene, not retinol (vitamin A). The body converts β-carotene into retinol during the transfer of the provitamin through the intestinal wall. Night blindness is prevented by regeneration of rhodopsin (visual purple) from retinol. **Vitamin A** aids in the prevention of infection by barring bacteria from entering

Eat polar bear liver sparingly. Thirty grams contain 450,000 IU of retinol; continued ingestion causes peeling of the skin from head to foot.

and passing through cell membranes. The vitamin performs its sentinel duty by producing and maintaining mucus-secreting cells. Bacteria stick to the mucus and are thus trapped.

The function of **vitamin E** as an antioxidant has been well established. Vitamin E is particularly effective in preventing the oxidation of polyunsaturated fatty acids, which readily form peroxides. Perhaps this is why vitamin E is always found distributed among fats in nature. The fatty acid peroxides are particularly damaging because they can lead to runaway oxidation in the cells. Vitamin E protects the integrity of cell membranes, which contain considerable fat. In addition, vitamin E helps maintain the integrity of the circulatory and central nervous systems, and it is involved in the proper functioning of the kidneys, lungs, liver, and genital structures. Vitamin E also detoxifies poisonous materials absorbed into the body.

Vitamin E is the only vitamin destroyed by the freezing of food.

According to some theories that view aging as the cumulative effects of the action of free radicals running wild, vitamin E, with its antioxidant properties, is considered a good candidate as an agent to inhibit aging or at least to help avoid premature aging.

Is vitamin E the fountain of youth?

A free radical has one or more unpaired valence electrons. An example is the methyl group, $CH_3\cdot$.

The B group of vitamins (the **B-complex vitamins**) work together, primarily as coenzymes in biochemical reactions leading to growth and to energy production. Their place of action is in the mitochondria of the cells. Being water-soluble, the B vitamins are easily eliminated during the processing and cooking of food. The effectiveness of vitamins B_3 and B_6 is diminished in the presence of light, especially if the food is hot.

Pyridoxine (B_6), considered the "master vitamin," is involved in 60 known enzymatic reactions, mostly in the metabolism and synthesis of proteins.

Vitamin C is involved in the destruction of invading bacteria, in the synthesis and activity of interferon, which prevents the entry of viruses into cells, and in decreasing the ill effects of toxic substances, including drugs and pollutants. The question of whether vitamin C will decrease the incidence of the common cold has been studied for many years. Results of the studies show an average decrease of about 30% in illness (particularly upper respiratory infection) as a result of ingestion of vitamin C supplements. Not as well publicized or studied, vitamin A in large doses also decreases colds and the effects of colds. In avoiding or breaking colds, some persons respond better to vitamin A than to vitamin C, others respond better to vitamin C than to vitamin A, and still others respond to neither. In any case, for either vitamin to be effective, it must be taken preferably before but no later than at the early onset of a cold. It is recommended that the vitamins not be taken in combination, since this seems to prolong the cold symptoms.

English sailors are called Limeys because the British admiralty ordered a daily ration of lime juice (vitamin C) to prevent scurvy.

Two sources of vitamins: green leafy vegetables and laboratory-synthesized tablets.

DIET AND CANCER

Along with heredity and exposure to carcinogens, diet is now established as a cause of certain kinds of cancer. The National Cancer Institute asserts that at least 35% of the cancer diagnosed in the United States is caused by dietary

Figure 20–7   Areas of the female and male human bodies where food components are linked with cause of cancer.

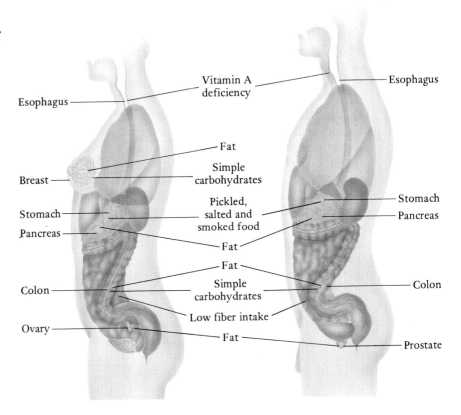

factors. Some of the correlations between certain excesses or wrong kinds of food and cancer in various areas of the human body are summarized in Figure 20–7. For example, deaths from breast cancer correlate proportionally with the dietary fat intake of the average American. Colon cancer is more prevalent with a fat and refined-carbohydrate-rich diet than with a diet rich in fiber. This has been known since 1972, when Dennis P. Burkett observed that certain African populations were protected from alimentary tract diseases (appendicitis, diverticulitis, and colon cancer) by high dietary fiber intake. Recently, some experts have concluded that moderate amounts of vitamin A and perhaps vitamins C and E reduce cancer occurrence in the esophagus.

On the positive side, some components of the diet prevent or decrease the spread of cancer. Natural enzymes for proteins (proteases) in soybeans have been shown to prevent the transformation of cells in cancer, and hence to prevent the spread of cancerous cells of various kinds. Beta-carotene in carrots can cut the incidence of epithelial (skin and covering tissue) cancer. Soil rich in selenium (Se) produces food that inhibits cancer. As with other constituents of the diet, some selenium is beneficial; too much Se is poisonous.

With over 400,000 deaths per year from cancer in the United States, what is the low-risk course to take for a cancer-free diet? The National

Cancer Institute and most experts agree that the following measures are unlikely to do harm and may be beneficial.

1. Eat less fat but enough to produce at least 20% of your calorie requirements. For the fats that are eaten, eat fats associated with the lowest risk of cancer: olive oil, rapeseed (canola) oil, and fish oils as found in fatty fish like salmon, mackerel, and bluefish. Avoid fried foods and fat-rich dressings. Use nonstick cookware to reduce the use of cooking fat.
2. Eat more fiber-rich foods such as whole-grain breads and cereals, whole grains like brown rice and whole wheat, fresh fruits and vegetables, preferably with the peel. The types of fibers believed to offer cancer-preventing benefits are the insoluble ones like bran although soluble fibers in oats, barley, apples, and grapes help to lower cholesterol in the blood. Look for labels saying 4 g of fiber per ounce in breakfast cereals.
3. Eat vegetables rich in cancer-blocking indoles frequently. These include asparagus, cabbages of all kinds, broccoli, cauliflower, kale, brussels sprouts, mustard greens, and bok choy.
4. Eat plenty of fruits and vegetables rich in β-carotene, which is converted in the body to vitamin A. Some foods rich in β-carotene are carrots, spinach, broccoli, kale, sweet potatoes, cantaloupe, cherries, papaya, apricots, tomatoes, and bell peppers. Supplements of vitamin A are not recommended because it is too easy to take a toxic overdose.
5. Eat more legumes, especially dried peas and beans, which contain substances believed to have cancer-preventing properties. They also help lower cholesterol levels in the blood.

Indole

SELF-TEST 20-C

1. Our mineral needs can be obtained by eating a _____ of _____ foodstuffs grown in _____.
2. Anemia can be caused by a deficiency of the mineral _____, which is used to make _____.
3. What mineral is required in melanin? Melanin is involved in _____.
4. A deficiency of the mineral _____ causes stretch marks in the skin.
5. Finicky eaters may have a deficiency of the mineral _____.
6. Minerals involved in transmitting a nerve impulse are _____ and _____.
7. For proper balance, the K/Na weight ratio in the body should be slightly ( ) greater than ( ) less than 1.
8. The mineral obtained from chlorophyll in green vegetables is _____.
9. The mineral with the largest weight in the body is _____, which is used mostly in the _____ and _____.
10. Calmodulin, a protein, collects excess calcium from the bloodstream and helps prevent _____ stones.

Whole grain wheat, oats, and barley supply fiber and vitamins, particularly the B-vitamins, to the diet.

11. Goiter is caused by a deficiency of _____ .

12. Vitamins are synthesized by cells of the body. True ( ) False ( )

13. Vitamins A, D, E, F, and K are _____ -soluble, whereas vitamins B-complex and C are _____ -soluble.

14. The relationship between β-carotene and vitamin A is that β-carotene is a _____ of vitamin A.

15. A good vegetable source of vitamin A is _____ ; a good animal source of the vitamin is _____ .

16. Polar bear liver is exceptionally rich in vitamin _____ .

17. Vitamin E is effective as a(n) _____ , particularly in preventing the deterioration of polyunsaturated fatty acids.

18. What vitamin is destroyed in the freezing of foods?

19. Rickets is caused by a deficiency of vitamin _____ .

20. Because it is involved in so many biochemical reactions, vitamin _____ is considered the master vitamin.

21. The most toxic vitamin is _____ .

22. The nickname "Limeys" for English sailors came from the sailors' having been given limes to prevent _____ .

23. B-complex vitamins function generally as _____ involved in the process of _____ .

24. A deficiency of thiamine, vitamin _____ , causes the disease _____ .

25. Pellagra is caused by a deficiency of the vitamin called _____ .

Figure 20–8   Between the harvested and the consumer-ready food, one often finds a large variety of food additives.

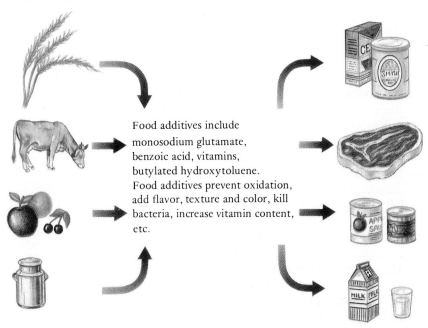

Food additives include monosodium glutamate, benzoic acid, vitamins, butylated hydroxytoluene. Food additives prevent oxidation, add flavor, texture and color, kill bacteria, increase vitamin content, etc.

FOOD ADDITIVES

Many chemicals with little or no nutritive value are added to food for a variety of reasons (Fig. 20–8). The chemicals are added during the processing and preparation of food for the purpose of preserving the food from oxidation, microbes, and the effects of metals. Food additives add and enhance flavor. They color the food, control pH, prevent caking, stabilize, thicken, emulsify, sweeten, leaven, and tenderize among other effects.

The GRAS List

The Food and Drug Administration lists about 600 chemical substances **"generally recognized as safe" (GRAS)** for their intended use. A small portion of this list is given in Table 20–11. It must be emphasized that an

TABLE 20–11 A Partial List of Food Additives Generally Recognized as Safe*

Anticaking Agents
  Calcium silicate
  Iron ammonium citrate

Acids, Alkalies, and Buffers
  Acetic acid
  Calcium lactate
  Citric acid
  Lactic acid
  Phosphates, Ca^{2+}, Na^+
  Potassium acid tartrate
  Sorbic acid
  Tartaric acid

Surface Active Agents
(Emulsifying Agents)
  Glycerides: mono- and diglycerides
    of fatty acids
  Sorbitan monostearate

Polyhydric Alcohols
  Glycerol
  Mannitol
  Propylene glycol
  Sorbitol

Preservatives
  Benzoic acid
    Sodium benzoate
  Propionic acid
    Propionates, Ca^{2+}, Na^+
  Sorbic acid
    Sorbates, Ca^{2+}, K^+, Na^+
  Sulfites, Na^+, K^+

Antioxidants
  Ascorbic acid
    Ascorbates, Ca^{2+}, Na^+
  Butylated hydroxyanisole (BHA)
  Butylated hydroxytoluene (BHT)
  Lecithin
  Sulfur dioxide and sulfites

Flavor Enhancers
  Monosodium glutamate (MSG)
  5′-Nucleotides
  Maltol

Sweeteners
  Aspartame
  Mannitol
  Saccharin
  Sorbitol

Sequestrants
  Citric acid
  EDTA, Ca^{2+}, Na^+
  Pyrophosphate, Na^+
  Sorbitol
  Tartaric acid
    NaK (tartrate)

Stabilizers and Thickeners
  Agar-agar
  Algins
  Carrageenin

Flavorings (1700)
  Amyl butyrate (pearlike)
  Bornyl acetate (piney, camphor)
  Carvone (spearmint)
  Cinnamaldehyde (cinnamon)
  Citral (lemon)
  Ethyl cinnamate (spicy)
  Ethyl formate (rum)
  Ethyl vanillin (vanilla)
  Geranyl acetate (geranium)
  Ginger oil (ginger)
  Menthol (peppermint)
  Methyl anthranilate (grape)
  Methyl salicylate (wintergreen)
  Orange oil (orange)
  Peppermint oil (peppermint)
  Wintergreen oil (wintergreen)
    (methyl salicylate)

* For precise and authoritative information on levels of use permitted in specific applications, consult the regulations of the U.S. Food and Drug Administration and the Meat Inspection Division of the U.S. Department of Agriculture.

Food additives and other contents of a popular granola bar. Note sorbitol, BHA, and citric acid, which are discussed in this chapter.

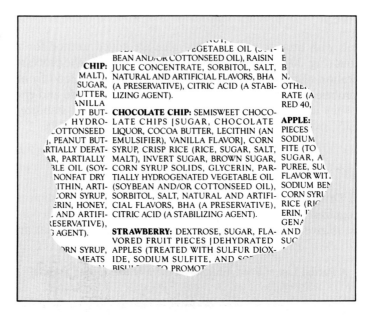

CHIP: ...NUT, ...EGETABLE OIL (...T- BEAN AND/OR COTTONSEED OIL), RAISIN MALT), JUICE CONCENTRATE, SORBITOL, SALT, SUGAR, NATURAL AND ARTIFICIAL FLAVORS, BHA ...UTTER, (A PRESERVATIVE), CITRIC ACID (A STABI- ...NILLA LIZING AGENT).

...UT BUT- , HYDRO- **CHOCOLATE CHIP:** SEMISWEET CHOCO- ...OTTONSEED LATE CHIPS [SUGAR, CHOCOLATE], PEANUT BUT- LIQUOR, COCOA BUTTER, LECITHIN (AN ...RTIALLY DEFAT- EMULSIFIER), VANILLA FLAVOR], CORN ...R, PARTIALLY SYRUP, CRISP RICE (RICE, SUGAR, SALT, ...3LE OIL (SOY- MALT), INVERT SUGAR, BROWN SUGAR, ...NONFAT DRY CORN SYRUP SOLIDS, GLYCERIN, PAR- ...ITHIN, ARTI- TIALLY HYDROGENATED VEGETABLE OIL ...CORN SYRUP, (SOYBEAN AND/OR COTTONSEED OIL), ...ERIN, HONEY, SORBITOL, SALT, NATURAL AND ARTIFI- ...AND ARTIFI- CIAL FLAVORS, BHA (A PRESERVATIVE), ...RESERVATIVE), CITRIC ACID (A STABILIZING AGENT). ...3 AGENT).

STRAWBERRY: DEXTROSE, SUGAR, FLA- VORED FRUIT PIECES [DEHYDRATED ...RN SYRUP, APPLES (TREATED WITH SULFUR DIOX- ...MEATS IDE, SODIUM SULFITE, AND S... ...BISU... TO PROMOT...

...E ...B N... OTH... RATE (A RED 40,

APPLE: PIECES ... SODIUM FITE (TO SUGAR, A PUREE, SU... FLAVOR WIT... SODIUM BE... CORN SYR... RICE (RI... ERIN, ... GEN... AND SU...

additive on the GRAS list is safe *only if it is used in the amounts and in the foods specified.* The GRAS list was published in several installments in 1959 and 1960. It was compiled from the results of a questionnaire asking experts in nutrition, toxicology, and related fields to give their opinions about the safety of various materials used in foods. Since its publication, few substances have been added to the GRAS list, and some, such as the cyclamates, carbon black, safrole, and Red Dye No. 2, have been removed.

There are more than 2500 known food additives, and many more chemicals than those that appear on the GRAS list are approved (or at least, not banned) for use as food additives by the FDA. It is quite expensive to introduce a new food additive with the approval of the FDA. Allied Chemical Corporation began research in 1964 on a new synthetic food color, Allura Red AC, which was approved by the FDA and went on the market in 1972. The cost of introducing this product was $500,000, and about half of this amount was spent on safety testing.

Preservation of Foods

Foods generally lose their usefulness and appeal a short time after harvest. Bacterial decomposition and oxidation are the prime reasons steps must be taken to lengthen the time that a foodstuff remains edible. Any process that prevents the growth of microorganisms or retards oxidation is generally an effective preservation process. Perhaps the oldest technique is the drying of grains, fruits, fish, and meat. Water is necessary for the growth and metabolism of microorganisms, and it is also important in oxidation. Dryness thus thwarts both the oxidation of food and the microorganisms that feed on it.

Chemicals may also be added as preservatives. Salted meat, and fruit preserved in a concentrated sugar solution, are protected from microorganisms. The abundance of sodium chloride or sucrose in the immediate

FOOD SPOILAGE

Food is shipped across America and around the world. As a natural consequence the rate of food spoilage is a vital factor for our global economy. As a matter of fact, the kinds of reactions that cause food to spoil are not very different from those the chemists study in the laboratory.

How do scientists study food spoilage and preservation?—We have asked Dr. Theodore Labuza, a food chemist.

The way I always like to talk about food, it's the study of messy chemistry. And the reason I say that is that, in a food which has so many different organic compounds and inorganic compounds together, there are lots and lots of different reactions that could have caused spoilage. What we can do, however, is narrow them down to several classes. For example, when you bite into an apple, you see it start to brown. That's an enzyme reaction. There are reactions that make food go rancid. Potato chips, for example, if they sit around for a long time, the fat goes rancid.

Modern techniques of food preparation and refrigeration have greatly reduced spoilage. Still, the shopper might like further proof that reactions have been retarded, that food is as fresh as possible.

One of the more interesting things that we're doing in our laboratory here, which is an application of chemical kinetics, is the study of little devices that can be used to monitor time-temperature when a food goes through a distribution cycle. Foods deteriorate at a faster rate when the temperature goes higher, and at a slower rate when the temperature goes lower. If you had a device that could be placed on a package that would essentially integrate the time-temperature exposure and show a color change that could be related to the loss of quality of the food, then, by simply picking up a package and looking at a device that may be on the package, you could tell how much more shelf life is left. Then you'd know whether to consume it.

The World of Chemistry (Program 13), "The Driving Forces."

A time-temperature monitor of food. (*The World of Chemistry*, **Program 13, "The Driving Forces."**)

environment of the microorganisms forms a **hypertonic** condition in which water flows by **osmosis** from the microorganism to its environment. Salt and sucrose have the same effect on microorganisms as does dryness; both dehydrate them.

A hypertonic solution is more concentrated than solutions in its immediate environment.

The canning process for preserving food, developed around 1810, involves first heating the food to kill all bacteria and then sealing it in bottles or cans to prevent access of other microorganisms and oxygen. Some canned meat has been successfully preserved for over a century. Newer techniques for the preservation of food include vacuum freezing, pasteurization, cold storage, irradiation, and chemical preservation.

Osmosis is the flow of water from a more dilute solution through a membrane into a more concentrated solution.

Antimicrobial Preservatives

Food spoilage caused by microorganisms is a result of the excretion of toxins. A preservative is effective if it prevents multiplication of the microbes during the shelf life of the product. Sterilization by heat or radiation, or inactivation

A preservative must interfere with microbes but be harmless to the human system—a delicate balance.

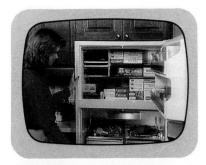

Cooling food slows down the rate of oxidation and enzyme action and thus is one way to preserve food longer. (*The World of Chemistry*, Program 17, "The Precious Envelope.")

by freezing, is often undesirable, since it impairs the quality of the food. Chemical agents seldom achieve sterile conditions but can preserve foods for considerable lengths of time.

Antimicrobial preservatives are widely used in a large variety of foods. For example, in the United States sodium benzoate is permitted in nonalcoholic beverages and in some fruit juices, fountain syrups, margarines, pickles, relishes, olives, salads, pie fillings, jams, jellies, and preserves. Sodium propionate is legal in bread, chocolate products, cheese, pie crust, and fillings. Depending on the food, the weight of the preservative permitted ranges up to a maximum of 0.1% for sodium benzoate and 0.3% for sodium propionate.

Sodium Benzoate   Sodium Propionate

Postulated mechanisms for the action of food preservatives may be grouped into three categories: (1) interference with the permeability of cell membranes of the microbes in foodstuffs, so the bacteria die of starvation; (2) interference with bacterial genetic mechanisms, so the reproduction processes are hindered; and (3) interference with intracellular enzyme activity, so that metabolic processes such as the Krebs cycle cease.

Sulfites have been used for sanitizing and preserving foods for over 2,000 years. Recently, sulfites have been identified as the cause of certain allergic reactions in asthmatics. In 1986, as a result of these toxicity findings, the FDA limited the use of sulfites on fruits and vegetables. In 1987, a regulation was introduced that requires the label of prepared foods to list the amount of sulfites if the amount is greater than 10 ppm.

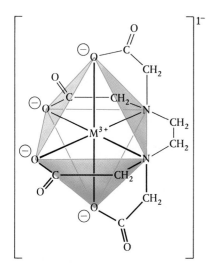

Figure 20–9   The structural formula for the metal chelate of ethylenediaminetetraacetic acid (EDTA).

Atmospheric Oxidation

Microbial activity results in oxidative decay of food, but it is not the only means of oxidizing food. The direct action of oxygen in the air, **atmospheric oxidation,** is the chief cause of the destruction of fats and fatty portions of food. Foods kept wrapped, cold, and dry are relatively free of air oxidation. An antioxidant added to the food can also hinder oxidation. Antioxidants most commonly used in edible products contain various combinations of butylated hydroxyanisole (BHA) and butylated hydroxytoluene (BHT).

BHA   BHT

To prevent the oxidation of fats, the antioxidant can donate the hydrogen atom ($H\cdot$) in the —OH group to reactive species; this effectively stops the reaction between fats and oxygen. If antioxidants are not present, the oxidation of fats leads to a complex mixture of volatile aldehydes, ketones, and acids, which cause a rancid odor and taste.

Sequestrants

Metals get into food from the soil and from machinery during harvesting and processing. Copper, iron, and nickel, as well as their ions, catalyze the oxidation of fats. However, molecules of citric acid bond with the metal ions, thereby rendering them ineffective as catalysts. With the competitor metal ions tied up, antioxidants such as BHA and BHT can accomplish their task much more effectively.

Citric acid belongs to a class of food additives known as **sequestrants.** For the most part sequestrants react with trace metals in foods, tying them up in complexes so the metals will not catalyze the decomposition or oxidation of food. Sequestrants such as sodium and calcium salts of EDTA (ethylenediaminetetraacetic acid) are permitted in beverages, cooked crab meat, salad dressing, shortening, lard, soup, cheese, vegetable oils, pudding mixes, vinegar, confections, margarine, and other foods. The amounts range from 0.0025 to 0.15%. The structural formula of EDTA bonded to a metal ion is shown in Figure 20–9.

To sequester means "to withdraw from use." The sequestering ability of EDTA accounts for its use in treating heavy-metal poisoning (Chapter 16).

Another example of chelation by EDTA is the Fe-EDTA chelate in soil nutrient application.

Flavor in Foods

Flavors result from a complex mixture of volatile chemicals. Since we have only four tastes—sweet, sour, salt, and bitter—much of the sensation of taste in food is smell (Fig. 20–10). For example, the flavor of coffee is determined largely by its aroma, which in turn is due to a very complex mixture of over 500 compounds, mostly volatile oils. These compounds are of undetermined toxicity.

Some 1700 natural and synthetic substances are used to flavor foods, making flavors the largest category of food additives.

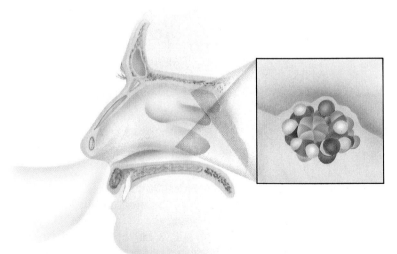

Figure 20–10 **A stereochemical interpretation of the sensation of smell. The substance fits a cavity in the back of the oral cavity. If the atoms are properly spaced, they sensitize nerve endings that transmit impulses to the brain. The brain identifies these sensations as a particular smell. A complete explanation of smell is certainly more involved than the simple idea presented here.**

Most flavor additives originally came from plants. The plants were crushed and the compound extracted with various solvents such as ethanol or carbon tetrachloride. Sometimes a single compound was extracted; more often the residue contained a mixture of several compounds. By repeated efforts, relatively pure oils were obtained. Oils of wintergreen, peppermint, orange, lemon, and ginger, among others, are still obtained in this way. These oils, alone or in combination, are then added to foods to produce the desired flavor. Gradually, analyses of the oils and flavor components of plants have revealed the active compounds responsible for the flavors. Today synthetic preparations of the same flavors actively compete with natural extracts.

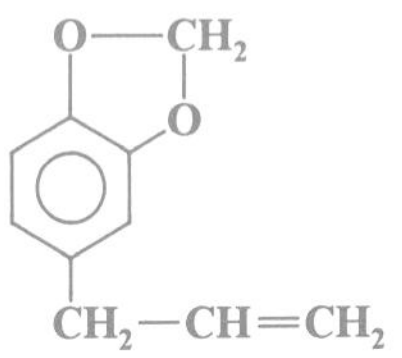

The FDA has banned some of the naturally occurring flavoring agents that used to be used, including safrole, the primary root beer flavor, found in the root of the sassafras tree.

Safrole

Flavor Enhancers

Flavor enhancers have little or no taste of their own but amplify the flavors of other substances. They exert synergistic and potentiation effects. **Synergism** is the cooperative action of discrete agents such that the total effect is greater than the sum of the effects of each used alone. **Potentiators** do not have a particular effect themselves but exaggerate the effects of other chemicals. The 5′-nucleotides, for example, have no taste but enhance the flavor of meat and the effectiveness of salt. Potentiators were first used in meat and fish but now are also used to intensify flavors or cover unwanted flavors in vegetables, bread, cakes, fruits, nuts, and beverages. Three commonly used flavor enhancers are **monosodium glutamate (MSG), 5′-nucleotides** (similar to inosinic acid), and **maltol.**

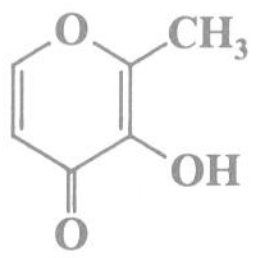

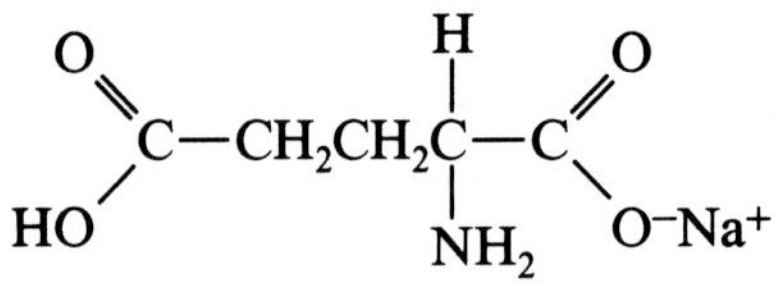

Maltol
(from pine needles)

In some people MSG causes the so-called Chinese restaurant syndrome, an unpleasant reaction characterized by headaches and sweating that usually occurs after an MSG-rich Chinese meal. Tomatoes and strawberries affect some individuals in the same way.

Monosodium Glutamate

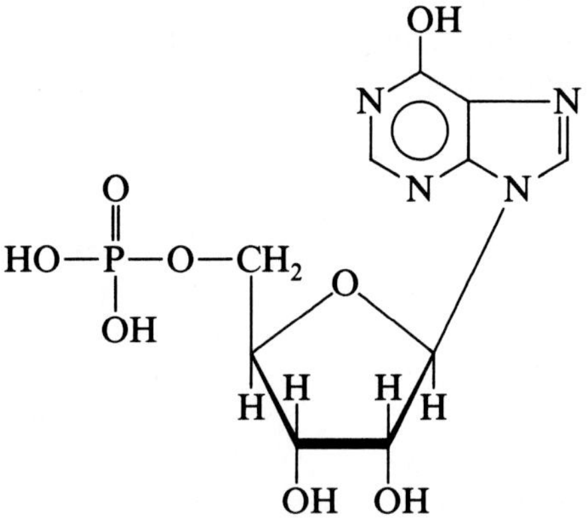

Inosinic Acid
(A 5′-nucleotide)

When MSG is injected in very high doses under the skin of 10-day-old mice, it causes brain damage. When these laboratory results were reported, considerable discussion ensued concerning the merits of MSG. National investigative councils have suggested that MSG be removed from baby foods, since infants do not seem to appreciate enhanced flavor. However, in the absence of hard evidence that MSG is harmful in the amounts used in regular food, no recommendations were made relative to its use.

MSG is a natural constituent of many foods, such as tomatoes and mushrooms.

Sweeteners

Sweetness is characteristic of a wide range of compounds, many of which are completely unrelated to sugars. Lead acetate [$Pb(CH_3COO)_2$] is sweet but poisonous. A number of **artificial sweeteners** are allowed in foods. These are primarily used for special diets such as those of diabetics. Artificial sweeteners have no known metabolic use in the body and do not need to be offset by insulin.

Insulin is a hormone that regulates glucose metabolism.

Saccharin

The most common artificial sweetener is saccharin:

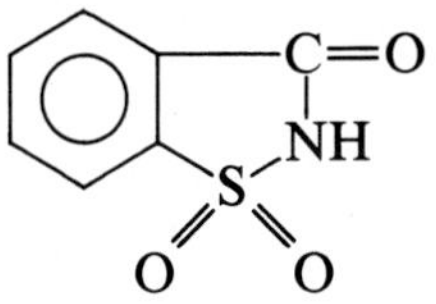

Saccharin

Saccharin is about 300 times sweeter than ordinary sugar (sucrose). When ingested, saccharin passes through the body unchanged; it therefore has no food value other than to render an otherwise bland mixture more tasty. Saccharin has a somewhat bitter aftertaste, which renders it unpleasant to some users. Glycine, the simplest amino acid, which is also sweet, is often added to counteract this bitter taste.

Saccharin is a synthetic chemical.

Both saccharin and another artificial sweetener, *dulcin,* were accidentally discovered about 100 years ago. Saccharin became the dominant artificial sweetener of the 20th century; dulcin never received much recognition and was removed from the market in 1951 by the Food and Drug Administration.

In 1878, Constantine Fahlberg was working in Ira Ramsen's laboratory at The Johns Hopkins University and was oxidizing a series of substituted compounds of toluene. In the process, he synthesized a sweet-tasting *o*-toluenesulfonamide. Although we are told today not to taste anything in the lab, it was quite common a hundred years ago for chemists to report the taste of their compounds. Saccharin was reported in an article published in 1879, later tested for toxicity, and patented in 1894.

In 1884, Joseph Berlinerblau prepared 4-ethoxyphenylurea while carrying out a series of organic reactions in the laboratory of Professor Schmitt at the University of Bern in Switzerland. In the paper reporting his work, Berlinerblau noted that his compound had a very sweet taste. He patented the synthesis of the compound in 1891. In 1893, A. Kossell, who tested the compound for toxicity, named it dulcin, meaning sweet. From the tests, it was noted that dulcin had no bitter aftertaste.

Laboratory studies have shown that high doses of saccharin cause cancer in mice. After months of consideration, the Institute of Medicine of the National Academy of Science joined the FDA in its 1978 statement that saccharin should be banned in U.S. foods. The U.S. Congress passed the ban and then suspended it pending the results of some ongoing studies in Canada. As of 1987, saccharin was still being sold, but with a warning label, "Use of this product may be hazardous to your health. This product contains saccharin which has been determined to cause cancer in laboratory animals."

Aspartame

A new entry in the sweetener market, aspartame is chemically an ester of a two-amino acid peptide with the name *N-L-α*-aspartyl-*L*-phenylalanine methyl ester and trade names of NutraSweet, Equal, and Tri-Sweet. The sweetener was approved by the FDA in 1974 and subsequently withdrawn by its maker, G. D. Searle Co., when toxicity questions were raised. When these questions were resolved in 1981, aspartame again received FDA approval.

Aspartame is about 180 times sweeter than table sugar (sucrose). The caloric value of aspartame is similar to that of proteins. The caloric intake of consumers using this product is reduced, since much smaller amounts of aspartame are needed to produce the same sweetening effect as sugar. Aspartame does not have the bitter aftertaste associated with other artificial sweeteners. It is normally metabolized in the body as a peptide.

From Aspartic Acid   From Phenylalanine Ester

Aspartame

Mannitol and sorbitol, both polyhydric alcohols, are sweeteners used in such products as sugarless gum. These sweeteners have some caloric value in the body.

Food and Esthetic Appeal

Food Colors

There are about 30 chemical substances used to color food. All are under investigation by the FDA, and some may be prohibited as the investigations progress. About half of the food colors are laboratory synthesized, and half are extracted from natural materials. Most food colors are large organic molecules with several double bonds and aromatic rings. The electrons of these conjugated structures can absorb certain wavelengths of light and pass the rest; the wavelengths passed give the substances their characteristic colors. Beta-carotene, an orange-red substance in a variety of plants that gives carrots their characteristic color, has a conjugated system of elec-

Colored organic substances often are conjugated molecules, having alternating double and single bonds in the carbon chain or ring.

trons and is used as a food color. Beta-carotene is a precursor (provitamin) of vitamin A.

Because one of the food colors, Yellow No. 5, causes allergic reactions (mainly rashes and sniffles) in an estimated 50,000 to 90,000 Americans, the FDA has required manufacturers to list Yellow No. 5 on the labels of any food products containing it.

pH Control in Foods

Weak organic acids are added to such foods as cheese, beverages, and dressings to give a mild acidic taste. They often mask undesirable aftertastes. Weak acids and acid salts, such as tartaric acid and potassium acid tartrate, react with bicarbonate to form CO_2 in the baking process.

Some acid additives control the pH of food during the various stages of processing as well as in the finished product. In addition to single substances, there are several combinations of substances that will adjust and then maintain a desired pH; these mixtures are called **buffers.** An example of one type of buffer is potassium acid tartrate, $KHC_4H_4O_6$.

Adjustment of fruit juice pH is allowed by the FDA. If the pH of the fruit is too high, it is permissible to add acid (called an **acidulant**). Citric acid and lactic acid are the most common acidulants used, since they are believed to impart good flavor, but phosphoric, tartaric, and malic acids are also used. These acids are often added at the end of the cooking time to prevent extensive hydrolysis of the sugar. In the making of jelly they are sometimes mixed with the hot product immediately after pouring. To raise the pH of a fruit that is unusually acid, buffer salts such as sodium citrate or sodium potassium tartrate are used.

The versatile acidulants also function as preservatives to prevent the growth of microorganisms, as synergists and antioxidants to prevent rancidity and browning, as viscosity modifiers in dough, and as melting-point modifiers in such food products as cheese spreads and hard candy.

Buffer solutions resist change in acidity and basicity; pH remains constant.

Small amounts of certain acids are allowed to be added to some foods.

Anticaking Agents

Anticaking agents are added to hygroscopic foods — in amounts of 1% or less — to prevent caking in humid weather. Table salt (sodium chloride) is particularly subject to caking unless an anticaking agent is present. The additive (magnesium silicate, for example) incorporates water into its structure as water of hydration and does not appear wet as sodium chloride does when it absorbs water physically on the surface of its crystals. As a result, the anticaking agent keeps the surface of sodium chloride crystals dry and prevents crystal surfaces from codissolving, and joining together.

Hygroscopic substances absorb moisture from the air.

Stabilizers and Thickeners

Stabilizers and thickeners improve the texture and blends of foods. The action of carrageenan (a polymer from edible seaweed) is shown in Figure 20–11. Most of this group of food additives are polysaccharides (Chapter 15), which have numerous hydroxyl groups as a part of their structure. The hydroxyl groups form hydrogen bonds with water to prevent the segregation of water from the less polar fats in the food and to provide a more even blend of the water and oils throughout the food. Stabilizers and thickeners are

Stabilizers and thickeners are types of emulsifying agents.

Figure 20–11   The action of carrageenan to stabilize an emulsion of water and oil in salad dressing. An active part of carrageenan is a polysaccharide, a portion of which is shown here. The carrageenan hydrogen-bonds to the water, which keeps it dispersed. The oil, not being very cohesive, disperses throughout the structure of the poly- saccharide. Gelatin (a protein) undergoes similar action in absorb- ing and distributing water to prevent the formation of ice crystals in ice cream.

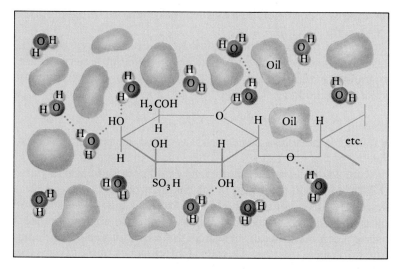

particularly effective in icings, frozen desserts, salad dressings, whipped cream, confections, and cheeses.

Surface Active Agents

Surface active agents are similar to stabilizers, thickeners, and detergents in their chemical action. They cause two or more normally incompatible (non- polar and polar) chemicals to disperse in each other. If the chemicals are liquids, the surface active agent is called an **emulsifier.** If the surface active agent has a sufficient supply of hydroxyl groups, as does cholic acid, the groups form hydrogen bonds to water. Cholic acid and its associated group of water molecules are distributed throughout dried egg yolk in a manner quite similar to that of carrageenan and water in salad dressing.

Some surface active agents have both hydroxyl groups and a relatively long nonpolar hydrocarbon end. Examples are diglycerides of fatty acids, polysorbate 80, and sorbitan monostearate. The hydroxyl groups on one end of the molecule are anchored by hydrogen bonds in the water, and the nonpolar end is held by the nonpolar oils or other substances in the food. This provides tiny islands of water held to oil. These islands are distributed evenly throughout the food.

Polyhydric Alcohols

Polyhydric alcohols are allowed in foods as humectants, sweetness con- trollers, dietary agents, and softening agents. Their chemical action is based on their multiplicity of hydroxyl groups that hydrogen-bond to water. They thus hold water in food, soften it, and keep it from drying out. Tobacco is also kept moist by the addition of polyhydric alcohols such as glycerol. An added feature of polyhydric alcohols is their sweetness. The two polyhydric alco- hols mentioned earlier for their sweetness are mannitol and sorbitol. The structures of these alcohols are strikingly similar to the structure of glucose (Chapter 15), and all three have a sweet taste.

Cholic Acid

Hydrogen bonding plays a major role in stabilizers, thickeners, surface active agents, and humectants.

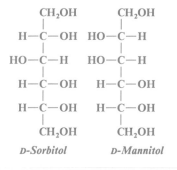

D-Sorbitol   D-Mannitol

Kitchen Chemistry

Leavened Bread

Sometimes cooking causes a chemical reaction that releases carbon dioxide gas, and the trapped carbon dioxide causes breads and pastries to rise. Yeast has been used since ancient times to make bread rise, and remains of bread made with yeast have been found in Egyptian tombs and the ruins of Pompeii. The metabolic processes of the yeast furnish gaseous carbon dioxide, which creates bubbles in the bread and makes it rise:

$$C_6H_{12}O_6 \xrightarrow[\text{from Yeast}]{\text{Zymase}} 2\ CO_2 + 2\ C_2H_5OH$$
$$\text{Glucose} \qquad\quad \text{(Gas)} \qquad \text{Ethanol}$$

When the bread is baked, the CO_2 expands even more to produce a light, airy loaf (Fig. 20–12).

Carbon dioxide can be generated in cooking by other processes. For example, baking soda (which is simply sodium bicarbonate ($NaHCO_3$), a base) can react with acidic ingredients in a batter to produce CO_2:

$$NaHCO_3 + H^+ \longrightarrow Na^+ + H_2O + CO_2(g)$$

Baking powders contain sodium bicarbonate and an added acid salt or a salt that hydrolyzes to produce an acid. Some of the compounds used for this purpose are potassium hydrogen tartrate ($KHC_4H_4O_6$), calcium dihydrogen phosphate monohydrate ($Ca[H_2PO_4]_2 \cdot H_2O$), and sodium acid pyrophosphate ($Na_2H_2P_2O_7$). The reactions of these white, powdery salts with sodium bicarbonate are similar, although the compounds all have somewhat different appearances. For example:

$$KHC_4H_4O_6 + NaHCO_3 \xrightarrow{\text{Water}} KNaC_4H_4O_6 + H_2O + CO_2(g)$$

Cooking and Precooking: "Preliminary Digestion"

The cooking process involves the partial breakdown of proteins or carbohydrates by means of heat and hydrolysis. The polymers that must be degraded if cooking is to be effective are the carbohydrate cellular wall materials in vegetables and the collagen, or connective tissue, in meats. Both types of polymers are subject to hydrolysis in hot water or moist heat. In either case, only partial depolymerization is required.

In recent years several precooking additives have become popular; the **meat tenderizers** are a good example. These are simply enzymes that catalyze the breaking of peptide bonds in proteins via hydrolysis at room temperature. As a consequence, the same degree of "cooking" can be obtained in a much shorter heating time. Meat tenderizers are usually plant products such as papain, a proteolytic (protein-splitting) enzyme from the unripe fruit of the papaw tree. Papain has a considerable effect on connective tissue, mainly collagen and elastin, and shows some action on muscle fiber proteins. On the other hand, microbial protease enzymes (from bacteria, fungi, or both) have considerable action on muscle fibers. A typical formulation for the surface treatment of cuts of beef consists of 2% commercial papain or 5% fungal protease, 15% dextrose, 2% monosodium glutamate (MSG), and salt.

Leavened bread is as old as recorded history.

About 350 million pounds of phosphates are added to foods in the United States each year: This is 20 to 25% of our phosphorus intake. Some phosphates are used as leavening agents. Sodium phosphate thickens puddings, retains juices, makes hams tender, and prevents canned milk from thickening on standing.

Figure 20–12 Yeast cells provide the catalyst zymase that produces CO_2 and causes bread to rise. (*The World of Chemistry*, Program 13, "The Driving Forces.")

Cooking starts the digestive process, although it is rarely needed for this purpose.

SELF-TEST 20-D

1. Flavor enhancers exert a(n) _____ or _____ effect on the flavors of foods.
2. Name two of the oldest means for preserving food.
3. Antimicrobial preservatives make foods sterile. True ( ) False ( )
4. Flavors result from ( ) volatile ( ) nonvolatile compounds.
5. Antioxidants are ( ) more ( ) less easily oxidized than the food into which they are placed.
6. Citric acid is an example of a(n) _____. Such compounds tie up metals in stable complexes.
7. A flavor in a food can usually be traced to a single compound. True ( ) False ( )
8. Monosodium glutamate is a(n) _____.
9. Salt is effective in preserving foods because it kills microorganisms by _____ them.
10. GRAS is an acronym for _____.
11. Which has the sweetest taste when an equal weight of each is tasted: table sugar, aspartame, or saccharin?
12. What molecular characteristic do most food colors have?
13. What acids are added to foods to lower the pH?
14. The gas released by leavening agents is _____.
15. Cooking _____ some chemical bonds.
16. Hydrogen bonding generally plays a very important role in the action of surface _____ agents.

MATCHING SET I

Match each nutrient with the disease(s) caused by a *deficiency* in the nutrient. Use all of the diseases.

Nutrient	*Disease*
_____ 1. Cobalamin	a. Xerophthalmia
_____ 2. Iodine	b. Kwashiorkor
_____ 3. Vitamin A	c. Goiter
_____ 4. Vitamin D	d. Scurvy
_____ 5. Vitamin C	e. Pernicious anemia
_____ 6. Thiamine	f. Rickets
_____ 7. Riboflavin	g. Anemia
_____ 8. Niacin	h. Finickiness in appetite, stretch marks
_____ 9. Protein	i. Pellagra
_____ 10. Fat	j. Hypoglycemia
_____ 11. Sugar	k. Eczema
_____ 12. Iron	l. Night blindness
_____ 13. Zinc	m. Beriberi
	n. Shark skin

MATCHING SET II

_____ **1.** Beta-carotene
_____ **2.** Monosodium glutamate
_____ **3.** Copper, nickel, and iron
_____ **4.** Sodium benzoate
_____ **5.** Potentiator
_____ **6.** Mineral
_____ **7.** Mannitol

a. Nutrient supplement in food
b. Food color
c. Flavor enhancer
d. Catalyze oxidation of fats
e. Antimicrobial preservative
f. Exaggerates some chemical effects
g. Sequestering agent
h. Sweetener
i. pH adjuster

QUESTIONS

1. What two factors influenced the U.S. government to pass laws on food? What problem was caused by the government regulations?
2. If all of our blood cells are renewed each 120 days, must new nutrients be ingested to make the new cells? Explain.
3. Distinguish between RDA and USRDA recommendations on the basis of the following:
   a. Which has the higher and which has the lower recommendations?
   b. Which has the greater breakdown with respect to age, sex, and so on?
   c. Which is used on package labels?
4. What are two sources of dietary fiber, and why is fiber important in the diet?
5. What problem is caused
   a. By hydrogenation of polyunsaturated fats (oils)?
   b. By heating, especially of unsaturated fats, to temperatures at which they smoke?
6. What is the name of the essential fatty acid? What vitamin is designated as this fatty acid?
7. Give two examples of dietary components in which balance is especially important.
8. **a.** What activities go on during a determination of the basal metabolic rate (BMR)?
   b. What three factors affect BMR in addition to weight, height, and age?
9. What are the functions of fat, protein, and carbohydrates in the body?
10. Based on their solubilities, what are the two classes of vitamins? An excess of which class of vitamins causes fewer problems?
11. Why is vitamin B_6 called the "master vitamin"?
12. Why should one eat the whole fruit rather than, for example, sucking the juice out and throwing the rest away?
13. What are some good food sources of complex carbohydrates?
14. What is the cause of kwashiorkor?
15. What foods are good sources of thiamine, vitamin B_{12}, niacin, ascorbic acid, vitamin A, vitamin D, and vitamin E?
16. What diseases or symptoms are caused by a deficiency of niacin, thiamine, proteins, calciferol, and ascorbic acid?
17. What group of vitamins is most easily destroyed or removed during food processing and cooking?
18. **a.** Distinguish between refined sugar and complex carbohydrates with respect to how they are assimilated by the body.
   b. What problems may arise from consuming too much
   (1) refined sugar?
   (2) refined grains (white flour)?
19. Use Table 20–4 to calculate the dietary calories in some fast-food hamburgers, French fries, and milkshakes. Which chain of fast-food restaurants offers the _lowest_ calorie total for a hamburger, milkshake, and an order of French fries? What is the total number of kcal?

	Burgers	Shakes	Fries
Chain A			
Protein (g)	11	10	2
Carbohydrates (g)	29	72	20
Fat (g)	9	9	19
Chain B			
Protein (g)	12	10	3
Carbohydrates (g)	30	66	26
Fat (g)	10	9	12
Chain C			
Protein (g)	13	11	2
Carbohydrates (g)	29	55	25
Fat (g)	11	7	10

20. Should salt tablets be taken after the body is acclimated to the heat? Why should or why should they not be taken?

21. a. What is the typical K/Na weight ratio in the body?
  b. What is a problem in maintaining this ratio?
  c. What can be done to maintain the proper K/Na ratio?

22. Why may a calcium deficiency occur in some women after menopause?

23. Why do we need each specific mineral and each specific vitamin? List one function or one deficiency effect for each.

24. Which vitamins are produced by bacteria in the intestine?

25. Does vitamin C decrease the symptoms of the common cold? Does vitamin A?

26. How does salt preserve food?

27. a. Why does it take less time to cook food in a pressure cooker than in an open pot of boiling water?
  b. Why does cooking aid digestion?

28. A label on a brand of breakfast pastries lists the following additives: dextrose, glycerin, citric acid, potassium sorbate, vitamin C, sodium iron pyrophosphate, and BHA. What is the purpose of each substance?

29. What is a common flavor enhancer? How do flavor enhancers work?

30. Choose a label from a food item, and try to identify the purpose of each additive.

31. Describe some of the chemical changes that occur during the cooking of
  a. A carbohydrate   **c.** A fat
  b. A protein

32. What causes bread to rise?

33. What do the letters *FDA* represent, and what does this government agency do?

34. What are the pros and cons of eating "natural" foods, as opposed to foods containing chemical additives?

35. What is the GRAS list?

36. What foods have you eaten during the past week that did not have chemicals added or applied to them?

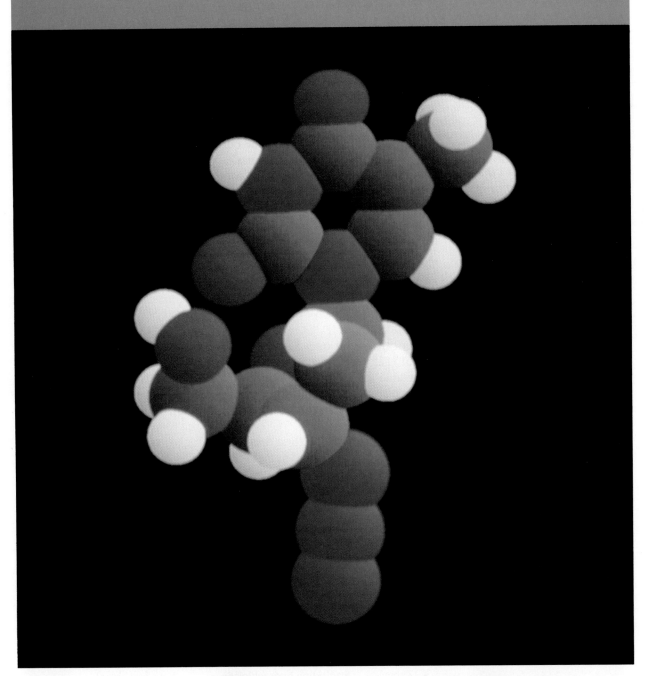

Chemistry and Medicine

Computer-generated model of AZT molecule. (Courtesy of Phillips Petroleum.)

The average life expectancy for men in the United States has risen from 53.6 years in 1920 to 71.3 years in 1986, a rise of 33.0%. During this same period, the life expectancy for women has risen from 54.6 years to 78.3 years, a rise of 43.4% (Table 21 – 1). What role has chemistry and chemical technology played in raising life expectancy and in improving the quality of life? The primary focus has been on the discovery of new and improved medicines and therapeutic drugs for treating both the mind and the rest of the body.

Within the past decade, remarkable progress has been made in understanding how chemical reactions control and regulate biological processes. As knowledge about the chemistry of biological processes and the chemical mechanism of drug action improves, drug design will receive increased emphasis over the earlier trial-and-error methods used for screening chemicals for drug use. This chapter will emphasize the role chemistry plays in improving mental and physical health.

TABLE 21 – 1 Life Expectancy in the United States for Men and Women (1920 – 1986)*

Year	Men	Women
1920	53.6	54.6
1930	58.1	61.6
1940	60.8	65.2
1950	65.6	71.1
1960	66.6	73.1
1970	67.1	74.8
1980	70.0	77.5
1986	71.3	78.3

* Data from U.S. Bureau of the Census: *Statistical Abstract of the United States,* 108th ed. Washington, D.C., 1988.

MEDICINES

Americans spend over $30 billion a year on medicines. This amounts to about $125 per person. As a result, the drug industry ranks as the fourth most profitable in the U.S. economy, following soft drinks, tobacco, and the communications media. The top ten prescription drugs, based on the number of prescriptions written (Table 21 – 2), show an interesting cross section of medicinal uses.

The generic name for a drug is the drug's generally accepted chemical name. The trade or brand name is the name used by the drug manufacturer. For example, the antibiotic amoxicillin (number 1 drug) is sold under brand names such as Amoxil, Amoxidall, Amoxibiotic, Infectomycin, Moxaline, Utimox, and Wymox. If the generic name is used, the prescription is often cheaper, particularly if the drug is not protected by patents and can be manufactured and marketed competitively by several companies.

TABLE 21 – 2 Ten Most Prescribed Drugs in 1988*

Trade Name	Generic Name	Use
1. Amoxil	Amoxicillin trihydrate	Antibiotic
2. Lanoxin	Digoxin	Heart disease
3. Xanax	Alprazolam	Tranquilizer
4. Zantac	Ranitidine hydrochloride	Antiulcer agent
5. Premarin	Mixture of estrogens	Menopausal symptoms
6. Dyazide	Triamterene/hydrochlorothiazide	Blood pressure (diuretic/antihypertensive)
7. Tagamet	Cimetidine	Antiulcer agent
8. Tenormin	Atenolol	Heart disease (beta blocker)
9. Naprosyn	Naproxen	Anti-inflammatory agent
10. Cardizem	Diltiazem hydrochloride	Heart disease (calcium blocker)

* Data from *American Druggist,* Vol. 199, no. 2, p. 40, February, 1989.

LEADING CAUSES OF DEATH

In 1900, five of the ten leading causes of death were infectious diseases (Fig. 21 – 1a). By 1985, only one of the ten was an infectious disease (Fig. 21 – 1b). This dramatic decrease can be attributed to the successful development of a variety of **antibiotics** that reduced mortality due to infectious diseases from 668 per 100,000 population in 1900 to 21 per 100,000 population in 1982. However, deaths from AIDS have caused the number to increase to 28 per 100,000 between 1982 and 1985. The lack of a cure and the experimental nature of present drugs being used to treat AIDS patients is likely to cause a significant increase in the number of deaths due to infectious diseases during the next decade. Public Health Service officials have predicted that by the end of 1992, 365,000 Americans will have developed AIDS and 263,000 of these will have died.

Heart Disease

Heart disease is the number one killer of Americans, claiming about 771,000 lives in 1985, or 37.0% of the total deaths. However, a variety of surgical techniques along with many new drugs are being used to decrease the death rate and improve the quality of life for persons suffering from heart disease. Since 1970, the number of deaths due to heart disease has dropped from 369 to 323 per 100,000 population. Medical observers attribute progress against heart disease to reduction of **atherosclerosis** (plaque buildup on artery walls), the most common cause. Risk factors contributing to cardiovascular diseases include smoking, eating too much saturated fat and cholesterol, and stress. These factors together with genetic influences may lead to **hypertension** (high blood pressure), and **hypercholesterolemia** (high concentrations of blood cholesterol).

Heart disease is an assortment of diseases, and death rates remain high despite the progress in recent years. Of about 1.5 million persons who experience heart attacks in the United States each year, approximately one third die. About 98% of all heart attack victims have atherosclerosis.

Atherosclerosis

Atherosclerosis is the buildup of fatty deposits called **plaque** on the inner walls of arteries. Cholesterol, a lipid, is a major component of atherosclerotic plaque. Many scientists believe that a high level of cholesterol in the blood, along with high blood levels of triglycerides, which are also lipids, contribute to the buildup of this plaque. The plaque buildup reduces the flow of blood to the heart. If a coronary artery is blocked by plaque, a heart attack occurs as a result of the reduced blood flow carrying oxygen to the heart. In the vast majority of cases the reduction in the blood supply can be traced to the formation of a blood clot in the plaque of a coronary artery. If prolonged, such an attack can cause part of the heart muscle to die.

Lipids include fats, oils, and other substances whose solubility characteristics are similar to those of fats and oils. See Chapter 15.

Review the discussion of cholesterol in Chapter 20.

Hypertension

Hypertension, or high blood pressure, is a common medical problem in the United States as evidenced by the number six drug in Table 21 – 2, dyazide. Approximately 60 million adults, or one in four, have high blood pressure.

Figure 21–1   The ten leading causes of death as a percentage of all deaths. (a) 1900. (b) 1985. (Source: National Center for Health Statistics, Division of Vital Statistics, National Vital Statistics System.)

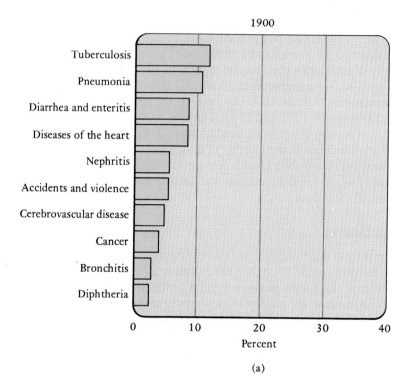

(a)

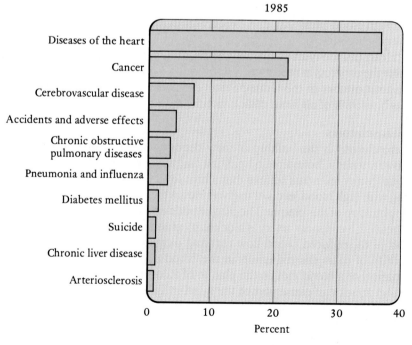

(b)

Hypertension, along with atherosclerosis, are the major causes of the 500,000 strokes and 1,500,000 heart attacks each year. Factors that contribute to high blood pressure are family history of hypertension, age, race, diabetic conditions, heavy salt and/or alcohol consumption, obesity, chronic stress, and atherosclerosis.

A person's blood pressure provides information about the extent of hypertension. Blood pressure is represented by two numbers, for example, 120/80. The 120 represents the **systolic pressure** (at the heartbeat) and the 80 represents the **diastolic pressure** (between heartbeats). The diastolic pressure is often used to diagnose hypertension. Normal diastolic pressure is considered to be 80 and diastolic pressures of 90 to 104 are arbitrarily classed as mild hypertension, 105 to 114 as moderate, and over 115 as severe. Each increase of 5 units in diastolic pressure increases the risk of heart attack by 20 to 25%.

Drugs Used to Treat Heart Disease

Atherosclerotic heart disease results in angina (chest pain on exertion), ischemia (partial deprivation of oxygenated blood), and myocardial infarction (heart attack). Accumulated damage to heart muscle by ischemia and heart attacks can lead to arrhythmias (abnormal heart rhythms) and heart failure. Hypertension and hypercholesterolemia accelerate atherosclerosis, and hypertension is also a cause of heart failure independent of atherosclerosis.

Some of the same drug types that control hypertension are also effective against angina and heart failure. Therefore, decisions about which antihypertensive drugs to prescribe must be weighed against their effects on arrhythmias, blood cholesterol, and heart failure.

Treatment of hypertension begins with loss of excess weight and restrictions on dietary Na^+. If these measures do not lower blood pressure sufficiently, drugs are required. The first choice is a **diuretic** such as dyazide, which stimulates the production of urine and excretion of Na^+. Since diuretics can also cause excess excretion of K^+, which is important in nerve conduction and muscle contraction, patients whose hearts are subject to arrhythmia must receive either a potassium supplement or a diuretic that does not cause K^+ excretion (Fig. 21–2).

Clinical studies have shown that increased levels of Na^+ raise the blood pressure in some individuals. See Chapter 20.

Figure 21–2 Diuretics that do not cause excretion of K^+.

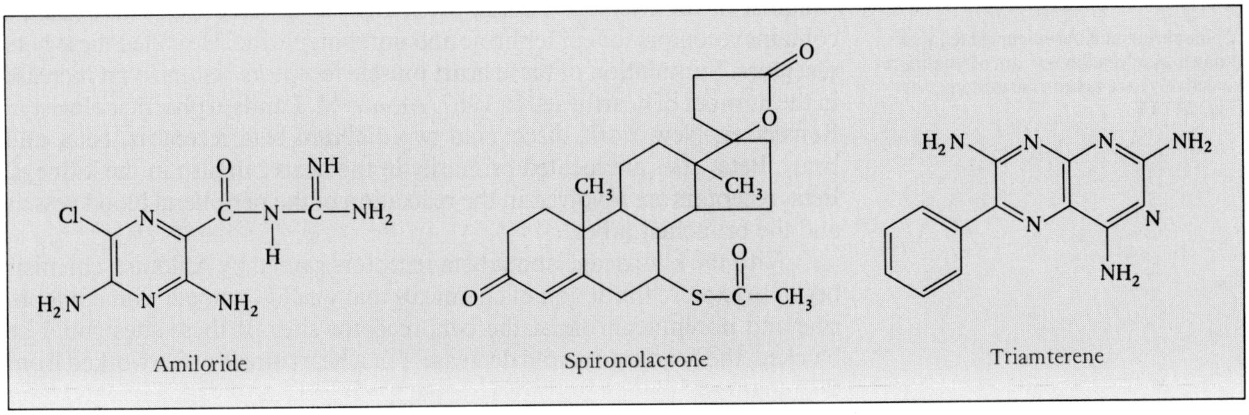

Amiloride   Spironolactone   Triamterene

The ultimate method of prevention of most forms of heart disease would be prevention of the buildup of atherosclerotic plaque in the arteries. In 1954, John W. Gofman of the University of California at Berkeley discovered four separate density categories of lipoproteins in the blood. About 65% of the cholesterol in the blood is carried by low-density lipoproteins (LDL), whereas only 25% of the cholesterol in the blood is carried by high-density lipoproteins (HDL). In 1968, John A. Glomset of the University of Washington showed that HDLs are effective in removing cholesterol from arterial walls and transporting it to the liver, where it is metabolized. This discovery opened up the possibility that already-formed atherosclerotic plaque might be dissolved by using a normal blood component. Scientists also have been attempting to develop drugs that either raise the level of HDLs or lower the level of LDLs for patients who cannot control these lipoproteins by proper diet.

Cholesterol-lowering drugs include niacin (nicotinic acid, a component of vitamin B complex), lovastatin, and cholestyramine resin (Questran). Niacin and lovastatin lower cholesterol levels in the blood by interfering with cholesterol synthesis in the liver. Questran lowers blood cholesterol by lowering the concentration of bile acids in the intestines, which causes the liver to convert more cholesterol into bile acids. Of the three drugs, lovastatin is the most recent on the market and is more effective for treating patients with very high blood cholesterol and/or LDL levels. Early indications are that lovastatin causes a drop in LDL levels of 19 to 39%.

Vasodilator drugs are used to relieve the pain from angina attacks. Vasodilators dilate, or open, veins, thus reducing the blood pressure against which the heart must work. Angina occurs because of insufficient oxygen delivery to the heart muscle. Angina attacks are brought on by exercise or anxiety, which increase the work of the heart and thus its oxygen demand. Symptoms include a crushing sensation in the chest, abdominal pain, and/or pain radiating from the chest to the left arm, throat and jaw, and sometimes to the right arm. Treatment of an attack centers on easing the work of the heart and thus its oxygen demand. Vasodilators include nitroglycerin and amyl nitrate, Lanoxin (number two drug in Table 21–2), as well as various beta blockers and calcium channel blockers, which are described in the next section.

Beta Blockers and Calcium Channel Blockers   In 1948, Raymond P. Ahlquist of the Medical College of Georgia discovered that heart muscle contains receptors for epinephrine and norepinephrine. He called these **beta receptors.** Stimulation of these heart muscle receptors results in an increase in the number of heartbeats. In 1967, Alonzo M. Lands, a pharmacologist in Rensselaer, New York, discovered two different beta receptors, beta$_1$ and beta$_2$. Beta$_1$ sites are located primarily in the heart but also in the kidneys. Beta$_2$ receptors are involved in the relaxation of the peripheral blood vessels and the bronchial tubes.

With the knowledge about beta receptors gained by Ahlquist, chemists began to explore the action of chemicals that would compete with epinephrine and norepinephrine at the beta receptor sites. If these sites could be blocked, the heart rate would decrease. For a heart already overworked from the buildup of plaque in the arteries supplying heart tissue with blood (and

Epinephrine and norepinephrine (also known as adrenalin and noradrenalin, respectively) are neurotransmitters. See Fig. 21–16.

oxygen), this might just produce enough relaxation to allow recovery from an impending attack. In addition, these drugs might be able to relieve high blood pressure and migraine headaches.

The first drugs of this type, called **beta blockers** because of their action of blocking beta receptor sites, came into use in the late 1950s and early 1960s but were later withdrawn because of undesirable side effects. In 1967, the beta blocker propranolol (trade name Inderal) was first prescribed. Its first use was for cardiac arrhythmias, but now it has been approved by the FDA for the treatment of angina, hypertension, and migraine headaches. Propranolol has high lipid solubility, so it passes through the blood-brain barrier (see Fig. 21–18) and builds up in the central nervous system, where it is more slowly metabolized. This buildup of the chemical in the central nervous system can cause side effects of fatigue, lethargy, depression, and confusion. In spite of these possible side effects, propranolol is a widely prescribed drug, and ranked 18th in prescribed drug sales in 1988.

A second beta blocker, metoprolol (trade name Lopressor) was introduced in the United States in 1978 for the treatment of hypertension. Because metoprolol is selective in its beta-blocking effects, blocking only $beta_1$ sites and not the $beta_2$ sites of the peripheral blood vessels or the bronchial tube, it is safe for asthma sufferers and for patients with severe blood vessel disorders. Metoprolol is not as soluble in lipids as propranolol and does not accumulate in the central nervous system. It is also more slowly metabolized by the liver, which allows more widely spaced doses, usually twice a day. Metoprolol ranked 26th in prescribed drug sales in 1988. A third beta blocker, atenolol (trade name Tenormin) was number 8 in prescribed drug sales in 1988 (Table 21–2).

Other heart disease drugs are the **calcium channel blockers.** Research has found that calcium ions move into the heart muscle by means of holes in the phospholipid membrane surrounding the muscle. In the muscle cells the calcium ions cause an interaction between the parallel protein filaments myosin and actin, and this interaction causes the cell to contract (Fig. 21–3). In addition, the double positive charge on the calcium ion neutralizes some of the negative charge of the muscle cell, also causing the muscle to contract. Movement of the calcium ions out of the cell restores the negative charge, and the cell relaxes. Blocking the flow of calcium ions into the cell causes the muscles to relax. When the smooth muscles in the walls of the coronary arteries are relaxed, these arteries expand and increase the supply of blood to the heart. Some calcium blockers also decrease the force of contraction and thus decrease the oxygen requirements of the heart.

Calcium blockers, unlike beta blockers, can prevent spasms of the coronary arteries. These spasms cause a blockage of blood flow to the heart and the intense pain of the angina attack. The causes of these spasms are poorly understood, but the calcium blockers do dilate the arteries and lessen the possibility of an angina attack.

After 19 years of use in Europe, verapamil (trade names Isoptin and Calan) became available in the United States and was first used to treat angina in 1981. In the same year a second calcium channel blocker, nifedipine (trade name Procardia) was introduced. Nifedipine is a powerful dilator of coronary arteries that has an immediate effect on patients suffering from angina. Although calcium blockers have fewer side effects than beta

Figure 21-3 (a) Calcium ions flowing into a muscle cell. (b) Contracted muscle cell caused by the presence of calcium ions in contractile protein fiber bundles. (c) Relaxed muscle cell with calcium channels blocked by drug molecules.

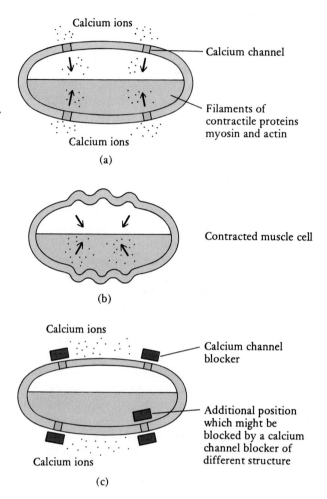

blockers, they may cause headaches, dizziness, flushing of the skin, and light-headedness. FDA-approved uses for calcium blockers now include angina, hypertension, and arrhythmias. Diltiazem (trade name Cardizem) has fewer side effects than the other calcium blockers, and as a result is the top-ranked calcium blocker, being ranked number 10 among prescribed drugs in 1988 (Table 21-2).

ACE Inhibitors An increased understanding of blood pressure regulation has resulted in the development of drugs to inhibit an enzyme, which regulates the synthesis of angiotensin II, a plasma protein that constricts blood vessels. These are referred to as **ACE inhibitors;** ACE stands for angiotension-converting enzyme. Three ACE inhibitors are Captopril, Enalapril, and Lisinopril.

Emergency Treatment of Heart Attack Victims

There are also new "clot-dissolving" drugs that are given to heart attack victims in the emergency room (or even in the ambulance), which show promise in reducing the death rate. Plasmin helps dissolve the blood clot naturally by catalyzing the breaking of the chain in the blood-clotting protein fibrin. Plasmin is generated from the circulating precursor plasminogen by enzymes called **kinases.** In recent years, drug companies have developed such kinases as **urokinase, streptokinase,** and **tissue-plasminogen activator (TPA)** to treat heart attacks. Intravenous injection of TPA is often enough to halt a heart attack within minutes and save heart tissue from damage. The only TPA now approved in this country is produced by Genentech, who use biotechnology to produce it from hamster ovary cells. Controversy has erupted over the relative costs of clot-dissolving kinases. Costs per treatment are $200 for streptokinase and $2200 for TPA or urokinase. Genentech has offered to reimburse hospitals for the cost of its TPA used to treat indigent patients.

One problem in reducing deaths by heart attack is the delay in seeking treatment—more than 350,000 heart attack victims die every year before reaching the hospital—even though an average of 3 hours elapses between the initial appearance of symptoms and death.

Does Aspirin Cut the Risk of a Heart Attack?
Recent studies have shown that aspirin can cut the risk of a heart attack by almost half even for those who have no overt signs of cardiovascular disease, apparently by working to reduce the incidence of clot formation. The results of a 6-year study covering 22,000 male physicians, ages 40 to 84, with no history of prior heart disease or stroke showed that taking a 325-mg tablet of aspirin every other day is associated with a 47% reduction in the risk of an initial heart attack. Aspirin works by blocking the manufacture of prostaglandins that are instrumental in the formation of blood clots. But there are potentially serious side effects. Persons susceptible to ulcers or bleeding should not take aspirin frequently.

Preventative Measures Against Heart Disease
Ultimately, the most effective measures are those that depend on an individual's own initiatives. Preventative measures taken by individuals are as important as having drugs for treating heart disease. Persons who exercise, don't smoke, have a low-fat, low-sodium diet, and avoid consuming more than 2 oz of 100-proof whiskey, 8 oz of wine or 24 oz of beer per day are less likely to develop heart disease.

Anticancer Drugs

Cancer is not one but perhaps 100 different diseases, caused by a number of factors. The sites of attack for major types of cancer are shown in Figure 21–4. A cancer begins when a cell in the body starts to multiply without restraint and produces descendants that invade tissues in the vicinity. It seems reasonable, then, that some drugs might exist that would be able either to stop this undesirable spreading of cancer cells or to prevent cancer from happening at all.

Cancers are treated by (1) surgical removal of whole areas affected by them, as well as the cancerous growths themselves; (2) irradiation to kill the cancer cells; and (3) chemicals to kill the cancer cells (**chemotherapy**). These

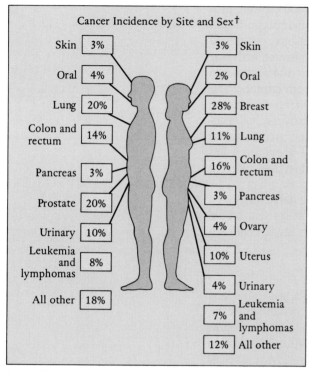

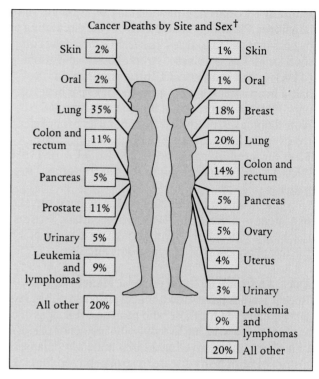

†Excluding non-melanoma skin cancer and carcinoma in situ.

Figure 21–4   1988 estimates of cancer incidence and deaths by site and sex.

treatment methods have resulted in dramatic improvements in the rates of survival of patients with certain cancers. A group of cancer patients can be considered cured if, after their treatment, they die at about the same rate as the general population. Another way of judging success in cancer therapy is by the number of patients who survive for 5 years after the treatment. Estimates of current survival rates are shown in Figure 21–6. In the 1930s less than one cancer patient in five was alive at least 5 years after treatment; in the 1940s it was one in four; in the 1960s it was one in three; and today it is about one in two.

Since 1930, the number of cancer deaths per 100,000 population has increased, and the major cause of this increase has been lung cancer (Fig. 21–5). Most lung cancers are caused by cigarette smoking, which has been implicated as the cause of 83% of lung cancers as well as many cancers of the mouth, pharynx, larynx, esophagus, pancreas, and bladder. Not only does smoking account for about 30% of all cancer deaths, it is also a major cause of heart disease and is linked to respiratory conditions ranging from colds to emphysema. As Figure 21–6 shows, only a few major types of cancer have low survival rates. These include lung cancer (13%) and pancreatic cancer (3%). Obviously, the high incidence of lung cancer and its low survival rate highlight the importance of not smoking.

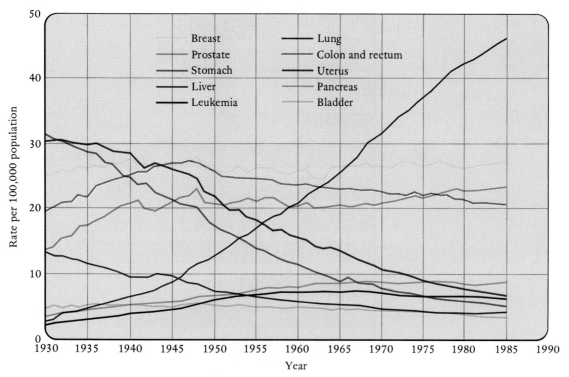

Figure 21–5 **U. S. cancer death rates by site, 1930–1985. The rate for the population is standardized for age based on the 1970 U. S. population. Rates are for both sexes combined except breast and uterus (female population only) and prostate (male population only). (Source: National Center for Health Statistics and Bureau of the Census, United States.)**

Chemotherapy

In World War I the toxic effects of a class of the chemical-warfare gases called mustard gases were recognized. These gases were found to cause damage to the bone marrow and to be mutagenic. In these ways they were acting like X rays, which are also toxic to cells and cause mutations.

$$Cl—CH_2CH_2—S—CH_2CH_2—Cl$$

Mustard Gas

The name *mustard gas* comes from its mustardlike odor; mustard gas, however, is not a gas but a high-boiling liquid that was dispersed as a mist of tiny droplets.

Beginning around 1935, other mustards of the nitrogen family were synthesized. They, too, caused mutations in some laboratory animals. In addition, they caused cancers in some animals.

Nitrogen Mustard
(general formula)

A Nitrogen Mustard

After World War II the secrecy surrounding the mutagenic nature of these chemicals was lifted, and it occurred to cancer researchers that cancers might be treated with chemicals that selectively destroy unwanted cells.

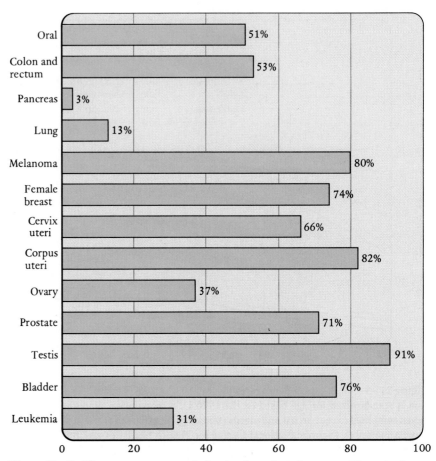

Figure 21–6   Five-year cancer survival rates for selected cancers. The rates have been adjusted for normal life expectancy, and the figure is based on cases diagnosed in 1979–1984. (Source: National Center for Health Statistics.)

One of the most widely used anticancer drugs is cyclophosphamide, a compound that contains the nitrogen mustard group (shown in color).

Cyclophosphamide

Compounds such as cyclophosphamide belong to the alkylating class of anticancer drugs. **Alkylating agents** are reactive organic compounds that transfer alkylating groups in chemical reactions. Their effectiveness as anticancer agents is due to the transfer of alkyl groups to the nitrogen bases in DNA, particularly guanine. The presence of the alkyl group in the guanine molecule blocks base pairing and prevents DNA replication, which stops cell division. Although alkylating agents attack both normal cells and cancer cells, the effect is greater for rapidly dividing cancer cells.

Figure 21-7 The structural arrangement of atoms when a *cis*[Pt(NH_3)$_2$] fragment binds to two nitrogen atoms in the guanine rings of a dinucleotide.

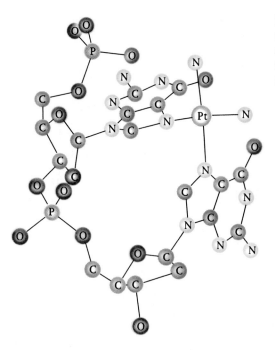

Another widely used anticancer drug, **cisplatin,** blocks DNA replication by a similar mechanism.

The discovery of anticancer properties of cisplatin was discussed in Chapter 1.

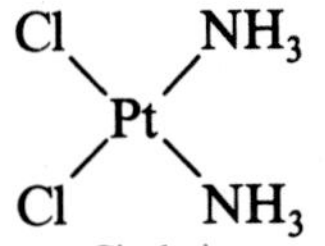

Cisplatin

Inside the cell, the Cl^- ions are displaced, and the unit binds to the nitrogen sites on the guanine bases in DNA (Fig. 21-7). As with alkylating agents, the blocking action prevents base pairing and DNA replication.

Another group of chemotherapeutic agents, called **antimetabolites,** interferes with DNA synthesis. One of these chemicals, 5-fluorouracil, gets involved in the synthesis of a nucleotide, which inhibits the formation of a thymine-containing nucleotide necessary for DNA synthesis. Lack of proper DNA synthesis slows cell division. 5-Fluorouracil has proved useful in the treatment of cancers of the breast.

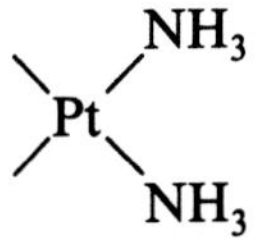

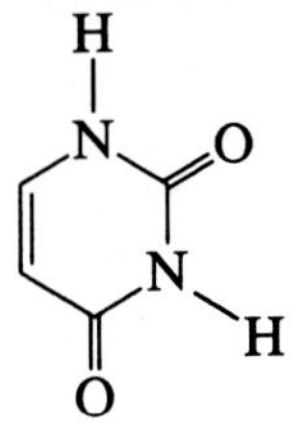

Uracil

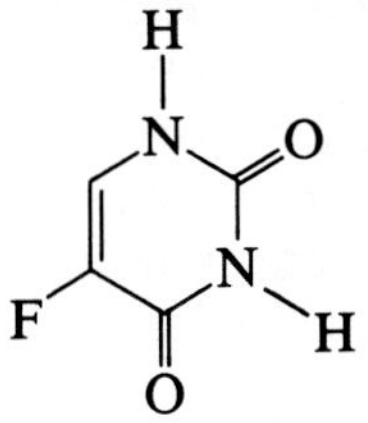

5-fluorouracil

Another antimetabolite is methotrexate, which has a structure similar to that of folic acid.

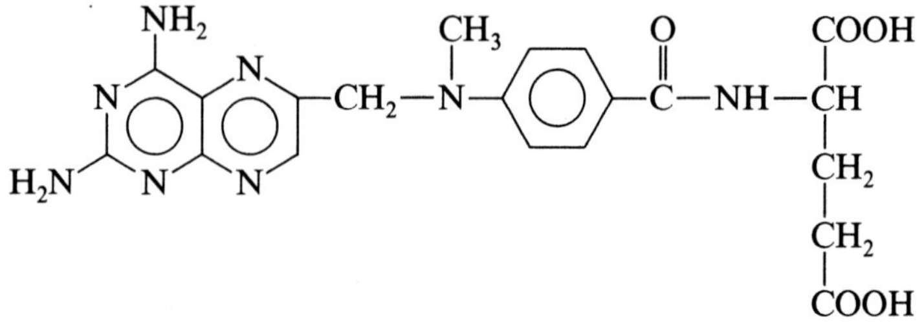

Methotrexate

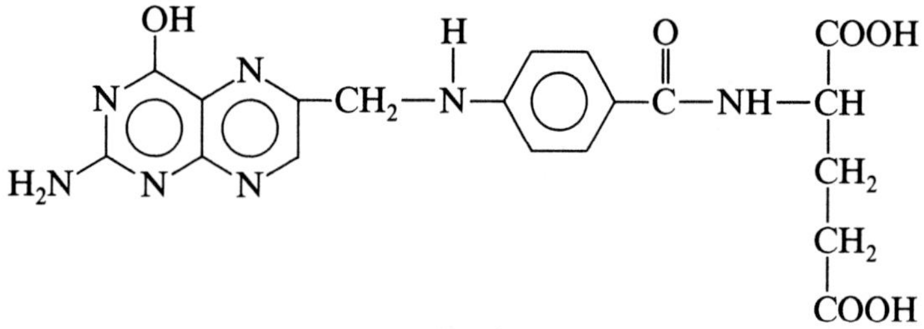

Folic Acid

A major route to the synthesis of nucleic acid in the body begins with folic acid from the diet. Methotrexate binds to the enzyme dihydrofolic reductase 10,000 times more strongly than folic acid, and this prevents the reduction of folic acid in the first step of the synthesis of nucleic acid. This results in a slowdown of cell growth. Leukemia is treated with methotrexate.

All cancer chemotherapy is tedious and has its risks. In addition to being highly toxic, most of the useful chemotherapy agents are themselves carcinogenic (cancer-causing). Often very high doses are necessary to effect treatment. As a result, single-agent chemotherapy has largely given way to combination chemotherapy because of the success of additive, or even synergistic, effects when two or more anticancer drugs are used. For example, a combination that is used in the treatment of several cancers is cisplatin and cyclophosphamide. Because of their synergistic action, lower doses of each compound can be used when they are given together than if they were each given alone, and this reduces the harmful side effects of chemotherapy.

Synergism is the working together of two things to produce an effect greater than the sum of the individual effects.

Childhood cancers respond most favorably to chemotherapy. Most children with leukemia who are treated with chemotherapy drugs enter a period of relapse-free survival. In terms of the definition given earlier, they are cured. During the early 1950s about 1900 children under the age of 5 years died of cancer per year in the United States. Today the number is fewer than 700 per year. A few other cancers, such as Hodgkin's disease, have shown similar increases in survival rates due to chemotherapy. Not all cancers have been so treatable, however, because cancer is so many different diseases.

SELF-TEST 21–A

1. The accepted chemical name for a drug is called its _____ name.
2. The most widely prescribed drug in the United States is used as a(an) _____ .
3. The two ingredients of atherosclerotic plaque are _____ and _____ .
4. When beta sites in the heart are stimulated by epinephrine, the heart beats ( ) faster ( ) slower.
5. When propranolol passes through the blood-brain barrier, it causes the side effects of _____ and _____ .
6. A drug that blocked calcium ions from flowing into heart muscles would have the effect of ( ) exciting ( ) relaxing the heart.
7. HDL stands for _____ and LDL stands for _____ .
8. Cholesterol is carried in the blood by both HDLs and LDLs. Which carries the greater percentage of cholesterol?
9. Most anticancer drugs can also cause cancer. True ( ) False ( )
10. The nitrogen mustards act on cancer cells by blocking _____ replication.
11. Chemicals that interfere with DNA synthesis are called _____ .
12. Which cancer has shown the greatest response to chemotherapeutic drugs?
13. _____ , a "clot-dissolving" drug produced by biotechnology, is given to heart attack victims to halt heart attacks and reduce heart damage.

ANTIBACTERIAL DRUGS

In 1904, the German chemist Paul Ehrlich (1854–1915; 1908 Nobel-Prize recipient) realized that infectious diseases could be conquered if toxic chemicals could be found that attacked parasitic organisms within the body to a greater extent than they did host cells. Ehrlich achieved some success toward his goal; he found that certain dyes that were used to stain bacteria for microscopic examination could also kill the bacteria. This led to the use of dyes against organisms causing African sleeping sickness and arsenic compounds against those causing syphilis. Ehrlich also introduced the term "receptor" in 1907 and proposed concepts of receptor binding, bioactivation, the therapeutic index, and drug resistance that are still valid in principle.

After experimenting with several drugs, Gerhard Domagk, a pathologist in the I. G. Farbenindustrie Laboratories in Germany, found in 1935 that Prontosil, a dye, was effective against bacterial infection in mice. Actually, Domagk discovered that Prontosil does not have antibacterial activity *in vitro* (outside the living body) but is metabolized *in vivo* (inside the body) to **sulfanilamide,** the actual antibacterial agent.

Many drugs in use today are not active themselves but are metabolized to the active agent. Drugs of this type are often referred to as prodrugs.

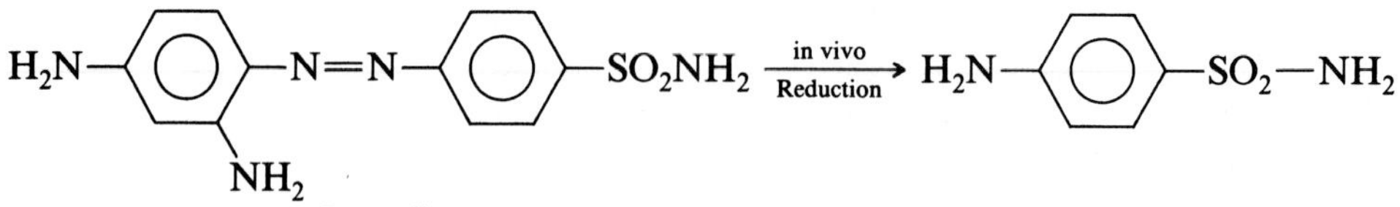

Prontosil Sulfanilamide

Sulfa Drugs

Sulfanilamide destroys bacteria that cause pneumonia, diphtheria, and gonorrhea.

Sulfanilamide, the first of a class of drugs referred to as sulfa drugs, is effective against streptococci, staphylococci, pneumococci, gonococci, meningococci, and dysentary bacteria. However, sulfanilamide has harmful side effects, and when used in high doses, can cause kidney damage. The success of sulfanilamide led to the synthesis and testing of over 5000 derivatives in an attempt to find more effective drugs with fewer side effects. All of these compounds are referred to as sulfonamides or **sulfa drugs** and have the general formula

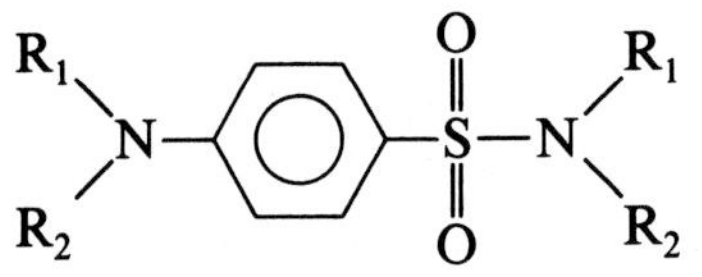

R_1 and R_2 may be hydrogen or an organic group.

Of thousands of compounds synthesized by drug companies as potential drugs, only a few actually make it to the drug market.

A sulfa drug mimics an essential compound.

This attempt to find better drugs by synthesizing and testing derivatives of a proven drug provides an excellent example of the strategy often used by the pharmaceutical industry in the search for better drugs. Structure-activity correlations derived from the testing results of sulfonamides are shown in Figure 21-8. Only about 30 of the more than 5000 known sulfonamides actually reached the point of being used as drugs. Some of the sulfa drugs still in use today are shown in Figure 21-9.

Sulfa drugs inhibit bacteria by preventing the synthesis of folic acid, a vitamin essential to their growth. The drugs' ability to do this lies in their structural similarity to *para*-aminobenzoic acid, a key ingredient in the folic acid synthesis in bacteria.

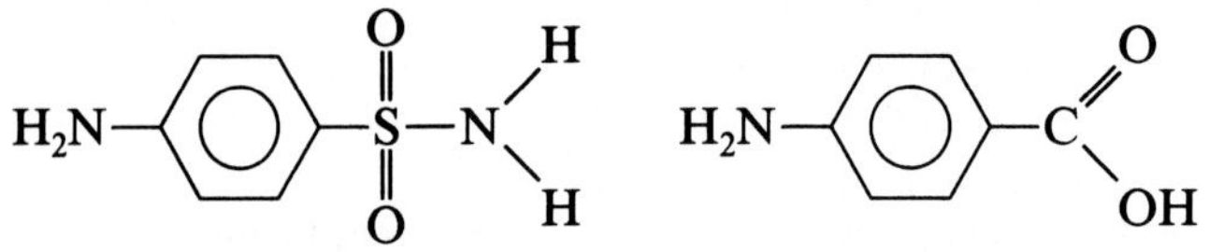

Sulfanilamide
(A Typical Sulfa Drug)

p-Aminobenzoic Acid

Recall that methotrexate is an effective anticancer agent because it takes the place of folic acid on an enzyme that promotes cell growth.

The close structural similarity of sulfanilamide and *p*-aminobenzoic acid permits sulfanilamide instead of *p*-aminobenzoic acid to be incorporated into the enzymatic reaction sequence. By bonding tightly, sulfanilamide shuts off the production of the essential folic acid, and the bacteria die of vitamin deficiency. Since humans and other higher animals obtain folic acid from their diet, *p*-aminobenzoic acid is not necessary for folic acid synthesis, and sulfa drugs do not interfere with normal cell growth.

Sulfa drugs were the miracle drugs during the late 1930s and early 1940s. During World War II, thousands of lives were saved by the effective use of sulfa drugs to prevent infection in wounds. However, in the 1940s another group of drugs, the antibiotics, were found to be more effective antibacterial drugs.

Antibiotics

An **antibiotic** is a substance produced by a microorganism that inhibits the growth of other organisms. Their job generally is to aid the white blood cells by stopping bacteria from multiplying. When a person falls victim to or is killed by a bacterial disease, it means that the invading bacteria have multiplied faster than the white blood cells could devour them and that the bacterial toxins increased more rapidly than the **antibodies** could neutralize them. The action of the white blood cells and antibodies plus an antibiotic is generally enough to repulse an attack of disease germs. Penicillins were the first class of antibiotics to be discovered.

The Penicillins

Penicillin was discovered in 1928 by Alexander Fleming, a bacteriologist at the University of London, who was working with cultures of *Staphylococcus aureus,* a germ that causes boils and some other types of infections. In order to examine cultures with a microscope, Fleming had to remove the covers of the culture plates for a while. One day as he started work he noticed that one culture was contaminated by a blue-green mold. For some distance around

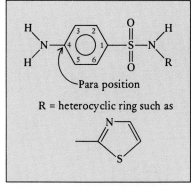

Figure 21–8   5000 compounds showed the highest activity for compounds that fit the following structural limitations: (1) the amino groups must be in the *para* position to the sulfonamide group and be unsubstituted; (2) the benzene ring can be substituted in positions 1 and 4 only; (3) the sulfonamide amine nitrogen can be monosubstitued only and heteroaromatic substituents increase the activity.

Figure 21–9   Some of the sulfa drugs still in use today.

Sulfathiazole

Sulfamerazine

Sulfadiazine

Sulfisoxazole

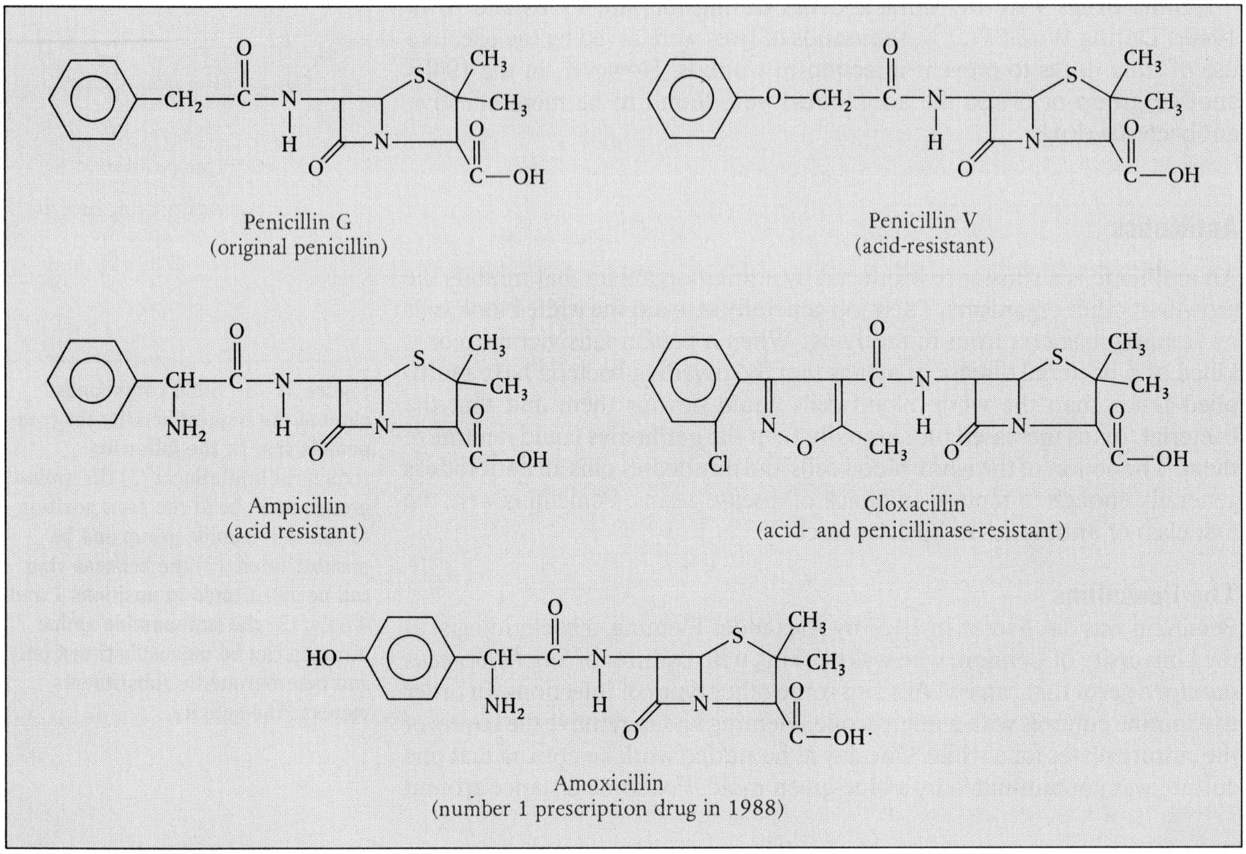

Figure 21–10   Structures of several types of penicillin.

the mold growth, the bacterial colonies were being destroyed. Upon further investigation, Fleming found that the broth in which this mold had grown also had an inhibitory or lethal effect on many pathogenic (disease-causing) bacteria. The mold was later identified as *Penicillium notatum* (the spores sprout and branch out in pencil shapes; hence the name). Although Fleming showed that the mold contained an active antibacterial agent which he called **penicillin,** he was not able to purify the active substance. In 1940, Howard Florey and Ernst Chain of Oxford University succeeded in isolating a product from the mold, called **penicillin G,** and by 1943 penicillin G was available for clinical use. By the end of World War II, penicillin G was saving many lives threatened by pneumonia, bone infections, gonorrhea, gangrene, and other infectious diseases. For their work, Fleming, Florey, and Chain received the Nobel Prize for medicine and physiology in 1945. The structures of several types of penicillin are shown in Figure 21–10.

Penicillin G had to be injected because it was destroyed by stomach acid when taken orally. Bacterial strains also became resistant to penicillin G. This led to the testing of other penicillins such as those shown in Figure 21–10. Note that amoxicillin was the number one prescribed drug in 1988. It can be taken orally and is effective against a wide variety of bacteria.

Another *Penicillium* strain that proved to be an excellent source of the new antibiotic was discovered on a moldy cantaloupe in a Peoria, Illinois market.

All penicillins kill growing bacteria by preventing them from making their normal cell walls. Since animal cells do not have cell walls, bacteria can be destroyed without damaging animal cells.

Tetracyclines

Antibiotics other than penicillin work in a variety of ways. Many interfere with the making or functioning of DNA in bacteria. The tetracyclines, streptomycin and erythromycin, prevent bacteria from making proteins from their DNA.

In 1937, following collaboration with René Dubos, Selman Waksman isolated a compound from a soil organism, *Streptomyces griseus;* this compound, which came to be known as **streptomycin,** was released to physicians in 1947. It was quite successful in controlling certain types of bacteria but was later withdrawn because of adverse side effects.

In 1945, B. M. Duggan discovered that a gold-colored fungus, *Streptomyces aureofaciens,* produced a new type of antibiotic, **aureomycin,** the first of the **tetracyclines.** Research then stepped up to a fever pitch. Pfizer Laboratories tested 116,000 different soil samples before they discovered the next antibiotic, **terramycin.**

Streptomycin has many undesirable side effects.

Chlortetracycline
(Aureomycin)

Oxytetracycline
(Terramycin)

Compounds of the tetracycline family are so named because of their four-ring structure. One side effect of these drugs is diarrhea, caused by the killing of the patient's intestinal flora (the bacteria normally residing in the intestines).

Tetracyclines get their names from their four-ring structures.

Cephalosporins

Cephalosporins, another class of antibiotics, were discovered in the 1950s. They are widely used because of their broad antimicrobial activity and low risk of serious adverse effects. The structures of some of the most important cephalosporins are shown in Figure 21–11. In 1988, Cefaclor (brand name Ceclor) was the top-ranked cephalosporin in sales, ranking 14th in total prescriptions in the United States.

Problems with Antibiotics

Although antibiotics have saved millions of lives by controlling infectious diseases, they also have some disadvantages. Repeated use of broad-spectrum antibiotics such as tetracyclines kills beneficial and harmless bacteria in the digestive tract and causes diarrhea. Some people are allergic to antibi-

Figure 21–11   Structures of several cephalosporins.

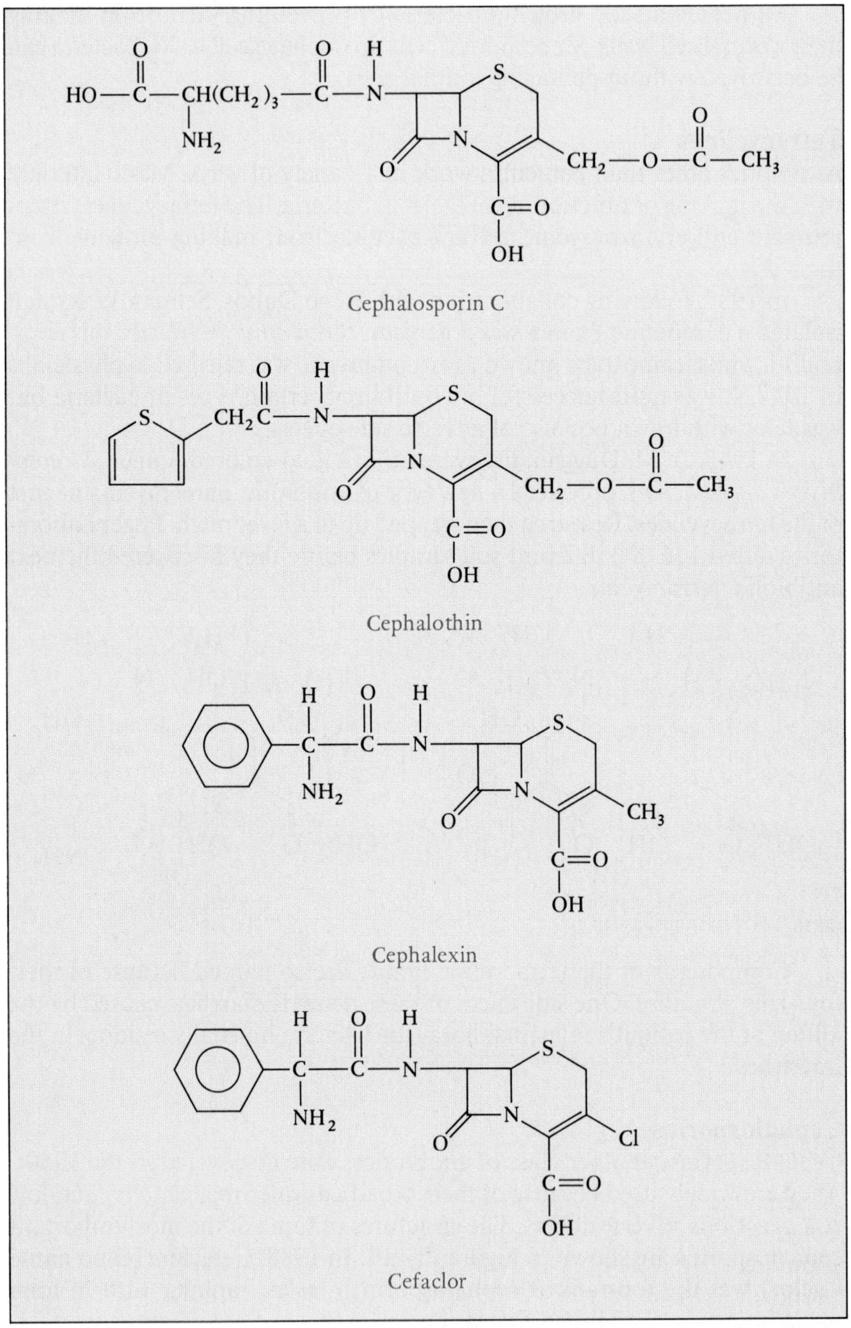

Cephalosporin C

Cephalothin

Cephalexin

Cefaclor

otics, especially penicillin, and suffer allergic responses occasionally severe enough to cause death. Hypersensitivity to penicillin is the most common side effect. Adverse reactions to penicillin may occur immediately or within 20 min. More delayed reactions are not as severe, and the risk of hypersensitivity reactions increases with prolonged therapy.

FOLK MEDICINE

There's a long history of folk medicine based on plants, especially in the Near and Far East. Now the question is: which of these things really works and how much of it is just a rumor? First the chemist checks out the plant rumored to be good and analyzes its extracts to see if there is any activity. One such substance is called fredericamycin. Fredericamycin is an antibiotic of some interest as a possible antitumor compound. It comes from a soil organism, a bacterium, found in the soil in Frederick, Maryland.

Dr. Kathlyn Parker, of Brown University, and other researchers like her, first analyze the naturally occurring substance in the lab. If they find an active component, they look further. As Dr. Parker says:

Then, if you are really interested in drug development, you have to isolate the active component, purify it, determine its structure, and then it gets handed over to the pharmaceutical people who decide how to package it. For some pharmaceuticals, what you really need is to be able to make a large amount of stuff really cheap. One solution is that you would develop methods so that it was so cheap to make something that you could distribute it to people in a way that they could afford it.

The World of Chemistry (Program 21), "Carbon."

Kathlyn Parker (*The World of Chemistry,* Program 21, "Carbon.")

Another problem is the development of genetic resistance. As antibiotics are used more and more for both people and livestock, resistant bacteria result. Strains of malaria, typhoid fever, gonorrhea, and tuberculosis have emerged that are resistant to various penicillins, streptomycin, and a number of tetracyclines. Certain bacteria can gain the ability to produce enzymes that inactivate the antibiotic. Other bacteria can become resistant to antibiotics by preventing the antibacterial agent from entering the bacterial cell.

ALLERGENS AND ANTIHISTAMINES

A person may have an unpleasant physiological response to poison ivy, pollen, mold, food, cosmetics, penicillin, aspirin, and even cold, heat, and ultraviolet light. In the United States about 5000 people die yearly from bronchial asthma, at least 30 from the stings of bees, wasps, hornets, and other insects, and about 300 from ordinary doses of penicillin. The reason is **allergies.** About one person in ten suffers from some form of allergy; more than 16 million Americans suffer from hay fever.

An allergy is an adverse response to a foreign substance or to a physical condition that produces no obvious ill effects in most other organisms, including humans. An **allergen** (the substance that initiates the allergic reaction) is in many cases a highly complex substance — usually a protein. Some allergens are polysaccharides or compounds formed by the combination of a

An allergy is a physiological response — such as sneezing, runny nose, coughing, or dermatitis — to a foreign substance. This foreign substance is called an allergen.

Figure 21–12   A postulated mechanism for the cause of and relief from hay fever. The details are described in the text.

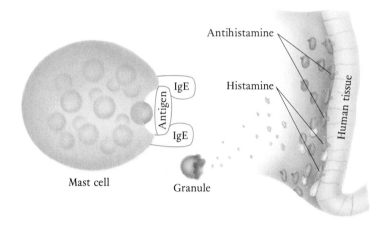

Mast cell   Granule

Antihistamine

Histamine

Antigen   IgE

IgE

Human tissue

Most allergens are high-molecular-weight substances.

10^{-12} is 0.000000000001
A picogram is 1×10^{-12} g.

Antibodies are high-molecular-weight proteins, called immunoglobulins, that attack foreign proteins such as allergens.

protein and polysaccharide. Usually allergens have a molecular weight of 10,000 or more.

Scientists have succeeded in isolating the principal allergen of ragweed pollen, a major allergy producer, named ragweed antigen E. It is a protein with a molecular weight of about 38,000; it represents only about 0.5% of the solids in ragweed pollen but contributes about 90% of the pollen's allergenic activity. A mere 1×10^{-12} g (1 picogram) of antigen E injected into an allergic person is enough to induce a response.

The allergens come in contact with special cells in the nose and breathing passages to which is attached the IgE antibody, which has a molecular weight of about 196,000. Allergic persons have 6 to 14 times more IgE in their blood serum than nonallergic persons. The IgE is formed in the nose, bronchial tubes, and gastrointestinal tract and binds firmly to specific cells, called **mast cells,** in these regions.

Antigen E from ragweed reacts with the IgE antibody attached to the mast cells, forming antigen–antibody complexes. The formation of these antigen–antibody complexes leads to the release of so-called allergy mediators from special granules in the mast cells (Fig. 21–12). The most potent of these mediators found so far is **histamine.** Although it is widely distributed in the body, histamine is especially concentrated in the 250 to 300 granules of the mast cells. Histamine accounts for many, if not most, of the symptoms of hay fever, bronchial asthma, and other allergies.

$$H_2NCH_2CH_2 \underset{\underset{H}{N}}{\overset{N}{\rule{0pt}{0pt}}}$$

Histamine

Histamine causes runny noses, red eyes, and other hay fever symptoms.

Chemical mediators such as histamine must be released from the cell to cause the symptoms of allergy. The release mechanism is an energy-requiring process in which the granules may move to the outer edge of the living cell and, without leaving the cell, discharge their histamine contents through a temporary gap in the cell membrane. This sends the histamine on its way to produce the toxic effects of hay fever.

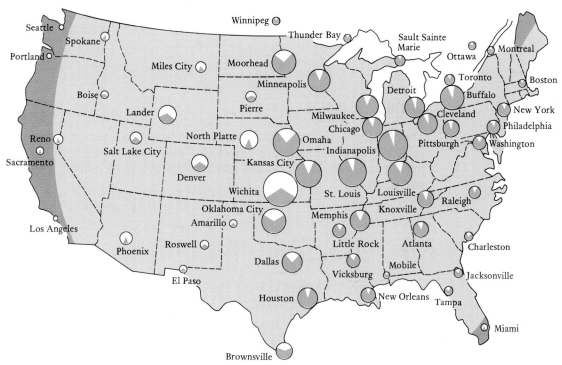

Figure 21–13   In this Abbott Laboratories' map, the size of each circle represents the amount of all late-summer and fall pollens found in the air in each city. Dark portions show the amount of ragweed pollen. Shaded areas are regions of low pollen count.

Treatment consists of three procedures: avoidance (Fig. 21–13), desensitization, and drug therapy. Desensitization therapy is costly and inconvenient, since 20 or more injections are required to achieve what is usually only a partial cure. One possible form of chemical desensitization is injection of a blocking antibody that preferentially reacts with the allergen so that it cannot react with the IgE allergy-sensitizing antibody. This breaks the chain of events leading to the release of histamine or other allergy-producing mediators. Many small injections, spaced over time, are required to build up a sufficient level of the blocking antibody.

Epinephrine (adrenalin), steroids, and antihistamines are effective drugs in treating allergies. The first two are particularly effective in treating bronchial asthma; the **antihistamines,** introduced commercially in the United States in 1945, are the most widely used drugs for treating allergies. More than 50 antihistamines are offered commercially in the United States. Many of these contain, as does histamine, an ethylamine group ($-CH_2CH_2N\zigzag$):

Pyribenzamine
(An Important Antihistamine)

These drugs act competitively by occupying the receptor sites on cells that are normally occupied by histamine, effectively blocking the action of histamine.

ANTIULCER DRUGS

The principal antiulcer drug is Tagamet (number seven in total prescriptions in 1988). Its generic name is **cimetidine.** This compound is used for the treatment of both duodenal and gastric ulcers; it acts by inhibiting gastric acid secretion.

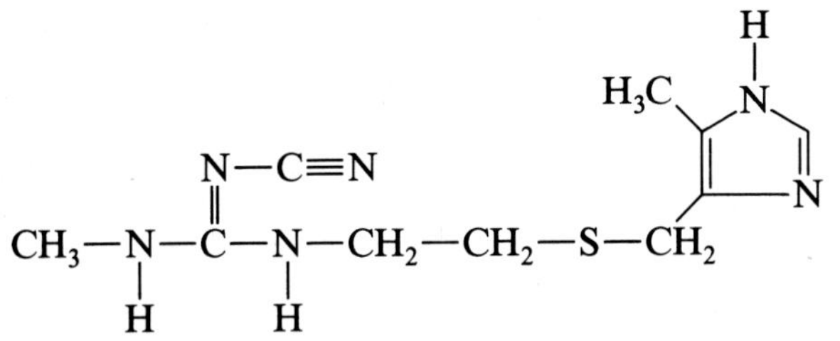

Cimetidine
(Tagamet)

ANTISEPTICS AND DISINFECTANTS

An **antiseptic** is a compound that prevents the growth of microorganisms. It now is legally considered a **germicide,** or a compound that kills microorganisms. A **disinfectant** is a compound that destroys pathogenic bacteria or microorganisms, but usually not bacterial spores. Disinfectants are generally poisonous and therefore suitable only for external use, as on the skin or a wound.

Common germicides, such as the halogens, sodium hypochlorite, and hydrogen peroxide, are effective because they can oxidize any kind of cell, including human cells. For this reason they are used mostly as disinfectants on nonliving objects. The quaternary ammonium compounds are surface-active agents, and their bactericidal effect seems to be related to their ability to weaken the cell wall so the cell contents cannot be contained.

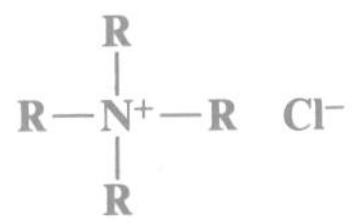

Iodophors, commonly used as disinfectants for restaurant glassware, are a complex of iodine with polyvinylpyrrolidone.

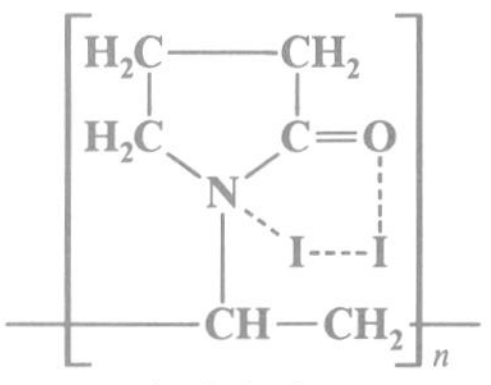

Because antiseptics and disinfectants are generally toxic, it is necessary to use dilute solutions that are applied only to the skin. Although they help prevent the spread of disease, germicides are practically useless in the treatment of disease because they act nonspecifically against all cells. However, there are a number of antibiotics that are used as germicides because they act more selectively against infecting bacteria. The most common one is **Neosporin,** which is applied topically to treat a number of skin conditions. Neosporin is a combination of three antibiotics: polymyxin B, bacitracin, and neomycin.

A Quaternary
Ammonium
Chloride

An Iodophor

HORMONES

Hormones are substances produced by glands to serve as chemical messengers in regulating biological processes. They are chemically diverse but are primarily proteins or steroids. Hormones are synthesized by specific glands —hypothalamus, pituitary, thymus, thyroid, parathyroid, pancreas, adrenals, and gonads—and are secreted directly into the blood.

Insulin

Diabetes mellitus, a disease that affects several million people, arises from a deficiency of the hormone insulin. Insulin is a polypeptide hormone that promotes the entry of glucose, some other sugars, and amino acids into muscle and fat cells. Diabetes is characterized by an elevated level of glucose in the blood and in the urine because the deficiency of insulin prevents sufficient transfer of glucose to muscle and fat cells.

Review the discussion of diabetes in Chapter 20.

Steroid Hormones

A large and important class of naturally occurring compounds is derived from the following tetracyclic structure:

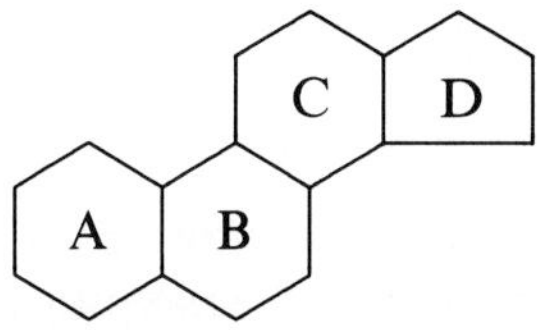

These compounds, known as **steroids,** occur in all plants and animals. The most abundant animal steroid is cholesterol. The human body synthesizes cholesterol and readily absorbs dietary cholesterol through the intestinal wall. An adult human contains about 250 g of cholesterol. Although cholesterol receives a lot of attention in connection with the correlation of blood cholesterol levels with heart disease, it is important to realize that proper amounts of cholesterol are essential to our health because cholesterol undergoes biochemical alteration or degradation to give milligram amounts of many important hormones, such as cortisone.

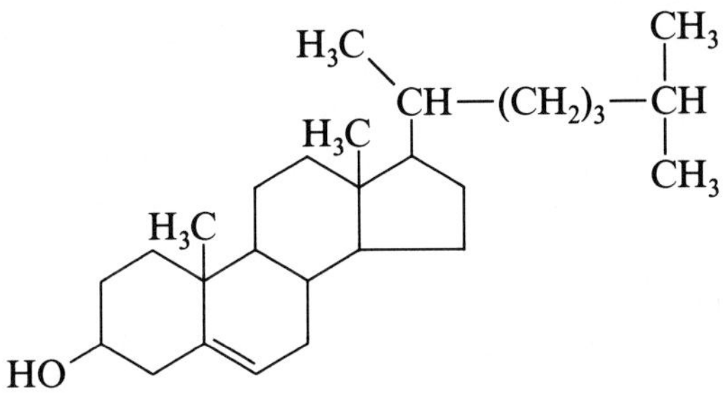

Cholesterol

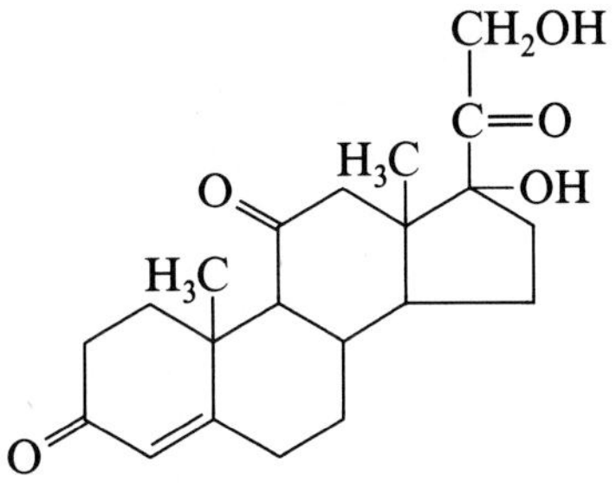

Cortisone

Cortisone is used as an antiinflammatory agent when applied topically or injected into a diseased joint, and it is used in treating acute cases of arthritis.

Structurally related to cholesterol and cortisone are the sex hormones. One female sex hormone, **progesterone,** differs only slightly in structure from an important male hormone, **testosterone.**

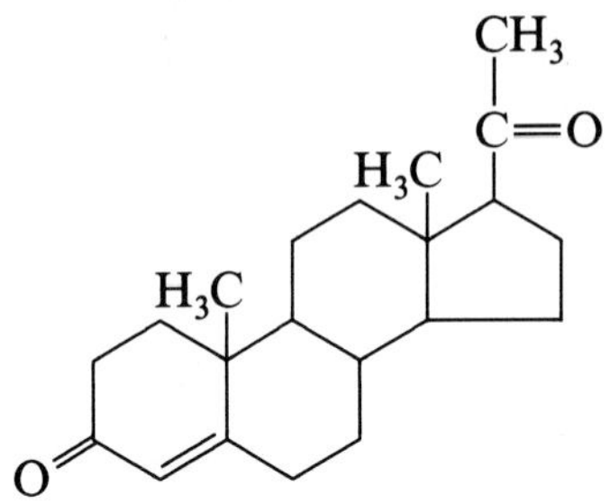

Progesterone

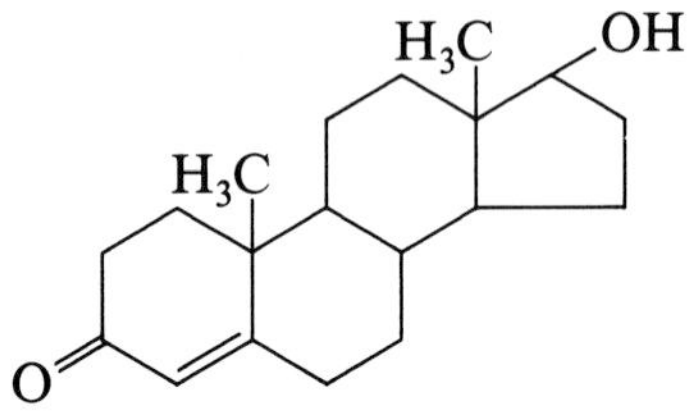

Testosterone

Other female hormones are estradiol and estrone, together called **estrogens.** The estrogens differ from the steroids discussed earlier in that they contain an aromatic ring.

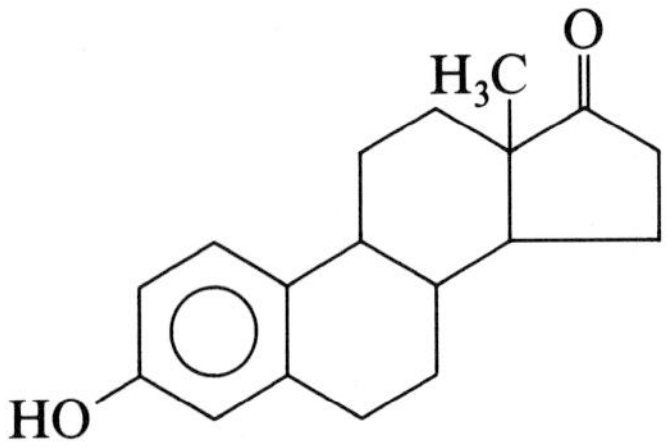

Estrone

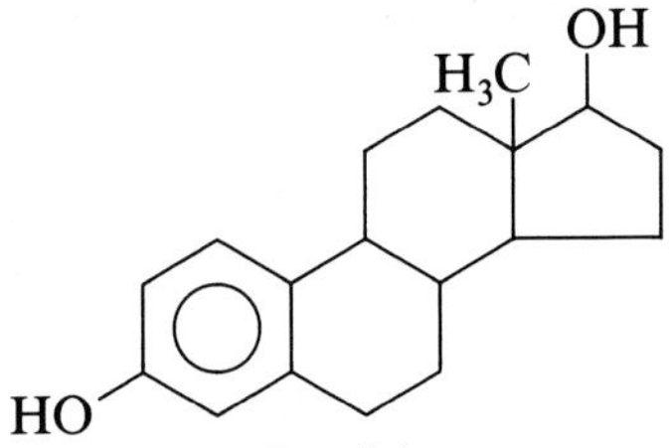

Estradiol

The estrogens and progesterone are produced by the ovaries. Estrogens are important to the development of the egg in the ovary, whereas progesterone causes changes in the wall of the uterus and after pregnancy prevents release of a new egg from the ovary (ovulation). Birth control drugs use derivatives of estrogens and progesterone to simulate the hormonal processes resulting from pregnancy and thereby prevent ovulation.

Birth Control Pills

One of the most revolutionary medical developments of the 1950s was the worldwide introduction and use of "the pill." Now there are two types of oral contraceptives, "the pill" and "the minipill."

The pill contains small amounts of synthetic analogs of estrogens and progesterone. The common ones are mestranol, a synthetic estrogen derivative, and norethindrone, a synthetic progesterone derivative. The estrogen derivative regulates the menstrual cycle, and the progesterone derivative establishes a state of false pregnancy resulting in the prevention of ovulation.

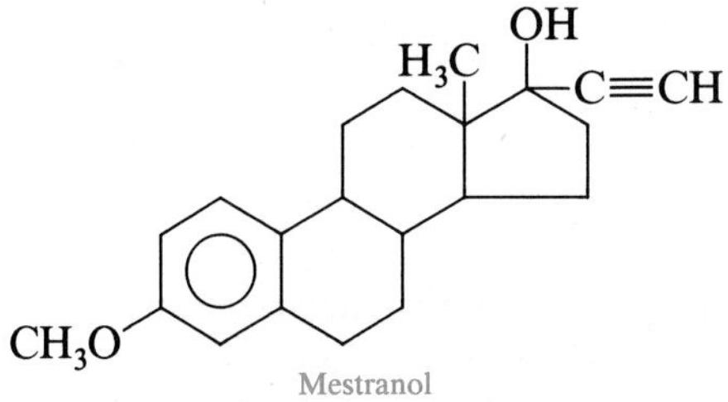

Mestranol
(synthetic estrogen)

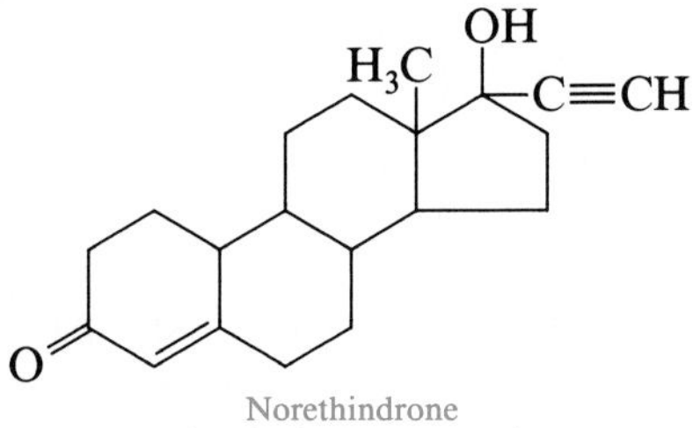

Norethindrone
(synthetic progesterone)

Early versions of the pill contained larger doses of the estrogen and progesterone derivatives. However, since the 1960s, researchers have succeeded in reducing the steroid content of the pill from about 200 mg to as little as 2.6 mg. In addition, "biphasic" and "triphasic" products were developed to reflect more accurately a woman's natural hormonal changes. These products are formulated to provide the lowest effective dosages of both progesterone and estrogen on different days of the menstrual cycle. An example of a biphasic oral contraceptive is Ortho-Novum 10/11, which provides 0.5 mg of norethindrone for 10 days, then 1.0 mg of norethindrone for 11 days; 35 μg of ethynyl estradiol is administered consistently for 21 days. Triphasic oral contraceptives change the amount of norethindrone three times instead of twice.

All oral contraceptives are prescription drugs and should be taken only after a checkup by a doctor. There are risks associated with the use of the pill. The risk of blood clots, the most common of the serious side effects, increases with age and with heavy smoking (more than 15 cigarettes per day). The chance of a fatal heart attack is about 1 in 10,000 in women between the ages of 30 and 39 who use oral contraceptives and smoke, compared with about 1 in 50,000 in users who do not smoke and 1 in 100,000 in nonusers who do not smoke.

Women are warned not to use oral contraceptives if they have or have had a heart attack or stroke, blood clots in the legs or lungs, angina pectoris, known or suspected cancer of the breast or sex organs, or unusual vaginal bleeding. In addition, they should not use the pill if they are pregnant or suspect they are pregnant.

The "minipill," developed in the 1970s, contains much smaller amounts of the synthetic progesterones (0.1–0.2 mg) and no synthetic estrogen. The minipill was introduced in the hope that its users would experience fewer side effects than users of the pill, since the estrogen component of the pill is regarded as the cause of many of the serious side effects. However, there is not sufficient information available to support this concept. Although the mechanism of minipills is not fully understood, these contraceptives are thought to stop conception by preventing release of the egg, by keeping the sperm from reaching the egg, and by making the uterus unreceptive to any fertilized egg that reaches it.

Steroid Drugs in Sports

The steroid testosterone is responsible for the muscle building that boys experience at puberty, in addition to the development of adult male sexual characteristics. Synthetic steroids have been developed in part to separate the masculinizing (androgenic) effects and muscle-building (anabolic) effects of testosterone. These steroids have been prescribed by physicians to correct hormonal imbalances or to prevent the withering of muscle in persons who are recovering from surgery or starvation.

Healthy athletes discovered that synthetic steroids appeared to have an anabolic effect on them as well. Initially these anabolic steroids were used by weight lifters and by athletes in track-and-field events like the shot-put and hammer throw. Later, some inconclusive evidence surfaced suggesting that anabolic steroids increased endurance, and this caused runners, swimmers, and cyclists to begin using them.

$1\ \text{mg} = 1 \times 10^{-3}\ \text{g}$

$1\ \mu\text{g} = 1 \times 10^{-6}\ \text{g or } 1 \times 10^{-3}\ \text{mg}$

Ethynyl estradiol

Such sports organizations as the National Collegiate Athletic Association and the International Olympic Committee have banned the use of anabolic steroids and other drugs by athletes. Although few human studies have been carried out on the use of anabolic steroids by healthy individuals, a number of harmful side effects have been identified.

The side effects of anabolic steroid use include acne, baldness, and changes in sexual desire. Some men experience enlargement of the breasts. Accompanying these noticeable changes are testicular atrophy and decreased sperm production. This is caused by an imbalance among the testes, pituitary, and hypothalamus due to the increased concentration of these male sex hormones in the bloodstream. High levels of male sex hormones cause the hypothalamus to signal the pituitary gland to lower production of two other hormones, luteinizing hormone and follicle-stimulating hormone, which stimulate sperm production in the testes. Although these changes appear to be reversible, additional testing is needed. In women the use of anabolic steroids produces facial hair, male-pattern baldness, deepening of the voice, and changes in the menstrual cycle. Most of these changes are not reversible.

In addition to these problems, oral-dose anabolic steroids are toxic to the liver. Testosterone taken orally is not very effective, since most of it is rapidly metabolized by the liver before it reaches the bloodstream. However, several of the common anabolic steroids are active when taken orally, in part because of an alkyl group in addition to the hydroxyl group at the carbon-17 position of the steroid nucleus (Fig. 21–14). This alkyl structure slows metabolism in the liver and thus allows more of the dose to reach the bloodstream, but it also increases liver toxicity. Some liver cancer has been reported in anabolic steroid users.

A small percentage of athletes continue to use anabolic steroids in spite of the publicity about harmful side effects and the ban on their use. David L. Black, director of the Athletic Drug Testing Laboratory at Vanderbilt University, reports that 1 to 2% of athletes in international competitions now test positive for steroids; in college athletics, 5% test positive; and in professional football, 7 to 8% test positive. Although there is no firm estimate of the number of people who take anabolic steroids, the percentages above together with the results of a 1988 survey of teenaged boys that indicated 7% used anabolic steroids suggests at least a million users in the United States, and the concern is that the demand is growing.

Use of Sex Hormones in Fattening Cattle

Sex hormones are used to promote more rapid growth in cattle being fattened for market. The hormones used include three natural ones (estradiol, progesterone, and testosterone) and two synthetic anabolic steroids (trenbolone acetate and zeranol, shown in Fig. 21–14). A controversy has developed over the use of sex hormones in cattle with the ban on hormone-treated cattle by the European Community on January 1, 1989. Since most U.S. cattle are given hormone implants to accelerate growth, the ban effectively blocks imports of beef from the United States valued at $100 million a year. The ban was imposed in spite of two reports—one from the World Health Organization and one commissioned by the European Community—that conclude that no residue standards are needed for the three natural hor-

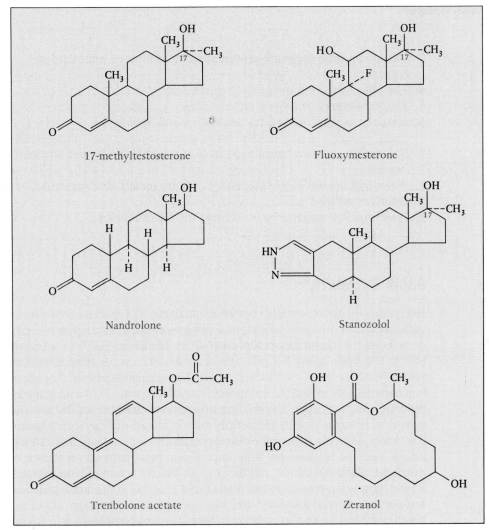

17-methyltestosterone

Fluoxymesterone

Nandrolone

Stanozolol

Trenbolone acetate

Zeranol

Figure 21–14 Structure of some anabolic steroids. The carbon-17 position in some oral-dose steroids is occupied by an alkyl group.

mones "when used under the appropriate conditions as growth promoters in farm animals." Limits of 1.4 and 2 μg/kg in beef muscle were suggested for trenbolone acetate and zeranol, respectively. Although these limits are stricter than those set by the U.S. Food and Drug Administration, U.S. officials have said that these limits would not be difficult to meet. However, U.S. officials argue that the problem with the recently imposed absolute ban on the use of hormones is the impossibility of certifying that an animal has never received the synthetic hormones.

SELF-TEST 21–B

1. The drug Prontosil is converted to sulfanilamide, which resembles _____ used for folic acid synthesis.
2. Penicillin is produced from cultures of what kind of organism?

3. The tetracycline drugs are produced from cultures of what kind of organism?
4. The male sex hormone is called _____.
5. The female sex hormone is called _____.
6. Anabolic as used in the term *anabolic steroid* means

   _____.

7. Name three undesirable side effects in male athletes who use anabolic steroids.
8. Drugs that are not active themselves but are metabolized to the active agent are called _____.
9. Penicillins kill bacteria by preventing them from making

   _____.

BRAIN CHEMISTRY

Everything we do is controlled by nerve signals racing between our brain and various parts of our nervous system. Estimates of the number of nerve cells, or **neurons,** in the human brain range from 10 billion (10^{10}) to a trillion (10^{12}). The brain along with the spinal cord make up the central nervous system (CNS) (Fig. 16 – 10). Other parts of the nervous system are the peripheral nervous system and the autonomic nervous system. The brain receives, processes, and acts upon information originating within the central nervous system or brought to it by the peripheral and autonomic nervous systems. The brain controls both our voluntary actions such as walking, talking, eating, and the involuntary functions of our body such as regulation of heartbeat, gland secretions, and the smooth-muscle action of blood vessels. Research in brain chemistry has made rapid progress in the last decade and most of the material discussed here was not known 30 years ago. However, much is still unknown about the brain because of the complexity and variety of actions controlled by the brain.

   The brain functions through its billions of neurons and millions of neuron networks. Neurons are like tiny circuits that pass electrical impulses

Figure 21 – 15   Two associated neurons. An electrical impulse is transmitted to another neuron across the synapse by a neurotransmitter. Neurotransmitter synthesis occurs in the axon terminals.

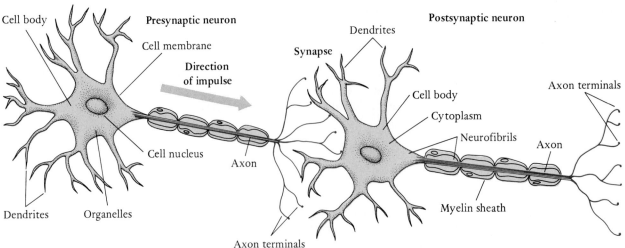

(messages) along a network of other neurons that result in a specific action (Fig. 21–15). The electrical impulse is transmitted in less than a thousandth of a second along the axon of one neuron across a small gap (synapse) to a dendrite on an adjacent neuron. The electrical impulse is based on changes in the concentration of Na^+ and K^+ inside and outside the neuron. A typical neuron may have ten thousand or more dendrites, and each dendrite can receive the input of numerous neurons.

The important thing to keep in mind is that these neuron networks are not physically connected. Chemicals known as **neurotransmitters** are released, which cross the synapse and bind to the dendrites of the adjacent neuron and trigger the flow of Na^+ and K^+ that allows the electrical impulse to travel to the adjacent neuron.

The importance of the K/Na ratio was discussed in Chapter 20. The K/Na ratio for normal brain cells is 1.7.

Neurotransmitters

Different neurons use different transmitters. Some of the principal neurotransmitters are shown in Figure 21–16. Some excite the receiving neuron to send the electrical impulse on to another neuron, and others halt the electrical impulse.

Figure 21–16   Some of the known neurotransmitters.

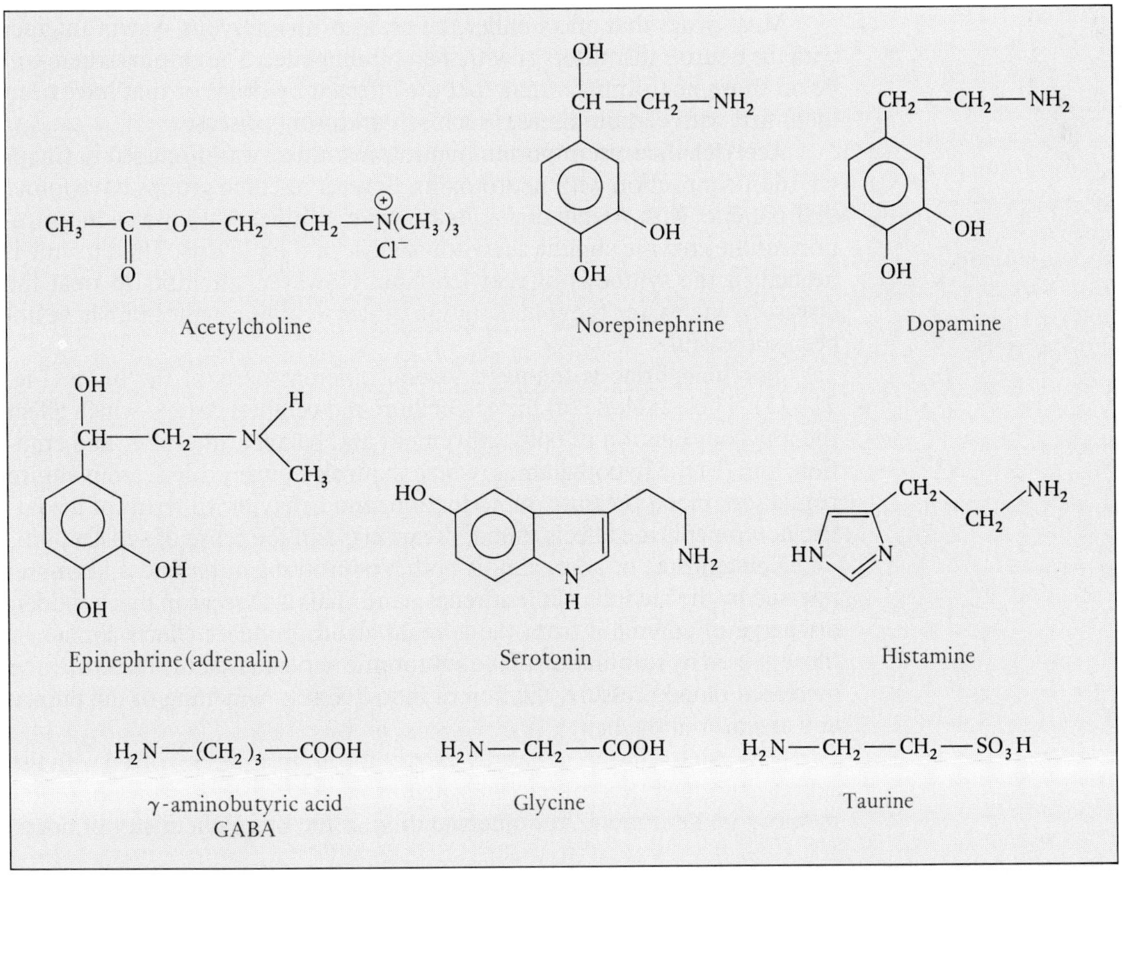

Figure 21–17 Diagram of major parts of the brain.

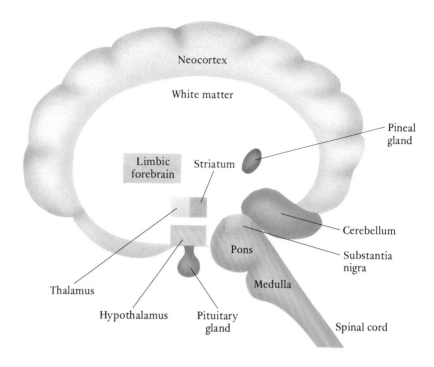

Most drugs that affect either the brain or the nervous system interact with the neurotransmitters or with their binding sites. The emphasis here will be on those neurotransmitters that are affected by drugs or that have been identified with certain diseases such as Parkinson's disease.

Acetylcholine, an important neurotransmitter, was discussed in Chapter 16 in connection with neurotoxins. Several research groups have found that patients with Alzheimer's disease have significantly lower concentrations of the enzyme choline acetyltransferase in their brains. This enzyme is needed in the synthesis of acetylcholine. However, attempts to treat the disease by increasing the concentration of acetylcholine in the brain have not been successful.

Norepinephrine is found in several different parts of the brain (Fig. 21–17). These include (a) the cerebellum and cerebral cortex, which affect the fine coordination of body movement and balance, alertness, and emotion; and (b) the hypothalamus, which controls hunger, thirst, temperature regulation, blood pressure, reproduction, and behavior. In terms of the latter, norepinephrine affects mood, dreaming, and the sense of satisfaction.

Epinephrine, or adrenalin, is both a neurotransmitter and a hormone released by the medulla of the adrenal gland. This is the reason that a sudden discharge of adrenalin from the adrenal gland produces effects similar to those caused by stimulation of the autonomic nervous system. These include increased blood pressure, dilation of blood vessels, widening of the pupils, and erection of the hair.

Dopamine is found in several areas of the brain and is involved with the action and integration of fine muscular movement as well as the control of memory and emotion. An understanding of the brain chemistry of dopa-

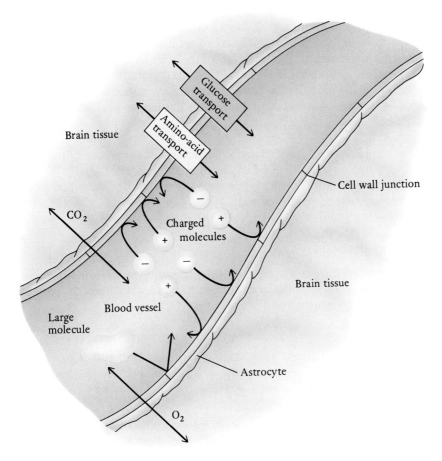

Figure 21-18   Blood-brain barrier.

mine led to the development of an effective treatment for Parkinson's disease. Patients with Parkinson's disease experience trembling and muscular rigidity, among other symptoms, because of a deficiency of dopamine. Dopamine will not cross the blood-brain barrier (Fig. 21-18), but research indicated that L-dopa is an effective drug because it crosses the blood-brain barrier and then reacts to produce dopamine.

The L in L-dopa indicates the left-handed optical isomer is used.

L-dopa

*asymmetric carbon

Since the adrenal gland also produces dopamine, a recent surgical procedure for patients with Parkinson's disease has involved transplanting dopamine-producing tissue from the patient's adrenal glands into the patient's brain. The hope is that the transplant will enable patients to produce sufficient dopamine to reduce the severity of their symptoms. Preliminary indications are that the procedure is effective, but the duration of the effect is not known.

Current research indicates that tissue from dead fetuses offers extraordinary possibilities for the treatment of diabetes, Alzheimer's disease, Parkinson's disease, and Huntington's chorea. Scientists have reported reversing the effects of Parkinson's disease in monkeys by implanting cells from the substantia nigra area of the midbrain of monkey fetuses. Other scientists have observed a reduction in chemically induced symptoms of Huntington's chorea, a fatal genetic brain disorder, in rats implanted with fetal nerve tissue. A moratorium has been placed on the use of fetal tissue in any federally funded research projects while the ethical dilemmas of its use are discussed.

Schizophrenia has been related to an excess of dopamine, and drugs that block dopamine receptor sites are used to treat this condition.

Serotonin controls sensory perception, the onset of sleep and body temperature, and may affect mood.

Drugs and the Brain

Drugs that affect the brain or nervous system generally work by inhibiting or promoting the production of neurotransmitters or by blocking the receptor sites occupied by neurotransmitters.

Analgesics

Analgesics are painkillers. They include both narcotic and nonnarcotic drugs. The **opiates** are analgesics with a strong narcotic action, producing sedation and even loss of consciousness.

Opiates Opium, obtained from the unripened seed pods of opium poppies contains at least 20 different compounds called **alkaloids** (organic nitrogenous bases described in Chapter 13). Some of the more important opiates are given in Figure 21–19. About 10% of crude opium is **morphine,** which is primarily responsible for the effects of opium.

Two derivatives of morphine are of interest. **Codeine,** a methyl ether of morphine, is one of the alkaloids found in opium and is used in cough syrup and for relief of moderate pain. Codeine is less addictive than morphine, but its analgesic activity is only about one fifth that of morphine. Another derivative of morphine is **heroin,** the diacetate ester of morphine, which does not occur in nature but can be synthesized from morphine. Heroin is much more addictive than morphine and for that reason has no legal use in the United States.

One of the most effective substitutes for morphine is **meperidine,** first reported in 1931 and now sold as Demerol. It is less addictive than morphine. Two other relatively strong pain relievers used today are **pentazocine** (Talwin) and **propoxyphene** (Darvon). Talwin is slightly addictive, whereas Darvon has not been shown to be addictive. However, Darvon has been much abused, and critics argue that Darvon is more dangerous and less effective in killing pain than other available opiates.

Considerable progress has been made in understanding the drug action of opiates. For many years scientists speculated about the action of opiates in the brain and the possible relationship to the human response to pain.

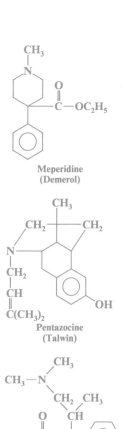

Meperidine
(Demerol)

Pentazocine
(Talwin)

Propoxyphene
(Darvon)

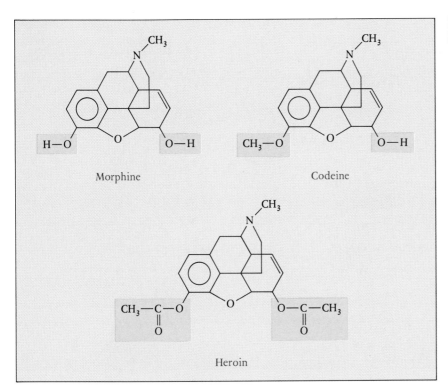

Figure 21–19   Some of the more important opiates.

Solomon Snyder and coworkers at Johns Hopkins University discovered in 1973 that the brain and spinal cord contain specific bonding or receptor sites that the opiate molecules fit as a key fits into a lock. This enhanced the search for opiate-like neurotransmitters. In 1975, John Hughes and Hans Koster-litz, of the University of Aberdeen, Scotland, isolated two peptides with opiate activity from pig brains. They decided to call these peptides **enkepha-lins** (from the Greek *en kephale,* meaning "within the head"); specifically, the two pentapeptides they isolated are known as methionine-enkephalin and leucine-enkephalin (Fig. 21–20). A year later, Roger Guillemin and coworkers at the Salk Institute isolated a longer peptide, called **beta-endor-phin,** from extracts of the pig hypothalamus. Beta-endorphin is 50 times more potent than morphine. Since this early work other enkephalins and endorphins have been isolated. In addition, the relationship of these natural opiates to pain has been studied.

Natural opiates include enkephalins (pentapeptides) and endorphins (peptides with about 30 amino acids).

Our bodies synthesize enkephalins and endorphins to moderate pain, and our pain threshold is related to levels of these neuropeptides in our central nervous system. Individuals with a high tolerance for pain produce more neuropeptides and consequently tie up more receptor sites than normal; hence, they feel less pain. A dose of heroin temporarily bonds to a high percentage of the sites, resulting in little or no pain. Continued use of heroin causes the body to reduce or cease its production of enkephalins and endorphins. If use of the narcotic is stopped, the receptor sites become empty and withdrawal symptoms occur.

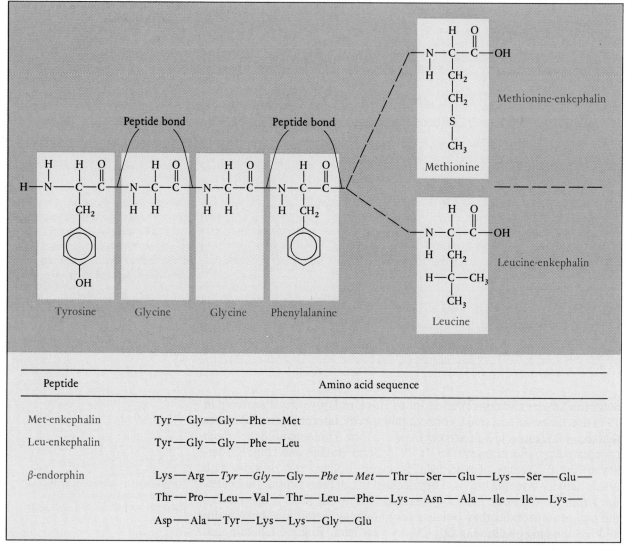

Figure 21–20 Some enkephalins and endorphins.

Peptide	Amino acid sequence
Met-enkephalin	Tyr—Gly—Gly—Phe—Met
Leu-enkephalin	Tyr—Gly—Gly—Phe—Leu
β-endorphin	Lys—Arg—*Tyr*—*Gly*—Gly—*Phe*—*Met*—Thr—Ser—Glu—Lys—Ser—Glu— Thr—Pro—Leu—Val—Thr—Leu—Phe—Lys—Asn—Ala—Ile—Ile—Lys— Asp—Ala—Tyr—Lys—Lys—Gly—Glu

Local Analgesics

Local analgesics block transmission of pain by plugging the Na^+ and K^+ channels of the neuron membranes (Fig. 21–21). Local analgesics include the naturally occurring **cocaine** (Table 21–3) derived from the leaves of the coca plant of South America, and the familiar **Novocain**.

Aspirin

When milder general analgesics are required, few compounds work as well for many people as **aspirin**. Not only is aspirin an analgesic, but it is also an antipyretic, or fever reducer. Each year about 40 million lb of aspirin are manufactured in the United States. Aspirin is thought to inhibit cyclooxy-

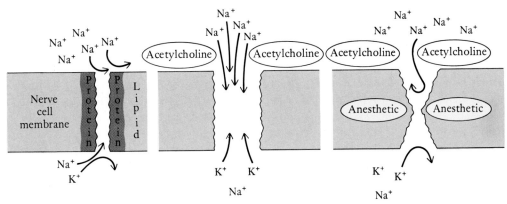

Figure 21–21 Action of acetylcholine and anesthetics in depolarizing the membrane of a nerve cell. Acetylcholine makes it possible for sodium and potassium ions to neutralize the negative charge associated with a nerve impulse so another impulse can be transmitted. Anesthetics block the action of acetylcholine and do not allow repetitive impulses to travel along the nerve.

genase, the enzyme that catalyzes the reaction of oxygen with polyunsaturated fatty acids to produce **prostaglandins.** Excessive prostaglandin production causes fever, pain, and inflammation—just the symptoms aspirin relieves.

Prostaglandins are a group of more than a dozen related compounds with potent effects on physiological activities such as blood pressure, relaxation and contraction of smooth muscle, gastric acid secretion, body temperature, food intake, and blood platelet aggregation. Their potential use as drugs is currently under widespread investigation. Two of the prostaglandins

TABLE 21–3 Some Local Analgesics

Name (Trade name)	Structural Formula	Use
Cocaine		Probably the first local analgesic used
Procaine (Novocain)		Often used in dental work
Lidocaine (Xylocaine)		More potent than procaine; can be applied to the skin

that have been characterized are prostaglandin E_1 (used to induce labor to terminate pregnancy) and prostaglandin E_2.

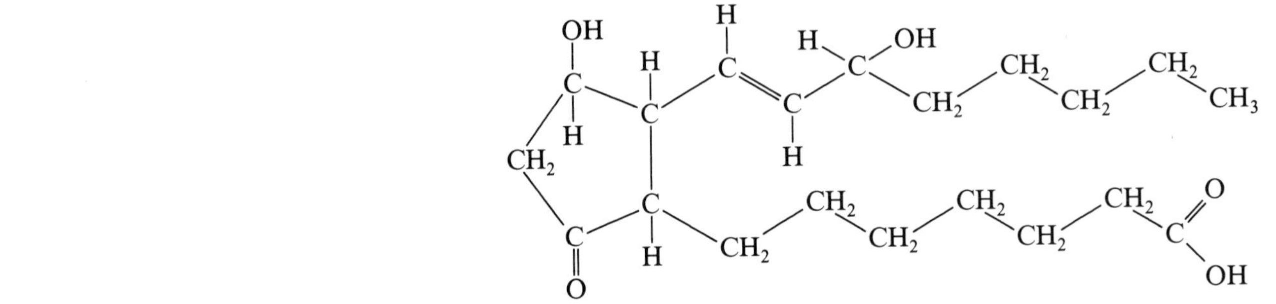

Prostaglandin $E_1 (C_{20}H_{34}O_5)$

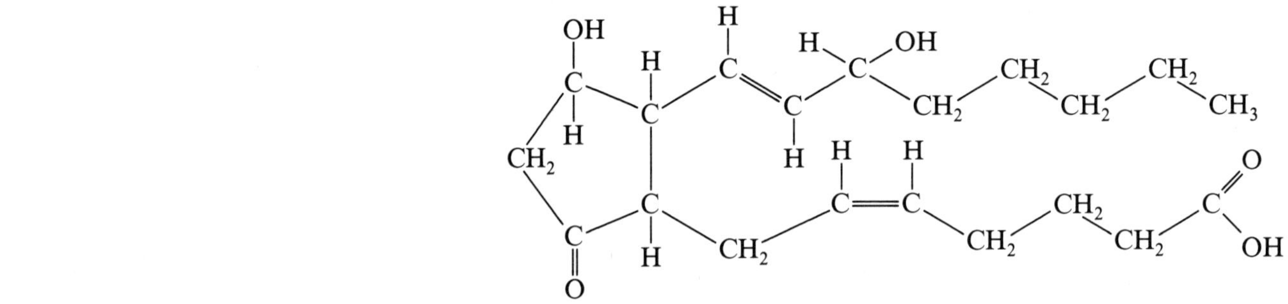

Prostaglandin $E_2 (C_{20}H_{32}O_5)$

A danger presented by aspirin is stomach bleeding, caused when an undissolved aspirin tablet lies on the stomach wall. As the aspirin molecules pass through the fatty layer of the mucosa, they appear to injure the cells, causing small hemorrhages. The blood loss for most individuals taking two 5-grain tablets is between 0.5 mL and 2 mL. However, some persons are more susceptible than others. Early aspirin tablets were not particularly fast dissolving, which aggravated this problem greatly. Today's aspirin tablets are formulated to disintegrate quickly, and the process can be sped up by crushing the tablet in a little water before ingestion.

Many aspirin tablets contain starch to hasten their disintegration in the stomach.

A greater potential danger of aspirin is its possible link to **Reye's syndrome,** a brain disease that also causes fatty degeneration in organs such as the liver. Reye's syndrome can occur in children recovering from the flu or chicken pox. Vomiting, lethargy, confusion, and irritability are the symptoms of the disease. Studies have shown a strong correlation between aspirin ingestion and the onset of Reye's syndrome. About one quarter of the 200 to 600 cases per year have proved fatal. Beginning in 1982, aspirin products were required to contain a warning about the possible link between aspirin and Reye's syndrome. As of now, there is no explanation for the relationship between aspirin and this disease.

Aspirin Substitutes

Several over-the-counter alternatives are now available for pain sufferers who have trouble with aspirin. The two principal ones are acetaminophen (Tylenol) and ibuprofen (Advil). Like aspirin, acetaminophen and ibuprofen are both analgesics and antipyretics. Of course, no drug should be taken without proper caution.

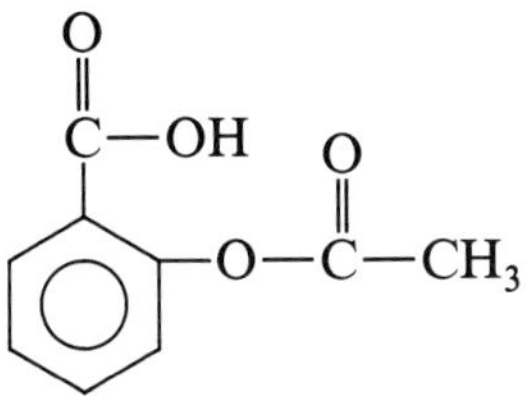

Acetylsalicylic Acid
(Aspirin)

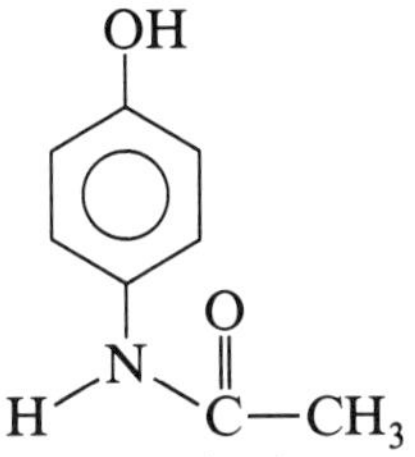

Acetaminophen
(Tylenol)

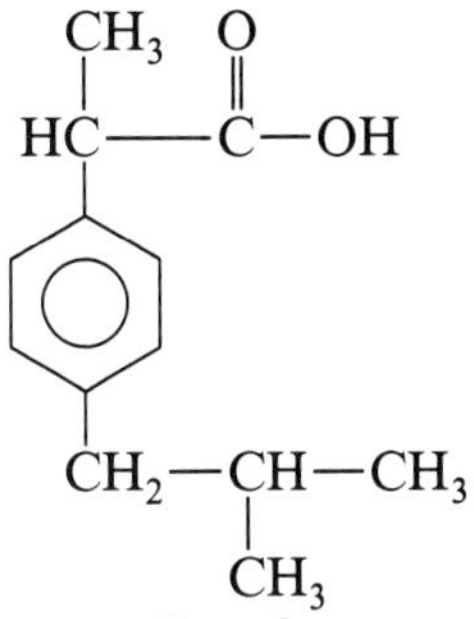

Ibuprofen
(Advil)

Depressants

Depressant drugs are either **sedatives,** which cause relaxation, or **hypnotics,** which induce sleep. The best-known depressants are the **barbiturates,** and structures of some of the common barbiturates are shown in Figure 21–22. Barbiturates are especially dangerous when ingested along with ethyl alcohol, another depressant, since the two together have a synergistic effect. This synergism has been the cause of many deaths.

Tranquilizers range from the mild diazepams such as Valium to the stronger promazines such as chlorpromazine, which is used to treat psychotic disorders. The structures are given in Figure 21–23 for some of the common tranquilizers.

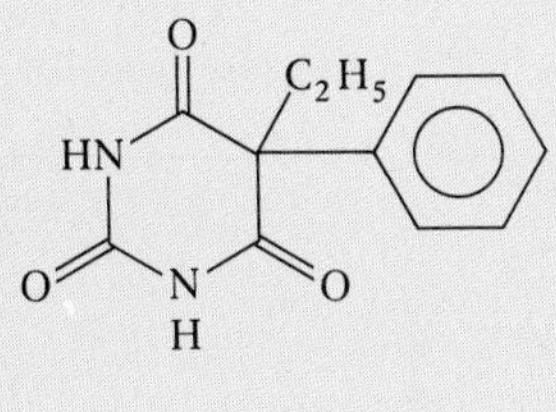

Phenobarbital
(Luminal)

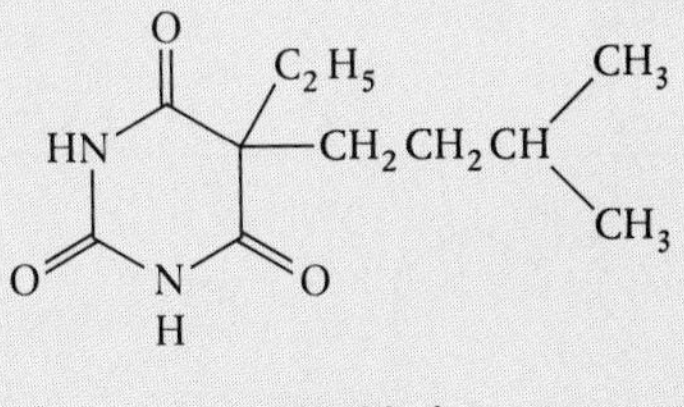

Amobarbital
(Amytal)

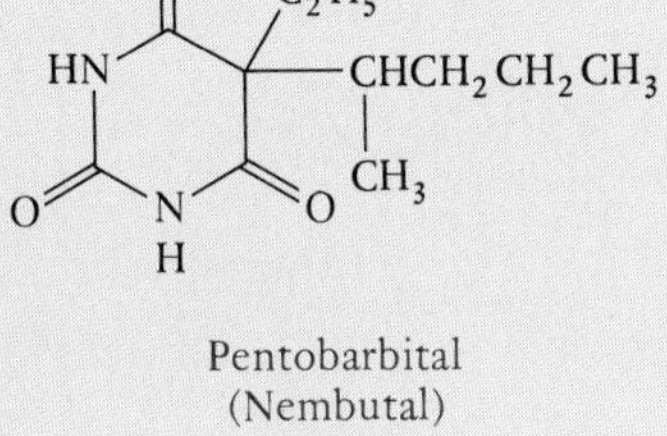

Pentobarbital
(Nembutal)

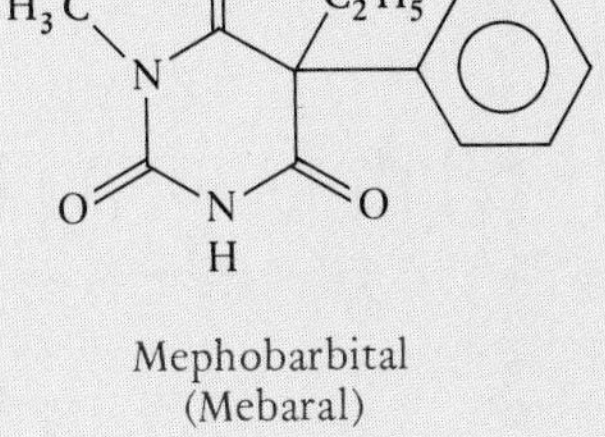

Mephobarbital
(Mebaral)

Figure 21–22   Some common barbiturates.

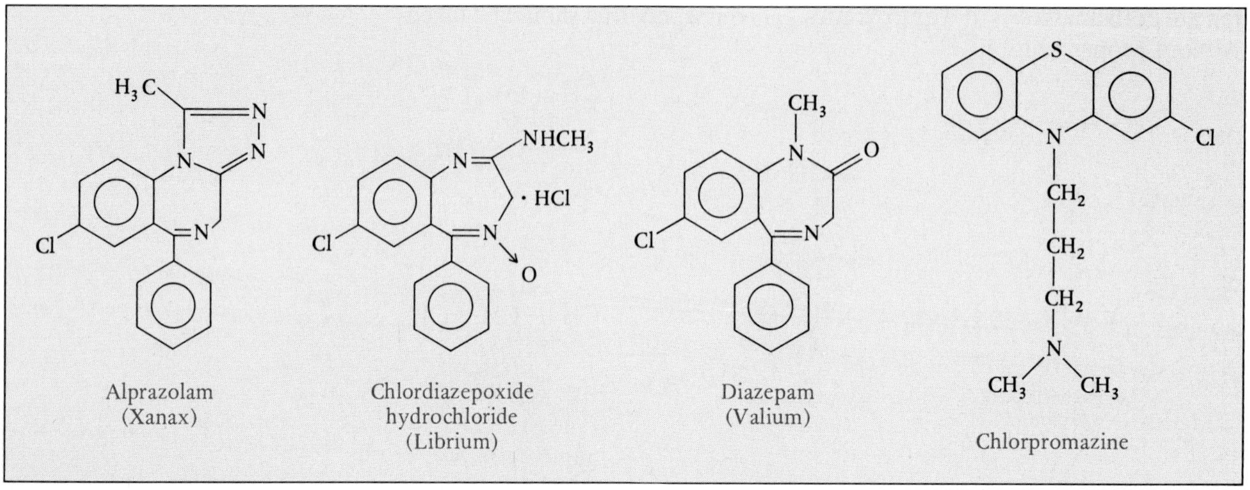

Figure 21–23   Some common tranquilizers.

Antidepressants and Stimulants

Antidepressants, or stimulants, increase the concentration of neurotransmitters such as norepinephrine and serotonin in the brain. The most well known are the **amphetamines.** Amphetamines are derivatives of phenylethylamine (Fig. 21–24). Note the similarity of the structure of amphetamine and methamphetamine to norepinephrine. This similarity helps us understand the stimulant activity of amphetamines. They have been used for years as a mood elevator and stimulant by persons who must stay awake. They are still used as an appetite suppressant. The two optical isomers of amphetamine have a different activity. The right-handed isomer, dextro-amphetamine (Dexedrine) is four times more active than the equal mixture of the two isomers, sold under the trade name Benzedrine.

Natural stimulants include the alkaloids **caffeine** and **nicotine** (Fig. 13–8).

Figure 21–24   Amphetamines are derivatives of phenylethylamine.

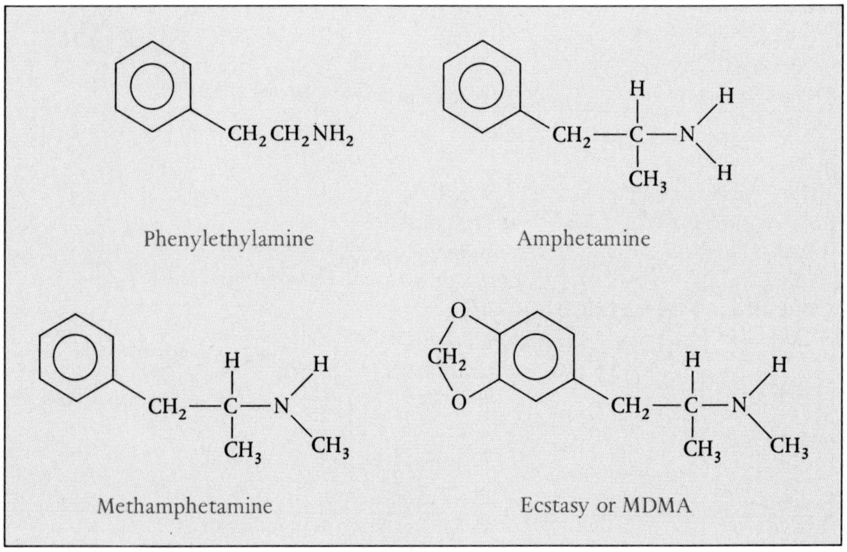

Drugs Used for Treating Mental Illness

Mood swings are caused by changes in levels of norepinephrine. An excess causes elation; a deficiency causes depression. The difference between normal persons and manic-depressive mental patients is the extent of excess or deficiency and the degree of difference between the two extremes. Because most drugs will only be able to control one problem—either excess or deficiency—there are limitations to the use of drugs in the treatment of this condition. Lithium salts seem to have the ability to control both mania and depression and to control the mood swings. Lithium carbonate has been used for 30 years in the treatment of manic-depressives. The reason for its action is more likely tied to the electrical balance since it seems to regulate both excess and deficiency.

Strong tranquilizers such as chlorpromazine calm manic-depressive patients and treat some of the symptoms of schizophrenia. Schizophrenia is believed to be caused by an excess of dopamine, and drugs that either reduce the production of dopamine or that block dopamine receptor sites are likely to be effective in treating schizophrenia. Chlorpromazine and the other strong tranquilizers work by blocking dopamine receptor sites, but proper dosage is important to avoid interfering with motor activity.

Drug Abuse

Although the list of drugs of abuse is wide ranging, a central theme is their relationship to brain chemistry and the effect of the drug on neurotransmitters or their receptor sites. Many of the drugs mentioned above are widely abused. These include amphetamines, narcotics, stimulants, and ethyl alcohol. Before discussing these, it is important to have an understanding of the extent of legal control of abused drugs. Table 21–4 lists the various classifications. Note that examples of controlled substances include drugs that are narcotics, hallucinogens, depressants, and stimulants.

TABLE 21–4 Classification of Drugs

Designation	Description	Examples
Over-the-counter (OTC)	Available to anyone	Antacids, aspirin, cough medicines
Prescription drugs	Available only by prescription	Antibiotics
Unregulated nonmedical drugs	Available in beverages, foods, or tobacco	Ethanol, caffeine, nicotine
Controlled substances*		
Schedule 1	Abused drugs with no medical use	Heroin, ecstasy, LSD, mescaline
Schedule 2	Abused drugs that also have medical uses	Morphine, amphetamines
Schedule 3	Prescription drugs that are often abused	Valium, phenobarbitol

* Drugs the sale, distribution, and possession of which are controlled by the Drug Enforcement Administration of the U.S. Department of Justice.

Problems with drug abuse and drug addiction are as old as civilization itself. However, a new dimension has been added in the 1980s as a result of three developments, "crack," "ice," and "designer drugs."

Crack

Review the reaction of sodium bicarbonate with acids in Chapter 9.

Crack is a purified form of cocaine obtained by heating a mixture of cocaine with sodium bicarbonate for 15 min. The reaction is an acid–base reaction since the base, sodium bicarbonate, is neutralizing cocaine hydrochloride, the normal form of cocaine. Since the reaction releases carbon dioxide gas, the term "crack" came about as a result of the crackling sound of the heated mixture during the release of carbon dioxide.

More than 20 million Americans have used cocaine, and about 4 million are using or abusing it now. The appearance of crack has caused an increase in the number of cocaine addicts, since crack is much more addictive than cocaine. The reason for its addictiveness is that crack is more potent than cocaine and is smoked rather than sniffed, giving the user a much quicker, more intense high. Because the high lasts less than 10 min, users have a tendency to use crack repeatedly over a short period, and most users thus become addicted after only one try. The problem is enhanced by the ready availability and low cost of crack ($10–15 per fix).

Ice

The newest drug abuse problem involves a smokable form of methamphetamine, nicknamed "ice." Ice is regarded as the number one drug problem in Hawaii and will likely become a serious national problem in the next few years.

Designer Drugs

Designer drugs are chemical substances that are structurally similar to legal drugs. Because of their action, they are potential drugs of abuse. All the designer drugs that have been discovered so far are either narcotics or hallucinogens. For example, fentanyl (Fig. 21–25) is a powerful narcotic marketed under the trade name Sublimaze. Fentanyl is about 150 times more potent than morphine and just as addictive, but very short-acting. Fentanyl is used in up to 70% of all surgical procedures in the United States. The derivatives of the fentanyl molecule are also potent narcotics. These drugs were called designer drugs when they first appeared on the streets because they had obviously been designed by some unscrupulous chemists for consumption by drug addicts. These fentanyl derivatives were every bit as potent as heroin, but because they were not listed on the U.S. Drug Enforcement Administration (DEA) list of controlled substances, they could be sold legally. Until a compound is recognized as being abused and is classified as a dangerous drug—a process known as *scheduling*—no laws apply to it.

In the past few years several fentanyl derivatives (Fig. 21–25) have appeared in California. Samples ranged from pure white powder, sold as China White, to a brown material. First came α-methyl fentanyl, then p-fluoro fentanyl, then α-methyl acetyl fentanyl, and in early 1984 3-methyl fentanyl, a compound 3000 times more potent than morphine. Because of this potency and because heroin addicts can use the fentanyl derivatives

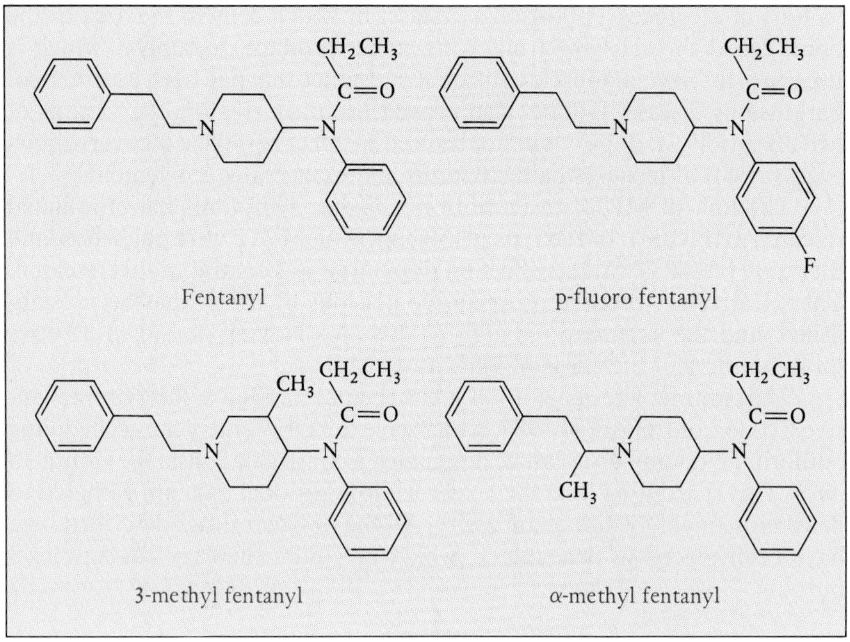

Figure 21-25 Fentanyl and several of its derivatives.

interchangeably with heroin, fentanyl derivatives, mostly 3-methyl fentanyl, have been responsible for over 100 overdose deaths in California.

Another group of designer drugs is derived from meperidine (Demerol). One of these is MPPP, which is short for 1-methyl-4-phenyl-4-propionoxy-piperidine. This compound was first synthesized in 1947, never used commercially, and never scheduled as a controlled substance. MPPP is about 3 times more potent than morphine and 25 times more potent than meperidine. It is structurally so close to meperidine that one has to look closely at the structures to see the difference (hint: look at the ester linkage). If the synthesis of MPPP is carried out at too high a temperature or at too low a pH, the product is MPTP, 1-methyl-4-phenyl-1,2,3,6-tetrahydropyridine.

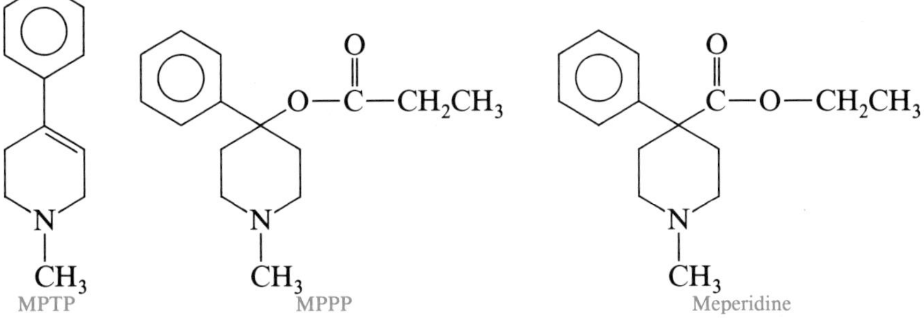

In 1982 a batch of MPTP-tainted MPPP, sold in San Jose, California, as "synthetic heroin," produced terrible side effects. It seems that MPTP causes the symptoms of Parkinson's disease, which include stiffness, impaired speech, rigidity, and tremors. Users of this batch of synthetic heroin became victims of advanced Parkinson's disease, in which cells in the area of the brain called the substantia nigra no longer produce dopamine, which is necessary for normal muscle control. A substance that had been used to treat Parkinson's disease, L-dopa, also proved useful in treating the victims of MPTP toxicity. L-dopa could not be used to effect complete recovery, however, since it also causes hallucinations and exaggerated movements.

The link of MPTP to Parkinson's disease symptoms has stimulated research in this area. In 1987, over 150 papers on MPTP were published, and the study of MPTP and its effect on dopamine was central to this research. The selective destruction of dopamine neurons by MPTP has been established, and the extensive research in this area is likely to aid in a better understanding of the cause of Parkinson's disease.

The control of designer drugs is easier since passage of the Comprehensive Crime Control Act of 1984, which gave the DEA emergency scheduling authority. Now any drug can be designated a controlled substance within 30 days. This scheduling lasts for 1 year while additional data are gathered to determine final scheduling authority. All the designer drugs described here have been placed in Schedule 1, which precludes their use for any legal purpose.

Hallucinogens

Hallucinogens are chemicals that cause vivid illusions, fantasies, and hallucinations. The most common hallucinogens obtained from natural sources are mescaline, which comes from the fruit of the peyote cactus, and lysergic acid diethylamide (LSD), which is made from lysergic acid derived from either the morning glory or ergot, a fungus that grows on wheat, rye, and other grasses. Marijuana is a mild hallucinogen and sedative made from the flowering tops, seeds, leaves, and stems of the female hemp plant, *Cannabis sativa.* The structures of the common hallucinogens are shown in Figure 21–26.

Although marijuana is not physically addicting and thus gets less attention than addicting drugs such as crack and cocaine, persons can become psychologically dependent on the drug. An estimated 6 million Americans between the ages of 18 and 25 smoke marijuana every day and another 16 million use it occasionally. Although these users may regard marijuana as a "safe" drug, recent studies indicate otherwise. Besides the dangers associated with psychological dependence, the smoke from marijuana and its active ingredient, tetrahydrocannabinol(THC), cause damage to the lungs, impede brain function, and hamper the immune system.

A hallucinogenic designer drug that appeared in the late 1970s is Ecstasy, also called XTC, Adam, or MDMA (Fig. 21–24). Before its appearance on the streets as a designer drug, ecstasy was used by some psychiatrists who found it useful in the treatment of schizophrenia, depression, and anxiety. By the early 1980s, this drug was being widely used as a recreational substance and was sold openly in bars, often accompanied by advertisements in windows. On July 1, 1985, the drug was placed in the Schedule I classification, which prevented its legal use by anyone. Nevertheless, there is evidence that the drug continues to be popular on college campuses. Results of surveys of college students at several campuses in 1987 indicated that 17 to 30% of students had tried the drug at least once.

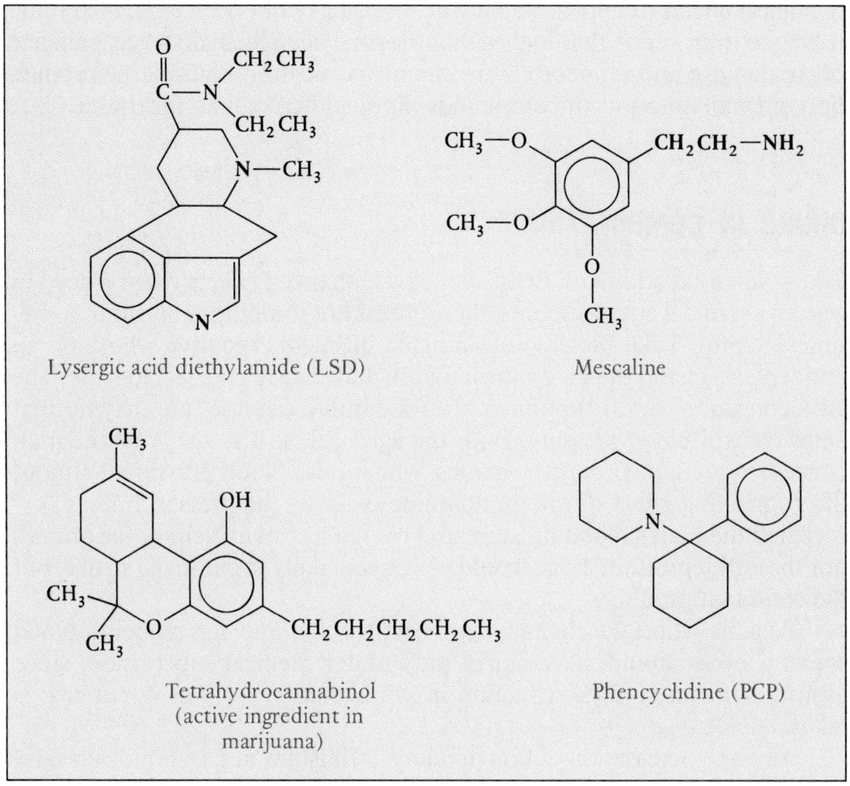

Figure 21-26 Some hallucinogens.

Lysergic acid diethylamide (LSD)

Mescaline

Tetrahydrocannabinol
(active ingredient in
marijuana)

Phencyclidine (PCP)

MDMA works by increasing the release of serotonin. Although the neurotoxicity is unknown, tests with rats show the MDMA destroys serotonin nerve terminals and results in long-term depletions of brain serotonin levels. Since the Drug Enforcement Agency and the Food and Drug Administration have reaffirmed MDMA's status as a Schedule 1 drug in 1988, studies of human neurotoxicity are presently not possible.

Alcoholism

Alcoholism is one of the largest health problems in the United States, where there are at least 10 million alcoholics and an estimated 200,000 deaths per year are attributed to alcohol abuse. A metabolic change that accompanies detoxification of ethanol in the liver is the synthesis of fat, which is deposited in liver tissue. Excessive drinking causes deterioration of the liver, known as **cirrhosis.** Cirrhosis of the liver is eight times more common among alcoholics than among nonalcoholics. Since 1974, cirrhosis of the liver has surpassed arteriosclerosis, influenza, and pneumonia to become the seventh leading cause of death. Alcoholics also tend to suffer from malnutrition and cardiovascular disease.

Genetic factors appear to play an important role in alcoholism. Research indicates that alcoholics metabolize acetaldehyde less effectively than nonalcoholics, probably because of a deficiency in the enzyme alcohol dehydrogenase. As a result, blood acetaldehyde levels are higher in alcoholics than in nonalcoholics for the same amount of alcohol intake. Although the biological effects of high acetaldehyde levels have not been explained fully, there are indications that higher than normal acetaldehyde levels enhance organ damage and influence brain chemistry, possibly causing the production of small amounts of compounds more addictive than morphine.

DRUGS IN COMBINATIONS

Like some food additives, drugs can have enhanced effects when placed in certain chemical environments; these effects are sometimes harmful, sometimes helpful. Take the case of an aging business executive who took an antidepressant and then ate a meal that included aged cheese and wine. The antidepressant was an inhibitor of monoamine oxidase, an enzyme that helps control blood pressure. Both the aged cheese and the wine the man consumed contained pressor amines, which raise blood pressure. Without the controlling effect of the monoamine oxidase, the pressor amines skyrocketed the man's blood pressure and caused a stroke. Neither the amines nor the antidepressant alone would have been likely to cause the stroke, but the combination did.

Pressor amines tend to increase blood pressure.

Persons who take digitalis for heart trouble and for reducing blood sodium levels should take aspirin only under medical supervision, since aspirin can cause a 50% reduction in salt excretion for 3 or 4 h after it is taken.

Alcohol increases the action of many antihistamines, tranquilizers, and drugs such as reserpine (for lowering blood pressure) and scopolamine (contained in many over-the-counter nerve and sleeping preparations), making such combinations extremely dangerous. Staying away from dangerous alcohol–drug combinations is not as easy as it may seem. Many people fail to realize that a large number of over-the-counter preparations, such as liquid cough syrup and tonics, contain appreciable amounts of alcohol.

Not all drug combinations are bad. The synergistic effect observed in the treatment of cancer by drugs in combination has already been mentioned. Doctors have been highly successful in prolonging the lives of leukemia and other cancer victims with combinations of drugs that individually could not do the job. Resistant kidney disease has also responded to drug combinations in cases in which single drugs were ineffective.

Perhaps the best advice is to take medicine only when you are seriously ill, making sure that a physician knows what you are taking.

THE ROLE OF THE FDA

Since 1940, more than 1200 new drugs have been introduced into the U.S. market. Included in this number are almost all of the drugs you probably recognize on your medicine shelf at home. Drug companies must petition the U.S. Food and Drug Administration (FDA) with data showing that a new drug is safe and effective for its intended use. Usually this involves extensive animal tests. If the FDA gives its approval, the drug undergoes limited controlled testing on healthy human test subjects. A second phase involves testing on research subjects who have the disease that the drug is intended to treat. Further tests are then carried out on larger groups to gauge the drug's effectiveness and safety. After all of this testing, only about one in ten drugs passes. The FDA has in the past been accused of being overly cautious, but the thalidomide incident (see Chapter 16) showed that caution can often prevent tremendous human anguish and needless suffering.

Dr. Frances O. Kelsey, a new drug investigator at the FDA in Washington, refused to allow thalidomide to be listed as safe and effective in light of some evidence she read in the data supplied concerning the drug and its effects on laboratory animals. Only after her refusal to certify thalidomide did the teratogenic effects of the drug become known. Thousands of young people born in the early 1960s in the United States owe Dr. Kelsey gratitude for their good health.

The FDA often gets embroiled in controversy regarding drugs that have claimed effectiveness in other countries. However, if a drug is not tested properly, the FDA has no recourse but to withhold approval. One drug that falls into this category is laetrile, a cyanide-containing compound found in the seeds of apples and peaches. This compound has been reputed to cure certain cancers. Yet the FDA refuses to approve its use as a chemotherapy agent because of a lack of sufficient evidence. Laetrile clinics are run in a number of countries, where U.S. citizens travel and receive treatments without FDA approval.

Another chemical involved in controversy is cyclosporin A, a cyclic peptide consisting of 11 amino acid residues with a molecular weight of 1202. Cyclosporin was discovered in 1970 by J. F. Borel of Switzerland as a metabolite in a culture broth of the fungus, *Tolypocladium inflatum Gams.* It was soon discovered that this peptide had powerful immunosuppressive effects, and by 1977 the chemical was being used in organ transplants in animals. A year later cyclosporin A was being used to suppress rejection of human liver transplants and marrow transplants in patients with leukemia. Because cyclosporin A so effectively involves itself with the immune system, it is not surprising to see it tried for the treatment of diseases in which the immune system is out of control (after all, the most effective anticancer drugs are themselves carcinogens). Cyclosporin A has recently been tested in the treatment of acquired immune deficiency syndrome (AIDS), although not in the United States, where it has not been approved for testing for that purpose. Whether cyclosporin A will ever be used in the treatment of AIDS in the United States depends on the data available to the FDA regarding its intended use.

If the FDA continues to do its job, we can be reasonably certain the drugs we take have been thoroughly tested before being approved for widespread human use.

EXPERIMENTAL DRUGS WITH PROMISE

The search for new drugs that are effective against diseases is a continuing process. Many of the drugs discussed in this section are still regarded as experimental, particularly those being used or considered for treatment of AIDS.

AZT

In 1987, the FDA approved the use of AZT for the treatment of AIDS.

AZT (azidothymidine, also called zidovudine) is the only drug presently approved for use in the United States in the treatment of AIDS patients. AZT is a derivative of thymidine, a nucleoside. Nucleosides contain a base and a ribose unit, whereas nucleotides (described in Chapter 15) contain a base, a ribose unit, and a phosphate group. The structures of thymine, thymidine, and thymidine phosphate are shown below:

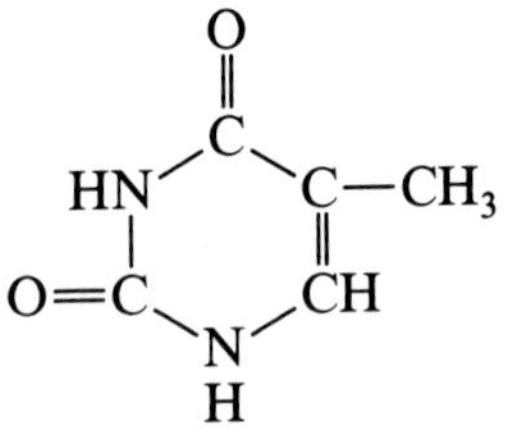

Thymine

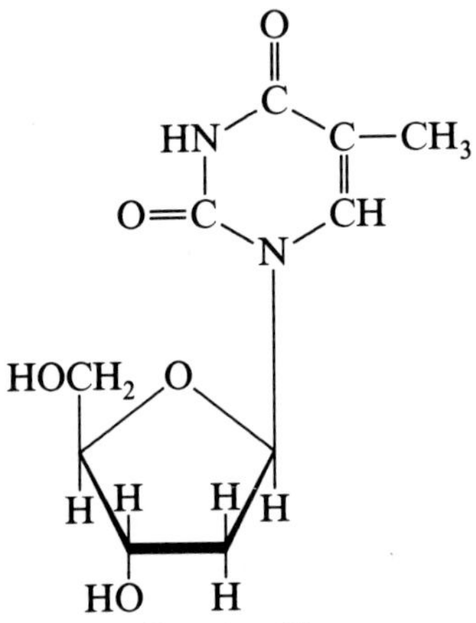

Deoxythymidine (Thymidine)

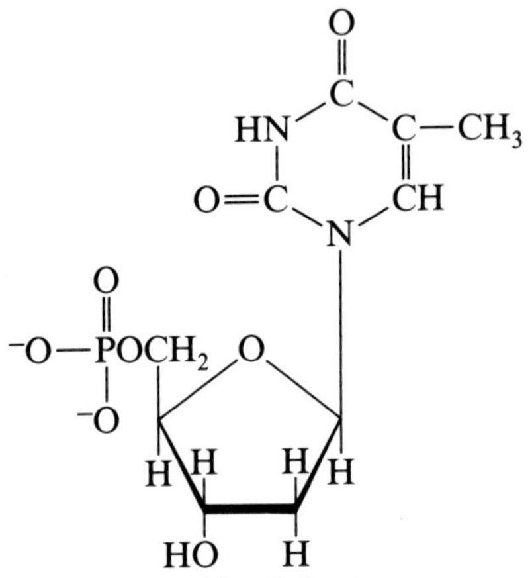

Thymidine 5′-phosphate

AIDS is caused by a **retrovirus,** a virus with an outer double layer of lipid material that acts as an envelope for several types of proteins, an enzyme called reverse transcriptase, and RNA. The term *retrovirus* is used because the virus enzyme carries out RNA-directed synthesis of DNA rather than the usual DNA-directed synthesis of RNA (see Fig. 15–32).

T cells are white blood cells that play a crucial role in controlling the immune response of the body.

The AIDS retrovirus penetrates the T cell, a key cell in the immune system of the body. Once the retrovirus is inside the T cell, the reverse transcriptase of the AIDS virus translates the RNA code of the virus into the T cell's double-stranded DNA, directing the T cell to synthesize more AIDS viruses. Eventually the T cell swells and dies, releasing more AIDS viruses to attack other T cells (Fig. 21–27).

AZT apparently works by being accepted by reverse transcriptase in place of thymidine. After AZT has become a part of the DNA chain, its structure prevents additional nucleosides from being added onto the DNA chain.

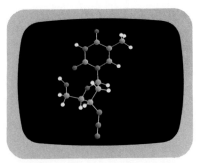

Computer-generated ball-and-stick model of AZT. (Courtesy of Phillips Petroleum.)

Azidothymidine was first synthesized in the 1960s by Jerome Horwitz, who was looking for a compound that would stop cancer cells from multiplying. He reasoned that the incorporation of a "fake nucleoside" into a DNA chain would prevent additional nucleosides from being added and thus cause cell division to stop. The idea didn't work because tumor cells recognized that AZT was not thymidine and didn't incorporate AZT into DNA chains. However, it appears the enzyme in the AIDS virus is fooled by the fake nucleoside.

Although AZT has been shown effective in retarding the progression of AIDS, it is not a cure. In addition, there are a number of problems with its

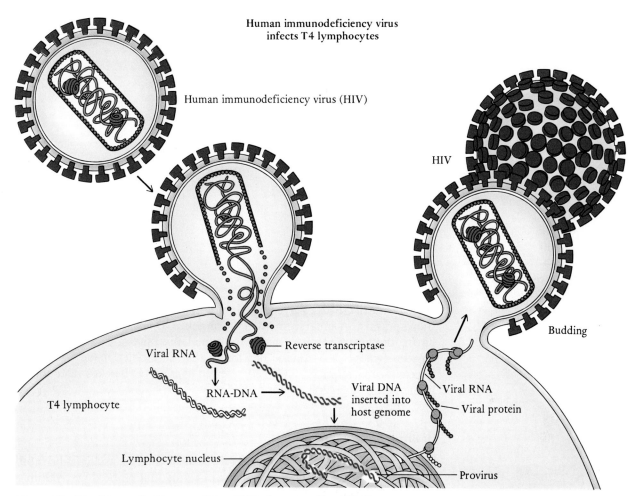

Human immunodeficiency virus
infects T4 lymphocytes

Human immunodeficiency virus (HIV)

HIV

Budding

Viral RNA

Reverse transcriptase

RNA-DNA →

Viral DNA
inserted into
host genome

Viral RNA

Viral protein

T4 lymphocyte

Lymphocyte nucleus

Provirus

Figure 21-27   Schematic diagram of the AIDS virus attacking T cells.

use. AZT has toxic side effects that many patients cannot tolerate; the cost of the drug is about $10,000 per year; and the AIDS virus can become resistant to AZT.

Preliminary tests with several other dideoxynucleosides (Fig. 21-28) indicate that they have fewer side effects and with further testing they may prove to be more effective than AZT. Another approach is to try to prevent the entry of the AIDS virus into cells. For example, the protein CD4, which is made using recombinant DNA technology, is a soluble form of the receptor to which the AIDS virus must bind in order to infect cells (Fig. 21-27). The idea is to inject CD4 into the bloodstream to tie up the AIDS virus and prevent it from spreading throughout the body.

Although these drugs offer hope for the treatment of AIDS, there has been little progress in developing a cure for it. A cure would require eliminating from the body the human immunodeficiency virus (HIV) that causes AIDS. The problems of developing a cure are associated with the latency period before AIDS develops and the genetic variability of the virus. How-

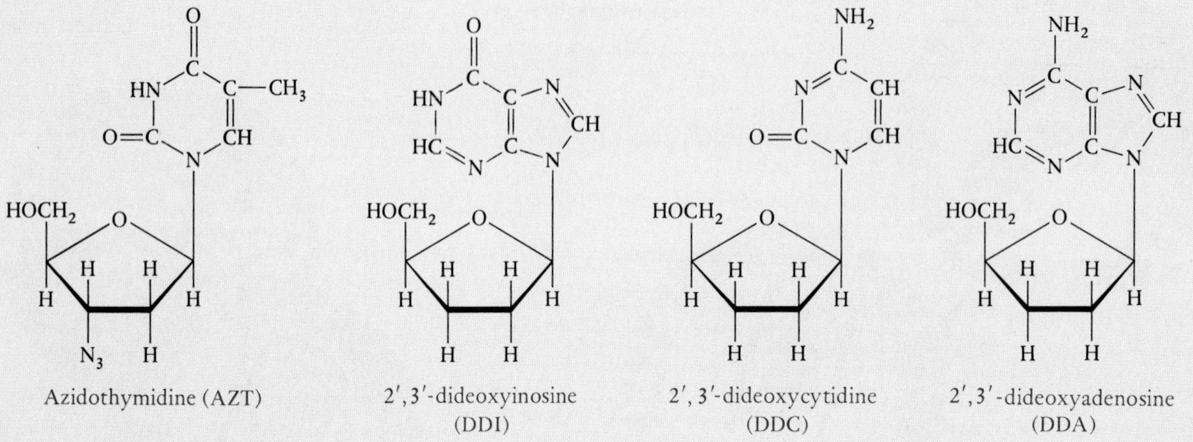

Figure 21–28   Structures of nucleoside derivatives that are being tested for treatment of AIDS. Note the similarity of their structures.

ever, the use of GP120, a glycoprotein that has a molecular weight of 120,000 shows promise. A small segment of the protein, CD4, was described above as having therapeutic value. David Ho of the University of California at Los Angeles School of Medicine has found that antibodies made by rabbits injected with a segment of GP120, referred to as C21E, blocked infection of human cells by several strains of HIV.

Bradykinin Blockers

Bradykinin is produced whenever body tissue is injured, and it is the most potent pain-producing chemical known. A drug that blocked bradykinin's action would act before the pain signal reached the central nervous system where morphine and other opiate drugs work. This would avoid the problem of using addictive narcotics such as morphine to ease the pain. Bradykinin is a polypeptide that contains nine amino acid residues in the following sequence

Arg-Pro-Pro-Gly-Phe-Ser-Pro-Phe-Arg

Arg = arginine, Pro = proline, Gly = glycine, Phe = phenylalanine, Ser = serine.

where the three letters stand for the particular amino acids present (See Chapter 15). Since bradykinin works by fitting into a receptor site, the strategy for developing blockers of bradykinin's action is to substitute one or more amino acids in the sequence. Over 200 potential blockers have been synthesized in which one or more amino acids have been changed. For example, swapping D-phenylalanine for proline in the number seven spot gives a polypeptide that has been shown to be effective in blocking pain.

Abortion Pill

A drug known as RU 486 was introduced in France in September, 1988 for use in terminating pregnancy up to the tenth week. The drug blocks progesterone so that the system stops its support of the ovum or fetus, thereby inducing spontaneous abortion. After opposition from antiabortion groups, the French company that developed the drug, Roussel Uclaf, abruptly withdrew it on October 26, 1988. However, two days later the French government in effect ordered Roussel to reintroduce it. Although prospects for the pill's sale in the United States are years away, extensive discussion of the ethical and moral issues by prochoice and antiabortion groups is already taking place.

SELF-TEST 21–C

1. Neurons signal the production of chemicals known as _____, which cross the synapse and bind to dendrites of adjacent neurons.
2. Patients with Parkinson's disease have a deficiency of the neurotransmitter _____ .
3. _____ is a derivative of morphine that is not found in nature.
4. _____ are pentapeptides found in the brain that are referred to as natural opiates.
5. Crack is a purified form of _____ .
6. Fentanyl derivatives that have been sold on the streets tend to act like what other drug of abuse?
7. AZT is used to treat _____ .
8. The difference between nucleotides and nucleosides is a _____ group.
9. The AIDS retrovirus penetrates _____, which are white blood cells that play a crucial role in controlling the immune response of the body.
10. An illegal drug, which may have pronounced psychological or addictive effects and differs little in a molecular sense from a legal drug, is often called a _____ .

MATCHING SET

_____ 1. Histamine
_____ 2. Cortisone
_____ 3. Tetracycline
_____ 4. Penicillin
_____ 5. Sulfa drug
_____ 6. Hallucinogenic designer drug
_____ 7. Propranolol
_____ 8. Procaine (Novocain)
_____ 9. Opium poppy
_____ 10. 3-Methyl fentanyl

a. 3000 times more powerful than heroin
b. Source of morphine
c. Interferes with cell wall synthesis
d. Analgesic used in dentistry
e. Related to Reye's syndrome
f. Lowers blood cholesterol
g. Causes symptoms of hay fever
h. Causes heart muscles to contract
i. Sulfanilamide
j. Muscle producing
k. Cyclosporin A
l. Methotrexate

_____ **11.** Anabolic

_____ **12.** Alkylating drug

_____ **13.** Antimetabolite drug

_____ **14.** Immune suppressant

_____ **15.** Calcium ion

_____ **16.** Nicotinic acid

m. Female sex hormone

n. Cyclophosphamide

o. Heart muscle relaxant

p. Antibiotic containing a four-ring structure

q. Ecstasy

r. Steroid

QUESTIONS

1. Discuss why women on the average live longer than men.
2. Why does the stomach not dissolve itself?
3. Name two alkaloid narcotics that are derived from morphine.
4. What is Reye's syndrome? What common drug is associated with Reye's syndrome?
5. Name three ions that neutralize negative charges associated with nerve and muscle cells.
6. Nitrogen mustards are alkylating agents. These interfere with DNA replication. Explain.
7. Explain how the antimetabolite methotrexate works to kill cells.
8. If the chemotherapeutic agents kill living cells, why do they have a preferential effect on cancer cells?
9. What is one of the dangers of chemotherapy using an alkylating agent or an antimetabolite?
10. Describe the action of antihistamines.
11. Describe how the sulfa drug sulfanilamide works.
12. How was penicillin discovered?
13. What happens when beta receptor sites in heart muscles are stimulated?
14. What can happen to a patient when a drug passes through the blood-brain barrier?
15. What is atherosclerotic plaque? What are its two major ingredients?
16. Explain how a beta blocker can lower blood pressure.
17. What effect does the calcium ion have on the heart?
18. How do the estrogens differ structurally from the other female hormone progesterone?
19. Name an anti-inflammatory steroid produced in the body.
20. Anabolic steroids that can be taken orally are unlike testosterone with respect to what structural feature?
21. What designer drug is toxic to the part of the brain called the substantia nigra? What disease does this drug cause?
22. The Drug Enforcement Administration has three different classifications of controlled substances. Explain the differences among these and give an example of each.
23. Look up the drugs scheduled by the Drug Enforcement Administration as Schedule 1 drugs. How many are there?
24. What is a retrovirus? How does the AIDS retrovirus attack the immune system of the body?
25. What is hypertension and how can a person's blood pressure provide information about the extent of hypertension?
26. What do HDL and LDL stand for? Which one should be at high levels and which one should be at low levels to reduce danger of heart attack? Explain.
27. Why is aspirin thought to cut the risk of a heart attack?
28. Explain the function of acetylcholine, norepinephrine, dopamine, and serotonin in the brain.
29. What is bradykinin and why are drug companies making derivatives of it?
30. What neurotransmitter affects mood changes in an individual? Explain how this information is used in treating mental illness.

22

Consumer Chemistry — Our Money for Chemical Mixtures

Marilyn Monroe. (Reprinted with permission from *Today's Chemist,* August 1989, 2(4), p. 21. Copyright 1989 American Chemical Society.)

We spend our money for food, clothing, housing, beauty and health products, transportation, and entertainment. To reap the full benefit of the things we purchase, we should know something of the types and the availability of the raw materials used, the physical and chemical modifications made in them, the properties that give the desired effects, and the precautions to be taken in their use. For example, it is common to find products with the same formulation from the same raw materials that differ considerably in cost due to differences in packaging, appearance, and advertising. Whether it is the selection of a nonpoisoning material for the water pipes in your home, a cleansing medium for costly apparel, or the preservation of a work of art, your knowledge of chemistry can help you make better choices.

A considerable number of consumer products, such as building materials in Chapter 11, have been discussed in the previous chapters. In this chapter we shall give attention to items that command consumer attention in the areas of cosmetics, cleansing agents, automotive products, and photography. The authors believe that you, using the approach presented here, can apply it to other products of interest to you; and, with a minimum of library research, base your consumer spending on chemical knowledge and understanding rather than on advertising hype. We live at a time when it is more and more necessary to **"read and understand the label."**

COSMETICS

The use of chemical preparations, which are applied to the skin to cleanse, beautify, disinfect, or alter appearance or smell, is older than recorded history. Such preparations are known as **cosmetics.** There is at best a fuzzy distinction between cosmetics and drugs. Traditionally drugs alter body functions, but cosmetics do not. However, antiperspirants are considered as cosmetics but do stop the secretions of the sweat glands. The distinction is becoming even more difficult with the introduction of creams to stop the aging process in wrinkling skin (Retin A) and to promote the growth of hair (minoxidil). Perhaps the best distinction is the level of governmental control in the introduction of new products. Drugs require elaborate safety testing prior to receiving approval by the U.S. Food and Drug Administration, but cosmetics do not.

Cosmetics are being expanded from their traditional use of adornment for esthetic reasons to scientific applications for skin care, protection from ultraviolet light, conditioning hair, and beautification with increasingly milder ingredients. Japan's Shiseido brand of cosmetics advertises the molecular formula of its humectant (sodium hyaluronate) right on the product label for a $50 tube of skin cream — an indication of the movement from an art to a science. Although there is considerable debate over the level of scientific understanding in cosmetics, there is no doubt about public appetite for cosmetics and the profit potential in the manufacture and sale of them. Approximately $25 billion is presently being spent each year for cosmetics in the United States.

Annual retail sales for cosmetics in Western Europe are approximately $30 billion; in Japan, $15 billion.

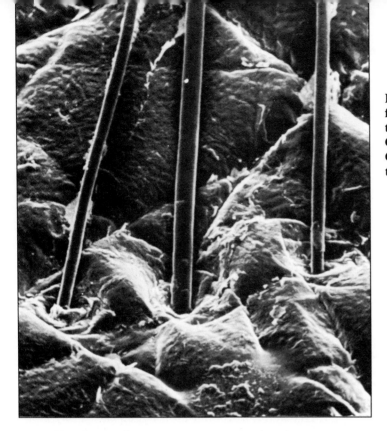

Figure 22–1 Replica of the surface of human forearm skin, showing three hairs emerging from the skin (× 225). (Courtesy of E. Bernstein and C. B. Jones: *Science*, Vol. 166, pp 252–253, 1969. Copyright 1969 by the American Association for the Advancement of Science.)

Skin, Hair, Nails and Teeth — A Chemical View

The skin, hair and nails are protein structures. Skin (Figs. 22–1 and 22–2), like other organs of the body, is not composed of uniform tissue and has several functions made possible by its structure; they are protection, sensation, excretion, and body temperature control. The exterior of the epidermis is called the stratum, corneum, or **corneal layer,** and is where most cosmetic

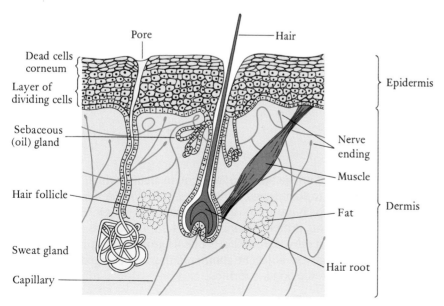

Figure 22–2 Cross section of the skin.

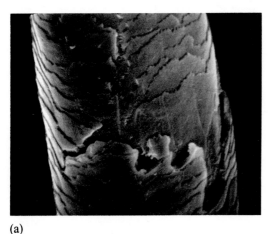

(a)

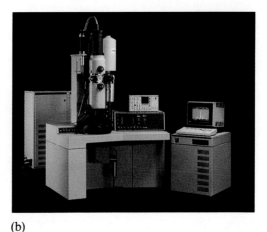

(b)

Figure 22–3 (a) Electron micrograph of human hair. Note the layers of keratinized cells. (b) An electron microscope. (Courtesy of Philips Electronic Instruments, Inc.)

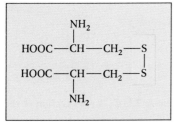

Figure 22–4 Molecular bonding in cystine. Note the disulfide bond.

The structures of protein tissues are due in part to disulfide cross-links and to ionic bonds between "molecules."

preparations for the skin act. The corneal layer is composed principally of dead cells with a moisture content of about 10% and a pH of about 4. The principal protein of the corneal layer is **keratin,** which is composed of about 20 different amino acids. Its structure renders it insoluble in, but slightly permeable to, water. Dry skin is uncomfortable, but an excessively moist skin is a good host for fungus organisms. An oily secretion, sebum, is secreted by the sebaceous glands to protect from excessive moisture loss. In order to control the moisture content of the corneal layer so that it does not dry out and slough off too quickly, moisturizers may be added to the skin.

Hair is also composed principally of keratin (Fig. 22–3). An important difference between hair keratin and other proteins is its high content of the amino acid cystine (Fig. 22–4). About 16 to 18% of hair protein is cystine, but only 2.3 to 3.8% of the keratin in corneal cells is cystine. This amino acid plays an important role in the structure of hair.

The toughness of both skin and hair is due to the bridges between different protein chains, such as hydrogen bonds and —S—S— linking bonds, called disulfide bonds.

Another type of bridge between two protein chains, which is important in keratin as well as in all proteins, is the ionic bond. Consider the interaction between a lysine —NH_2 group and a carboxylic group —COOH of glutamic acid on a neighboring protein chain. At pH 4.1, protons are added to the —NH_2 groups and removed from the —COOH groups, resulting in —NH_3^+ and —COO^- groups on adjacent chains. If the two charged groups approach closely, an ionic bond is formed.

$$\underset{\text{Lysine}}{\overset{|}{\text{H}}\overset{|}{\text{C}}\text{CH}_2\text{CH}_2\text{CH}_2\text{CH}_2\text{CH}_2\text{NH}_2} + \underset{\text{Glutamic Acid}}{\text{HOOCCH}_2\text{CH}_2\overset{|}{\underset{|}{\text{CH}}}} \xrightarrow{\text{at pH 4.1}}$$

$$\overset{|}{\text{H}}\overset{|}{\text{C}}\text{CH}_2\text{CH}_2\text{CH}_2\text{CH}_2\text{CH}_2\text{NH}_3^+ \quad {}^-\text{OOCCH}_2\text{CH}_2\overset{|}{\underset{|}{\text{CH}}}$$

$$\underset{\text{Ionic Bond}}{\uparrow}$$

As the pH rises above 4, keratin will swell and become soft as these cross-links are broken. This is an important aspect of hair chemistry since the pH of most shampoos and even water is above 4. Finger and toe nails are composed of hard keratin, a very dense type of this protein. These epidermal cells grow from epithelial cells lying under the white crescent at the growing end of the nail. Like hair, the nail tissue beyond the growing cells is dead.

The mineral content, or the hard part, of bones and teeth consists of two compounds of calcium. Calcium carbonate ($CaCO_3$) is present in bones and teeth in the crystalline form, known to mineralogists as aragonite. The second calcium compound found in teeth is calcium hydroxyphosphate [$Ca_5(OH)(PO_4)_3$], or apatite (Fig. 22–4).

Creams and Lotions

To remain healthy, the moisture content of skin must stay near 10%. If it is higher, microorganisms grow too easily; if lower, the corneal layer flakes off. Washing skin removes fats that help retain the right amount of moisture. If dry skin is treated with a fatty substance after washing, it will be protected until enough natural fats have been regenerated.

An **emollient** is a skin softener. Lanolin, an excellent emollient, is a complex mixture of esters from hydrated wool fat. The esters are derived from 33 different alcohols of high molecular weight and 37 fatty acids. Cholesterol is a common alcohol in lanolin and is found both free and in the esters. Cholesterol has hydrophilic properties due to the hydroxyl groups (—OH) and causes the fat mixtures to hydrogen-bond with water. Any preparation that holds moisture in the skin is also termed a **moisturizer** (Fig. 22–5). Emollients also include other natural and synthetic esters, ethers, fatty alcohols, hydrocarbons, and silicones.

Creams are generally emulsions of either an oil-in-water type or a water-in-oil type. An **emulsion** is simply a colloidal suspension of one liquid in another. The oil-in-water emulsion has tiny droplets of an oily or waxy substance dispersed throughout a water medium; homogenized milk is an example. The water-in-oil emulsion has tiny droplets of a water solution dispersed throughout an oil; examples are natural petroleum and melted butter. An oil-in-water emulsion can be washed off the skin surface with tap water, whereas a water-in-oil emulsion gives skin a greasy, water-repellent surface when rubbed under running water. The product is a **lotion** if the oil or water content is increased to provide fluidity.

(a)

(b)

Graphic of hair structure. (a) The structure of hair begins with the spiral proteins, which are covered in strands (b) as they are bound together. This allows for hair to be fine or coarse in texture.

You may wish to review the concept of colloidal sizes of particles in Chapter 2.

Figure 22–5 The hydroxyl groups of lanolin form hydrogen bonds with water and keep the skin moist. The fat parts of the molecule are "soluble" in the protein and fat layers of the skin.

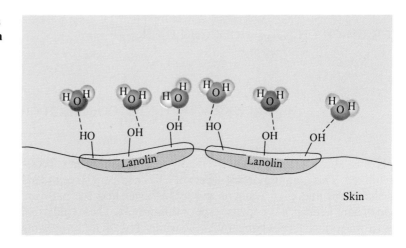

Spermaceti oil is chiefly cetyl palmitate, an ester of cetyl alcohol and palmitic acid. Identify the ester group in its formula: $CH_3(CH_2)_{14}COO(CH_2)_{15}CH_3$.

Cold cream originally was an emulsion of rose water in a mixture of almond oil and beeswax. Subsequently, other ingredients were added to get a more stable emulsion. An example of a modified cold cream composition is: almond oil, 35%; beeswax, 12%; lanolin, 15%; spermaceti (from whale oil), 8%; and strong rose water, 30%. With the wide variety of oils, waxes, emollients, and perfumes available, compositions for skin creams appear limitless. There are already many tested products on the market. Japan's Shiseido Company claims to have duplicated the moisture-binding substance plentiful in infants' skin, hyaluronic acid (Fig. 22–6), and they market the resulting cream for more than $100 for two 1-oz bottles, a 0.12% mixture for daytime and a 0.15% mixture for night. Retinoic acid (a derivative of vitamin A) has been used as an acne treatment for many years, but now Johnson and Johnson will test this cream ingredient to see whether it will reduce wrinkle formation in human skin. And the search goes on.

Figure 22–6 Hyaluronic acid, claimed to be the moisture-binding substance in infants' skin, represents the current effort to understand the molecular basis of moisture control in skin. (Shiseido Company.)

Creams offer an excellent application base for other cosmetic preparations or medical applications as in the cases of cream deodorants and antifungal foot products. The cream is the vehicle for chemical delivery into the skin structure.

Lipstick

The skin on our lips is covered by a thin corneal layer that is free of fat and consequently dries out easily. A normal moisture content is maintained from the mouth. In addition to being a beauty aid, lipstick, with or without color, can be helpful under harsh conditions that tend to dry lip tissue.

Lipstick consists of a solution or suspension of coloring agents in a mixture of high-molecular-weight hydrocarbons or their derivatives, or both. Consistency of the mixture over a wide temperature range is all important in a product that must have even application, holding power, and a resistance to running of the coloring matter at skin temperature. Lipstick is perfumed and flavored to give it a pleasant odor and taste. The color usually comes from a dye, or "lake," from the eosin group of dyes. A **lake** is a precipitate of a metal ion (Fe^{3+}, Ni^{2+}, Co^{3+}) with an organic dye. The metal ion modifies the natural color of the dye and usually produces a more intense color; the metal also keeps the dye from dissolving in the oil medium, thus preventing the color from running.

Two commonly used dyes are dibromofluorescein (yellow-red) and tetrabromofluorescein (purple) [Fig. 22–7].

The ingredients in a typical formulation of lipstick are given in Table 22–1.

Face and Body Powder

Powder is used to give the skin a smooth appearance and a dry feel. Face powders often contain dyes to impart color or shading to the skin. The principal ingredient in body powder is talc ($Mg_3(OH)_2Si_4O_{10}$), a natural mineral able to absorb both water and oil. The absorptive properties of talc are due to the electronegative oxygen atoms, which can hydrogen-bond to water, and to the extensive amount of surface area resulting from the

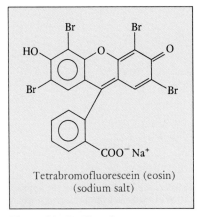

Tetrabromofluorescein (eosin)
(sodium salt)

Figure 22–7 Tetrabromofluorescein, a purple dye used in lipstick.

A lake is a coloring agent made up of an organic dye adhering to an inorganic substance called a mordant. Some lakes are also approved as food colors.

TABLE 22–1 A Typical Lipstick Formulation

Ingredient	Purpose	Percentage
Dye	Furnishes color	4–8
Castor oil, paraffins, or fats	Dissolves dye	50
Lanolin	Emollient	25
Carnauba and/or beeswax	Makes stick stiff by raising the melting point	18
Perfume	Gives pleasant odor	Small amount
Flavor	Gives pleasant taste	Small amount

Note: Carnauba wax and beeswax are mixtures of high-molecular-weight esters.

TABLE 22–2 General Formula for Body Powder

Ingredient	Purpose	Percentage
Talc	Absorbent	56
Precipitated chalk ($CaCO_3$)	Absorbent	10
Zinc oxide	Astringent	20
Zinc stearate	Binder	6
Dye	Color	Trace
Perfume	Odor	Trace

Powders are less effective for the application of medicines but preferred when dryness is desired, as in foot powders.

fineness of the ground powder. A binder, such as zinc stearate, is necessary to increase adherence to the skin. Zinc oxide or another astringent is added to shrink tissue and reduce fluid flow in oily skins. A general formulation for body powder is given in Table 22–2.

Eye Makeup

Petrolatum, or petroleum jelly, is a semisolid mixture of hydrocarbons (saturated, $C_{16}H_{34}$ to $C_{32}H_{66}$; and unsaturated, $C_{16}H_{32}$; etc.; melting point, 34–54°C.

Eye makeup consists of emollients, solvents, preservatives, and colors. Mixtures of fats, oils, petrolatum, lanolin, beeswax, and paraffin can be blended to give the desired consistency and melting point. Eyebrow pencils can be colored black with lampblack (carbon), brown with a mixture of iron oxide and lampblack, or a variety of colors with other dyes.

The molding and sticky qualities of mascara are obtained by increasing the amounts of soap and waxes in the mixture. Chromic oxide imparts a dark green color to mascara while ultramarine (a sodium and aluminum silicate admixed with sodium sulfide) gives a blue color. Mascara may be made water-soluble or water-resistant depending on the emollients and solvents used.

Glycerides are triesters of glycerine and fatty acids. Refer to the section on soap making later in this chapter.

A typical eye shadow base is composed of 60% petroleum jelly, 6% lanolin, 10% fats and waxes (beeswax, spermaceti oil, and cocoa butter are commonly used) and the balance zinc oxide (white) plus tinting or coloring dyes. Cocoa butter is composed of glycerides of stearic, palmitic, and lauric acids.

Suntan Lotions

Ultraviolet light darkens skin as the skin responds to the light by increasing the concentration of the natural pigment **melanin.** The melanin absorbs some of the ultraviolet photons and changes the energy to heat, thus protecting the molecular structure of the skin from damage by these photons. However, even with melanin's protection, ultraviolet light in the short-wavelength region causes a general degradation of the skin and, in extreme cases, skin cancer. The problem is more acute for fair-skinned people, whose skin has smaller amounts of melanin.

Since chemicals selectively absorb particular wavelengths of light, it is possible to screen the skin from the harmful radiation by placing an absorbing chemical in a lotion to be applied to the skin. The chemical must function in a manner similar to melanin. The first popular **sunscreens** contained *p*-aminobenzoic acid (Fig. 22–8) and some of its derivatives as the active

Sunbather (*The World of Chemistry*, Program 10, "Signals from Within.")

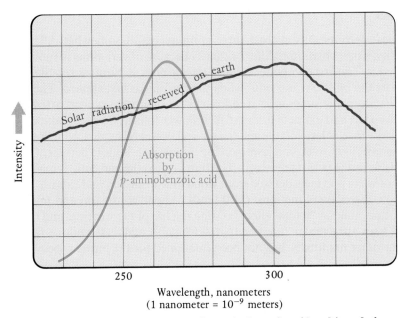

Coppertone assembly line. (*The World of Chemistry*, Program 10, "Signals from Within.")

Figure 22–8   Absorption spectrum of *p*-aminobenzoic acid and its relationship to solar ultraviolet radiation received on Earth. Maximum absorption occurs at 265 nm, although it absorbs at other wavelengths as shown. The maximum of the deep-burning ultraviolet radiation received on Earth is about 308 nm.

ingredients. The absorption spectrum for *p*-aminobenzoic acid is shown in Figure 22–8, showing that it absorbs an appreciable amount of the radiation in the dangerous portion of the spectrum.

There is a growing concern relative to possible toxic effects with the *p*-aminobenzoic acid compounds. In Europe, the market is moving toward cinnamates, derivatives of cinnamic acid, and in the United States the compounds on the rise in this market are the benzophenones.

Sunbathers refer to *p*-aminobenzoic acid and its close derivatives as PABA.

Sunscreen testing (*The World of Chemistry*, Program 10, "Signals from Within.")

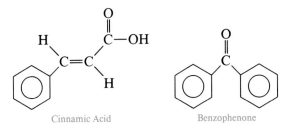

Cinnamic Acid   Benzophenone

The sun protection factor (SPF) for a sunscreen gives the ratio of the protection of the screen to that in natural skin. An SPF of 4, then, would provide four times the skin's natural sunburn protection. While numbers above 30 are advertised, the FDA has questioned whether numbers above 15 are realistic.

Some suntan products contain a dye, which dyes the skin. Local anesthetics such as benzocaine are added to some suntan preparations to overcome the pain associated with overexposure to the sun.

A recent application of sunscreening is the introduction to the market of eyedrops that claim protection from ultraviolet light for several hours.

Perfume

A typical perfume has at least three components of somewhat different volatility. Their volatilities differ primarily because of variations in molecular weights. The first component, called the **top note,** is the most volatile and therefore the most obvious odor when the perfume is first applied. The top note in natural perfumes is derived from citrus oils, herbs such as rosemary and lavender, and extracts from crushed flower foliage. The second, called the **middle note,** is less volatile and is a natural flower oil extract from flowers such as rose, lily, violet, and jasmine. The middle note is more persistent and noticeable as the perfume is worn. The third, called the **end note,** is made of resins or waxy polymers derived from wood, musk, amber, and balsam. The end note serves two functions: (1) it tends to bond to the top and middle notes, thus making "a total effect" during the life of the wearing, and (2) it acts as a sex attractant, a function that is debatable and that, while certainly true for other mammals, has not been quantified for the human species.

Figure 22–9   Synthetic compounds with jasminelike odor. Although some of the natural compounds have been isolated, analyzed, and then synthesized, the natural perfume essence is a complicated mixture of compounds.

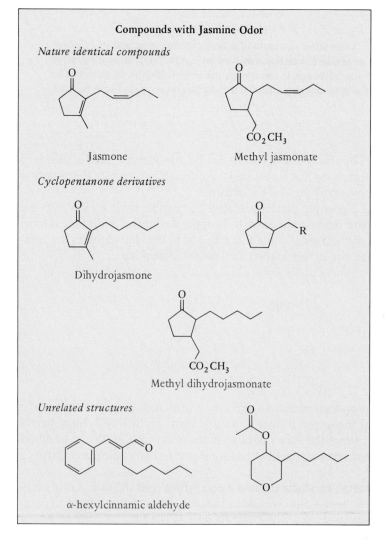

Compounds with Jasmine Odor

Nature identical compounds

Jasmone   Methyl jasmonate

Cyclopentanone derivatives

Dihydrojasmone

Methyl dihydrojasmonate

Unrelated structures

α-hexylcinnamic aldehyde

TABLE 22–3 Approximate Costs of Jasmine Note (1988)

Component	Cost (per lb)
Absolute natural jasmine	$2500–4000
Synthetic jasmone and methyl jasmonate mix	$250–400
Simpler derivatives with "same" effect	$8–40
Unrelated molecules with similar effects	$1.60–4.00

Note: Less than 10 tons of the natural and nature-identical materials are sold each year on a worldwide basis, whereas over 10,000 tons of the cheaper materials are used annually.

One hundred years ago perfumes were only available to the very rich. Now they are so common they are included in soaps, detergents, household cleaners, and even bleaches. This change resulted from advances in analytical techniques, such as the identification of over 200 chemicals in jasmine extract, and in organic synthesis, which made possible the synthesis of the natural compounds and related compounds that sometimes have better qualities than the natural ones. The famous Chanel No. 5 was the first fine perfume to employ synthetic chemicals in 1921. Since then essentially all of the natural chemicals in the top, middle, and end notes have been replaced by synthetics in the perfume industry. The natural products are still prepared for those who prefer and can afford them.

Figure 22–9 gives the molecular structures for the two most active ingredients in jasmine, jasmone and methyl jasmonate; and Table 22–3 gives the approximate costs for the natural products and for the various substitutions that can be made.

Typical perfumes are 10 to 25% perfume essence and 75 to 90% alcohol and a fixative to retain the essential oils. After-shave lotions and colognes are diluted perfumes, about one-tenth (or less) as strong. In addition to giving cosmetics a desirable odor, perfumes are added to mask the odors of other constituents that are unpleasant. Perfumes are mildly bactericidal and antiseptic because of the alcohol content.

The relationship between molecular structure and odor is beginning to be understood. There are seven primary odors for the human; they are camphorous, musky, floral, pepperminty, ethereal, pungent, and putrid. Although substances in a given class do not share similarity in empirical and structural formulas, they do share roughly the same molecular shape and size. Seven kinds of receptor sites exist in the olfactory cells of the nose. When a molecule of the correct size and shape fits into a complementary receptor site, a particular smell impulse is initiated (See Fig. 20–10). As this relationship between molecular shape and the sense of smell is further elucidated, people will surely use this knowledge to modify their surroundings with a more pleasing environment.

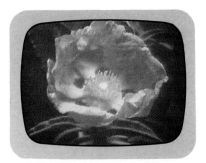

Perfume science seeks to explain scent in terms of molecular structure. (*The World of Chemistry*, Program 13, "The Driving Forces.")

SELF-TEST 22–A

1. The surface of the skin epidermis is known as the _____ layer.
2. The oily secretion of skin is _____.

3. Keratin is a protein found in _____, _____, and
_____.

4. Bridges between protein chains may be _____ linkages or
_____ bonds.

5. What is an ideal moisture content for human skin?

6. Lanolin is a skin softener or an _____.

7. A skin cream is either an oil-in-water or a water-in-oil _____.

8. A cosmetic that is colored with "a lake" is _____.

9. The major mineral ingredient in face or body powder is
_____.

10. An end note is likely to be the attracting odor in a perfume. (   ) True
(   ) False

11. An example of an active chemical that will absorb ultraviolet light in
a sunscreen lotion is _____.

12. Skin darkens because of an increased concentration of the skin
pigment _____.

CURLING, COLORING, CONDITIONING, GROWING, AND REMOVING HAIR

The curl, color, texture, and presence of human hair are matters of personal
choice that vary considerably from person to person. Chemical means are
available to alter these conditions, except for the growing of hair on skin
where none is present. The interest in hair care is evidenced by the fact that
more money is spent in the United States on hair care than any other group
of cosmetics and toiletries.

Changing the Shape of Hair

When hair is wet, it can be stretched to one and a half times its dry length
because water (pH 7) weakens some of the ionic bonds and causes swelling of
the keratin. Imagine the disulfide cross-links remaining between two protein
chains in hair as in Figure 22 – 10. Winding the hair on rollers causes tension

Annual U. S. sales of cosmetics and toiletries in billions of dollars.

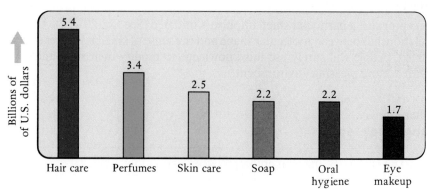

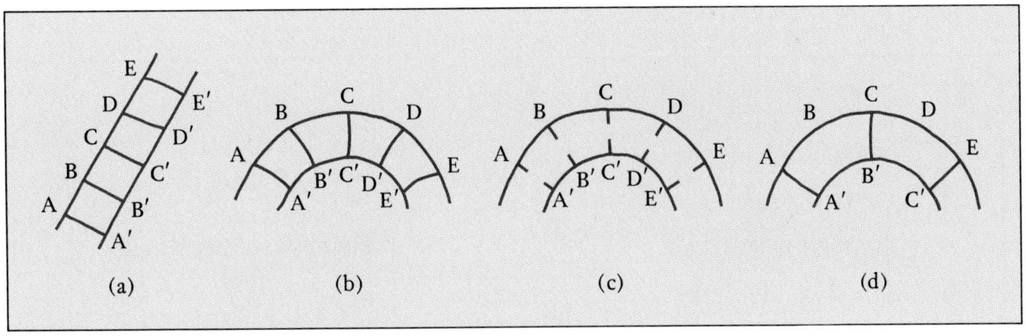

Figure 22-10 A schematic diagram of a permanent wave.

to develop at the cross-links (b). In "cold" waving, these cross-links are broken by a reducing agent (c), relaxing the tension. Then, an oxidizing agent regenerates the cross-links, (d) and the hair holds the shape of the roller. The chemical reactions in simplified form are shown in Figure 22-11.

The most commonly used reducing agent is thioglycolic acid (Fig. 22-12). The common oxidizing agents used include hydrogen peroxide, perborates ($NaBO_2 \cdot H_2O_2 \cdot 3H_2O$), and sodium or potassium bromate ($KBrO_3$). A typical neutralizer solution contains one or more of the oxidizing agents dissolved in water. The presence of water and a strong base in the oxidizing solution also helps to break and re-form hydrogen bonds between adjacent protein molecules. However, too-frequent use of strong base causes hair to become brittle and "lifeless."

Various additives are present in both the oxidizing and the reducing solutions in order to control pH, odor, and color, and for general ease of application. A typical waving lotion contains 5.7% thioglycolic acid, 2.0% ammonia, and 92.3% water.

Hair can be straightened by the same solutions. It is simply "neutralized" (or oxidized) while straight (no rolling up).

Coloring and Bleaching Hair

Hair contains two pigments: brown-black melanin and an iron-containing red pigment. The relative amounts of each actually determine the color of the hair. In deep black hair melanin predominates and in light-blond, the iron pigment predominates. The depth of the color depends upon the size of the pigment granules.

Melanin: brown-black; Iron pigment: red.

Formulations for dyeing hair vary from temporary coloring (removable by shampoo), which is usually achieved by means of a water-soluble dye that acts on the surface of the hair, to semipermanent dyes, which penetrate the hair fibers to a great extent (Fig. 22-13). The more permanent dyes often consist of cobalt or chromium complexes of dyes dissolved in an organic solvent and are generally "oxidation" dyes. They penetrate the hair, and then are oxidized to give a colored product that either is permanently attached to the hair by chemical bonds or is much less soluble in shampoo water than the reactant molecule. Permanent hair dyes generally are derivatives of phenylenediamine. Phenylenediamine dyes hair black. A blond dye can be formulated with *p*-aminodiphenylaminesulfonic acid or

Some hair dyes are suspected of being carcinogenic.

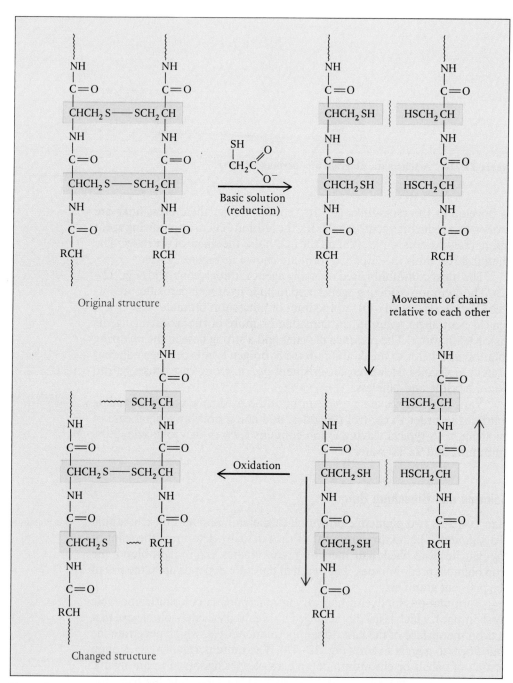

Figure 22–11.   Structural changes at the molecular level that occur in hair during a permanent wave.

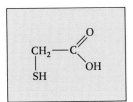

**Figure 22–12   Thioglycolic acid.
Note the relationship of the
molecular structure to acetic acid
and the basic structure in amino acids.**

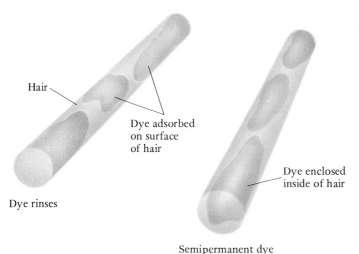

**Figure 22–13   Methods of dyeing
hair.**

p-phenylenediaminesulfonic acid. The structures of these three dye molecules are given in Figure 22–14.

The active compounds are applied in an aqueous soap or detergent
solution containing ammonia to make the solution basic. The dye material is
then oxidized by hydrogen peroxide to develop the desired color. The amine
groups in the dye molecules are oxidized to nitro compounds.

$$-NH_2 + 3\ H_2O_2 \xrightarrow{\text{Oxidation}}\ -NO_2\ + 4\ H_2O$$
$$\text{Amine} \qquad\qquad\qquad \text{Nitro Compound}$$

Hair can be bleached by a more concentrated solution of hydrogen
peroxide, which destroys the hair pigments by oxidation. The solutions are
made basic with ammonia to enhance the oxidizing power of the peroxide.
Parts of the chemical process are given in Figure 22–15. This drastic treatment of hair does more than just change the color. It may destroy sufficient
structure to render the hair brittle and coarse.

Hair Control — Sprays, Conditioners, and Mousses

Wet hair can be shaped better than dry hair since some of the ionic bonds
maintaining the protein shape are broken as water hydrates the ionic centers
and thereby isolates the ionic charges from each other. Hair sprays are
solutions of resins in a volatile solvent. When the solvent evaporates, the hair

Just about any shade of hair color can be
prepared by varying the modifying group
on certain basic dye structures.

There are several dangers in breathing
the vapor of hair sprays, such as
possible harm from chemicals acting on
delicate lung tissue and the danger of
asphyxiation by the plastic coating the
lining of the lungs.

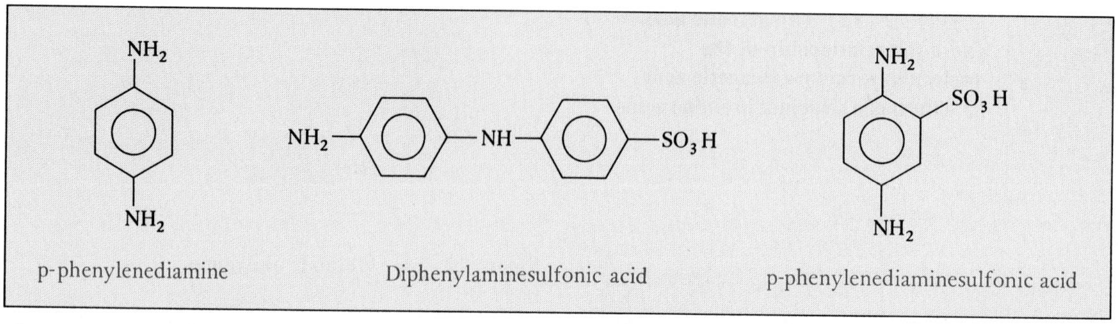

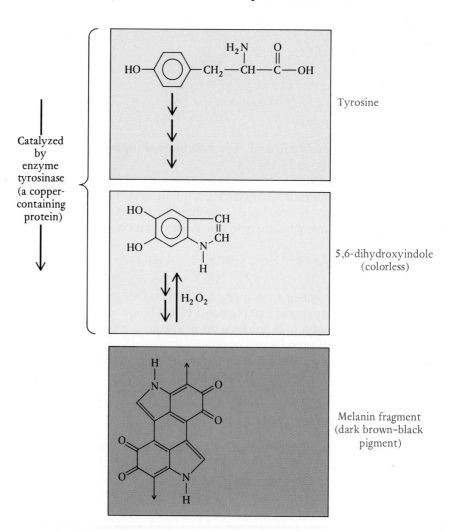

p-phenylenediamine   **Diphenylaminesulfonic acid**   **p-phenylenediaminesulfonic acid**

Figure 22–14   Hair dyes.

is coated with a film of sufficient strength to hold the hair in place (Fig. 22–16). A suitable resin is a copolymer of vinylpyrrolidone and vinyl acetate (Fig. 22–17). In addition to the resin, solvent, and propellant (the liquid that gasifies when the valve is opened), a hair spray is likely to have a plasticizer to enhance elasticity and a silicone oil to impart a sheen to the hair.

Figure 22–15   Bleaching of the hair by hydrogen peroxide. There are several chemical intermediates between the amino acid, tyrosine, and the hair pigment, melanin, which is partly protein. Hydrogen peroxide oxidizes melanin back to colorless compounds, which are stable in the absence of the enzyme tyrosinase (found only in the hair roots). Melanin is a high-molecular-weight polymeric material of unknown structure. The structure shown here is only a segment of the total structure.

Catalyzed by enzyme tyrosinase (a copper-containing protein)

Tyrosine

5,6-dihydroxyindole (colorless)

H_2O_2

Melanin fragment (dark brown–black pigment)

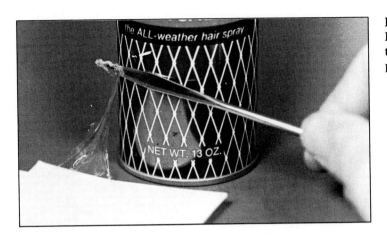

Figure 22–16   Film of hair spray. Hair spray was allowed to dry on the white surface and was then pulled up to reveal film.

An after-shampoo conditioner attempts to manage hair without the film. Added to the water–alcohol dispersing medium, the conditioner contains emollients, oils, waxes, resins, and proteins which adhere to and penetrate into the hair to produce a more pliable and elastic fiber that is not as likely to dry and be affected by atmospheric conditions. Holding the correct amount of moisture is the key to control since too much water causes the hair to be limp and too little causes it to be fly-away.

The mousse has recently become a very popular technique for the application of chemicals to the hair. The French word means "froth, foam, lather, or whipped cream." A mousse foam can deliver any haircare chemical with a delightful advantage. They aren't messy, which is very important in coloring and curling hair, and they give pinpoint accuracy with no overspray or runoff. While the technique is as old as whipped cream from a bottle pressurized with carbon dioxide, the mousse in haircare products has only become popular since 1983.

Since the permanent-waving process depends upon breaking and reforming cystine bonds present in hair keratin, it appears possible to actually

Figure 22–17   A copolymer of vinylpyrrolidone and vinyl acetate is a resin used in hair sprays. Recalling the formula for acetic acid, see if you can identify the vinyl acetate unit in the polymeric chain.

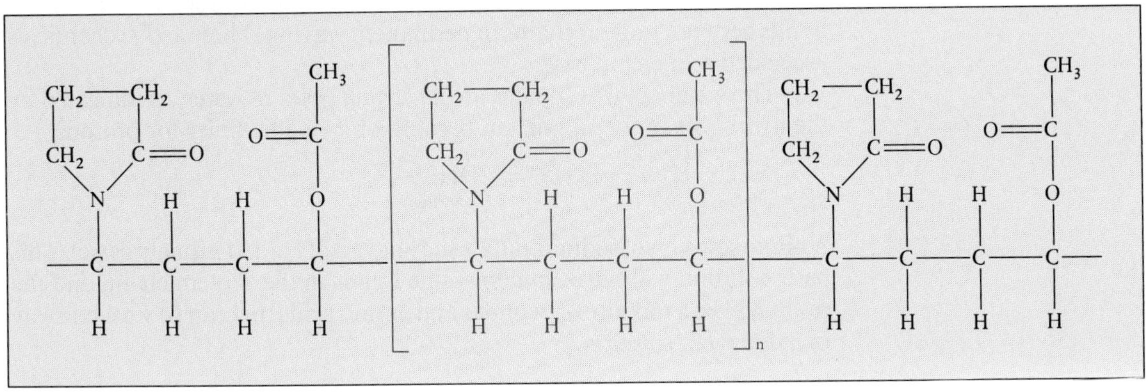

add protein structures to the hair by covalent bonding. Products are now on the market that make this claim; in addition to chemically bonding protein bulk to the hair, the process is thought to restore deteriorated hair by adding protein bridges between broken fragments within a strand.

Growing Hair

Minoxidil is a peripheral vasodilator, an oral medication that enlarges or dilates the blood vessels in the extremities of the body. Blood pressure is reduced on taking minoxidil as the volume for the same amount of blood is increased. Applied to the skin, this chemical causes the growth of fine hair anywhere hair follicles exist on the skin. Oral doses produce the effect over the entire body. The effect, though temporary during the application period of the drug, offers some possibilities for the controlled growth of human hair. The drug is sold under the tradename Regaine in Canada and Europe. The Upjohn pharmaceutical company is making a major effort to supply minoxidil to the U.S. market as a prescription drug, a market that some have estimated to be as high as $250 million per year. The individual cost of minoxidil, meant to be used on a permanent basis, is expected to be in excess of $800 per year.

Removing Hair — Depilatories

The purpose of a depilatory is to remove hair chemically. Since skin is sensitive to the same kind of chemicals that attack hair, such preparations should be used with caution and, even then, some damage to the skin is almost unavoidable. Because of this, depilatories should be used only weekly, and should never be used on skin that is infected or when a rash is present. Depilatories should not be used with a deodorant with astringent action. If the sweat pores are closed by the deodorant, the caustic chemicals are retained and can do considerable harm.

Hair is also removed by electrical cauterization, commonly called "electrolysis." The hair follicle is destroyed by a high-voltage electric spark.

The chemicals used as depilatories include sodium sulfide, calcium sulfide, strontium sulfide (water-soluble sulfides), and calcium thioglycolate [$Ca(HSCH_2COO)_2$], the calcium salt of the compound used to break S—S bonds between protein chains in permanent waving. These active chemicals are added to a cream base.

The water-soluble sulfides are all strong bases in water, as indicated by the hydrolysis of the sulfide ion because of its high affinity for protons.

$$S^{2-} + H_2O \longrightarrow HS^- + OH^-$$
Sulfide   Hydroxide

A dilute solution of sodium sulfide may have a pH of 13 (strongly basic). This basic solution will break some peptide bonds in the protein chain, and the result will be a mixture of peptides and amino acids that can be washed away in a detergent solution.

DEODORANTS

The two million sweat glands on the surface of the body are primarily used to regulate body temperature via the cooling effect produced by the evaporation of the water they secrete. This evaporation of water leaves solid constituents, mostly sodium chloride, as well as smaller amounts of proteins and other organic compounds. Body odor results largely from amines and hydrolysis products of fatty oils (fatty acids, acrolein, etc.) emitted from the body and from bacterial growth within the residue from sweat glands. Sweating is both normal and necessary for the proper functioning of the human body; sweat itself is quite odorless, but the bacterial decomposition products are not.

Body odor is promoted by bacterial action.

There are three kinds of deodorants: those that directly "dry up" perspiration by acting as astringents, those that have an odor to mask the odor of sweat products, and those that remove odorous compounds by chemical reaction. Among those that have astringent action are hydrated aluminum sulfate, hydrated aluminum chloride ($AlCl_3 \cdot 6H_2O$), aluminum chlorohydrate [actually aluminum hydroxychloride, $Al_2(OH)_5Cl \cdot 2H_2O$ or $Al(OH)_2Cl$ or $Al_6(OH)_{15}Cl_3$], and alcohols. Those compounds that act as deodorizing agents include zinc peroxide, essential oils and perfumes, and a variety of mild antiseptics to stop the bacterial action. Zinc peroxide removes odorous compounds by oxidizing the smelly amines and fatty acid compounds. The essential oils and perfumes absorb or otherwise mask the odors.

An astringent closes the pores, thus stopping the flow of perspiration.

SELF-TEST 22–B

1. What is the effect of water on the natural protein bonding in hair?
2. What is the purpose of the residue film left by hair sprays?
3. What acid is used to reduce hair chemically (break the cross-links between protein molecules)?
4. Is the last step in the production of a permanent wave in hair oxidation or reduction?
5. What color is imparted to hair by phenylenediamine dyes?
6. What is the purpose of the protein included in hair conditioners?
7. Name three chemicals used as depilatories.
8. What is the purpose of aluminum chlorohydrate in deodorants?
9. What advantage does a mousse have over a liquid hair conditioner?
10. What is a danger associated with the use of hair sprays?

CLEANSING AGENTS

Dirt can be defined as matter in the wrong place. Tomato soup may be tasty food, but on your shirt it is dirt. There are over 1200 commercial cleansing, or surface-active compounds (**surfactants**) capable of removing dirt with

Surfactants stabilize suspensions of nonpolar materials in polar solvents or vice versa. Over 6 billion lb are produced in the U.S. per year. Examples include soaps, detergents, wetting agents, and foaming agents.

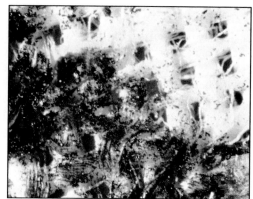

Figure 22–18   Photomicrograph of clean cotton cloth (left) and soiled cotton cloth (right). The proper application of surface-active agents should return the soiled cloth to its original state.

results similar to that shown in Figure 22–18. The classic surfactant, soap, dates back in recorded history to the Sumerians in 2500 B.C., in what is now Iraq and Iran. Soap has always been made from the reaction of a fat with an alkali. The Greek physician Galen referred to this recipe and stated further that soap removed dirt from the body as well as serving as a medicament. What is now new is that soap can be made in a very pure state and many other compounds, both natural and synthetic, have been found to be excellent surfactants and are commercially employed in a variety of applications.

Soap

Review Chapter 13 for the molecular structure of fats and oils.

The reaction of fats and oils in strongly basic solutions is a hydrolysis reaction that produces glycerol and salts of the fatty acid. Such hydrolysis reactions are called **saponification** reactions: the sodium or potassium salts of the fatty acids formed are **soaps.** Pioneers prepared their soap by boiling animal fat with an alkaline solution obtained from the ashes of hardwood. The resulting soap could be "salted out" by adding sodium chloride, making use of the fact that soap is less soluble in a salt solution than in water. Inventory of cleaning materials in a typical modern household might include half a dozen or more formulated products designed to be the most suitable for a specific job, whether cleaning clothes, floors, or the family car.

Principal fats and oils for soap making: tallow from beef and mutton, coconut oil, palm oil, olive oil, bone grease, and cottonseed oil.

$$CH_3(CH_2)_{16}COO-CH_2$$
$$CH_3(CH_2)_{16}COO-CH + 3\ NaOH \longrightarrow 3\ CH_3(CH_2)_{16}COO^-Na^+ + HO-CH$$
$$CH_3(CH_2)_{16}COO-CH_2 \qquad\qquad HO-CH_2$$

Tristearin (Glyceryl Tristearate)   Sodium Stearate   Glycerol
                                        (A Soap)

Floating soaps float because of trapped air.

The cleansing action of soap can be explained in terms of its molecular structure. Substances that are water soluble can be readily removed from the skin or a surface by simply washing with an excess of water. To remove a sticky sugar syrup from one's hands, the sugar is dissolved in water and rinsed away. Many times the material to be removed is oily, and water will merely run over the surface of the oil. Since the skin has natural oils, even

Toilet soaps are generally pure soaps to which dyes and perfumes are added.

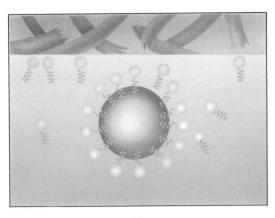

(a)

Figure 22–19 The cleaning action of soap. (a) Soap molecules in water interact strongly with the water through electrical interaction at the salt end of the molecule. The hydrocarbon end of the soap molecule, having more attraction for other hydrocarbons than water, is "pulled" along into solution by the water-salt interaction. (b) The soap molecule, with its oil-soluble and water-soluble ends, will become oriented at an oil-water interface such that the hydrocarbon chain is in the oil (with molecules that are electrically similar, nonpolar) and the salt group ($-COO^-Na^+$) is in the water. When greasy dirt is broken up in soapy water, a process that is aided by mechanical agitation, the oily particles are surrounded and insulated from each other by the soap molecules.

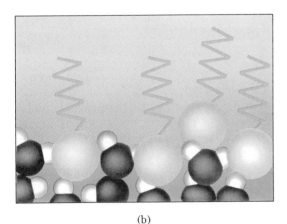

(b)

substances such as ordinary dirt that are not oily themselves can adhere to the skin quite strongly. The cohesive forces (forces between molecules tending to hold them together) within the water layer are too large to allow the oil and water to intermingle (Fig. 22–19). When present in an oil-water system, soap molecules such as sodium stearate

> Soap, water, and oil together form an emulsion, with the soap acting as the emulsifying agent.

$$CH_3CH_2CH_2CH_2CH_2CH_2CH_2CH_2CH_2CH_2CH_2CH_2CH_2CH_2CH_2CH_2CH_2C \overset{\displaystyle O}{\underset{\displaystyle O^-Na^+}{\Big\backslash}}$$

will move to the interface between the two liquids. The hydrocarbon chain, which is a nonpolar organic structure, will mix readily with the nonpolar grease molecules, whereas the highly polar $-COO^-Na^+$ group enters the water layer because the polar groups become hydrated (Fig. 22–19b). The soap molecules will then tend to lie across the oil-water interface. The grease is then broken up into small droplets by agitation, each droplet surrounded by hydrated soap molecules. The surrounded oil droplets cannot come together again since the exterior of each droplet is covered with $-COO^-Na^+$ groups that interact strongly with the surrounding water. If enough soap and

> Stearates: sodium — hard soap; potassium — soft soap; ammonium — liquid soap.

> Pure soap is a mildly basic material because it is the salt (salts are ionic) of a strong base and a weak acid (Stearate ions + H₂O → Hydrogen stearate + OH⁻).

water are available, the oil will be swept away, leaving a clean and water-wet surface.

Shampoos

Shampoos are generally more complex than a simple soap solution, with a number of ingredients to satisfy different requirements for maintaining clean and healthy-looking hair. Condensation products like that obtained from diethanolamine and lauric acid are often used because good surfactant properties are obtained without the alkaline properties characteristic of soap.

$$HN(CH_2CH_2OH)_2 + CH_3(CH_2)_{10}COOH \longrightarrow CH_3(CH_2)_{10}\overset{\overset{\displaystyle O}{\|}}{C}-N(CH_2CH_2OH)_2 + H_2O$$

Diethanolamine   Lauric Acid   An Amide Detergent

Some shampoos contain anionic detergents, which are less damaging to the eyes than cationic detergents (see p. 735 for typical formulas). Sodium lauryl sulfate is an example of an anionic detergent.

$$CH_3(CH_2)_{11}OSO_3^-Na^+$$

Sodium Lauryl Sulfate

The hair is more manageable and has a better sheen if all the shampoo is removed. An anionic detergent can be removed by using a **rinse** containing a dilute solution of a cationic detergent, which electrically attracts the anions and facilitates their removal. Caution should be taken with the cationic rinse because of the possible irritation to the eyes.

Shampoos also contain compounds to prevent the calcium or magnesium ions in hard water from forming a precipitate. Ethylenediaminetetraacetic acid (EDTA), a metal complexing agent, added to a shampoo, will tie up the calcium, magnesium, and iron and avoid the sticky precipitate.

Lanolin and mineral oil (or their substitutes) are often added to shampoos to replace the natural oils in the scalp, thus preventing it from drying out and scaling. The presence of oil additives and stabilizers gives the shampoo a pearlescent appearance.

Calcium, magnesium, and iron soaps are insoluble in water, forming a sticky precipitate, a common problem when using soap in "hard water." Soap and rain water—clean hair; soap and hard water—a sticky mess.

Clear shampoos are preferred in the U.S., 70% of the market; the Europeans buy 80% pearlescent products.

There is a synthetic detergent for almost every type of cleaning problem.

A typical laundry detergent might contain: surfactants, builders, ion exchangers, alkalies, bleaches, fabric softeners, anticorrosion materials, antiredeposition materials, enzymes, optical brighteners, fragrances, dyes, and fillers.

$$R-O-\overset{\overset{\displaystyle O}{\uparrow}}{\underset{\underset{\displaystyle O}{\downarrow}}{S}}-O^-\qquad R-\overset{\overset{\displaystyle O}{\uparrow}}{\underset{\underset{\displaystyle O}{\downarrow}}{S}}-O^-$$

Sulfate Group   Sulfonate Group

Synthetic Detergents

Synthetic detergents ("syndets") are derived from organic molecules designed to have the same cleansing action but less reaction than soaps with the cations found in hard water. As a consequence, synthetic detergents are more effective in hard water than soap. Soap leaves undesirable precipitates that have no cleansing action and tend to stick to laundry.

There are many different synthetic detergents on the market. The molecular structure of a detergent molecule consists of a long oil-soluble (hydrophobic) group and a water-soluble (hydrophilic) group. The hydrophilic groups include sulfate ($-OSO_3^-$), sulfonate ($-SO_3^-$), hydroxyl ($-OH$), ammonium ($-NH_3^+$), and phosphate [$-OPO(OH)_2$] groups.

Cationic (positively charged) detergents are almost all quaternary ammonium halides with the general formula:

$$R_1 - \underset{\underset{R_4}{|}}{\overset{\overset{R_2}{|}}{N^+}} - R_3 \ X^-$$

where one of the R groups is a long hydrocarbon chain and another frequently includes an —OH group. X^- represents a halogen ion such as chloride. In these the water-soluble portion is positively charged; so they are sometimes called invert soaps (in soaps the water-soluble portion is negatively charged).

Cationic detergents frequently exhibit pronounced bactericidal qualities. Cationic detergents are incompatible with anionic detergents. When they are brought together, a high-molecular-weight insoluble salt precipitates out, and this has none of the desired detergent properties of either detergent.

Some detergents are nonionic. They have a polar, but not an ionic, grouping attached to a large organic grouping of low polarity. Consider the following formula that has a long hydrocarbon chain attached through an ester group to a carbon chain containing multiple ether links.

$$CH_3(CH_2)_{11}COO(CH_2)_2O(CH_2CH_2O)_2CH_2CH_2OH$$
Carbon Chain   Ester Group (—COO—)   Ether Links (—O—)

The carbon chain is oil-soluble and the rest of the molecule is hydrophilic, the properties needed for the molecule to be a detergent.

The nonionic detergents have several advantages over ionic detergents. Since nonionics contain no ionic groups, they cannot form salts with calcium, magnesium, and iron ions and consequently are unaffected by hard water. For the same reason, nonionic detergents do not react with acids and may be used even in relatively strong acid solutions.

In general, the nonionic detergents foam less than ionic surface-active agents, a property that is desirable where nonfoaming detergents are required, as in dishwashing. Nonionics tend to be viscous liquids with melting points below room temperature. Although scarcely used in 1970, liquid detergents will hold one third or more of the detergent market in the early 1990s, based on presently measured increases.

In spite of the major impact of synthetics on the detergent industry, soap is still the number 1 surfactant, holding approximately 39% of the market.

Fillers or Builders

A number of materials are added to soap powders for laundry purposes. These materials are often quite basic, and their addition gives the soap a greater detergent action. Commonly added materials include sodium carbonate, sodium phosphates, sodium polyphosphates, and sodium silicate. Rosin neutralized with sodium hydroxide is also commonly added to laundry soaps in large amounts. The rosin is mostly abietic acid (Fig. 22–20). The neutralized acid has the nonpolar (hydrocarbon) part and polar end required for a soap. Such soaps are not to be recommended for use on the

Cationic detergents act as disinfectants.

In commercial production, approximately 62% of the surfactants are anionic, 29% nonionic and 9% cationic, with nonionics increasing their share faster.

Hydrophilic substances have an attraction for water usually through dipole–dipole attraction or hydrogen bonding. Cotton, for example, develops a moist feeling in humid air.

Figure 22–20 (a) Molecular formula for abietic acid. (b) Rosin, important to the violinist in chunk form but to the baseball pitcher in powdered form, is mostly abietic acid. The sodium salt of abietic acid has surfactant qualities that make it suitable as an additive in laundry detergents.

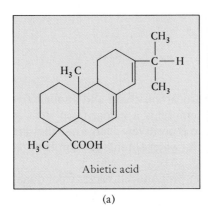

Abietic acid

(a)

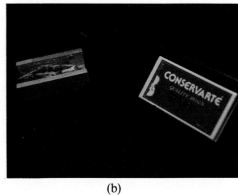

(b)

human skin. Phosphates, carbonates, and silicates hydrolyze to give OH^- ions, which, in turn, react with grease to make soaps.

$$PO_4^{3-} + H_2O \longrightarrow HPO_4^{2-} + OH^-$$

$$CO_3^{2-} + H_2O \longrightarrow HCO_3^- + OH^-$$

$$SiO_3^{2-} + H_2O \longrightarrow HSiO_3^- + OH^-$$

Builders also assist in negating the effect of the ions that cause water to be hard. Since the phosphate, carbonate, and hydroxide compounds of these ions are insoluble, these ions are precipitated.

See the discussion of water-softening in Chapter 17.

$$HCO_3^- + OH^- \longrightarrow CO_3^{2-} + H_2O$$

$$Ca^{2+} + CO_3^{2-} \longrightarrow CaCO_3$$

$$Mg^{2+} + 2\,OH^- \longrightarrow Mg(OH)_2$$

It is fortunate that these precipitates are powdery and easily rinsed away. In contrast, the soap precipitates form scum that sticks to the material being washed.

Other Cleansers

Soaps containing pumice (finely powdered volcanic ash) will wash out ground-in dirt.

A very large number of special cleaners or cleansing agents is available. Simple abrasive cleansers contain a large percentage of an abrasive such as silica (SiO_2) or pumice (65 to 75% SiO_2, 10 to 20% Al_2O_3), a variable amount of soap, and generally some polyphosphates to raise the pH. They may also contain one or more of a variety of salts that react with water to produce a basic solution: trisodium phosphate, sodium carbonate, sodium bicarbonate, sodium pyrophosphate, or sodium tripolyphosphate, plus a detergent and perhaps pine oil to give an attractive odor. Metal cleansers may contain strong acid or strong base to dissolve oxide films. Many cleaning liquids contain organic solvents such as perchloroethylene, 1,1,1-trichloroethane, and the like. The vapors of these are quite toxic so the cleansers must be used in ventilated areas.

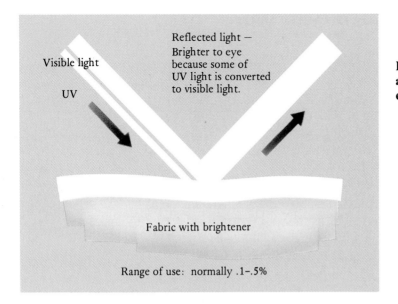

Visible light

UV

Reflected light —
Brighter to eye
because some of
UV light is converted
to visible light.

Fabric with brightener

Range of use: normally .1–.5%

Figure 22–21 Optical brighteners absorb unseen ultraviolet light and emit the energy as visible light.

Whiter Whites and Bleaching Agents

Bleaching agents are compounds used to remove color from textiles. Most commercial bleaches are oxidizing agents such as sodium hypochlorite. Optical brighteners are quite different, since they act by converting a portion of the incoming invisible ultraviolet light into visible blue or blue-green light, which is emitted. Together or separately, these two classes of compounds find their way into commercial laundry and cleaning preparations, since they seem to make clothes cleaner.

In earlier times textiles were bleached by exposure to sunlight and air. In 1786, the French chemist Berthollet introduced bleaching with chlorine, and subsequently this process was carried out with sodium hypochlorite, an oxidizing agent prepared by passing chlorine into aqueous sodium hydroxide:

$$2\,Na^+ + 2\,OH^- + Cl_2 \longrightarrow Na^+ + OCl^- + Na^+ + Cl^- + H_2O$$
Sodium Hypochlorite

Shortly thereafter, hydrogen peroxide was introduced as a textile bleach. Later, a number of other oxidizing agents based on chlorine were developed and introduced.

One way to decolorize materials is to remove or immobilize those electrons in the material activated by visible light.

Colored (or stained) material − electrons ⟶ White material

The hypochlorite ion, because it is an oxidizing agent, is capable of removing electrons from many colored materials. In this process, the hypochlorite ion is reduced to chloride and hydroxide ions.

$$ClO^- + H_2O + 2\,e^- \longrightarrow Cl^- + 2\,OH^-$$

Optical brighteners are fluorescent compounds. A fluorescent material absorbs light of a shorter wavelength and emits light of a longer wavelength (Fig. 22–21). When optical brighteners are incorporated into textiles or

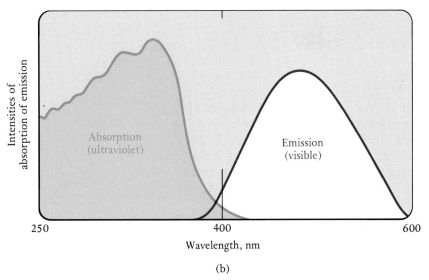

(a)

(b)

Figure 22–22 (a) Optical brighteners are composed of complex organic molecules containing numerous double bonds between the atoms. (b) A typical absorption and emission spectral pattern for an optical brightener. The energy of the ultraviolet is "rendered visible" as it is shifted to visible wavelengths. Compare this to special glasses that allow you to see better under night-darkness conditions.

Many stains can be removed by an appropriate solvent or chemical reagent.

paper, they make the material appear brighter and whiter. An example of such a brightener is represented by the complex formula in Figure 22–22 along with its absorption and emission spectra in the ultraviolet and visible regions, respectively.

Spot and Stain Removers

To a large extent, stain removal procedures are based on solubility patterns or chemical reactions. Many stains, such as those due to chocolate or other fatty foods, can be removed by treatment with the typical dry-cleaning solvents such as tetrachloroethylene, $Cl_2C=CCl_2$.

Stain removers for the more resistant stains are almost always based on a chemical reaction between the stain and the essential ingredients of the stain remover. A typical example is an iodine stain remover, which is simply a concentrated solution of sodium thiosulfate. The stain removal reaction is:

$$\underset{\text{Iodine}}{I_2} + 2\,Na_2S_2O_3 \longrightarrow \underset{\text{(Soluble in Water and Colorless)}}{2\,NaI + Na_2S_4O_6}$$

Other examples are given in Table 22–4.

TABLE 22–4 Some Common Stains and Stain Removers*

Stain	Stain Remover
Coffee	Sodium hypochlorite
Lipstick	Isopropyl alcohol, isoamyl acetate, Cellosolve (HOCH$_2$CH$_2$OCH$_2$CH$_3$), chloroform
Rust and ink	Oxalic acid, methyl alcohol, water
Airplane cement	50/50 amyl acetate and toluene or acetone
Asphalt	Benzene or carbon disulfide
Blood	Cold water, hydrogen peroxide
Berry, fruit	Hydrogen peroxide
Grass	Sodium hypochlorite in alcohol
Nail polish	Acetone
Mustard	Sodium hypochlorite or alcohol
Antiperspirants	Ammonium hydroxide
Perspiration	Ammonium hydroxide, hydrogen peroxide
Scorch	Hydrogen peroxide
Soft drinks	Sodium hypochlorite
Tobacco	Sodium hypochlorite

* Before any of these stain removers are used on clothing, the possibility of damage should be checked on a portion of the cloth that ordinarily is hidden. Some stain removers, such as benzene and chloroform, are suspected carcinogens, and some are toxic, such as methanol and carbon disulfide.

TOOTHPASTE

The structure of tooth enamel is essentially that of a stone composed of calcium carbonate and calcium hydroxy phosphate (apatite), Figure 22–23. Such structures are readily attacked by acid. Since the decay of some food particles produces acids and since bacteria will convert plaque, a deposit of dextrins, to acids, it is important to keep teeth clean and free from prolonged contact with these acids if the hard, stonelike enamel is to be preserved.

The two essential ingredients in toothpaste are a detergent and an abrasive. The abrasive serves to cut into the surface deposits, and the detergent

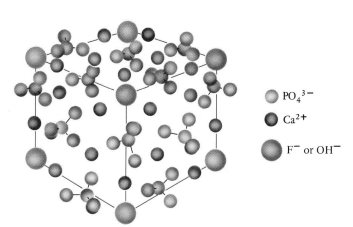

PO$_4$$^{3-}$

Ca^{2+}

F$^-$ or OH$^-$

Figure 22–23 Structure of apatite and fluoroapatite. The grey circles denote Ca^{2+} ions; purple, phosphorus; red, oxygen; and the brown circles represent OH$^-$ or F$^-$ groups.

Figure 22–24   Apatite, a natural mineral as well as being a major constituent of teeth and bones. (Bill Tronca/Tom Stack and Associates.)

assists in suspending the particles in a water medium to be carried away in the rinse. Abrasives commonly used in toothpaste formulations include hydrated silica (a form of sand, $SiO_2 \cdot nH_2O$); hydrated alumina ($Al_2O_3 \cdot nH_2O$); and calcium carbonate ($CaCO_3$). It is difficult to select an abrasive that is hard enough to cut the surface contamination yet not so hard as to cut the tooth enamel. The choice of detergent is easier; any good detergent such as sodium lauryl sulfate will do quite well.

Since the necessary ingredients in toothpaste are not very palatable, it is not surprising to see the inclusion of flavors, sweeteners, thickeners, and colors to appeal to our senses.

One addition to the toothpaste mixture has caused a dramatic decrease in the amount of tooth decay in our population; it is the addition of stannous fluoride (SnF_2) to provide a low level of fluoride ion concentration in the brushing medium. Some of the fluoride ions actually replace the hydroxide ions in the hydroxyapetite structure [$Ca_{10}(PO_4)_6(OH)_2$] (Fig. 22–23), to form fluoroapatite [$Ca_{10}(PO_4)_6F_2$]. The fluoride ion forms a stronger ionic bond because of its high concentration of negative charge in the crystalline structure, and as a result, the fluoroapatite is harder and less subject to acid attack than the hydroxyapatite. Hence, there is less tooth decay. The fluoride ion is also introduced into essentially all of the public water supplies for this same purpose.

Most teeth are now lost as a result of gum disease, which results from the lack of proper massage, irritating deposits below the gum line, bacterial

Figure 22–25   "No cavities!" (*The World of Chemistry*, Program 3, "Measurement: The Foundation of Chemistry.")

(a)

(b)

infection and poor nutrition. More attention is being given to toothpastes containing disinfectants such as peroxides in addition to the other ingredients.

SELF-TEST 22-C

1. A fat is a triester of glycerol and _____ acids.
2. To make soap, a fat is treated with _____.
3. Is the acid or salt group in a soap molecule more soluble in water or in oil?
4. What makes floating soap float?
5. Which is more likely to precipitate the hard-water ions (Ca^{2+}, Mg^{2+}, Fe^{3+}) as a sticky precipitate, traditional soaps or synthetic detergents?
6. Surfactant is a short term for _____ _____
   _____.
7. Why should a cationic detergent not be mixed with an anionic detergent in a laundry blend?
8. Which foams the least: ( ) cationic, ( ) anionic ( ) nonionic detergents?
9. Give two purposes for adding a detergent-builder to a laundry product.
10. Optical brighteners transform ultraviolet light into _____ light.
11. What are the two fundamental ingredients in a toothpaste?
12. A compound of what element is added to toothpaste to replace some of the hydroxide ions in apatite?

(c)

Figure 22–26   Many living organisms (a) concentrate calcium carbonate, which collects in massive geological formations such as this limestone bed in Verde River, Arizona (b). It is no surprise then that teeth contain calcium carbonate and are attacked by acid as are carbonate deposits in general. (c) The action of strong acid on carbonate chips. (a: Dick Georges/ Tom Stack and Associates; b: James Cowlin; c: Charles Steele.)

$$\left(CH_3 - \bigcirc - O \right)_3 PO$$

Tricresyl phosphate

Tricresyl phosphate is a gasoline additive that prevents preignition.

AUTOMOTIVE PRODUCTS

Proper fuels, lubricants, cooling solutions, and protective dressings must be selected and used by the consumer to keep an automobile in top condition.

Gasoline Additives

The production of gasoline from petroleum and the antiknock properties of gasoline were discussed in Chapter 13. In addition to antiknock additives (aromatic compounds are replacing tetraethyllead because of lead pollution), numerous other chemicals are added to gasoline to improve its properties (Table 22–5).

Lubricants and Greases

Lubricants have been used to separate moving surfaces, and thus minimize friction and wear, for a long time. Even before 1400 B.C., animal tallow was used to lubricate chariot wheels. Petroleum lubricating oils and greases came into widespread use after the famous Drake oil well was drilled at Titusville, Pennsylvania, in 1859. The use of lubricants has progressed rapidly since about 1930; synthetic lubricants have been developed largely since World War II.

Lubricating oils from petroleum consist essentially of complex mixtures of hydrocarbons. These generally range from low-viscosity oils, having molecular weights as low as 250 amu to very viscous lubricants with molecular weights as high as about 1000 amu. The viscosity of an oil can often determine its use. For example, if the oil is too viscous, it offers too much resistance to the metal parts moving against each other. On the other hand, if the oil is not viscous enough, it will be squeezed out from between the metal surfaces, and consequently offer insufficient lubricating power. For these

Recall the atomic weight scale in atomic mass units (amu's) from Chapter 2. The carbon-12 isotope of carbon has a mass of exactly 12 atomic mass units.

TABLE 22–5 Gasoline Additives

Additive	Chemical (Example)	Purpose
Deposit modifiers	Tricresyl phosphate (TCP)	Quenches one explosion to prevent preignition of next explosion.
Antioxidants	Phenylenediamine	Prevents peroxide formation that leads to engine knock.
Complexing agents	Ethylenediamine	Masks metals such as copper that catalyze gum formation.
Antirust agents	Trimethylphosphate	Coats metal surfaces in engine to prevent moisture corrosion.
Anti-icing agents	Ethylene glycol	Prevents ice crystals from forming as gasoline is vaporized.
Detergents	Alkylammonium dialkyl phosphates	Helps to prevent gum deposits in carburetor through detergent cleaning.

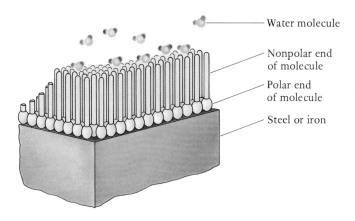

Water molecule

Nonpolar end
of molecule

Polar end
of molecule

Steel or iron

Figure 22-27   The action of
surface-active agents, such as rust
inhibitors, mild antiwear agents, and
some deicing agents depends on a
surface coating of polar molecules.
Compare this with the molecular
action of soap at the interface of oil
and water.

and other reasons, motor oil is often a mixture of oils with varying viscosities.
The common 10W-40 oil, for instance combines the low-temperature 10W
viscosity classification established by the Society of Automotive Engineers
(SAE) for easy low-temperature starting with SAE 40 high-temperature viscosity for better load capacity in bearings at the normal running temperature.

The SAE scale for rating motor oils is based on the viscosity of the oils.
The viscosity criteria for the SAE scale are given in Table 22-6.

Additives to motor oils include antiwear agents for extreme pressure
applications, oxidation inhibitors, rust inhibitors, detergents, viscosity im-
provers, and foam inhibitors.

Greases are essentially lubricating oils thickened with a gelling agent
such as fatty acid soaps of lithium, calcium, sodium, aluminum, or barium.
The soaps form a network of fibers that entrap the oil molecules within the
interlacing fiber structure. Carbon black, silica gel, and clay are also used to
thicken petroleum greases. Chemical additives similar to those used in lubri-

**TABLE 22-6 Viscosity Data at $-18°C$ and $99°C$ for the SAE Method of Rating
Motor Oils**

	Viscosity (SUS)*			
	$-18°C$		$99°C$	
Motor Oil	*Min*	*Max*	*Min*	*Max*
5W		4,000	39	
10W	6,000	12,000	39	
20W	12,000	48,000	39	
20W			45	58
30W			58	70
40W			70	85
50W			85	110

* SUS is the Saybolt universal second, which is the time in seconds required for 50 ml of oil to
empty out of the cup in a Saybolt viscometer through a carefully specified capillary opening.
Note the extremely shortened time for the outflowing of the hotter oil. Gaps in the table
represent improper uses of a particular weight of oil in an internal combustion engine. Oils with
low viscosity offer relatively little protection at high temperatures and viscous oils tend not to
flow at low temperatures.

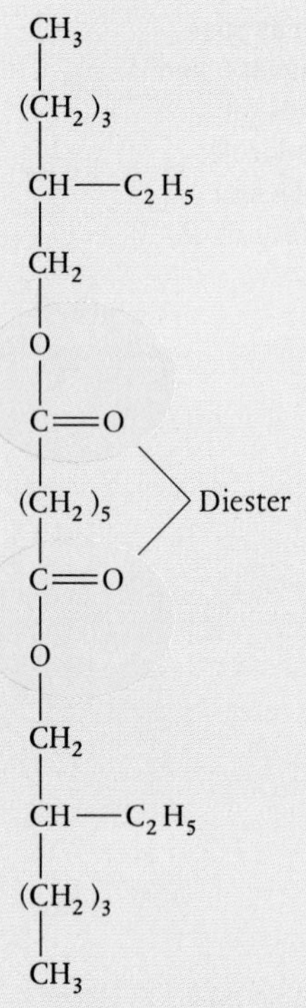

CH₃ structure — Di(2-ethylhexyl) sebacate:

$$CH_3$$
$$(CH_2)_3$$
$$CH-C_2H_5$$
$$CH_2$$
$$O$$
$$C=O$$
$$(CH_2)_5 \quad \rangle \; Diester$$
$$C=O$$
$$O$$
$$CH_2$$
$$CH-C_2H_5$$
$$(CH_2)_3$$
$$CH_3$$

Di(2-ethylhexyl) sebacate

Di(2-ethylhexyl) sebacate is a grease.

The principal ingredient in commercial automotive antifreeze is ethylene glycol. (Charles Steele.)

cating oils and gasolines are added to greases to improve oxidation resistance, rust protection, and extreme pressure properties. Synthetic greases are being developed that deteriorate so slowly that longer intervals between grease jobs are now possible. Silicone greases have a useful life of up to 1000 h at 450°F (232°C). Unfortunately, silicone greases provide relatively poor lubrication for gears and other sliding devices. Diester greases such as di(2-ethylhexyl)sebacate have found extensive use among synthetic greases. Lithium soaps dissolve well in the diester oil and form a grease with equal or better lubrication characteristics and have a considerably longer useful life than petroleum greases. Blends of silicone oil and diester oil provide greases with good low-resistance lubrication power even at low temperatures (−73°C).

Antifreeze

An antifreeze is a substance that is added to a liquid, usually water, to lower its freezing point. Although various substances have been used as antifreezes in the past, nearly all of the current market is supplied by ethylene glycol and methyl alcohol.

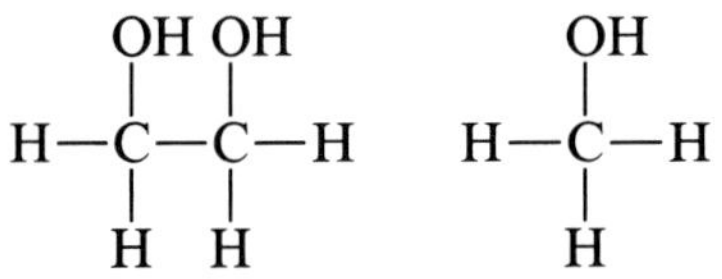

$$H-\overset{\displaystyle OH}{\underset{\displaystyle H}{C}}-\overset{\displaystyle OH}{\underset{\displaystyle H}{C}}-H \qquad H-\overset{\displaystyle OH}{\underset{\displaystyle H}{C}}-H$$

Ethylene Glycol   Methyl Alcohol

More than 95% of the antifreeze on the market is "permanent" antifreeze, having ethylene glycol as the major constituent. The lower-boiling methyl alcohol will boil away in a hot radiator and leave the system unprotected against freezing. Water as a coolant has two serious disadvantages. First, it has a relatively high freezing point and, second, under normal operating conditions, it is corrosive. Modern antifreeze mixtures effectively counteract these problems. The commercial preparation of antifreeze is presented in Chapter 13.

Figure 22–28 displays the relationship between the freezing points and concentrations for ethylene glycol–water and methyl alcohol–water mixtures. Since the density of the coolant solution is a function of its antifreeze concentration, you can keep up with your antifreeze protection with a simple hydrometer available at your auto supply store.

Commercial antifreeze contains various additives to prevent corrosion, leaks, damage to rubber, and foaming.

The prevention of corrosion is the second most important job of antifreeze solutions. Metals in the cooling system that are subject to corrosion are copper, steel, cast iron, aluminum, solder (lead and tin), and brass (copper and zinc). Reducing agents, such as nitrite solutions, which are easier to oxidize than the metals, protect the metals until they are depleted. Phosphates are added as ion scavengers since they complex the metal ions that catalyze the oxidation of metal surfaces. Detergents (e.g., triethanolamine) are added to coat the internal metal surfaces. All inhibitors are depleted with use. Corrosion is greatly increased if the pH of the coolant drops much below

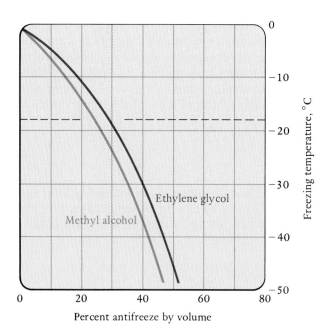

7 (acid solution). A rust inhibitor that contains an alkali can be added to raise the pH back into the alkaline range. Many of the commercial products are color-coded with indicators to signal failure of the antifreeze solution.

Any fast-moving fluid exposed to a gas will tend to foam, and any leak into the coolant system will accentuate the foaming problem. Gas bubbles in the coolant dramatically reduce the cooling efficiency. Antifoam additives include silicones, polyglycols, mineral oil, high-molecular-weight alcohols, organic phosphates, alkyl lactates, castor oil soaps, and calcium acetate. These substances interfere with the molecular bonding in the bubble films causing their breakdown.

SELF-TEST 22–D

1. Name a deposit modifier for internal combustion engines.
2. What is the benefit of a detergent in gasoline?
3. What is necessary to form a grease?
4. What chemical is the main ingredient in antifreeze solutions?
5. Name a nonpermanent antifreeze.
6. Which oil weight represents the higher viscosity, 10W or 40W?

PHOTOGRAPHIC CHEMISTRY

The recording of photographic images is approximately 150 years old. Starting with the recording of a fuzzy black-and-white image focused onto a surface with a magnifying glass, we have progressed to the point where the

For 150 years cameras have been used to graph latent light images in photosensitive films. Pictures resulted with development. At least a part of the future of photography will increasingly involve the recording of the light image on magnetic surfaces as in the case of the video recorder and the computer floppy or hard disc. (Courtesy of Canon U.S.A., Inc.)

entire cover of an issue of the *National Geographic Magazine,* December, 1988 (and popular charge cards as well) presents a holograph that, not only records the color in a three dimensional view, but varies the view depending on the angle at which you are looking at the picture. The fall anniversary issue, 1988, of *Life Magazine* presents a pictorial (with some commentary) history of the 150th anniversary of the birth of photography, the graphing of light pictures.

Black and White Photography

J. H. Schulze observed in 1727 that a mixture of silver nitrate ($AgNO_3$) and chalk darkened on exposure to light. The first permanent images were obtained in 1824 by Nicéphore Niepce, a French physicist, by means of glass

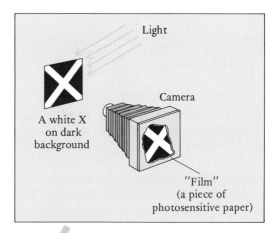

Figure 22–29   The Talbot process.

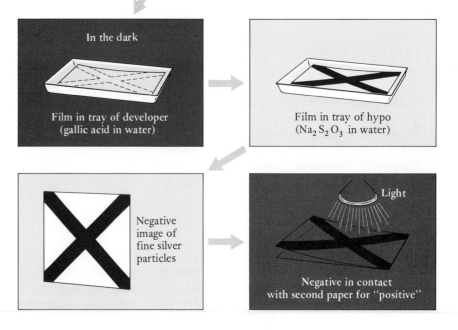

plates coated with a coal derivative (bitumen) containing silver salts. In the early 1830s, Niepce's partner, Louis Daguerre, discovered that mercury vapor was capable of developing an image from a silver-plated copper sheet that had been sensitized by iodine vapor. The daguerreotype image was rendered permanent by washing the plate with hot concentrated salt solution. In 1839, Daguerre demonstrated his photographic process to the Academy of Sciences in Paris. The process was improved by using sodium thiosulfate to wash off the unexposed silver salts.

In 1841, an Englishman, William Henry Fox Talbot, announced the calotype process. The Talbot process (Fig. 22–29) involved a paper made sensitive to light by silver iodide. The light-sensitive paper could be developed into a negative image with gallic acid in a development process essentially the same as is used today. When made with semitransparent paper, Talbot's negatives could be laid over another piece of photographic paper which, when exposed and developed, yielded a "positive," or direct copy of the original. Although the Talbot process required less time than the Daguerre process, the Talbot images were not sharp. It was obvious that some way of holding the silver halides on a transparent material would have to be devised.

At first, the silver salts were held on glass with egg white as a binder. This provided sharp, though easily damaged, pictures. By 1871, the problem had been solved by an amateur photographer and physician, R. L. Maddox. He discovered a way to make a gelatin emulsion of silver salts and apply it to glass. In 1887, George Eastman introduced the Kodak, a camera using film made by attaching a gelatin emulsion to a plastic (cellulose nitrate) base (Fig. 22–30). The camera could take 100 pictures and then camera and film had to be sent to Rochester, New York, for processing. The age of modern photography had arrived.

Photochemistry of Silver Salts

A typical photographic film contains tiny crystallites called grains (Fig. 22–31), which are composed of a slightly soluble silver salt, such as silver bromide (AgBr). The grains are suspended in gelatin, and the resulting gelatin emulsion is melted and applied as a coating on glass plates or plastic film.

Civil War photographers actually processed some of their pictures in the field in portable laboratory equipment in order to get clearer pictures.

Cellulose acetate replaced easily combustible cellulose nitrate as the film support in 1951.

Stereographic photos became popular in the latter part of the 19th century; required were two cameras making simultaneous pictures from the points of view of the two eyes and a viewer, thus allowing each eye to look at its own picture at one time.

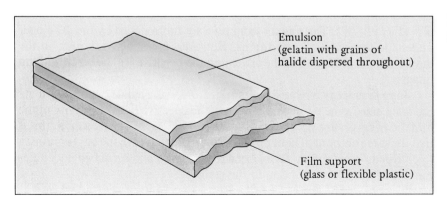

Emulsion
(gelatin with grains of halide dispersed throughout)

Film support
(glass or flexible plastic)

Figure 22–30 A photosensitive gelatin emulsion.

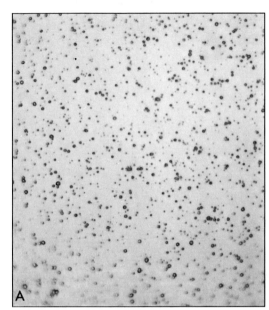
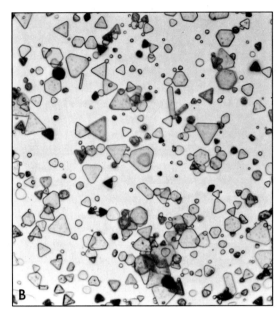

Figure 22–31   (a) Photomicrograph of the grains in a slow positive emulsion. (b) The grains in a high-speed negative emulsion at the same magnification are much larger.

When light of an appropriate wavelength strikes one of the grains, a series of reactions begins that leaves a small amount of free silver in the grain. Initially, a free bromine atom is produced when the bromide ion absorbs the photon of light:

$$Ag^+Br^- \xrightarrow{\text{Light Absorption}} Ag^+ + Br^\circ + e^-$$

Then the silver ion can combine with the free electron to produce a silver atom.

$$Ag^+ + e^- \longrightarrow Ag^\circ$$

For an exposed AgBr grain to be developable, it will need a minimum of four silver unoxidized atoms as Ag_4°. The superscript $^\circ$ indicates unoxidized silver atoms.

Association within the grains produces species such as Ag_2^+, Ag_2°, Ag_3^+, Ag_3°, Ag_4^+, and Ag_4°. The presence of this free silver in the exposed silver bromide grains provides the **latent image,** which is later brought out by the development process. The unreduced silver in the grains containing the free silver in the form of Ag_4° is readily reduced by the developer to form relatively massive amounts of free silver; hence a dark area appears at that point on the film (Fig. 22–32). The unexposed grains are not reduced by the developer under the same critical conditions. However, continued exposure to the developer will reduce all of the silver and the negative will be black with no image.

Film sensitivity is rated on the American Standards Association (ASA) scale. The larger the number, the more sensitive the film is to light.

Film sensitivity is related to grain size and to the halide composition. As the grain size in the emulsion increases, the effective light sensitivity of the film increases up to a point of saturation (Fig. 22–32). The reason for this is that the same amount of light produces more free silver in the larger grains; a free silver atom in a grain having been reduced by a photon of light facilitates the rest of the silver reduction in the grain.

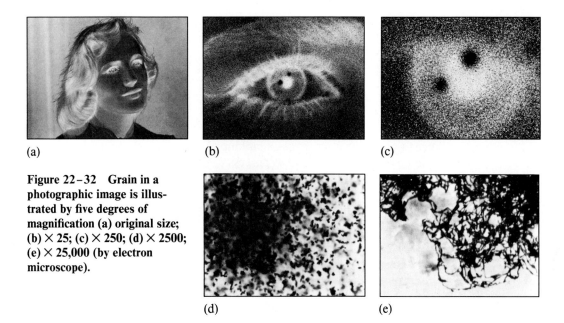

(a)   (b)   (c)

Figure 22–32   Grain in a photographic image is illustrated by five degrees of magnification (a) original size; (b) × 25; (c) × 250; (d) × 2500; (e) × 25,000 (by electron microscope).

(d)   (e)

Amplification of the Latent Image — Development

Silver halides are not the most photosensitive materials known. Why, then, are they effective image producers? The answer lies in the fact that the impact of a single photon on a silver halide grain produces a nucleus of at least four silver atoms, and this effect is amplified as much as a billion times by the action of a proper reducing agent **(developer)**. When an exposed film is placed in developer, the grains that contain silver atom nuclei are reduced faster than those grains that do not. The more nuclei present in a given grain, the faster the reaction. The reduction reaction is:

$$Ag^+_{\text{(in grains containing } Ag_4^\circ)} + e^- \longrightarrow Ag^\circ$$

Factors such as temperature, concentration of the developer, pH, and the total number of nuclei in each grain determine the extent of development and the intensity of free silver (blackness) deposited in the film emulsion in a given time.

The blackness on the negative is due to free silver atoms, Ag°.

Not only must the developer be capable of reducing silver ions to free silver, but it must be selective enough not to reduce the unexposed grains, a process known as "fogging." Figure 22–33 gives the formulas for some compounds used as photographic developers.

Most developers used for black-and-white photography are composed of hydroquinone and Metol or hydroquinone and Phenidone. A typical developer consists of a developing agent (or two), a preservative to prevent air oxidation, and an alkaline buffer to prevent the actual reduction reaction from being retarded (Table 22–7). When hydroquinone acts as a developer, quinone is formed. Two hydrogen ions are also produced for every two silver atoms.

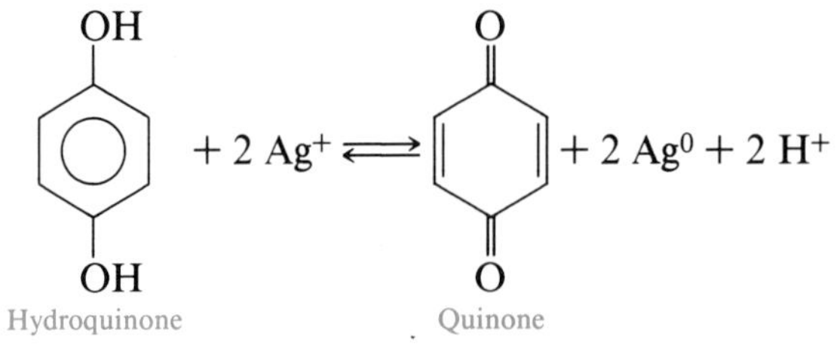

Since this reaction is reversible, a buildup of either hydrogen ions or quinone could impede the development process. The sodium sulfite reacts with quinone to form a sulfite derivative of hydroquinone, thus preventing the reverse reaction back to hydroquinone. The hydrogen ions formed in the developing reaction are neutralized by the alkaline buffer:

$$H^+ + OH^- \longrightarrow H_2O$$

If development proceeds either too long or at a higher temperature than recommended, sufficient fogging occurs to render a negative useless. Since the rates of the development reactions increase with increasing temperature, the photographer usually controls the temperature of the development bath very carefully.

The development process is terminated by a **stop bath.** The stop bath usually contains a weak acid such as acetic acid, which decreases the pH. Since hydrogen ions are a product of the developing reaction, the developing

Figure 22–33 Some developers.

Some Compounds Used as Photographic Developers

NAME	FORMULA
Gallic acid	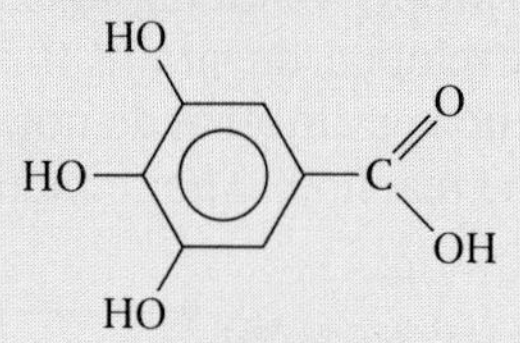
o-aminophenol	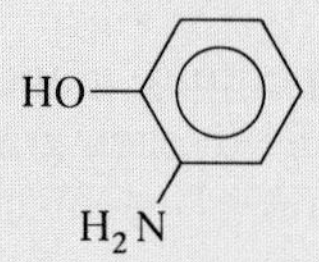
Hydroquinone	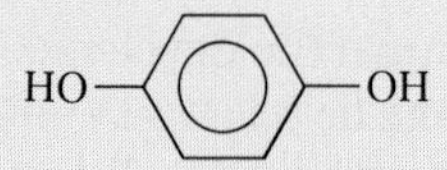
p-methylaminophenol (metol)	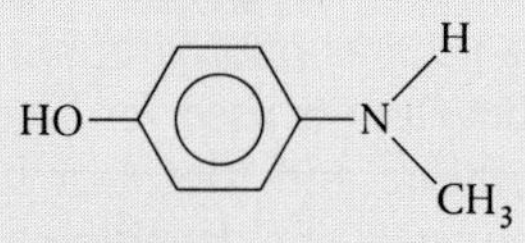
1-phenyl-3-pyrazolidone (phenidone)	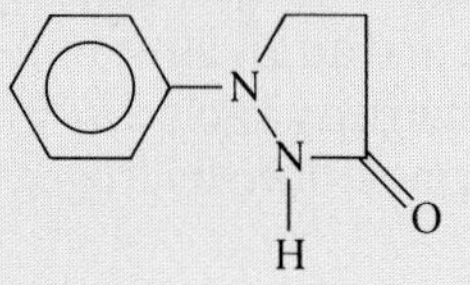

TABLE 22–7 Formula for a Typical Developer for Black-and-White Films

Dissolve in 750 mL water at 50°C	
Metol	2.0 g
Hydroquinone	5.0
Sodium sulfite (Na_2SO_3)	100.0
Borax ($Na_2B_4O_7 \cdot 10H_2O$)	2.0
Add cold water to make 1 L of solution.	

process can be stopped by a sudden increase in a reaction product, the hydrogen ion.

Fixing

If development only produces free silver where the light intensity was greatest and nothing further is done to the negative, the undeveloped silver halide will be exposed the instant the negative is taken into the light. After that, almost any reducing agent will completely fog the negative. In order to overcome this problem, a suitable substance had to be found to remove the unreduced silver halides after development. The most commonly used fixing agent in black-and-white photography is the thiosulfate ion ($S_2O_3^{2-}$). Thiosulfate ions form stable, water-soluble complexes with silver ions, and consequently are easily washed from the gelatin medium on the film.

Solutions of sodium thiosulfate ($Na_2S_2O_3$), known as "hypo," were first used by J. W. F. Herschel to "fix" negatives.

$$\underset{\text{Insoluble Salt}}{AgBr(s)} + \underset{\text{From Hypo Solution}}{2\,S_2O_3^{2-}} \longrightarrow \underset{\text{Water-Soluble Complex}}{Ag(S_2O_3)_2^{3-}} + Br^-$$

Spectral Sensitivity

Probably the most important ingredients in a black-and-white photographic emulsion, other than the silver halide salts themselves, are the spectral-sensitizing dyes. Silver halides are most sensitive to blue light or higher energy per

Figure 22–34 The Polaroid process made "instant" pictures popular. Utilizing the same basic chemistry this process holds the chemicals in layers until they are used. The chemicals are released by the pressure between rollers.

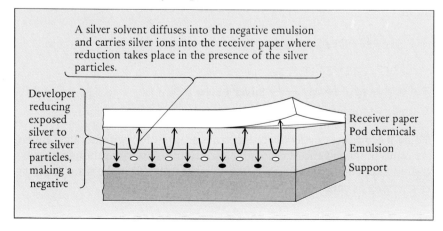

A silver solvent diffuses into the negative emulsion and carries silver ions into the receiver paper where reduction takes place in the presence of the silver particles.

Developer reducing exposed silver to free silver particles, making a negative

Receiver paper
Pod chemicals
Emulsion
Support

Figure 22–35   Spectral sensitivity of a typical AgBr emulsion.

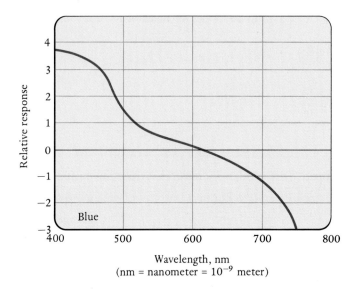

Relative response

Wavelength, nm
(nm = nanometer = 10^{-9} meter)

quantum electromagnetic radiation such as ultraviolet light (Fig. 22–35). A film manufactured with only silver halides as the photosensitive agents will be only blue-sensitive and will not "see" reds, yellows, greens, and so on, as ordinary colors.

In 1873, while trying to eliminate light-scattering problems in photographing the solar spectrum, W. H. Vogel, a German chemist, added a yellow dye to his emulsion. To his surprise he discovered that he could now record black-and-white images in the green region of the visible spectrum. Later, in 1904, another German, B. Homolka, discovered a dye, Pinacyanol (Fig. 22–36), which when added to silver halide emulsion rendered it sensitive to the entire visible spectrum (Fig. 22–37). Films of this type are called **panchromatic** or "pan" films.

The mechanism by which a dye molecule can impart spectral sensitivity to silver halide grains seems to involve initially the absorption of a photon of light by the dye molecule. Next, the excited molecule ejects an electron into the silver halide grain, where a free silver atom is formed. The electron-deficient dye molecule then oxidizes the bromide ion, producing a bromine atom:

$$\text{Dye} \xrightarrow{\text{Light}} \text{Dye}^{*}_{\text{(excited dye molecule)}}$$

$$\text{Dye}^{*} \longrightarrow \text{Dye}^{+} + e^{-}$$

$$\text{Ag}^{+} + e^{-} \longrightarrow \text{Ag}^{\circ} \longrightarrow \text{Silver nuclei}$$

$$\text{Dye}^{+} + \text{Br}^{-} \longrightarrow \text{Dye} + \text{Br}^{\circ}$$

Thus, the process is effectively the same as a photon with enough energy (photons in the red end of the visible spectrum are not large enough to cause the reduction of the silver directly) striking the bromide in the grain itself, the dye serving as catalyst.

By adding spectral-sensitizing dyes, photographic emulsions can be made that are sensitive to selected regions of the spectrum with wavelengths from 100 nm (ultraviolet) to 1300 nm (infrared).

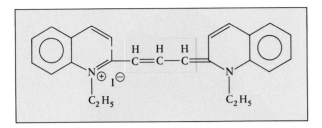

Figure 22–36 Pinacyanol, a cyanine dye. The conjugated group (double bond, single bond, etc.—one sequence in the highlighted block) produces the color of the dye. Other cyanine dyes have more CH groups, absorb longer wavelengths of light, and shift the film sensitivity toward the red end of the spectrum.

Color Photography

Color photography dates from 1861, and it was James Clerk Maxwell, a pioneer in the understanding of electromagnetic radiation, who first photographed an object in color. He photographed the object through three color filters, resulting in three black-and-white negatives made with the three different regions of the visible spectrum. The negative made with blue light darkened where the blue light, passing through the blue filter, exposed it. Now a negative of the blue-light negative would pass blue light at the point where blue light struck the original negative and darkened it. He then superimposed the projections of the three images using the three different color filters and observed the colored picture on the screen. Due to a fortunate peculiarity of his experiment, Maxwell's film was panchromatic even though panchromatic film was not reported in the literature and explained at this time.

Additive and Subtractive Primary Colors

As early as 1611, De Dominis showed that the visible spectrum could be approximated by three fundamental colors: red, green, and blue (known as **additive primaries**), Figure 22–38. This concept has since proved useful in the development of color vision theory and in color photography. After 1861, the idea slowly evolved that in order to reproduce color images, a film would have to be made with three different layers, each layer sensitive to one

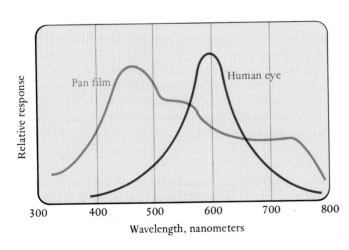

Figure 22–37 Spectral sensitivity of a panchromatic film compared with that of the human eye.

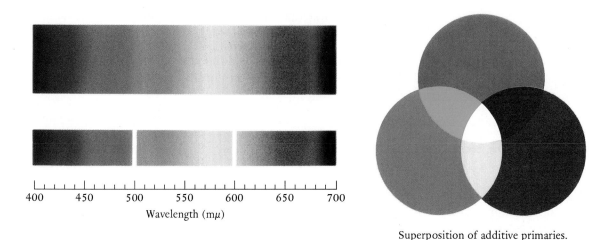

Superposition of additive primaries.

Figure 22–38 Additive primary colors. Red and green light together produce yellow, green and blue mix to form cyan light, and blue and red blend into magenta. All three primaries together produce white light.

When mixing light (not pigments), blue + green = cyan; green + red = yellow; and blue + red = magenta.

of the three primary colors. After the discovery of color-sensing dyes and panchromatic film, several different techniques for color photography were developed; but it was not until 1935, when Kodachrome was placed on the market, that the products reached the consumer. The Kodachrome process produces transparencies (or slides) that are viewed by projected light.

Combinations of additive primary colors can be used to reproduce any visible color on a projection screen. However, if the primary color filters are used, all of the light will be subtracted and there will be no light projected; the projection will be black (black is the absence of light). Another system of primary-color filters was developed, known as **subtractive primary** colors (Fig. 22–39). These colors are produced by dyes that absorb a primary color and pass the rest of the visible spectrum. A dye that transmits or reflects red light produces the additive primary color red. A dye that transmits or reflects all of the visible spectrum *except red light* (it absorbs red light) produces a subtractive primary color (cyan, a greenish blue color). Absorption of blue light from white leaves yellow, and absorption of green light results in bluish red, known as magenta. So the three subjective primary colors are cyan, yellow, and magenta.

When the proper mixture of subtractive primary dyes is layered in a photographic emulsion, an image is developed that will transmit the "true" color of the object photographed. For example, a mixture of magenta and cyan dyes would appear blue, since the magenta dye absorbs green light and the cyan dye absorbs red light, leaving only blue to be transmitted out of the three primary components of white light. White is produced by the absence of all three subtractive primaries and black is produced by an equal balance of the three subtractive primaries.

Color Film
Generally, a color film consists of a support and three color-sensitive emulsion layers. The blue-sensitive layer is usually on the top since silver halides are inherently blue-sensitive. Next, a yellow-colored filter layer is added. This layer absorbs blue light (transmits green and red, which together appear yellow) and serves to protect the lower emulsion layers from blue light. A

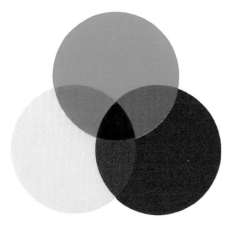

Superposition of subtractive primaries.

Figure 22–39 Subtractive primary colors. Since a dye absorbs only some of the white light that shines on it, it is a subtractive process. Three primary dye colors (cyan, magenta, and yellow) can be used not only to produce these three colors but can be mixed to produce red, green, and blue as well. The primary student quickly learns that if he/she colors yellow over cyan, the result on the paper will be green. The cyan dye reflects green and blue light to produce the cyan color, but the yellow dye absorbs the blue light, leaving only the green light to be reflected. All three subtractives absorb all of the light so the result is no light or black.

green-sensitive layer is next, and is followed by a red-sensitive layer on the supporting film (Fig. 22–40). These layers are rendered color-sensitive by dyes similar to those in the cyanine class, which render black-and-white film panchromatic. It should be realized that the color-sensitizing dyes in the film are not the color-producing dyes in the developed product; their job is to render each layer of the film sensitive to the appropriate portion of the visible spectrum. It is the final processing of the color film that yields the color image.

The thickness of the entire emulsion of color film is only about 0.0254 mm (0.001 in.).

Color Development

In the development process for the layered color film, each layer must be developed in sequence; hence time, temperature, and chemical concentrations are critical in each step to be sure that the reaction is taking place in the appropriate layer.

Most color films are developed with the aid of a dye-forming color process first introduced by a German chemist, R. Fischer, in 1912. The basis

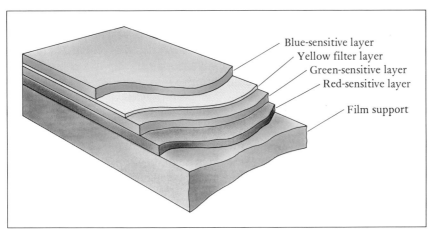

Blue-sensitive layer
Yellow filter layer
Green-sensitive layer
Red-sensitive layer

Film support

Figure 22–40 A typical arrangement of color-sensitive emulsion layers in a color film.

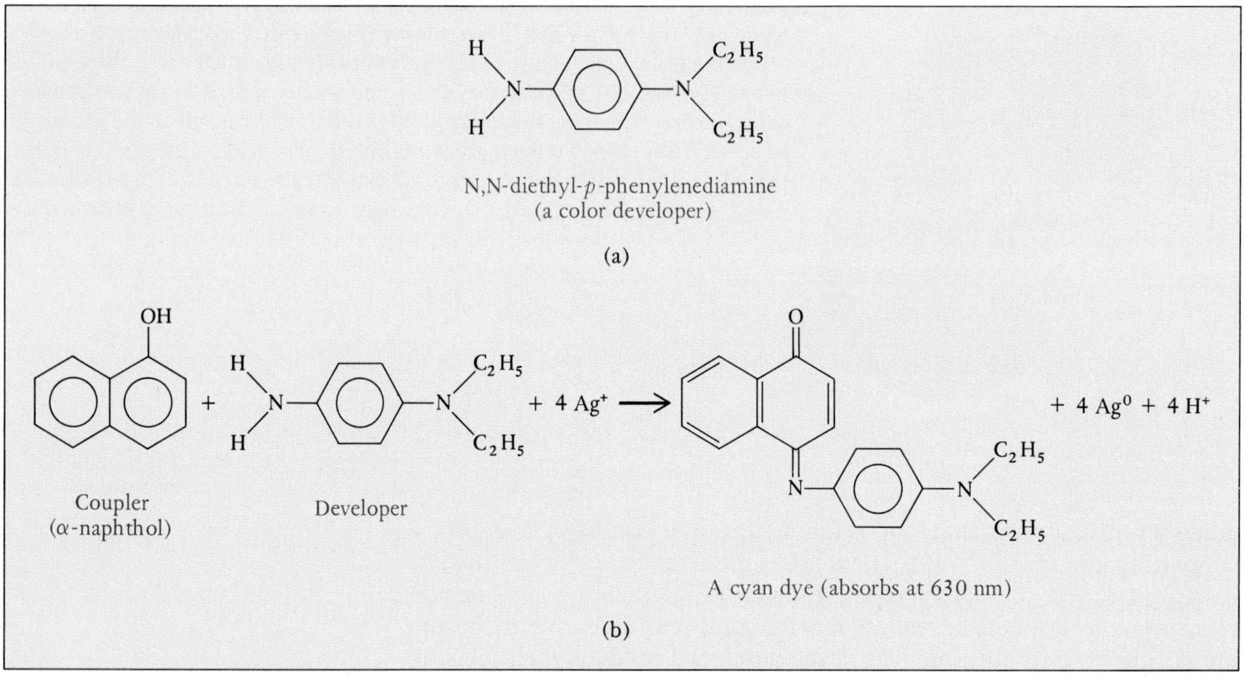

Figure 22–41 (a) A color developer: *N,N*-diethyl-*p*-phenylenediamine (b) A developer reacts with a film coupler to produce a cyan dye.

for this process is the oxidation of the developer to a dye-forming substance, which is then allowed to react with a molecule called a **coupler** to form the dye in the emulsion layer.

Color developers are generally substituted amines, and as such are reducing agents that will reduce the undeveloped silver in a film emulsion. An example is *N,N*-diethyl-*p*-phenylenediamine (Fig. 22–41).

To form a cyan dye during the development process, the developer reacts with a coupler, such as α-naphthol, and the silver ions in the emulsion layer to form the colored material. The reaction in words is:

Developer + Coupler + Oxidized silver $\longrightarrow$

Dye + Reduced silver + Hydrogen ions

The formulas for the reactants and products in this reaction are given in Figure 22–41. Thus, in the development of an exposed silver halide grain in the red-sensitive emulsion layer, a small amount of cyan dye is produced. The free silver must be bleached out prior to finishing.

Light with a wavelength of 630 nm is red.

The Kodachrome Process
An interesting example of a widely used color photography system is the Kodachrome process of the Eastman Kodak Company. The Kodachrome process is a **reversal** process, meaning that colors are reproduced in terms of their correct values and not their negative or complementary colors. The first developer in the Kodachrome process is a black-and-white developer, and in its application the reduction of the exposed silver halide is essentially complete.

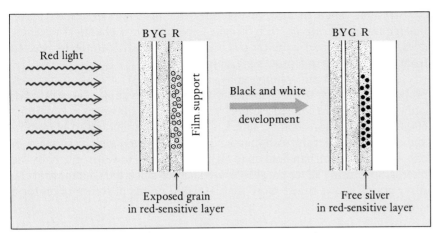

Figure 22–42 Simplified color image-forming process. B, blue-sensitive layer; Y, yellow filters; G, green-sensitive layer; R, red-sensitive layer.

The remaining unexposed silver halide in the three color-sensitive emulsions is a positive record of the original exposure. For example, red light striking the film would, upon black-and-white development, leave free silver in the red-sensitive layer (Fig. 22–42). Since no other color-sensitive layers were exposed by the original image, they contain no information. Now, selective reexposure and color development will produce free silver throughout the emulsion layers, along with the colored dyes, *except* where the red light originally struck the film. No dye forms there since the silver was previously reduced with a black-and-white developer.

Next, all the silver in the three emulsion layers, as well as the yellow-colored protective layer, is bleached with an oxidant such as the cyanoferrate ion [$Fe(CN)_6^{3-}$]

$$Ag° + Fe(CN)_6^{3-} \longrightarrow Ag^+ + Fe(CN)_6^{4-}$$

Once oxidized, the silver is treated with hypo and washed from the emulsion. The resulting emulsion is transparent except for the dyes that it contains. Considering that red light originally exposed the film, we see that the transmitted light will appear red (Fig. 22–43).

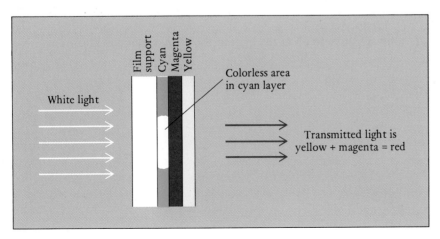

Figure 22–43 White light passes through a three-layer transparency. Since blue light and green light are absorbed, the transmitted light is red.

"Instant" color pictures involve the same principles described above, but require a *delicate balance* of light exposure, photochemical reagents, dyes, developers, and couplers since excess chemicals cannot be washed from the emulsion in a processing step.

The future of still photography, as well as that of motion pictures, probably lies in the use of cameras that use no film at all. The still video camera can record a colored picture by using a single silicon chip containing 600,000 tiny photoreceptors, called pixels, each responding to brightness and color. The "picture" can then be dumped onto a floppy disk, which in one working version can hold up to 50 photographs. However, currently the photographs using silver emulsions are sharper since the smallness of the silver grains allows much more than 600,000 points of information for a single picture.

SELF-TEST 22–E

1. What chemical causes the dark regions on a black-and-white negative?
2. Name a chemical that can be used as a black-and-white developer.
3. In the development process, silver is ( ) oxidized ( ) reduced?
4. Complete the following equation for the fixing process.

$$Ag^+ + 2\ S_2O_3^{2-} \longrightarrow \underline{\hspace{2cm}}$$

5. Which are more sensitive to light: ( ) very small ( ) somewhat larger grains of AgBr?
6. Photograph developers in black-and-white photography are ( ) oxidizing ( ) reducing agents?
7. Hypo is another name for _____.
8. **a.** Cyan + magenta = _____
   b. Cyan + yellow = _____
   c. Magenta + yellow = _____
9. The primary additive colors are _____, _____, and _____.
10. A Kodachrome color transparency has no silver in it. ( ) True ( ) False
11. In the formation of a dye in color film, silver is ( ) oxidized ( ) reduced.
12. Film sensitive to all of the light in the visible region is known as _____ film.

MATCHING SET I

_____ **1.** Keratin
_____ **2.** Melanin
_____ **3.** Sodium lauryl sulfate
_____ **4.** SPF-15
_____ **5.** Alcohol
_____ **6.** Hydrated aluminum chloride
_____ **7.** Soap
_____ **8.** Fat
_____ **9.** Cholesterol
_____ **10.** Thioglycolic acid
_____ **11.** _p_-Aminobenzoic acid
_____ **12.** Glycerol
_____ **13.** Sodium tripolyphosphate
_____ **14.** Whiteners
_____ **15.** Sodium hypochlorite
_____ **16.** Calcium carbonate
_____ **17.** Tin (II) fluoride (stannous fluoride)

a. Salt of fatty acid
b. Reducing agent in wave lotion
c. Reasonable limit for sunscreen
d. Abrasive in toothpaste
e. Holds moisture in skin
f. Common alcohol in lanolin
g. Ultraviolet absorber in suntan lotion
h. Alcohol produced in saponification of fat or oil
i. Detergent builder
j. Radiate a different wavelength of light than that absorbed
k. Synthetic detergent
l. Laundry bleach
m. Dark pigment
n. Deodorant component
o. Furnishes fluoride for stronger teeth
p. Dehydrates skin microbes
q. Skin and hair protein

MATCHING SET II

_____ **1.** Grease
_____ **2.** Ethylene glycol
_____ **3.** Negative
_____ **4.** Larger grain size in film
_____ **5.** Deguerreotype
_____ **6.** George Eastman
_____ **7.** Hydroquinone
_____ **8.** Acid solution
_____ **9.** Sodium thiosulfate
_____ **10.** Panchromatic
_____ **11.** Red, green, and blue
_____ **12.** Cyan, magenta, and yellow

a. Developer
b. Additive primary colors
c. Stop bath
d. Fixer
e. Subtractive primary colors
f. Permanent antifreeze
g. Exposed areas are dark
h. Oil with soap added
i. More light-sensitive film
j. Early photographic film
k. Introduced Kodak
l. Sensitive to broad band of wavelengths

QUESTIONS

1. **a.** What is the purpose of an emulsifier?
   b. In which of the following cosmetics is an emulsifier important: suntan lotion, hair spray, cold cream?
2. What is the purpose of each of the following?
   a. Detergent in toothpastes
   b. Polyvinlypyrrolidone in hair sprays
   c. Aluminum chloride in deodorants
   d. *p*-Aminobenzoic acid in suntan lotion
3. What specific substance is broken down during the bleaching of hair?
4. Hydrogen bonding is a very handy theoretical tool. Name three applications of hydrogen bonding in cosmetics and cleansing agents.
5. If you were going to formulate a suntan lotion, what particular spectral property would you look for in choosing the active compound?
6. Why are synthetic detergents better cleansing agents than soaps in regions where the water supply contains calcium or magnesium salts?
7. Why is a soap made from coconut oil more soluble in water than a soap made from palm oil?
8. Name and give the functions of four chemical ingredients of perfume.
9. Explain how an optical brightener in a detergent works.
10. What do skin, hair, and finger nails have in common?
11. What is the major difference between keratin in the hair and other proteins?
12. Why do the lips dry so easily?
13. Describe in chemical terms what happens when a person gets a permanent.
14. What three types of chemical bonds hold hair proteins together?
15. If a substance is an astringent, what is its action?
16. What is the purpose of talc in face powder?
17. Name an astringent widely used in deodorants.
18. What is the major ingredient in lipstick?
19. What is the chemical action of hydrogen peroxide on hair?
20. Is a fat an acid, an alcohol, an ester, or an alkane?
21. Can vegetable oils be used as effectively as animal oils to make soap?
22. Is the hydrocarbon end of the soap molecule polar or nonpolar? Explain.
23. Which is more soluble in water, calcium stearate or sodium stearate?
24. What would be the effect of lower pH on a typical developer?
25. Explain the term *fixing*. What is a fixer, and why is it important in photography?
26. Write a chemical equation for the development of silver ions by Metol.
27. Explain what the stop bath solution does in the development process.
28. What are the subtractive primary colors?
29. Is all of the silver reduced in the development of a black-and-white film? Explain.
30. What is the purpose of sodium sulfite in black-and-white photography?
31. What was done to increase the spectral range of black-and-white film?

A
Appendix

The International System of Units (SI)

A coherent system of units known as the Système International (SI system), bearing the authority of the International Bureau of Weights and Measures, has been in effect since 1960 and is gaining increasing acceptance among scientists. It is an extension of the metric system that began in 1790, with each physical quantity assigned a unique SI unit. An essential feature of both the older metric system and the newer SI is a series of prefixes that indicate a power of 10 multiple or submultiple of the unit.

UNITS OF LENGTH

The standard unit of length, the **meter,** was originally meant to be 1 ten-millionth of the distance along a meridian from the North Pole to the equator. However, the lack of precise geographical information necessitated a better definition. For a number of years the meter was defined as the distance between two etched lines on a platinum-iridium bar kept at 0°C (32°F) in the International Bureau of Weights and Measures at Sèvres, France. However, an inability to measure this distance as accurately as desired prompted a recent redefinition of the meter as being a length equal to 1,650,763.73 times the wavelength of the orange-red spectrographic line of $^{86}_{36}$Kr.

The meter (39.37 in.) is a convenient unit with which to measure the height of a basketball goal (3.05 meters), but it is unwieldy for measuring the parts of a watch or the distance between continents. For this reason, prefixes are defined in such a way that, when placed before the meter, they define distances convenient for our particular purposes. Some of the prefixes with their meanings are:

nano— 1/1,000,000,000 or 0.000000001 (or 10^{-9})
micro— 1/1,000,000 or 0.000001 (or 10^{-6})
milli— 1/1,000 or 0.001 or 10^{-3})
centi— 1/100 or 0.01 (or 10^{-2})
deci— 1/10 or 0.1 (or 10^{-1})
deka— 10 (or 10^{1})
hecto— 100 (or 10^{2})
kilo— 1,000 (or 10^{3})
mega— 1,000,000 (or 10^{6})

The corresponding units of length with their abbreviations are the following:

nanometer (nm)— 0.000000001 meter
micrometer (μm)— 0.000001 meter
millimeter (mm)— 0.001 meter
centimeter (cm)— 0.01 meter
decimeter (dm)— 0.1 meter
meter (m)— 1 meter
dekameter (dam)— 10 meters
hectometer (hm)— 100 meters
kilometer (km)— 1,000 meters
megameter (Mm)— 1,000,000 meters

Since the prefixes are defined in terms of the decimal system, the conversion from one metric length to another involves only shifting the decimal point. Mental calculations are quickly accomplished.

How many centimeters are in a meter? Think: Since a centimeter is the one-hundredth part of a meter, there would be 100 cm in a meter.

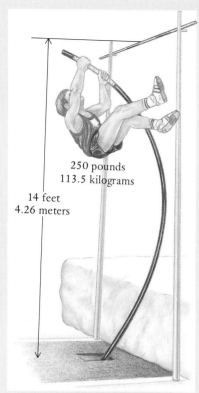

250 pounds
113.5 kilograms

14 feet
4.26 meters

Figure A–1   The pole-vaulter is easily recognized as hefty when described as weighing 250 lb and making a jump something less than 14 ft. To persons in the habit of using the system of international measurements, 113.5 kg and 4.26 m produce a similar conceptualization.

Figure A–2   A meter equals 1.094 yd.

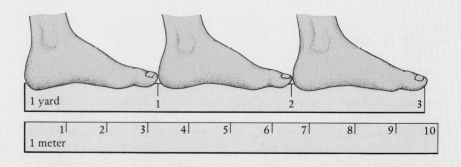

TABLE A–1 Conversion Factors*

Length:	1 inch (in.)	= 2.54 centimeters (cm)
	1 yard (yd)	= 0.914 meter (m)
	1 mile (mi)	= 1.609 kilometers (km)
Volume:	1 ounce (oz)	= 29.57 milliliters (mL)
	1 quart (qt)	= 0.946 liter (L)
	1.06 quart (qt)	= 1 liter (L)
	1 gallon (gal)	= 3.78 liters (L)
Mass (weight)†:	1 ounce (oz)	= 28.35 grams (g)
	1 pound (lb)	= 453.6 grams (g)
	1 ton (tn)	= 907.2 kilograms (kg)

* Common English units are used.
† Mass is a measure of the amount of matter, whereas weight is a measure of the attraction of the earth for an object at the earth's surface. The mass of a sample of matter is constant, but its weight varies with position and velocity. For example, the space traveler, having lost no mass, becomes weightless in earth orbit. Although mass and weight are basically different in meaning, they are often used interchangeably in the environment of the earth's surface.

Conversion of measurements from one system to the other is a common problem. Some commonly used English–SI equivalents (conversion factors) are given in Table A–1.

UNITS OF MASS

The primary unit of mass is the **kilogram** (1000 g). This unit is the mass of a platinum-iridium alloy sample deposited at the International Bureau of Weights and Measures. One pound contains a mass of 453.6 g (a five-cent nickel coin contains about 5 g).

Conveniently enough, the same prefixes defined in the discussion of length are used in units of mass, as well as in other units of measure.

UNITS OF VOLUME

The SI unit of volume is the **cubic meter** (m^3). However, the volume capacity used most frequently in chemistry is the liter, which is defined as 1 cubic decimeter (1 dm^3). Since a decimeter is equal to 10 cm, the cubic decimeter is equal to 10 cm^3 or 1,000 cubic centimeters (cc). One cc, then, is equal to one milliliter (the thousandth part of a liter). The mL (or cc) is a common unit that is often used in the measurement of medicinal and laboratory quantities. There are then 1,000 liters (L) in a kiloliter or cubic meter.

UNITS OF ENERGY

The SI unit for energy is the **joule** (J), which is defined as the work performed by a force of one newton acting through a distance of one meter. A newton is defined as that force which produces an acceleration of one meter per second per second when applied to a mass of one kilogram. Conversion units for energy are:

$$1 \text{ calorie} = 4.184 \text{ J}$$
$$1 \text{ kilowatt-hour} = 3.5 \times 10^6 \text{ J}$$

OTHER SI UNITS

Other SI units are listed below.

Time	second (s)
Temperature	Kelvin (K)
Electric current	ampere (A) = 1 coulomb per second
Amount of molecular substance	mole (mol) = 6.023×10^{23} molecules
Pressure	pascal (Pa) = 1 newton per square meter
Power	watt (W) = 1 joule per second
Electric charge	coulomb (C) = 6.24196×10^{18} electron charges
	= 1.036086×10^{-5} faradays

Further information on SI units can be obtained from "SI Metric Units —An Introduction," by H. F. R. Adams, McGraw-Hill Ryerson Ltd., Toronto, 1974.

B
Appendix

Calculations with Chemical Equations

The bases for calculations with chemical equations were presented in Chapter 6. Problems of a more complex nature and a systematic approach to their solution are presented in the following examples. Finally, a list of exercise problems is given for further study.

EXAMPLE 1

Balanced Equations Express Number Ratios for Particles

In the reaction of hydrogen with oxygen to form water, how many molecules of hydrogen are required to combine with 19 oxygen molecules?

Solution: A chemical equation can be written for a reaction only if the reactants and products are identified and the respective formulas determined. In this problem, the formulas are known and the unbalanced equation is:

$$H_2 + O_2 \longrightarrow H_2O$$

It is evident that one molecule of oxygen contains enough oxygen for two water molecules and the equation, as written, does not account for what happens to the second oxygen atom. As it is, the equation is in conflict with the conservation of atoms in chemical changes. This conflict is easily corrected by balancing the equation:

$$2 H_2 + O_2 \longrightarrow 2 H_2O$$

Now all atoms are accounted for in the equation and it is obvious that two hydrogen molecules are required for each oxygen molecule. In other words, two hydrogen molecules are equivalent to one oxygen molecule in their usage. This can be expressed as follows:

2 hydrogen molecules {are equivalent to} 1 oxygen molecule;

or,

$2 H_2$ molecules $\approx O_2$ molecule;

or,

$$\frac{2 H_2 \text{ molecules}}{O_2 \text{ molecule}},$$

which can be read as two hydrogen molecules per one oxygen molecule. Using the factor-label approach, the solution is readily achieved.

? H_2 molecules $= 19 O_2$ molecules

$$? H_2 \text{ molecules} = 19 \, \cancel{O_2 \text{ molecules}} \times \frac{2 H_2 \text{ molecules}}{\cancel{O_2 \text{ molecule}}}$$

$$= 38 H_2 \text{ molecules}$$

Note: The reader is likely to say at this point that the method is cumbersome and that he can quickly see the answer to be 38 H_2 molecules without "the method." However, problems to follow are made much easier if a systematic method is used.

EXAMPLE 2

Laboratory Mole Ratios Identical with Particle Number Ratios

How many moles of hydrogen molecules must be burned in oxygen (the reaction of Example 1) to produce 15 mol of water molecules (about a glassful)?

Solution: The balanced equation

$$2 H_2 + O_2 \longrightarrow 2 H_2O$$

tells us that two molecules of hydrogen produce two molecules of water; or,

2 molecules hydrogen $\approx$ 2 molecules water,

EXAMPLE 3   **A-7**

and therefore,

    1 molecule hydrogen ≈ 1 molecule water.

It is obvious then that the number of water molecules produced will be equal to the number of hydrogen molecules consumed regardless of the actual number involved. Therefore,

    6.02×10^{23} molecules of hydrogen ≈ 6.02×10^{23} molecules of water

Since 6.02×10^{23} is a number called the mole, it follows that 1 mol of hydrogen molecules will produce 1 mol of water molecules. The general conclusion, then, is the following: the ratio of particles in the balanced equation is the same as the ratio of moles in the laboratory. The solution to the problem logically follows:

$$\left.\begin{array}{c}? \text{ moles of hydrogen} \\ \text{molecules}\end{array}\right\} = \left\{\begin{array}{c} 15 \text{ mol} \\ \text{water molecules}\end{array}\right\} \times \frac{1 \text{ mol hydrogen molecules}}{1 \text{ mol water molecules}}$$

$$= 15 \text{ mol hydrogen molecules}$$

Note: Again the solution to the problem looks simple enough without resorting to the factor-label method. However, in Examples 3 and 4, the numbers become such that a quick mental solution is not readily achieved by most students.

EXAMPLE 3

Mole Weights Yield Weight Relationships

How many grams of oxygen are necessary to react with an excess of hydrogen to produce 270 g of water?

Solution: From the balanced equation

    $2 H_2 + O_2 \longrightarrow 2 H_2O$

the mole ratio between oxygen and water is immediately evident and is 1 mol of oxygen molecules per 2 mol of water molecules, or

$$\frac{1 \text{ mol oxygen molecules}}{2 \text{ mol water molecules}}$$

This mole ratio can be changed into a weight ratio since the mole weight can be easily calculated from the atomic weights involved. One molecule of oxygen (O_2) weighs 32 amu (16 amu for each oxygen atom). Therefore, a mole of oxygen molecules weighs 32 g. Similarly, 2 mol of water weigh 36 g [2(16 + 1 + 1)]. Therefore, the weight ratio is:

$$\frac{1 \text{ mol oxygen molecules} \times \dfrac{32 \text{ g oxygen}}{\text{mole oxygen molecules}}}{2 \text{ mol water molecules} \times \dfrac{18 \text{ g water}}{\text{mole water molecules}}}$$

or,

$$\frac{32 \text{ g oxygen}}{36 \text{ g water}}$$

This weight relationship is exactly the conversion factor needed to answer the original question:

$$? \text{ grams oxygen} = 270 \text{ g water} \times \frac{32 \text{ g oxygen}}{36 \text{ g water}}$$

$$= 240 \text{ g oxygen}$$

Note: It should be observed that a weight relationship could be established between any two of the three pure substances involved in the reaction, regardless of whether they are reactants or products.

EXAMPLE 4

How many molecules of water are produced in the decomposition of eight molecules of table sugar? The unbalanced equation is as follows:

$$C_{12}H_{22}O_{11} \longrightarrow C + H_2O$$

Solution: Balance the equation

$$C_{12}H_{22}O_{11} \longrightarrow 12\,C + 11\,H_2O$$

$$? \text{ molecules of water} = 8 \text{ molecules of sugar} \times \frac{11 \text{ molecules water}}{1 \text{ molecule sugar}}$$

$$= 88 \text{ molecules of water}$$

EXAMPLE 5

How many grams of mercuric oxide are necessary to produce 50 g of oxygen? Mercuric oxide decomposes as follows:

$$2\,HgO \longrightarrow 2\,Hg + O_2$$

Solution:

Weight of 2 mol of HgO $= 2(201 + 16) = 2(217) = 434$ g

Weight of 1 mol of $O_2 = 2(16) = 32$ g

$$? \text{ g HgO} = 50 \text{ g oxygen} \times \frac{434 \text{ g mercuric oxide}}{32 \text{ g oxygen}}$$

$$= 678 \text{ g mercuric oxide}$$

EXAMPLE 6   **A-9**

EXAMPLE 6

How many pounds of mercuric oxide are necessary to produce 50 lb of oxygen by the reaction:

$$2 \, HgO \longrightarrow 2 \, Hg + O_2$$

Solution: Note that the problem is the same as Example 5 except for the units of chemicals. Also note that the conversion factor of Example 5

$$\frac{434 \text{ g mercuric oxide}}{32 \text{ g oxygen}}$$

can be converted to any other units desired:

$$\frac{434 \text{ g mercuric oxide} \times \dfrac{1 \text{ lb}}{454 \text{ g}}}{32 \text{ g oxygen} \times \dfrac{1 \text{ lb}}{454 \text{ g}}} = \frac{434 \text{ lb mercuric oxide}}{32 \text{ lb oxygen}}$$

It is evident that the ratio, 434/32, expresses the ratio between weights of mercuric oxide and oxygen in this reaction regardless of the units employed.

$$? \text{ pounds mercuric oxide} = 50 \text{ lb oxygen} \times \frac{434 \text{ lb mercuric oxide}}{32 \text{ lb oxygen}}$$

$$= 678 \text{ lb of mercuric oxide}$$

Problems

1. What weight of oxygen is necessary to burn 28 g of methane (CH_4)? The equation is:

$$CH_4 + 2 \, O_2 \longrightarrow CO_2 + 2 \, H_2O \qquad \textit{Ans.} \; 112 \text{ g oxygen}$$

2. Potassium chlorate ($KClO_3$) releases oxygen when heated according to the equation:

$$2 \, KClO_3 \longrightarrow 2 \, KCl + 3 \, O_2$$

What weight of potassium chlorate is necessary to produce 1.43 g of oxygen? 　　　　　　　　　　　　　　　　*Ans.* 3.65 g $KClO_3$

3. Fe_3O_4 is a magnetic oxide of iron. What weight of this oxide can be produced from 150 g of iron? 　　　*Ans.* 207 g oxide

4. Steam reacts with hot carbon to produce a fuel called water gas; it is a mixture of carbon monoxide and hydrogen. The equation is:

$$H_2O + C \longrightarrow CO + H_2$$

What weight of carbon is necessary to produce 10 g of hydrogen by this reaction? 　　　　　　　　*Ans.* 60 g carbon

5. Iron oxide (Fe_2O_3) can be reduced to metallic iron by heating it with carbon.

$$2\,Fe_2O_3 + 3\,C \longrightarrow 4\,Fe + 3\,CO_2$$

How many tons of carbon would be necessary to reduce 5 tons of the iron oxide in this reaction?   *Ans.* 0.56 ton carbon

6. How many grams of hydrogen are necessary to reduce 1 lb (454 g) of lead oxide (PbO) by the reaction:

$$PbO + H_2 \longrightarrow Pb + H_2O$$ *Ans.* 3.91 g hydrogen

7. Hydrogen can be produced by the reaction of iron with steam.

$$4\,H_2O + 3\,Fe \longrightarrow 4\,H_2 + Fe_3O_4$$

What weight of iron would be needed to produce 0.5 lb of hydrogen?   *Ans.* 10.5 lb iron

8. Tin ore, containing SnO_2, can be reduced to tin by heating with carbon.

$$SnO_2 + C \longrightarrow Sn + CO_2$$

How many tons of tin can be produced from 100 tons of SnO_2?   *Ans.* 79 tons tin

Answers to Self-Test Questions and Matching Sets

CHAPTER 1

Self-Test 1–A

1. observed experimental facts
2. the same
3. the integrated circuit
4. *E. coli*
5. (a) theories, (b) laws, (c) facts
6. cadmium
7. Clean Air Act
8. tissue-plasminogen-activator

CHAPTER 2

Self-Test 2–A

1. dirt, wood, dusty air, salt water, etc.
2. operational definition
3. rain water, gold, quartz, diamond, etc.
4. false
5. physical, chemical, nuclear
6. solution
7. true

Matching Set

1. c
2. a
3. d
4. b

Self-Test 2–B

1. (a) metals: iron, copper, gold, silver, chromium, magnesium, etc.
   (b) nonmetals: oxygen, silicon, carbon, nitrogen, chlorine, fluorine, etc.
2. 89 found in nature and 20 claimed artificially
3. false
4. cutting diamond, blowing glass, molding plastic, slicing bread, etc.
5. burning coal, dissolving iron ore in an acid, making steel from iron ore, making aspirin, etc.
6. chemical substance
7. Submicroscopic
8. (a) macroscopic, (b) microscopic, (c) molecular
9. nuclear
10. (a) oxygen, (b) iron, (c) hydrogen

11. chromatography, distillation, recrystallization, filtration
12. A student should be able to list a dozen or more elements.

Self-Test 2-C
1. (a) oxygen, an oxygen atom or a mole of oxygen atoms
   (b) two oxygen molecules, each containing two atoms, or a mole of oxygen molecules
   (c) a molecule of methane or a mole of methane molecules
   (d) reacts to form
   (e) a water molecule or a mole of water molecules
   (f) two water molecules or two moles of water molecules
2. the element, an atom of the element, or a mole of the elemental atoms
3. the elements present and the relative number of each type of atom
4. coefficient
5. false
6. seven

Matching Set

1. b	7. i	12. o
2. e	8. k	13. l
3. d	9. g	14. m
4. a	10. h	15. n
5. c	11. j	16. p
6. f		

CHAPTER 3

Self-Test 3-A
1. Leucippus, Democritus
2. (b) philosophy
3. gained, chemical
4. CO, CO_2
5. (a) new compound, (b) 2:4:1, (c) law of multiple proportions
6. (d) Atoms are recombined into different arrangements.
7. (a) the same, (b) atoms
8. repel, attract
9. alpha (α), beta (β), gamma (γ), gamma (γ)
10. scanning tunneling

Self-Test 3-B
1. protons, neutrons
2. small
3. nucleus
4. electrons, protons, neutrons
5. electrons, protons
6. atomic

7. about 1836
8. 33, 33, 42
9. different, identical
10. electrons
11. electrons, protons
12. false

Self-Test 3-C
1. particles, waves
2. spectrum
3. farther from, closer to
4. (d) Wave nature of the electron
5. wavelength
6. 18, 2
7. (a) No, not the paths
   (b) In a way, yes. They are representations of the space in which we can expect to find the electron with 90% certainty.
   (c) Of a given type, yes, with 90% certainty

Matching Set

1. e	6. n	11. h ·
2. a	7. b	12. c
3. j	8. f	13. k
4. i	9. m	14. l
5. d	10. g	15. o

CHAPTER 4

Self-Test 4-A
1. IIA, VIII, VIIA, IIIA, IB
2. 2, 5, 6, 7, 4
3. R, R, T, N, I
4. Mendeleev
5. atomic numbers
6. periodic
7. metals
8. nonmetals

Self-Test 4-B
1. 3, $GaCl_3$; 2, $BaCl_2$; 2, $SeCl_2$, 1, ICl
2. metalloid, metal, nonmetal, metal, metal, nonmetal, nonmetal, metalloid
3. 1, 2, 7, 7, 6, 3, 4
4. ionization
5. He, F, Br, S
6. K, Br, Na, In
7. IA, IVA, IIA

Matching Set

1. i	5. d	8. h
2. c	6. g	9. f
3. e	7. b	10. k
4. a		

CHAPTER 5

Self-Test 5-A
1. ions
2. ionic
3. one
4. CaI_2
5. Cl^-
6. valence
7. losing
8. smaller
9. gaining
10. larger
11. Rb, one electron lost; Ca, two electrons lost; K, one electron lost; S, two electrons gained; Mg, two electrons lost; Br, one electron gained

Self-Test 5-B
1. (a) H_2, (b) HF
2. six
3. three
4. fluorine
5. (a) eight, (b) octet, (c) most of the time
6. fluorine
7. decrease
8. H_2O

Self-Test 5-C
1. nonbonding, bonding
2. linear, trigonal planar, tetrahedral, octahedral
3. two, two
4. three, one
5. bending, stretching, rotating
6. Infrared
7. hydrogen bonding

Matching Set-I
1. l	6. g	10. c
2. d	7. b	11. h
3. j	8. e	12. a
4. f	9. k	13. m
5. i		

Matching Set-II
1. a	4. e	6. g
2. c	5. b	7. f
3. d		

CHAPTER 6

Self-Test 6-A
1. (a) 6, 6, 1, 6 (The 1 is understood.)
   (b) 6
   (c) 6
   (d) 44, 180
   (e) 264 g
2. 136 kcal
3. (a) 2, 1, 2
   (b) 1, 2, 1
   (c) 4, 3, 2
4. (a) 2, 2, 1
   (b) 18 g
   (c) 16 g
   (d) 18 tons

Self-Test 6-B
1. temperature, because the rate of bacterial growth slows with decreasing temperatures
2. hydrogen, oxygen
3. hemoglobin and oxygen, and calcium oxide and water
4. (a) Catalysts are used to increase reaction rates.
   (b) Catalysts remain in original amounts.
5. no
6. Equilibrium constants vary with temperature.
7. (b) the same flour in dust form
8. hemoglobin uptake and release of oxygen
9. the formation of any very stable compound such as water or table salt
10. Ordinary chemical change occurs with a minimum amount of change in the structure of the atoms, molecules, or ions involved.

Matching Set
1. g	5. c	9. h
2. f	6. a	10. l
3. e	7. j	11. i
4. b	8. k	

CHAPTER 7

Self-Test 7-A
1. cloud chamber
2. $^{87}_{36}Kr$
3. $^{212}_{82}Pb$
4. 0.125 g
5. $^{8}_{4}Be$
6. James Chadwick; alpha-ray bombardment of beryllium
7. lead-206
8. 5730 years

Self-Test 7-B
1. one billion (10^9)
2. cyclotron
3. neptunium (Np)
4. 109
5. roentgen equivalent man

6. less than 0.1 rad per year
7. Dallas
8. Energies in the billions or trillions of electron volts are required.
9. genetic

Self-Test 7 – C
1. sterilization, tracers, medical diagnosis
2. (a) ^{60}Co
3. (d) all of these
4. (b) imaging
5. (c) metastable
6. (a) one eighth of the original dose
7. (b) 6-h half-life isotope
8. radium-226
9. lead-206

Matching Set

1. b	6. c	11. h
2. i	7. l	12. m
3. a	8. d	13. f
4. k	9. o	14. g
5. e	10. n	15. j

CHAPTER 8

Self-Test 8 – A
1. petroleum and coal
2. natural gas
3. 33%
4. coal, petroleum, natural gas
5. oxygen, water
6. true
7. work
8. quantitatively
9. disorder (entropy)
10. quality
11. calorie, joule, Btu, kilowatt-hour
12. watt, kilowatt, joule/sec
13. aluminum, aluminum

Self-Test 8 – B
1. petroleum
2. 50%
3. 42
4. CO, H_2, N_2
5. by a primary source of energy
6. electricity
7. coal
8. no, yes

Self-Test 8 – C
1. fission
2. fusion
3. containment of reactants at high temperature

4. tritium
5. amorphous
6. plutonium ($^{239}_{94}$Pu), uranium ($^{233}_{92}$U)
7. water, air, greenhouse
8. natural gas
9. arsenic, boron (or gallium, see Chapter 11)

Matching Set

1. f	6. e	10. h
2. l	7. g	11. k
3. b	8. a	12. m
4. j	9. p	13. o
5. c, e		

CHAPTER 9

Self-Test 9 – A
1. nonelectrolyte
2. base, acid
3. base, acid
4. amphiprotic
5. neutral
6. water
7. 2 mol
8. 0.50 M
9. acid, base (either order)
10. 12

Self-Test 9 – B
1. 10, (b) basic; 7, (c) neutral; 3, (a) acidic
2. 1400 g, or 1.4 kg
3. NH_4^+, OH^-
4. H_3O^+, Cl^-
5. pH of 2
6. yes
7. buffers
8. Strong acids: H_2SO_4 (sulfuric), HCl (hydrochloric), HNO_3 (nitric)
   Weak acids: CH_3COOH (acetic), citric, ascorbic (vitamin C)
   Strong bases: NaOH (sodium hydroxide), KOH (potassium hydroxide)
   Weak bases: NH_4OH (ammonium hydroxide), NH_3 (ammonia)
9. 7
10. 10^{-2} or 0.01 times as large
11. hydroxide, constant value (10^{-14} at 25°C)
12. hydrogen phosphate (HPO_4^{2-}), dihydrogen phosphate ($H_2PO_4^-$)
13. low
14. high

Matching Set

1. k	6. l	10. e
2. k	7. j	11. g

3. b	8. h	12. i
4. a	9. c	13. d
5. f		

13. a	15. j	17. m
14. d	16. b	18. o

CHAPTER 10

Self-Test 10-A
1. burning/combustion
2. carbon dioxide, water
3. oxide
4. rust
5. water, carbon dioxide
6. carbon monoxide
7. oxidizing agent (oxidant)
8. acetaldehyde
9. Na^+ ion

Self-Test 10-B
1. reduced
2. reducing agent
3. reduced
4. reducing agent
5. sodium metal
6. magnesium metal
7. electrolysis
8. reduction
9. oxidation
10. hydrogen, oxygen

Self-Test 10-C
1. electrolysis
2. oxidation
3. iron, water, oxygen
4. magnesium
5. yes
6. fluorine
7. hydrogen, oxygen
8. lithium

Self-Test 10-D
1. fluorine
2. lithium
3. Copper plates out on the zinc and zinc goes into solution.
4. sulfuric acid
5. cathodic
6. secondary
7. primary
8. fuel

Matching Set
1. h	5. f	9. e
2. o	6. l	10. p
3. k	7. g	11. c
4. q	8. n	12. i

CHAPTER 11

Self-Test 11-A
1. oxygen
2. aluminum
3. Minnesota
4. limestone
5. copper
6. copper, aluminum, magnesium
7. slag
8. reduced
9. cathode
10. magnesium
11. positive ions

Self-Test 11-B
1. nitrogen, oxygen
2. liquid oxygen
3. cold materials in the temperature range of liquid air
4. oxygen: used in making steel, as an oxidizing agent, in welding, and in controlled atmospheres; nitrogen: used in cryosurgery, in an inert atmosphere, as a welding gas blanket, and in freezing food

Self-Test 11-C
1. silicon, oxygen
2. lead
3. silicon
4. clay, sand, feldspar
5. cement
6. sulfur
7. chlorine
8. air
9. sodium carbonate, sodium hydrogen carbonate (sodium bicarbonate)

Matching Set
1. i	5. h	8. a
2. j	6. e	9. c
3. b	7. f	10. d
4. g		

CHAPTER 12

Self-Test 12-A
1. organic
2. tetrahedral
3. 13, no
4. false
5. two (a, b, or d; and e or f)
6. structural
7. ethene (ethylene)

8. ethyne (acetylene)
9. double bond
10. 2,4-dimethylhexane
11. $-C_2H_5$
12. geometric

13.
$$H-\overset{\overset{\displaystyle H}{|}}{\underset{\underset{\displaystyle H}{|}}{C}}-\overset{\overset{\displaystyle H}{|}}{C}=\overset{\overset{\displaystyle C_2H_5}{|}}{C}-\overset{\overset{\displaystyle H}{|}}{\underset{\underset{\displaystyle H}{|}}{C}}-\overset{\overset{\displaystyle H}{|}}{\underset{\underset{\displaystyle H}{|}}{C}}-H$$

Self-Test 12 – B

1. cyclopropane
2. hydrogen
3. alkyne class
4. delocalized
5. false
6. four
7. rotation of polarized light
8. (a) alcohol, (b) carboxylic acid, (c) aldehyde, (d) ketone
9. (a) C_2H_6, C_2H_5OH, CH_3CHO, CH_3COOH, $(C_2H_5)_2O$, $C_2H_5NH_2$
   (b) ethane, ethyl alcohol, acetaldehyde, acetic acid, diethyl ether, ethyl amine
   (c) $-C_2H_5$ or ethyl group in ethanol, diethyl ether, and ethyl amine; $-CH_3$ or methyl in acetaldehyde and acetic acid
10. (a) diethyl ether
    (b) ethanol
    (c) acetic acid
    (d) acetone
11. 12 (6 C, 6 H)
12. 1,2,4-trimethylbenzene

Matching Set

1. c	5. b	8. h
2. i	6. f	9. g
3. a	7. e	10. d
4. j		

CHAPTER 13

Self-Test 13 – A

1. fractional distillation
2. methane
3. methyl tertiary-butyl ether (MTBE)
4. catalytic re-forming
5. petroleum
6. cracking
7. pyrolysis
8. aromatics
9. c

Self-Test 13 – B

1. 42%
2. acetaldehyde

3. denatured
4. isopropyl (rubbing)
5. fatty
6. acetic acid
7. formaldehyde
8. methanol
9. New Zealand

Self-Test 13 – C

1. glycerol
2. esters
3. aldehydes, carboxylic acids, alcohols
4. alkaloids, caffeine
5. c
6. $CH_3CH_2CH_2O\overset{\overset{\displaystyle O}{\|}}{C}CH_2CH_3$ (an ester) $+ H_2O$
7. (a) A fat is a solid at room temperature, whereas an oil is a liquid. The fat molecule has fewer double bonds.
   (b) by hydrogenation (addition of hydrogen to the $C=C$ double bonds).

Matching Set

1. i	5. h	9. e
2. g	6. b	10. d
3. k	7. a	11. m
4. j	8. f	12. l

CHAPTER 14

Self-Test 14 – A

1. monomers
2. thermoplastic
3. (a) $H_2C=\overset{\overset{\displaystyle }{|}}{\underset{\underset{\displaystyle CH_3}{|}}{CH}}$

   (b) $HC=CH_2$

   (c) $F_2C=CF_2$
   (d) $H_2C=CHCl$
4. (a) $\left(\overset{\overset{\displaystyle H}{|}}{\underset{\underset{\displaystyle H}{|}}{C}}-\overset{\overset{\displaystyle H}{|}}{\underset{\underset{\displaystyle H}{|}}{C}}\right)_n$

   (b) $\left(\overset{\overset{\displaystyle H}{|}}{\underset{\underset{\displaystyle H}{|}}{C}}-\overset{\overset{\displaystyle H}{|}}{\underset{\underset{\displaystyle CN}{|}}{C}}\right)_n$
5. isoprene
6. copolymer

Self-Test 14–B
1. condensation or polyamide
2. water
3.

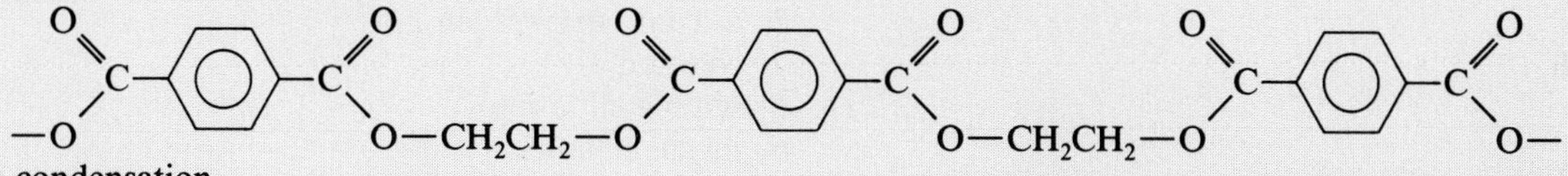

4. condensation
5. polyester
6. thermosetting
7. iodine
8.

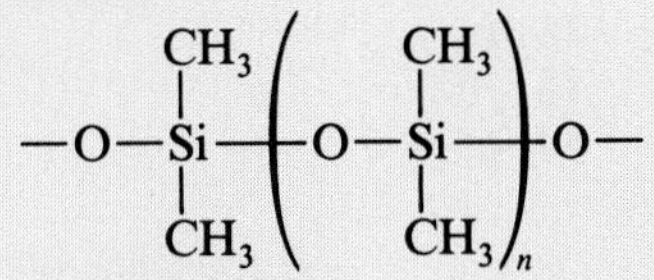

9. ultraviolet light
10. alcohols, acids
11. plasticizer
12. O (oxygen); it is a polymer held together by a network of Si—O bonds
13. HCl
14. ultraviolet
15. polyethylene

Matching Set

1. p	6. g	11. o
2. d	7. b	12. m
3. j	8. c	13. e
4. k	9. f	14. n
5. a	10. i	15. h

CHAPTER 15

Self-Test 15–A
1. carbon, hydrogen, oxygen
2. monosaccharides
3. D-glucose, D-fructose
4. D-glucose
5. α-D-glucose
6. hydrogen bonding
7. hydrogen bonding

Self-Test 15–B
1. amino acids
2. essential amino acids
3.

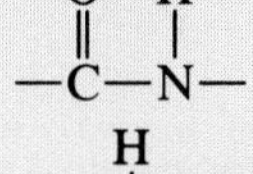

4.

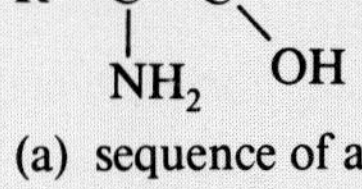

5. (a) sequence of amino acids
   (b) hydrogen bonded structures to form helices or sheets

(c) arrangements of helices into super helices or balls (globs)

(d) positioning of the tertiary structures (globs)
6. (a) 27, (b) 6
7. Hydrogen bonds form the helices (H-bonds between each third amino acid in the chain) and the sheets (H-bonds between chains).
8. catalyst
9. key, lock
10. enzyme
11. enzyme
12. active site
13. hydrogen bonding, London forces, disulfide bridges

Self-Test 15–C
1. the Sun
2. ATP
3. ADP, phosphate or phosphoric acid, energy
4. CO_2, H_2O, energy
5. chlorophyll
6. hydrolysis
7. emulsifying agents
8. ATP, lactic acid
9. CO_2, H_2O, ATP
10. rhodopsin
11. rods, cones
12. rods, cones
13. β-carotene

Self-Test 15–D
1. (a) DNA, (b) transfer RNA
2. hydrogen
3. ATP
4. adenine; thymine or uracil; cytosine: guanine; guanine: cytosine; thymine: adenine; uracil: adenine
5. false
6. false (complement, not duplicate)
7. ribose, deoxyribose
8. phosphoric acid, a sugar (ribose or deoxyribose), a nitrogenous base
9. double helix

Matching Set

1. g	5. c	9. l
2. f	6. b	10. o
3. a	7. d	11. q
4. e	8. k	12. i

13. j   15. r   17. p
14. n   16. s   18. m

CHAPTER 16

Self-Test 16-A
1. dehydration, hydrolysis
2. oxidizing
3. hemoglobin, oxygen
4. false
5. CN^-, cytochrome oxidase, oxygen
6. fluoroacetic acid
7. heavy-metal poisons, complex
8. true
9. water, food, paint

Self-Test 16-B
1. neurotoxins
2. synapses
3. acetylcholine
4. genes, chromosomes
5. false
6. chimney sweeping
7. heart disease, cancer
8. teratogen

Matching Set

1. b	5. j	9. c
2. f	6. d	10. k
3. h	7. g	11. a
4. i	8. e	

CHAPTER 17

Self-Test 17-A
1. 70
2. oceans
3. 90, 230 million, 20.7 billion
4. 10 gal
5. metal ions, pesticides, organic solvents
6. groundwater
7. aquifer
8. runoff

Self-Test 17-B
1. biochemical oxygen demand, BOD
2. 5°C water
3. batteries, petroleum
4. solvents, pesticides
5. a. auto battery; lead, sulfuric acid
   b. paints; solvents
   c. motor oil; metals
6. aluminum, paper, plastics, glass

Self-Test 17-C
1. Na^+, Mg^{2+}, Ca^{2+}, K^+
2. toxic
3. calcium, magnesium
4. chlorine
5. settling, filtration
6. metal, organics

Matching Set

1. l	7. n	13. g
2. o	8. q	14. r
3. p	9. d	15. i
4. a	10. m	16. j
5. c	11. e	17. k
6. b	12. f	

CHAPTER 18

Self-Test 18-A
1. ppm, 10,000, 0.001
2. decrease
3. ozone
4. adsorb
5. absorb
6. particulates
7. warm, cool
8. coal burning
9. oxides
10. $NO\cdot$ and $O\cdot$
11. $O\cdot$
12. aldehydes, ketones
13. peroxyacetyl nitrate (PAN)

Self-Test 18-B
1. carbon dioxide, sulfur dioxide
2. sulfur trioxide, sulfuric acid
3. lime
4. sulfuric acid, nitric acid
5. 5.6, carbon dioxide
6. 1872
7. C—Cl bond
8. $O_2 + O \longrightarrow O_3$
9. ClO
10. C—Br bond
11. Antarctica
12. ultraviolet

Self-Test 18-C
1. clear-cutting, automobiles, fossil fuel burning to produce electricity
2. photosynthesis, dissolving in oceans
3. carbon dioxide, water vapor, ozone. carbon dioxide
4. true
5. 350 ppm

6. Texas
7. toluene
8. smoking

Matching Set

1. b	6. c	11. o
2. o, g	7. e	12. a
3. l	8. m	13. k
4. j	9. i	14. h
5. n	10. d	

CHAPTER 19

Self-Test 19–A
1. clays, silts, sandy soils, loams
2. sour
3. acidic, basic
4. size of soil particles, chemical composition of soils
5. a trivalent ion like Fe^{3+} (hydrolyzes more than Na^+)
6. humus
7. silicate
8. calcium, magnesium, sulfur
9. element
10. true

Self-Test 19–B
1. false
2. nitrogen, phosphate, potash
3. potassium nitrate for potassium
4. yes
5. gas
6. nitric acid (HNO_3), nitrous acid (HNO_2)
7. denitrification
8. K_2CO_3
9. magnesium
10. 33%
11. DDT
12. chlordan
13. selective herbicide
14. 2,4-D

Matching Set

1. g	7. h	12. a
2. b	8. m	13. i
3. l	9. k	14. r
4. p	10. o	15. f
5. c	11. q	16. d
6. e		

CHAPTER 20

Self-Test 20–A
1. proteins, fats, carbohydrates, vitamins, minerals, water
2. oxygen, calcium
3. USRDA

4. partially true
5. fat
6. 1, 1
7. basal metabolic rate, weight in pounds
8. fats

Self-Test 20–B
1. apoenzyme
2. protein
3. false
4. urea
5. lecithin, cholesterol
6. triglycerides
7. linoleic acid
8. pork lard, chicken fat
9. fat, carbohydrate
10. fat
11. peroxides, polymers, free radicals
12. false
13. sugar, flour
14. *trans*

Self-Test 20–C
1. variety, whole, different places
2. iron, hemoglobin
3. manganese, skin pigmentation
4. zinc
5. niacin (B_3)
6. sodium, potassium
7. greater than 1
8. magnesium
9. calcium, bones, teeth
10. kidney
11. iodine
12. false
13. fat, water
14. provitamin
15. carrots, liver
16. A
17. antioxidant
18. vitamin E
19. D
20. B_6
21. vitamin D
22. scurvy
23. coenzymes, protein production
24. B_1, beriberi
25. niacin (B_3)

Self-Test 20–D
1. synergistic, potentiation
2. drying, salting
3. false (preserve for shelf life)
4. volatile compounds

5. more easily oxidized
6. complexing agent (sequestrant)
7. false
8. flavor enhancer
9. dehydrating
10. generally recognized as safe
11. saccharin
12. conjugated chains with aromatic rings
13. citric acid, lactic acid
14. carbon dioxide (CO_2)
15. breaks
16. active

Matching Set – I

1. e	6. m	10. k
2. c	7. n	11. j
3. l, a	8. i	12. g
4. f	9. b	13. h
5. d		

Matching Set – II

1. b	4. e	6. a
2. c	5. f	7. h
3. d		

CHAPTER 21

Self-Test 21 – A
1. generic
2. antibiotic
3. cholesterol, triglycerides
4. faster
5. fatigue, lethargy, depression, confusion
6. relaxing
7. high-density lipoprotein, low-density lipoprotein
8. low-density lipoprotein
9. true
10. DNA
11. antimetabolites
12. leukemia
13. tissue-plasminogen-activator (TPA)

Self-Test 21 – B
1. p-aminobenzoic acid
2. mold
3. fungus
4. testosterone
5. progesterone
6. muscle building
7. liver damage, liver cancer, acne, baldness, changes in sexual desire, enlargement of breasts
8. prodrugs
9. cell walls

Self-Test 21 – C
1. neurotransmitters
2. dopamine
3. heroin
4. enkephalins
5. cocaine
6. heroin
7. AIDS
8. phosphate
9. T cells
10. designer drug

Matching Set

1. g	7. o	12. n
2. r	8. d	13. l
3. p	9. b	14. k
4. c	10. a	15. h
5. i	11. j	16. f
6. q		

CHAPTER 22

Self-Test 22 – A
1. corneal
2. sebum
3. skin, hair, nails
4. disulfide, ionic
5. 10%
6. emollient
7. emulsion
8. lipstick
9. talc
10. False
11. p-aminobenzoic acid
12. melanin

Self-Test 22 – B
1. stretches because it weakens ionic bonds and causes swelling of keratin
2. to hold hair in shape and place against its natural tendency
3. thioglycolic acid
4. oxidation
5. black
6. to penetrate hair in order to help produce a more pliable and elastic fiber
7. sulfides of sodium, calcium, strontium
8. acts as an astringent
9. makes it easier to hold chemical at desired spot
10. breathing spray chemicals into lungs

Self-Test 22 – C
1. fatty
2. alkali or lye

3. water
4. air
5. traditional soaps
6. surface-active agents
7. produces a precipitate
8. nonionic detergents
9. increases soap action, precipitates hard-water ions
10. visible
11. detergent, abrasive
12. fluorine

Self-Test 22–D
1. tricresyl phosphate (TCP)
2. prevents high-boiling residues on walls of carburetor
3. lubricating oils and gelling agents such as fatty acid soaps
4. ethylene glycol
5. methyl alcohol
6. 40W

Self-Test 22–E
1. silver
2. hydroquinone
3. reduced

4. $Ag(S_2O_3)_2^{3-}$
5. somewhat larger
6. reducing
7. sodium thiosulfate
8. (a) blue, (b) green, (c) red
9. red, blue, green
10. true
11. reduced
12. panchromatic

Matching Set–I

1. q	7. a	13. i
2. m	8. e	14. j
3. k	9. f	15. l
4. c	10. b	16. d
5. p	11. g	17. o
6. n	12. h	

Matching Set–II

1. h	5. j	9. d
2. f	6. k	10. l
3. g	7. a	11. b
4. i	8. c	12. e

Index